Lowell Wakefield Fisheries Symposium

ECOSYSTEM APPROACHES FOR FISHERIES MANAGEMENT

Proceedings of the Symposium on Ecosystem Considerations in Fisheries Management, September 30-October 3, 1998, Anchorage, Alaska

University of Alaska Sea Grant College Program
AK-SG-99-01
1999

Price $40.00

Elmer E. Rasmuson Library Cataloging-in-Publication Data

Symposium on Ecosystem Considerations in Fisheries Management (1998 : Anchorage, Alaska)

Ecosystem approaches for fisheries management : Proceedings of the Symposium on Ecosystem Considerations in Fisheries Management, September 30-October 3, 1998, Anchorage, Alaska. — Fairbanks : University of Alaska Sea Grant Program, 1999.

x, 756 p. cm. — (Lowell Wakefield Fisheries Symposium ; [16th]), (University of Alaska Sea Grant College Program ; AK-SG-99-01)

The symposium met jointly with the annual (1998) meeting of the Alaska Chapter of the American Fisheries Society.

Includes bibliographical references and index.

ISBN 1-56612-061-6

1. Fishery management, International—Congresses. 2. Fishery management—Alaska—Congresses. 3. Fishes—Ecology—Congresses. I. Title. II. Series: Lowell Wakefield Fisheries Symposia Series ; 16th. III. Series: Alaska Sea Grant College Program report ; AK-SG-99-01.

SH328.S847 1998

Citation for this volume is: 1999. Ecosystem Approaches for Fisheries Management. University of Alaska Sea Grant, AK-SG-99-01, Fairbanks. 756 pp.

Acknowledgments

This book is published by the University of Alaska Sea Grant College Program, which is cooperatively supported by the U.S. Department of Commerce, NOAA National Sea Grant Office, grant no. NA86RG-0050, project A/75-01; and by the University of Alaska Fairbanks with state funds. The University of Alaska is an affirmative action/equal opportunity institution.

Sea Grant is a unique partnership with public and private sectors combining research, education, and technology transfer for public service. This national network of universities meets changing environmental and economic needs of people in our coastal, ocean, and Great Lakes regions.

University of Alaska Sea Grant
P.O. Box 755040
203 O'Neill Bldg.
Fairbanks, Alaska 99775-5040
(907) 474-6707 Fax (907) 474-6285
http://www.uaf.edu/seagrant/

Contents

Species Interactions

Habitat and Spatial Considerations

Whole Ecosystem Approaches

About the Symposium

Alaska Sea Grant invited the Alaska Chapter of the American Fisheries Society to hold its 1998 annual meeting jointly with a Wakefield symposium. At the time discussions began, the Alaska Chapter of AFS had proposed to host the 1998 annual meeting of the Western Division of AFS. The proposal was successful and the three meetings were held jointly. The topic, Ecosystem Considerations in Fisheries Management, was jointly chosen by representatives of the three entities.

The meeting was held September 30 to October 3, 1998, in Anchorage, Alaska. Nearly 250 presentations were made, with several plenary speakers to introduce the topic and discuss the partnership of Sea Grant and the American Fisheries Society. Due to the dedication of the following committee members, the meeting was deemed successful and attracted nearly 500 attendees.

Wakefield Program Committee

Milo Adkison, University of Alaska Fairbanks, Fisheries Division
Brenda Baxter, University of Alaska Fairbanks, Alaska Sea Grant
Cindy Hartmann, American Fisheries Society Alaska Chapter
Gordon Kruse, Alaska Department of Fish and Game
Patricia Livingston, U.S. National Marine Fisheries Service
Clarence Pautzke, North Pacific Fishery Management Council

Local Arrangements Committee

Brenda Baxter, University of Alaska Fairbanks, Alaska Sea Grant
Bill Bechtol, Alaska Department of Fish and Game
Allen Bingham, Alaska Department of Fish and Game
Jane DiCosimo, North Pacific Fishery Management Council
Valerie Elliott, U.S. Minerals Management Service
Steve Fried, Alaska Department of Fish and Game
Lee Ann Gardner, RWJ Consulting
Jeanne Hanson, U.S. National Marine Fisheries Service
Pat Hansen, Alaska Department of Fish and Game
Cindy Hartmann, U.S. National Marine Fisheries Service, Alaska Region
Bill Hauser, Alaska Department of Fish and Game
Gail Heineman, Alaska Department of Fish and Game
Carol Kerkvliet, Alaska Department of Fish and Game
Steve Miller, U.S. Fish and Wildlife Service
Larry Peltz, Alaska Department of Fish and Game
Joe Sullivan, Alaska Department of Fish and Game
Bill Wilson, LGL Alaska Research Associates
Carol Ann Woody, U.S. National Biological Service
Steve Zemke, U.S. Forest Service

American Fisheries Society Program Committee

Jeff Barrett, Pacific Lumber Co.
Steve Bauer, U.S. Environmental Protection Agency
Robert Bilby, Chair, National Marine Fisheries Service, Northwest Fisheries Science Center
Pete Bisson, U.S. Forest Service
Pete Hagen, Alaska Department of Fish and Game
Jeff Cederholm, Washington Dept. of Natural Resources
Charlie Corrarino, Oregon Department of Fish and Wildlife
Cindy Hartmann, U.S. National Marine Fisheries Service, Alaska Region
Kim Hyatt, Department of Fisheries and Oceans Canada
Carol Kerkvliet, Alaska Department of Fish and Game
Steve Klosiewski, U.S. Forest Service
Eric Knudsen, U.S. Geological Survey, Alaska Science Center
K. Koski, U.S. National Marine Fisheries Service
Tom Kline, Prince William Sound Science Center
Gino Lucchetti, King County, Washington
Alan Maki, Exxon Corporation
Hal Michael, Washington Department of Fish and Wildlife
Duane Neitzel, Battelle Laboratories
Steve Ralph, U.S. Environmental Protection Agency
Dana Schmidt, R.L.&L. Environmental Services
Ramona Schreiber, U.S. National Oceanic and Atmospheric Administration
Kate Wedemeyer, DOD, Elmendorf Air Force Base
Alex Wertheimer, U.S. National Marine Fisheries Service
Mark Wipfli, U.S. Forest Service

Symposium Sponsors

Alaska Sea Grant College Program
Alaska Department of Fish and Game
  Commercial Fisheries Division
  Division of Sport Fish
North Pacific Fishery Management Council
U.S. National Marine Fisheries Service
  Alaska Fisheries Science Center
Alaska Chapter of the American Fisheries Society
Colorado-Wyoming Chapter of the American Fisheries Society
North Pacific International Chapter of the American Fisheries Society
Western Division of the American Fisheries Society

Contributing sponsors

Alaska Crab Coalition
Alaska Longline Fishermen's Association
Aleutian Pribilof Island Community Development Association
Alyeska Seafoods
Anchorage Convention and Visitors Bureau
Kaladi Brothers Coffee Company
Kodiak Vessel Owners Association
Once in a Blue Moon Gift Shop
ProFish International
Sealaska Corporation
U.S. Fish and Wildlife Service.

The Lowell Wakefield Symposium Series

The University of Alaska Sea Grant College Program has been sponsoring and coordinating the Lowell Wakefield Fisheries Symposium series since 1982. These meetings are a forum for information exchange in biology, management, economics, and processing of various fish species and complexes as well as an opportunity for scientists from high latitude countries to meet informally and discuss their work.

Lowell Wakefield was the founder of the Alaska king crab industry. He recognized two major ingredients necessary for the king crab fishery to survive—ensuring that a quality product be made available to the consumer, and that a viable fishery can be maintained only through sound management practices based on the best scientific data available. Lowell Wakefield and Wakefield Seafoods played important roles in the development and implementation of quality control legislation, in the preparation of fishing regulations for Alaska waters, and in drafting international agreements for the high seas. Toward the end of his life, Lowell Wakefield joined the faculty of the University of Alaska as an adjunct professor of fisheries where he influenced the early directions of the university's Sea Grant Program. Three Wakefield symposia are planned for 2000-2002.

About This Proceedings

This publication, the 16th in the Lowell Wakefield Fisheries Symposium series, has 49 papers presented in the Wakefield sessions of the combined meeting.

Layout, format, and proofing are by Brenda Baxter, Sue Keller, and Carol Kaynor, and cover design is by David Brenner of University of Alaska Sea Grant. Copy editing is by Catherine W. Mecklenburg, Point Stephens Research Associates, Auke Bay, Alaska.

Opening Comments

R.K. Dearborn
University of Alaska Fairbanks, Alaska Sea Grant College Program, Fairbanks, Alaska

Welcome on behalf of the University of Alaska Fairbanks and its School of Fisheries and Ocean Sciences. Welcome on behalf of the Alaska Sea Grant College Program.

Most importantly I wish to welcome you on behalf of a partnership that for 18 years has brought together scientists on topics chosen by the fishery scientists and managers in this region through the mechanism of the Lowell Wakefield symposia. Although the symposium series is hosted by Alaska Sea Grant, and symposium manager Brenda Baxter works for Alaska Sea Grant, the Lowell Wakefield symposia are really the result of a partnership between the University of Alaska Fairbanks and its Sea Grant Program, the Alaska Department of Fish and Game, the National Marine Fisheries Service, and the North Pacific Fishery Management Council. The program committee always includes members from these five partners. This year's symposium committee includes: Milo Adkison of our School of Fisheries and Ocean Sciences in his first committee assignment; Gordon Kruse of the Alaska Department of Fish and Game, who has served on many of the Lowell Wakefield committees and to whom we all owe a significant debt of gratitude; Patricia Livingston of the National Marine Fisheries Service (NMFS) Alaska Fisheries Science Center; Clarence Pautzke, Executive Director of the North Pacific Fishery Management Council; and this year, because of our special partnership with the American Fisheries Society (AFS) for this meeting, Cindy Hartman of AFS and the NMFS Alaska Region. As for every Wakefield Symposium, this communal effort has been guided by the tender buggy whip of Brenda Baxter. There is no such thing as a non-working assignment to one of Brenda's committees. You will note that every one of these committee members, plus Bob Bilby of NMFS, is also serving as a session chair.

Because the Lowell Wakefield Symposium series is driven by a partnership of university and agency scientists, a retrospective look at the symposium topics also provides a look at our growth and the growth of the science. There are three striking signals in the data:

1. The early symposia addressed the biology of single commercial species—crab, herring, sablefish—but more recently the symposia have addressed topics like the role of forage fishes, combining multiple information sources in stock assessment, and this year, ecosystem considerations in fisheries management. The series has not abandoned

single-species topics, but I think there is a clear signal of the complexities that have entered fisheries management.

2. If one reads the proceedings of one of the very early symposia, and especially the discussions of "where do we go from here," it is clear that the learning that takes place at these symposia and the resulting discussions influence work over subsequent years. The tremendous progress that is reported in a subsequent proceedings volume is clearly a rewarding extension of the priorities discussed in the early symposium on a related topic. I trust that this year will be no exception to our trek up the learning curve.

3. Last, the symposia reflect an increasingly multinational flavor. We have long recognized that even though Alaska has more than half of our nation's shoreline and lands half the nation's fish, we do not have as residents half of the nation's, and certainly not half the world's, fishery or ocean scientists. Over the past decades, leaders of the fishing industry have pushed the industry beyond national boundaries and often ahead of our scientific knowledge base. Responsible leaders in the industry, such as Lowell Wakefield, recognized that the development of the industry has moved at a faster pace than knowledge can be developed and encouraged the kind of scientific information exchange that takes place at these meetings. This region's fisheries are of world scale. Our Canadian colleagues to our south have for many years been contributors to our discussions, and we have valued that tremendously. As the questions addressed by these symposia have become increasingly complex, we have been blessed by a continuing enrichment of talent from around the globe. We welcome with greed this international participation.

Thank you for joining us, both for the Wakefield symposium and for the meetings of the Alaska and Western Divisions of the American Fisheries Society.

Let me introduce you to your next speaker. Those of us from Alaska who think of lakes like Becherof and Iliamna as vast and productive lakes are reminded by people like our next speaker, a former executive secretary of the Canada–U.S. Great Lakes Fishery Commission, that there are larger lakes. I have had the privilege of working with him over recent years as a member of the National Review Panel to the Sea Grant College Program. You may know him better as a past president of the American Fisheries Society. At any rate, he is a long and faithful friend of us all. Let us welcome Mr. Carlos Fetterolf.

Sea Grant, the American Fisheries Society, and Essential Fish Habitat: A Developing Relationship

Carlos Fetterolf
National Sea Grant Review Panel, Ann Arbor, Michigan

Carlos Fetterolf is a past president of the American Fisheries Society and a current member of the National Sea Grant Review Panel. This paper contains his unique perspective on these two organizations and how they interact.

Abstract

Dr. Ronald Baird, Director of the National Sea Grant College Program, and Paul Brouha, then Executive Director of the American Fisheries Society (AFS), created an AFS/Sea Grant Intern Program in 1997. A guidance committee decided on aquatic habitat as the focus for the intern, Lee Benaka, thus fitting with the Congressional reauthorization of the Magnuson Fishery Conservation and Management Act. The act calls for designation and special management of "those waters and substrates necessary to fish for spawning, breeding, feeding, or growth to maturity" as essential fish habitat ("fish" includes other aquatic organisms, e.g., lobsters, clams). The National Marine Fisheries Service (NMFS) will carry out the provisions of the act. Discussions with Rolland Schmitten, NMFS director, resulted in agreement to develop a symposium on essential fish habitat for the 1998 annual meeting of AFS in Hartford, Connecticut. NMFS and Sea Grant agreed to publish the symposium proceedings (25 papers) in the AFS book program with intern Benaka as editor.

Following the symposium, a dozen Sea Grant and NMFS personnel met to decide how to continue their symbiotic relationship, and how the resources of the 29 individual Sea Grant college programs could be brought to bear on the identification, evaluation, and protection of essential fish habitat. In the words of Dr. Baird at this meeting, "Sea Grant has a ready cadre of researchers, educators, communicators, and outreach experts."

A foundation has been established for continued and expanded AFS intern programs with not only Sea Grant, but with NMFS, the National Oceanic and Atmospheric Administration, the Fish and Wildlife Service, the Geological Survey and its Biological Resources Division, the Environmental Protection Agency, the Department of Agriculture (aquaculture), the Forest Service, and others.

Introduction

Many of the attendees at today's plenary session from the United States and Canada are either members of, or familiar with the American Fisheries Society, but we also have attendees and authors from Australia, China, Denmark, Italy, Jakarta, Malaysia, New Zealand, the Philippines, Russia, South Africa, Sweden, and the United Kingdom. While the American Fisheries Society has more than 9,000 members in over 70 countries, it is unlikely our guests are familiar with AFS, the oldest and largest society of fisheries and aquatic scientists in the world.

American Fisheries Society

AFS and its members are dedicated to:

- The advancement of fisheries and aquatic science;
- Communication of scientific and technical information to other professionals, decision makers, and the public;
- Encouraging the teaching of fisheries and aquatic sciences in colleges and universities; and
- Continuing education and development for fisheries and aquatic science professionals.

AFS takes reasonable and informed positions on resource issues that affect aquatic environments and resources. To this end, AFS:

- Supports research to identify and understand fishery resources and ecosystems;
- Publishes four peer-reviewed journals focused on fisheries and aquatic sciences, fish health, aquaculture, and management;
- Supports managing aquatic resources and habitats for sustained yield;
- Supports regulation of aquatic pollutants;
- Supports strong, scientifically defensible government policies and effective interjurisdictional management structures; and
- Organizes science seminars, workshops, and technical meetings.

AFS is structured into 51 chapters, four geographic divisions, and 21 discipline-oriented sections ranging from fisheries management, to fisheries law, to physiology, to socioeconomics. In short, AFS has something for everyone involved with the fisheries profession. AFS is not just about fish.

When elected to the AFS presidential succession in 1988, I had served for many years on the Michigan Sea Grant Policy Committee, and had often thought of the potential win-win relationship if the two organizations could be brought together. Upon retiring from the Canada–U.S. Great Lakes Fishery Commission, which, by the way, is a sister organization to the commissions you are more familiar with on our Northwest Coast—the Pacific Halibut Commission and the Pacific Salmon Commission—I was appointed by the Secretary of Commerce to the National Sea Grant Review Panel in 1993, and the win-win relationship concept returned strongly.

National Sea Grant College Program

Sea Grant is a partnership and a bridge among federal and state government, academia, industry, scientists, and citizens to help Americans understand and sustainably use our precious coastal, ocean, and Great Lakes resources. Through the 29 Sea Grant institutions, which draw on faculty from over 200 colleges and universities, Sea Grant addresses such areas as aquaculture, aquatic nuisance species, coastal economic development, coastal habitat enhancement, coastal hazards, education, fisheries, marine biotechnology, and seafood technology. In the early days of Sea Grant, the programs very often dealt with fisheries and Sea Grant was colloquially known as Fish Grant, but now, as with AFS, it is not just about fish.

The Sea Grant effort is three-pronged: research, outreach, and education. Scientists tackle the important questions and the research. Extension specialists take the information to the users. Communication specialists build public understanding for science-based resource management. Educators, on a fourth front, bring the discoveries to the nation's students.

Bringing AFS and Sea Grant together

In June 1996, Dr. Ron Baird became Director of the National Sea Grant College Program, and he began building partnerships and working relationships. My AFS/Sea Grant win-win concept came to the front and I wrote a combined letter to Ron Baird and Paul Brouha, then Executive Director of AFS, urging them to meet and consider the potential for a symbiotic relationship which would benefit both organizations and the fishery resource. Both leaders quickly seized the opportunity; a memorandum of understanding was drafted; a guidance committee was formed (Dr. Jeffrey Reutter, Ohio Sea Grant; Dr. Robert Stickney, Texas Sea Grant; and the author); a position description was agreed upon; and AFS/Sea Grant intern Lee Benaka was hired. The committee met at the 1997 AFS annual meeting in Monterey, California, and settled on aquatic habitat as the intern's focus. This fit with the 1996 Sustainable Fisheries Act (P.L. 104-297) which amended the 1976 Magnuson Fishery Conservation and Management Act, in part, by adding provisions for the care of essential fish habitat. These provisions require the designation and protection of essential fish habitat defined as "those waters and substrates necessary to fish for spawning, breeding, feeding, or growth to maturity." Essential fish habitat (EFH) is destined to be the fisheries buzz phrase of the decade as we work to define it, refine it, and fully understand it. EFH is not just about fish. It's about other aquatic organisms as well, and living systems. The National Marine Fisheries Service is the agency responsible for carrying out the provisions of the act, which is landmark environmental legislation. The NMFS is to provide the national fishery management councils with ecologically sound guidance that is both feasible and scientifically defensible as required by the act, and has strived valiantly to do so. Much work remains, however, to expand what we know about marine habitats and the interactions of aquatic organisms with them. NMFS clearly needs help to support this breakthrough in marine legislation which adds huge new responsibilities and duties. Early discussions with Rollie Schmitten, NMFS director, resulted in agreement for Lee Benaka and the guidance committee to develop a symposium on EFH at the 1998 AFS annual meeting in Hartford, Connecticut. Further, NMFS and Sea Grant agreed to provide equal funding to publish the symposium proceedings as an AFS book edited by Benaka.

The Resulting Symposium

We titled the symposium "Fish Habitat: Essential Fish Habitat and Rehabilitation" and addressed various viewpoints on EFH by government, industry, and nongovernmental organizations; EFH identification; protecting habitat from fishing impacts; protecting habitat from nonfishing impacts; fish habitat rehabilitation; and socioeconomic issues. The 25 oral presentations plus posters showcased the habitat-related research capabilities of both NMFS personnel and Sea Grant–funded investigators which demonstrated what we know and the vast amounts we do not know. The symposium brought together researchers and policy makers from the Sea Grant network and NMFS under the umbrella of the AFS and further strengthened relations in pursuit of filling in the information gaps.

Despite the facts that the essential fish habitat mandate represents a new recognition by Congress, that Congress overwhelmingly endorsed reauthorization of the act, and Senator Hollings stated that "habitat protection has become a greater concern because coastal development and pollution threaten the environment and subsequently the effect of fish stocks," the impact of the EFH provisions have not been fully recognized by government, fishermen, developers, and industry—yet. But wait until a few enforcement actions are under way. Essential fish habitat is not glamorous—yet, but it should ultimately shift to those who would alter the habitat the burden of proof to show they will not damage its production potential.

I hope it is not the same old story as when I joined the Great Lakes Fishery Commission staff in 1975. When the sea lamprey control persons talked, the fish folks slept. When the fish folks talked, the lamprey guys slept. And when the habitat guys talked, everyone's eyes glazed over. The situation in the Great Lakes is very different now, but how different it was then from the theme of today's conference: "Ecosystem Considerations in Fisheries Management." Essential fish habitat will suddenly become glamorous when the dollars begin to flow, but what will be the consequences if we begin to call all waters EFH, instead of concentrating on the "essential essential fish habitat"? Will there be a backlash? Has it already started? What will be the outcome?

American Fisheries Society and Future Internships

In Hartford, a dozen ranking Sea Grant and NMFS personnel considered how Sea Grant might undertake a National Strategic Investment for Research, Education, and Outreach on EFH, how the resources of the two agencies could be brought together; how research priorities could be coordinated; how Sea Grant researchers could provide the "scientifically defensible data base" the act calls for; and how Sea Grant communicators and extension agents could be effective in educating and informing stakeholders with regard to compliance, intent, value, and legal aspects.

As Ron Baird stated at the meeting, Sea Grant has "a ready cadre of researchers, educators, communicators, and outreach experts," and as some NMFS regional folks said following the meeting, "It seems like a perfect fit."

Given a successful ending to this initial internship, there is potential for continued and expanded AFS intern programs not only with Sea Grant, but with NMFS, the Fish and Wildlife Service, the Geological Survey and its Biological Resources Division, the Department of Agriculture for aquaculture issues, the Forest Service, the Corps of Engineers, the Environmental Protection Agency, and others in a variety of areas. I am confident NMFS and Sea Grant are pleased with their exposure and the information accrued on a topic of critical current concern under the umbrella of the American Fisheries Society. AFS benefited from the services of a bright young person, and Lee Benaka, the intern, benefited. Most important of all, living marine resources and their habitats will benefit in the future. The internship was a good fit and a good start.

Taking the Next Step in Fisheries Management

Richard J. Beamish
Department of Fisheries and Oceans, Pacific Biological Station, Nanaimo, British Columbia, Canada

Conrad Mahnken
National Marine Fisheries Service, Port Orchard, Washington

Abstract

For several decades fisheries biologists focused their attention on the taxonomy, life history, and population dynamics of single species of fishes. Most attention went to the preferred commercial species, but less popular fishes were not totally ignored. There was always the intention to piece together the species relationships into some integrated big picture, but the incentive was never strong enough to bring enough people together for a long enough time to begin to understand how whole ecosystems could be understood and protected. Several recent events have now provided this incentive. The most serious is the recognition that climate impacts must be understood both for fisheries management and for the detection of global warming impacts. The lessons from recent fisheries management issues such as East Coast cod, Atlantic salmon management, and coho problems, clearly have shown that there are some problems with previous concepts and that it is cost effective to study marine ecosystems. It is also good politics. It is easier to make difficult decisions when people are well informed about what is known and what is not known.

It is time to manage and protect whole ecosystems. This will not be a linear extension of single-species thinking. A more abstract concept is needed in which the single species is seen in relation to the processes that affect ecosystems and less in terms of numbers of individuals. The timing of copepod production, the condition of juveniles at certain times of the year, and the abundance trends of associated species may all become ways of assessing fishing impacts. Ecosystem management requires an understanding of the influences that regulate species naturally. For salmon, we propose a new concept of natural regulation that we call the critical size–critical period hypothesis. According to this hypothesis, the abundance of salmon is determined both in the early marine period and in the first fall and winter of ocean residence. The amount of mortality late in the first marine year is related to the rate of growth during the summer. Like all difficult but essential tasks, it is important to get started with ecosystem management. It is also important to recognize that the communication and coordination of relevant information for ecosystem management may be as challenging as acquiring the understanding of how to do it.

Introduction

In the 1960s and 1970s we first heard about ecology. I think most of us liked what we heard. If you remember, we considered ecology to be the interaction of living things and nonliving things. In fisheries, the interest in ecology always seemed to be curiosity-based. It was hard to see how the complexities of food chains and environmental factors would ever find their applications in fisheries management. It reminded us of this little poem called The Purist.

> I give now Professor Twist
> A conscientious scientist
> Trustees exclaimed, "He never bungles"
> And sent him off to distant jungles
> Camped on a tropic riverside
> One day he missed his loving bride
> She had, the guide informed him later
> Been eaten by an alligator
> Professor Twist could not help but smile
> You mean he said, "a crocodile."

The point, of course, is that we do not want to miss the message in the detail of the event.

Ecosystem interactions are complex, but we can no longer avoid dealing with the complexities. We do have to consider the impacts of associated species and we do have to consider the impacts of the environment. As fisheries science developed, we had to simplify the complexities of ecosystems, but it is time to understand how to incorporate ecosystem dynamics into management. This is an improved understanding, not a complication of the simplicities we had to use. We must always be careful not to elevate our intelligence above the species relationships we do not understand. We are, in reality, stewards, and not managers.

Two of the most frequent news topics in recent years have been fisheries and climate. They are newsworthy because the changes we observed were extreme or unexpected or both. Climate will continue to be an important item in the news as we begin to detect the impacts of global warming, but fisheries may become less newsworthy as we stabilize our expectations through an improved understanding of the interrelationships among species and their ecosystems.

For example, our forecasts on the west coast of Canada a decade ago bear little resemblance to the fisheries we have today, not because of poor management, but because of a poor understanding of how ecosystems function. Our rather embryonic understanding of processes that affect fish abundance is a product of single species thinking. Our well-intended emphasis on interpreting numbers without the associated emphasis on understanding how the system works was a response to a rapidly developing fishery and economy that needed quotas. The lesson from our experiences with Pacific sardines, northern cod, Atlantic salmon, Pacific coho salmon, and a number of other species is that there is a complexity that must be understood if we are to have fisheries and be good stewards of our resources. In the next decade we will recognize the need to separate natural abundance changes from fishing effects. The fluctuations in abundance will continue, but our response to these changes needs to be different. We need to be able to forecast the shifts in our marine ecosystems that will change the trends in abundance of salmon, herring, halibut, and other species. We need to assess the health of a stock in relation to its carrying capacity rather than its historic abundance or its demand by fishermen. I think that a decade from now all of us will be more comfortable with the state of our west coast fisheries, not because there will be more fish, but because we will better understand why changes occur. It requires a major change in the thinking of biologists, fishermen, managers, educators, and the public, but we have the resources, the people, and the policies to make it happen if we can just get started.

It may be fine to preach all this ecosystem stuff, but how do we use it in management? Does an Atlantic cod with a skinny body mean anything? Of course it does. Does a shift in the intensity of winter winds in the North Pacific mean anything? Of course it does. In this paper, we try to show why and how we think ecosystem management will make us better stewards.

Regimes

A few years ago the word *regime* was not in the vocabulary of many fisheries biologists. It is today, even though we are not entirely sure how to define regimes. In general, regimes are large, linked climate-ocean ecosystems that shift in states over 10- to 30-year periods. In response to these changes, there may be changes in the dynamics of the resident fishes and other biota. A number of authors have written about regimes, but the best-known studies are studies by Hare and Francis (1995) and Mantua et al. (1997). Both these studies clearly demonstrated the persistence of trends and the abruptness of the periods of change. Most researchers studying trends in climate agree that there was a regime shift in 1977 and some feel that another change occurred in 1989 (Watanabe and Nitta 1998, Beamish et al. 1999). Beamish et al. (1998a) think that a recent change occurred in 1995. They use a number of indices to study regimes and regime shifts

(Beamish et al. 1999) and here we show how three indices changed about 1995.

The length of day (LOD) is the time it takes the solid earth to complete one rotation. The measure of LOD and the relevance of the measurement are discussed by Beamish et al. (1999). Changes in the annual trend of the LOD represent changes in the angular momentum of the solid earth and a redistribution of energy on the planet. The recent LOD time series shows changes in trends about 1973, 1987, and recently in 1994 (Fig. 1). The North Atlantic Oscillation Index (NAOI) is the normalized difference in the normalized winter sea level pressures between Lisbon, Portugal, and Iceland (van Loon and Rogers 1978). The NAOI clearly changed in 1989 and 1995. The changes in the 1970s were less clear (Fig. 2). The southern Oscillation Index (SOI) is commonly known on the "El Niño index." It is analogous to the NAOI but in an east-west direction. It is the standardized series of the standardized Tahiti data minus the standardized Darwin data (Climate Prediction Center, National Center for Atmospheric Research, Boulder, Colorado). The trend in the pattern in the SOI changed in 1977 (Fig. 3) from an oscillation of positive (La Niña) and negative (El Niño) anomalies to persistent El Niño conditions up to 1989 and then to extreme, persistent El Niño conditions to 1995. Beginning in 1996, the oscillations were more like the pre-1977 state but more extreme.

Critical Size–Critical Period Hypothesis

It is now clear to salmon biologists that abundance changes in salmon can be related to natural changes in climate and in the ocean. We propose that the linkage of the climate/ocean environment to final abundance is through summer growth rates. We call our hypothesis the critical size–critical period hypothesis.

We propose that the marine carrying capacity for Pacific salmon species is regulated naturally by the requirement to grow at a critical rate and thus achieve a critical size by a critical period. Regulation occurs in two major stages (Fig. 4). In the first stage, marine mortality is predation-based and occurs immediately after entry into salt water. Mortality at this stage is partly related to the size as well as to the density of smolts and the density and type of predators. The second period of major mortality occurs in the fall and winter when coho that are not at a critical size, are unable to maintain minimum metabolic requirements and enter a growth trajectory that leads to death. The second mortality occurs in the late fall and winter and is not predation-based, rather it is a function of the condition of the juvenile. The "culling" impact of winter on land animals such as muskoxen (*Ovibos moschatus*) is a recognized mechanism of regulating abundance (Bartmann 1984, Gunn et al. 1989) and we suggest that an analogous mechanism occurs in the ocean for coho salmon (*Oncorhynchus kisutch*) and possibly juveniles of other Pacific salmon species. The amount of fall and winter mortality is related to the condition of the fish at a

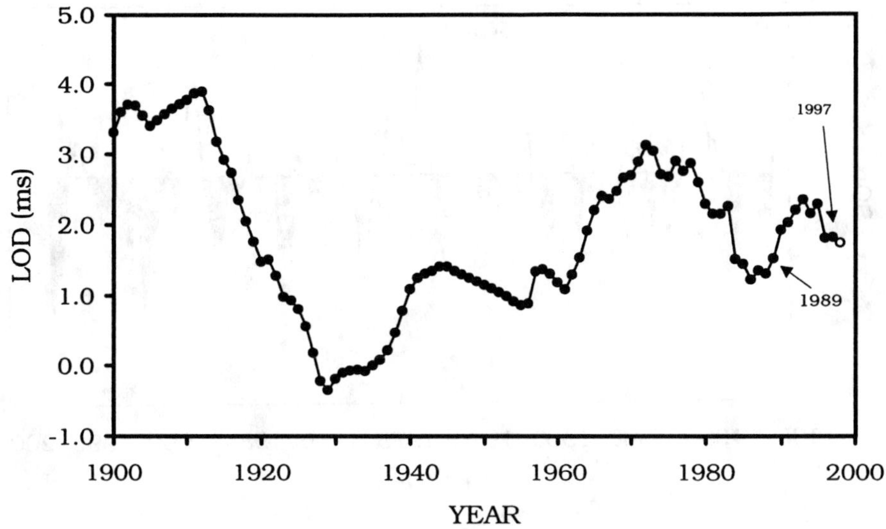

Figure 1. The length of day (LOD) time series from 1900 to 1997 in milliseconds (ms). Changes occur about 1974, 1989, and 1996.

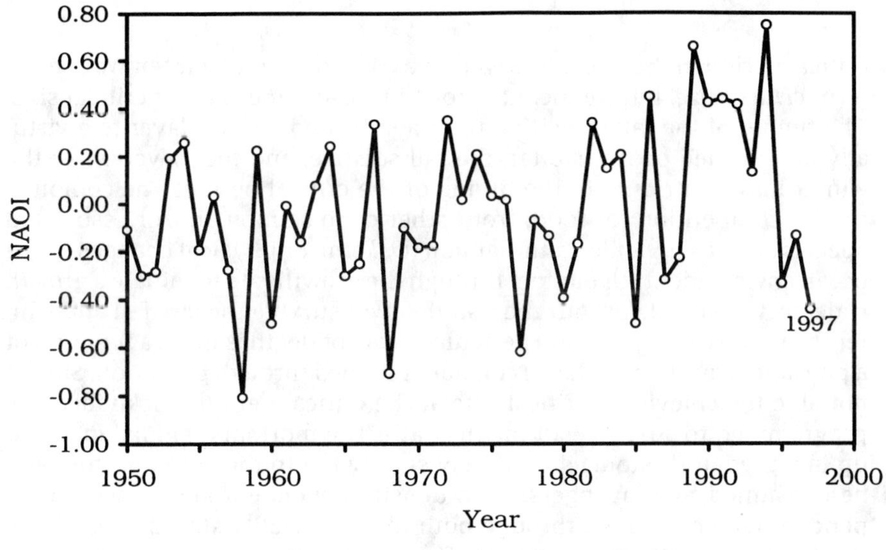

Figure 2. The North Atlantic Oscillation Index (NAOI) from 1950 to 1997 showing the change to positive anomalies in 1989 and a major change to negative anomalies in 1995.

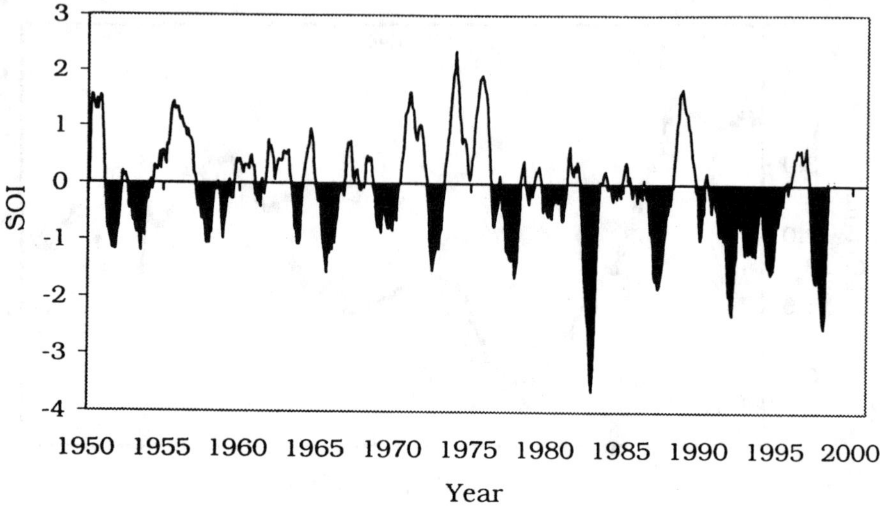

Figure 3. The southern oscillation Index (SOI) from 1951 to 1997, smoothed with a running average of 5s. The extreme negative anomalies are El Niño events and the extreme positive anomalies are La Niña events. Trends exist from 1951 to 1976 and from 1977 to 1989 and 1990 to 1995. The pattern may have changed again in 1996. The actual values are not shown.

critical period in the late fall as well as to the stresses of winter. We use the term *critical size*, but we include growth rate as a measure of critical size. The timing of the fall transition from a stratified surface layer to a virtually unstratified ocean habitat, the fall solstice, and the "severity" of the winter may contribute to the timing of the critical period. The combination of competition for food, from other coho salmon and/or associated species such as juvenile chum salmon (*O. keta*), or physical changes to the ocean environment, all interact through the growth rate to initiate a growth trajectory that will sort out the fish that will survive the late fall and winter from those that will die. The actual cause of death is more a function of a physiological change, than a constant removal through predation. Smolts not able to achieve the critical size at the critical period quickly become programmed to a path resulting in death. The moribund animal may end up in a predator's stomach or it may simply fall to the ocean bottom and be consumed by scavengers. Both density-dependent and density-independent factors interact through both these mortality stages to establish the brood-year strength after the first winter in the ocean. Interannual variability in abundances at the stock and regional levels will contribute to the variations about a mean carrying capacity that is characteristic of a regime. Shifts in the mean carrying capacity occur when there are shifts in a regime.

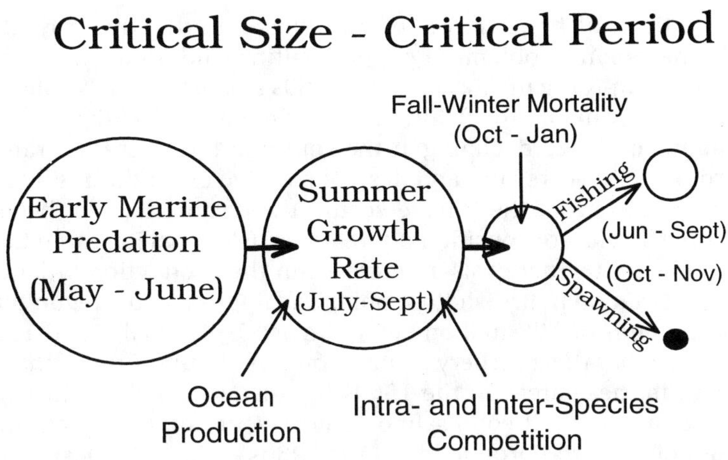

Figure 4. Diagram of the marine phase of critical size-critical period hypothesis. The areas of the circles are proportional to the speculated numbers of individuals in a stock. The early marine predation period reduces the number of juveniles to the abundance represented by the circle for the July to September growth period. It is during this time that the rate of growth in the summer determines the size of the stock after the first marine winter (October-January). The abundance after the marine winter will be the approximate brood-year size and during June to September fishing will remove a large percentage, leaving the spawning escapement in October and November.

According to this hypothesis, the freshwater stage of Pacific salmon is a period of relatively safe refuge for reproduction, which results in the production of a large number of young with diverse genetic traits. The large number of young greatly exceeds the available ocean habitat, thus high marine mortalities will occur. The diverse, genetically based life history traits help optimize survival in the ocean, which is a more hostile environment than fresh water. Thus, even though egg to smolt mortalities are extremely high, the marine environment regulates the final abundance in a manner analogous to the regulation of abundance of any plant or animal that produces very large numbers of young. It is a fundamental principle in ecology that in such cases the final abundance is a function of available habitat and not the large number of seeds or babies (Colinvaux 1978).

In North America, coho salmon are perhaps the salmon species that has the broadest general recognition. They are both a sought-after sport and commercial species and an indicator of the health of freshwater streams and rivers. They also are relatively easy to rear in hatcheries, making the additions of hatchery-reared juveniles an attractive way of addressing management problems or supplementing wild production with the expectation of producing more. In the 1960s and early 1970s coho salmon were readily available for fishing but by the late 1970s there were signs that abundance was declining at the southern limits of their range. It is now roughly 20 years later and despite massive expenditures by governments as well as from the private sector, the decline has not been halted and some stocks are considered threatened (National Research Council 1996). In the 1960s there was an increase in the production of hatchery-reared coho salmon in British Columbia, Washington, and Oregon that reached a maximum of 198 million smolts in 1981 (Beamish et al. 1997). We estimated that this hatchery production was about 1.5 times the wild production in these areas in the 1960s. By the late 1980s and early 1990s, the total numbers of coho salmon smolts entering the ocean probably were double the wild production of the 1960s and 1970s. Despite this increase in hatchery-reared smolt releases, the total catch of coho salmon continued to decline, reaching astonishingly low levels in the mid-1990s (Fig. 5). In the states of Washington and Oregon, and the Province of British Columbia, total government expenditures over the past 20 years have probably exceeded 1 billion U.S. dollars. The exact amount is less important than the inability to manipulate the population dynamics of coho salmon despite the good intentions of a large number of very concerned individuals. We propose that the explanation is that except for very low abundances, coho salmon abundance is less a function of the number of smolts entering the ocean and more of a function of the ocean conditions. We suggest that the final size of the population or brood-year strength is determined in the fall and winter. This hypothesis explains how carrying capacity is affected by ecosystem changes. Carrying capacity can be defined in several ways, but we consider carrying capacity to be the mean biomass that can be supported in an ecosystem in a particular state or regime. Within a regime, variation in abundance will occur. However, we consider carrying capacity to be the mean level. For example, the production of Pacific salmon stocks in Alaska has been shown to switch from one mean level to another (Hare and Francis 1995). We would consider that each mean level represents a distinct carrying capacity.

Ricker (1945) defined critical size as the mean length or weight at which the growth of a year class just balances mortality, i.e., the average size of the year class when the total weight is neither increasing nor decreasing. For exploited populations, it is the mean size at which the instantaneous rate of growth equals the instantaneous rate of natural mortality. The concept of a critical period is frequently associated with Hjort (1914) and has been interpreted to relate to larval survival in relation to predation and food. According to these earlier definitions, therefore, the critical period implies that a time in the larval stage is critical for stock size and the critical size identifies a stage in the growth and death of a year class. The critical size, according to our theory, is the minimum size (or growth rate) of the fish at the critical period, that will ensure it survives the first winter in the ocean.

Evidence to Support Hypothesis

In this report we describe results from three studies that support the hypothesis that growth during the first marine summer is an important factor in the determination of brood-year strength. Additional studies are currently in progress.

Swept Volume Abundance Estimates

The results of abundance estimates of coho salmon in the Strait of Georgia (between Vancouver Island and mainland British Columbia) provided the initial evidence that relatively large numbers of juveniles were still alive by the fall of their first marine year. A number of surveys have been completed, but here we report the results from the September 1997 cruise as an example of the methods used to estimate abundance.

Surveys were conducted in the Strait of Georgia using a large rope trawl (Beamish and Folkes 1998). The net was fished at a speed of approximately 5 knots, with an average mouth opening of 18 m × 15 m, and was effective at catching all species and sizes of Pacific salmon in most weather conditions. In the Strait of Georgia in September 1997, over 75% standardized catch of coho salmon occurred in the top 15 m and virtually all coho were in the top 45 m.

Swept volume estimates were made according to the procedures described in Beamish et al. (1998a). The estimates were for the top 45 m only and the net was assumed to have a catchability of 1. This means that all juvenile coho salmon in front of the net opening would be captured. We doubt that this is correct, which means that the abundance estimates are minimal estimates. In September 1997, 96 sets were completed throughout the Strait of Georgia. A total of 2,280 ocean-age-0 coho salmon were captured, with an average, standardized catch of 44.0 coho salmon per hour in the top 45 m. The abundance estimate was 2,840,000 with 95% error limits of 1,583,000-4,097,000. This is a minimal abundance and if the catchability is smaller (as it probably is), the estimates would be larger. In Russia, studies of the catchability of larger rope trawls (mouth opening 45 m × 45 m) fishing at the surface has been estimated to be 0.3 (Shuntov et al. 1988, 1993).

In 1997, Canadian hatcheries released approximately 10 million coho salmon smolts into the Strait of Georgia. Our studies in September 1997 showed that approximately 77% of the ocean-age-0 coho salmon were from hatcheries (Beamish et al. 1998c). Therefore, the remaining 23% were either wild fish from Canada or hatchery and wild fish from the United States. We estimated the hatchery and wild percentage from the United States to be 7.3% based on the marked and unmarked coho salmon released into Puget Sound and samples we collected in our surveys in Puget Sound in 1997 (Beamish et al. 1998b). Assuming that there was no selective

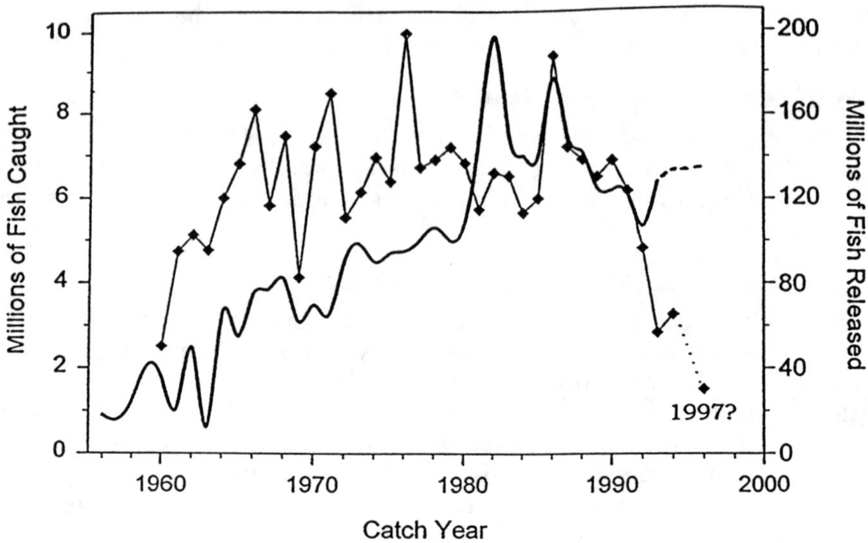

Figure 5. The estimated catch of coho salmon (line with diamonds) from British Columbia, Washington and Oregon (International North Pacific Fish Commission, statistical year books). Data from 1974 to the present include sports catch. The total hatchery production from British Columbia, Washington, and Oregon is shown as a solid line to 1992 and an estimate (dashed line) to 1996. We estimated smolt production in the 1960s by assuming that the combined harvest of sport and commercial catches would be approximately 6 million with a harvest rate of 50%. Assuming an equal sex ratio, 3,000 eggs per female, and a 2% survival from egg to smolt, there would be about 180 million smolts produced annually. Hatchery production averaged about 60 million (Beamish et al. 1997), leaving about 120 million wild smolts. Hatchery release year is lagged +1 year to match catch year.

movement of rearing types (Beamish et al. 1998c), the number of wild Canadian coho salmon entering the Strait of Georgia could be determined using the known number of hatchery releases and would be approximately 2.2 million. The total hatchery and wild smolt production, therefore, would be approximately 13 million fish. The minimal abundance estimate of 2.8 million represents 22% of this total smolt production or 19% if only smolts of Canadian origin are counted. If the actual abundance approximately is double because of a catchability of 0.5 rather than 1.0, the percentage of coho salmon remaining in the Strait of Georgia could approach about 40% of the number that entered salt water in the spring. In 1997, the total returns of coho salmon that entered the Strait of Georgia in 1996 (catch and spawning abundance) was less than 500,000 fish. The abundance estimate in September 1996 was 3.6 million (with 95% error limits of 5.1 million-2.1 million). Again, the actual number could be higher if the true catchability were known. Even at 3.6 million, the total return of less than 500,000 in 1997 indicates that the mortality in the fall and winter is substantial. It is too early to estimate the total returns in 1998 to compare with the 1997 abundance estimate, but preliminary estimates are that the total returns will be less than in 1997. This indicates that the marine mortality that occurs after September of the first year in the ocean is of major importance in the determination of the final brood-year strength.

Marine Pen Rearing Studies

Mahnken et al. (1982) studied coho salmon from 12 hatcheries in Washington State to determine the size that would optimize their survival in net-pens in salt water. In the net-pens, coho salmon were in a smolt, transitional, or parr-like stage (Fig. 6). The parr-like stage resulted from some fish that did not develop into the other stages and fish that reverted to the parr stage. These changes occurred despite the addition of artificial food. The parr-like fish did not feed or grow normally and eventually died. The fish in reversion fed and grew, but eventually became parr-like and died. It was possible for some of the fish in reversion to survive through the winter and resmolt the following spring (Folmar et al. 1982). The reversion process accelerated after the summer solstice and continued to the winter solstice. The association with photoperiod suggested that reversion was associated with genetically controlled osmoregulatory requirements that were linked to the time of year and a requirement to grow at a minimum rate to survive. The different growth rates resulted in reversion occurring throughout the year (Fig. 6B). In the net-pens, the size of the largest parred coho salmon (Fig. 6B) was a good indicator of the minimum size required for the fish to continue in a growth trajectory leading to survival rather than reversion. Mahnken et al. (1982) called this size the critical size (not our definition), which increased throughout the year. In fresh water during the smoltification of coho salmon it was possible to predict survival after transfer to net-pens based on size and condition (Folmar and Dickhoff 1981). Mahnken et al. (1982) expanded this technique to salt water and used the percentage of the population in salt water smaller than the largest parred individual to predict saltwater mortality. The observations of Mahnken et al. (1982) identified a possible ecosystem and trophic relationship with growth that could result in death in the late fall and winter through osmoregulatory dysfunction. Although these studies were in net-pens, we propose that a similar response could occur for hatchery and wild fish during their first marine year.

We reanalyzed the data of Mahnken et al. (1982) to compare ration size with cumulate mortality (Fig.7). In all experiments there was an abrupt increase in the trend between mid-September and early October in the low-ration experiments, with abrupt increases in mortalities ranging from 13 to 17%. The ration size was also related to the cumulative number of stunted fish (Fig. 7), although there was not an abrupt change in the trend in late September. The rate of stunting (Fig. 7A) was greatest immediately following summer solstice. By winter solstice the commutative percentage of stunted fish was large (60-75%) in the populations fed the two lower rations, and low in fish fed the two high rations. There was a large, but delayed, increase in mortality (Fig. 7B) between September and November, with highest mortality in populations containing large numbers of stunted fish. Stunting continued to produce significant mortality at low-ration levels from November through termination of the experiment in March/April.

It is apparent from the feeding experiments that ration size is related to marine survival in controlled environments. The mechanism causing death is associated with physiological changes that are detectable, both as reversions to parr-like stages and stunting. As predation is not a factor in net-pens, it is clear that coho salmon require specific growth conditions in order to survive the changes that occur in their environment beginning in the fall of their first marine year.

Growth and Survival of Sockeye Salmon in the First Marine Year

One of the most extensive studies of marine growth and survival of Pacific salmon has been for Fraser River sockeye salmon (*O. nerka*). A number of investigators have produced some excellent data and analyses (Foerster 1954; Ricker 1962, 1982, 1995; Henderson and Cass 1991; Cox 1997). However, the relationship between size of smolts and total return remains unclear. Henderson and Cass (1991) showed that the larger smolts entering the ocean had higher marine survival. However, the years in which the average size of the smolts entering the ocean was larger did not produce higher average marine survivals; i.e., producing more larger smolts at the time of ocean entry would not necessarily produce a larger return. Henderson provided us with an extensive set of scale measurements of distance to the first ocean annulus (Fig. 8) for the 1949-1985 brood years, (1951-1987 years-to-sea). The scale measurements showed that the distance to the first marine annulus remained relatively constant, indicating that there could be a critical size of Chilko Lake sockeye salmon at the time of the first annulus formation in the ocean. Only the 1975 brood year, or the smolts that went to sea in 1977, were significantly different in size (much smaller). Despite this absence of a trend in sizes at the time of first annulus formation, others had observed a trend in adult sizes Ricker (1995) that changed with the regime change in 1976-1977 (Cox 1997).

According to our hypothesis, the explanation would be related to the two types of mortality in the first 12 months of ocean residence. The early marine mortality is size-related and predation-based, but the final regulation is size- or growth-related, which is a function of the carrying capacity. The numbers of juveniles reaching the critical size will vary, but the critical size would be more stable as it is a function of the intrinsic physiology of the sockeye salmon juvenile. The trend in adult size, however, is re-

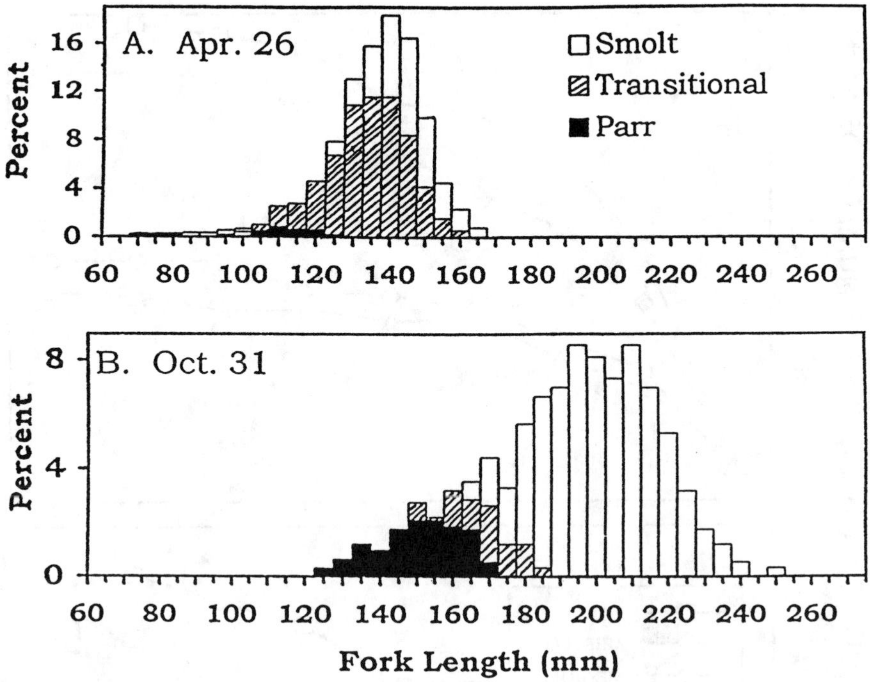

Figure 6. Fork length frequency distribution of coho salmon maintained in net-pens in salt water. (A) Early in the year (April 26) when coho salmon from 12 hatchery stocks were added, most fish were in a transitional state between parrs and smolts. (B) By October 31 few transitional fish were present. These fish would not survive winter conditions. The largest parred fish (i.e., 170 mm on October 31) was considered to be the size below which coho salmon would not survive after the critical period. From Mahnken et al. (1982).

lated to the available food after the carrying capacity was established. We emphasize that at this stage in the development of this hypothesis, critical size may also be a critical growth rate which would allow for more variation in the size at the time of the formation of the first marine annulus. Our hypothesis, therefore, appears to explain the observations on marine growth and survival of Chilko Lake sockeye salmon. One problem is the anomalous 1975 brood year. Why would the critical size be different only for 1 year out of 27 years? These juveniles entered the ocean in 1977 and exceptional survival was recorded for many species in this year (Beamish 1993). It is possible that the winter conditions were exceptionally favorable and smaller juveniles were able to survive the winter according to the mechanisms reported previously by Folmar et al. (1982).

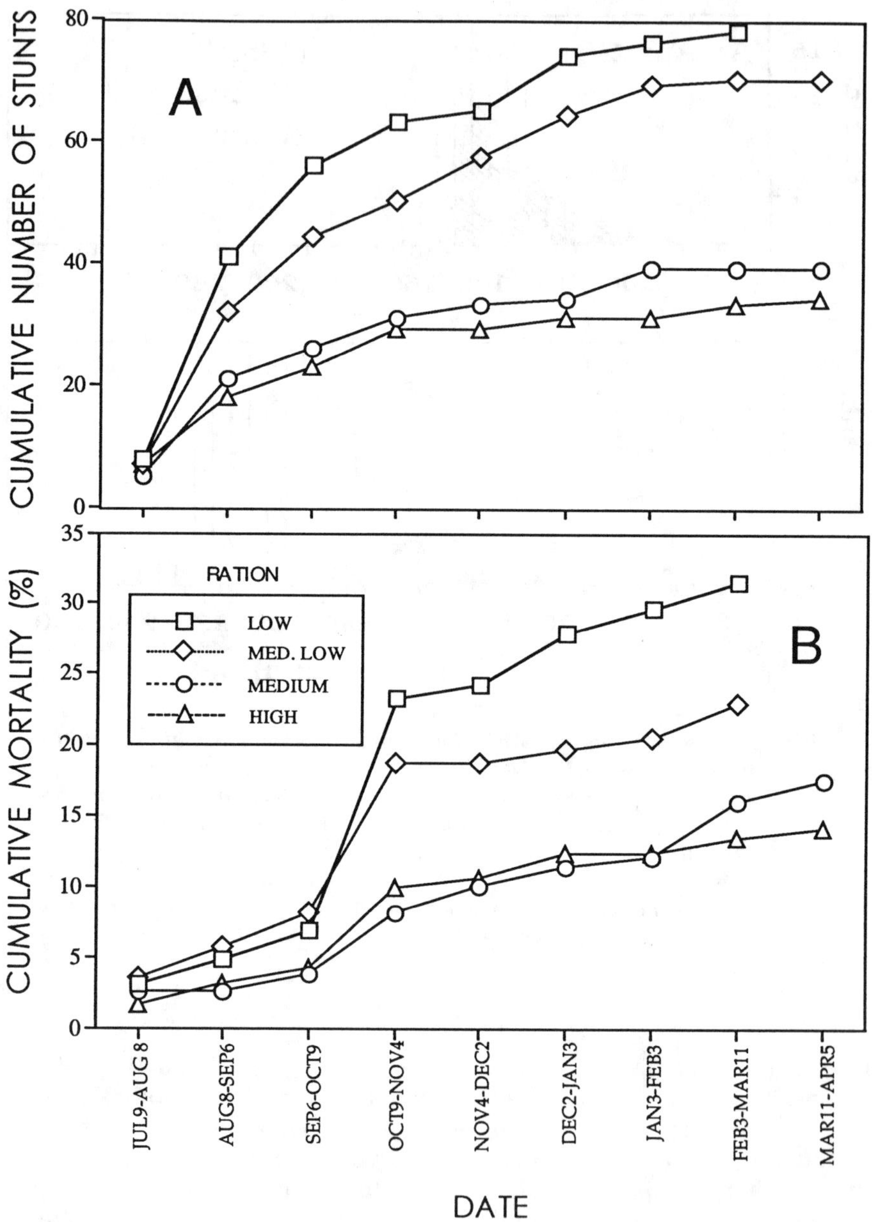

Figure 7. The data used in Fig. 5. (A) There was a larger number of stunted fish in the two experiments with the lowest rations that continued to increase in numbers throughout the study. (B) If aggregated by ration size there was a large increase in mortality between October 19 and November 4 for the two experiments with the lowest ration.

Climate Change and the Critical Period–Critical Size Hypothesis

There is an increasing amount of evidence indicating that Pacific salmon abundance is closely linked to long-term trends in climate (Beamish and Bouillon 1993, Hare and Francis 1995, Mantua et al. 1997). The linkage between the ocean environment and salmon abundance may be thought of in terms of carrying capacity. According to our hypothesis, changes in climate that affect ocean productivity and temperatures change carrying capacity by affecting the ability of juveniles to achieve a critical size in their first ocean year. According to this view, carrying capacity would be affected by the amount of food available in the summer months, competition for this food by other species, density-dependent competition from other juveniles of the same species, and physical conditions that affect the amount of energy required to find and metabolize food. It is apparent that a number of situations could interact to affect carrying capacity. For example, very high early marine predation-based mortality could reduce competition for food and improve summer growth and survival over the winter even if total food production was low. Warmer sea surface temperature can have the confounding effect of being associated with lower ocean productivity and higher metabolic rates, resulting in fewer juveniles surviving the winter. Adding more smolts to the ocean in a less favorable regime would probably reduce returns as competition for food and intrinsic population dynamics of a species may reduce the numbers of fish that achieve a critical size. Climate changes that alter the timing of food production could favor the survival of competing species. Again, the addition of more smolts should not be expected to improve returns, as it is the amount of food available to each fish in the summer that regulates returns and not the number of fish feeding. In periods of exceptional ocean productivity, as apparently occurred after the 1976-1977 regime shift (Beamish 1993), survivals tend to be better because competition for food is reduced. Perhaps the most dramatic impacts of climate impacts on carrying capacity are the changes in survival that occurred for coho salmon off Oregon and chinook salmon (*O. tshawytscha*) in the Strait of Georgia after the 1976-1977 climate change. Pearcy (1992) and Beamish et al. (1995a) both showed that abrupt declines in abundance and survival occurred even though more smolts were added into the ocean from hatcheries. In both cases, there was no evidence of large increases in predators and no explanation for the reduced returns. Clearly something changed, and we suggest that the change was increased winter mortalities resulting from reduced summer growth.

Density-Dependent and Density-Independent Effects

Traditional interpretations of the mechanisms regulating population size incorporate natural mortality that is density-dependent and density-independent. In simple language, density-dependent mortality is death from

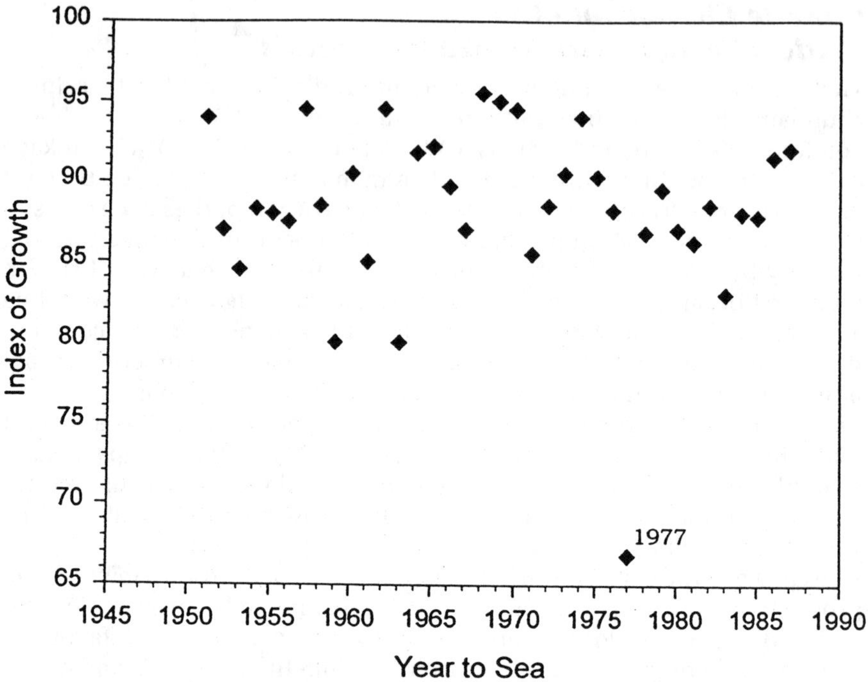

Figure 8. Index of growth of Chilko Lake sockeye salmon during the first year of marine growth. Index is the scale distance between the last freshwater circulus and first marine annulus, magnified 87 times, using 100 scales each year collected from the spawning ground.

overcrowding. The causes are seldom known. Density-independent mortality is mortality associated with physical changes in the environment and is independent of the number of animals in a population. In a classical Ricker type of population response, these two factors affect the shape of the stock and recruitment curve (Ricker 1975).

Our critical size–critical period hypothesis identifies the ocean environment as having a powerful impact on the final stock size. In the Strait of Georgia, the conditions were clearly unfavorable for chinook salmon immediately and for coho salmon later (Beamish et al. 1995a, 1995b). At issue is the relationship to density. Because mortality is measured as a percentage, some biologists feel that if mortality is independent of density, the reduced percentage of survival caused by a change in the environment can be mitigated by producing more juveniles. Our hypothesis considers that the environmental effect is most important in the conditioning of juveniles for late fall and overwinter survival and is related to the density of competitors for the common food resource, including other fish in the population.

At some theoretical level of smolt abundance, the population must stop increasing and the mechanisms must involve density-dependent interactions within the population, otherwise the populations would not be limited in abundance. Our hypothesis accounts for this theoretical control by proposing that competition for food in the summer is both external and internal to a population. As density increases, the internal processes become increasingly active.

Discussion

Ecosystem management of our fisheries is a logical progression from single-species management. There is growing concern about overfishing, a developing appreciation of biodiversity, and an understanding that we must protect our marine plants and animals. This interest has fostered a desire to broaden our protection of aquatic resources and the concept of stewardship of whole ecosystems has become a priority of many management agencies. Ecosystem management is not a new concept but it has not advanced very far because the urgency of single-species management overwhelmed the work required to understand associations with other species and with the environment. It is now time to establish a practical set of guidelines and begin to manage on a more holistic basis. There is a wealth of literature on the concept so it remains to find ways to apply the theories. Most importantly it is *time to begin the process,* which will be experimental in the beginning.

We recognize that it has been difficult to model marine ecosystems. We also recognize that it will not be possible to maintain annual sampling programs for all the appropriate parameters for all of the key species in the various ecosystems. However, we believe it is important to maintain a process that looks at general or larger-scale relationships. It is easy to be pessimistic about the possibility of managing ecosystems. It is equally easy to be optimistic about the information that can be obtained by participating in processes that practice this approach. Several authors have suggested that ecosystem management depends upon the good effort, good judgment and good faith of those involved.

Ecosystem management is a philosophical change in our management approach. Ecosystem management is an exercise in developing realistic expectations for our fisheries. Realistic expectations are a function of improved understanding of how the system works and ensuring that the public are informed about what we know and what we do not know. Ecosystem management is an exercise in long-term, precautionary thinking. It is acceptable not to know things. It is not acceptable to place self-interest ahead of a resource. In some respects we may have to start over again. Native species and natural ecosystems need to be recognized and protected.

We do not have to defend our inability to understand the impacts of everything humans want to do to ecosystems. We do need to ensure that we are not preventing species from replenishing themselves. Be careful about promising stability. We need to recognize that extreme events may be rare but they are a normal factor in the natural selection process. One of the most difficult aspects of ecosystem management is communication among scientists. Experts in the private or public sectors are going to have to find ways to communicate their knowledge so that a collective wisdom is provided to managers. You have to be able to talk to your colleagues. You do not have to like them: just be able to work together.

We suggest that our hypothesis of the natural regulation of salmon abundance can be used as a simple example of ecosystem management. Abundance of salmon is more than a function of the abundance of spawning females. Climate is important. Our hypothesis identifies food limitation as the reason for the mortality that occurs after the critical period. Food limitation or availability may appear to be an unlikely factor controlling mortality because it can be shown that there is a large amount of food in the ocean relative to the amount eaten by some salmon (Walters et al. 1978). However, we propose that availability of food is not measured by total abundance. Laevastu and Favorite (1977) consider that partial starvation is an important component of marine mortality and this is consistent with our hypothesis of programmed death, if the critical size is not achieved by the critical period. This concept of minimal growth rates required to maintain a genetically determined rate of growth or perish is a familiar concept for larval fish survival (Iles 1980).

Some biologists believe that studies of the mechanisms that regulate salmon abundance in the ocean are not useful in management because we cannot control the ocean. First, it certainly is debatable if our perceived "control" is real. Second, if our hypothesis is correct, the value of understanding the mechanisms that produce final stock sizes is the wise use of funds and in the careful identification of expectations of yield. If our hypothesis is correct, it is clear that we can neither expect to change the carrying capacity of the ocean, nor should we want to attempt such an intervention into the marine ecosystem with our rather dismal understanding of how they work. We may expect that a better understanding of the mechanisms regulating carrying capacity will provide a better understanding of both how many fish should spawn and the importance of maintaining diversity within the life history types. We would no longer equate the straightforward production of smolts with the expectation of ever-increasing returns until some desirable historic catch is achieved. We also would understand why it is so difficult to identify stock and recruitment relationships using recruitment estimates determined before smolts enter the ocean. With such a complex set of factors in the ocean that regulate the final returns, the process of estimating the returns before any marine mortality has occurred is particularly difficult. Improving our understanding of the mechanisms that regulate abundance of salmon in fresh water and in salt water improves the credibility of management for those that either harvest salmon or value knowing that stocks are healthy. Improving our understanding of ecosystems improves everyone's appreciation of the impact of our own intervention into our ecosystems.

References

Bartmann, R.M. 1984. Estimating mule deer winter mortality in Colorado. J. Wildl. Manage. 48:262-266.

Beamish, R.J. 1993. Climate and exceptional fish production off the west coast of North America. Can. J. Fish. Aquat. Sci. 50:2270-2291.

Beamish, R.J., and D.R. Bouillon, 1993. Pacific salmon production trends in relation to climate. Can. J. Fish. Aquat. Sci. 50:1002-1016.

Beamish, R.J., and M. Folkes. 1998. Recent changes in the marine distribution of juvenile chum salmon off Canada. Proceedings of the Conference on Pacific Salmon. North Pacific Anadromous Fish Commission Bull. No. 1:443-453. In press.

Beamish, R.J., M. Folkes, and C. Mahnken. 1998b. Intra-annual shifts in abundance of coho, chinook, and chum salmon in Puget Sound in 1997. Proc. Puget Sound Research 1998. In press.

Beamish, R.J., C. Mahnken, and C.M. Neville. 1997. Hatchery and wild production of Pacific salmon in relation to large scale, natural shifts in the productivity of the marine environment. ICES J. Mar. Sci. 54:1200-1215.

Beamish, R.J., R. Sweeting, and Z. Zhang. 1998c. Estimating the percentage of hatchery-reared juvenile coho salmon in the Strait of Georgia in 1997. PSARC working paper S98-8, Pacific Biological Station, Nanaimo, B.C., Canada. Cited with permission.

Beamish, R.J., C. Neville, J. Rice, and Z. Zhang. 1995b. Factors affecting the marine survival of coho salmon in the Strait of Georgia. Pacific Stock Assessment Review Committee (PSARC) Working Paper S95-4. Pacific Biological Station, Nanaimo, B.C., Canada. Cited with permission.

Beamish, R.J., J. King, D. Noakes, G.A. McFarlane, and R. Sweeting. 1998a. Evidence of a new regime starting in 1995 and the relation to Pacific salmon catches. Noth Pacific Anadromous Fish Commission Doc. No. 321. 21 pp.

Beamish, R.J., B.E. Riddell, C.M. Neville, B.L. Thomson, and Z. Zhang. 1995a. Declines in chinook salmon catches in the Strait of Georgia in relation to shifts in the marine environment. Fish. Oceanogr. 4:243-256.

Beamish, R.J., D. Noakes, G.A. McFarlane, L. Klyashtorin, V.V. Ivanov, and V. Kurashov. 1999. The regime concept and natural trends in population of Pacific salmon. Can. J. Fish. Aquat. Sci. 56. In press.

Colinvaux, P. 1978. Why big fierce animals are rare. Princeton University Press. 256 pp.

Cox, S.P. 1997. The effects of physical and biological oceanographic factors on marine growth of Fraser River sockeye salmon. M.Sc. thesis, University of British Columbia, Vancouver, B.C.

Foerster, R.E. 1954. On the relation of adult sockeye salmon (*Oncorhynchus nerka*) returns to know smolt seaward migrations. J. Fish Res. Board Can. 11:339-350.

Folmar, L.C., and W.W. Dickhoff. 1981. Evaluation of some physiological parameters as predictive indices of smoltification. Aquaculture 23:309-324.

Folmar, L.C., W.W. Dickhoff, C.V.W. Mahnken, and F.W. Waknitz. 1982. Stunting and parr-reversion during smoltification of coho salmon (*Oncorhynchus kisutch*). Aquaculture 28:91-104.

Gunn, A., F.L. Miller, and B. McLean. 1989. Evidence for the possible causes of increased mortality of bull musk-oxen during severe winter. Can. J. Zool. 67:1106-1111.

Hare, S.R., and R.C. Francis. 1995. Climate change and salmon production in the northeast Pacific Ocean. In: R.J. Beamish (ed.), Climate change and northern fish populations. Can. Spec. Publ. Fish. Aquat. Sci. 121:357-372.

Henderson, M.A., and A.J. Cass. 1991. Effect of smolt size on smolt-to-adult survival for Chilko Lake sockeye salmon (*Oncorhynchus nerka*). Can. J. Fish. Aquat. Sci. 48:988-994.

Hjort, J. 1914. Fluctuations in the great fisheries of northern Europe viewed in the light of biological research. Rapp. P.-V. Reun. Cons. Int. Explor. Mer. 20:1-228.

Iles, T.D. 1980. Environmental pressure and intra- and inter-year-class competition as determinants of year-class size. In: A. Saville (ed.), The assessment and management of pelagic fish stocks. A symposium held in Aberdeen, 3-7 July 1978. International Council for the Exploration of the Sea, Copenhagen, pp. 315-331.

Laevastu, T., and F. Favorite 1977. Minimum sustainable biomasses of marine ecological groups off central and northern California, Oregon, Washington, and Vancouver Island coasts. National Marine Fisheries Service, Washington, D.C.

Mahnken, C., E. Prentice, W. Waknitz, G. Monan, C. Sims, and J. Williams. 1982. The application of recent smoltification research to public hatchery releases: An assessment of size/time requirements for Columbia River hatchery coho salmon (*Oncorhynchus kisutch*). Aquaculture 28:251-268.

Mantua, N.J., S.R. Hare, Y. Zhang, J.M. Wallace, and R.C. Francis. 1997. A Pacific interdecadal climate oscillation with impacts on salmon production. Bull. Am. Meteorol. Soc. 78:1069-1079.

National Research Council. 1996. Upstream: Salmon and society in the Pacific Northwest. Committee on Protection and Management of Pacific Northwest Anadromous Salmonids. National Academy Press, Washington, D.C. 452 pp.

Pearcy, W.G. 1992. Ocean ecology of North Pacific salmonids. University of Washington Press, Seattle.

Ricker, W.E. 1945. A method of estimating minimum size limits for obtaining maximum yield. Copeia 1945(2):84-94.

Ricker, W.E. 1962. Comparison of ocean growth and mortality of sockeye salmon during their last two years. J. Fish. Res. Board Can. 19:531-547.

Ricker, W.E. 1975. Computation and interpretation of biological statistics of fish populations. Bull. Fish. Res. Board Can. 191. 382 pp.

Ricker, W.E. 1982. Size and age of sockeye salmon (*Oncorhynchus nerka*) in relation to environmental factors and the fishery. Can. Tech. Rep. Fish. Aquat. Sci. 1115. 117 pp.

Ricker, W.E. 1995. Trends in the average size of Pacific salmon in Canadian catches. In: R.J. Beamish (ed.), Climate change and northern fish populations. Can. Spec. Publ. Fish. Aquat. Sci. 121:593-602.

Shuntov, V.P., A.F. Volkov, and A.Y. Efimkin. 1988. Composition and present state of pelagic fish communities in the western Bering Sea. Biol. Morya 2:3-41. (In Russian.)

Shuntov, V.P., V.I. Radchenkov V.V. Lapko, and Y.N. Poltev. 1993. Distribution of salmons in the west part of the Bering Sea and adjacent waters of the Pacific Ocean during anadromous migration. Vopr. Ikhtiol. 33(3):337-347. (In Russian.)

van Loon, H., and J.C. Rogers. 1978. The seesaw in winter temperatures between Greenland and Northern Europe. Part 1: General description. Monthly Weather Rev. 106:296-310.

Walters, C.J., R. Hilborn, R.M. Peterman, and M.J. Staley. 1978. Model for examining early ocean limitation of Pacific salmon production. J. Fish. Res. Board Can. 35:1303-1305.

Watanabe, M., and T. Nitta. 1998. Abrupt shifts in the atmospheric circulation and associated decadal climate variations in the Northern Hemisphere winter: A diagnostic study. J. Climate. In press.

Climate Variation, Ecosystem Dynamics, and Fisheries Management in the Northwestern Hawaiian Islands

Jeffrey J. Polovina
National Marine Fisheries Service, Honolulu Laboratory, Honolulu, Hawaii

Wayne R. Haight
University of Hawaii, Joint Institute for Marine and Atmospheric Research, Honolulu, Hawaii

Abstract

A dramatic ecosystem shift from high to low carrying capacity occurred in the late 1980s in the northern portion of the Hawaiian Archipelago. This shift was observed for a range of trophic levels and species including seabirds, monk seals, spiny lobsters, and reef fishes. Concurrent with the ecosystem shift were physical changes associated with weakening of the Aleutian Low Pressure System and the Subtropical Counter Current. A trap fishery operating at Maro Reef rapidly depleted the spiny lobster population after the ecosystem shift. At an adjacent bank, Laysan Island, closed to commercial lobster fishing, the spiny lobster population declined much more gradually but ultimately was depleted as well. Harvest from the lobster fishery during much of the 1990s averaged about 25% of the harvest in the 1980s. Fishermen adjusted to the reduced harvest by moving into the pelagic longline fishery which was apparently not impacted by the ecosystem shift.

Introduction

The Northwestern Hawaiian Islands (NWHI) comprise a 1,500 km chain of islands, atolls, coral reefs, and banks in the northern portion of the Hawaiian Archipelago (Fig. 1). The region has no permanent human population but supports a rich marine ecosystem. In the mid-1970s a comprehensive research effort began to study various components of this ecosystem which

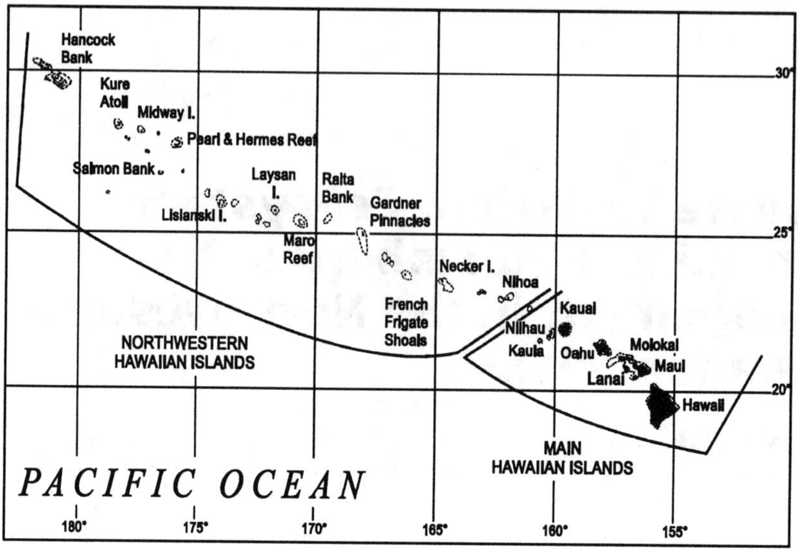

Figure 1. The Hawaiian Archipelago.

led, in subsequent years, to standardized time series of indices of ecosystem productivity and population abundance for some components of this ecosystem. These indices and other survey data will be used to describe the sudden ecosystem shift which occurred concurrently with a climate shift in the late 1980s. One of the resources we will examine is the spiny lobster which is targeted by a commercial trap fishery and hence provides an example of interactions between both fishery and climate impacts.

The Climate Shift

A North Pacific climate event characterized by an intensification and southward shift in the Aleutian Low Pressure System occurred during 1977-1988 (Trenberth and Hurrell 1994). This climate shift impacted the northern portion of the Hawaiian Archipelago, probably through a southward shift of winter storm tracks and increased wind speed and stress, resulting in a deepening of the mixed layer depths by 30-80% during the winter (Polovina et al. 1994, 1995). An ecosystem model of nitrate, phytoplankton, and zooplankton driven by mixed layer depth was used to evaluate the impacts to the plankton ecosystem from the lower light but greater nutrient levels caused by the deeper mixed layer depth during the 1977-1988 period. The model suggested that due to nutrient limitation in the NWHI, the deeper vertical mixing observed during this climate shift could increase both phytoplankton and zooplankton production by 50% (Polovina et al. 1995).

Changes in vertical mixing are only one physical response; another is a change in horizontal transport. Data from tide gauges along the archipelago suggest that the Subtropical Counter Current (SCC), which crosses the northern portion of the Hawaiian Archipelago, was stronger during 1977-1985 and weaker subsequently (Polovina and Mitchum 1992). A positive correlation between an index of the strength of the SCC and recruitment to the fishery 4 years later at Maro Reef has been documented (Polovina and Mitchum 1992). It is not known whether the stronger SCC was the result of the more intense Aleutian Low or just coincidental, but both features were strong from 1977 until the mid-1980s and then shifted to weaker states which have persisted at least through the mid-1990s. Concurrent with this shift in physical features, a drop in ecosystem carrying capacity in the NWHI was observed, which we believe resulted from this climate shift.

The Ecosystem Shift

Time series of indices of productivity for spiny lobster *(Panulirus marginatus)*, the Hawaiian monk seal *(Monachus schauinslandi)*, and two species of seabirds, the red-footed booby *(Sula sula)* and the red-tailed tropicbird *(Phaethon ruricauda)*, all show declines between the 1980s and 1990s (Fig. 2). In the case of spiny lobsters, indices of recruitment of 3-year-olds at two banks, Necker Island and Maro Reef, are measured as the catch rates from an annual standardized research survey. At Maro Reef 3-year-old recruitment completely collapsed in the late 1980s and has not shown any recovery since, while at Necker Island, 750 km southeast of Maro, recruitment dropped about 50% in the late 1980s but has shown some subsequent recovery (Fig. 2). Monk seal pup survival, measured as the fraction of pups born which survive to age 2, sharply declined at French Frigate Shoals in the late 1980s with no subsequent recovery, while at Laysan Island, north of French Frigate Shoals, the decline was more gradual (Fig. 2). For both species of seabirds at French Frigate Shoals reproductive success, measured as the fraction of eggs laid which result in fledgling chicks, shows a sharp drop of about 50% between the early 1980s and the period since the mid-1980s (Fig. 2). When lobster and monk seal time series are shifted backwards 2 or 3 years to account for the appropriate time lags (lobster data are for 3-year-olds, monk seal data are based on 2-year-olds), temporal coherence between the time series is seen. While the causes of these declines have not been completely determined, in the case of monk seals and seabirds field observations suggest reduced food availability during their winter fledgling season is a likely factor. A reduction in food availability has been cited as the reason for declines in reproductive success in other seabird populations (Baird 1992). In the NWHI in the 1990s, the higher egg mortality appears to be the result of longer exposure to the sun and higher chick mortality to be the result of starvation, all of which are consistent with lower food availability for adults

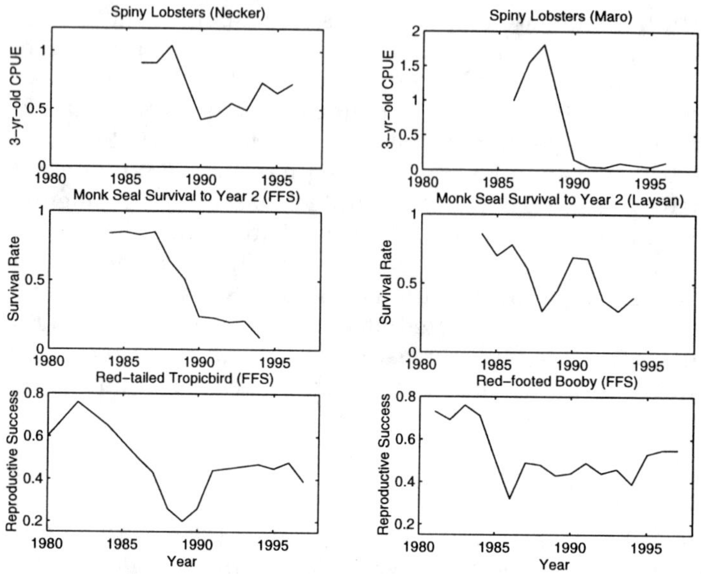

Figure 2. *Time series of indices of productivity for seabirds, monk seals, and spiny lobsters in the Northwestern Hawaiian Islands: Necker, Maro, and Laysan islands and French Frigate Shoals.*

resulting in more time spent away from the nest foraging and less food for the chicks. These seabirds forage primarily on flying fishes and squids which are linked to oceanic rather than reef productivity.

The decline in monk seal pup survival has also been inferred to be the result of a decline in prey, based on widespread observations of emaciated pups and the lack of other identifiable causes including direct human impacts and disease (Gilmartin and Ragen 1992). The diet of monk seal pups is not known but generally the areas and depths where they are seen foraging suggest they forage on prey in the benthic coral reef and shelf ecosystem. Like the seabirds, winter appears most critical to monk seal pup survival since the pups first forage then.

The reef fishes are a major component of the NWHI ecosystem, and given the remote location of the region, are not impacted by human exploitation. While a continuous time series of reef fish densities is not available, comparisons between densities at the same reefs at French Frigate Shoals and Midway Atoll between the early 1980s and early 1990s show that densities in the early 1990s are about one-third lower than in the early 1980s, providing still another indication of the drop in carrying capacity seen in other elements of the ecosystem (DeMartini et al. 1996).

Climate and Fishery Interactions

The catch rate of 3-year-old spiny lobsters at Maro Reef, obtained from standardized research cruises and used as an index of recruitment to the fishery, showed a dramatic collapse in 1990 (Fig. 2). Age composition of the trap catches from these research cruises shows an abundance of 3-year-olds along with some older and younger lobsters in 1987 and 1988, but since 1990 and persisting through the present, the spiny lobster population at Maro Reef appears to be severely depleted (Fig. 3). Unfortunately, 1989 age-frequency data which might show a transition between 1988 and 1990 levels are not available due to the lack of a research cruise in 1989.

Maro Reef and Necker Island were the two primary fishing grounds for the lobster trap fishery during the 1980s, and annual harvest at Maro Reef during that period ranged from about 200,000 to 450,000 lobsters (Fig. 4). The picture from Figs. 2-4 suggests a collapse in recruitment of 3-year-old spiny lobsters to the fishery in 1989, as a result of a sudden weakening of the SCC during the mid-1980s (Polovina and Mitchum 1992). The fishery was largely harvesting a single age group, 3-year-olds, and with the collapse in recruitment to the fishery in 1989, the remaining population of older lobsters was quickly depleted by the fishery.

Laysan Island, about 110 km north of Maro Reef, has been closed to commercial lobster trapping since the beginning of the fishery to protect the large population of monk seals on Laysan from becoming entangled in lobster traps and line. Limited research trapping at Laysan shows an absence of 2- and 3-year-olds in 1991 compared to 1977 and 1986, suggesting that the recruitment collapse which impacted Maro Reef also impacted Laysan by 1991 (Fig. 5). However, apparently due to the absence of a fishery at Laysan, unlike at Maro Reef, there was still a significant population of older animals at Laysan. Unfortunately, by 1996, even in the absence of a fishery, the lobster population at Laysan was severely depleted, apparently the result of the collapse of recruitment in the mid-1980s and the subsequent impact of natural mortality (Fig. 5).

Response of Fisheries Management

Since 1983, the fishery has been regulated under a fishery management plan by the Western Pacific Regional Fishery Management Council, primarily with a minimum size set slightly above the size at onset of maturity, escape vents in traps to allow escapement of sublegal lobsters, and some areas closed to fishing. These closed areas, although not specifically designed to protect lobsters but rather to reduce fishery interactions with protected species, did close perhaps as much as 10% of the lobster habitat to trapping, including all of Laysan Island and all portions of banks shallower than about 20 m.

The sudden collapse of the spiny lobster population in 1990 caught fisheries managers by surprise. In 1989 a dynamic production model was

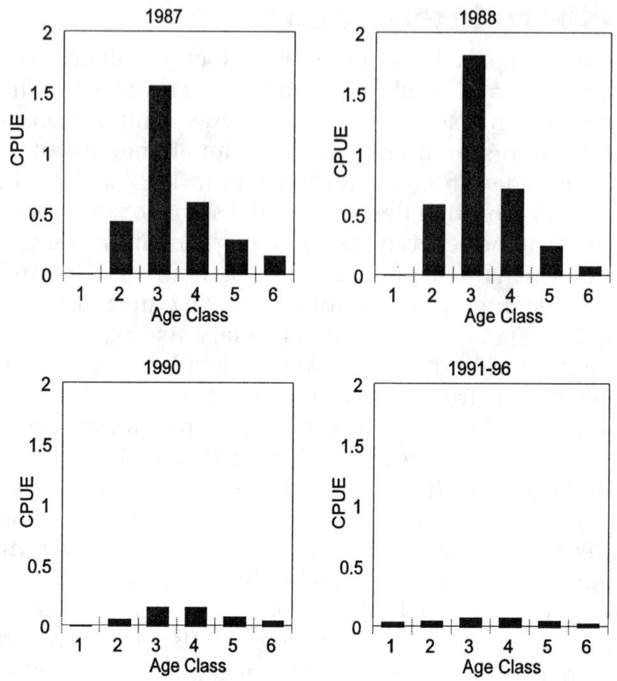

Figure 3. Age-frequency of spiny lobsters from Maro Reef research cruises.

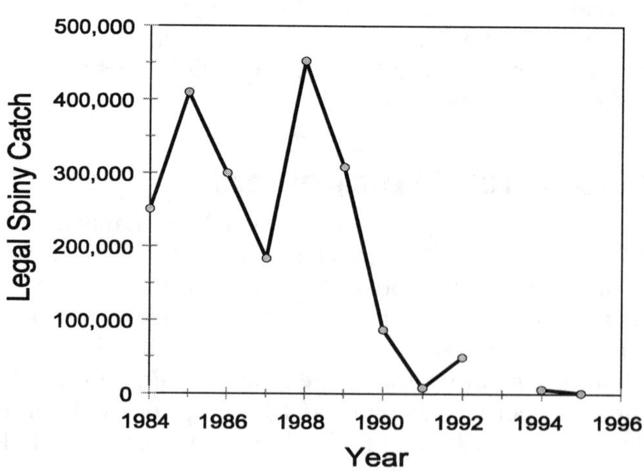

Figure 4. Commercial landings of spiny lobsters from Maro Reef.

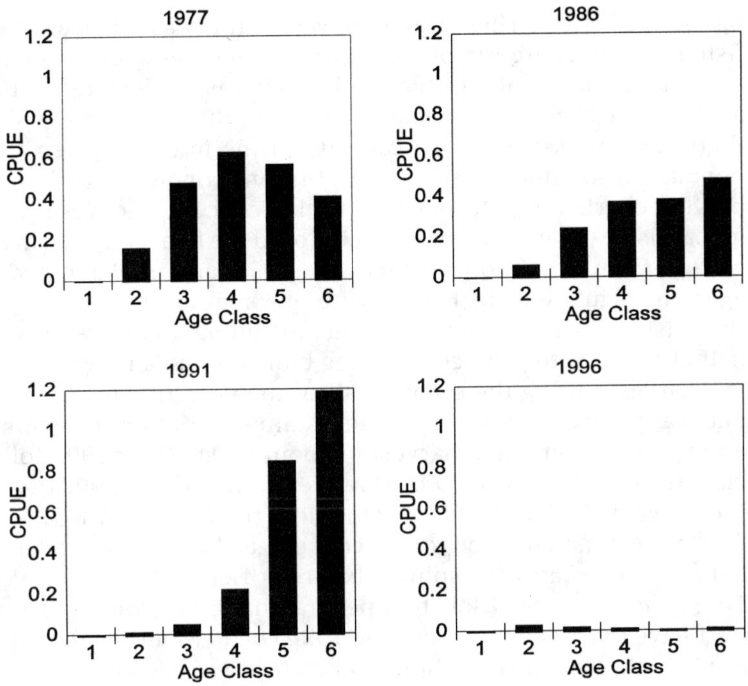

Figure 5. Age-frequency of spiny lobsters from Laysan Island research cruises.

used with historical commercial fishery data to estimate a maximum sustainable yield (MSY) for the NWHI fishery (slipper and spiny lobsters combined) of 1 million lobsters, with an effort of 1 million trap-hauls (Polovina 1990). The 1989 fishery catch and effort of about 1.2 million lobsters and 1.1 million trap-hauls suggested the fishery was at about MSY levels. However, by 1990 it was obvious that the population of spiny lobsters at Maro Reef and other northern banks had collapsed. The report on status of the stocks in 1990 (Polovina 1991) noted that the spiny lobster spawning biomass was estimated at 22% of the pre-fishery level and recommended both a seasonal closure from January through August to protect spawning biomass, and a reduction of annual effort to 200,000 trap-hauls.

Catch and effort were not restricted during 1990, and about 775,000 lobsters were caught with 1.2 million trap-hauls. The collapse of a portion of the spiny lobster population in 1990 without any reduction in effort, meant that the performance of the fishery measured as the catch rate (the ratio of total catch divided by total effort) dropped 40% between 1989 and 1990. Prior to 1990, industry was not generally supportive of restrictions on catch or effort. Their argument was that it was unnecessary since the fishing grounds are remote and effort would cease once catch rates fell below an economically viable level which was sufficiently high to protect against recruitment overfishing. However, the 1990 season showed that there was excessive fishing capacity in the industry given the reduced population size and raised concern that an economic threshold might not prevent overfishing. Responding to this concern, the Western Pacific Regional Fishery Management Council closed the fishery under an emergency closure from May 8 to November 11, 1991, and implemented new management regulations in 1992. These new measures added regulations to the existing management measures, including a closure during the first half of the year to protect spawning biomass and a harvest quota to control harvest during the second half of the year. Initially the new harvest quota approach resulted in high interannual variation: a combined slipper and spiny lobster NWHI harvest of about 850,000 for 1992 followed by a closure of the fishery in 1993. However, from 1994 to 1997 quotas were in the range 101,000-310,000, well below the long-term MSY of 1 million lobsters estimated prior to the collapse at Maro Reef. Under this lower harvest, catch rates for spiny lobsters at Necker Island recovered about 50%, from 0.5 to 0.75 lobsters per trap-haul, but spiny lobsters at Maro Reef have not recovered since the 1990 collapse, although there has been a substantial increase in slipper lobsters at Maro Reef.

Recent research suggests that the source of spiny lobster larvae at Maro Reef is the spawning population at Maro Reef and hence the recovery of the spiny lobster population at Maro Reef may be a long, slow process based on rebuilding the resident spawning biomass (Polovina et al. 1999). The industry has been generally supportive of the management action reducing harvest levels. One important reason for this support was that all 15 permit holders were also able to obtain permits for the pelagic longline or bottomfish fishery, so their economic condition was not tied solely to the lobster fishery.

In the absence of the evidence of a climate and ecosystem shift, management might have a goal of rebuilding the fishery to achieve an MSY of 1 million lobsters. This would require closing the fishery for an extended period, but it is unlikely this recovery is achievable under the current less productive regime. Current management action recognizes there is a much lower MSY under the new physical and ecosystem regime.

Summary

A shift in climate occurred in about 1989 in the central North Pacific characterized by a weakening of the SCC and Aleutian Low Pressure System. The NWHI ecosystem shifted, apparently triggered by the climate shift, from a high carrying capacity to a low one. The shift occurred rapidly and nearly simultaneously over a range of trophic levels. There are significant spatial patterns; for example, the drop in spiny lobster recruitment was greatest in the northern portion of its range while the drop in pup survival for monk seals was greatest at the southern end of its range. It is likely that there is not one physical cause for this ecosystem shift. In the case of lobster, evidence suggests that a reduction in lobster larval retention at Maro Reef occurred with the weakening of the SCC. In the case of seabird reproductive success and monk seal pup survival, evidence suggests a decline in prey arising from lower primary productivity attributed to weaker vertical mixing.

At Maro Reef, fishing prior to the drop in productivity reduced that population age structure, making it less resilient to interannual recruitment variation. After the recruitment collapse, a few years of unrestricted fishing at Maro Reef quickly eliminated most of the remaining spawning biomass.These results illustrate interactions between fishing and climate shifts: fishing alters the age-structure and can rapidly deplete a stock when the ecosystem shifts to a regime of lower carrying capacity. At Laysan, protection from fishing preserved more of the age structure but after a decade of poor recruitment, the spawning biomass was apparently depleted from natural mortality. Thus marine reserves may not always provide the protection many proponents assume.

One consequence of the collapse in the spiny lobster fishery at Maro Reef is that it raised concern about the ability of open access management to protect the spiny lobster population and provided the impetus to managers and industry to move to a limited access fishery with harvest controls. The ability of management to reduce catch and effort to about one-fourth the level prior to the stock collapse at Maro Reef was greatly facilitated by the permit holders who all had permits for the pelagic longline fishery and were able to move into another fishery not impacted by the ecosystem shift. Therefore, one step fisheries management can take to prepare for the impacts of climate variation is to encourage fishermen to diversify between fisheries that are not likely to fluctuate together. Unfortunately, the current trend toward limited access fisheries may restrict the ability of fishermen to diversify.

It has been a decade since the last climate and ecosystem shift. An issue of great interest to all those involved with the NWHI ecosystem is whether these states alternate on a decadal cycle so we can look forward to a return to a more productive period in the near future or whether these shifts are rare and irregular events.

References

Baird, P.H. 1992. Seabirds as indicators of the oceanic environment. In: P. Grifman and S. Yoder (eds.), Perspectives on the marine environment. Sea Grant Program, University of Southern California, Los Angeles, pp. 91-104.

DeMartini, E.E., F.A. Parrish, and J.D. Parrish. 1996. Interdecadal change in reef fish populations at French Frigate Shoals and Midway Atoll, Northwestern Hawaiian Islands: Statistical power in retrospect. Bull. Mar. Sci. 58(3):804-825.

Gilmartin, W.G., and T.J. Ragen. 1992. Collection of juvenile monk seals at French Frigate Shoals for rehabilitation and release at Midway Islands. SWFSC unpublished report. NMFS, Southwest Fisheries Science Center, Honolulu, HI 96822-2396.

Polovina, J.J. 1990. Status of lobster stocks in the Northwestern Hawaiian Islands, 1989. SWFSC Admin. Rep. H-90-02. NMFS, Southwest Fisheries Science Center, Honolulu, HI 96822-2396. 11 pp.

Polovina, J.J. 1991. Status of lobster stocks in the Northwestern Hawaiian Islands, 1990. SWFSC Admin. Rep. H-91-04. NMFS, Southwest Fisheries Science Center, Honolulu, HI 96822-2396. 16 pp.

Polovina, J.J., and G.T. Mitchum. 1992. Variability in spiny lobster, *Panulirus marginatus*, recruitment and sea level in the Northwestern Hawaiian Islands. Fish. Bull., U.S. 90:83-493.

Polovina, J.J., P. Kleiber, and D.R. Kobayashi. 1999. Application of TOPEX-POSEIDON satellite altimetry to simulate transport dynamics of larvae of spiny lobster, *Panulirus marginatus*, in the Northwestern Hawaiian Islands, 1993-1996. Fish. Bull., U.S. 97:132-143.

Polovina, J.J., G.T. Mitchum, and G.T. Evans. 1995. Decadal and basin-scale variation in mixed layer depth and the impact on biological production in the central and North Pacific, 1960-88. Deep-Sea Res. 42(10):1701-1716.

Polovina, J.J., G.T. Mitchum, N.E. Graham, M.P. Craig, E.E. DeMartini, and E.N. Flint. 1994. Physical and biological consequences of a climate event in the central North Pacific. Fish. Oceanogr. 3:5-21.

Trenberth, K.E., and J.W. Hurrell. 1994. Decadal atmosphere-ocean variations in the Pacific. Climate Dynamics 9:303-319.

Spatial and Temporal Patterns of Covariation in Components of Recruitment of British Columbia and Alaska Sockeye Salmon

Randall M. Peterman, Brian J. Pyper, Michael F. Lapointe, and Milo D. Adkison
Simon Fraser University, School of Resource and Environmental Management, Burnaby, British Columbia, Canada

Carl J. Walters
University of British Columbia, Fisheries Centre, Vancouver, British Columbia, Canada

Extended Abstract

Sockeye salmon *(Oncorhynchus nerka)* populations typically show large year-to-year variation in survival rates, body size at maturity, and age at maturity. To better manage these stocks, it is important to understand the spatial and temporal scales over which environmental processes drive this variation in each component of recruitment. We used a multi-stock comparison to identify these spatial and temporal characteristics of variability using data from the late 1940s through brood year 1992 returns for 29 sockeye stocks from a wide geographical area across British Columbia and Alaska: 16 Fraser River stocks (southern British Columbia), the Skeena and Nass River stocks (central British Columbia), the Copper River and Upper Cook Inlet stocks (central Alaska), and 9 Bristol Bay stocks (western Alaska). These stocks overlap to varying degrees during their marine life stage.

Specifically, we examined patterns of covariation among these stocks using time-series indices of (1) survival rate (residuals from the best-fit

Current address for M.F. Lapointe is Pacific Salmon Commission, 600-1155 Robson St., Vancouver, B.C., Canada, V6E 1B5. Current address for M.D. Adkison is School of Fisheries and Ocean Sciences, University of Alaska, 11120 Glacier Highway, Juneau, AK 99801.

stock-recruitment curve); (2) length at a given adult age; and (3) mean age at maturity. For each component of recruitment, we computed pairwise correlation coefficients among the various stocks. In addition, we used (step) intervention analysis to quantify changes in survival-rate indices coinciding with the regime shift in climatic and oceanographic conditions that occurred in the North Pacific in the mid-1970s (Graham 1994). For indices of body size and mean age, we used principal components analysis (PCA) to further identify the temporal patterns of variability shared among stocks.

For indices of survival rate, we found strong positive covariation among the nine Bristol Bay stocks (all 36 correlations were positive; average $r = 0.44$), and weaker but predominantly positive covariation among the 16 Fraser River sockeye stocks (102 of the 120 correlations were positive; average $r = 0.16$). However, there was no evidence of positive covariation between Bristol Bay and Fraser River stocks (average $r = -0.05$) or with stocks of other regions in British Columbia and Alaska. These results suggest that regional-scale environmental processes influence survival rates of sockeye stocks within each region (i.e., Bristol Bay or Fraser River), and that these processes are distinct for each of these two regions. Furthermore, intervention analysis showed that there were large and persistent increases in the mean productivity of all nine Bristol Bay stocks coinciding with the mid-1970s climate shift. In contrast, survival rates of the 16 Fraser River stocks did not show a consistent response to the climate shift: some increased, some decreased, and many remained largely unchanged. Thus, these results and others presented in Peterman et al. (1998) and Adkison et al. (1996) suggest that environmental processes driving variability among survival rates of Fraser River sockeye stocks differ from those processes affecting Bristol Bay sockeye, and that only the latter were strongly influenced by the mid-1970s climate shift.

For body size at maturity, we found widespread positive covariation across all ages and stocks from British Columbia and Alaska (2,335 of the 2,556 correlations were positive; average $r = 0.37$). There was evidence of both distinct, within-region patterns of covariation, as well as positive covariation between regions. For example, for lengths of age 1.2 fish, there was a strong tendency toward positive covariation among Bristol Bay stocks (all 28 correlations were positive; average $r = 0.50$) and among Fraser River stocks (all 190 correlations were positive; average $r = 0.60$), with weaker but predominantly positive covariation between these two regions (136 of 160 correlations were positive; average $r = 0.23$). The stronger covariation within regions indicates that regional-scale processes are important determinants of final body size of sockeye salmon. In addition, the positive covariation between regions suggests that, to some extent, either fish of different regions share similar distributions during periods critical to growth, or that these fish have different distributions at such times but are influenced similarly by large-scale environmental processes. In addition, we found that much of the covariation in body size across ages and stocks was due to shared declining trends over time. The dominant principal component for body size, which accounted for 58% of the total variation, had a significant decreasing time trend ($P < 0.001$) over the period 1971-1997. These declining trends in body size of British Columbia and Alaska sockeye salmon may be related to physical oceanographic variables and ocean abundance of salmon (Cox and Hinch 1997).

Indices of age at maturity also showed positive covariation within and between regions, although the covariation was weaker (394 of the 496 correlations were positive; average $r = 0.17$). The dominant principal component (accounting for 33% of the total variation) for mean age had a significant increasing time trend ($P < 0.001$) from 1971 to 1994. These patterns of covariation and time trends in mean age suggest similar hypotheses about important environmental processes as discussed above for body size. Pyper et al. (1999) provides more detailed results on these analyses of body size and age at maturity.

References

Adkison, M.D., R.M. Peterman, M.F. Lapointe, D.M. Gillis, and J. Korman. 1996. Alternative models of climatic effects on sockeye salmon *(Oncorhynchus nerka)* production in Bristol Bay, Alaska, and the Fraser River, British Columbia. Fish. Oceanogr. 5:137-152.

Cox, S.P., and S.G. Hinch. 1997. Changes in size at maturity of Fraser River sockeye salmon *(Oncorhynchus nerka)* (1952-1993) and associations with temperature. Can. J. Fish. Aquat. Sci. 54:1159-1165.

Graham, N.E. 1994. Decadal-scale climate variability in the tropical and North Pacific during the 1970s and 1980s: Observations and model results. Clim. Dynam. 10:135-162.

Peterman, R.M., B.J. Pyper, M.F. Lapointe, M.F. Adkison, and C.J. Walters. 1998. Patterns of covariation in survival rates of British Columbian and Alaskan sockeye salmon *(Oncorhynchus nerka)* stocks. Can. J. Fish. Aquat. Sci. In press.

Pyper, B.J., R.M. Peterman, M.F. Lapointe, and C.J. Walters. 1999. Patterns of covariation in length and age at maturity of British Columbia and Alaska sockeye salmon *(Oncorhynchus nerka)* stocks. Can. J. Fish. Aquat. Sci. 56. In press.

Predictability of Returns of Sockeye Salmon *(Oncorhynchus nerka)* to Bristol Bay, Alaska, 1-4 Years in the Future

Milo D. Adkison
U.S. Geological Survey, Biological Resources Division, Anchorage, Alaska

Randall M. Peterman
Simon Fraser University, School of Resource and Environmental Management, Burnaby, British Columbia, Canada

Extended Abstract

A variety of forecasting methods has been used to forecast Bristol Bay sockeye salmon (*Oncorhynchus nerka*) returns 1 year into the future (Fried and Yuen 1987, Rogers 1997, Cross 1998). However, none of the methods has performed exceptionally well, particularly recently. Our purpose in this study was to improve the accuracy of preseason forecasts of sockeye salmon returns to Bristol Bay by selecting the most appropriate set of independent variables for forecasting. In addition, we wanted a reliable estimate of the uncertainty of such forecasts, a goal that may be even more important for some purposes.

We used cross-validation to select among several candidate forecasting models and to estimate the probable error in forecasting (Efron and Tibshirani 1993). Cross-validation estimates the error in forecasting by repeatedly fitting a model to subsets of the data and using these model fits to forecast the left-out data (the validation data). Because the parameters of the model are chosen as if the validation data were not yet known, forecasting these data is analogous to forecasting future observations. Thus, the errors in forecasting the validation data should be of similar magnitude to those of future forecasts.

Current address for Milo Adkison is Juneau Center, School of Fisheries and Ocean Sciences, University of Alaska Fairbanks, 11120 Glacier Highway, Juneau, AK 99801.

We constructed forecasts and confidence intervals for sockeye salmon returns to each of the nine principal Bristol Bay drainages 1 to 4 years in the future. We used cross-validation to select among a wide class of models (Adkison et al. 1996) that could include as independent variables the number of spawners, sibling returns, air and sea surface temperatures, and the previous year's deviation from the expected return. We found that no combination of predictor variables provided highly accurate forecasts. A typical 80% confidence interval for predicting returns to a drainage 1 year in advance spanned a four-fold range. Confidence intervals for forecasts 2-4 years in the future were even wider.

Using cross-validation to estimate the forecast error apparently resulted in an accurate representation of the uncertainty in future returns. The coverage probability of the confidence intervals about the forecasts was generally what it should have been. However, the returns in 1996, 1997, and 1998 were well below the forecast and outside of the 80% confidence intervals.

References

Adkison, M.D., R.M. Peterman, M.F. Lapointe, D.M. Gillis, and J. Korman. 1996. Alternative models of climatic effects on sockeye salmon, *Oncorhynchus nerka*, productivity in Bristol Bay, Alaska, and the Fraser River, British Columbia. Fish. Oceanogr. 5:137-152.

Cross, B.A. 1998. A synopsis and critique of forecasts of sockeye salmon returning to Bristol Bay, Alaska, in 1997. Alaska Department of Fish and Game, Division of Commercial Fisheries Management and Development. Regional Information Report 2A98-02. Anchorage, Alaska.

Efron, B., and R.J. Tibshirani. 1993. An introduction to the bootstrap. Chapman and Hall, New York. 436 pp.

Fried, S.M., and H.J. Yuen. 1987. Forecasting sockeye salmon (*Oncorhynchus nerka*) returns to Bristol Bay, Alaska: A review and critique of methods. In: H.D. Smith, L. Margolis, and C.C. Wood (eds.), Sockeye salmon (*Oncorhynchus nerka*) population biology and future management. Can. Spec. Publ. Fish. Aquat. Sci. 96, pp. 273-279.

Rogers, D.E. 1997. Forecasts of the 1998 sockeye salmon runs to Bristol Bay. Fisheries Research Institute, University of Washington, FRI-UW-9710. Seattle. 24 pp.

Comparison of Methods for Detecting Climate-Induced Changes in Productivity of Pacific Salmon Stocks

Randall M. Peterman, Jeff Grout, and Brian J. Pyper
Simon Fraser University, School of Resource and Environmental Management, Burnaby, British Columbia, Canada

Extended Abstract

Recent research has shown that Pacific salmon populations can experience abrupt and persistent changes in mean productivity that appear to be closely related to climatic changes. However, it is difficult to forecast such non-stationarity in salmon productivity because of large uncertainties about mechanisms that link climatic changes, freshwater and oceanographic conditions, and salmon. Furthermore, inseason monitoring of salmon abundance and changes in regulations have not usually been able to make up for such errors in forecasts when attempting to meet management objectives. Therefore, management agencies need methods to reliably detect and respond to such changes in productivity in a timely manner to avoid costly, suboptimal harvests or depletion of stocks. The purpose of our research was to compare the effectiveness of various parameter estimation methods at tracking such changes in a timely, accurate, and precise manner.

To meet this objective, we evaluated the relative effectiveness of several methods of annually updating estimates of parameters of Ricker stock-recruitment relationships: standard regression, Walters' (1990) bias correction to that regression, and a Kalman filter. The latter is a method often used by electrical engineers to more accurately estimate the underlying signal amidst noise. However, with a few exceptions, this method has rarely been used in fisheries. We used Monte Carlo simulations in which we hypothesized a wide range of possible future climate-induced changes in productivity of a salmon stock. We represented such changes by a time-varying *a* parameter in the Ricker model because Adkison et al. (1996) documented that this parameter, rather than the Ricker *b*, showed consistent changes across Bristol Bay sockeye salmon stocks in the 1970s in response to major climatic changes. We analyzed three types of scenarios in *a*: a sine wave with a 60-year period, a "step" function with a 60-year period, and an autoregressive *a*.

We then used each parameter estimation method on the stochastic simulated stock-and-recruitment data to estimate the population's parameters each year and set escapement targets and harvest goals accordingly. Because errors in parameter estimates translate into deviation from the true optimal escapement, we compared parameter estimation methods in terms of their resulting cumulative harvests and their accuracy and precision. We compared these methods to a fixed harvest rate policy that has been advocated previously for such nonstationary situations (Walters and Parma 1996). We also compared the resulting harvests to what could be obtained if there had been perfect knowledge of the true *a* parameter each year (i.e., expected value of perfect information).

The Kalman filter resulted in greater catch than the standard least squares approach for the scenarios that used an underlying 60-year sine wave or step function in the Ricker *a* parameter. However, for the scenario with an autoregressive *a*, these two methods gave roughly the same catches. Walters' (1990) bias correction method consistently performed more poorly than the Kalman filter and least squares methods. All of these methods also generated greater catch than the best fixed-harvest-rate strategy.

The results above were derived from analyses where the model assumed "perfect control," i.e., if a particular escapement goal was estimated, it was achieved exactly as long as recruitment was greater than that goal, and catch was simply the excess of recruitment over the goal. When recruitment was too low, catch was zero. However, this assumption of perfect control generated catches that were almost as large as the expected value of perfect information. Therefore, in a second set of analyses, we assumed "implementation error," whereby the target was never achieved exactly, which reflects the field situation in which preseason and inseason knowledge of abundance is imperfect, as is control over harvest rate. When implementation error was added, the cumulative catches from all parameter estimation methods dropped dramatically and there was a smaller difference between catches from different parameter estimation methods.

The Kalman filter did better than least squares in some scenarios because it generally had more precise parameter estimates and therefore more estimated target escapements fell near the true optimum than with least squares. Analogously, Walters' (1990) bias correction method did not perform as well as the other approaches because, although its parameter estimates were less biased, they were also less precise (Korman et al. 1995), which caused estimated optimal escapements to deviate further from the true optimum than with the other two approaches. However, the potential advantage of a Kalman filter in tracking changes in productivity can only be realized if there also is drastic reduction in implementation error.

References

Adkison, M.D., R.M. Peterman, M.F. Lapointe, D.M. Gillis, and J. Korman. 1996. Alternative models of climatic effects on sockeye salmon productivity in Bristol Bay, Alaska, and the Fraser River, British Columbia. Fish. Oceanogr. 5:137-152.

Korman, J., R.M. Peterman, and C.J. Walters. 1995. Empirical and theoretical analyses of correction of time-series bias in stock-recruitment relationships of sockeye salmon. Can. J. Fish. Aquat. Sci. 52(10):2174-2189.

Walters, C.J. 1990. A partial bias correction factor for stock-recruitment parameter estimation in the presence of autocorrelated environmental effects. Can. J. Fish. Aquat. Sci. 47:516-519.

Walters, C.J., and A. Parma. 1996. Fixed exploitation rate strategies for coping with effects of climate change. Can. J. Fish. Aquat. Sci. 53:148-158.

Influence of Physical Processes on the Early Life History Stages of Walleye *(Stizostedion vitreum)* in Western Lake Erie

Edward F. Roseman, William W. Taylor, and Daniel B. Hayes
Michigan State University, Department of Fisheries and Wildlife, East Lansing, Michigan

Robert C. Haas
Michigan Department of Natural Resources, Mt. Clemens Fisheries Station, Mt. Clemens, Michigan

David H. Davies
Ohio Department of Natural Resources, Division of Wildlife, Sandusky Fisheries Station, Sandusky, Ohio

Scudder D. Mackey
Ohio Department of Natural Resources, Division of Geological Survey, Lake Erie Geology Group, Sandusky, Ohio

Abstract

Physical processes create the environmental settings in which biological processes occur. Therefore, physical processes can have strong effects on fish populations, especially during vulnerable early life history stages. Much of the variability in recruitment of Lake Erie walleye is thought to be caused by variable survival during the egg and larval stages. Recent research on Lake Erie walleye indicates that density-independent physical processes directly affect the vital rates of these early life history stages. In western Lake Erie, large scale storm events in spring were related to high mortality of walleye eggs and larvae on reefs and subsequently produced poor recruitment. In addition, variability in environmental conditions (e.g., water warming rate) prolongs life history stage durations over which high mortality rates operate. Slower water warming rates in western Lake Erie prolonged walleye egg incubation and pelagic larval stages extending the period of vulnerability to disease and predation for these early life history stages and negatively impact year-class strength. Physical processes rarely function independently but facilitate the underlying ecological mechanisms that drive fish population dynamics. Therefore, it is essential to examine physical processes in relation to biological processes in order to gain a more complete understanding of the linkages between them. By doing so, we gain a stronger basis for predicting the response of fish populations to environmental variability.

Introduction

Physical processes and other abiotic factors can have powerful, although usually indirect, effects on fish population and community dynamics by creating the environmental conditions in which many important biological processes operate (Dunson and Travis 1991). While many physical processes are difficult to predict (e.g., weather), we can gain insight into the response of fish populations to such factors by understanding how physical processes influence fish and their habitat. Therefore, it is important to identify and understand the various physical processes that structure fish populations and determine how physical processes influence the biological processes involved in fish population dynamics.

Lake Erie supports the world's largest naturally reproducing population of walleye (*Stizostedion vitreum*), producing valuable sport and commercial fisheries in the United States and Canada. Like most natural populations, Lake Erie walleye have exhibited significant variation in year-class strength over the past 20 years. The abundance of age-2 fish entering the stock has fluctuated nearly 60-fold since 1977 (Fig. 1) and the causes of this recruitment variability have been unclear (Henderson and Nepszy 1994, Madenjian et al. 1996). Recent field research in western Lake Erie, however, reveals the importance of density-independent physical processes in walleye year-class formation with the strongest year classes occurring in years with fast water warming rates and few storm events during egg incubation and larval dispersal periods (Mion 1996, Roseman et al. 1996). These studies concluded that variability in environmental conditions influenced the timing of life history events and caused differential survival rates during the egg and larval stages which translated into variability in recruitment to the adult stock. The goal of this paper is to review the effects of physical processes occurring in western Lake Erie on the vital rates (development, growth, and survival) of the egg and larval stages of walleye to gain a better understanding of how these processes temper the environmental setting for year-class development.

Because walleye are the dominant predator in Lake Erie and support important sport and commercial fisheries, fluctuations in their population size and structure can have direct implications for the entire Lake Erie ecosystem and local economies. For example, Knight and Vondracek (1993) found that increases in the abundance of walleye in the late 1970s played

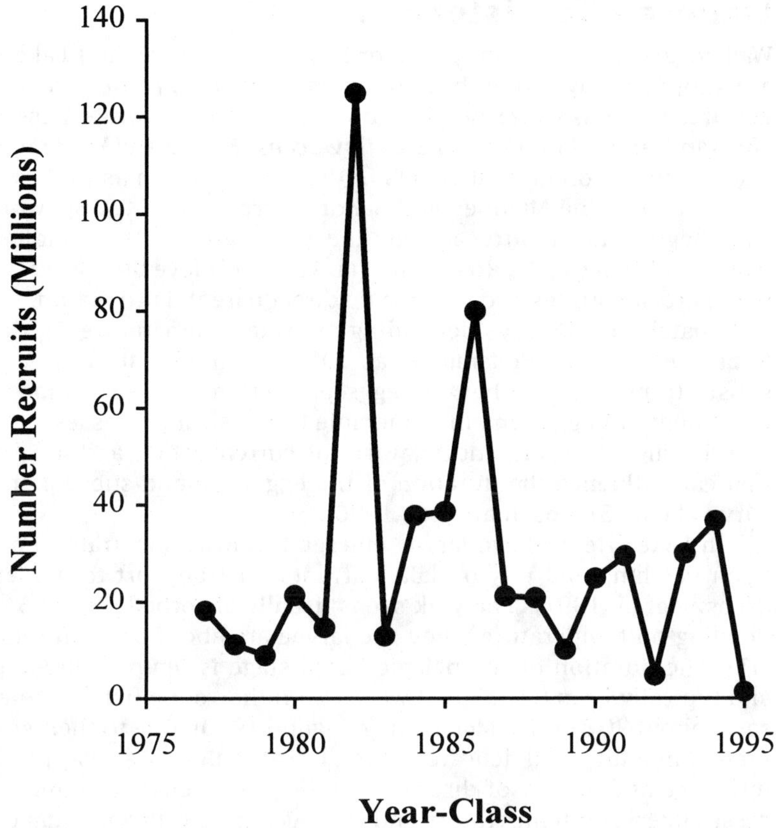

Figure 1. Number of age-2 walleye entering the Lake Erie stock as determined by Ohio Division of Wildlife fall gillnet surveys (Turner et al. 1998).

a major role in the decline in abundance of soft-rayed species such as emerald shiner (*Notropis atherinoides*), spottail shiner (*N. hudsonius*), and alewife (*Alosa pseudoharengus*). They concluded that management goals focusing primarily on walleye affected not only the targeted species but the entire fish community of western Lake Erie. During this same time period, abundant walleye populations attracted interest throughout the midwestern United States and led to the development of a substantial sportfishing industry in the states bordering the lake as well as a rejuvenated commercial gillnet fishery in Canada. Consequently, fluctuations in the walleye population have resulted in direct economic consequences to businesses and families relying on these fisheries (Knight 1997, Lichtkoppler 1997).

Walleye Life History

Walleye are the dominant predator in the western basin of Lake Erie and are supported by a prey base of gizzard shad (*Dorosoma cepedianum*), alewife, shiners, white perch (*Morone americana*), white bass (*Morone chrysops*), and yellow perch (*Perca flavescens*) (Knight and Vondracek 1993). Walleye spawn on mid-lake reefs in the western basin as well as in tributaries such as the Maumee and Sandusky rivers (Fig. 2). Spawning typically begins shortly after ice-out and peaks around the middle of April (Baker and Manz 1971, Roseman et al. 1996). Walleye broadcast their eggs over hard substrates and provide no direct parental protection. Eggs typically hatch in 7-15 days depending on water temperature (Hurley 1972, Nepszy et al. 1991, Roseman et al. 1996). Small interstitial spaces in the substrate provide the benthic eggs protection from predation and displacement but eggs remain vulnerable to physical processes such as water warming rate, wind-induced wave and current action, and sedimentation that can influence the duration of the egg stage and subsequently their survival (Fig. 3) (Roseman et al. 1996).

In Lake Erie, walleye larvae emerge from the substrate immediately upon hatching and rely on lake currents for transport to nursery areas (Nepszy et al. 1991). The yolk sac is usually absorbed after 3-5 days depending on temperature when the larvae are about 9.5 mm total length (TL). The duration of the pelagic larval stage is temperature-dependent and typically lasts from 3 to 4 weeks until the young fish become demersal at about 30 mm TL (McElman and Balon 1979). Due to their small size, limited mobility, and delicate nature, larval walleye are susceptible to the influence of a variety of direct and indirect influence from physical processes like water temperature and wind-generated currents that can influence the timing and magnitude of larval development, growth, and survival (Fig. 3).

Physical Processes Influencing the Early Life History Stages of Walleye

Water temperature has been shown to be one of the most influential physical factors in walleye egg development and survival (Allbaugh and Manz 1964, Hurley 1972). Laboratory experiments have demonstrated that optimal egg survival occurs when water temperatures increase from 5°C at a rate of 1°C per day (Smith and Koenst 1975). Incubation periods are lengthened when waters warm more slowly, thereby increasing the time eggs are exposed to predation, disease, and severe wind events (Wolfert et al. 1975, Roseman et al. 1996). In western Lake Erie, Busch et al. (1975) and Roseman et al. (1996) found a positive relationship between water warming rate and walleye year-class strength. The index of fall walleye abundance never exceeded 5 fish per hour of trawling when waters warmed at a rate less than 0.2°C per day but greatly exceeded that number at higher warming

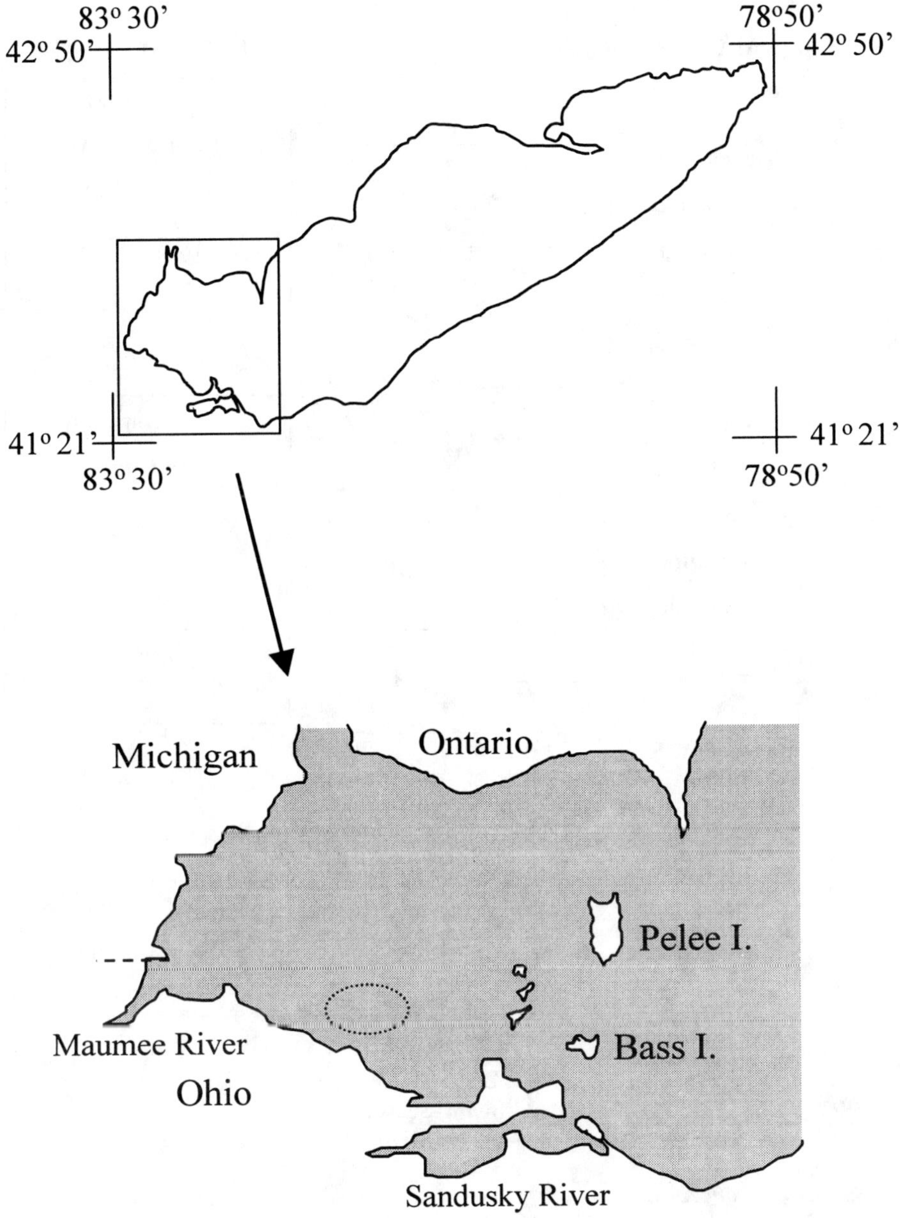

Figure 2. Map of Lake Erie detailing the western basin. Dotted ellipse represents mid-lake reef complex.

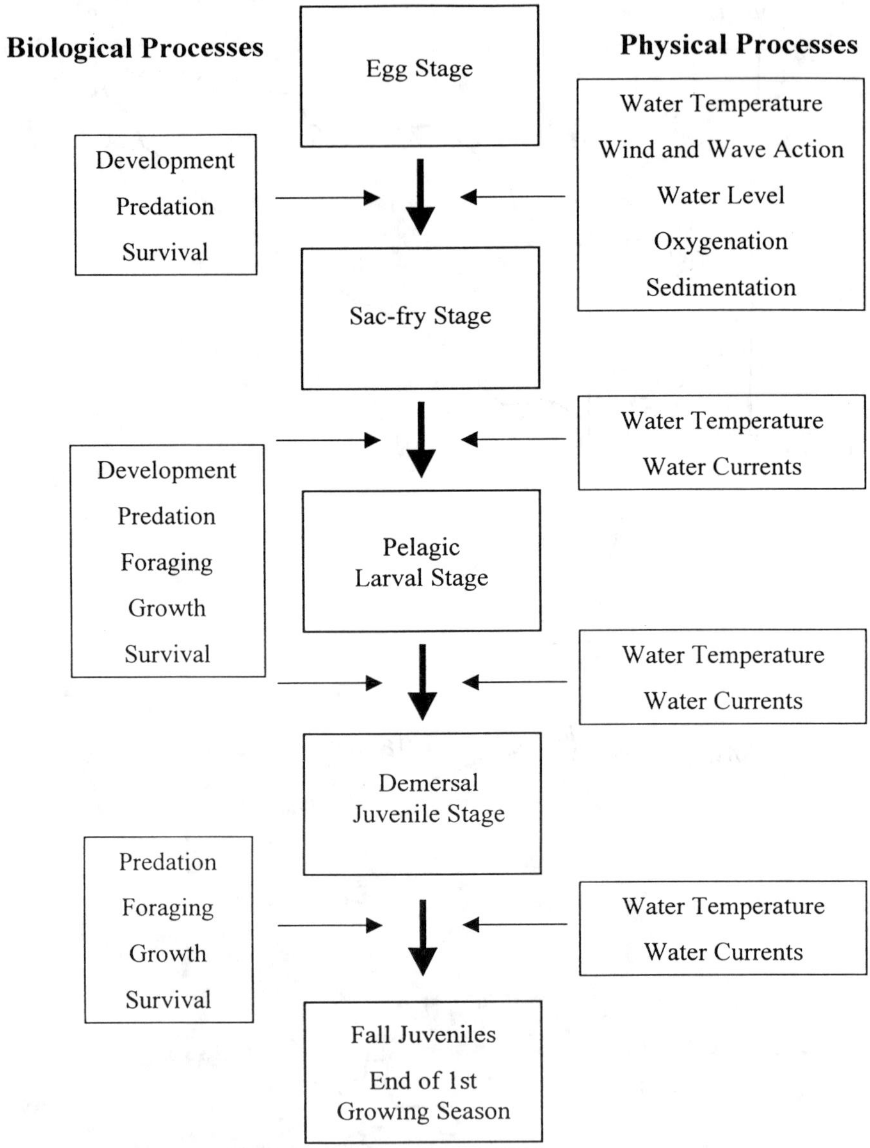

Figure 3. Conceptual model of physical and biological processes that influence walleye early life history stages.

rates (Fig. 4), indicating that 0.2°C per day may be a critical minimum threshold warming rate for walleye egg survival in western Lake Erie.

Water temperature mediates larval walleye activity including metabolism, feeding rate, growth rate, and ultimately, survival (Hokanson and Koenst 1986, Santucci and Wahl 1993, Mion 1996, Roseman 1997). Hokanson and Koenst (1986) reported that larval and juvenile walleye exhibited optimum growth in laboratory experiments at temperatures between 22° and 28°C. Growth rate and development of sac fry is temperature-dependent, as temperature determines the rate of yolk absorption (McElman and Balon 1979) while post-sac-fry larval walleye growth is a function of food consumption and metabolism, both of which are temperature-regulated (McElman and Balon 1979, Johnston and Mathias 1994). Warmer water temperatures also benefit production of plankton (Wetzel 1975), which translates to increased walleye growth. Zooplankton are important prey for larval walleye in western Lake Erie and ichthyoplankton become important as the young walleye grow, especially after they become demersal (Paulus 1969, Roseman 1997).

Low-intensity winds (<15 km per hour) serve to mix lake waters and create low-velocity currents important for distributing oxygen and nutrients throughout the water column (Bowden 1983). Higher-intensity winds over western Lake Erie can cause more extensive mixing of lake waters, which retards water warming and may create currents that dislodge walleye eggs from shallow reef substrates and deposit them in areas unsuitable for incubation (Busch et al. 1975, Roseman et al. 1996). Roseman et al. (1996) documented that the cumulative effect of 21 days of intense wind events in 1995, especially winds from the east-northeast which have long fetches, contributed to slow water warming rates and created strong currents and wave action that displaced walleye eggs from reefs.

Although the effects of moderate wind events may be cumulative, short-term high-velocity wind events may have catastrophic consequences for egg survival. In early April of 1998, Roseman (Unpubl. data) documented the removal of over 80% of walleye eggs from reefs in western Lake Erie by wave and current action generated during a single intense storm (Fig. 5). This storm originated from the northeast and produced 36 hours of gale-force winds and wave heights in excess of 4 m in western Lake Erie. They hypothesized that displaced eggs were likely deposited in deeper waters where silt substrates and reduced bottom dissolved-oxygen levels (<3.0 ppm) suffocated the embryos (Seifert and Spoor 1974, Balon et al. 1977, McMahon et al. 1984).

As described earlier, winds create waves and current patterns that have both positive and negative effects on early life history stages of walleye. Prolonged strong wind events create wave and current conditions that can transport fish large distances (Martin et al. 1992) as well as cause damage to delicate larvae (Cordone and Kelley 1961). Further, heavy wave action is known to stir up sediments causing increased turbidity associated with sediment transport (Gedney and Lick 1972). Increased turbidity

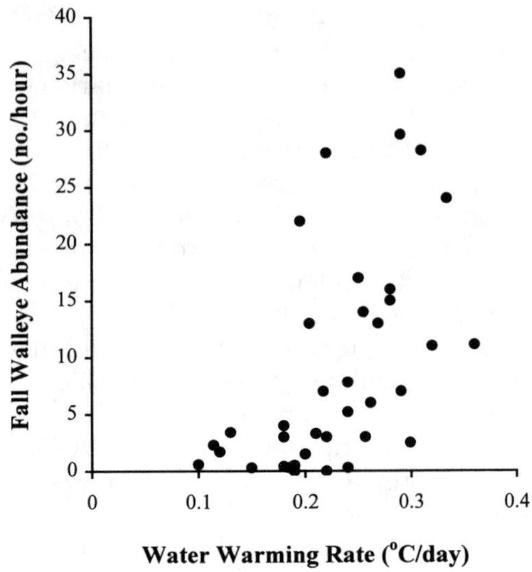

Figure 4. Relationship between spring water warming rate and abundance of age-0 walleye collected in Ohio Division of Wildlife bottom trawls. Data from Busch et al. (1975), Roseman (1997), and Ohio Division of Wildlife (1997).

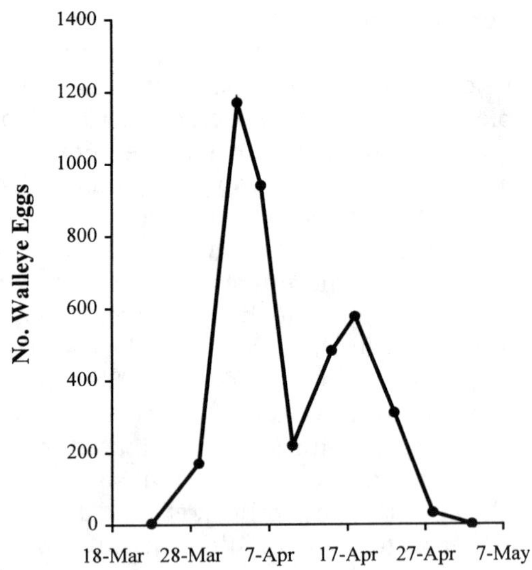

Figure 5. Numbers of walleye eggs collected per 2-minute tow on reefs in western Lake Erie in spring of 1998 showing removal of eggs caused by 8-9 April storm. From Roseman et al. (1999, In prep.).

Table 1. Relative peak walleye egg abundance (number per tow), walleye egg survival, water warming rate during egg incubation period (1 April-15 May), number of intense wind events (as determined by criteria in Busch et al. 1975), peak walleye larval abundance (number per 1,000 cubic meters), and index of August walleye year-class strength (number of fish per hour of bottom trawling) in western Lake Erie, 1994-1998.

	1994	1995	1996	1997	1998
Egg abundance	11,300	3,610	7,230	5,450	695
Egg survival	0.09	0.01	0.08	0.07	0.02
Warming rate	0.22	0.16	0.22	0.19	0.24
Winds	15	21	15	17	15
Larval abundance	28.4	2.0	8.1	13.1	16.8
August YCS	18.2	1.9	28.3	11.4	–

Data from Roseman et al. (1996, 1998), Roseman (1997), and Turner et al. (1998).

reduces the phototrophic zone thereby reducing primary productivity (Wetzel 1975).

Pelagic walleye larvae are poor swimmers (Houde 1969) and consequently are subject to transport by water currents, which is the mechanism by which larvae move from spawning areas to nursery areas both in rivers and on reefs (Houde and Forney 1970, Mion 1996, Roseman 1997). Walleye spawned on reefs in western Lake Erie rely on current patterns in the spring to transport the larvae to the warmer and more productive nearshore nursery habitats (Roseman 1997). Ichthyoplankton samples collected from 1994 through 1998 showed high densities of walleye larvae and other ichthyoplankton in nearshore areas of western Lake Erie, suggesting that current patterns during this time of year act to transport pelagic larvae in these areas (Mion 1996; Roseman 1997; E.F. Roseman, Unpubl. data). These nearshore areas warm quickly in the spring and are known to have higher densities of prey than offshore areas (Roseman 1997) providing habitat conditions favorable to larval walleye growth and survival.

Conclusions

In this paper we have identified specific factors and mechanisms that influence the vital rates of egg and larval stages of walleye, providing insight to the ecological functions mediated by physical processes. While we do not propose that physical processes are more important than biological processes in determining the vital rates of walleye eggs and larvae or structuring year-class strength, we do contend that physical processes create the environment in which biological processes occur. Further, we agree with Walters and Collie (1988) that research efforts should not focus solely on examining physical processes but rather should involve multidisciplinary approaches that examine the ecological linkages through which environmental variability influences the vital rates of fish populations.

Both episodic events (weather-related events) and subtle variability (temperature-dependent development) in early life history vital rates can significantly affect recruitment of walleye in western Lake Erie (Busch et al. 1975; Mion 1996; Roseman et al. 1996, 1998). Episodic events are usually catastrophic in nature, removing large numbers of individuals from the population (Houde 1989). Such events may relieve survivors of density-dependent regulatory mechanisms (i.e., competition) and shorten the duration of vulnerable early life history stages allowing good growth and survival of the remaining cohort. Conversely, decelerated development and larval growth rates precipitated by unfavorable temperatures can lead to substantially longer early life history stage durations over which high mortality rates can operate (Houde 1987), resulting in fewer recruits.

Walleye in western Lake Erie are recruited from two different spawning stocks: those that spawn in tributaries and those that spawn on the mid-lake reefs. Because walleye life stages most vulnerable to storm effects overlap in time (larvae in rivers, eggs on reefs), spring storms may ultimately drive survival in both habitats. This hypothesis is consistent with previous observations of synchrony of fish year-class strengths over large geographic areas in freshwater (Koonce et al. 1977) and marine (Ruzecki et al. 1976, Koslow 1984) systems as generated by large-scale climatic events. Linkage in survival between river- and reef-spawned fish should produce greater variability in total walleye recruitment than if survival was determined independently in the two habitats.

While we can modify the physical habitat in which fish populations exist, we do not have the capabilities to manage or manipulate most physical processes. For example, we can construct artificial reefs with deep crevices to shelter eggs from strong wave and current action, but we cannot alter the actual wave and current action to any appreciable degree that would benefit egg survival. Instead, we must rely on developing a sound understanding of the linkages between physical processes and the ecological processes that regulate fish production and recruitment. By doing so, we will be able to better predict the response of fish populations to variability in their habitat and respond with appropriate management strategies.

Acknowledgments

This work was funded in part by grants from the Michigan Sea Grant College Program (Project R/GLF 43), Ohio Sea Grant College Program, Michigan Department of Natural Resources, Ohio Division of Wildlife, and the Department of Fisheries and Wildlife at Michigan State University. The authors acknowledge the efforts of Kenneth Baker, Roger L. Knight, Kristine Dawson-Lynch, Joseph B. Mion, Kurt R. Newman, Edward O. Roseman, and David A. Thomas who assisted with the preparation of this manuscript.

References

Allbaugh, C.A., and J.V. Manz. 1964. Preliminary study of the effects of temperature fluctuations on developing walleye eggs and fry. Prog. Fish-Cult. 26:175-180.

Baker, C.T., and J.V. Manz. 1971. Walleye spawning area study in western Lake Erie. Ohio Department of Natural Resources, Division of Wildlife, Final Rep. Res. Proj. F-35-R-10. 23 pp.

Balon, E.K., W.T. Momot, and H.A. Regier. 1977. Reproductive guilds of percids: Results of the palaeogeographical history and ecological succession. J. Fish. Res. Board Can. 34:1910-1921.

Bowden, K.F. 1983. Physical oceanography of coastal waters. Ellis Horwood Limited, Chichester. 302 pp.

Busch, W.D.N., R.L. Scholl, and W.L. Hartman. 1975. Environmental factors affecting the strength of walleye (*Stizostedion vitreum vitreum*) year-classes in western lake Erie, 1960-70. J. Fish. Res. Board Can. 32:1733-1743.

Cordone, A.J., and D.W. Kelley. 1961. The influences of inorganic sediment on the aquatic life of streams. Calif. Fish Game 47:189-228.

Dunson, W.A., and J. Travis. 1991. The role of abiotic factors in community organization. Am. Nat. 138:1067-1091.

Gedney, R.T., and W. Lick. 1972. Wind-driven currents in Lake Erie. J. Geophys. Res. 77:2714-2723.

Henderson, B.A., and S.J. Nepszy. 1994. Reproductive tactics of walleye (*Stizostedion vitreum*) in Lake Erie. Can. J. Fish. Aquat. Sci. 51:986-997.

Hokanson, K.E.F., and W.M. Koenst. 1986. Revised estimates of growth requirements and lethal temperature limits of juvenile walleye. Prog. Fish-Cult. 48:90-94.

Houde, E.D. 1969. Sustained swimming ability of larval walleyes, *Stizostedion vitreum vitreum*, and yellow perch, *Perca flavescens*. J. Fish Res. Board Can. 26:1647-1659.

Houde, E.D. 1987. Fish early life dynamics and recruitment. Am. Fish. Soc. Symp. 2:17-29.

Houde, E.D. 1989. Subtleties and episodes in the early life of fishes. J. Fish Biol. 35:29-38.

Houde, E.D., and J.L. Forney. 1970, Effects of water currents on distribution of walleye larvae in Oneida Lake, New York. J. Fish. Res. Board Can. 27:445-456.

Hurley, D.A. 1972. Observations on incubating walleye eggs. Prog. Fish-Cult. 34:49-54.

Johnston, T.A., and J.A. Mathias. 1994. The effects of temperature on feeding in zooplanktivorous walleye, *Stizostedion vitreum*, larvae. Environ. Biol. Fishes 40:189-198.

Knight, R.L. 1997. Successful interagency rehabilitation of Lake Erie walleye. Fisheries (Bethesda) 22(7):16-17.

Knight, R.L., and B. Vondracek. 1993. Changes in prey fish populations in western Lake Erie, 1969-88, as related to walleye, *Stizostedion vitreum*, predation. Can. J. Fish. Aquat. Sci. 50:1289-1298.

Koonce, J.F., T.B. Bagenal, R.F. Carline, K.E.F. Hokanson, and M. Nagiec. 1977. Factors influencing year-class strength of percids: A summary and a model of temperature effects. J. Fish. Res. Board Can. 34:1900-1909.

Koslow, J.A. 1984. Recruitment patterns on Northwest Atlantic fish stocks. Can. J. Fish. Aquat. Sci. 41:1722-1729.

Lichtkoppler, F.R. 1997. Ohio's Lake Erie charter fishing industry: 1985-1994. Fisheries (Bethesda) 22(1):14-21.

Madenjian, C.P., J.T. Tyson, R.L. Knight, M.W. Kershner, and M.J. Hansen. 1996. First-year growth, recruitment, and maturity of walleyes in western Lake Erie. Trans. Am. Fish. Soc. 125:821-830.

Martin, T.J., D.P. Cyrus, and A.T. Forbes. 1992. Episodic events: The effects of cyclonic flushing on the ichthyoplankton of St. Lucia estuary on the southeast coast of Africa. Netherlands J. Sea Res. 30:273-278.

McElman, J.F., and E.K. Balon. 1979. Early ontogeny of walleye, *Stizostedion vitreum*, with steps of saltatory development. Environ. Biol. Fishes 4:309-348.

McMahon, T.E., J.W. Terrell, and P.C. Nelson. 1984. Habitat suitability information: Walleye. U.S. Fish and Wildlife Service, Washington, D.C. 43 pp.

Mion, J.B. 1996. River discharge drives differential survival of larval walleye. M.S. thesis, Ohio State University, Columbus.

Nepszy, S.J., D.H. Davies, D. Einhouse, R.W. Hatch, G. Isbell, D. MacLennan, and K.M. Muth. 1991. Walleye in Lake Erie and Lake St. Clair. In: P.J. Colby, C.A. Lewis, and R.L. Eshenroder (eds.), Status of walleye in the Great Lakes: Case studies prepared for the 1989 workshop. Great Lakes Fish. Comm. Spec. Publ. 91-1, pp. 145-168.

Ohio Division of Wildlife. 1997. Ohio's Lake Erie fisheries. Annual report prepared for the Lake Erie Committee, March 1997. 20 pp.

Paulus, R.D. 1969. Walleye fry food habits in Lake Erie. Ohio Department of Natrual Resources, Ohio Fish Monogr. 2. 45 pp.

Roseman, E.F. 1997. Factors influencing the year-class strength of reef-spawned walleye in western Lake Erie. M.S. thesis, Michigan State University, East Lansing. 140 pp.

Roseman, E.F., W.W. Taylor, and D.B. Hayes. 1999. Removal of walleye eggs from reefs in western Lake Erie by a catastrophic storm. In preparation.

Roseman, E.F., W.W. Taylor, D.B. Hayes, R.C. Haas, R.L. Knight, and K.O. Paxton. 1996. Walleye egg deposition and survival on reefs in western Lake Erie. Ann. Zool. Fenn. 33:341-351.

Ruzecki, E.P., and six co-authors (eds.). 1976. The effects of tropical storm Agnes on the Chesapeake Bay estuarine system. The Chesapeake Research Consortium, Publ. 54. 639 pp.

Santucci, V.J., and D.H. Wahl. 1993. Factors influencing survival and growth of stocked walleye (*Stizostedion vitreum*) in a centrarchid-dominated impoundment. Can. J. Fish. Aquat. Sci. 50:1548-1558.

Seifert, R.E., and W.A. Spoor. 1974. Effects of reduced oxygen on embryos and larvae of the white sucker, coho salmon, brook trout, and walleye. In: J.H.S. Blaxter (ed.), The early life history of fish. Springer-Verlag, New York, pp. 487-495.

Smith Jr., L.L., and W.M. Koenst. 1975. Temperature effects on eggs and fry of percoid fishes. Ecological Research Services, Report EPA-660/3-75-017. 91 pp.

Turner, M., D. Einhouse, R. Haas, T. Johnson, R. Kenyon, R. Knight, D. MacLennan, and M. Thomas. 1998. Report of the Lake Erie Walleye Task Group. Great Lakes Fisheries Commission, Ann Arbor, Michigan. 24 pp.

Walters, C.J., and J.S. Collie. 1988. Is research on environmental factors useful to fisheries management? Can. J. Fish. Aquat. Sci. 45:1848-1854.

Wetzel, R.G. 1975. Limnology. W.B. Saunders Co., Philadelphia. 743 pp.

Wolfert, D.R., W.-D.N. Busch, and C.T. Baker. 1975. Predation by fish on walleye eggs on a spawning reef in western Lake Erie, 1969-71. Ohio J. Sci. 75:118-125.

Managing the Baltic Sea Cod (*Gadus morhua*): Stable Catches in an Unstable Environment

Olle Hjerne and Sture Hansson
Stockholm University, Department of Systems Ecology, Stockholm, Sweden

Extended Abstract

Overcapacity is one of the most serious problems in the world fisheries, frequently resulting in overfishing (FAO 1995). The risk of attaining an overcapacity is particularly large in fisheries for fluctuating species (Ludwig et al. 1993). Constant harvest rate strategies, with catches proportional to the target species abundance, stimulate investments when stock sizes are large which can result in overcapacity. Constant catch strategies, on the other hand, lack this incentive. Therefore we explored the possibilitiy of a management strategy based on constant catches by modeling the eastern Baltic Sea cod (*Gadus morhua*) population. The eastern Baltic cod stock is naturally fluctuating due to highly varying spawning conditions in the brackish Baltic Sea (Ojaveer et al. 1981, Bagge et al. 1994, Nissling 1994).

The model used is single species and cohort based, with mortality caused by the fishery, cannibalism, and "other causes." To model cod recruitment, we used the empirical relationship between year-class strength, spawning volume (the volume of water with an oxygen content > 2 mL/L and salinity > 11 PSU), and spawning stock biomass found by Sparholt (1996). A Monte Carlo model with actual data from 1966 to 1992 simulates the variation in spawning conditions.

The risk of overfishing within 5, 20, and 50 years for different constant catch levels was calculated. If we do not accept any risk of overfishing within 50 years the long-term yield will, according to the model, be less than half the yield resulting from the constant harvest rate strategy which corresponds to the maximum sustainable yield. To increase the long-term yield we have to accept a higher risk of overfishing.

To avoid the low long-term yield and the high risk of overfishing resulting from the strict constant catch strategy, we developed a more dynamic constant catch strategy (the DCC strategy) based on alternative constant catches and two overfishing criteria. Normal catches were allowed when the spawning stock biomass was above a certain level. When the spawning stock biomass was below this level, catches were reduced by 50%. If the spawning stock biomass fell below a second and lower threshold, fishing was stopped totally. According to the model, the long-term yield with the DCC strategy was almost as large as the long-term yield with the constant harvest rate strategy. The DCC approach had the advantage of smaller variation in catches between years and a better degree of utilization of the fishery capacity.

In conclusion, we find the DCC approach to be an interesting and competitive management strategy, probably able to decrease the problems of overcapitalization and overfishing for long-lived and highly fluctuating species like the Baltic Sea cod.

References

Bagge, O., F. Thurow, E. Steffensen, and J. Bay. 1994. The Baltic cod. Dana 10:1-28.

FAO. 1995. The state of world fisheries and aquaculture. FAO, Rome. 57 pp.

Ludwig, D., R. Hilborn, and C. Walters. 1993. Uncertainty, resource exploitation, and conservation: Lessons from history. Science 260:17,36.

Nissling, A. 1994. Survival of eggs and yolk-sac larvae of Baltic cod (*Gadus morhua* L.) at low oxygen levels in different salinities. ICES Mar. Sci. Symp. 198:626-631.

Ojaveer, E., A. Lindroth, O. Bagge, H. Lehtonen, and J. Toivonen. 1981. Fishes and fisheries. In: A. Voipio (ed.), The Baltic Sea. Elsevier Oceanography Series 30. Elsevier Scientific Publishing Company, Amsterdam, The Netherlands, pp. 275-350.

Sparholt, H. 1996. Causal correlation between recruitment and spawning stock size of central Baltic cod? ICES J. Mar. Sci. 53:771-779.

Retrospective Projection Using Monte Carlo Simulation: An Application of a Length-Based Model to Kachemak Bay Pink Shrimp

Caihong Fu, Terrance J. Quinn II, and Milo D. Adkison
University of Alaska Fairbanks, School of Fisheries and Ocean Sciences, Juneau, Alaska

Abstract

Pink shrimp (*Pandalis borealis*), the formerly abundant and commercially important species in Kachemak Bay, Alaska, has declined in population abundance since the mid-1980s. No sign of population recovery has been detected in recent years. This paper is a retrospective look at what the risk to a population would have been under various levels of harvest rate, incorporating uncertainty in natural mortality and recruitment. A length-based population dynamics model (Quinn et al., In press) was employed and Monte Carlo simulation was conducted to estimate the probabilities that various harvest strategies would have caused the shrimp population to fall below a threshold. Our results indicated that natural mortality was the most important factor controlling population dynamics, and the increasing trend in natural mortality in the 1980s resulted in the population crash. Because the pink shrimp population had a very high background risk of extinction, all harvest strategies led to a similar population decline. More effort should be expended in studying the relationship between pink shrimp and its predators. The common assumption of constant natural mortality in stock assessment and population projection should be revisited for forage species like pink shrimp.

Introduction

Traditional fish stock assessment focuses on point estimates of population parameters such as recruitment and fishing intensity with a common assumption of constant and known natural mortality. This assumption is vital in that it often leads to underestimation of uncertainty in population dynamics and, most important, in the risk of population collapse. Risk, according to Burgman et al. (1993), is the potential or probability of an adverse event, e.g., population abundance going below some defined level. For many fish populations there exist threshold levels below which the population will have difficulty in recovering (Quinn et al. 1990, Zheng et al. 1993). This is especially true for some crustacean populations such as pink shrimp, in which copulation is necessary in the reproduction process. Animal populations that live in a highly variable and unpredictable environment can go extinct at some point even free from human disturbance. This is termed the "Background Risk of Extinction" (Burgman et al. 1993). Human activities such as harvesting can increase this risk of extinction. The impacts on population dynamics from natural and fishing mortality should be translated into the language of risk for proper understanding.

Pink shrimp (*Pandalus borealis* Krøyer), also known as northern shrimp or northern pink shrimp, was once a commercially important species in Kachemak Bay, constituting more than half of the harvests in the pandalid fishery (Davis 1982). While landings of pandalid shrimp from the otter trawl fisheries in Alaska have declined dramatically since the record high yield of 58,600 metric tons (t) in 1976, trawl catches in Kachemak Bay were relatively constant from 1970 to 1983. One important reason for the sustained yields is that conservative management of seasons and guideline harvest levels had been carried out based on the survey estimation of population biomass. Despite this conservative management, the catch during 1983 decreased to nearly $^1/_{10}$ of that in 1980 due to reduced population abundance. The fishery was closed after 1986 due to continued low stock size (Gustafson 1994) and has not reopened.

The depletion of pandalid shrimp populations in Alaska was synchronous with the collapses of many other crustacean stocks. After simultaneously examining historical data from several populations, Orensanz et al. (1998) concluded that the sequential patterns of rise-peak-demise in total catch shown by many crustacean fisheries represented a serial depletion of resources. The serial depletion cannot be explained simply as a result of overfishing. Some lightly or seldom-fished pink shrimp populations experienced similar population declines (Anderson 1991). As proposed by Orensanz et al. (1998), climate change may have reduced reproductive success and increased natural mortality.

The serial depletion of crustacean populations in Alaska regardless of fishing pressure underlines the necessity of seriously considering temporal variations in both recruitment and natural mortality. One of the purposes in this study is to understand what roles recruitment and natural mortality played in the collapse of pink shrimp population in Kachemak Bay using estimates of recruitment and natural mortality trends from a length-based model (Quinn et al., In press) in conjunction with other assumptions. We also use retrospective analysis to examine how different harvest rates might have altered the population's fate. Monte Carlo simulation together with probabilistic risk analysis provide the essential tools for accomplishing these goals. This approach not only captures variability present in real population fluctuations but also presents results in terms of risks, which are useful in decision-making. Finally, we discuss the possibility of incorporating major environmental factors such as predator abundance into population projection and harvest recommendations.

Methods

Data

For doing retrospective simulations, two components are important: i.e., parameters, some of which are random variables, and a model to describe the population in a mathematical language. To obtain population parameters, historical data need to be analyzed. Data utilized in this study come from the commercial fishery and research surveys. Annual commercial yield data from the shrimp fishery in Kachemak Bay between 1971 and 1986 were provided by the Alaska Department of Fish and Game (ADF&G). A trawl survey for pandalid shrimp in Kachemak Bay was initiated by the National Marine Fisheries Service (NMFS) in 1971, and continued once every year in May until 1974 using a 66-foot Nordby trawl net with assumed 50% net efficiency. From 1975 to 1990, the trawl survey was conducted twice every year (October and May) by ADF&G using 61-foot NMFS-designed net with assumed 100% net efficiency. Since 1991, the survey has been conducted only once every 2 years in May due to a depressed shrimp population. Data available from spring surveys include absolute biomass estimates from 1971 to 1991 and length frequency data during the same period with data missing from 1975 to 1977. For the purpose of doing simulations, we filled in length frequencies for the 3 years using those observed from 1971 to 1973 to stabilize the estimation. Survey data after 1991 are not yet available due to processing lags. During fall surveys (mid-October), females are nearly 100% ovigerous, which makes identifying sex easy. Female proportion at length was therefore calculated from length frequency data obtained in fall surveys.

Model and Parameter Estimation

A length-based age-structured population model (Quinn et al., In press) was applied to the population to obtain parameter estimates. This model follows individual cohorts through time and has flexibility to incorporate general forms of selectivity and natural mortality (Appendix 1). Parameter estimation was accomplished using the software Autodif Model Builder (ADMB) which estimates parameters in a step-by-step fashion, known as a multi-phase procedure (Otter Research Ltd. 1996). Parameters estimated include growth parameters: L_∞, κ, σ, σ_r and mean length at recruitment $\mu_{r,t}$;

annual recruitments R_t and α, β, and ϕ in the spawner-recruitment (S-R) relationship (described in Appendix 1); full recruitment fishing mortality F_t and selectivity parameters γ and $L_{50\%}$; and natural mortality M_t. Interannual variation of M_t was modeled as a time series structure of a random walk $\ln(M_t) = \ln(M_{t-1}) + \delta_t$, where δ_t is a normal error term. The initial value of natural mortality was set at 0.2 per year for 1970, which gives a better fit than larger values. Parameters were estimated by minimizing an objective function (Appendix 1).

Retrospective Projection Procedure

Examining the Impact of M and R

To reduce the complexity of the retrospective projection procedure, we only took account of the uncertainties in recruitment and post-recruitment natural death. All other parameters were fixed at the estimated values from the parameter estimation procedure. Temporal variations in both recruitment (R) and natural mortality (M) were simulated with log-normal error: $\Theta' = \Theta e^{\mathrm{cv}(\Theta)Z}$, where Θ' is the random value of M or R, Θ is the deterministic value, and $Z \sim N(0,1)$.

To carry out the projection, we employed three ways of generating a deterministic M for each year. First, because traditional methods of stock assessment and population projection often make the assumption of a constant and known survival rate over time, we investigated the validity of this assumption and its effect on the population analyses. The constant mean was assumed to be 0.2 per year. Second, the estimates of M from the model were simulated with lognormal errors of equal variability across years. Third, a linearly increasing trend in M was also imposed for years between 1980 and 1990 together with a constant M of 0.2 per year for the years before 1980. Parameters for this linear function were obtained from a linear regression of estimated M for years between 1980 and 1990 versus year so that the linear function would approximate the estimated M. The intercept and slope were −10.559 and 0.134, respectively. We assumed a coefficient of variation (cv) of 30% in M for all the M sets.

Temporal variation in R has been the focus of considering uncertainties in population projections (Francis 1992). Four ways were used to generate deterministic recruitment for the projections because the Beverton-Holt (B-H) model with autocorrelated errors fitted the estimated recruitment best, this relationship was first considered for projecting. The *S-R* parameters were obtained from the parameter estimation procedure (Table 1). Second, because density-independent *S-R* models are more conservative and therefore desirable when the population level is low (Ferson 1993), a linear increasing function of spawners and recruitment with autocorrelation was also applied. The slope for the linear *S-R* function was estimated to be 0.228 by linearly regressing the estimated R versus S through the origin. Third, the common assumption of constant R was considered with the constant mean calculated as the average value of the

Table 1. Estimates and coefficients of variation (cv) for scalar parameters.

Parameter	Estimate	cv
$S\text{-}R\ \alpha$	1.826	73.87%
$S\text{-}R\ \beta$	0.002	90.11%
$S\text{-}R\ \phi$	0.267	34.15%
L_∞	24.698	0.36%
κ	0.367	1.06%
σ	0.529	2.20%
σ_r	1.127	0.88%
L_{50}	16.720	0.58%
γ	0.572	0.66%

estimated *R*. Fourth, for gaining a better understanding of recruitment impact on population dynamics, the estimated values of *R* were utilized. The cv for *R* was chosen as 20%, which was close to the average cv of 12% estimated from the model.

Combinations of each *M* and *R* model gave rise to 12 scenarios. For each of these 12 scenarios, annual *F* was fixed at the estimated value from the model in order to examine the effects on projection of uncertainties in *M* and recruitment *R*. Each scenario used the same random seeds, and 5,000 replicates were implemented.

Projection was initiated using the estimated abundance at ages and lengths in 1971 from the parameter estimation procedure and continued through 1991, covering the period when historical survey indices for biomass were available. The projection procedure produced estimates of population biomass, yield, and probabilities of going below the quasi-extinction threshold level (Ferson et al. 1989) over years. The quasi-extinction threshold level was arbitrarily defined as 10% of the maximum historical biomass from the surveys.

Evaluating Harvest Strategies

To examine how alternative fishery management might have affected the pink shrimp population, nine different harvest strategies relative to the estimated fishing mortality (F_{est}) were investigated. They were 0, $0.25F_{est}$, $0.5F_{est}$, $0.75F_{est}$, F_{est}, $1.5F_{est}$, and F_{est} with earlier fishery closures in 1980, 1982, and 1984.

For these analyses the scenario with the estimated *M* and a B-H spawner recruitment relationship was employed. The comparison criteria among various harvest strategies were cumulative yield over the entire fishing period and the risk of quasi-extinction in years 1987, when the fishery was closed, and 1991, the last year when survey data were available.

Results

Model Parameter Estimation

Estimated scalar parameters had relatively small values of cv except for the *S-R* relationship parameters (Table 1). Estimated recruitment displayed a dramatic drop in 1978 and stayed relatively low until a peak occurred in 1984 and 1985 (Fig. 1). Field observations did indicate a rebound of biomass in 1985 and 1986, but mainly with small male shrimp (Gustafson 1994) which is in concordance with our recruitment estimates. No strong spawner-recruitment relationship was discernable (Fig. 2). In some years, low spawner biomass resulted in high recruitment, for instance, 1984 and 1985, while in some other years, such as 1978 and 1981, spawner biomass of median size did not produce reasonable recruitment as we would expect. Anderson (1991) found with a pink shrimp population in Pavlof Bay that a relatively small spawning stock may give rise to an outstandingly large year class in some years, with the majority of years conforming to a Ricker relationship. For our estimated spawners and recruitment, the B-H model with autocorrelation resulted in a better fit.

The estimates of *M* displayed an increasing trend, which started in 1980 (Fig. 3). In accordance with the change in *M* over time, there was an increasing trend in the relative abundance of Pacific cod, considered to be the major predator, in the late 1980s and through the 1990s (Bechtol 1997). The mean frequencies of occurrence for Pacific cod were obtained during shrimp trawl surveys in Kachemak Bay. In contrast, estimated *F* was relatively stable except for that in 1983, when the fishery was partially closed, and in 1986, the last year with a fishery operation.

The overall fit of biomass and yield from the parameter estimation procedure was good after 1979. However, there were some discrepancies between the predicted and observed values in the 1970s (Fig. 4), which might partially be due to the missing length frequency data from 1975 to 1977.

Comparison of Projections with Different Models of M and R

Monte Carlo simulations produced 5,000 replicates from which statistical summaries and the probabilities of extreme events were drawn. We used the scenario with the estimated *M* and *R* to illustrate the possible ways of summarizing results. Although each individual trajectory from the 5,000 replicates has a different path, 50% of the projected biomass values fall into the range between the 25th and 75th percentiles (Fig. 5). This range followed the pattern of the observed biomass trend. The mean line of the trajectories appeared to be a good representative of the observed bio-

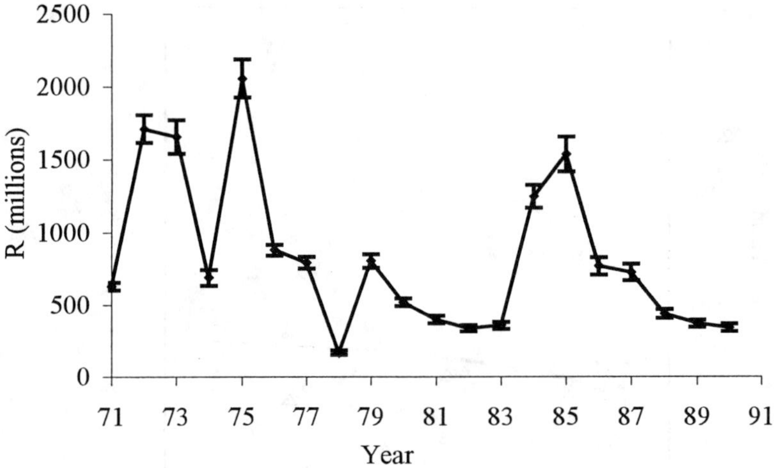

Figure 1. Estimated recruitment in number (± 1 SE) from the length-based model.

mass. Therefore, we chose the trajectory mean to demonstrate the impact of *M* and *R* models on population projections.

Comparisons were made among the *R* models with each set of *M*. With constant *M*, the trajectory means were driven by the fluctuations in *R* and *F*. No matter what recruitment scenario was examined, there were big discrepancies between the projections and the observations after 1983 when *F* began to decrease, and they got particularly larger after 1986, when the fishery was completely closed (Fig. 6). This discrepancy between the observed decline in biomass and a projected increasing trend in biomass indicated that *M* was greatly understated in the 1980s with the constant *M* model. In addition, there was one biomass peak in 1976 when the estimated R was used, which may have resulted from a combination of high recruitment anomalies in 1972, 1973, and 1975, and a peak M in 1975 (Fig. 3).

In contrast, when we used the estimated *M* values as the deterministic mean, all *R* models resulted in a population decline similar to what was actually observed (Fig. 7). The best projection came from the scenario with the estimated *R*. Once again the underestimation of biomass for years with peak biomass may have resulted from high recruitment anomalies. One prominent result from this exercise was that a constant *R* did not project the population any worse than the B-H or linear model did. Precise prediction of recruitment anomalies would help the projection, but for most years, recruitment predicted from general models is likely to be adequate.

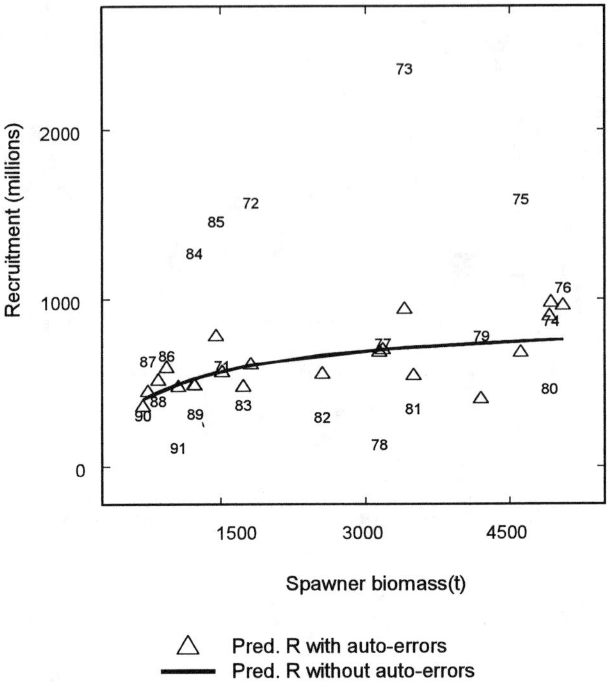

Figure 2. Fits of estimated recruitment in number versus spawner biomass using the Beverton-Holt (B-H) S-R model with and without autocorrelation.

The approximation of *M* to the estimated values by using the linearly increasing function resulted in projections consistent with the observed biomass values in the 1980s, similar to what were obtained from the scenario with the estimated *M* (Fig. 8). This also implies that the natural mortality must have had some increasing trend in the 1980s, in order for the pink shrimp population to crash. In conclusion, an increasing trend in *M* appears to be the most likely explanation of the decline of the pink shrimp population in Kachemak Bay. Recruitment did not seem to be as critical in this case.

Comparison of Harvest Strategies

Harvest strategies are commonly evaluated in terms of fishery or economic gain and their impact on the exploited populations as well. We compared the consequences of different harvest strategies based on the scenario with the estimated *M* and B-H *R*. The best harvest level was 75% of the F_{est}, which resulted in not only the highest cumulative yield, but

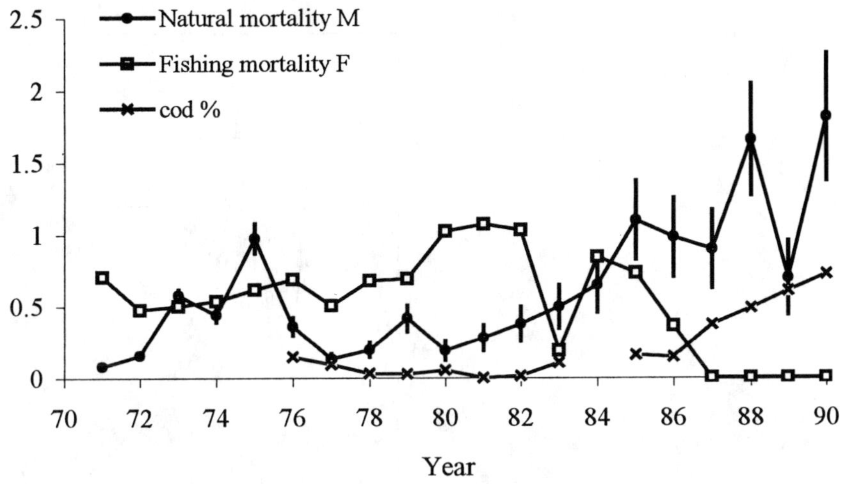

Figure 3. Estimated natural mortality (±1 SE) from 1970 to 1990 and fishing mortality estimates from 1970 to 1986 from the length-based model and relative frequencies of occurrence of Pacific cod obtained from the same shrimp trawl surveys.

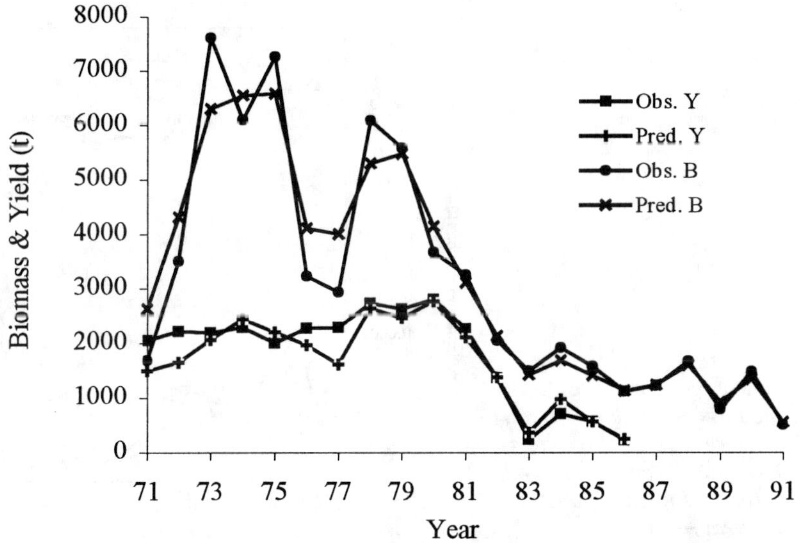

Figure 4. Comparisons between observed and predicted biomass and yield from the length-based model.

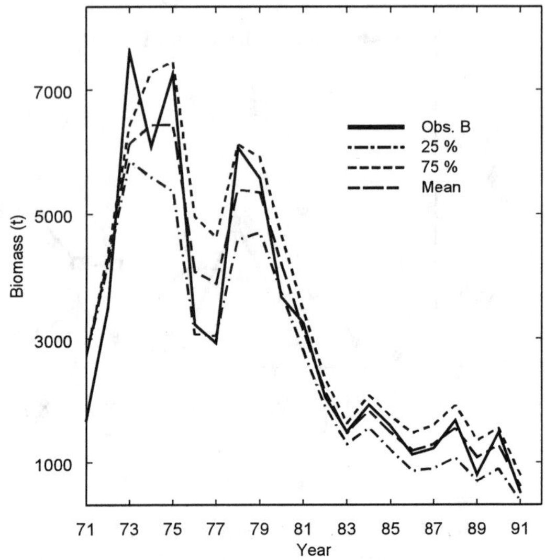

Figure 5. Projected biomass at the mean level and 25th and 75th percentiles from the scenario with estimated natural mortality and recruitment.

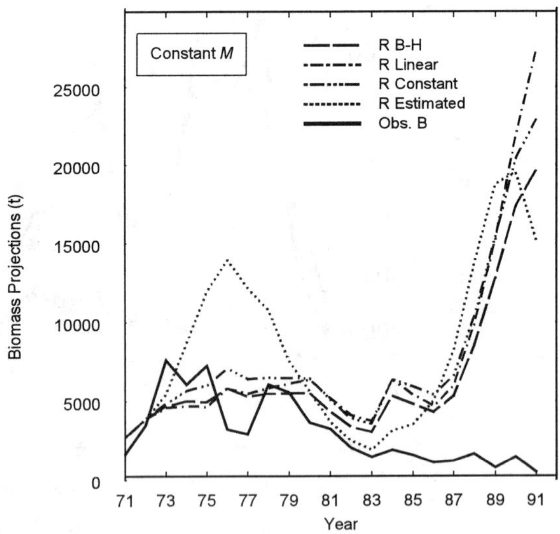

Figure 6. Comparisons of biomass projections among four recruitment models: B-H, linear, constant R, and estimated R using constant natural mortality.

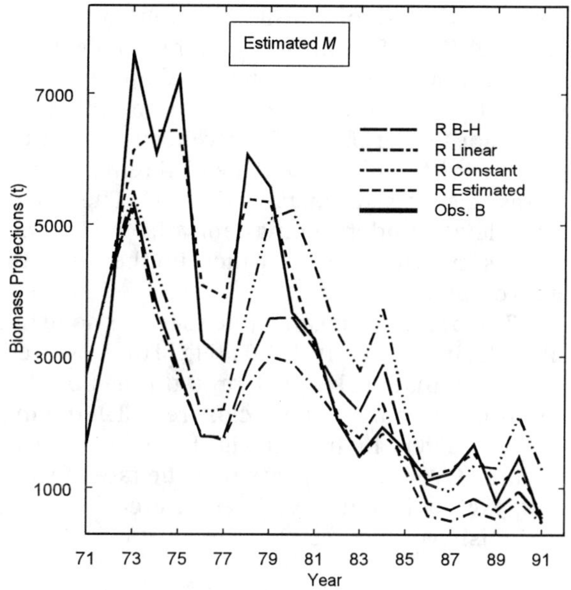

Figure 7. Comparisons of biomass projections among four recruitment models: B-H, linear, constant R, and estimated R using estimated natural mortality.

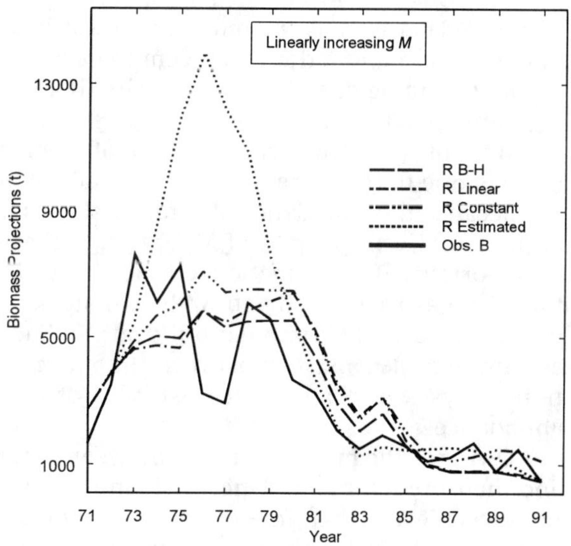

Figure 8. Comparisons of biomass projections among four recruitment models: B-H, linear, constant R, and estimated R using constant natural mortality before 1980, and a linearly increasing function for years after 1980.

also a lower risk of quasi-extinction compared with those under harvest levels of F_{est} or 50% of the F_{est} (Table 2).

The population had a high background risk of quasi-extinction in 1991 (56%). The risk of quasi-extinction in 1991 became larger as fishing mortality got higher or the fishery closure was implemented later. Nevertheless, the difference among the different harvest strategies was not large in terms of the risk of quasi-extinction. The risk of going below various biomass levels under fishing mortality of F_{est} and no fishing mortality had great similarity (Fig. 9), which implies that the population decline was unavoidable.

The projected mean biomass under no fishing mortality was extremely high during the early 1980s (Fig. 10). However, there was a dramatic decline after that peak, which brought the population down to a level similar to the observed. A fishery closure in 1980 temporarily brought the population to a very high level (Fig. 11), which was also followed by a dramatic decline. Overall, it appears that the fate of the pink shrimp population to collapse could not have been altered by lessening fishing or executing early fishery closures.

Discussion

Because of parameter confounding, measurement error, and/or process error, the accuracy of *M* and *R* estimates is unknown. Nevertheless, the retrospective simulation with various sets of *M* and *R* values indicated that the conclusion from the simulation is robust to the assumed *M* or *R* patterns. That is, *M* is the most important factor controlling the population dynamics and *R* does not seem to be as critical in this case.

Both parameter estimation and simulations indicated that an increasing trend in natural mortality in the 1980s had led the pink shrimp population to collapse. The serial depletion of other crustacean stocks in Alaska is likely due to the same reason: high natural mortality.

Because the abundance of forage fishes is highly correlated with that of predators, a major part of *M* for forage species may come from predation mortality. The dramatic increase in the abundance of shrimp predators, such as Pacific cod and walleye pollock, may have directly resulted in the increase in shrimp natural death in Kachemak Bay. Therefore, the assumption of constant post-recruitment *M* will inevitably bias population analyses under the circumstances of dramatic change in predator abundance.

Owing to the high correlation between natural death of forage species and their predator abundance, the temporal variation in *M* is likely to be more visible and therefore easier to study than that in *R*, which is generally driven by many physical environmental factors. This study leads us to conclude that more effort is desirable to investigate dynamics of predator populations in order to obtain more realistic population analyses and projections for forage species. Applying this conclusion to other pink

Table 2. Projected total yield and risk of quasi-extinction using estimated *M* and B-H *R*.

Scenario	*F*	Total yield	1987 risk[a]	1991 risk[a]
1	0	0.0	5.38%	56.02%
2	0.25 F_{est}	31.66	11.90%	60.22%
3	0.5 F_{est}	43.14	24.58%	75.26%
4	0.75 F_{est}	45.90	43.04%	71.68%
5	F_{est}	44.86	66.40%	78.34%
6	1.5 F_{est}	38.95	96.68%	89.84%
7	Closure 1980	31.60	6.90%	57.08%
8	Closure 1982	39.64	11.32%	60.88%
9	Closure 1984	43.19	25.08%	67.74%

[a] Risk is measured by the percentage of simulations in which biomass is less than 10% of the historical maximum.

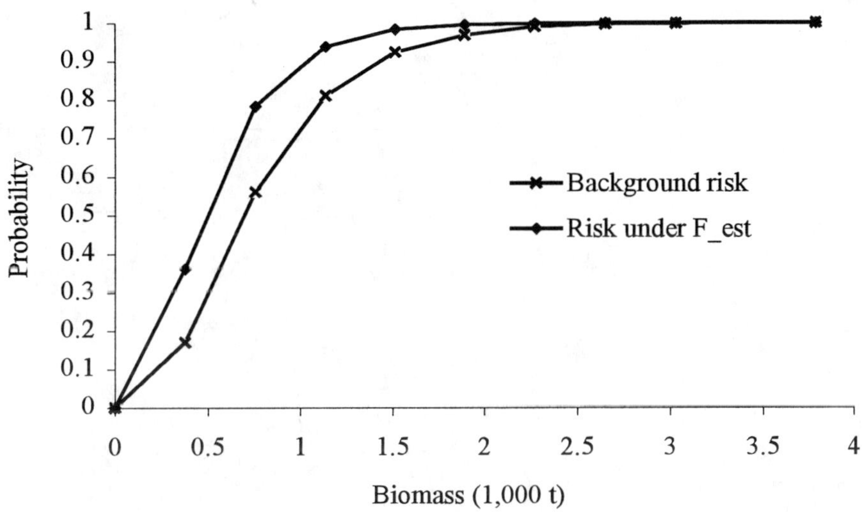

Figure 9. Comparisons between the background risk of biomass going below various levels and the risk when fishing at level F_{est}, using the scenario with the estimated M and predicted recruitment from the B-H model with autocorrelation in 1991.

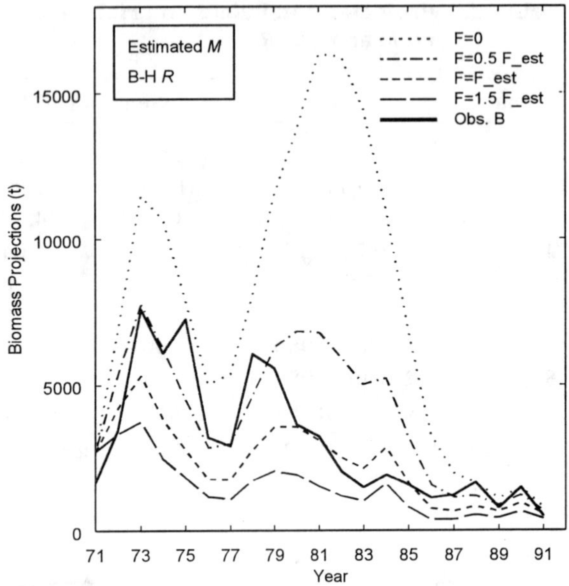

Figure 10. Comparisons of biomass projections under fishing mortality F = 0, 0.5 F_{est}, F_{est}, and 1.5 F_{est} using the scenario with estimated natural mortality and predicted recruitment from B-H model with autocorrelation.

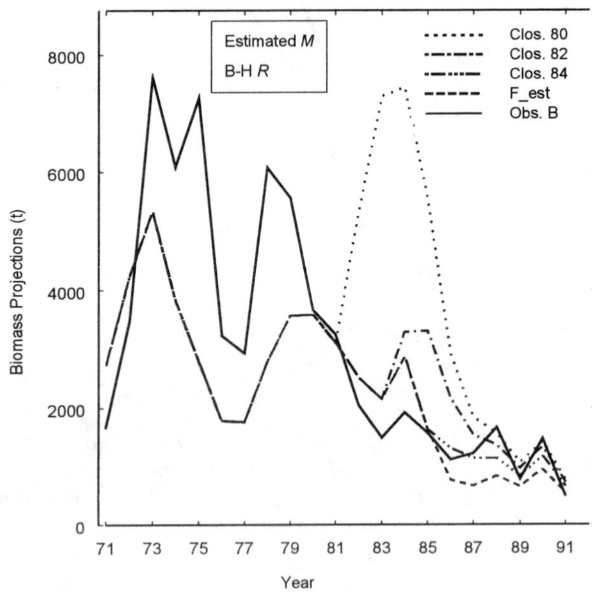

Figure 11. Comparisons of biomass projections under fishery closure in 1980, 1982, and 1984 using the scenario with estimated natural mortality and predicted recruitment from B-H model with autocorrelation.

shrimp populations, particularly those that are still under commercial fishing, prudent study of *M* is important. Models incorporating the relationship between *M* and the abundance of predators should be helpful for projecting the dynamics of forage species.

Acknowledgments

We appreciate the comments and support of Dr. Gordon Kruse, Alaska Department of Fish and Game. This publication is the result of research sponsored by Alaska Sea Grant with funds from the National Oceanic and Atmospheric Administration, Office of Sea Grant, Department of Commerce, under grant number NA90AA-D-SG066, project numbers R/07-21 and R/31-03, from the University of Alaska with funds appropriated by the State of Alaska. The U.S. Government is authorized to produce and distribute reprints for government purposes, notwithstanding any copyright notation that may appear hereon.

References

Anderson, P.J. 1991. Age, growth, and mortality of the northern shrimp *Pandalus borealis* Krøyer in Pavlof Bay, Alaska. Fish. Bull., U.S. 89:541-553.

Bechtol W.R. 1997. Changes in forage fish populations in Kachemak Bay, Alaska, 1976-1995. In: Forage fishes in marine ecosystems. Proceedings of the International Symposium on the Role of Forage Fishes in Marine Ecosystems. University of Alaska Sea Grant, AK-SG-97-01, Fairbanks, pp. 441-455.

Burgman, M.A., S. Ferson, and H.R. Akcakaya. 1993. Risk assessment in conservation biology. Chapman and Hall, London. 309 pp.

Cohen, M.D., and G.S. Fishman. 1980. Modeling growth-time and weight-length relationships in a single year-class fishery with examples on North Carolina pink and brown shrimp. Can. J. Fish. Aquat. Sci. 37:1000-1011.

Davis, A.S. 1982. The commercial otter trawl shrimp fishery of Cook Inlet. October 1982. Alaska Department of Fish and Game Report sheets.

Ferson, S. 1993. Using fuzzy arithmetic in Monte Carlo simulation of fishery populations. In: G.H. Kruse, D.M. Eggers, R.J. Marasco, C. Pautzke, and T.J. Quinn II (eds.), Proceedings of the International Symposium on Management Strategies for Explored Fish Populations. University of Alaska Sea Grant, AK-SG-93-02. Alaska Fairbanks, pp. 595-608.

Ferson, S., L. Ginzburg, and A. Silvers. 1989. Extreme event risk analysis for age-structured populations. Ecol. Modell. 47:175-187.

Francis, R.I.C.C. 1992. Use of risk analysis to assess fishery management strategies: A case study using orange roughy (*Hoplostethus atlanticus*) on the Chatham Rise, New Zealand. Can. J. Fish. Aquat. Sci. 49:922-930.

Gustafson, R. 1994. Trawl shrimp index fishing in the southern district of the Cook Inlet management area (Spring 1992 and 1993). Alaska Department of Fish and Game, Regional Information Report 2A94-23.

Orensanz, J.M., J. Armstrong, D. Armstrong, and R. Hilborn. 1998. Crustacean resources are vulnerable to serial depletion: The multifaceted decline of crab and shrimp fisheries in the Greater Gulf of Alaska. Rev. Fish Biol. Fish. 8:117-176.

Otter Research Ltd. 1996. An introduction to AD Model Builder: For use in nonlinear modeling and statistics. Otter Research Ltd., Nanaimo, B.C. 51 pp.

Quinn II, T.J., R. Fagen, and J. Zheng. 1990. Threshold management policies for exploited populations. Can. J. Fish. Aquat. Sci. 47:2016-2029.

Quinn II, T.J., C.T. Turnbull, and C. Fu. 1998. A length-based population model for hard-to-age invertebrate populations. In: F. Funk, T.J. Quinn II, J.N. Ianelli, J.E. Powers, J.F. Schweigert, P.J. Sullivan, and C.-I. Zhang (eds), Fishery stock assessment models. University of Alaska Sea Grant, AK-SG-98-01, Fairbanks, pp. 531-556.

Zheng, J., T.J. Quinn II, and G.H. Kruse. 1993. Comparison and evaluation of threshold estimation methods for exploited fish populations. In: G.H. Kruse, D.M. Eggers, R.J. Marasco, C. Pautzke, and T.J. Quinn II (eds.), Proceedings of the International Symposium on Management Strategies for Exploited Fish Populations. University of Alaska Sea Grant, AK-SG-93-02, Fairbanks, pp. 267-290.

Appendix 1. Details of the Length-Based Model

Assume that the length frequency distribution of shrimp population at recruitment, before entering the fishery, is a simple discrete normal. The probability density function (PDF) for the lengths x at recruiting age r and time t is:

$$f_{r,t}(x) \sim N_D(\mu_{r,t}, \sigma_r^2) = \frac{e^{-\frac{1}{2\sigma_r^2}(x-\mu_{r,t})^2}}{\xi_{r,t}}, \quad \text{where } \xi_{r,t} = \sum_x e^{-\frac{1}{2\sigma_r^2}(x-\mu_{r,t})^2}.$$

The abundance of the newly recruited shrimp at length x is: $N_{r,t}(x) = N_{r,t} \times f_{r,t}(x)$. To obtain abundance $N_{a+1,\,t+1}(x)$ from abundance $N_{a,t}(x)$ at age a (starting at age r) and time t, the process of growth was modeled first followed by mortality. The standard von Bertalanffy growth model is utilized to express the growth pattern from one age to next, assuming that growth is constant over time:

$$L_{a+1} = L_\infty(1-\rho) + \rho L_a + \varepsilon, \text{ where } \rho = e^{-\kappa}$$
is the Brody coefficient, and $\varepsilon \sim N(0, \sigma^2)$.

The expected length and variance at age a+1 for an individual of length x at age a are from Cohen and Fishman (1980):

$$\mu_{a+1}(x) = L_\infty(1-\rho) + \rho x, \quad \text{and} \quad \sigma_{a+1} = \sigma^2 \frac{1-\rho^{2(a+1-r)}}{1-\rho^2} + \rho^{2(a+1-r)}\sigma_r^2.$$

After one growth increment, the relative distribution of lengths $p_{a+,t+}(L)$ is:

$$p_{a+,t+}(L) = \sum_x f_{a,t}(x) e^{-\frac{1}{2\sigma_{a+1}^2}[L-\mu_{a+1,t+1}(x)]^2}.$$

Assume full recruitment fishing mortality at time t is proportional to the ratio of observed yield and estimated biomass from the survey at time t with a deviation term e^{vt}:

$$F_t = \frac{Y_t^{obs}}{B_t^{obs}} e^{v_t}.$$

The selectivity function follows the common assumption of a logistic curve:

$$S_x = \frac{1}{1+e^{-r(x-L_{50\%})}}$$

where γ is the shape parameter and $L_{50\%}$ is the length at which 50% of the individuals are vulnerable to fishing.

Absolute abundance, $N_{a+1,\ t+1}(L)$, at the start of time t+1 after fishing mortality $F_t s_x$ and natural mortality M_t is:

$$N_{a+1,t+1}(L) = N_{a,t} p_{a+,t+}(L) e^{-(M_t+F_t s_x)},$$

where $M_t + F_t s_x = Z_{t,x}$ is the total mortality at time t, and natural mortality M_t is assumed to be constant for all length intervals after recruitment.

Abundance is summed over all ages for length L at time t to get length distributions of the population at time t:

$$N_{t,x} = \sum_a N_{a,t,x}.$$

Length frequency distributions in the sample at time t are:

$$LF_{t,x} = n_t \frac{N_{t,x} s_x}{\sum_x N_{t,x} s_x},$$

where n_t is the sample size for measuring length.

The expected survey biomass at time t is:

$$B_t = \sum_x N_{t,x} s_x W_x,$$

where weight at length is modeled as the allometric relationship: $W_x = \alpha^* x^{\beta^*}$. The parameters α^* and β^* were obtained independently from the survey conducted in spring 1997, the same time of year when survey biomass indices and length frequency distributions were collected. We assume the length-weight relationship has no interannual variation. Estimated yield from the model at time t is:

$$Y_t = \sum_x N_{t,x} \frac{F_{t,x}}{Z_{t,x}} (1 - e^{-Z_{t,x}}) w_x.$$

Expected female abundance at time t and length x is: $FN_{t,x} = N_{t,x} \chi_{t,x}$, where $\chi_{t,x}$ is the female proportion at length x and time t. This proportion is modeled as a logistic function:

$$\chi_x = \frac{1}{1+e^{-\delta(x-P_{50\%})}},$$

where $P_{50\%}$ is the length at which 50% individuals are composed of females. Note that the female proportion parameters were estimated from the survey data outside the model to avoid complications.

Parameters were estimated by minimizing the objective function:

$$f = \lambda_B \sum_t (\sqrt{B'_t} - \sqrt{B_t})^2 + \lambda_L \sum_t \sum_l (\sqrt{LF'_{t,l}} - \sqrt{LF_{t,l}})^2 + \lambda_Y \sum_t (\sqrt{Y'_t} - \sqrt{Y_t})^2$$
$$+ \lambda_R \sum_t [\ln(R'_t) - \ln(R_t)]^2 + \lambda_M \sum_t \delta_t^{\,2},$$

where variables B'_t, LF'_t, and Y'_t denote observed values, while those without primes denote corresponding predicted values. Square root transformation of data is more desirable than a logarithm transformation (Quinn et al., In press). The component $\lambda_R \sum_t [\ln(R'_t) - \ln(R_t)]^2$ is for estimating the spawner-recruitment parameters α, β and ϕ in the Beverton-Holt (B-H) *S-R* relationship with first-order autocorrelation, which fitted the recruitment estimates best:

$$\ln(R'_t) = \ln(\alpha S_{t-1}) - \ln(1 + \beta S_{t-1}) + \phi\varepsilon_{t-1} + \upsilon_t = \ln(R_t) + \upsilon_t,$$

where R'_t are observed values, $\phi\varepsilon_{t-1}$ is the first-order autocorrelated error term, and $\upsilon_t \sim N(0, \sigma^2)$. The last part of the objective function, $\lambda_M \sum_t \delta_t^2$, is a penalty term controlling the degree of variability in the time series of M from year to year.

The weighting factors λs for the observations were determined based on their accuracy and variance ratio ($\lambda_B = \lambda_Y = 1$, λ_L (1971 to 1974 and 1978 to 1991) = 4, and λ_L (1975 to 1977) = 1). Weights λ_R and λ_M were determined based on the assumptions on the variation in the spawner-recruitment parameters and M (λ_R=5, and λ_M=10).

Variations in Abundance of Year Classes of Threadfin Hakeling (*Laemonema longipes*, Moridae) in the Northwestern Pacific Corresponding to Oceanographic Conditions

Vadim P. Pavlychev and Andrew B. Savin
Pacific Scientific Research Fisheries Centre (TINRO-Centre), Vladivostok, Russia

Extended Abstract

This paper gives data on the biology of threadfin hakeling (*Laemonema longipes*) and compares year-class abundance and subarctic water regimes at 149°E for the period 1965-1990. A quasi-seven-year period, caused by the "pole" tide, is present in the variations of threadfin hakeling year-class abundance and the subarctic water regime. It is assumed that high year-class abundance of threadfin hakeling will occur in the late 1990s and early 2000s.

The threadfin hakeling is an important commercial bathyal fish of the northwest Pacific. It was first mentioned by Taranets (1937) and described by Shmidt (1938). Matsubara (1938) also described threadfin hakeling.

Significant aggregations of threadfin hakeling were found during the 1974 TINRO expedition to Alaid Hollow (Okhotsk coast of Paramushir Island) and waters off the east coast of Honshu. Russian fishermen began harvesting it in the same year. In the late 1980s and early 1990s the threadfin hakeling catch ranged between 17,000 t and 87,000 t, composing 66-88% of the general world catch of all species of Moridae for that time (FAO 1995).

The amount of water with temperatures less than 5°C at 149°E longitude between 43° and 33°N latitude at 0-1,000 m of depth during the summer is used as an indicator of interannual fluctuations of the water regime. Threadfin hakeling year-class abundance was estimated by adding the number of fish of the same year class caught in different years from one unit of effort.

Threadfin hakeling inhabit the continental slope and mesopelagial of the northwestern Pacific to the north of 33°N, the Sea of Okhotsk, the Bering Sea, and the Gulf of Alaska, but are most abundant near Japan and the Kuril Islands (Shmidt 1950, Rass 1954, Pautov 1980, Kodolov and Pautov 1986, Savin 1993). Catching threadfin hakeling near eastern Kamchatka, in the Bering Sea, and the Gulf of Alaska is rare and those regions are zones of its emigration paths. Such vast threadfin hakeling areas with different habitation conditions are caused by the presence of well-expressed seasonal and yearly migrations.

Temperature and salinity characteristics of threadfin hakeling habitat correspond in most cases to the warm intermediate layer of the subarctic water mass structure. According to Burkov and Arsenyev (1958) and Moroshkin (1964), temperature and salinity of this layer increase from 0.8°C and 33.3‰ in the Sea of Okhotsk to 4-6°C and 34.5‰ in the spawning area near southeastern Honshu (Fig. 1). The peak of threadfin hakeling density is usually observed at 400-800 m of depth.

On the basis of data obtained, the migration cycle of threadfin hakeling is presented in the following description (Savin 1993). In spring adult threadfin hakeling from the spawning grounds off the Idzu Archipelago migrate along the continental slope to the north at the foraging areas near Hokkaido and the Kuril Islands. In autumn after foraging they migrate back to the south and reach the spawning grounds in midwinter, completing the yearly migration cycle (Fig. 2).

After spawning, eggs and larvae are transported to the north in the northeastern branch of the Kuroshio that includes several gyres of an anticyclonic character-the peculiar "vortex" track. In their second year juveniles enter the Sea of Okhotsk and feed there until 3 years of age. They then enter through deepwater straits to the oceanic side of the Kuril Ridge and as they mature they migrate south toward Japan, recruiting to the spawning stock.

Spawning grounds of the threadfin hakeling are located near the southeastern coast of Honshu in the regions of the Boso Peninsula and the Idzu Archipelago (Fig. 3). They localize at depths of 400-800 m between the coast and the Kuroshio current (Yasui et al. 1979, Savin 1993). The Kuroshio, moving along Honshu, agitates the upwelling of relatively cold and low-salinity deep waters enriched by organic elements near the coast. Consequently, an environment favorable for threadfin hakeling spawning is created. The higher temperatures in the intermediate layers, compared with other areas, are apparently a required condition for final maturing and spawning. The Kuroshio, that carries more warm water masses from the south, limits the entrance of *L. longipes* to the south and southwest along the continental slope of the Idzu Archipelago and Honshu.

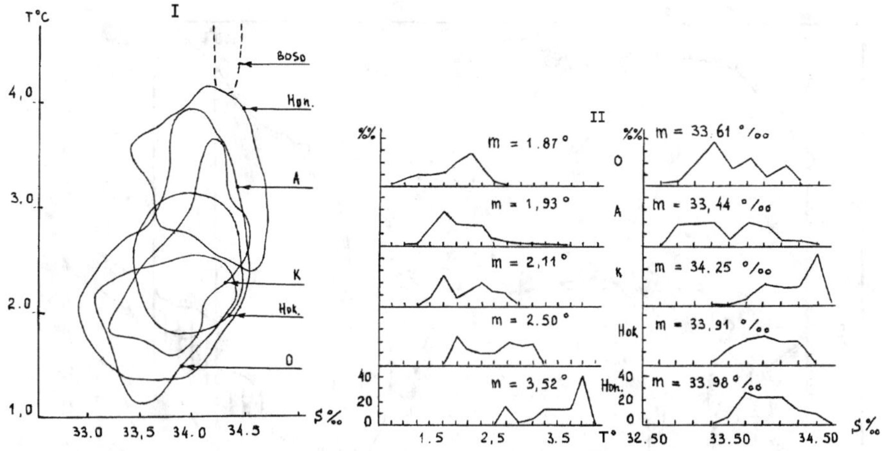

Figure 1. *Distribution of threadfin hakeling catch in the different regions corresponding to the bottom temperature and salinity for 1976-1989.*
I – Boundaries of catch in TS-plane
II – Change of average catch as % of their total
O – Sea of Okhotsk
A – Alaid Hollow
K – Kuril Ridge
Hok – Hokkaido Island
Hon – Honshu Island

It is obvious that variability in position of the Kuroshio and Oyashio currents has considerable effect on threadfin hakeling year-class abundance.

Threadfin hakeling is a fluctuating species. As investigations have shown, relatively high abundance of year classes of threadfin hakeling occurred in 1963, 1971, and 1978. Trawl survey data for 1984-1991 showed high abundance for the 1987 year class (Fig. 4). Judging from the amount of water of temperatures less than 5°C at 149°E during these years, the subarctic waters expanded to the east from Japan (Fig. 4). Moreover, from data for 1965-1990 the absolute maximum was observed in 1987. In spring 1963 a distinct cooling of the water near the southeastern coast of Honshu was observed as a consequence of the extreme southerly position of the coastal Oyashio branch (Fujimori 1964). As an analysis showed, a quasi-seven-year period, caused by the "pole" tide (Maximov 1970), is present in the variations of threadfin hakeling year-class abundance and subarctic water regimes. In addition, data resulting from spectrum analysis show the presence of a similar cycle in interannual dynamics of the Kuroshio

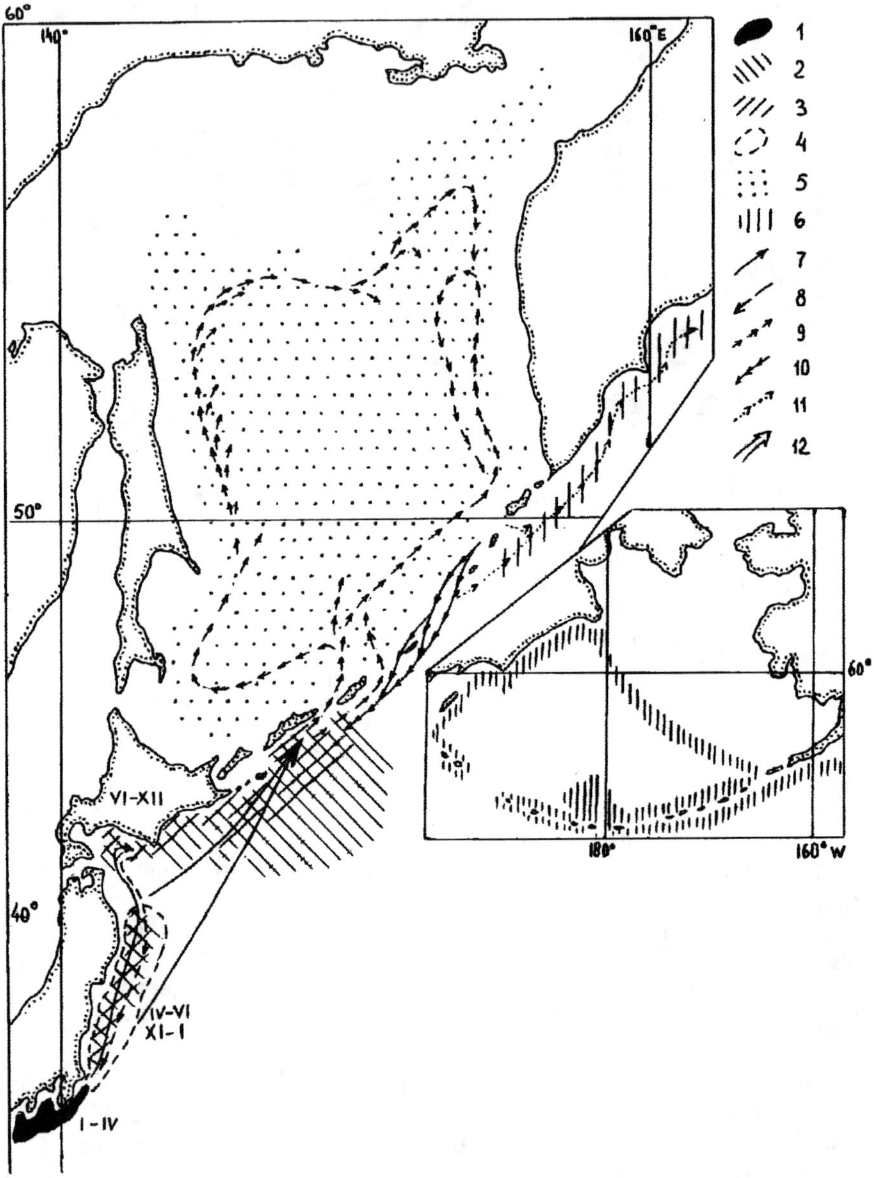

Figure 2. Scheme of threadfin hakeling distributions and migrations. Distributions: 1 – spawning grounds; 2 – mature females; 3 – mature males; 4 – region of primary gravid females and larvae; 5 – region of juvenile habitat; 6 – zone of emigration paths. Migrations: 7 – postspawning; 8 – prespawning; 9 – juvenile; 10 – first maturing young; 11 – emigration; 12 – direction of juvenile transport. Roman numerals show the number of months of adult occupancy.

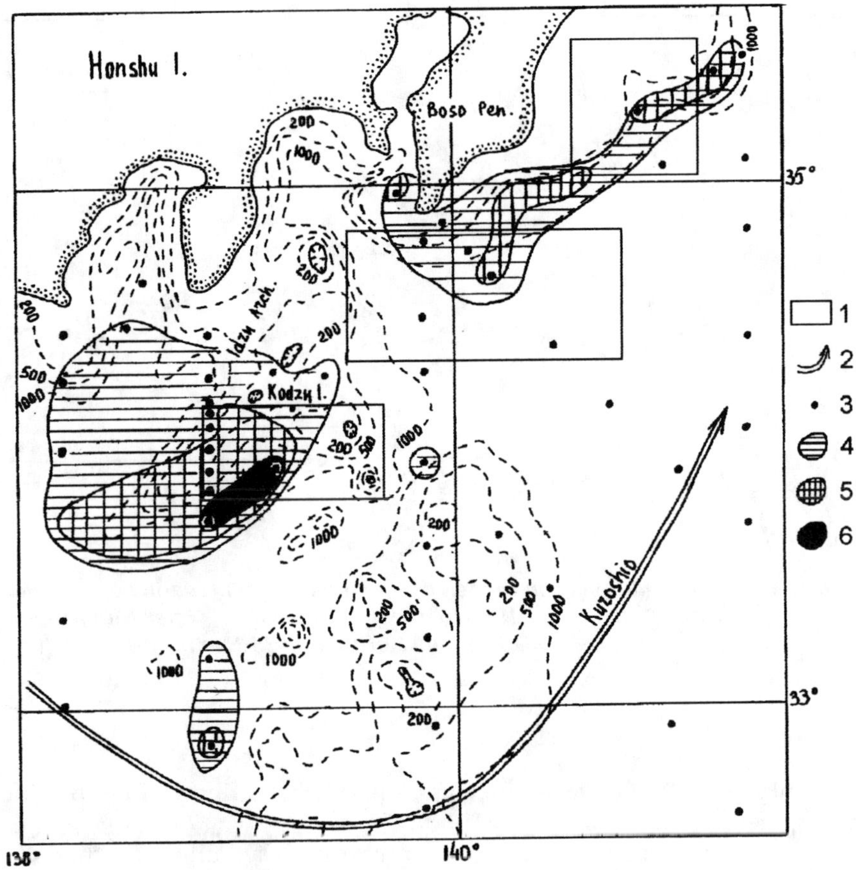

Figure 3. Distribution of threadfin hakeling eggs in March 1975.
1 – regions of threadfin hakeling removal
2 – Kuroshio axis
3 – egg station
Egg catch per one set (Yasui et al. 1979):
4 – 1-10 eggs
5 – 10-100 eggs
6 – >100 eggs

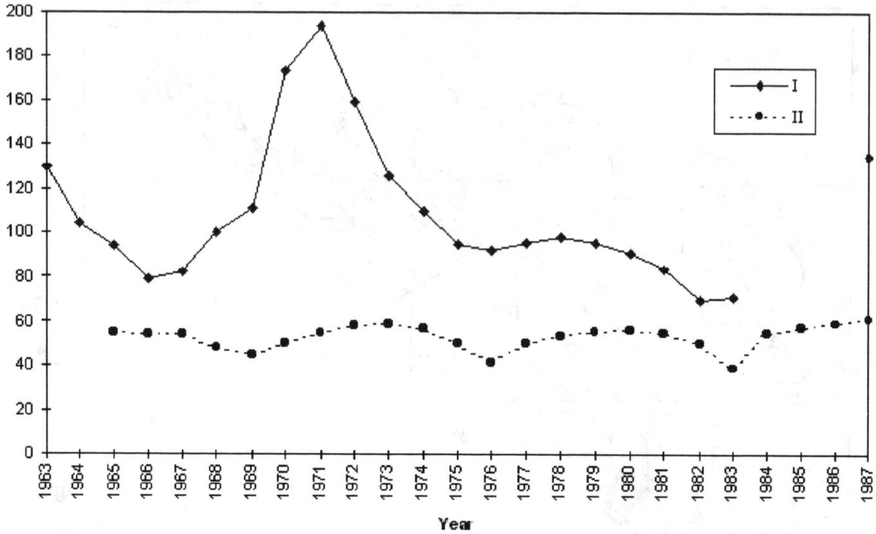

Figure 4. Annual variations in the relative abundance of threadfin hakeling year classes (%) to level of 1968 (I) and in the amount of water with temperature less than 5°C at 149°E between 43° and 33°N in the 0-1,000 m layer (II).

cyclonic meander between 135° and 140°E where a powerfully productive zone is formed (Pavlychev 1980).

Penetration of subarctic waters increases the suitability for threadfin hakeling spawning in this area and also has a favorable influence on the survival of early life stages.

The correlation between threadfin hakeling year-class abundance and subarctic water regime has great forecast importance, making it possible to predict threadfin hakeling year-class abundance. We predict that high abundance of year classes of threadfin hakeling will occur in the late 1990s and early 2000s.

References

Burkov, V.A., and V.S. Arsenyev. 1958. Experiment of water masses detaching in zone of contact of Kuroshio and Kuril currents. Trudy Inst. Okeanol. Akad. Nauk SSSR 27:5-11. (In Russian.)

FAO (Food and Agriculture Organization of the United Nations). 1995. Yearbook of fishery statistics, 1993. Catches and landings. Vol. 76. FAO, Rome, p. 187.

Fujimori, T. 1964. Preliminary report on the oceanographic anomaly in the early half of 1963 in the waters adjacent to Kanto District. Bull. Tokai Reg. Fish. Res. Lab. 38:77-98.

Kodolov, L.S., and G.P. Pautov. 1986. *Laemonema longipes.* Biological resources of the Pacific ocean. Moscow, pp. 181-186. (In Russian.)

Matsubara, K. 1938. Studies on the deep-sea fishes of Japan. IX. Description of a new gadid fish. Bull. Jpn. Soc. Sci. Fish. 7(2):61-62.

Maximov, I.V. 1970. Geophysical forces and waters of ocean. Hydrometeoizdat., Moscow, p. 447. (In Russian.)

Moroshkin, K.V. 1964. Water masses of the sea of Okhotsk. Moscow. 67 pp. (In Russian.)

Pautov, G.P. 1980. Distribution and biology of threadfin hakeling (*Laemonema longipes* Shmidt, 1938). Izv. TINRO 39:132-140. (In Russian.)

Pavlychev, V.P. 1980. Long-term fluctuations of the Kuroshio current. In: Kuroshio IV. Proceedings of the Fourth CSK Symposium, Tokyo, pp. 95-105.

Rass, T.S. 1954. To study of Pacific Ocean Moridae (Pisces, Gadiformes). Trudy Inst. Okeanol. Akad. Nauk SSSR 11:56-61. (In Russian.)

Savin, A.B. 1993. Distribution and migrations of threadfin hakeling (*Lamonema longipes*, Moridae) in the northwestern part of Pacific. Vopr. Ikhtiol. 33(2):190-197. (In Russian.)

Shmidt, P.Yu. 1938. Three new deep-sea fishes from the Sea of Okhotsk. Dokl. Akad. Nauk SSSR 19(8):653-656. (In Russian.)

Shmidt, P.Yu. 1950. Fishes of the Sea of Okhotsk. Akad. Nauk SSSR, Moscow. 370 pp. (In Russian.)

Taranets, A.Ya. 1937. Handbook for identification of fishes of Soviet Far East and adjacent waters. Izv. TINRO 11. 200 pp. (In Russian.)

Yasui, T., M. Mikawa, and M. Watanabe. 1979. Report on accurate investigations of unused resources at the shelf slope east of Honshu. In: Collection of reports on accurate investigations of unused resources at the shelf slopes. Tokyo, pp. 3-43.

Ecosystem Approaches for Fisheries Management
Alaska Sea Grant College Program • AK-SG-99-01, 1999

Monitoring Changes in Oceanographic Forcing Using the Carbon and Nitrogen Isotopic Composition of Prince William Sound Pelagic Biota

Thomas C. Kline Jr.
Prince William Sound Science Center, Cordova, Alaska

Abstract

Changes in the physical environment known to affect phytoplankton and zooplankton production can be linked to fish production through carbon and nitrogen stable isotope natural abundance measurements. Stable isotopic analyses of herbivorous copepods and juvenile fishes from Prince William Sound (PWS) and the northern Gulf of Alaska (GOA) were conducted as part of the Sound Ecosystem Assessment (SEA) project, a comprehensive multidisciplinary ecosystem study. The advective regime connecting the GOA with PWS was postulated by SEA to affect recruitment and nutritional processes in juvenile fishes. Herbivorous zooplankton, an indicator for pelagic production sources, had distinctive carbon isotope signatures when sampled in the GOA compared to those from PWS. PWS carbon had consistent spring bloom carbon isotopic signatures during 1994-1996 while GOA carbon differed in 1996. Nevertheless, PWS carbon was always distinctive from GOA carbon. This variation suggested that interannual fluctuations occurring at the food chain base are driven by processes in the gulf.

Analyses of nitrogen isotope ratios and C/N ratios of juvenile fishes were used to normalize their carbon isotope ratios, enabling determination of their relative affinity for GOA or PWS carbon. The data suggest a large affinity range, changing on annual time scales, consistent with observed oceanographic phenomena. For example, there was a shift to a greater dependency on GOA carbon in 1995 compared with 1994 and 1996. A parallel shift to increased GOA-originating copepods undergoing diapause (resting phase) in 1995 suggesting an influx of GOA zooplankton, provided a second line of evidence. Thus herring and other fishes partially dependent on GOA carbon are subject to vagaries of carbon flow that fall under the domain of physical oceanographic processes connecting the GOA with PWS as well as processes occurring on the GOA continental shelf adjacent to PWS.

Introduction

Cycles in the size of fish and shellfish populations have had dramatic effects on Gulf of Alaska coastal fisheries of Alaska (McDowell et al. 1989). Variations in natural stable isotope abundance provide evidence that the advective regime connecting the northern Gulf of Alaska (GOA) with Prince William Sound (PWS) can alter nutritional processes in PWS pelagic food webs (Kline 1999). Gradients of stable isotope ratios of carbon in the PWS study area serve as effective tracers of energy supply due to conservative transfer of carbon isotope ratios between the lower trophic levels (phytoplankton to zooplankton to forage fishes, etc.) up to the top consumers. Fishes acquire these isotope ratios in response to the importance of the food in bulk body tissues (muscle and fat). Isotope ratio analysis of these tissues can provide insight into the effects of fluctuations of GOA carbon sources on PWS fish populations (Kline 1999).

The large interzonal herbivorous copepods of the genus *Neocalanus*, which comprise the bulk of the zooplankton during vernal zooplankton blooms, have a yearly life history pattern. They undergo a prolonged stage copepodite-V diapause at depth following termination of feeding at the end of the bloom. The diapaused copepods provide the "seed" for the following generation since reproduction takes place at depth in late winter. A portion of PWS has depths great enough to accommodate the *Neocalanus* diapause stage; i.e., depths > 400 m (Kline 1999). The offspring of *Neocalanus* diapausing in PWS form a source of primary consumers of the PWS spring bloom. A secondary source of *Neocalanus* are populations diapausing beyond the continental shelf in the GOA. The Alaska Coastal Current flow in PWS may transport *Neocalanus* into PWS while feeding or during ontogenetic migrations in the water column (Kline 1999).

Diapausing samples were analyzed for $^{13}C/^{12}C$ for comparison with the regional isotope gradient obtained during the bloom to assess input of GOA copepods after feeding is complete. Concomitant shifts in the $^{13}C/^{12}C$ content of diapausing copepods and fishes in PWS during the 1994-1996 study period suggested a common system-wide cause.

Materials and Methods

Copepods were collected at the terminal feeding stage (Copepodite IV and V) throughout PWS and stations in the northern GOA and during diapause (Copepodite V and VI) at stations located in suitable habitat in PWS (Kline

1999). Juvenile Pacific herring and walleye pollock were collected throughout PWS during 1994-1997 from multivessel broad-scale surveys (Kline 1999). Laboratory handling procedures, isotopic analysis preparation, and data analysis were as described by Paul et al. (1998) and Kline (1999).

Normalization for lipid (DeNiro and Epstein 1977) and trophic level effects on $\delta^{13}C$ values of fish enabled "fingerprinting" of their carbon source (Kline 1997, 1999). The McConnaughey and McRoy (1979) lipid normalization based upon the C/N ratio using a C/N = 4 as the base level was used. The Kline (1997) method for the normalization for trophic level effects used: (1) the $\delta^{15}N$ value of the herbivore, *N. cristatus*, as the trophic level baseline (Vander Zanden et al. 1997); and (2) an assumed constant N and C trophic fractionation factor ratio of 3.4:1. In general, normalization reduces sources of ^{13}C variability, enabling comparisons among species without the confounding effects of trophic level and lipid content (Kline 1997). Thus, normalized $^{13}C/^{12}C$ content of fishes could be compared directly with *N. cristatus* at a consistent lipid storage level (Kline 1999).

The expressions $\delta^{13}C$, $\delta^{13}C'$, $\delta^{13}C_{TL}$, or $\delta^{13}C'_{TL}$ denote $^{13}C/^{12}C$ abundance in relation to the international standard (Vienna PDB), normalized for lipid content, normalized for trophic level, and normalized for lipid content and trophic level, respectively, using conventional delta notation (the per mil [‰] deviation from the international standard). Whereas $\delta^{13}C'$, $\delta^{13}C_{TL}$, and $\delta^{13}C'_{TL}$ are used in accordance to a particular data analysis context, "^{13}C" is used to reflect generic $^{13}C/^{12}C$ isotopic trends irrespective of normalization. Similarly, $^{15}N/^{14}N$ abundance is expressed relative to air N_2 as $\delta^{15}N$ values.

Results

Neocalanus cristatus sampled throughout the PWS and GOA were consistently different (Kline 1999). The mean values plotted versus time show a consistency for PWS carbon $\delta^{13}C'$ of −19 to −20‰ (Fig. 1). GOA carbon was usually ^{13}C-depleted having $\delta^{13}C'$ values of −22 to −23.5‰. However, in 1996, *Neocalanus cristatus* (as well as other *Neocalanus* species) shifted to an anomalous high $\delta^{13}C'$ of ~ −17.5‰ (Fig. 1).

The isotopic composition of *Neocalanus cristatus* diapausing in PWS include those with signatures consistent with those feeding in PWS (upward arrows, Fig. 2) and those feeding in the GOA (downward arrows, Fig. 2). While mean $\delta^{15}N$ was consistently ~ +8‰ each year, the relative proportion of high and low $\delta^{13}C'$ varied (Fig. 2.). While the contribution of PWS and GOA copepods was approximately the same in 1994, most of those in 1995 and 1996 originated from the GOA.

Juvenile herring and pollock < 100 mm long shifted in $\delta^{13}C'_{TL}$ content in concert during 1994-1997 (Fig. 3). The lowest $\delta^{13}C'_{TL}$ values measured in juvenile fishes in 1995 coincided with low $\delta^{13}C'_{TL}$ in diapausing *Neocalanus cristatus* (Fig. 2).

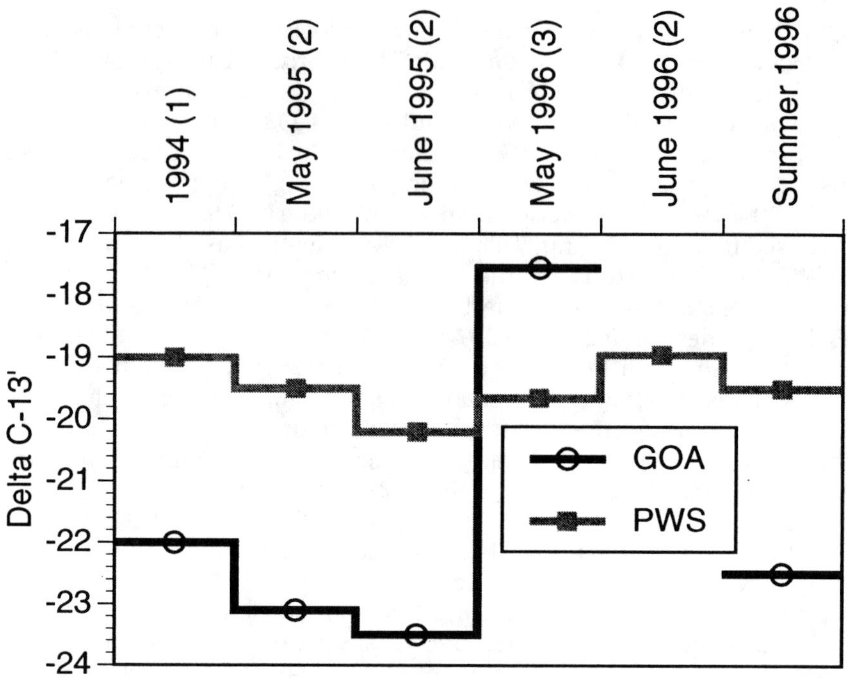

Figure 1. Time series of $\delta^{13}C'$ measured in feeding Neocalanus cristatus from Prince William Sound (PWS) and the Gulf of Alaska (GOA), 1994-1996. Points indicate mean values, standard deviations were 0.5-1‰. PWS and GOA values were consistently statistically different (t-test $P < 0.05$). (1) Mean values inferred from diapaused population modes. (2) Mean values of terminal feeding stages sampled with 0.5 m diameter plankton net. (3) Mean values of terminal feeding stages sampled with MOCNESS sampler. (4) Mean values inferred from diapaused population modes (Fig. 2: 1996).

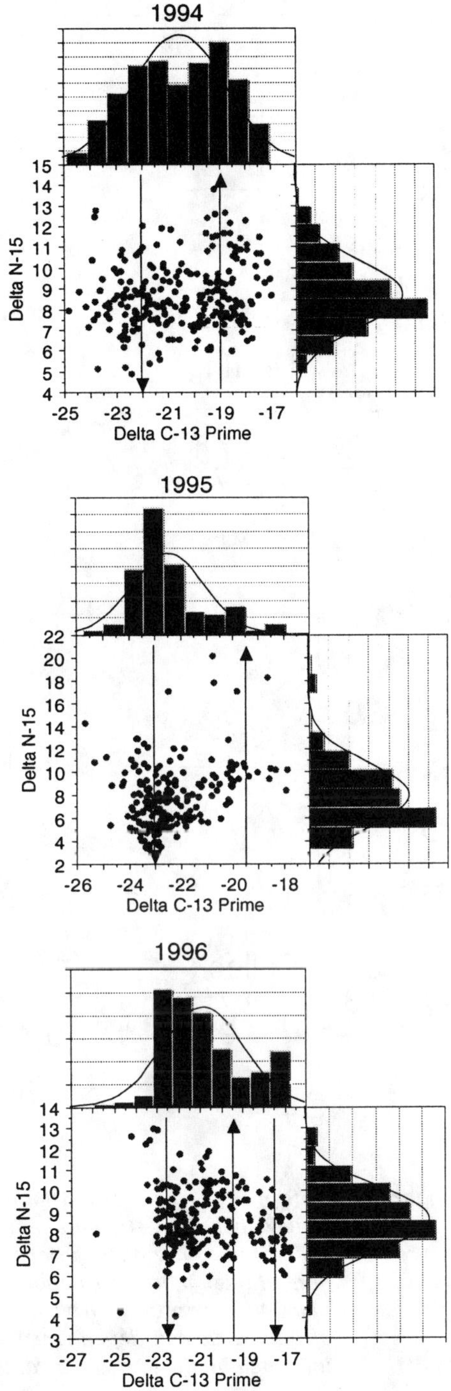

Figure 2. The $\delta^{13}C'$ composition of Neocalanus cristatus diapausing in Prince William Sound (PWS) from three cohorts. The upward-pointing arrows indicate $\delta^{13}C'$values of Neocalanus cristatus originating in PWS which were inferred from mode values (1994) or measured in spring sampling (1995 and 1996; Fig. 1). The downward-pointing arrows indicate $\delta^{13}C'$ values of Neocalanus cristatus originating in the Gulf of Alaska which were inferred from mode values (1994 and −22.5 value in 1996) or measured in spring sampling (1995 and −17.5 value; Fig. 1). The isotopic shifts indicate a greater extent of zooplankton transport into PWS in 1995 than 1994 and 1996.

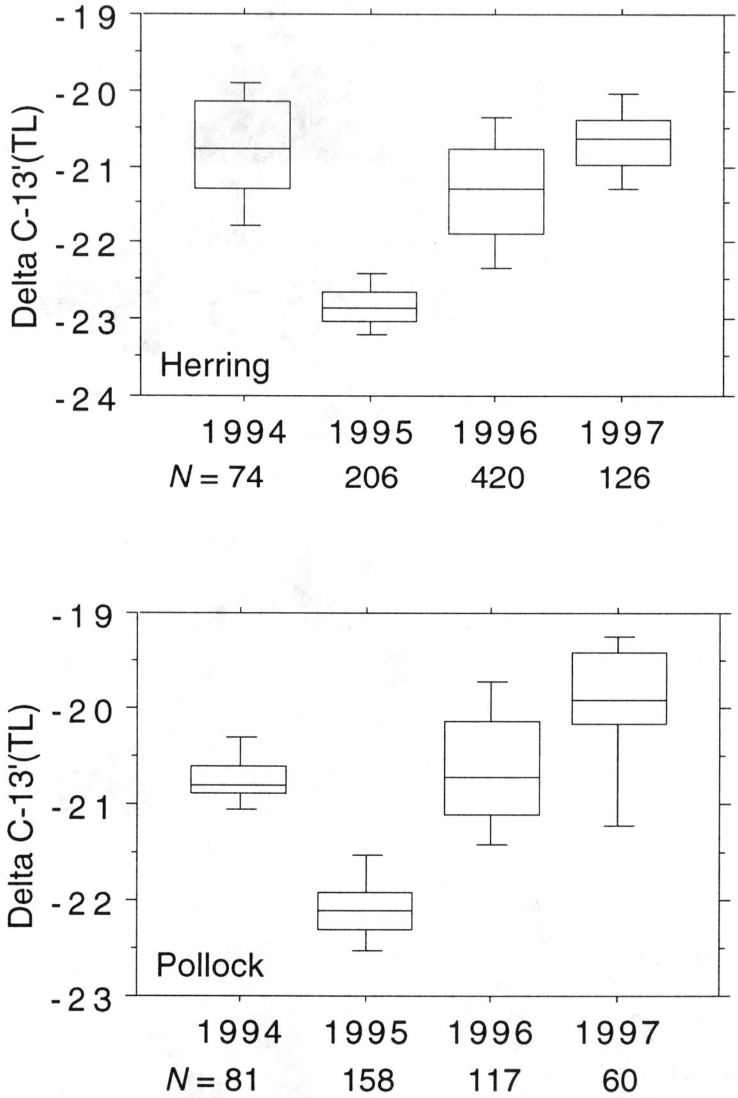

Figure 3. Shift in $\delta^{13}C'_{TL}$ for Prince William Sound Pacific herring and walleye pollock < 100 mm long in 1994-1997; sample sizes as indicated. The distributions of values are shown as box and whisker plots that denote the 10th, 25th, 50th, 75th, and 90th percentiles. There was a large shift to greater Gulf of Alaska carbon dependency in 1995 for both species as indicated by the low $\delta^{13}C'_{TL}$ values.

Discussion

Regional Isotopic Gradient and Interannual Variability

There was a consistent isotopic gradient between PWS and GOA in the 1994-1996 period except for May 1996 when the gradient reversed owing to a large magnitude change in the GOA signature (Fig. 2). Whereas PWS mean $\delta^{13}C'$ values ranged within 1‰, and the difference between PWS and GOA averaged 3‰, the GOA mean value shifted in spring 1996 by 5‰. This large shift probably reflected a change in phytoplankon fractionation during uptake of CO_2 which varies as a function of growth rate (Laws et al. 1995). Thus the productivity pattern during the spring bloom of 1996 was markedly different from other times. These isotopic values corroborate model predictions made for spring 1996 that suggest poor GOA primary and zooplankton productivity (Eslinger and Cooney 1997). If fluctuations in GOA productivity are typical, they could affect year-to-year inconsistencies in food availability for consumers. Thus the question arises: Are fluctuations in GOA spring bloom $\delta^{13}C'$, and inferred growth rate changes (Laws et al. 1995), typical?

Juvenile Fish Carbon Source Dependencies

A change in carbon source dependency for fishes can be inferred from their $\delta^{13}C'_{TL}$ shifts (Fig. 3). Juvenile herring and pollock are the dominant pelagic fishes in PWS and both consume zooplankton. Juvenile herring and pollock from PWS shifted in $\delta^{13}C'_{TL}$ content during this study from which a change in carbon source dependency can be inferred (Fig. 1). Although both species shifted in concert to lower $\delta^{13}C'_{TL}$ suggesting a greater GOA dependency in 1995 than 1994, pollock were generally 0-1‰ more positive in $\delta^{13}C'_{TL}$ suggesting a lower dependency on GOA carbon. Juvenile pollock and herring occupy different levels in the water column, have different schooling behavior, and recruit from the larval stage at different times, effecting access to a different forage base as confirmed by differences in the data. This difference may not be reflected in the species composition of diet, but instead, the where and when of the production cycle as integrated into the isotopic signature (Kline 1999), which reflects the assimilated carbon pool of the fish. The greater reliance on GOA-derived carbon in herring may reflect their dependence on carbon generated later in the season during the time when advection of GOA production was nearly the sole carbon source in 1995 (Kline 1999).

Significance of Parallel Isotopic Shifts

The concordant shift to greater GOA dependency by both fish species throughout PWS in 1995 implied that system-scale oceanographic process-driven bottom-up effects permeated PWS. Furthermore, *Neocalanus cristatus* diapausing in fall 1995 principally came from the GOA. A common process that explains these observations is the physical transport of water containing zooplankton (both copepods entering diapause and zooplankton fish forage) into PWS from the GOA sometime after the spring bloom via the Alaska Coastal Current or deepwater renewal (Kline 1999). The greater degree of GOA carbon dependency by fishes and the greater proportion of GOA copepods in diapause in 1995 than other years suggests that the influx of GOA water was stronger that year. Variations in zooplankton transport into PWS among years are thus evidenced. Such system-wide changes suggest that there are year-to-year inconsistencies in the food web base and fish production potential for PWS and thus should be a concern for management of exploited populations.

Ecosystem Considerations for Management

Fishes living in PWS but dependent on GOA carbon are subject to changes in carbon flow that result from physical oceanographic processes (e.g., currents and gradients of temperature and salinity) that connect the GOA with PWS. It is conjectured that affinities range from total dependency on PWS carbon to significant input of carbon from the GOA. Fishes wholly dependent on PWS are more likely to be directly affected by internal PWS processes. Increased competition for PWS carbon by all species, however, may occur if GOA carbon is less available to those that normally use it. Shifting to increased dependency on PWS carbon by species with normal affinity for GOA carbon during years of poor GOA carbon availability would provide evidence of competition for a limited carbon supply by the increasing overlap in their $^{13}C/^{12}C$ signature. Increased competition for PWS carbon by all species, however, may occur if GOA carbon is less available to those that normally use it. Time-series measurements of natural stable isotope abundance in fishes combined with data on fish populations and cogent oceanographic measurements will enable a new understanding of how basic environmental processes affect fish recruitment and interaction.

Acknowledgments

This project was funded by the *Exxon Valdez* Oil Spill Trustee Council through the Sound Ecosystem Assessment (SEA) and Pacific Herring Productivity Dependencies projects. However, the findings presented by the author are his own and not necessarily the Trustee Council's position. SEA colleagues collected fish specimens. J.M. Paul and P. Shoemaker did the initial fish sample preparation. Captain Mike Gaegel, the crew of the R/V *Bering Explorer*, E. Stockmar, J. Murphy, B. Barnett, and T. Cooney assisted with *N. cristatus* sampling. Kim Antonucci and John Williams did the laboratory processing of samples for stable isotope analysis. Bruce Barnett and Norma Haubenstock at the University of Alaska Fairbanks Stable Isotope Facility performed the stable isotope analysis.

References

DeNiro, M.J., and S. Epstein. 1977. Mechanism of carbon isotope fractionation associated with lipid synthesis. Science 197:261-263.

Eslinger, D.L., and R.T. Cooney. 1997. Biophysical modeling and validation through remote sensing. *Exxon Valdez* Oil Spill Trustee Council Annual Report (Restoration Project 96320-R), Alaska Department of Fish and Game, Anchorage.

Kline, T.C. 1997. Confirming forage fish food web dependencies in the Prince William Sound ecosystem using natural stable isotope tracers. In: Forage fishes in marine ecosystems. Proceedings of the International Symposium on the Role of Forage Fishes in Marine Ecosystems. University of Alaska Sea Grant, AK-SG-97-01, Fairbanks, pp. 257-269.

Kline, T.C. 1999. Temporal and spatial variability of $^{13}C/^{12}C$ and $^{15}N/^{14}N$ in pelagic biota of Prince William Sound, Alaska. Can. J. Fish. Aquat. Sci. 55(Supple. 2).

Laws, E.A., B.N. Popp, R.R. Bridigare, M.C. Kennicut, and S.A. Macko. 1995. Dependence of phytoplankton carbon isotopic composition on growth rate and $[CO_2]aq$: Theoretical considerations and experimental results. Geochem. Cosmochim. Acta 59:1131-1138.

McConnaughey, T., and C.P. McRoy. 1979. Food-web structure and the fractionation of carbon isotopes in the Bering Sea. Mar. Biol. 53:257-262.

McDowell, E., J. Calvin, and N. Gilbertsen. 1989. Alaska seafood industry: A technical report. Distributed by University of Alaska Sea Grant, Fairbanks. 96 pp.

Paul, A.J., J.M. Paul, and E. Brown. 1998. Fall and spring somatic energy content for Alaskan Pacific herring (*Clupea pallasi* Valenciennes 1847) relative to age, size and sex. J. Exp. Mar. Biol. Ecol. 223:133-142.

Vander Zanden, M.J., G. Cabana, and J.B. Rasmussen. 1997. Comparing trophic position of freshwater fish calculated using stable nitrogen isotope ratios ($\delta^{15}N$) and literature dietary data. Can. J. Fish. Aquat. Sci. 54:1142-1158.

Marine Survival of Puget Sound Coho Salmon: Deciphering the Climate Signal

William D. Pinnix
University of Washington, School of Fisheries, Seattle, Washington

Abstract

Each year the State of Washington forecasts the number of adult coho salmon (*Oncorhynchus kisutch*) that will return to Washington waters. The abundance forecast is conducted jointly by the Washington Department of Fisheries and Wildlife (WDFW) and the Northwest Indian Fisheries Commission in January and February of each year to aid the Pacific Fisheries Management Council in West Coast salmon allocation. This forecast relies on accurate predictions of marine survival combined with hatchery and wild smolt abundance estimates. Historical estimates of the number of hatchery and wild smolts have been quite accurate, whereas marine survival for Puget Sound hatchery coho has been underestimated in the recent past.

Recent studies have implicated localized oceanographic conditions at the time of ocean entry as the critical phase of marine survival for Pacific salmon (*Oncorhynchus*) species. Coded wire tag (CWT) data are used to calculate time series of coho salmon marine survival for Puget Sound, the Washington coast, west coast of Vancouver Island, and the Strait of Georgia. Several climatic variables related to Puget Sound's freshwater input are analyzed by principal component analysis.

A conceptual model is presented to visualize how interannual climatic variability affects marine survival of coho salmon in Puget Sound. Changes in climatic variables coincide with changes in marine survival at the decadal scale. The Strait of Georgia and Puget Sound show different patterns in marine survival of hatchery coho salmon due to differences in the freshwater input to volume ratio of these two estuarine systems.

Author's current address: 2030 Marine Science Center Dr., Newport, OR 97365.

Introduction

The Pacific Fishery Management Council conducts salmon fishery management off the coasts of Washington, Oregon, and California. This complicated and epic task relies upon preseason estimates of abundance provided by state-level fishery management agencies. Every year the Washington Department of Fish and Wildlife (WDFW) and the Northwest Indian Fisheries Commission jointly forecast the abundance of both natural and hatchery adult coho salmon (*Oncorhynchus kisutch*) that will return to Washington coastal waters and Puget Sound.

The current Puget Sound forecast methodology makes separate forecasts for wild and hatchery fish. Wild forecasts are separated into four management units (Skagit, Stillaguamish-Snohomish, South Sound, and Hood Canal), each with its own forecast methodology. The 1997 (1994 brood year) forecast of *wild* marine survival incorporated an average wild marine survival from the brood years 1985-1992 (Seiler 1997); 1997 *hatchery* returns were forecast in a similar manner, but using the hatchery average of the 1988-1992 (brood years) marine survival estimates. The weakness in the forecast is the application of the f-year average of marine survival because it under-forecasts large runs and over-forecasts low runs. Recent forecasts (1990-1995) have over-forecasted preseason natural coho salmon abundance by more than 250% in some river systems (PFMC 1997), due to over-estimating marine survival.

Spatial and temporal variations exist at the regional scale in coho salmon marine survival , a measure of salmon productivity. Coronado-Hernandez (1995) calculated marine survival from smolt to adult of hatchery coho salmon from coded wire tag (CWT) data dating to 1971. Historically, coho salmon from the Strait of Georgia and Puget Sound (long-term average ≈ 8.5%) had much higher marine survival than their coastal counterparts (long-term average ≈ 3.9%), the west coast of Vancouver Island and the Washington coast. Puget Sound coho salmon marine survival peaked at 17% in 1984 and Strait of Georgia coho salmon marine survival peaked at 26% in 1973. In addition, Coronado-Hernandez (1995) found no relationship between length of hatchery operation and marine survival of coho salmon.

An internal memo from the WDFW states "presently, no correlation with ocean environmental conditions has been found to explain the observed inter-annual variation in [Puget Sound] marine survival" (Seiler 1997). It is hoped that by carefully examining the oceanography and interannual climate variability in Puget Sound that a robust relationship between marine survival and ocean conditions will be found. Hatchery fish are chosen because of large release numbers (approximately 35 million smolts per year), and a long time series (1971-1995) stored in an easily accessible database (Pacific States Marine Fisheries Commission, Regional Mark Information System). Wild and hatchery coho salmon from the same system show remarkable similarity in patterns of marine survival, although wild survival is typically much higher than hatchery survival (Fig. 1).

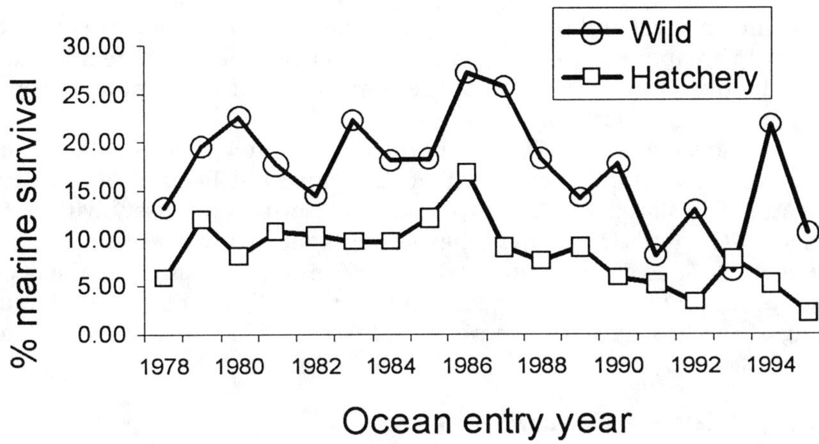

Figure 1. Puget Sound hatchery and wild coho salmon marine survival. Adapted from Seiler (1997).

Marine survival, a measure of salmon production, is established during *early marine residency* (Hare and Francis 1995). Evidence from the Oregon coast suggests that the majority of marine mortality occurs in the first 3 months of ocean residency (Pers. comm., William Pearcy, Oregon State University, Corvallis, OR 97331, February 1997). Beamish et al. (1995) found synchronous timing of an increase in mean temperature of the Strait of Georgia and marine survival of hatchery reared chinook salmon (*O. tshawytscha*) released into the Strait of Georgia. Francis and Hare (1994) found a relation between a climate index and subsequent returns of salmon in Alaska, adjusted for year of entry. This evidence suggests that the effects of climate variability on salmon are manifested in the first 3 months of ocean residency.

Marine survival is thought to be a function of the ability of coho salmon smolts to outgrow the initial suite of predators, and therefore a function of growth rate. Holtby et al. (1990) found that growth during the first few months of life in the ocean, was well correlated with total marine survival. It is thought that growth is related to availability of food.

Gargett (1997) proposed a hypothesis that juvenile salmon pass through a "stability window" on their seaward migration. This mechanistic hypothesis attempts to explain the apparent oscillation in production of Pacific salmon between Alaska and the Pacific Northwest. If oceanic conditions create an optimal stability, then in theory, conditions favoring higher than average plankton production will ultimately benefit coho salmon in higher growth rates, marine survival, and production. On the other hand, if oceanic conditions are less than optimal then juvenile salmon marine survival and production would be less than average. Thus a "bottom-up" response is proposed as the link between climate and salmon production, and is the basis of the North Pacific trophic model proposed by Francis et al. (1998).

Coho salmon from Puget Sound spend about 3 months in Puget Sound before entering the open ocean (Pers. comm., Bill Tweit, WDFW, Olympia, WA 98501, March 1997), so their critical period is entirely within Puget Sound. Gargett's (1997) hypothesis adapts nicely to Puget Sound, because Puget Sound is an estuarine system. Therefore, by examining the oceanographic components of Puget Sound we should gain a better understanding of the link between the physical forces of climate and the biology of coho salmon.

Puget Sound Oceanography

Exchange of water between Puget Sound and the Strait of Juan de Fuca is driven by the difference in density, or angle of the pycnocline (Ebbesmeyer et al. 1989). These different water masses (Puget Sound is estuarine and the Strait of Juan de Fuca is more oceanic) meet at Admiralty Inlet (Fig. 2). Differences in river runoff, changes in wind direction, and the gradient in temperature, regulate the density difference across Admiralty Inlet. Changes in the exchange and inflow depth of oceanic water are driven by decadal fluctuations (Fig. 3) in atmospheric circulation over the North Pacific Ocean (Ebbesmeyer et al. 1989). This atmospheric forcing, in turn, drives changes in climate over the North Pacific Ocean and Puget Sound.

The unique circulation properties of the main basin of Puget Sound regulate residence time and stratification properties of the sub-basins (Fig. 2). Residence time in the lower layer of the main basin is up to 3 weeks (Ebbesmeyer and Barnes 1980) but organisms riding the currents, such as plankton, may stay resident in Puget Sound much longer because of the recycling of surface waters that occurs at Admiralty Inlet. As much as 60% of the surface water is downwelled into the deeper waters of Puget Sound on every ebb tide (Strickland 1983:48).

Optimal conditions for phytoplankton blooms in Puget Sound occur when a balance between mixing (bringing nutrients to the surface), light, and stratification occurs (Strickland 1983:45). The coastal waters of Alaska are thought to be light limited (a function of stratification), whereas the coastal waters of the Pacific Northwest are thought to be mixing (and nutrient) limited (Gargett 1997). Mixing is important for bringing nutrients to the surface, and stratification is important for keeping phytoplankton near energy from the sun. Optimal phytoplankton conditions in Puget Sound lie somewhere between Ebbesmeyer et al.'s (1989) two different regimes of velocity profiles (Fig. 3), and therefore between heavily mixed and highly stratified.

This paper answers a number of specific questions. Does marine survival of hatchery coho salmon in Puget Sound show regime-like steps in

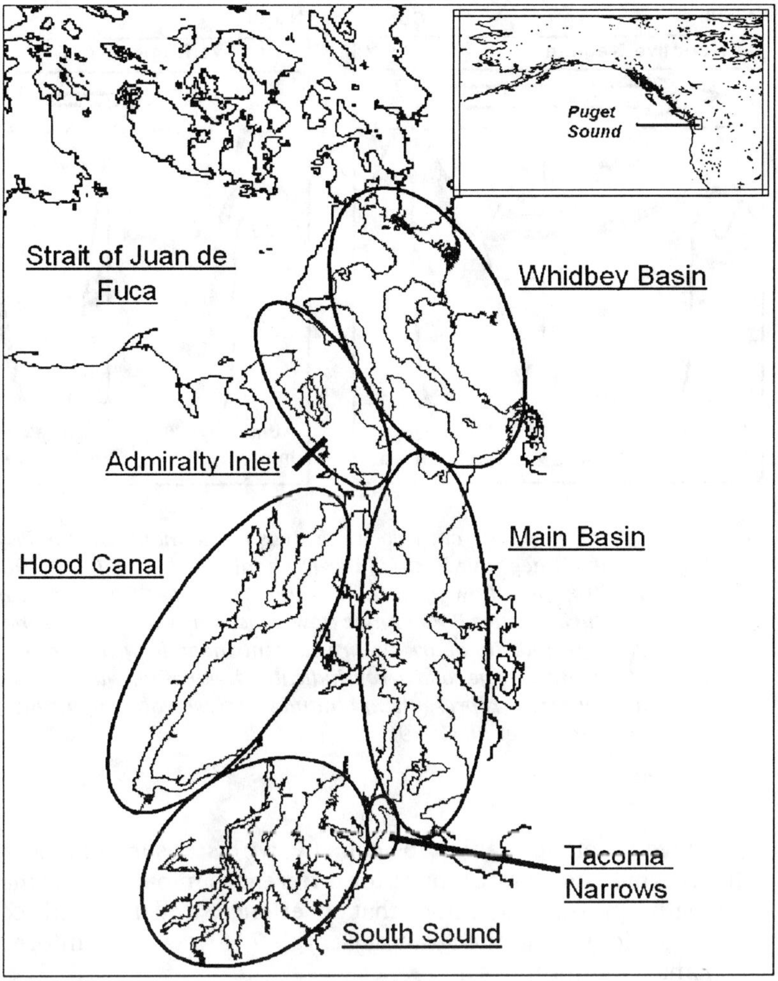

Figure 2. Sub-basins of Puget Sound.

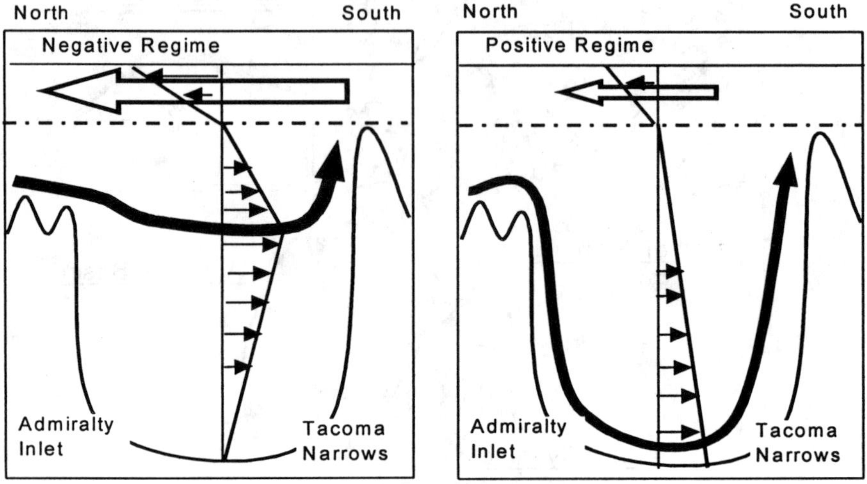

Figure 3. Schematic of current velocity profiles in the main basin of Puget Sound. During a negative Pacific inter-Decadal Oscillation, (PDO, Mantua et al. 1997) regime flow of oceanic water into Puget Sound is at mid-depth. During a positive PDO regime flow of oceanic water is near the bottom of Puget Sound. Small arrows are current shear, length representing speed of currents. Large solid arrow indicates depth of saltwater intrusion. Large hollow arrow represents magnitude of freshwater input. Adapted from Ebbesmeyer et al. (1989).

mean levels, and do these steps coincide with steps in climatic variables? What mechanisms may account for differences in patterns of marine survival between adjacent regions that experience similar climatic conditions (i.e., Puget Sound and the Strait of Georgia)? How can this information aid the management process?

Data and Methods

Coded wire tag data were obtained from the Pacific States Marine Fisheries Commission Regional Mark Processing Center. The processing center houses information pertaining to the mark, release, and catch of Pacific salmonids throughout the Pacific region involving the CWT marking project. Data exist from 1971. Each release group of smolts has a unique tag code. A hatchery in a given year may have a number of tag codes. Because coho salmon return as adults at different ages, it is necessary to standardize marine survival to one age class.

Marine survival by tag code is calculated as the ratio between the standardized number of tags recovered and the total number of smolts released with tags (equation 1). Tag recoveries are standardized to age 3 using virtual population analysis (VPA; Walters and Hilborn 1992) following Coronado-Hernandez (1995) (equation 2).

$$\text{Marine survival} = \frac{N_3}{N_{released}} \tag{1}$$

$$N_3 = C_1 s_1 s_2 + C_2 s_2 + C_3 + \frac{C_4}{s_4} + \frac{C_5}{s_4 s_5} + \frac{C_6}{s_4 s_5 s_6} + \frac{C_7}{s_4 s_5 s_6 s_7} \tag{2}$$

where

N_3 number of fish that would be alive if they were not caught, or spawned at an age different than age 3.

C_t number of fish recovered (caught + spawned [escaped]) at age t

S_t survival rate from age t–1 to t, 0.5 for all ages except age 3, which has no S component (Hilborn and Hare 1992).

The survival rates used in calculation of of numbers alive at age 3, do not make a large difference in the result of N_3 because over 90% of returning coho salmon are age 3 (Hilborn and Coronado 1997). The percent age composition of Puget Sound coho salmon from Coronado-Hernandez (1995) is: age 1, 0.01%; age 2, 6.68%; age 3, 91.20%; age 4, 1.91%; age 5, 0.17%; and age 6, 0.02%.

Yearly averages of marine survival estimates within Puget Sound were made for Hood Canal, South Sound, and Whidbey Basin (see Fig. 2). These sub-basins correspond to the management units in Puget Sound; the Skagit and Stillaguamish-Snohomish units both lie within Whidbey Basin. Standardized anomalies were computed for each sub-basin. Averages of the three sub-basin yearly standard anomalies were computed to form a Puget Sound marine survival average. Time series of marine survival were calculated for the Strait of Georgia, the west coast of Vancouver Island, and the Washington coast from CWT data.

Coded Wire Tag Selection Criteria

Coded wire tag codes were selected based on several quality and stock composition criteria. Tag codes were selected from hatcheries that are on or near marine waters. This was done to eliminate any freshwater influence that could possibly cloud the marine survival estimate. Survival estimates from these groups is thought to be representative of marine survival (Pers. comm., Bill Tweit, WDFW, Olympia, WA 98501, March 1997). Tag code notes were reviewed to eliminate any releases of fish that may have been diseased or impaired in any way that would inflate marine mortality. When possible, tag codes were selected for stocks of fish native to the region, because stocks of salmon are transported from their natal streams to be reared in hatcheries and released in non-natal waters. Hatchery managers in Puget Sound have begun releasing coho salmon smolts late in the summer (July-September) to encourage residency in the sound to enhance the sport fishery. Late release (after 1 July) tag codes were not included in the analyses.

Extended Coded Wire Tag Time Series

The coded wire tag time series for Puget Sound extends back to brood year 1970. The tagging program was in its infancy at that time, and recoveries of tags may have been underestimated (producing low marine survival estimates). In order to validate early tag code estimates, and to extend the marine survival time series back in time, the literature was reviewed for estimates of marine survival prior to 1975. Olson (1978) analyzed factors contributing to Puget Sound hatchery-reared coho salmon. I used survival estimates for brood years 1966-1974 from his table of smolt-to-adult survival estimates for a number of hatcheries in Puget Sound. Standardized anomalies were computed within these 9 years and are not included in rest of time series statistics because the estimates do not include a large portion of catch.

The time series for each sub-basin was extended another 3 years using data presented in Mathews (1980). The long-term mean (calculated from the CWT data) was removed from Mathew's marine survival estimates and was divided by the long-term standard deviation. The inclusion of Mathew's and Olsen's data allowed me to create a marine survival time series from broods 1964-1993 (ocean entry years 1966-1995). Mathew's data (1980) covers ocean entry years 1968-1970. Olsen's data (1978) covers ocean entry years 1970-1976. Coded wire tag data cover ocean entry years 1972-1995. Where time series overlapped, yearly mean anomalies were calculated.

The extended Puget Sound coho salmon marine survival index was regressed against total Puget Sound coho salmon run size. Run size was obtained from WDFW's run-reconstruction model (Pers. comm., Bill Tweit, WDFW, Olympia, WA 98501, March 1997).

Climate Variables

In this study a suite of variables, often used to describe the state of climate and the ocean, was examined. Each of the variables was summarized in terms of the water year (October of year *t*–1 through September of year *t*). For example, water year 1995 is October 1994 through September 1995. Seven environmental variables were collected from various sources. Table 1 presents the data gathered and the sources. All variables were averaged over the winter (October-March) and the summer season (April-September).

The time series of winter and summer standardized anomalies of these variables were calculated and a principal component analysis (PCA) was performed on this suite of climatic variables. The PCA analysis extracts vectors that explain the largest amount of variance in the combined time series. The significance of the components was calculated following North

Table 1. Monthly environmental variables collected and their sources.

	Dates	Source	Units
Air temperature	1895-1997	NCDC[a]	°F
Precipitation	1895-1997	NCDC	Inches
Sea surface	1950-1997	Neah Bay, SIO[b]	°C
temperature	1921-1998	Race Rocks, IOS[c]	
Snow pack	1926-1997	AWC, NWS[d]	Inches
Streamflow	1914-1997	USGS[e]	Ft^3/s

[a] National Climatic Data Center, Division 3 (Puget Sound Lowlands).
[b] Scripps Institution of Oceanography, Shore Station Database.
[c] Institute of Ocean Sciences, Victoria, British Columbia.
[d] Avalanche Warning Center, National Weather Service, Seattle, Washington.
[e] United States Geological Survey.

et al. (1982). The principal components represent the temporal structure of the variance of the combined time series while the loading vectors represent the spatial component, or contribution of each variable to the particular principal component.

Intervention Analysis

Intervention analysis (Francis and Hare 1994) was performed on the marine survival index and the first principal component of the climate variables. Principal component 1 was tested for the effects of interventions in the years 1942, 1977, and 1989, while the marine survival time series was tested for interventions in 1977 and 1989. Hare and Francis (1995) reported on major shifts in both climate and salmon stocks in the early 1940s and late 1970s, and Mantua et al. (1997) identified climatic shifts that occurred in 1925, 1942, and 1977. Recent changes in atmospheric circulation (Pers. comm., Tom Murphree, Naval Postgraduate School, Monterey, CA 93943, September 1998) and shifts in marine survival of Strait of Georgia coho salmon (Pers. comm., Richard Beamish, Pacific Biological Station, Nanaimo, B.C., Canada, V9R 5K6, June 1998) point toward significant biophysical shifts in 1989.

Results

Coded Wire Tag Data

The values of marine survival computed from coded wire tags, along with their associated standardized anomalies, are presented in Table 2. The extended Puget Sound coho salmon marine survival index expressed as standardized anomalies is presented in Fig. 4. Total Puget Sound coho salmon run size was plotted with the marine survival index (Fig. 5). The

Table 2. Percent marine survival of Puget Sound hatchery coho salmon.

Brood year	Hood Canal	South Sound	Whidbey Basin	Puget Sound average
1971	3.50 (–1.30)	1.35 (–1.26)	6.67 (–0.63)	4.04 (–1.11)
1972	4.05 (–1.17)	N/A	8.60 (–0.08)	6.33 (–0.62)
1973	6.62 (–0.56)	6.30 (–0.11)	9.53 (0.18)	8.88 (–0.16)
1974	9.35 (0.10)	N/A	15.86 (1.98)	13.95 (1.03)
1975	N/A	N/A	3.57 (–1.51)	3.57 (–1.51)
1976	N/A	3.70 (–0.72)	6.40 (–0.71)	5.86 (–0.72)
1977	N/A	12.10 (1.24)	12.12 (0.92)	11.95 (1.08)
1978	11.50 (0.60)	12.40 (1.31)	8.10 (–0.22)	8.20 (0.56)
1979	8.00 (–0.23)	12.70 (1.38)	11.18 (0.65)	10.67 (0.60)
1980	8.00 (–0.23)	7.23 (0.11)	11.96 (0.87)	10.31 (0.25)
1981	12.57 (0.86)	12.33 (1.29)	8.86 (–0.01)	9.67 (0.72)
1982	11.13 (0.52)	9.75 (0.69)	8.89 (0.00)	9.65 (0.40)
1983	15.27 (1.51)	15.08 (1.93)	11.49 (0.74)	12.02 (1.39)
1984	19.47 (2.51)	6.73 (–0.01)	16.23 (2.09)	16.84 (1.53)
1985	11.00 (0.49)	2.90 (–0.90)	11.49 (0.74)	8.96 (0.11)
1986	9.98 (0.24)	5.48 (–0.30)	8.63 (–0.07)	7.63 (–0.04)
1987	10.68 (0.41)	8.44 (0.39)	11.98 (0.88)	9.10 (0.56)
1988	4.07 (–1.17)	5.35 (–0.33)	7.45 (–0.41)	5.76 (–0.64)
1989	9.37 (0.10)	2.93 (–0.89)	5.49 (–0.97)	5.26 (–0.58)
1990	4.20 (–1.14)	4.18 (–0.60)	8.30 (–0.17)	3.31 (–0.63)
1991	10.01 (0.25)	1.57 (–1.21)	8.61 (–0.08)	7.74 (–0.35)
1992	7.56 (–0.33)	3.06 (–0.86)	5.20 (–1.05)	5.32 (–0.75)
1993	2.72 (–1.49)	1.88 (–1.14)	1.92 (–1.98)	2.19 (–1.54)

Values in parentheses are standardized anomalies. Values in the Puget Sound Average column are averages of the yearly values of the subbasins.

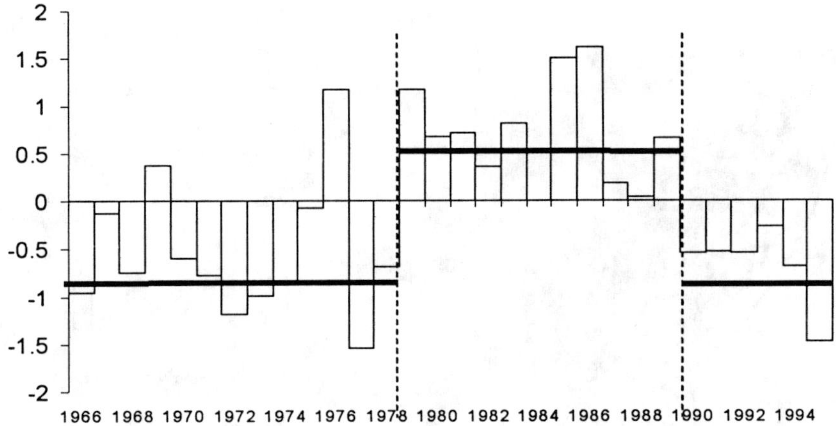

Figure 4. Puget Sound marine survival index, ocean entry years 1966-1995. Y axis is standard deviations from the mean. Dashed vertical lines represent significant interventions. Solid horizontal lines are mean levels for each regime.

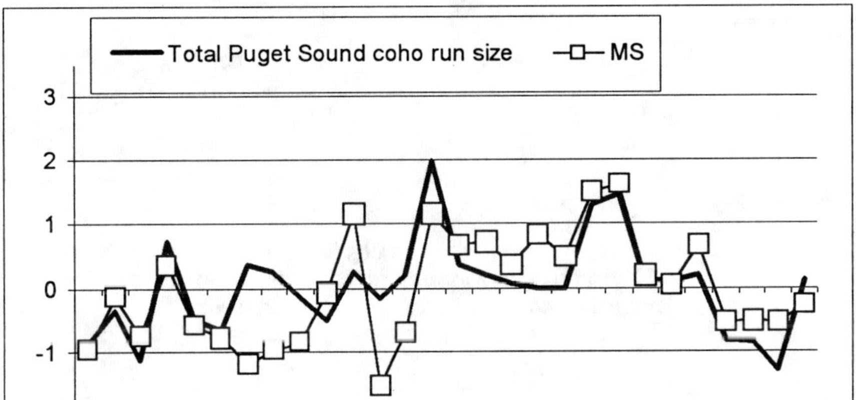

Figure 5. Puget Sound coho salmon run-size standard anomalies plotted with coded-wire-tag Puget Sound marine survival (MS) index. X axis is run-size year. Marine survival time series has been shifted forward 1 year to coincide with associated run size.

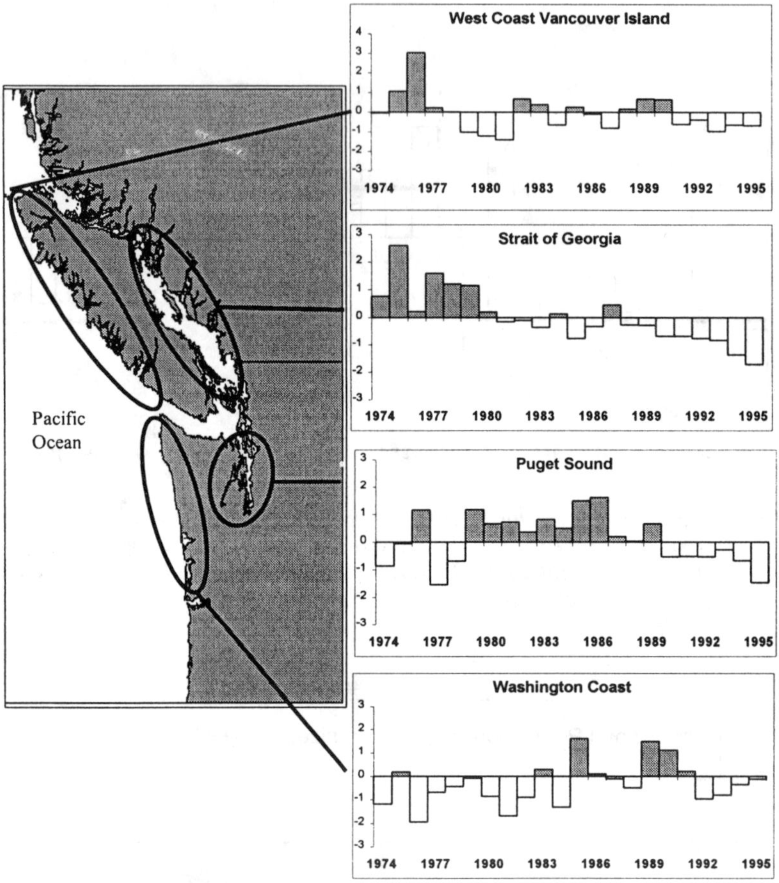

Figure 6. *Regional patterns of marine survival standard anomalies computed from coded wire tag data. Y axis represents number of standard deviations. X-axis is ocean entry year.*

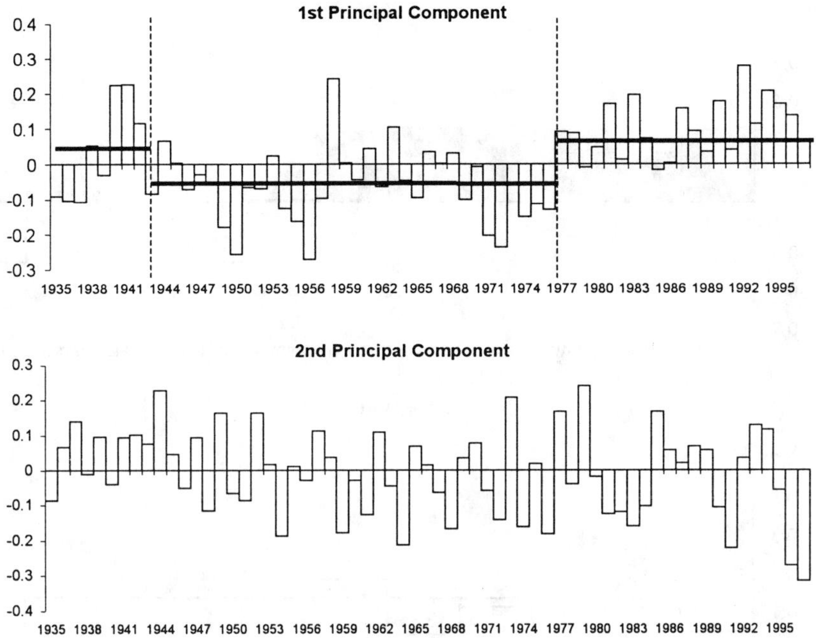

Figure 7. *First (PC1) and second (PC2) principal components of Puget Sound environmental variables. Dashed vertical lines represent significant interventions. Solid horizontal lines are mean levels for each regime.*

marine survival index accounted for 86% of the variance ($p < 0.001$) in total Puget Sound run size. Regional patterns in hatchery coho salmon marine survival are presented in Fig. 6.

Principal Component Analysis

The first two principal components are statistically significant and together account for 56.6% of the variance of the combined time series. Figure 7 presents the time series of the first two principal components and Fig. 8 presents the loading vectors of these components.

The first principal component (PC1) shows a strong interdecadal pattern similar to the Pacific inter-Decadal Oscillation (PDO, Mantua et al. 1997). The loading vector of PC1 suggests that the decadal pattern of climate in the Puget Sound region follows a warm-dry cold-wet pattern (Fig. 8, PC1). That is, when air and sea surface temperatures are warm, less rain and snowfall, making the region dry. In contrast when air and sea surface temperatures are cool, more precipitation falls and more snow accumulates.

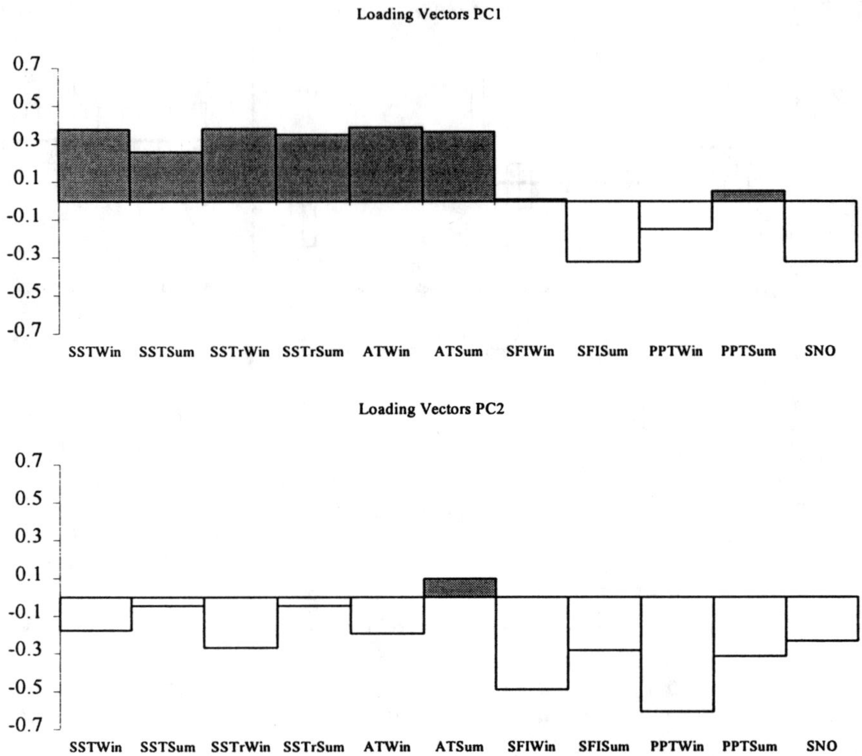

Figure 8. Loading vectors of Puget Sound environmental variables for PC1 and PC2. On the X axis, Win and Sum signify winter and summer mean values, respectively. SST is sea surface temperature at Neah Bay, Washington. SSTr is sea surface temperature at Race Rocks, British Columbia. AT is air temperature. SFI is streamflow at Concrete, Washington, on the Skagit River. PPT is precipitation. SNO is snowpack at Paradise Meadows, Mount Rainier, Washington. Y axis represents magnitude of loading vector.

The loading vectors of PC2 suggest that on an interannual time scale, the Puget Sound region follows a warm-wet and cool-dry pattern. That is, when air and sea surface temperatures are warm more precipitation falls, and the converse holds true.

Intervention Analysis

The extended Puget Sound marine survival index had significant ($p < 0.001$) interventions in ocean entry years 1978 and 1989. This differs from the interventions found by Mantua et al. (1997), and other studies that have identified a climatic shift occurring in 1977.

Intervention analysis shows that the PC1 time series shows significant ($p < 0.001$) steps in its sign in 1942 and 1977 calendar years. This is consistent with the regime shift concept put forth by Hare and Francis (1995) and supported by the findings of Mantua et al. (1997).

Discussion

The time series of Puget Sound coho salmon marine survival shows significant steps in mean levels in 1978 and 1989. These roughly coincide with changes in Puget Sound climatic data. The first principal component of Puget Sound climatic data shows significant steps in 1942, 1977, and 1989. So why the discrepancy in the late 1970s shift? The answer is twofold. First, although the first principal component captures a significant shift in climate, 1977 was a particularly dry year with very low streamflow with values comparable to the early 1990s. Second, since plankton production in Puget Sound is nonlinear, low as well as high values of streamflow will lead to lower plankton production, and therefore lower marine survival. With higher streamflow stability is greatly weakened because of enhanced estuarine circulation, and with lower than average streamflow, stability is enhanced because of retarded estuarine flow. This may seem counter-intuitive, but is supported by Ebbesmeyer (1989).

Gargett's (1997) optimal stability window (Fig. 9) that links oceanographic properties and salmon production is supported by Puget Sound climatic data. In Puget Sound this optimal stability window hypothesis can be framed in terms of the PDO. The first principal component of Puget Sound climatic data, PC1, has shifts in its sign that coincide with shifts in the PDO.

During the negative PDO of 1942-1976 more precipitation fell, the snowpack was higher, and temperatures were cooler. This would shift the frame of the stability window in Puget Sound to the left of optimal conditions (Fig. 9). With increases in runoff there is more vertical mixing in Puget Sound because of enhanced estuarine flow and it follows that stability would break down. Thus the period prior to 1977 would be a period where the conditions in Puget Sound were not met for optimal growth of the phytoplankton pathway that supports coho salmon, as reflected in lower marine survival during this period.

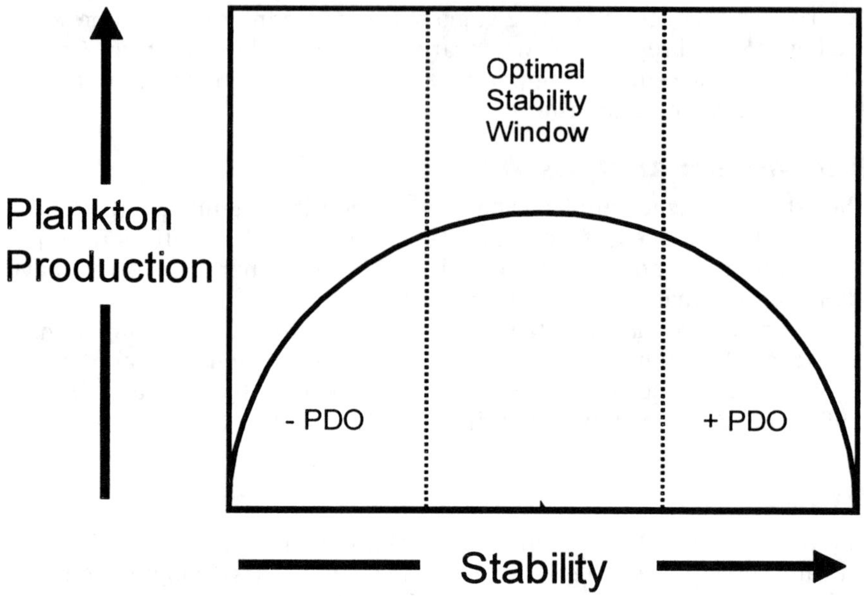

Figure 9. Schematic of Gargett's (1997) optimal stability window hypothesis, as it applies to hypothesized Puget Sound plankton production.

During the positive PDO of 1977-1984, the stability window would be framed to the right of the optimal window because of reduced runoff and retarded estuarine flow (Fig. 9). The period from 1976 to 1989 was a time of near normal streamflow, and following the hypotheses of Gargett (1997) and Strickland (1983), would be a period of optimal conditions necessary for the fish-enhancing phytoplankton pathway in Puget Sound.

The climate conditions in Puget Sound since 1990 seem to be an enhancement of the positive PDO signal. Air and sea surface temperatures are much higher than normal during both the winter and summer and higher than the period 1977-1989. Snowpack has been much lower than normal and streamflow has been sharply reduced during the summer months. These changes could be due to increases in mean global temperatures (Mann et al. 1998), or changes in global atmospheric circulation (Pers. comm., Tom Murphree, Naval Postgraduate School, Monterey, CA 93943, September 1998). Regardless of the drivers of these changes, impacts on the marine ecosystem of Puget Sound have been significant.

Why do we see different patterns in coho salmon marine survival in neighboring (Puget Sound and Strait of Georgia) estuarine systems? This optimal stability window should hold true for the Strait of Georgia. It too is an estuarine system similar in many properties to Puget Sound. Beamish et al. (1994) found a relationship between Fraser River discharge and interannual salmon production in the Strait of Georgia, relating to freshwater input into this estuarine system. A potential answer to the different patterns lies in the amount of freshwater input to volume ratio.

Puget Sound has a freshwater input (m^3/s) to volume (km^3) ratio of 13.0, while the ratio for the Strait of Georgia is 5.5 (Pers. comm., Rick Keil, University of Washington, Seattle, WA 98195, February 1997). Thus Puget Sound has more than twice the amount of freshwater input per volume. So in order to get to the optimal window in the Strait of Georgia, it would require very high streamflow, precisely the conditions prior to 1977. This explains why Strait of Georgia marine survival was highest prior to 1977. These same climatic conditions in Puget Sound would deluge the system with water shifting away from optimal plankton conditions

The optimal window hypothesis assumes that climate will force changes in absolute magnitude of plankton production. Brodeur (1990) showed that growth of Oregon coast coho salmon was related to plankton species assemblages and these assemblages were related to climatic conditions. The paucity of plankton data in Puget Sound prohibits testing either of these hypotheses at present. The optimal stability window is adequate in capturing the physical oceanography of Puget Sound, as supported by Ebbesmeyer's (1989) work. Comparing Puget Sound and Strait of Georgia coho salmon marine survival supported the optimal window hypothesis at least qualitatively.

How can fishery managers use this climate information to quantify coho salmon abundance? The impetus for this research was to improve the forecast of marine survival of Puget Sound coho salmon. This paper identified climate patterns of Puget Sound and presented a hypothesis on how these patterns alter the oceanography of this unique estuarine ecosystem. By examining regional differences and similarities in time series of marine survival, key components of the Puget Sound ecosystem (a linear combination of PC1 and PC2 which represents an index of freshwater input) were identified that were then used in a forecast model for Puget Sound. The description of the forecast model is reserved for a future publication.

Acknowledgments

The author would like to acknowledge his Master of Science thesis committee, Robert C. Francis, Nathan J. Mantua, and Ray Hilborn, who provided insightful ideas that greatly improved this research. Special thanks to Richard Beamish and Sandy McFarlane for helpful insights into Strait of Georgia coho, and to Bill Tweit for his efforts in selecting appropriate coded wire tag data. This work was funded by the University of Washington, Pacific Northwest Climate Impacts Group, a Joint Institute for the Study of Atmosphere and Oceans/School of Marine Affairs interdisciplinary research team. This paper is JISAO Contribution 638.

References

Beamish, R.J., C.E.M. Neville, P.J. Thomson, B.L. Harrison, and M. St. John. 1994. A relationship between Fraser River discharge and interannual production of Pacific salmon (*Oncorhynchus* spp.) and Pacific herring (*Clupea pallasi*) in the Strait of Georgia. Can. J. Fish. Aquat. Sci. 51(12):2843-2855.

Beamish, R.J., B.E. Riddell, C.E.M. Neville, B.L. Thomson, and Z. Zhang. 1995. Declines in chinook salmon catches in the Strait of Georgia in relation to shifts in the marine environment. Fish. Oceanogr. 4(3):243-256.

Brodeur, R.D. 1990. Feeding ecology and food consumption by juvenile salmon in coastal waters, with implications for early ocean survival. Ph.D. thesis, University of Washington, Seattle.

Coronado-Hernandez, M.C. 1995. Spatial and temporal factors affecting survival of hatchery-reared chinook, coho and steelhead in the Pacific Northwest. Ph.D. thesis, University of Washington, Seattle.

Ebbesmeyer, C.C., and C.A. Barnes. 1980. Control of a fjord basin's dynamics by tidal mixing in embracing sill zones. Estuarine Coastal Mar. Sci. 2:310-330.

Ebbesmeyer, C.C., C.A. Coomes, G.A. Cannon, and D.E. Bretschneider. 1989. Linkage of ocean and fjord dynamics at decadal period. Geophysical Monogr. 55:399-417.

Gargett, A.E. 1997. The optimal stability "window": A mechanism underlying decadal fluctuations in North Pacific salmon stocks? Fish. Oceanogr. 6:9.

Francis, R.C., and S.R. Hare. 1994. Decadal-scale regime shifts in the large marine ecosystems of the Northeast Pacific: A case for historical science. Fish. Oceanogr. 3:279-291.

Francis, R.C., S.R. Hare, A.B. Hollowed, and W.S. Wooster. 1998. Effects of interdecadal climate variability on the oceanic ecosystems of the NE Pacific. Fish. Oceanogr. 7:1-21.

Hare, S.R., and R.C. Francis. 1995. Climate change and salmon production in the Northeast Pacific Ocean. Can. Spec. Publ. Fish. Aquat. Sci. 121:357-372.

Hilborn, R.H., and C. Coronado. 1997. Changes in ocean survival of coho and chinook salmon in the Pacific Northwest. In: Estuarine and ocean survival of northeastern Pacific Salmon. Proceedings of NOAA workshop. NOAA Tech. Memo. NMFS-NWFSC-29.

Hilborn, R.H., and S.R. Hare. 1992. Hatchery and wild fish production of anadromous salmon in the Columbia River basin. University of Washington, Fisheries Research Institute Technical Report, FRI-UW-9207, Seattle. 34 pp.

Holtby, L.B., B.C. Anderson, and R.K. Kadowaki. 1990. Importance of smolt size and early ocean growth to interannual variability in marine survival of coho salmon (*Oncorhynchus keta*). Can. J. Fish. Aquat. Sci. 47:2181-2194.

Mann, M.E., R.S. Bradley, and M.K. Hughes. 1998. Global-scale temperature patterns and climate forcing over the past six centuries. Nature 192(23 April):779-787.

Mantua, N.J., S.R. Hare, and others. (1997). A Pacific interdecadal climate oscillation with impacts on salmon production. Bull. Am. Meteorol. Soc. 78(6):1-11.

Mathews, S.B. 1980. Trends in Puget Sound and Columbia River salmon. In: W.J. McNeil and D.C. Himsworth (eds.), Salmonid ecosystems of the North Pacific. Oregon State University Press, Corvallis, pp. 133-145.

Mysak, L.A. 1986. El Niño, interannual variability and fisheries in the northeast Pacific Ocean. Can. J. Fish. Aquat. Sci. 43:464-497.

North, G.R., T.L. Bell, R.F. Cahalan, and F.J. Moeng. 1982. Sampling errors in the estimation of empirical orthogonal functions. Mon. Wea. Rev. 110:699-706.

Olson, F.W. 1978. An evaluation of factors affecting the survival of Puget Sound hatchery-reared coho salmon (*Oncorhynchus kisutch*). Master of Science thesis, University of Washington.

PFMC. 1997. The Pacific Fishery Management Council Preseason Report 1. Stock abundance analysis for 1997. Ocean salmon fisheries. February 1997. Appendices B-2 and B-3.

Seiler, D. 1997. 1997. Wild coho run forecasts. Internal memo: State of Washington Department of Fish and Wildlife Fish Management Program. February 6, 1997.

Strickland, R.M. 1983. The fertile fjord. University of Washington Press, Seattle.

Walters, C.J., and R.W. Hilborn. 1992. Quantitative fisheries stock assessment: Choice, dynamics, and uncertainty. Chapman & Hall, New York.

First-Year Energy Storage Patterns of Pacific Herring and Walleye Pollock: Insight into Competitor Strategies

A.J. Paul and Judy M. Paul
University of Alaska, Institute of Marine Science, Seward Marine Center, Seward, Alaska

Abstract

Changes in length and whole body energy content for Pacific herring and walleye pollock from Prince William Sound, Alaska, were monitored in 1996 to describe growth patterns of these two pelagic competitors during their first year. Metamorphosed walleye pollock were first captured in June, when they averaged 34 mm in standard length (SL) and their whole body energy content (WBEC) was 2.7 kJ/g wet wt. In August they had grown to 69 mm and 3.4 kJ/g wet wt. In October they averaged 81 mm and 3.6 kJ/g wet wt. Metamorphosed Pacific herring were first captured in July, and averaged 28 mm and 2.5 kJ/g wet wt. In August they had grown to 38 mm and 3.1 kJ/g wet wt. In October they averaged 75 mm and 5.0 kJ/g wet wt.

Walleye pollock metamorphosed earlier than Pacific herring and in August and October they were on average 31 mm and 6 mm longer than their herring competitors. In August the WBEC of the two species was similar. In October the herring had stored on average ≈1.4 kJ/g more energy than pollock to enter the overwinter period. The survival strategy for first year pollock includes an early metamorphosis followed by rapid growth in length. For herring, metamorphosing later when conditions are warmer and storing more somatic energy for overwintering appear to be primary tactics.

Introduction

In the northern Gulf of Alaska two of the most common pelagic forage fishes are the Pacific herring *(Clupea pallasi* Valenciennes 1847) and the walleye pollock *(Theragra chalcogramma* [Pallas 1814]). Both species are mobile and vertically migrate in the water column, and during the first year of life they are competitors for the same prey resources (Willette et al. 1997). Both species are spring spawners; their larvae eat copepod nauplii (McGurk et al. 1993, Hillgruber et al. 1995) and then switch to consuming small copepods like *Pseudocalanus* (Willette et al. 1997). During their first summer both are pelagic by habit and exist in the same geographic regions although they do not always coexist in the same depth strata. As they grow beyond the first year of life, the herring relies on its gill rakers to continue to capture copepods while pollock tend to supplement their copepod diet with macrozooplankton and fish to a greater degree. Pollock may also exploit benthic prey as well as pelagic species as they mature.

The objective of this report is to compare standard length (SL), body weight, and whole body energy content (WBEC) of recruiting herring and pollock. This was done to see if these measurements provided insight into the similarities and differences in their respective survival strategies during their first feeding season.

Materials and Methods

Sampling

During the spring and summer of 1996 age-0 herring were captured from four sites in Prince William Sound, Alaska: Simpson Bay, Eaglek Bay, Whale Bay, and Zaikof Bay (Fig. 1). Collection dates were 10-13 May, 10-15 June, 3-11 July, 1-5 August, and 3-8 October of 1996. Juvenile herring were captured with 50 m diameter × 4 m deep purse seines with 3 mm stretch mesh. At each collection site the nets were set at least three times to capture specimens.

Sampling for juvenile pollock was done at the same sites, and dates with an otter trawl (head rope 21 m, foot rope 29 m, mouth 3 × 20 m) with a 32 mm mesh cod end. At each collection site one trawl was made to capture recruiting 1996 year-class pollock. The number of pollock in the collections appears in the data plots.

After capture all fish were immediately frozen in seawater aboard ship and kept frozen until processing. Because the number of individuals captured at some sites, on some dates, was small the WBEC measurements for all herring and pollock captured at the different sites were combined for this presentation.

The sampling sites are remote from any facilities so to portray the thermal conditions during the study, water temperatures from the Seward Marine Center Laboratory seawater system are shown (Fig. 2). The laboratory is about 100 km downstream from Prince William Sound and seawater temperatures at a depth of 75 m and at the surface are recorded daily.

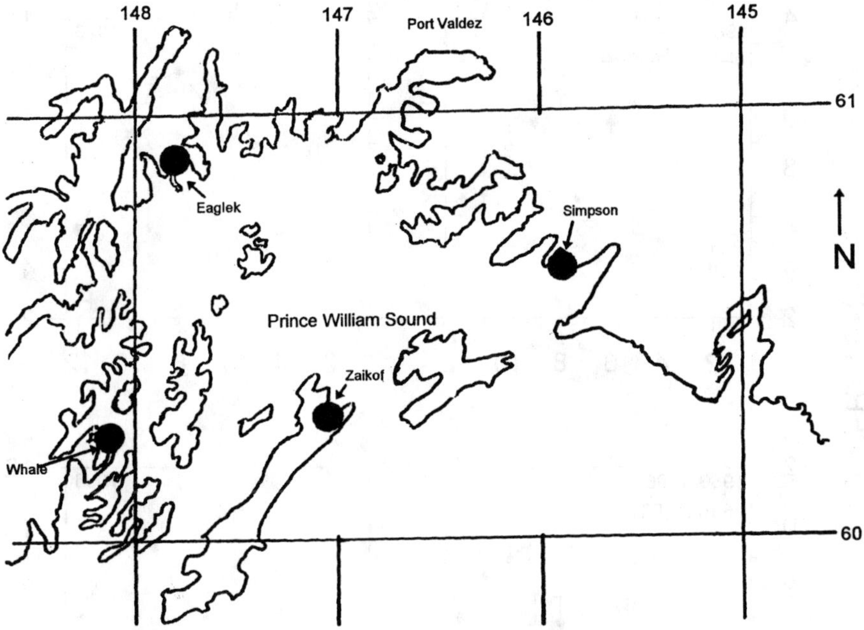

Figure 1. *Map of the study area in Prince William Sound, Alaska, where Pacific herring and walleye pollock were captured for analysis of somatic energy content.*

Whole Body Energy Content

In the laboratory the fish were partially thawed, just enough to handle but not enough so fluids were lost. All fish were measured for standard length to the nearest millimeter, then weighed to the nearest 0.1 g. Each fish was freeze-dried whole until there was no apparent moisture. After freeze drying, fish were placed in a convection oven at 60°C until they reached a constant weight. Individual wet and dry weight values were used to calculate the moisture content of every fish. Whole dried fish were ground in a mill and measurements of caloric content made by bomb calorimetry. All calorimetric samples were weighed to the 0.0001 g level with a single 0.5-1.0 g sample burned per fish. Whole body energy content is expressed as kJ/g wet wt.

Aging

Comparison of the test parameters were made for young-of-the-year (YOY) pollock and herring. Scales were removed from every herring just above

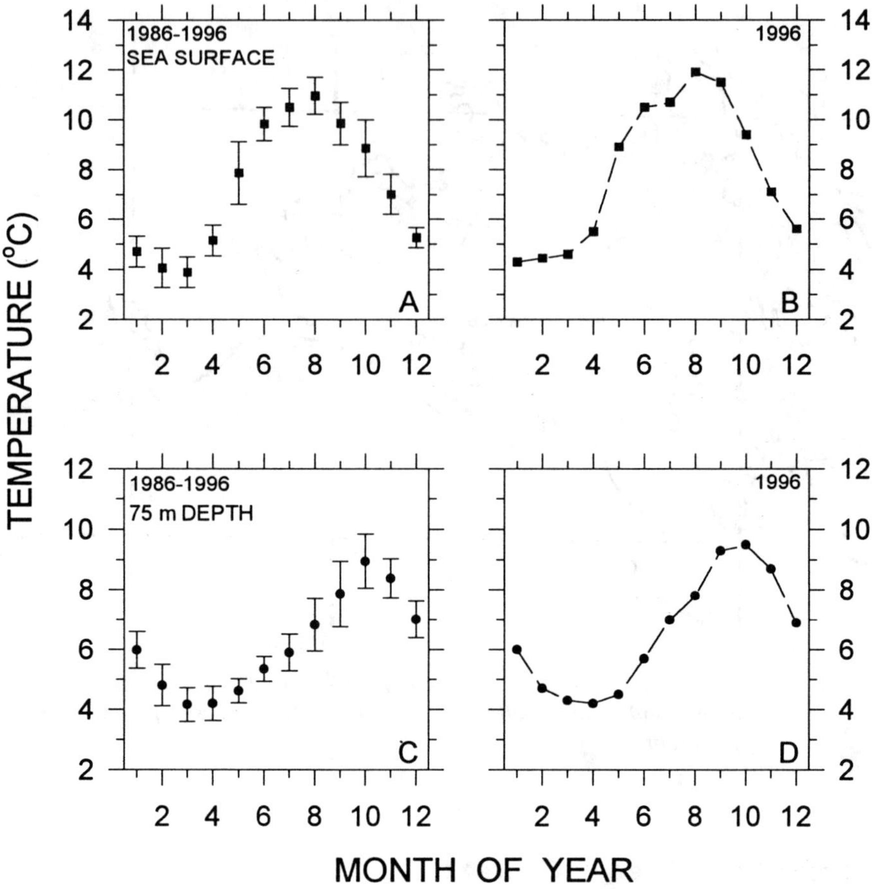

Figure 2. Seawater temperatures at the surface (upper) and 75 m depth (lower) near Seward, Alaska (date = monthly mean and SD) during 1986-1996 (left). Monthly mean temperature values for 1996 (right panels).

or below the lateral line, three rows behind the operculum for determining age. They were cleaned manually and mounted on glass slides. Mounted scales were placed in a microfiche reader and winter annuli were counted in the conventional manner. Age-0 herring are easy to identify both by SL and scale age since they are much smaller than other age groups and have no winter growth rings in fall collections and only one winter ring the following spring. Herring with more than one annuli on the scales were not assigned an age, but their WBEC was plotted against SL. Pollock were not aged, only fish less than 105 mm SL were processed with the assumption that most Gulf of Alaska pollock this size would be YOY (Brodeur and Wilson 1996).

The relationship between the percentage of whole body that is dry weight to WBEC in pollock has been described (Harris et al. 1986). The relationship for juvenile herring was established in this study.

Results and Discussion

Thermal Conditions

Based on temperatures at Seward (Fig. 2), the 1996 thermal conditions were close to the mean monthly values recorded in the upper 75 m of the water column during 1985-1996. Thus, from a thermal-metabolic perspective the YOY fish encountered temperatures that were typical of those seen during the last decade.

Walleye Pollock

The changes in WBEC of juveniles during the 1996 growing season are shown in Fig. 3. In May only individuals greater than 90 mm SL were captured (Fig. 3A); those fish had recruited in 1995 and overwintered. May specimens had an average WBEC of 4.0 (±0.4) kJ/g. On 13 June recently metamorphosed 1996 year-class pollock (Fig. 3B) were captured and they had a mean SL of 34 mm (±6 mm) and a mean WBEC of 2.7 (±0.2) kJ/g. By 3 August the 1996 recruits had increased their mean SL to 69 mm (±6 mm), which was a 109% increase in the mean SL (Fig. 3C). Those fish had an average WBEC of 3.4 (±0.3) kJ/g, which was a 26% increase in the mean for YOY fish since the June sampling. By 6 October YOY pollock had an average SL of 81 mm (±10 mm) and WBEC of 3.6 (±0.2) kJ/g (Fig. 3D). In October there were no fish with WBEC less than 3.0 kJ/g while in August they were relatively common (Fig. 3).

Herring

In May and June only herring that had overwintered were present in the catches (Fig. 4A,B). There was a tendency for WBEC to exhibit considerable variability relative to SL in all seasons. In May the average WBEC for all fish was 4.5 (±1.0) kJ/g wet wt, which increased to 5.6 (±1.6) kJ/g wet wt in

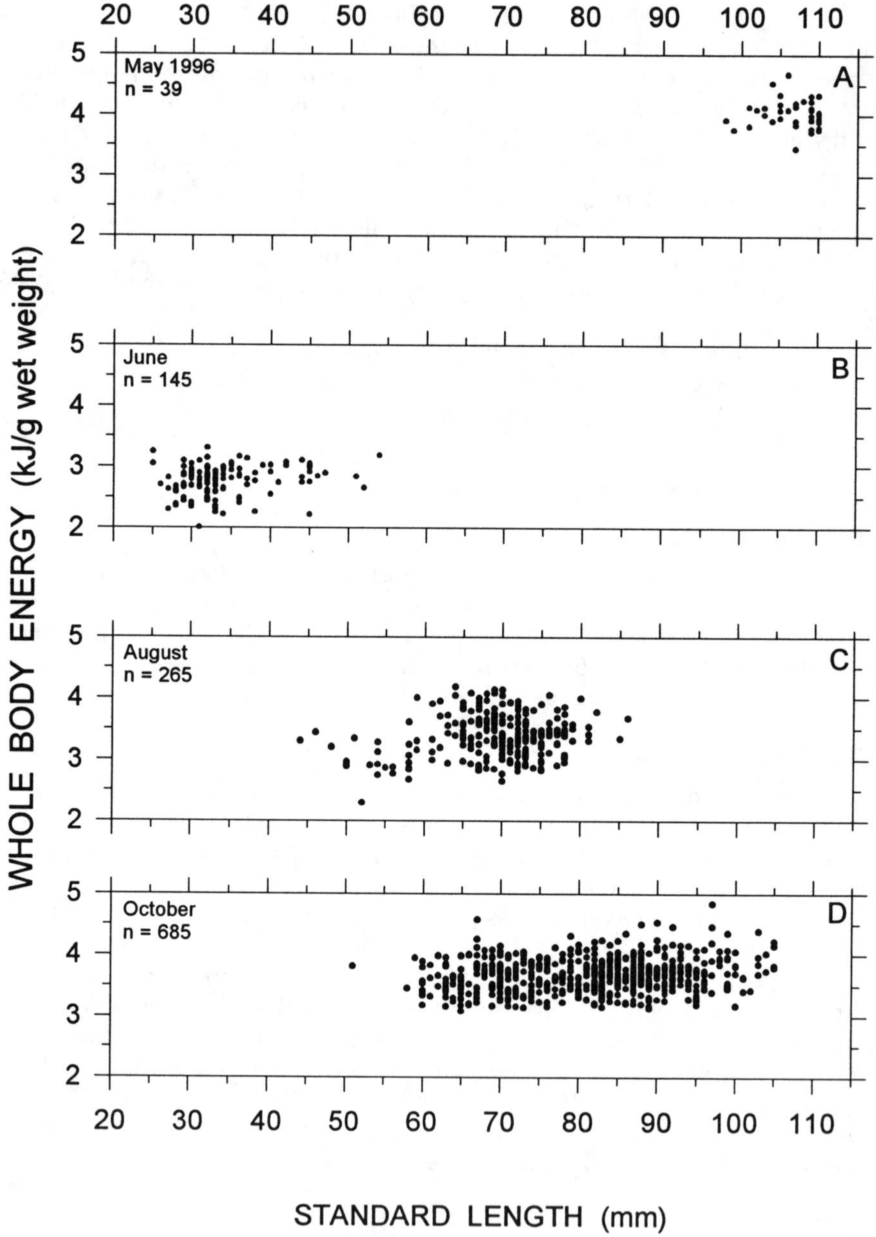

Figure 3. Whole body energy content (kJ/g wet weight) relative to standard length for juvenile walleye pollock captured in Prince William Sound, Alaska, during 1996.

June. In July the 1996 year class first appeared at about 25-30 mm SL (Fig. 4C). The 1996 recruits less than 30 mm SL had a WBEC of 2.5 (±0.3) kJ/g wet wt in July while fish greater than 70 mm SL that had overwintered had 4-8 kJ/g wet wt. By August the YOY fish measured 20-60 mm SL, and fish less than 60 mm had an average WBEC of only 3.1 (±0.6) kJ/g (Fig. 4D). In October (Fig. 4E) few fish greater than 100 mm SL were captured and they had a mean SL of 75 mm (±13 mm) and WBEC of 5.6 (±2.8) kJ/g wet wt.

The relationship between the percent dry weight of whole bodies and WBEC for YOY herring was linear in nature. The following equations describe these relationships:

$$\text{WBEC kJ/g dry wt} = 0.55\ (\%\ \text{dry wt}) + 9.47;\ r^2 = 0.86 \qquad (1)$$

$$\text{WBEC kJ/g wet wt} = 0.38\ (\%\ \text{dry wt}) - 3.44;\ r^2 = 0.96 \qquad (2)$$

Species Comparisons

Graphic representations of the mean (± SD) size and WBEC of both species occurs in Fig. 5. The months of August and October are the only periods when both YOY pollock and herring were captured. In August the pollock were markedly bigger (mean SL = 69 mm, mean wet wt = 2.8 g) than herring (mean SL = 38 mm, mean wet wt = 0.7 g). By October the mean SL was 81 mm for pollock vs. 75 mm for herring. In August the pollock WBEC averaged 3.4 kJ/g wet wt vs. 3.1 kJ/g for herring. A Mann-Whitney rank sum test indicated that these values were significantly different ($P < 0.0001$). Thus, during June through August YOY pollock are a more energy-rich prey than YOY herring. By October the herring had markedly higher WBEC (mean = 5.0 kJ/g wet wt) than the pollock (mean = 3.6 kJ/g).

The most obvious difference in the growth strategies of the two species was the much earlier metamorphosis of the pollock. They had nearly a month more to grow than the herring. The pattern for growth between Julian days 164 and 279 in SL for both species, based on mean values in Fig. 5 was linear, during June to October (Table 1). The growth rate for pollock was 0.4 mm per day vs. 0.5 mm per day for YOY herring. The change in WBEC between Julian days 164 and 279 was 8 J/g per day for pollock and 30 J/g per day for herring. The herring allocated over 3 times more of their energy intake to somatic storage than the pollock.

The survival strategy for YOY pollock includes an early metamorphosis followed by rapid growth in length. Presumably with rapid growth in length comes increased swimming speed useful in capturing prey and avoiding predators. Herring metamorphosed later than the pollock, but they exhibited a slightly faster growth in SL in the warmer conditons. They also stored more energy for overwintering. Pollock appear to feed and grow in length through the winter (Paul et al. 1998) while herring rely

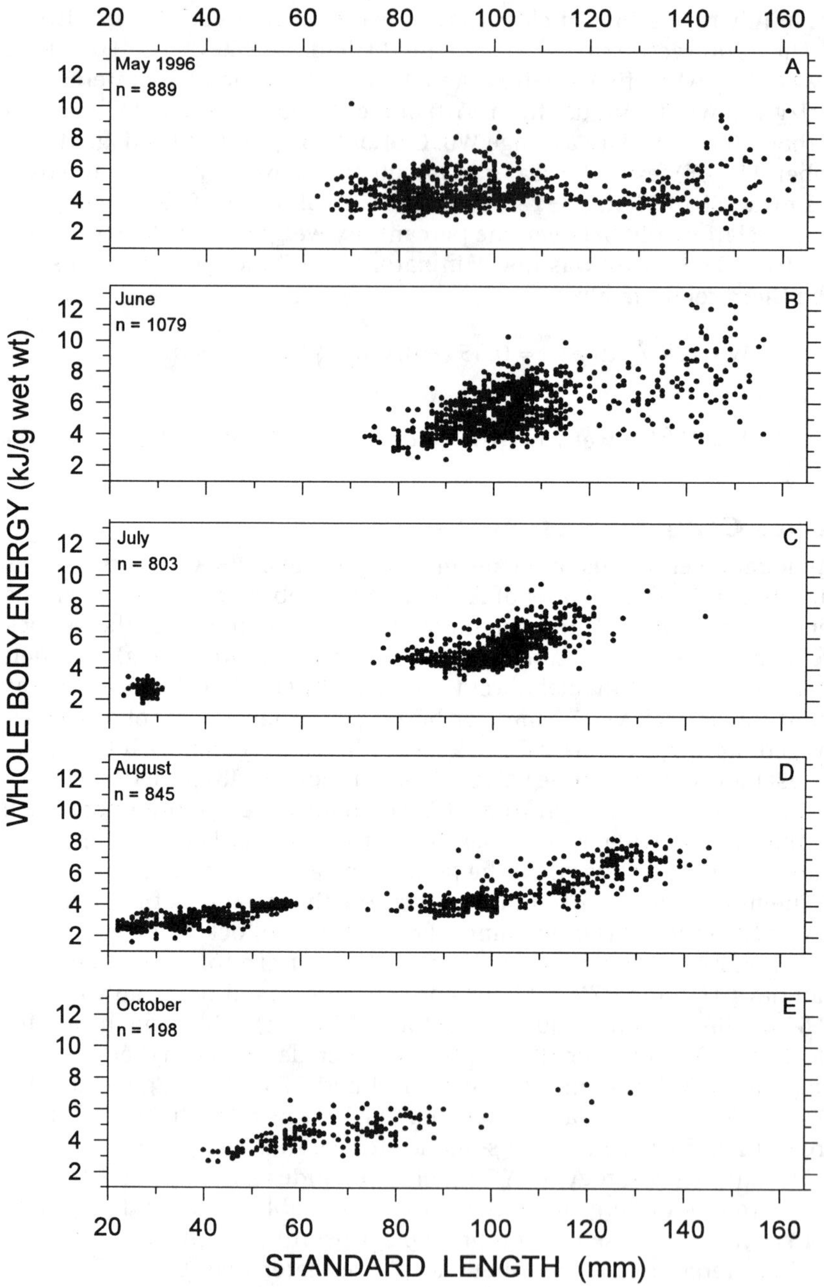

Figure 4. Whole body energy content (kJ/g wet weight) relative to standard length for juvenile Pacific herring captured in Prince William Sound, Alaska, during 1996.

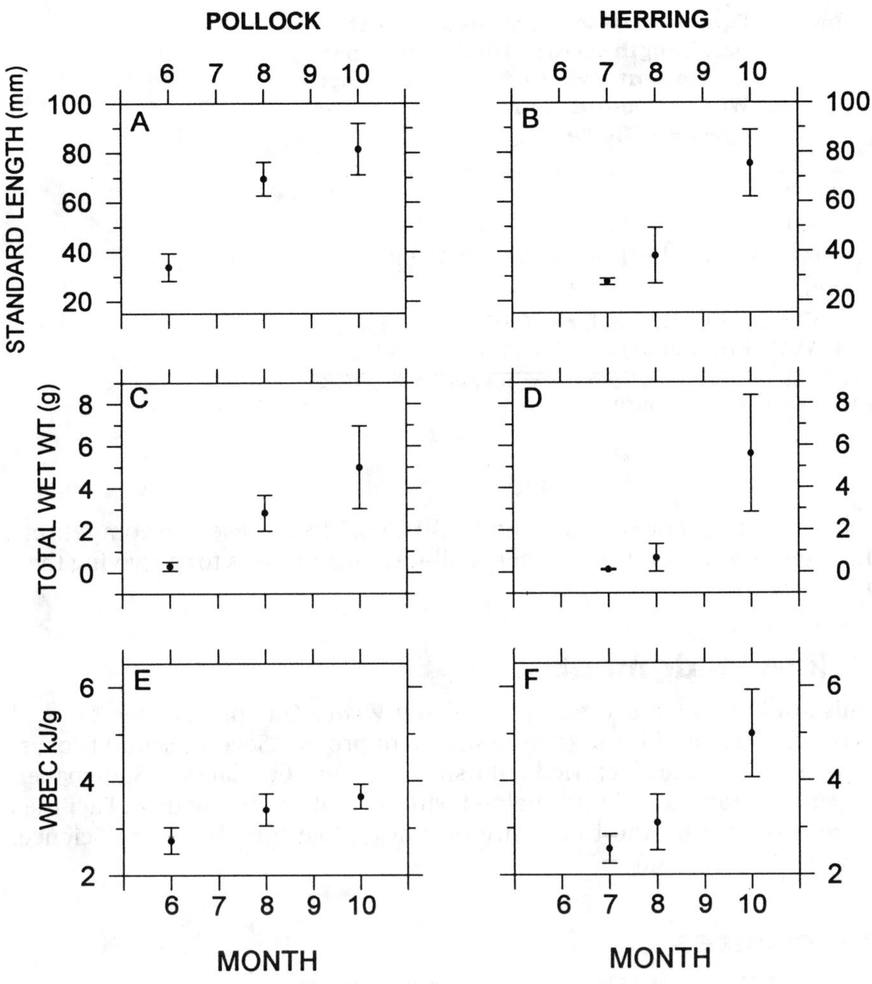

Figure 5. *Changes in standard length (mm; A, B), whole body wet weight (g; C, D) and whole body energy content (kJ/g wet wt; E, F) for age-0 walleye pollock and Pacific herring.*

Table 1. Equations describing changes in standard length (SL) and whole body energy content (WBEC) for age-0 Prince William Sound walleye pollock and age-0 Pacific herring.

Pollock
SL(mm) = $0.40X - 26.2$; $r^2 = 0.88$
WBEC (kJ/g wet wt) = $0.008X + 1.6$; $r^2 = 0.86$
Herring
SL(mm) = $0.52X - 73.0$; $r^2 = 0.99$
WBEC (kJ/g wet wt) = $0.02X - 1.3$; $r^2 = 0.94$

The X value is the Julian day between days 164 and 279 for pollock and days 188 and 279 for herring.

more on stored energy (Paul and Paul 1998). Thus, these two competitors have evolved very different energy allocation strategies for surviving their first year.

Acknowledgments

This study was sponsored by the *Exxon Valdez* Oil Spill Trustee Council through the Sound Ecosystem Assessment project. Several Sound Ecosystem Assessment staff carried out fish collections. C. Adams, P. Shoemaker, M. McEwen, and R. Schmidt helped with laboratory procedures. Facilities were provided by the University of Alaska, Institute of Marine Science, Seward Marine Center.

References

Brodeur, R.D., and M.T. Wilson. 1996 A review of the distribution, ecology and population dynamics of age-0 walleye pollock in the Gulf of Alaska. Fish. Oceanogr. 5:148-166.

Harris, R.K., T. Nishiyama, and A.J. Paul. 1986. Carbon, nitrogen and caloric content of eggs, larvae, and juveniles of the walleye pollock, *Theragra chalcogramma*. J. Fish Biol. 29:87–98.

Hillgruber, N., L. Haldorson, and A.J. Paul. 1995. Feeding selectivity of larval walleye pollock *Theragra chalcogramma* in the oceanic domain of the Bering Sea. Mar. Ecol. Prog. Ser. 120:1–10.

McGurk, M.D., A.J. Paul, K.O. Coyle, D.A. Ziemann, and L.J. Haldorson. 1993. Relationships between prey concentration and growth, condition and mortality of Pacific herring, *Clupea pallasi*, larvae in an Alaskan subarctic embayment. Can. J. Fish. Aquat. Sci. 50:163–180.

Paul, A.J., and J.M. Paul. 1998. Comparisons of whole body energy content of captive fasting age zero Alaskan pacific herring (*Clupea pallasi* Valenciennes) and cohorts over-wintering in nature. J. Exp. Mar. Biol. Ecol. 226:75-86.

Paul, A.J., J.M. Paul, and R.L. Smith. 1998. Seasonal changes in whole-body energy content and estimated consumption rates of age 0 walleye pollock from Prince William Sound, Alaska. Estuarine Coastal Shelf Sci. 47:251-259.

Willette, M., M. Sturdevant, and S. Jewett. 1997. Prey resource partitioning among several species of forage fishes in Prince William Sound, Alaska. In: Forage fishes in marine ecosystems. Proceedings of the International Symposium on the Role of Forage Fishes in Marine Ecosystems. University of Alaska Sea Grant, AK-SG-97-01, Fairbanks, pp. 11-29.

Feeding Behavior of Herring *(Clupea pallasi)* Associated with Zooplankton Availability in Prince William Sound, Alaska

Robert J. Foy and Brenda L. Norcross
University of Alaska Fairbanks, Institute of Marine Science, Fairbanks, Alaska

Extended Abstract

The goal of our project was to examine the feeding ecology of age-0 Pacific herring (*Clupea pallasi* Valenciennes 1847) within their nursery areas in Prince William Sound, Alaska. One of the many questions that led us to this goal was what defines the juvenile habitat and what influences within that habitat affect juvenile herring behavior and distribution. We were interested in the trophic interactions that juvenile herring encounter, in particular, their seasonal feeding behavior. Early hypotheses suggested that feeding dynamics in the spring and summer when zooplankton biomass is at its peak were critical to the success of the juveniles. Current work emphasizes fall and winter feeding behaviors as being important for the juveniles to survive to the next spring phytoplankton bloom (Paul and Paul 1998, Foy and Paul 2000).

Age-0 herring were collected from Simpson Bay in March, May, June, August, October, and November 1996, and February and March 1997 (Fig. 1). We focused on Simpson Bay, located in the eastern part of Prince William Sound, because of the high biomass of juvenile herring encountered during this study. Herring were collected from commercial seining vessels with 150-mm stretch mesh anchovy nets during dusk and daylight conditions. Fifteen fish were taken from each catch for the following diet analyses. Sample sizes for each month ranged from 35 to 135 fish. Fish were preserved in a 10% buffered formaldehyde solution. After 24 hours, samples were transferred to 50% isopropanol for diet analysis. Each fish was blotted dry, weighed to the nearest 0.01 g, and standard length (SL) measured to the nearest 1.0 mm. Energy density of assimilated food (kJ g^{-1} food wet wt) was determined from literature values of zooplankton

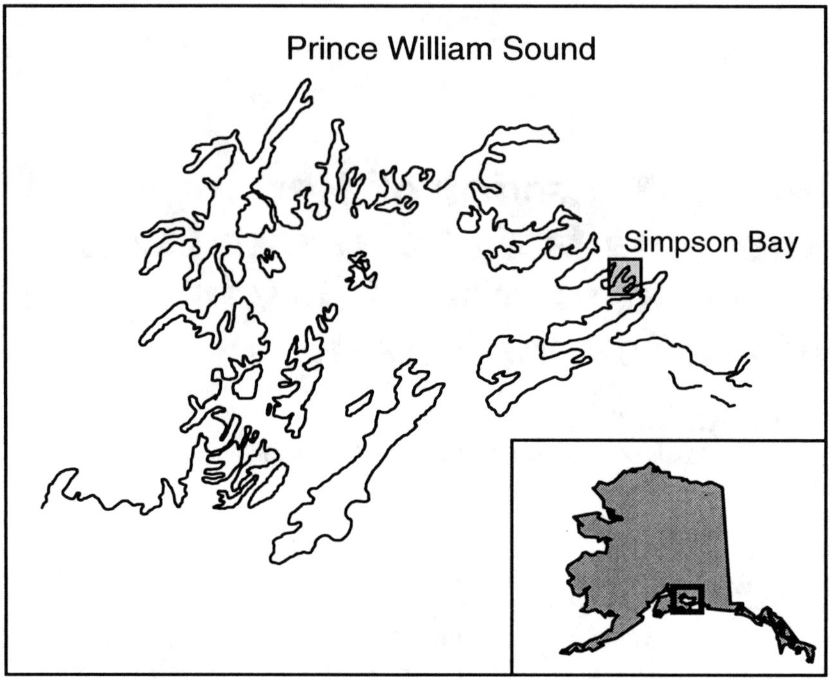

Figure 1. Map of the Simpson Bay study area in Prince William Sound, Alaska, where age-0 herring were captured for diet analysis.

energy density and the species composition in the herring diets (Foy and Norcross 1999, Foy and Paul 2000).

CTD measurements were taken in the water column in each month sampled. For March and July 1996, the average of multiple CTD sites at a depth of 5 m was used. In October 1996, loggers were placed in the bay at 5 m of depth and recorded temperature every 0.5 hour. Zooplankton were sampled with 0.5-m, 300-µm-mesh vertical ring nets to a depth of 30 m from multiple sites in the bay. Taxa were identified as close to species as possible. Weights were taken from multiple subsets and used for the biomass estimates.

Temperature was 11°C in July and declined through the winter until February at 4°C (Fig. 2a). Temperatures were consistently low in March 1996 and 1997. The response of the zooplankton community to the 1996 spring phytoplankton bloom occurred in June with the biomass peaking at over 1.5 g m^{-3} (Fig. 2b). Biomass was lowest in February and began to increase slightly in March 1997. Fish eggs, copepod nauplii, large Calanoida adults (*Metridia* sp.), and Euphausiacea dominated species composition in March. Small Calanoida were most abundant during the summer months

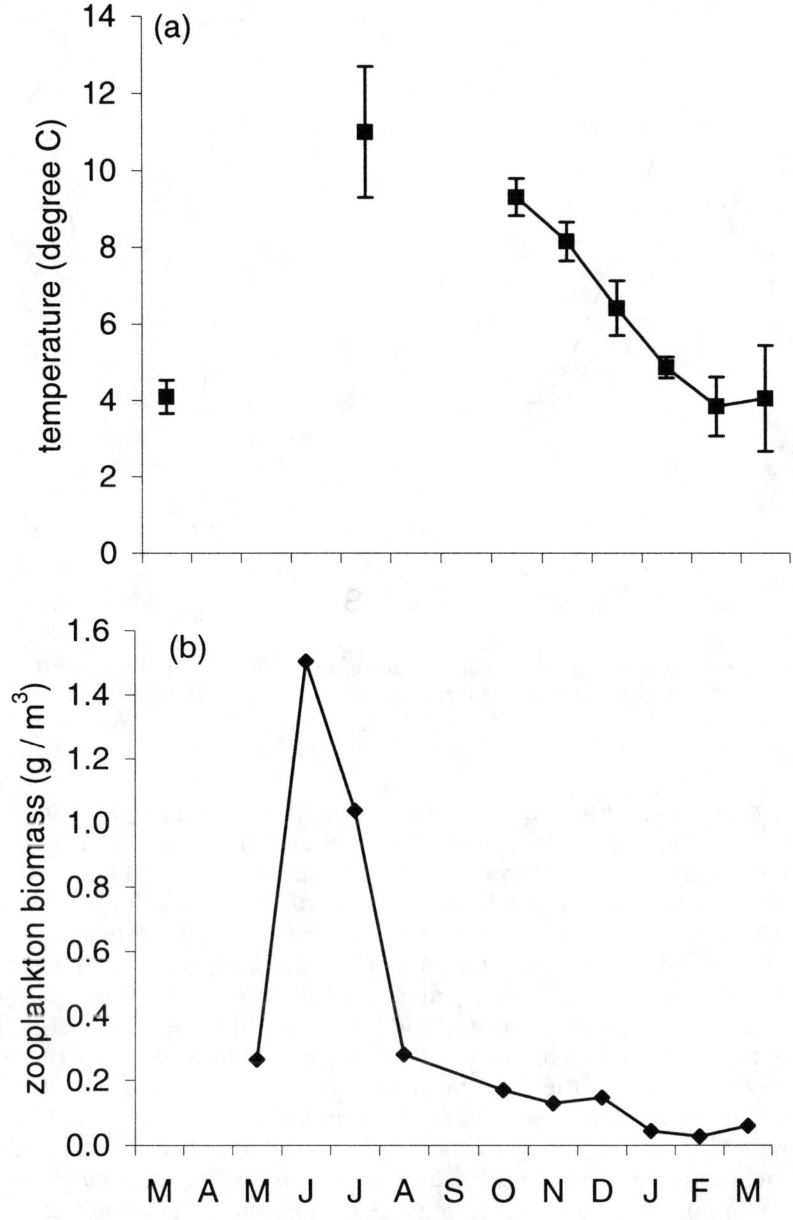

Figure 2. (a) Monthly 5 m mean (±SD) temperatures recorded in March, July, and October 1996 to March 1997. (b) Mean (SE) zooplankton biomass (g m^{-3}) from May 1996 to March 1997.

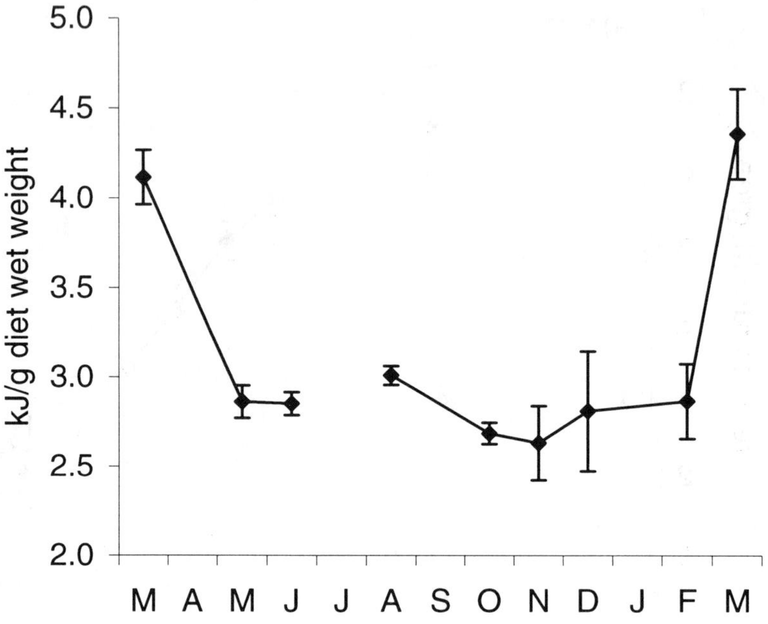

Figure 3. Mean (±SE) diet energy density (kJ g^{-1} food wet wt) for age-0 Pacific herring from March 1996 to March 1997.

and Larvacea were dominant in August and October samples. *Mesocalanus* sp. and Gastropoda larvae were more abundant in November and December. Based on this information it was expected that the age-0 herring in this bay would not put large amounts of effort into feeding until June when temperatures and zooplankton concentrations were high.

Age-0 herring were first encountered in Simpson Bay in August. The monthly mean of the energy density of the prey items found in the diets ranged from 2.6 to 4.4 kJ g^{-1} food wet weight (Fig. 3). The energy density of food items taken by the age-0 herring was highest in March of 1996 and 1997. The rest of the year the energy density of the prey was statistically similar regardless of species composition. Diets in March 1996 were dominated by fish eggs, Cirripedia nauplii, and Harpacticoida. In March 1997, diets contained mostly Euphausiacea, *Neomysis* sp., and *Metridia* sp. Small Calanoida were present in the diets throughout the summer and fall and Larvacea dominated the October diets. November and December diets were composed of Gastropoda larvae and Polychaeta juveniles. The species composition of the prey in November and December suggests that the fish are feeding near the bottom and may in fact be relatively inactive. Therefore, when the availability of zooplankton is low in March, the quality of prey may make feeding at this time worth while.

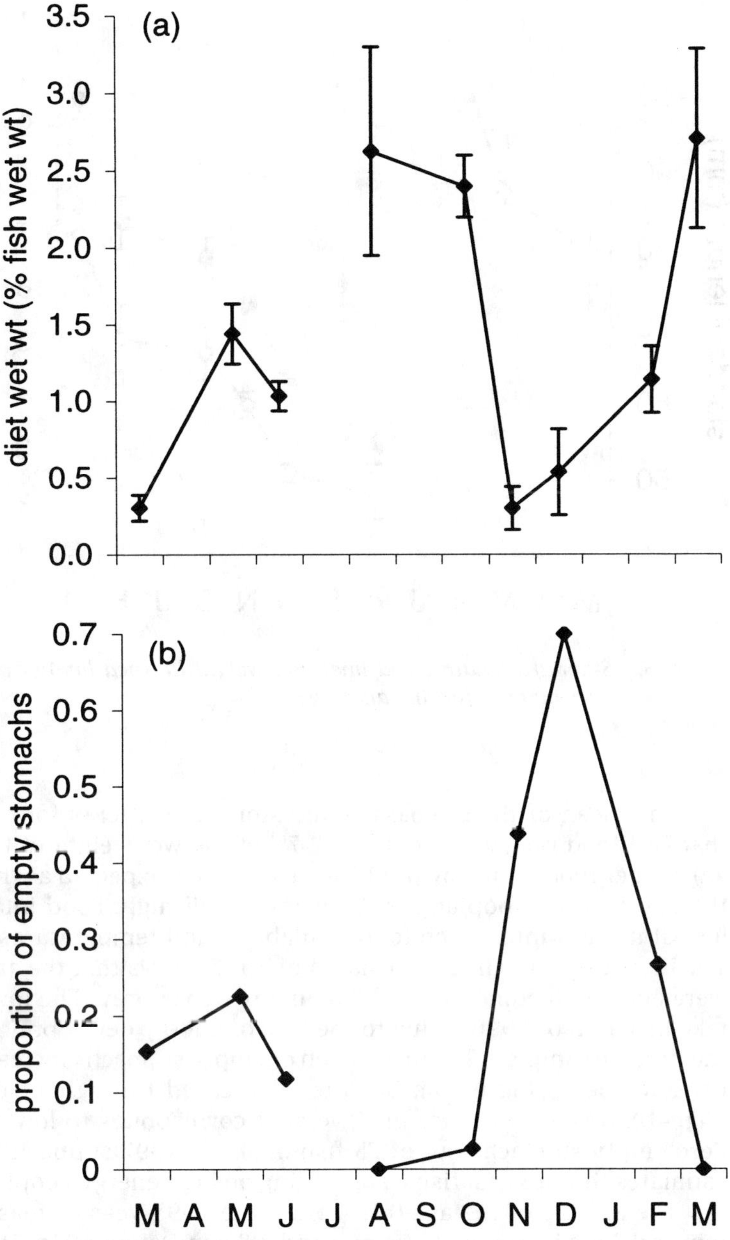

Figure 4. (a) Mean (±SE) diet wet weight (% fish wet wt) for age-0 herring from March 1996 to March 1997. Empty stomachs not included. (b) Proportion of herring empty stomachs encountered from March 1996 to March 1997.

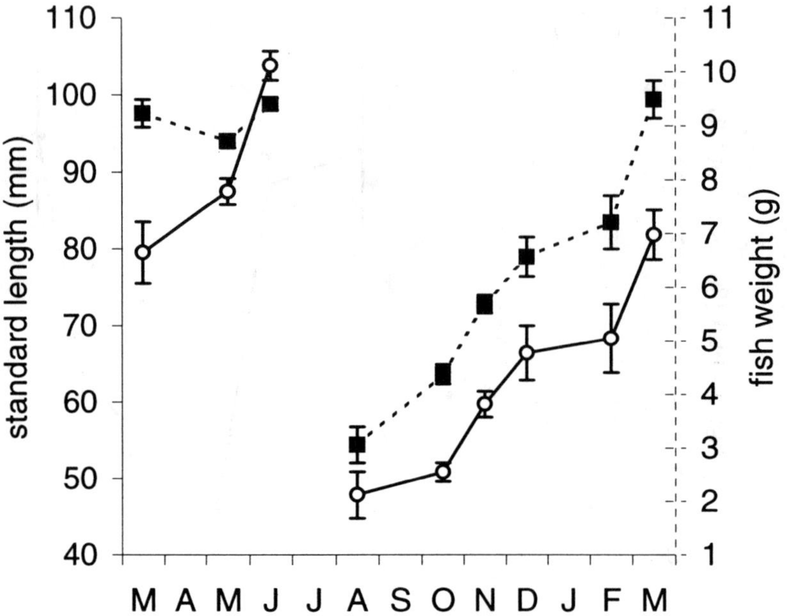

Figure 5. Standard length (solid line) and weight (dashed line) of age-0 herring used for diet analysis.

The mean of the biomass of the stomach contents for only the fish that had food ranged from 0.3 to 2.7% of the wet weight of the fish (Fig. 4a). Larger food intake by the 1996 year class is expected as the fish enter the bay and the zooplankton biomass is still high. Food intake is at its lowest in the winter when food availability and temperatures are lowest. The increased food intake in March of 1997 shows that the age-0 herring were actively feeding on the high energy dense prey. The lower food intake in March of 1996 is due to the lack of energy dense prey available at the time we sampled. The proportion of empty stomachs increased slightly prior to the spring bloom but then decreased throughout the summer (Fig. 4b). A sharp increase in November corresponds to low food intake. Zero empty stomachs out of 75 fish in March 1997 supports food intake estimates that suggest fish were feeding on high energy zooplankton taxa.

The SL size of the age-0 fish from the 1995 year increased from 54 mm and 2.6 g in August to 64 mm and 3.8 g in October (Fig. 5). Fish below 60 mm and 2.5 g were caught in August and November but were not found in the bay in subsequent sampling months. Increases in length and weight from October to March are hypothesized to be a result of mortalities of the smallest fish causing the increase in mean size of fish in the bay. This suggests that the size of the fish going into winter and therefore the feeding behavior in the summer and fall is critical to survival (Foy and Paul In prep.). No growth is expected during the winter and fish are assumed to remain in the bay between the sampling periods, both of which are realistic at this time of year due to low temperatures and prey availability. Age-0 herring from the 1995 year class began to grow between May and June because of warmer temperature and prey availability.

In conclusion, two important trends where found in relation to the feeding behavior of age-0 herring in Simpson Bay. First, increased age-0 herring feeding in response to the peak in prey availability in June and July is expected. The response of the herring to the availability of high-energy prey in March was not expected due to the low temperatures and total zooplankton biomass. This feeding in March suggests that the duration of winter and starvation for age-0 herring may be more dependent on photoperiod than the temperature of the water column. Second, achieving larger size and therefore higher energy stores prior to winter is critical in order to survive through the winter.

References

Foy, R.J., and B.L. Norcross. 1999. Spatial and temporal differences in the diet of juvenile Pacific herring (*Clupea pallasi*) in Prince William Sound, Alaska. Can. J. Zool. In press.

Foy, R.J., and A.J. Paul. 2000. Winter feeding and changes in somatic energy content for age-0 Pacific herring (*Clupea pallasi*) in Prince William Sound, Alaska. Trans. Am. Fish. Soc. In press.

Paul, A.J., and J.M. Paul. 1998. Comparisons of whole body energy content of fasting age zero Alaskan Pacific herring (*Clupea pallasi*). J. Exp. Mar. Biol. Ecol. 226:75-86.

Some Processes Affecting Mortality of Juvenile Fishes During the Spring Bloom in Prince William Sound, Alaska

T. Mark Willette
Alaska Department of Fish and Game, Cordova, Alaska

Ted Cooney
University of Alaska Fairbanks, School of Fisheries and Ocean Sciences, Fairbanks, Alaska

Karen Hyer
Alaska Department of Fish and Game, Cordova, Alaska

Extended Abstract

Piscivory among herring (*Clupea pallasi*) and walleye pollock (*Theragra chalcogramma*) during the spring bloom in Prince William Sound, Alaska, was examined using field data. A midwater wing trawl equipped with a net sounder was used to sample adult pollock (>30 cm). Herring, immature pollock (<30 cm), and various juvenile fishes were sampled with small-mesh purse seines and variable-mesh gillnets. The diet composition of herring and pollock was estimated from specimens collected in 1994 (n = 4,106), 1995 (n = 7,342), and 1996 (n = 3,958). Herring, immature pollock, and adult pollock diets were generally dominated by large calanoid copepods (mostly *Neocalanus*) during the bloom of these copepods in May, but both fish species switched to alternative prey in June (Willette et al. 1999). An analysis of the functional responses of herring and adult pollock feeding on large calanoid copepods was conducted to determine the copepod biomass at which these fishes switched to alternative prey as well as their probable feeding modes (Willette et al. 1999). Initially, nonlinear regression analyses were conducted to estimate handling times (H) using several possible values for the cross-sectional area of the reactive field (γ) and swimming speed (U) as constants. Then handling times estimated during the first stage were used as constants to estimate U using two values of γ

consistent with ram filter and particulate feeding. For filter feeders, γ was assumed to be the cross-sectional area of the mouth gape. Functional responses of herring and adult pollock were best described by type III models (Table 1) with parameters consistent with particulate and filter feeding modes, respectively (Willette et al. 1999). Herring began to switch to alternative prey as the biomass of large calanoid copepods in the upper 20 m of the water column declined below 0.2 g per m^3 Adult pollock switched to alternative prey as copepod biomass in high-density layers declined below about 1.0 g per m^3 (Willette et al. 1999). Analysis of covariance indicated that herring and pollock tended to switch from large copepods to nekton prey after the copepod bloom declined (Willette et al. 1999). During this period, herring and immature pollock largely consumed juvenile gadids and salmonids less than 6 cm in length, whereas adult pollock consumed squid and fish less than 15 cm (Table 2). Herring generally consumed fish 5-25% of their length, immature pollock 5-35%, and adult pollock 3-40% (Willette et al. 1999).

An analysis of covariance was conducted to test the null hypothesis that mortality of juvenile pink salmon (*Oncorhynchus gorbuscha*) was not related to the duration of the bloom of large calanoid copepods from ocean entry to the time the fish reached 6 cm in length (Willette et al. 1999). The analysis was structured to enable examination of the simultaneous effects of the zooplankton bloom on juvenile growth rate and predator switching behavior. Juvenile body weight at release, juvenile growth rate, number of juveniles released from each hatchery each day, and zooplankton bloom duration were included as independent variables in the model. Mortality of juvenile pink salmon from fry to adult was estimated for 147 coded-wire-tag (CWT) groups of pink salmon (Geiger 1990) released from four hatcheries in Prince William Sound (1989-1995). Only CWT groups with fish less than 0.35 g body weight at release were included in the analysis, due to small sample sizes for groups composed of larger fish. Growth of juvenile CWT salmon was estimated from an analysis of covariance function relating growth to juvenile release and environmental data (Willette 1996, Willette et al. 1999). Mortality of CWT pink salmon was significantly reduced when the zooplankton bloom was prolonged during the juvenile life-stage (Willette et al. 1999). This result was generally consistent with our analyses of predator stomach contents suggesting that predator switching behavior was the cause. However, mortality was also positively correlated with juvenile growth rate, juvenile body weight at release, and the number of juveniles released (Willette et al. 1999). Model predicted values exhibited some coherence with auto-correlated patterns of mortality by date of juvenile release (Fig. 1). Examination of temporal patterns in the independent variables indicated that these mortality patterns were probably caused by interactions among the independent variables rather than the effect of any single variable. Changes in predator abundance or size composition may be related to temporal trends in the residuals. Growth rate of CWT juvenile pink salmon was positively correlated (Table 3) with

Table 1. Nonlinear regression parameters for type II and III functional response models for herring and adult pollock feeding on large calanoid copepods.

Predator	Type	γ (m^2)	Estimated *U* (km/day)	*H* (day/g)	MS Regression	MS Error
Herring	II	0.0001	43.0 (7.3)	0.167	7.86	0.16
		0.0791	0.6 (0.5)	1.930	7.91	0.16
	III	0.0001	17.2 (2.0)	–1.913	2.99	0.33
		0.0791	10.0 (5.3)	1.247	9.40	0.10 *
Adult	II	0.0022	8.1 (3.8)	0.248	208.02	5.52
pollock		0.2788	0.1 (0.2)	0.524	156.47	6.96
	III	0.0022	16.7 (5.0)	0.142	243.46	4.54 *
		0.2788	0.4 (0.5)	0.438	185.24	6.16

Handling times (*H*) estimated during the first stage of the analysis were used to estimate swimming speed (*U*) using 2 values of γ associated with ram filter and particulate feeding. The best-fit model for each species is indicated with an asterisk, and the standard error of the estimated swimming speed is indicated in parentheses.

Table 2. Mean prey length (cm) and prey-predator length ratios (%) for various nekton in the diets of herring, immature pollock, and adult pollock. Data from 1994-1996 combined.

	Herring		Immature pollock		Adult pollock	
Prey species	Length	Ratio	Length	Ratio	Length	Ratio
Herring	–	–	–	–	11.5	20.9
Gadids	2.8	14.1	3.4	20.3	2.3	5.0
Salmonids	4.0	18.9	4.0	19.5	3.9	8.1
Capelin	–	–	–	–	9.2	19.1
Sand lance	–	–	5.7	22.9	9.9	17.6
Squid	–	–	–	–	6.9	13.1
Other fish	1.9	10.4	2.4	12.5	4.8	8.6
Unidentified fish	2.2	11.4	2.4	9.6	4.2	8.2

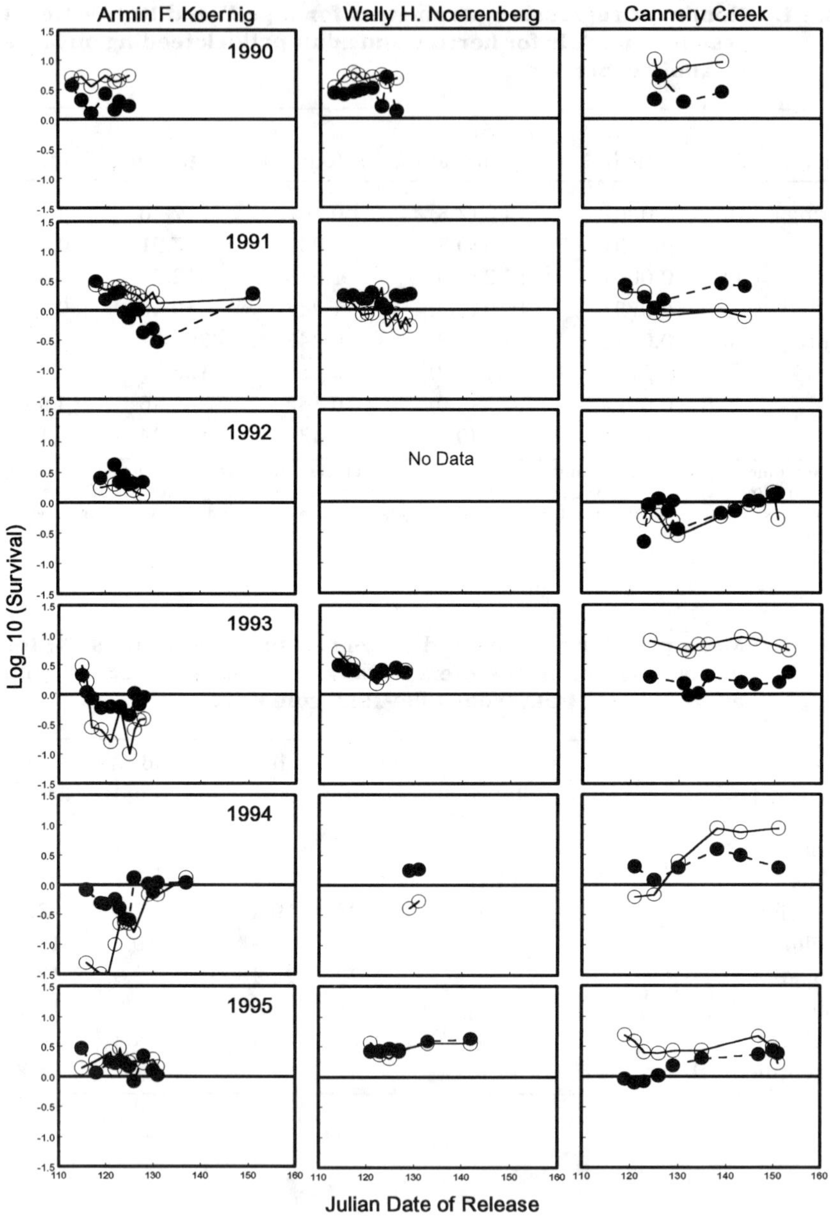

Figure 1. Actual (open circles) and predicted (closed circles) survival of coded-wire-tagged juvenile pink salmon released from the Armin F. Koernig, Wally H. Noerenberg, and Cannery Creek hatcheries in Prince William Sound, 1990-1995. Data are plotted by Julian date of release from each hatchery within each year of release.

Table 3. Parameters estimated from an analysis of covariance.

Parameter		Estimate	Standard error	*p*-value
Intercept		–5.762	1.468	<0.001
Mean zooplankton volume		3.201	0.445	<0.001
Mean ocean temperature		2.421	0.780	0.002
Number juveniles released:	low	4.850	0.676	<0.001
	high	0.000	–	–
Zooplankton × release number:	low	–2.729	0.480	<0.001
	high	0.000	–	–

Mean growth rate of coded-wire-tagged juvenile pink salmon was the dependent variable and mean zooplankton settled volume, mean ocean temperature, and number of juveniles released from each hatchery each day were the independent variables.

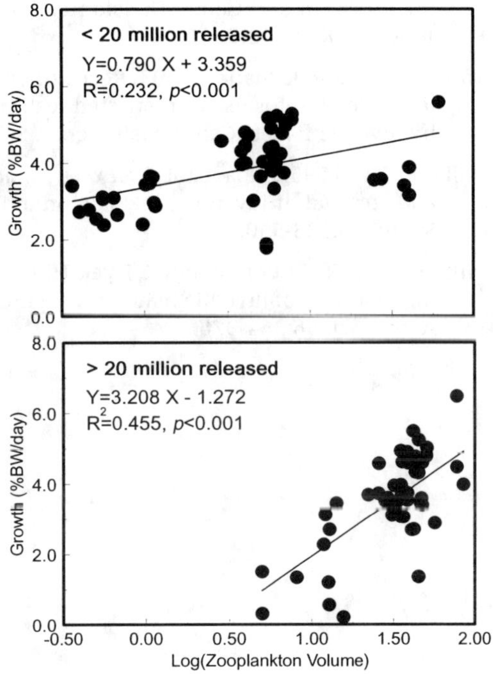

Figure 2. Relationships between zooplankton settled volume during juvenile life and growth rate of coded-wire-tagged juvenile pink salmon released in groups of less than and greater than 20 million.

ocean temperature and zooplankton settled volume during the period from juvenile release to recapture (Willette et al. 1999). The parameters of the relationship between growth and zooplankton settled volume were also significantly different between small and large release groups (Fig. 2), indicating that juvenile pink salmon probably compete for limited prey resources in nearshore nursery habitats (Walters and Juanes 1993). Our analyses indicate that bottom-up processes affect mortality of juvenile fishes during the spring bloom through effects of temperature and prey density on juvenile growth rate. However, bottom-up processes also modify top-down processes involving feeding-mode shifts toward piscivory among the later life-stages of herring and pollock, thus increasing mortality of juvenile fishes.

References

Geiger, H.J. 1990. Pilot studies in tagging Prince William Sound hatchery pink salmon with coded-wire tags. Fishery Research Bulletin No. 90-02, Alaska Department of Fish and Game, Division of Commercial Fisheries, P.O. Box 3-2000, Juneau, AK 99802-2000.

Walters, C.J., and F. Juanes. 1993. Recruitment limitation as a consequence of natural selection for use of restricted feeding habitats and predation risk taking by juvenile fishes. Can. J. Fish. Aquat. Sci. 50:2058-2070.

Willette, T.M. 1996. Impacts of the *Exxon Valdez* oil spill on the migration, growth, and survival of juvenile pink salmon in Prince William Sound. Am. Fish. Soc. Symp. 18:533-550.

Willette, T.M., R.T. Cooney, and K. Hyer. 1999. Predator foraging mode shifts affecting mortality of juvenile fishes during the subarctic spring bloom. Can. J. Fish. Aquat. Sci. 56:364-376.

Are Changes in Bering Sea Crab and Groundfish Populations Related?

Gordon H. Kruse and Jie Zheng
Alaska Department of Fish and Game, Division of Commercial Fisheries, Juneau, Alaska

Extended Abstract

Are decadal changes in abundance of crabs and groundfish in the Gulf of Alaska and eastern Bering Sea related? One popular opinion is that groundfish predation caused declines in crab populations. Research into crab-groundfish relationships off Alaskan coasts is scanty, and purported cause and effect relationships remain speculative. Routine examination of stomach contents of a few commercially important groundfish species in the eastern Bering Sea shows that Pacific cod *(Gadus macrocephalus)* are the primary crab consumer (Livingston 1991, Livingston et al. 1993, Livingston and deReynier 1996). Other species that eat much fewer crabs include yellowfin sole *(Pleuronectes asper)*, rock sole *(Pleuronectes bilineatus)*, Pacific halibut *(Hippoglossus stenolepis)*, Alaska plaice *(Pleuronectes quadrituberculatus)*, walleye pollock *(Theragra chalcogramma)*, and skates.

We examined three hypotheses about crab-groundfish relationships in the eastern Bering Sea: H_{01} (Predation hypothesis): Increased predation by groundfish caused declines in crab recruitment; H_{02} (Common forcing hypothesis): Physical forcing caused inverse responses in recruitment of crabs and groundfish; and H_{03} (Species replacement hypothesis): Declines in crab abundance preceded species replacement by groundfish. To test the predation hypothesis, time series of groundfish biomass were compared with indices of crab year-class strength lagged to age 1. To test the common forcing hypothesis, we compared annual crab and groundfish year-class strengths lagged to the common year of hatching or spawning. Finally, for the replacement hypothesis, time series of crab abundance were compared with groundfish year-class strength lagged to age 0.

Contribution PP-184 of the Alaska Department of Fish and Game, Division of Commercial Fisheries, Juneau.

Time series of abundance and recruitment were obtained for five crab stocks and three groundfish stocks in the eastern Bering Sea (Table 1). Length-based analyses provided abundance and recruitment estimates for red king crab, *Paralithodes camtschaticus* (Zheng et al. 1997), and Tanner crab, *Chionoecetes bairdi* (Zheng et al. 1998). Estimates for blue king crab, *Paralithodes platypus*, were based on catch-survey analyses (Zheng et al. 1997), and area-swept estimates were obtained for snow crab, *Chionoecetes opilio* (Stevens et al. 1998). Stock synthesis models provided biomass and recruitment estimates for Pacific cod (Thompson and Dorn 1997), yellowfin sole (Wilderbuer 1997), and rock sole (Wilderbuer and Walters 1997).

For each hypothesis and crab-groundfish species pair, a one-tailed t-test was conducted to determine whether the correlation coefficient, r, was significantly ($\alpha = 0.05$) less than zero. Following the advice of Pyper and Peterman (1998), significance tests were based on the effective number of degrees of freedom (df*) to account for autocorrelation in the time series of abundance and log-transformed recruitment.

Of the 45 correlations examined, only 4 were statistically significant. Most (26) correlation coefficients were of the wrong sign (+) contrary to the hypotheses. For the predation hypothesis, Bristol Bay red king crab year-class strength was significantly related to the biomass of Pacific cod ($r = -0.68$, df* = 7). For the common forcing hypothesis, significant correlation existed between year-class strengths of Bristol Bay red king crab and rock sole ($r = -0.75$, df* = 4) and between St. Matthew blue king crab and Pacific cod ($r = -0.45$, df* = 13). And for the replacement hypothesis, Bristol Bay Tanner crab abundance was significantly related to rock sole year-class strength ($r = -0.74$, df* = 6). Quite a few more correlations were statistically significant when autocorrelation was ignored with unadjusted degrees of freedom (df = n–2). However, such tests violate the assumption of serial independence of the observations (Pyper and Peterman 1998).

Statistical significance does not necessarily imply biologically meaningful relationships. We expect 2 out of 45 correlations to be significant at the 95% confidence level by chance alone. Multiple statistical tests increase the probability of Type I error. For instance, the negative relationship between red king crab year-class strength and abundance of Pacific cod may be spurious. In a detailed study of predation and population trends, Livingston (1989) concluded that cod predation was not responsible for declines of red king crab in Bristol Bay in the early 1980s. Estimates of red king crab consumed by cod during 1981 and 1983-1992 (Livingston 1991, Livingston et al. 1993, Livingston and deReynier 1996) constitute only 0-11% of the crab population. Most king crabs in cod stomachs are softshell females > 80 mm CL (Livingston 1989)—well beyond the size at which year class strength is determined. Interestingly, a good case for a predation mechanism exists for Tanner and snow crabs as juveniles of both species are heavily preyed upon by Pacific cod (Livingston 1989), yet strong year classes of both crab species co-occurred with high cod biomass resulting in positive correlations. Negative correlations between

Table 1. Abundance and year-class strength indices for crab and groundfish stocks in the eastern Bering Sea. Time lag is either the number of years from hatching to recruitment (crabs) or spawning to recruitment (fish). Recruitment is not necessarily recruitment to the fishery (see data source for definition).

Stock	Time series	Abundance measure	Year-class strength index	Time lag (yr)
Bristol Bay red king crab	1972-1997	Males ≥ 95 mm CL	Male recruits ≥ 95 mm CL	6
Pribilof Islands blue king crab	1975-1997	Males ≥ 120 mm CL	Newshell males 120-134 mm CL	7
St. Matthew I. blue king crab	1978-1997	Males ≥ 105 mm CL	Newshell males 105-119 mm CL	7
Bristol Bay Tanner crab	1975-1997	Males ≥ 93 mm CW	Male recruits ≥ 93 mm CL	7
EBS snow crab	1975-1997	Males ≥ 50 mm CL	Newshell males 50-60 mm CL	4
EBS Pacific cod	1978-1997	Biomass ≥ age 3	Age 3 recruits	3
EBS yellowfin sole	1965-1997	Biomass ≥ age 2	Age 5 recruits	5
EBS rock sole	1975-1997	Biomass ≥ age 2	Age 4 recruits	4

EBS = eastern Bering Sea. CL = carapace length. CW = carapace width.

year-class strengths of Bristol Bay red king crab and rock sole and between St. Matthew Island blue king crab and Pacific cod are consistent with apparent opposite responses of year-class strengths of groundfish (Hollowed and Wooster 1995) and red king crab (Tyler and Kruse 1996, Zheng and Kruse In press) to dynamics of Aleutian low pressure systems in winter. An inverse relationship between Tanner crab abundance and rock sole year-class strength seems plausible. There is considerable overlap in geographic distributions and diets, including polychaetes, bivalves, amphipods, ophiuroids, and echinoids (Tarverdieva 1976, Lang 1992). Depletion of Tanner crab could improve the growth and survival of rock sole competitors through a feeding response.

The lack of general inverse relationships between crabs and groundfish in the Bering Sea is largely attributable to the variety of patterns of year-class strengths among crab stocks. Many groundfish populations experienced common patterns. Strong year classes in the mid-1970s through the 1980s caused large increases in biomass throughout the 1980s, and moderate year classes in the 1990s caused a leveling off or decline in biomass in the 1990s. Year-class strength of most crab stocks declined from the early to mid-1970s—opposite the groundfish trend—but thereafter crab trends were stock specific. Year-class strength of St. Matthew Island blue king crab increased from the mid-1970s through 1990, whereas year-class strength of the Pribilof Islands stock continued its decline until the early 1980s and since then recruitment was moderate through 1990. Year-class strength of Tanner crab increased sharply in the late 1970s to early 1980s and then declined. Snow crab year-class strength was very high in the mid-1970s and throughout the 1980s but fell off in the early 1990s. Finally, year-class strength of red king crab continued to decline through the 1980s with a relatively strong one in 1990. Several red king and Tanner crab stocks in the Gulf of Alaska had patterns similar to Bristol Bay red king crab with the exception that the 1990 year class does not appear to be strong in the gulf. We suspect that common beliefs about inverse crab-groundfish relationships stem largely from the notoriety of the red king crab declines, the lack of knowledge about divergent trends for other crab species, and lack of appreciation for the statistical consequences of autocorrelation in time series of fish data.

Based on our exploratory data analysis, we reject the notion that groundfish predation and competition provide sweeping explanations for crab population fluctuations in the eastern Bering Sea. Most crab-groundfish relationships were positive, and we found few instances in which inverse relationships were statistically significant. Although we reject tenets such as "the groundfish ate all the crabs," we feel that more complex crab-groundfish relationships are certainly deserving of further study. It may be that inter-specific relationships are non-linear, that spatial distributions must be considered to uncover true underlying relationships, and that databases of groundfish stomach contents are not reflective of total crab consumption due to limited temporal and spatial sampling. We plan to explore these crab-groundfish relationships in greater detail in a forthcoming paper.

References

Hollowed, A.B., and W.S. Wooster. 1995. Decadal-scale variations in the eastern subarctic Pacific: II. Response of northeast Pacific fish stocks. In: R.J. Beamish (ed.), Climate change and northern fish populations. Can. J. Fish. Aquat. Sci. 121:373-385.

Lang, G.M. 1992. Food habits of three congeneric flatfishes: Yellowfin sole, *Pleuronectes asper*, rock sole, *P. bilineatus*, and Alaska plaice, *P. quadrituberculatus*, in the eastern Bering Sea, 1984-1988. Masters thesis, University of Washington, Seattle. 124 pp.

Livingston, P.A. 1989. Interannual trends in Pacific cod, *Gadus macrocephalus*, predation on three commercially important crab species in the eastern Bering Sea. Fish. Bull., U.S. 87:807-827.

Livingston, P.A. (ed.). 1991. Groundfish food habits and predation on commercially important prey species in the eastern Bering Sea from 1984 to 1986. NOAA Tech. Memo. NMFS F/NWC-207, Seattle. 240 pp.

Livingston, P.A., and Y. deReynier. 1996. Groundfish food habits and predation on commercially important prey species in the eastern Bering Sea from 1990 to 1992. NMFS AFSC Processed Rep. 96-04, Seattle. 214 pp.

Livingston, P.A., A. Ward, G.M. Lang, and M.-S. Yang. 1993. Groundfish food habits and predation on commercially important prey species in the eastern Bering Sea from 1987 to 1989. NOAA Tech. Memo. NMFS-AFSC-11, Seattle. 192 pp.

Pyper, B.J., and R.M. Peterman. 1998. Comparison of methods to account for auto correlation in correlation analyses of fish data. Can. J. Fish. Aquat. Sci. 55:2127-2140.

Stevens, B.G., J.A. Haaga, and R.A. MacIntosh. 1998. Report to industry on the 1998 eastern Bering Sea crab survey. NMFS, AFSC Processed Rep. 98-07, Kodiak, Alaska. 54 pp.

Tarverdieva, M.I. 1976. Feeding of the Kamchatka king crab *Paralithodes camtschatica* and Tanner crabs *Chionoecetes bairdi* and *Chionoecetes opilio* in the southeastern part of the Bering Sea. Sov. J. Mar. Biol. (Engl. Transl. Biol. Morya) 2(1):34-39.

Thompson, G.G., and M.W. Dorn. 1997. Pacific cod. In: Stock assessment and fishery evaluation report for the groundfish resources of the Bering Sea/Aleutian Islands regions. North Pacific Fishery Management Council, Anchorage, Alaska, pp. 126-158.

Tyler, A.V., and G.H. Kruse. 1996. Conceptual modeling of brood strength of red king crabs in the Bristol Bay region of the Bering Sea. In: High latitude crabs: Biology, management, and economics. University of Alaska Sea Grant, AK-SG-96-02, Fairbanks, pp. 511-543.

Wilderbuer, T.K. 1997. Yellowfin sole. In: Stock assessment and fishery evaluation report for the groundfish resources of the Bering Sea/Aleutian Islands regions. North Pacific Fishery Management Council, Anchorage, Alaska, pp. 160-185.

Wilderbuer, T.K., and G.E. Walters. 1997. Rock sole. In: Stock assessment and fishery evaluation report for the groundfish resources of the Bering Sea/Aleutian Islands regions. North Pacific Fishery Management Council, Anchorage, Alaska, pp. 246-269.

Zheng, J., and G.H. Kruse. In press. Recruitment patterns of Alaskan crabs and relationships to decadal shifts in climate and physical oceanography. ICES J. Mar. Sci.

Zheng, J., G.H. Kruse, and M.C. Murphy. 1997. Status of king crab stocks in the eastern Bering Sea in 1997. Alaska Department of Fish and Game, Reg. Inform. Rep. 5J97-13, Juneau, Alaska. 20 pp.

Zheng, J., G.H. Kruse, and M.C. Murphy. 1998. A length-based approach to estimate population abundance of Tanner crab, *Chionoecetes bairdi*, in Bristol Bay, Alaska. In: G.S. Jamieson and A. Campbell (eds.), Proceedings of the North Pacific Symposium on Invertebrate Stock Assessment and Management. Can. Spec. Publ. Fish. Aquat. Sci. 125:97-105.

Optimum Control in a Predator-Prey Model Based on a Two-Species Exploited Ecosystem

Tatiana Bulgakova
Russian Federal Research Institute of Fisheries and Oceanography (VNIRO), Moscow, Russia

Abstract

Various "biological reference points" are applied when choosing a rational regime of a fishery for a single species system. In the present paper an attempt has been made to illustrate the necessity of including multispecies considerations in precautionary fisheries management advice. As a starting point for such a precautionary approach it is possible to choose the simplest form of multispecies model which should have at least one steady equilibrium point.

The area of allowable management for a two-species system is constructed using as coordinates the fishing mortality rates of both species and some boundaries of the area that correspond to crash lines for this ecosystem. Trophic interactions of the "predator-prey" type reduce the allowable area of control as compared with areas that would be assumed if the interactions between fish populations were neglected. For example, the point corresponding to the MSY criterion for the prey species lies on the boundary of the allowable area and should be considered as a crash point with respect to the predator population.

The area of allowable control is also constructed for two competing population models. The problem of maximizing economic profit of a two-species fishery in an equilibrium state and during a set time period (for example, one year) inside the allowable area of control is touched upon.

Introduction

The theoretical basis of multispecies modeling in biology was set in the 1920s. The analytical models of community populations dynamics were worked out by Volterra (1931) and Lotka (1925) in the form of differential equation sets that took into consideration biological interactions between species. Qualitative analysis of those models showed that fluctuations in population abundance can be caused by the interaction between species. At the same time or a little earlier Baranov (1918, 1925) worked out the basis of the "fishery theory" having built the mathematical model for an exploited fish population. The main advantage of this model consisted of the fact that it was possible to estimate its parameters on the basis of biological and fishery data for specific important targets of the fishery. Later these two scientific schools developed separately. Modern catch-at-age analysis grew from Baranov's ideas. At the same time production models of the fishery are connected to Volterra's models.

Recently a new approach to selecting a rational management scheme for a fishery was detailed within the framework of ICES (International Council for the Exploration of the Sea) stock assessment working groups. This approach is based on the estimation of various "biological reference points." The algorithms of the definition of these points are described in various papers (Jakobsen 1992,1993; ICES 1996, 1997a), but they concern only isolated populations, i.e., an ideal situation. One of the objectives of the multispecies assessment working group (ICES 1997b) was to evaluate how the existence of multispecies interactions affect the estimates of the biological reference points.

The objective of this paper is to apply this approach to simple multispecies community management and to show the need to take into account the trophic interactions between fish populations when defining a rational fishing strategy as these interactions influence "a vector of rational management" of a fish community.

Methods

Often a single species precautionary approach is based on simplest single species considerations—production models. Thus as a starting point for a multispecies precautionary approach, it is possible to choose the simplest form of multispecies model (equation 1).

Since any real ecosystem persists in nature for a sufficiently long time, even the simplest model describing it should have at least one stable equilibrium point; for example:

$$\begin{aligned} \frac{dx_1}{dt} &= x_1(a_1 - bx_1 - cx_2) - F_1x_1 \\ \frac{dx_2}{dt} &= x_2(a_2x_1 - M) - F_2x_2 \end{aligned} \qquad (1)$$

where

x_1 is abundance of a prey species;

x_2 is abundance of a predator species;

a_1 and b are parameters of growth of a prey population in existence of limiting factor (if there are no predators in the system, the prey population grows according to logistics law);

cx_2 is a mortality rate attributable to the predators (described by a "Volterra term");

a_2x_1 is a predator population growth rate in terms of reproduction rate, proportional to the number of prey individuals;

M is the predator natural mortality rate; and

F_1 and F_2 are the fishing mortality rates for the prey and predator species respectively.

It follows from equation (1) that predator species go extinct in the absence of the prey species.

If the community is not exploited, equilibrium of equation (1) gives:

$$X_{10} = \frac{M}{a_2}; X_{20} = \frac{D}{a_2c}.$$ The solution is positive if:

$$D = a_1a_2 - bM > 0. \tag{2}$$

It is assumed that equation (2) is always true in this analysis.

If the fishing mortality rate is constant over time, the equilibrium point of system (1) (equilibrium stock abundance) has coordinates:

$$X_{10} = \frac{M = F_2}{a_2}; X_{20} = \frac{D - (a_2F_1 + bF_2)}{a_2c} \tag{3}$$

For point (X_{10}, X_{20}) to lie in a positive quadrant it is necessary that the fishing mortality rates (the parameters of control on our community) meet the condition

$$a_2F_1 + bF_2 < D \tag{4}$$

Research into the stability of an equilibrium point (X_{10}, X_{20}) on the I method by Ljapunov has shown that under condition (4) and $b > 0$ this point is asymptotically stable. As model (1) is a special case of a model by Kolmogoroff (1972), we obtain an additional feature of the equilibrium point: if the following condition is met

$$0.25b^2(M + F_2) + ba_2(a_2 + F_2) > a_2^2(a_1 - F_1)$$

it is a steady node; otherwise, a stable focal point. Thus, as parameters b, F_1 and F_2 increase (until condition (4) is broken) the system is damped. At

$b = 0$ a condition of stability is broken (system (1) is now the elementary set of V. Volterra's equations the solution of which is undamped oscillations).

Results

Using fishing mortality rates $\{F_1, F_2\}$ as coordinates, Fig. 1 shows the region (triangle MON together with its boundaries ON and OM) corresponding to an area of allowable management of the two-species community, as derived from condition (4) and $F_1 \geq 0$, $F_2 \geq 0$. The equation for line MN is obtained by changing (4) from an inequality to an equality and solving for F_1:

$$F_1 = \frac{D}{a_2} - \frac{F_2 b}{a_2} \tag{5}$$

Fishing Strategies for a Prey Population

Let us assume that in managing a fishery of a prey population we neglect any effort of the predator on the prey population; we consider it to be isolated. Then using accepted terminology (ICES 1996) $F_1 = a_1$ is a maximum (limiting) value of prey fishing mortality named F^*_{1crash}, and maximum sustainable yield (MSY) corresponds to a value $F^*_{1MSY} = 0.5a_1$ (point S in Fig. 1). The manager expects to obtain the catch $MSY^* = \frac{a_1^2}{4a_2}$. Sign * means that the formula concerns a model of an isolated population.

However, since the predators exist, the maximum intensity of fishing on the prey species is reduced. If the predators are not fished, $F_{1crash} = D/a_2 < a_1$ (on equation [2]), and with the increase of F_2, F_{1crash} linearly decreases according to equation (5). The line MN represents a scenario in which fishing mortality rates F_1 and F_2 are such that the fishery results in the extinction of the predator population (i.e., X_{20}, = 0). That is an example of limiting the crash line in a community management area corresponding to a collapse of the community structure.

Stable annual yield of the prey species in a year is given by

$$Y_{10} = F_1 X_{10} = \frac{F_1(M + F_2)}{a_2}$$

The function Y_{10} (F_1, F_2) has no maximum in a positive area of argument and increases with increasing F_1 and F_2 up to the intersection with a plane perpendicular to the plane $\{F_1,F_2\}$ and passing through the line MN.

There are two possible cases, as follows.

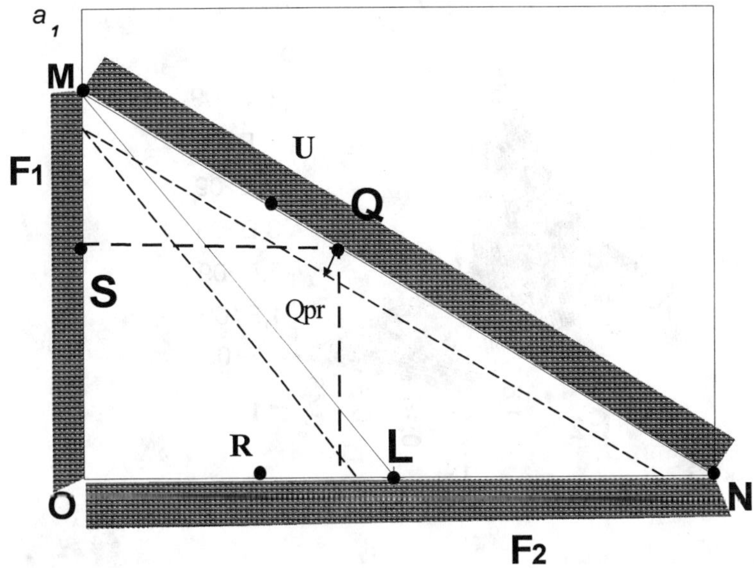

Figure 1. *The area of allowable management for a predator-prey system (unshaded triangle MON) built for Case 1; a point S lies inside the allowable area.*

Case I: $a_1/2 < D/a_2$

In this case point S lies inside the allowable area, maximum sustainable yield of the prey population on line MN is given by $MSY_1 = a_1^2 / 4b$, with the corresponding coordinates $F_1 = 0.5a_1$ and $F_2 = (a_1 a_2/2b - M)$. This is indicated as point Q on the boundary of allowable control area (Figs. 1 and 2a). Thus point Q is not a target point but rather a crash point with respect to the predator population. For prey species point Q could look like a target point, but fishing control corresponding to point Q is unallowable because it results in the extinction of the predator species. To manage the fishery using a precautionary approach, it is necessary to select the management strategy so as to save the predator species. It means, that instead of a point Q it is necessary to select a point lying inside area ONM. The precise choice of a management reference point depends on inter alia the relative value (biological or economic) of both species as well as on agreement between managers. For example, if a 90% control criterion were adopted the line MN would have to be shifted parallel to itself to pass through point $F_1 = 0.9D/A_2$. The shifted line (dashed line on Fig. 1) would obviously be included in the new allowable area of control. A point Q_{pr} on the this dashed line corresponds to the target point of the prey fishery.

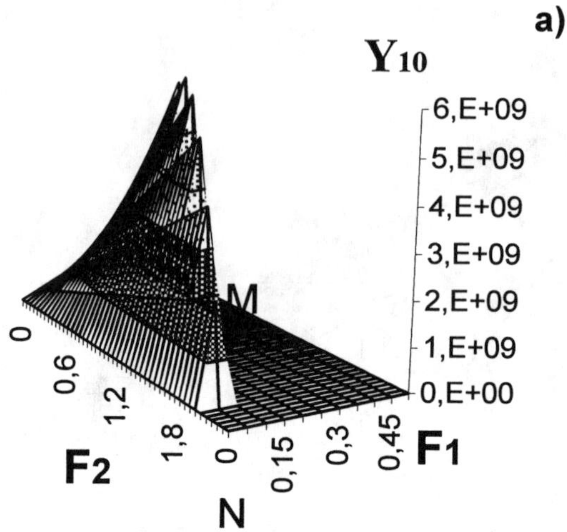

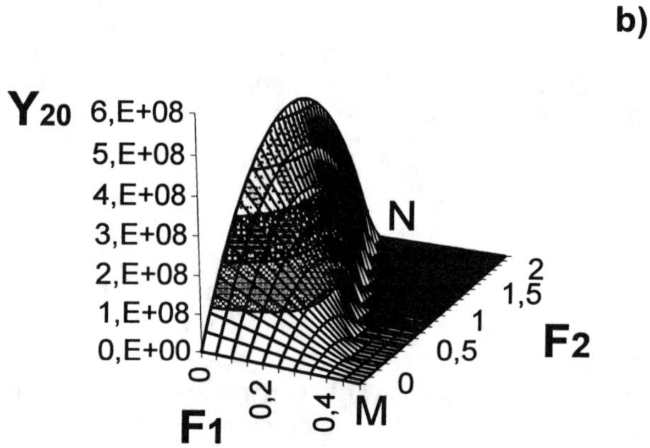

Figure 2. Sustainable yield for the prey species Y_{10} (a) and for the predator species Y_{20} (b) in the allowable area of management for Case 1: $a_1 = 0.5$; $b = E{-}11$; $c = 5E{-}10$; $a_2 = 5E{-}11$; $M = 0.2$; $D = 2.3E{-}11$.

Case II: $a_1/2 > D/a_2$ (i.e., $M < a_1 a_2/b < 2M$)

In this case point S (the point corresponding to a level MSY for an "isolated" population) lies outside the allowable area of control. It means that an attempt to manage the prey population using the conventional MSY criterion for an isolated population will result in the extinction of the predator population; i.e., the destruction of the community even in the complete absence of a predator fishery! This case is most dangerous for a multispecies fishery system. The greatest value of the yield of the prey species is equal to DM/a_2^2 and is reached in a point M in Fig. 1 ($F_1 = D/a_2$, $F_2 = 0$). For a precautionary approach a point corresponding to a smaller F_1 value should be chosen (with possible F_1 values up to $F_1 = 0.9D/a_2$).

Fishing Strategies for a Predator Population

In a steady state of community the yield of the predator species is given as

$$Y_{20} = F_2 X_{20} = \frac{F_2(D - a_2 F_1 - bF_2)}{a_2 c} \tag{6}$$

For fixed F_1, the predator yield is maximized at

$$F_2 = \frac{(D - a_2 F_1)}{2b} \tag{7}$$

represented by the line ML, which is inside the allowable control area and is a median of angle OMN. By substituting equation (7) into equation (6), we receive Y_{20} as a function of F_1 as shown below:

$$Y_2^* = \frac{(D - a_2 F_1)^2}{4a_2 bc}$$

This surface is shown in Fig. 2b.

For any fixed F_1 the line ML gives both levels of predator fishing mortality rate F_{2MSY} and of maximum predator yield. Predator yield is maximized at $F_1 = 0$ at point L$\{F_1 = 0,\ F_2 = D/2b\}$ on the boundary of the allowable area. We obtain an expected conclusion: the maximum yield of the predator population can be found in a lack of prey fishing. Point L is therefore a target reference point for the predator population. An example of a precautionary point may be taken as Lpr = $0.9D/2b$.

In Fig. 2 functions Y_{10} (F_1, F_2) and Y_{20} (F_1, F_2) are built for the model (equation 1) parameter values that satisfied the Case 1 condition ($a_1/2 < D/a_2$): $a_1 = 0.5$; $a_2 = 5\text{E}{-11}$; $b = \text{E}{-11}$; $c = 5\text{E}{-10}$; $M = 0.2$; $D = 2.3\text{E}{-11}$.

The Problem of Maximization of Economic Profit

Consideration of the problem of optimal control on a community as a whole warrants investigation of the problem of maximization of economic profit of a two-species fishery in an equilibrium state and during a set time period (for example, one year) inside the allowable area of control MON. In the simplest form, economic profit can be expressed as the function:

$$P(F_1, F_2) = A_1 F_1 (M + F_2) / a_2 + A_2 F_2 (D - a_2 F_1 - bF_2) / ca_2 - B_1 F_1 - B_2 F_2 \tag{8}$$

where parameters A_1 and A_2 are profit per unit yield of populations I and II, respectively; and B'_1 and B'_2 are the respective expenses per unit of fishing effort E. Assuming the fishing mortality rate is proportional to fishing effort, i.e., $F_i = q_i E_1$, (i = 1, 2) where the q_i is a constant factor, we have $B_1 = B'_1 / q_1$ and $B_2 = B'_2 / q_2$. Equation (8) may be written in a polynomial form

$$P(F_1, F_2) = k_1 F_1 + k_2 F_2 - k_3 F_1 F_2 - k_4 F_2^2 \tag{9}$$

where

$$k_1 = A_1 X_{10} - B_1$$
$$k_2 = A_2 X_{20} - B_2$$
$$k_3 = -A_1 / a_2 + A_2 / c$$
$$k_4 = A_2 b / ca_2$$

In this case, X_{10} and X_{20} are the coordinates of an equilibrium point of the unexploited system (equation 2) (at $F_1 = F_2 = 0$), and therefore k_1 and k_2 are profits per unit of fishing mortality rate in this "initial" equilibrium point. If the fishery is cost effective at this point, k_1 and k_2 are positive constants. The sign of k_3 can be either: for example, if $(A_2/c) >> (A_1/a_2)$, then the predator population is more valuable, then $k_3>0$, and vice versa.

Analysis of the function of two variables (equation 9) shows that it has neither a maximum nor a minimum, so that its greatest value is achieved on the boundary of the allowable area. The function P on each of the boundary lines is given by the following functions of one argument

P_1(on line ON) $= k_2 F_2 - k_4 F_2^2$

P_2(on line OM) $= k_1 F_1$

P_3(on line MN): substituting $F_1 = D/a_2 - F_2\, b/a_2$ in equation (9), we obtain the function of F_2

$$P_3(F_2) = F_2^2\, \alpha_2 + F_2\, \alpha_1 + \alpha_0 \tag{10}$$

where

$\alpha_2 = -A_1 b/a_2^2$

$\alpha_1 = A_1 (D - bM) / a_2^2 + B_1 b/a_2 - B_2$

$\alpha_0 = (A_1 M/a_2 - B_1) D/a_2$

The constant term α_0 is equal to a value of function *P* at point M. The value *P* at point N is equal $(-B_2D/b) < 0$, i.e., at this point the fishery is unprofitable. In a point $\{F_1 = 0, F_2 = 0\}$ $P = 0$.

It is necessary to consider four separate cases:

(1). $k_1 > 0, k_2 > 0$.

On line ON (with zero fishing on the prey species) $P_1(F_2)$ is maximized at the point R $\{0, k_2/2k_4)$ which lies left of point L $\{0, D/2b\}$, because $k_2/2k_4 = D/2b - B_2/2k_4$. The value of *P* at this point is $k_2^2/4k_4$.

On line OM, $P_2(F_1)$ increases with increasing F_1, so that on this line the greatest value of the profit occurs at point M and is equal to k_1D/a_2.

On line MN, $P_3(F_2)$ has a maximum at $F_2 = -\alpha_1/2\alpha_2$ (point U), provided $\alpha_1>0$, and max $P_3 = \alpha_0 - \alpha_1^2/4\alpha_2 > P$ (at point M).

If the condition $\alpha_1 > 0$ holds, it is therefore necessary to calculate and compare the value of the function *P* at each of the points U and R since profit is maximized at one of these two points. If this condition does not hold, the value of *P* at points M and R should be compared.

(2). $k_1 < 0, k_2 > 0$. The fishery for the prey species is not cost effective.

On line MN, $P < 0$. On line ON maximum occurs at point R.

(3). $k_1 > 0, k_2 < 0$. The fishery for the predator species is not cost effective, the greatest value *P* occurs at point M.

(4). $k_1 < 0, k_2 < 0$. The fishery for either of the two species is unprofitable.

The task of maximizing the summed yield of two species, given that they have different economic values, can easily be solved by substituting zeros in place of B_1 and B_2 in the profit function. From equation (9) it follows that the factors k_1 and k_2 are now positive, so that it is fair to assume that conclusions of case 1 hold. Now

$$\alpha_2 = -A_1 b/a_2^2; \ \alpha_1 = A_1 (D - bM) / a_2^2; \ \alpha_0 = A_1 M D/a_2^2.$$

Point U will therefore coincide with point Q, and point R with point L. Using 90% criterion for *F* rates, precautionary management points now correspond to points Q_{pr}, M_{pr} and R_{pr}, so that the values of *P* at these points should be compared. For example, point U will change to the corresponding precautionary point $U_{pr} = \{0.9F_1(Q), 0.9F_2(Q\}$ and so on. Thus

$$P(U_{pr}) \cong A_1 (D^2 - M^2 b^2)/(2ba_2^2);$$

$$P(R_{pr}) \cong A_2 D^2 / (4bca_2). \text{ (Sign } \cong \text{ arose from } 0.9^2 \cong 0.8).$$

The choice of optimal control strategy will depend on the relation between both the biological and the economic (A_1 and A_2) parameters.

Discussion

A simplified single predator-single prey model is considered in this paper, whereas the structure of real ecosystems is obviously much more complicated. At a shortage of prey species the predator switches to another prey; i.e., the risk of its extinction is not so high. There are more advanced models of community. For example, Collie and Spencer (1994) applied more complicated function in a predator-prey model to describe the predation: if in a set of equations (equation 1) the food consumption rate is proportional to prey number a_2x_1, these authors apply function $gx_1^2 / (A^2 + x_1^2)$, which takes into account the existence of threshold prey abundance A and level of saturation corresponding to maximum ration. These authors also described a natural mortality by quadratic function taking into account the interspecies competition of a predator and (or) cannibalism. The obtained model for different relations among parameters of a model has one or two steady equilibrium points in the positive quadrant. Outcomes of research of a model are very interesting: with changing parameter values or increasing F for prey population, the population number can "slide" from one equilibrium state to another.

Collie and Spencer (1994), to simplify the analysis, did not consider predator fishing; but if to the right side of the equation for a predator we add the term ($-F_2x_2$), it is simply to show (for example, with help of the analysis of isoclines) that the area of parameters for which there is the lower equilibrium state of a prey is narrowed; i.e., fishing for predators stabilizes prey population fluctuations and promotes the stabilization of its abundance near the top level of an equilibrium.

Incorporation of alternative prey into a two-species predator-prey model (Spencer and Collie 1995) does model more realistically but complicates an analysis.

Besides the prey-predator relation, it is interesting to consider other types of interspecies relations such as competition for some common resource. The task of optimizing a fish community with two competing species according to the Gause model (Gause 1935, Gause and Witt 1935) has previously been considered (Bulgakova 1970). Competition is manifest only in an environment with limited resources and hence a logistics growth model was used for each population. The set of equations describing an exploited community of this sort has the form:

$$\frac{dX_i}{dt} = \frac{X_i a_i}{K_i}(K_i - X_i - b_i X_j) - F_i X_i, \; i = 1,2 \tag{11}$$

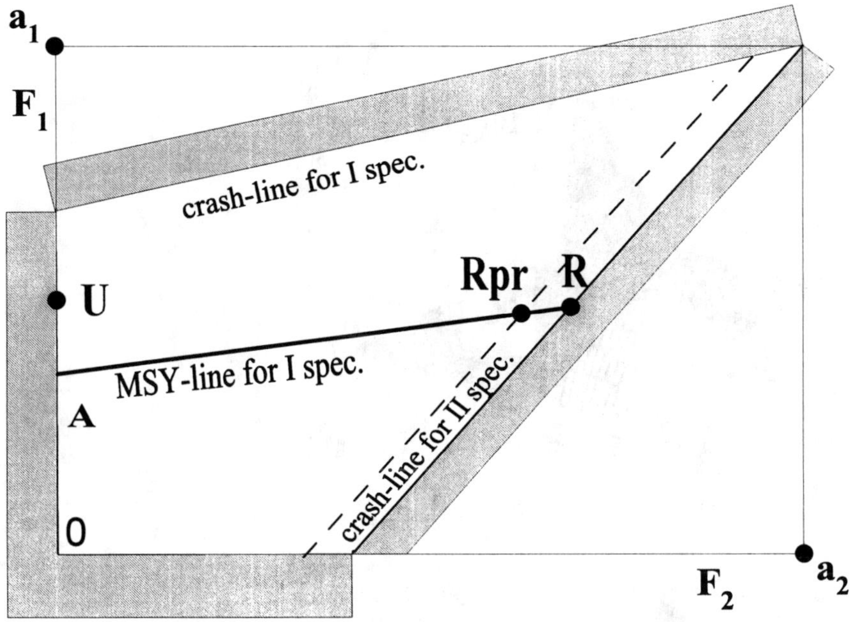

Figure 3. The area of available management for two competing species (unshaded area).

where K_i is the maximum abundance of species i in the absence of a competing species. Competition parameters b_i mean that in terms of the competition interaction X_j individuals of species j are equivalent to b_iX_j individuals of species i (i.e., they consume the same amount of resource limiting development of the species).

Further details pertaining to the analysis of this system are presented by Bulgakova (1970). We describe here only the main features of the results. The area of allowable control for the system (equation 11) is shown in Fig. 3. The line F_{1crash} is the line where $X_{10} = 0$; i.e., the prey population goes extinct. The line F_{2crash} is similarly a boundary where the predator species goes extinct. For this model, the allowable area is bounded by two crash lines! It is clear from this figure that an increase in F_2 allows an increase in the allowable F_1 value.

For an "isolated" prey population MSY_1 is reached at point U such that $F_1 = 0.5a_1$. In the presence of the competitor (the second population) MSY_1 increases with increasing F_2 and occurs on the line AR. The max Y_{10} corresponds to point R that lies on the crash line for the second population. It implies that control of this sort results in the displacement of the second

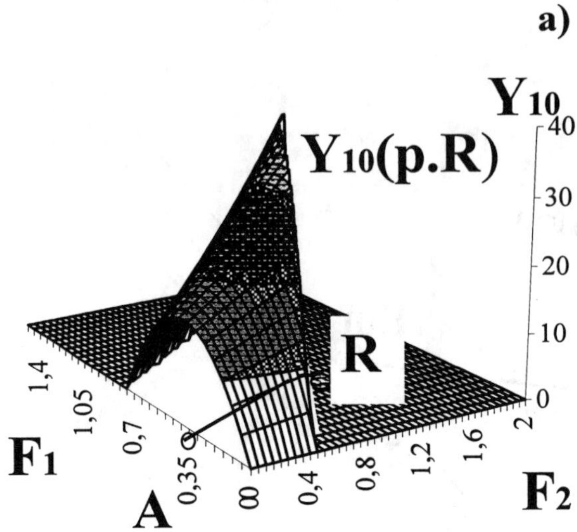

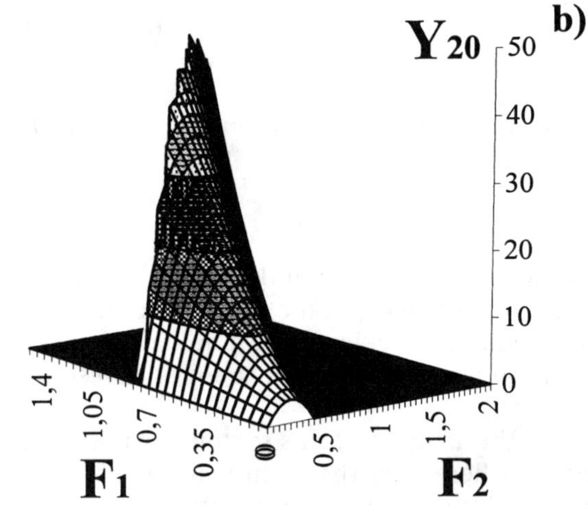

Figure 4. Sustainable yield for two competing species for $K_1 = K_2 = 100$; $a_1 = 1.5$; $a_2 = 2$; $b_1 = 0.5$; $b_2 = 0.8$.

species and therefore in a collapse of ecosystem structure. To avoid displacing the second species with the first species, it is necessary to select a point of control that lies inside the allowable area on the line AR (on the dashed line), i.e., point R_{pr}. The functions Y_{10} and Y_{20} for model (equation 11) are represented in Figs. 4a and 4b as functions of F_1 and F_2.

Conclusions

1. The management strategy chosen for an exploited prey-predator fish community should be selected on the basis of the relationship between the parameters of this community. It is important to determine whether this community corresponds to case 1 or case 2 as described in the text. A different precautionary approach has to be applied for different community types.

2. For a solution of the problem of optimal control it is useful to construct an area of allowable control using as coordinates the fishing mortality rates of the species in the community.

3. Consideration of interspecies interaction suggests that the effect of fishing can be very risky to a community because the fishing of one species can result, for example, in the extinction of the predators in the system. Certainly, we considered a simplified single-predator single-prey model. At a shortage of prey species the real predator switches to another prey; i.e., the risk of its extinction is not so high.

4. The model (equation 1) may serve to illustrate the need to elaborate on the theory of a precautionary approach to community management.

5. Fishing on a population corresponding to lower trophic level without considering the influence of predators on them is unacceptable. It is necessary to expect that there are a number of predators on the same prey population in the natural ecosystem that increase the pressure on the prey.

Acknowledgments

I am grateful to Eva Plaganyi (South Africa) who edited my English. Preparation of this paper was funded by INTAS/RFBR (proj. no. 95-1204).

References

Baranov, F.I. 1918. On the question of biological basis of fishery. In: F.I. Baranov, Izbrannye Trudy 3. Pischevaya Promyshlennost, Moscow, 1971(reprint):12-56. (In Russian.)

Baranov, F.I. 1925. On the question of the dynamics of fishery. In: F.I. Baranov, Izbrannye Trudy 3. Pischevaya Promyshlennost, Moscow, 1971(reprint):56-65. (In Russian.)

Bulgakova, T.I. 1970. On the problem of optimal catch of two competitive populations (mathematical model). Trudy Molodych Uchenych VNIRO 4, ONTI VNIRO, Moscow, pp. 143-156. (In Russian.)

Bulgakova, T.I. 1977. Mathematical investigation of the ecosystems stability and some aspects for the sea animals management. Ph.D.thesis. VNIRO, Moscow. 26 pp. (In Russian.)

Collie, J.S., and P.D. Spencer. 1994. Modelling predator-prey dynamics in a fluctuating environment. Can. J. Fish. Aquat. Sci. 51:2665-2672.

Gause, G.F. 1935. Investigation of the struggle for existence in mixed populations. Zool. J. Acad. Sci. USSR (Moscow) 14(2):243-270. (In Russian.)

Gause, G.F., and A.A. Witt. 1935. Behaviour of mixed populations and the problem of natural selection. Am. Nat. 69:596.

ICES. 1996. Report of the comprehensive fishery evaluation working group. ICES CM1996/Assess:20. 68 pp.

ICES. 1997a. Report of the study group on the precautionary approach to fisheries management. ICES CM 1997/Assess:7. 37 pp.

ICES. 1997b. Report of the multispecies assessment working group. ICES CM 1997 /Assess:16, Part 1. 136 pp.

Jakobsen, T. 1992. Biological reference points for north-east Arctic cod and haddock. ICES J. Mar. Sci. 49:144-166.

Jakobsen, T. 1993. The behaviour of Flow, Fmed and Fhigh in response to variation in parameters used for their estimation. In: Risk evaluation and biological reference points for fisheries management. Can. Spec. Publ. Fish. Aquat. Sci. 120:119-125.

Kolmogoroff, A.N. 1972. Qualitative investigation of mathematical models of population dynamics. Problems of Cybernetics 25:100-106. (In Russian.)

Lotka, A.J. 1925. Elements of physical biology. Williams and Wilkins, Baltimore.

Spencer, P.D., and J.S.Collie. 1995. A simple predator-prey model of exploited marine fish populations incorporating alternative prey. ICES J. Mar. Sci. 53:615-628.

Volterra, V. 1931. Leçons sur la theorie mathematique de la lutte pour la vie. Gauthier-Villars, Paris.

Consumption and Harvest of Pelagic Fishes and Squids in the Gulf of Maine–Georges Bank Ecosystem

William J. Overholtz, Jason S. Link, and Lynette E. Suslowicz
National Marine Fisheries Service, Northeast Fisheries Science Center, Woods Hole, Massachusetts

Abstract

The fish biomass on the continental shelf off the eastern United States and Canada has shifted toward the pelagic community in the 1990s. Biological interactions play a major role in the structure and function of this dynamic ecosystem and thus a greater understanding of these mechanisms can provide a basis for better fisheries management. Low-frequency monitoring of predator diet compositions, such as the current program at the Northeast Fisheries Science Center, can provide a basis for conclusions on the impact of biological interactions on this or similar ecosystems. This study provides estimates of consumption by 12 species of piscivorous fish on Atlantic mackerel (*Scomber scombrus*), Atlantic herring (*Clupea harengus*), butterfish (*Peprilus triacanthus*), sand lance (*Ammodytes* sp.), short-finned squid (*Illex illecebrosus*), and long-finned squid (*Loligo pealei*) during 1977-1997. Results suggest that total consumption by these predatory fish is significant when compared to landings, ranging from 1.5 million to 3.0 million t per year. Predation on pelagic fish and squids appears to be an important and large component of the overall system dynamics of this ecosystem. Consumption of pelagic fish and squids by predatory fish appears to equal or exceed landings in most years from 1977-1997. In several cases, notably for long-finned squid, herring, and butterfish, consumption also exceeds the current MSY for these stocks.

Introduction

The Georges Bank–Gulf of Maine ecosystem in the Northwest Atlantic found along eastern Canada and the United States has undergone significant changes in biomass and relative abundance of fishes during 1970-1997 (Clark and Brown 1977, Mayo et al. 1992, Fogarty and Murawski 1998). Traditionally important fisheries on groundfish (i.e., gadids and flounders), and pelagic species experienced major declines associated with large increases in foreign and domestic fishing effort throughout the 1970s and 1980s (Fig. 1). Following these reductions, formerly underutilized species such as goosefish, white hake, and spiny dogfish became important sources of landings in the 1990s, which then resulted in declines of these stocks. Concurrently, lack of interest by commercial fleets allowed for recovery of the herring and mackerel stocks present in this ecosystem; such that the community is now dominated by pelagic species.

Reported landings of herring and mackerel reached 470,000 t and 437,000 t in 1968 and 1973, respectively, and both stocks collapsed several years later (NEFSC 1996, 1998). Stock rebuilding of these species was complete by the mid-1990s and current biomass for both is on the order of 1-3 million t (NEFSC 1998). Previous research indicated that these species and several other pelagic fish and squids comprise major portions of the diets of piscivorous fish, seabirds, and marine mammals in this ecosystem (Cohen et al. 1982, Bowman and Michaels 1984, Overholtz et al. 1991a).

Understanding the interactions that occur in this pelagic complex should be important for understanding and managing this dynamic system. Future single species and multispecies management regimes for this region may benefit greatly from an increased understanding of the role that biological interactions play in the region (Larkin 1996). Previous research and analysis of diet composition data of piscivorous fish from the region of the Gulf of Maine to Cape Hatteras suggests that predation on pelagic fishes and squids may be one of the more important areas for investigation of biological interactions in this ecosystem. Prior studies suggest that piscivorous fish consumed more biomass than the current harvest of many pelagic species (Cohen et al. 1982, Overholtz et al. 1991a,b). Current fishery management in the United States is focused on sustainable levels of fishing as mandated in the recent revision of the Magnuson-Stevens Fishery Conservation and Management Act. Consumption of fish by predators may in many cases exceed the maximum sustained yield (MSY) of various stocks, with significant overlap of size distributions between commercial landings and predators.

The objective of this research was to utilize the time-series of diet compositions of 12 important predatory fish collected on bottom trawl surveys during 1977-1997 in the area from Maine to North Carolina, to produce estimates of consumption of the more important pelagic species, namely: Atlantic mackerel, Atlantic herring, butterfish, sand lance, short-

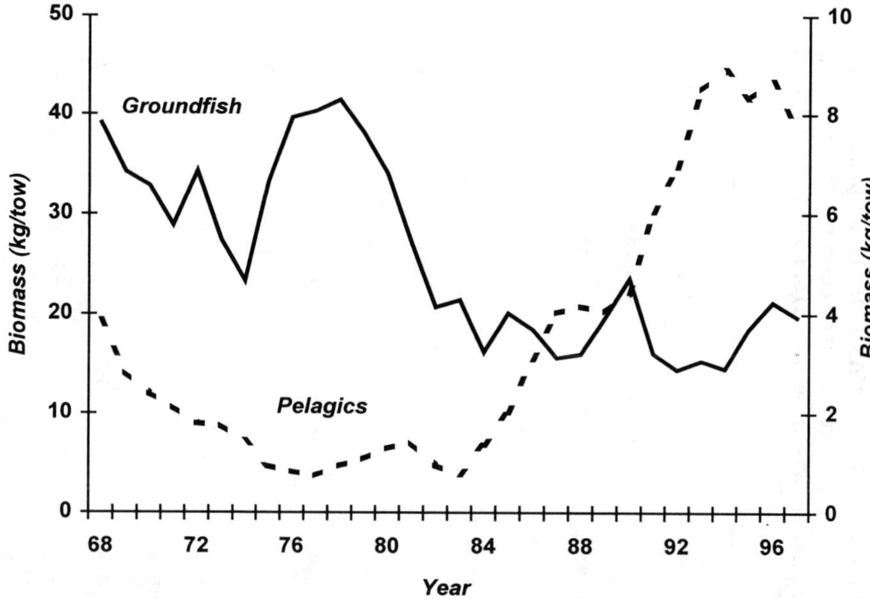

Figure 1. Aggregate catch per tow of principal groundfish and flounders (Groundfish) and herring and mackerel (Pelagics) from research bottom trawl surveys during 1968-1997 from Maine to Cape Hatteras, North Carolina.

finned squid, and long-finned squid. We contrasted fish consumption with commercial landings and current estimates of MSY. Estimates of individual and total consumption for the entire set of predators were produced. We also quantified and discussed various sources of uncertainty.

Methods

Input Data and Analysis

Diet composition data from spring and autumn research survey cruises for 1973-1990 were analyzed to determine a list of significant predators of Atlantic mackerel, Atlantic herring, sand lance, butterfish, and short-finned and long-finned squid. A set of 12 piscivorous fish was chosen based on percentage of pelagic prey in the diets and consistency over time (Table 1). Data were also analyzed in 5 cm intervals to determine appropriate lengths of predators where the fishes and squids began to comprise a significant portion of the diets. Cutoff lengths for predators averaged about 30 cm and ranged from 20 to 40 cm.

Sample sizes of stomachs were examined, and too few samples were available to use a calender year approach for most predators. Data from

Table 1. **Number of samples (N), minimum and maximum predator length (MIN_LEN, MAX_LEN), mean predator fork length (MEAN_LEN), standard error of predator length (SE_LEN), average total stomach content (AVE_STOM, g), and standard error of total stomach contents (SE_STOM) by species based on data from 1973-80 and 1981-1990 bottom trawl surveys from all areas.**

S NAME	N	MIN_LEN	MAX_LEN	1973-1980 MEAN_LEN	SE_LEN	AVE_STOM	SE_STOM
Spiny dogfish *Squalus acanthias*	2,368	22	109	67.669	0.4570	8.558	0.6155
Winter skate *Raja ocellata*	640	19	108	73.131	0.7359	7.079	0.6116
Silver hake *Merluccius bilinearis*	4,453	3	76	22.281	0.1816	1.899	0.1576
Atlantic cod *Gadus morhua*	1,977	3	150	51.564	0.5837	25.275	2.1104
Pollock *Pollachius virens*	601	16	120	63.160	0.9061	12.263	1.7951
Red hake *Urophycis chuss*	1,881	4	72	29.913	0.2785	1.759	0.1204
White hake *Urophycis tenuis*	668	7	113	48.419	0.7081	15.934	3.596
Spotted hake *Urophycis regia*	479	5	38	20.507	0.2988	1.144	0.2022
Summer flounder *Paralichthys dentatus*	632	15	82	36.161	0.43493	3.021	0.577
Bluefish *Pomatomus saltatrix*	536	9	83	35.604	0.81150	15.775	1.912
Weakfish *Cynoscion regalis*	367	7	75	26.071	0.64224	2.333	0.389
Goosefish *Lophius americanus*	745	7	124	59.110	0.8950	58.121	6.2823

Table 1. (Continued.)

S NAME	N	MIN_LEN	MAX_LEN	1981-1990 MEAN_LEN	SE_LEN	AVE_STOM	SE_STOM
Spiny dogfish *Squalus acanthias*	19,620	20	114	69.495	0.1395	14.981	0.347
Winter skate *Raja ocellata*	3,603	15	109	67.331	0.3594	9.527	0.459
Silver hake *Merluccius bilinearis*	13,766	6	66	27.407	0.0612	3.019	0.106
Atlantic cod *Gadus morhua*	6,943	9	140	55.035	0.2422	27.526	0.962
Pollock *Pollachius virens*	1,659	14	108	54.347	0.5961	18.993	1.257
Red hake *Urophycis chuss*	4,942	8	70	32.440	0.1148	4.037	0.180
White hake *Urophycis tenuis*	4,723	11	136	44.525	0.2156	17.732	0.9086
Spotted hake *Urophycis regia*	1,581	7	40	23.488	0.1226	2.281	0.141
Summer flounder *Paralichthys dentatus*	1,348	15	72	35.052	0.2436	2.966	0.283
Bluefish *Pomatomus saltatrix*	1,619	8	88	35.613	0.4808	20.490	1.6171
Weakfish *Cynoscion regalis*	763	8	79	26.388	0.3879	3.635	0.3750
Goosefish *Lophius americanus*	1,108	10	121	50.017	0.6885	49.440	5.1675

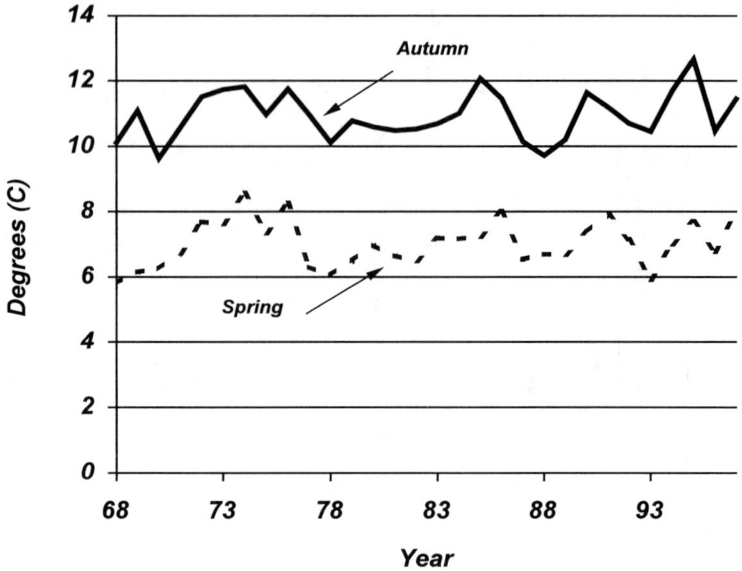

Figure 2. Average spring and autumn bottom temperature during 1968-1997 for the region from Maine to Cape Hatteras, North Carolina.

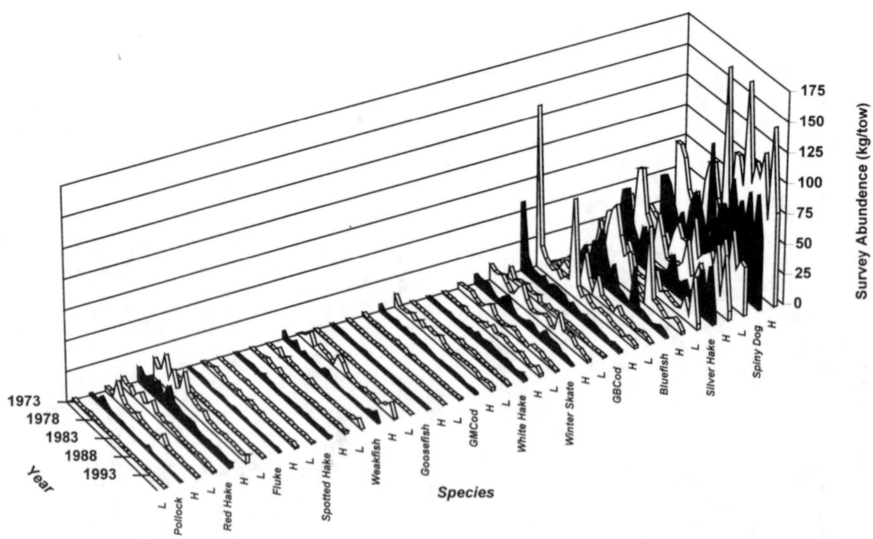

Figure 3. Mean number of fish per tow and 95% CI from research bottom trawl surveys for 12 species of predatory fish during 1973-1997.

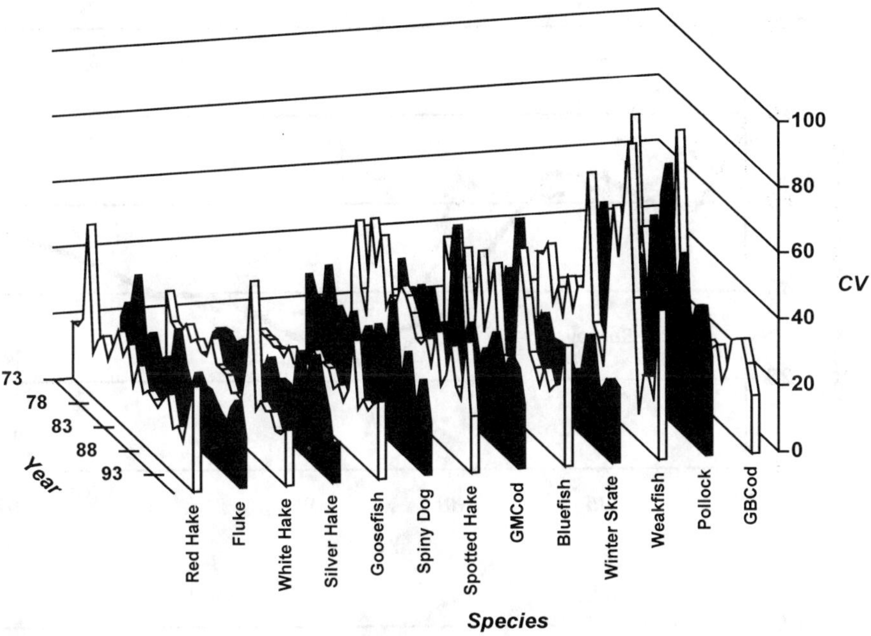

Figure 4. Coefficient of variation (CV) of mean survey abundance for 12 species of predatory fish during 1973-1997.

1977-1997 were aggregated on 2-year intervals for all predators except spotted hake, bluefish, and weakfish which were averaged over 5-year intervals. Interval length was based on having at least 50-75 stomachs available in each season (i.e., spring and autumn). We assumed that spring data would serve as a proxy for winter and autumn for the summer periods of the year.

Bottom temperature data were obtained for the same research survey cruises during 1977-1997. Average profiles for the entire region from the Gulf of Maine to Cape Hatteras, North Carolina, were calculated using methods developed by Holzwarth and Mountain (1992) to interpolate between stations and regions (Fig. 2).

Estimates of annual predator biomass and abundance were obtained from spring and autumn bottom trawl surveys during 1977-1997 (Fig. 3). Survey catch per tow in numbers were trimmed to account for each predator cutoff length and area swept abundance for the entire survey region was estimated. Coefficients of variation in the trawl survey abundance estimates averaged 20-40% for most species with occasional values at 50% or greater in specific years (Fig. 4). For species with virtual population analysis (VPA) available an average catchability coefficient (q) was calculated

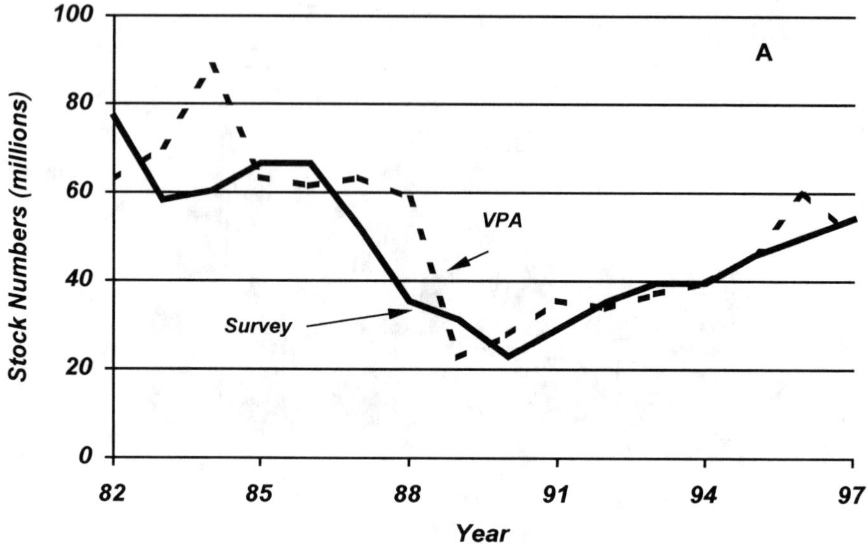

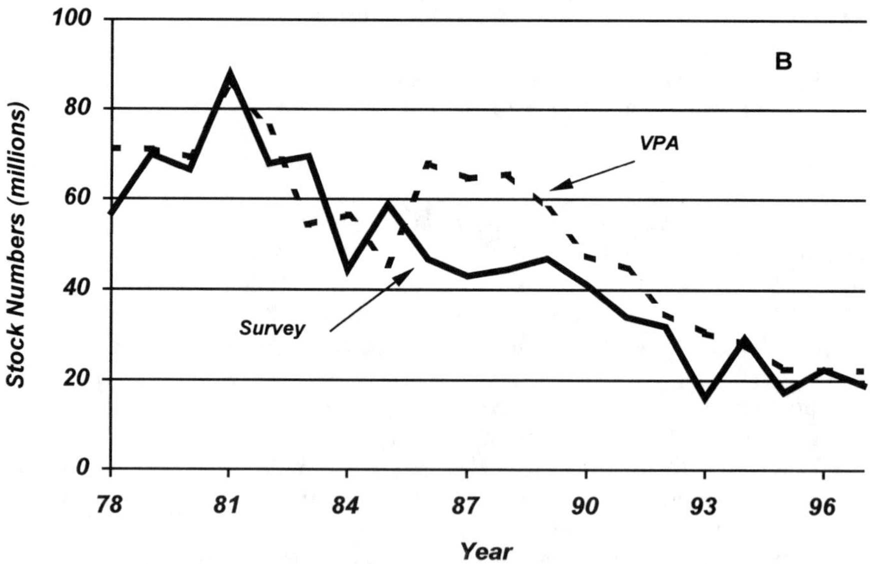

Figure 5. *A: Total stock numbers from virtual population analysis (VPA) and survey q method (Survey) for summer flounder during 1982-1997. B: Total stock numbers from VPA and survey q method for Georges Bank cod during 1978-1997.*

from age-specific results appropriate for each predator cutoff length. This average q was divided into area swept numbers to produce a virtual population estimate for each year. This method appears to produce useful results similar to current VPAs (Fig. 5), and is necessary for estimating predator numbers during years prior to the first year in the VPA. For species without a current VPA, weighting coefficients from Clark and Brown (1977) were applied in the same manner to produce estimates of population numbers. Exceptions to this were spiny dogfish and winter skate where area swept numbers were used. Spiny dogfish are very vulnerable to bottom trawl survey gear and following the same procedures used in the assessment for this stock (NEFSC 1998), an area swept estimate of numbers was used in this analysis. The estimate for winter skate was also based on an estimate of area swept numbers, but lacking any collaborative analysis, an estimate based on night tows (higher vulnerability) was used. Since availability to survey gear can be influenced by numerous factors in any given year, large, but not actual changes in average number and weight per tow may occur, resulting in very abrupt interannual variability in bottom trawl time series data. To compensate for this type of problem we used LOWESS (Cleveland 1979) smoothing methods to produce estimates of predator biomass. This approach is consistent with the current methodology applied in the spiny dogfish assessment (NEFSC 1998). Predator numbers used to estimate consumption are shown in Fig. 6.

The biomass of pelagic fish and squids varied considerably during 1973-1997. Butterfish, long-finned squid, and short-finned squid were generally more abundant during the 1970s, showed some decline in the 1980s, but did not fluctuate to any great degree during the entire period (Fig. 7). Sand lance on the other hand were very abundant during the late 1970s and early 1980s, but declined to very low abundances thereafter. Mackerel and herring, stocks that were heavily fished, were very low through the mid-1970s and early 1980s, but increased greatly during the 1990s (Fig. 7).

Estimation of Consumption

Many approaches to estimating consumption are available in the literature, but we chose the approach developed by Eggers (1977), Elliot and Persson (1978), and Pennington (1985) that utilizes the relationship between average stomach contents and gut evacuation rate as:

$$C = 24RS^{\gamma} \tag{1}$$

where C is consumption, 24 is the number of hours in a day, S is the mean stomach weight per 24-hour period per fish predator, γ is a derived constant (usually assumed equal to 1), and R is the evacuation rate, where:

$$R = \alpha\, e^{\beta T} \tag{2}$$

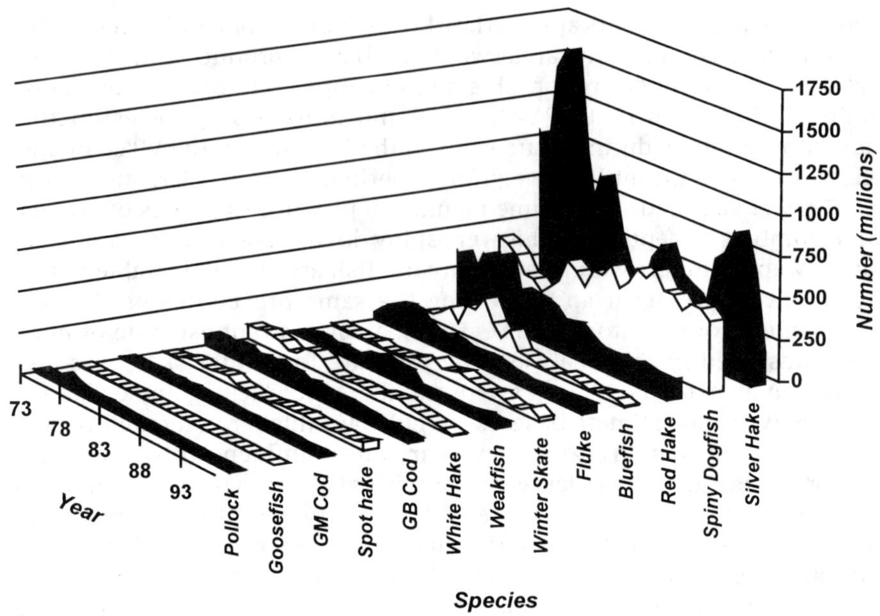

Figure 6. Total stock size of predatory fish used to estimate consumption during 1977-1997.

where α and β are fitted constants and T is temperature. Values for parameters in these two equations were obtained from the literature after a review of methods and estimates from various analyses (Tyler 1970, Jones 1974, Kioeboe 1978, Durbin et al. 1983). The alpha and beta parameters in equation 2 were assumed to equal 0.004 and 0.115, respectively, conservative values for teleost fishes from Durbin et al. (1983). Since results are tied closely to the choice of values for these two important parameters, a sensitivity analysis was completed for both the alpha and beta parameters over a range of temperature values. As expected, evacuation rates change minimally, but detectably, over a range of assumptions for the alpha parameter (Fig. 8 top). The beta parameter has a much more pronounced impact on evacuation rates resulting in large and dramatic increases in *R*, as expected due to the exponential effect of this parameter (Fig. 8 bottom).

Stomach content data for 1977-1980 were estimated in grams per stomach while data collected during 1981-1997 were estimated in volumetric units (cubic centimeters per stomach). We used a conversion of 1.1 g per 1.0 cc of stomach material in the current analysis. Data for estimating the

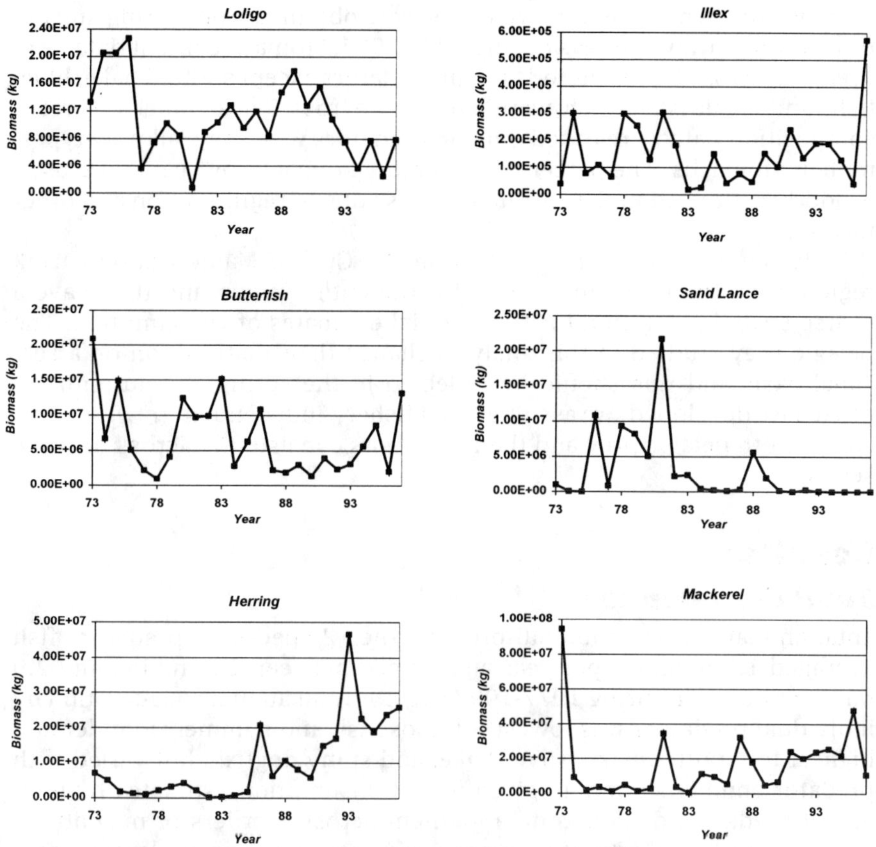

Figure 7. Survey biomass prey species 1973-1997. Biomass (kg) of Loligo, Illex, butterfish, mackerel, herring, and sand eels from research bottom trawl surveys during 1973-1997.

average stomach content in predators were obtained from spring and autumn bottom trawl surveys during 1977-1997. Stomach content data were averaged across 2-year periods for all predators except spotted hake, bluefish, and weakfish, which were averaged across 5-year blocks. Percent composition data were based only on predators with prey (all prey types) in their stomachs. The percentage of empty stomachs by year and season is poorly estimated in the current database due to regurgitation and other factors.

Although our primary focus is on the Gulf of Maine–Georges Bank region, we assumed homogeneity for the entire region and thus have a coarse spatial resolution for these initial estimates of consumption. The pelagic prey studied in this analysis change their distribution on a seasonal basis and this should be reflected in the spring and autumn approach we developed. We assumed that fishery-independent trawl surveys collect both pelagic prey and their predators consistently across the time series.

Results

Total Consumption

Total annual consumption (all prey) by the 12 species of predatory fish averaged 1.5 million t per year and ranged between 1.3 million and 2.9 million t per year during 1977-1997 (Fig. 9). Total annual consumption by individual predators was lowest for goosefish and summer flounder and highest for spotted hake, silver hake, and spiny dogfish, but varied with predator abundance in any given year. Consumption estimates by individual predators demonstrate a gradient across 3 orders of magnitude, mainly reflective of predator size and abundance. For example, total consumption by spiny dogfish averaged 619,000 t per year and ranged from 113,000 to 890,000 t per yer during 1977-1997 (Fig. 10). Total annual consumption for bluefish averaged 108,000 t per year and ranged from 16,000 to 167,000 t per year during this same period (Fig. 10). Goosefish total annual consumption averaged 14,000 t per year and ranged from 3,600 to 61,000 t per year during 1977-1997 (Fig. 10).

Consumption of Pelagics

Consumption of pelagic fishes and squids generally tracked levels of prey abundance, particularly for herring, mackerel, and sand lance (Fig. 11). Predation on sand lance reached very high levels in the late 1970s and early 1980s coincident with the collapse of mackerel and herring and the large biomass of sand lance present at this time (Fig. 11). As the mackerel stock began to recover in the early 1980s, predation increased, reaching 89,000 t in 1988 (Fig. 11). Consumption of herring averaged 200,000 t

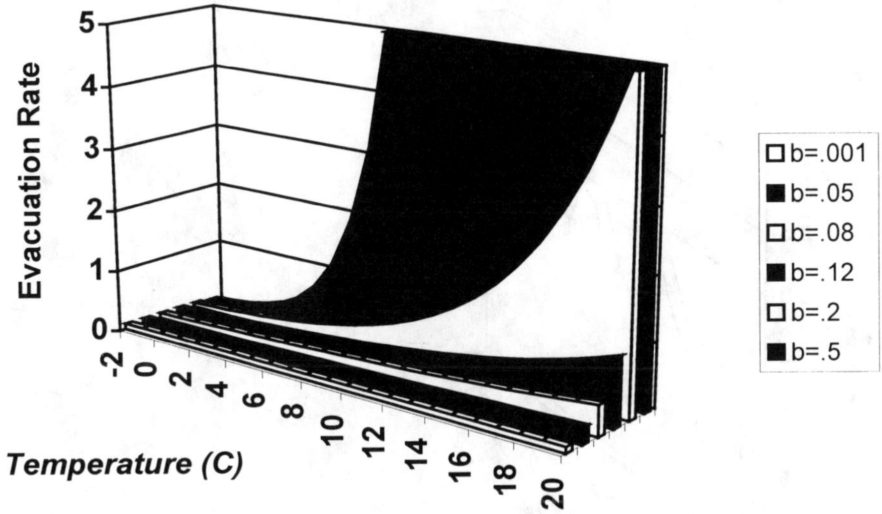

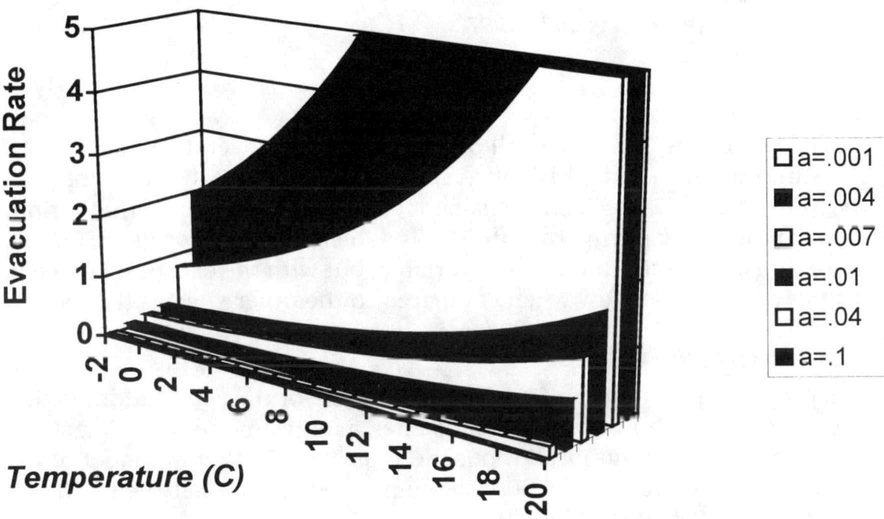

Figure 8. (Top) Results of sensitivity analysis with $\alpha = 0.004$ and β ranging from 0.001 to 0.5 over temperatures from -2° to 20° C. (Bottom) Results of sensitivity analysis with $\beta = -0.1$ and α ranging from 0.001 to 0.1 over temperatures from -2° to 20° C.

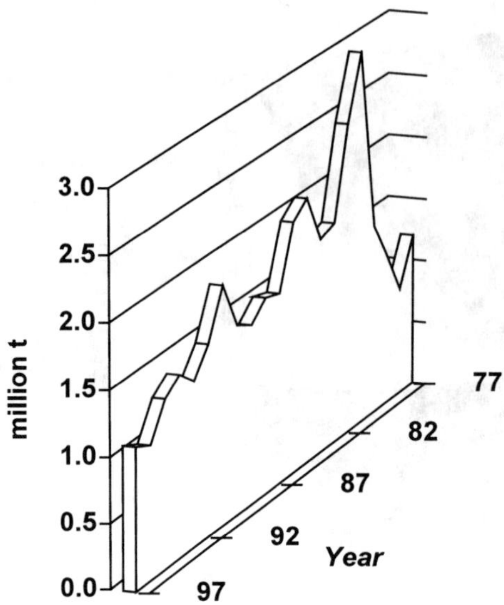

Figure 9. Total consumption for 12 predatory fish during 1977-1997.

during the 1990s, reaching well over 300,000 t in several years (Fig. 11). Consumption of *Illex* and *Loligo* averaged 24,000 t and 46,000 t, respectively, during 1977-1997, but remained relatively constant over this period, with a minor peak during the late 1970s and early 1980s for *Illex* (Fig. 11). Predation on butterfish was more variable, but with the exception of a few years, was relatively low when compared to the other species (Fig. 11).

Unidentified Fish

The proportion of unidentified fish in the diets of these predators varied over 1977-1997, but generally represented a large amount of biomass, on the order of several hundred thousand t per year. This component of the predator diet averaged 292,000 t annually and ranged between 170,000 and 436,000 t per year (Fig. 12).

Consumption vs. Available Biomass of Mackerel and Herring

Consumption of mackerel during 1977-1997 generally followed mackerel biomass until 1989, when a sharp decline occurred ($r = 0.21$) (Fig. 13 top). The reduction in mackerel consumption by piscivorous fishes continued through 1993 in spite of large increases in mackerel biomass (Fig. 13 top). Consumption of herring was relatively low during 1977-1988, and then

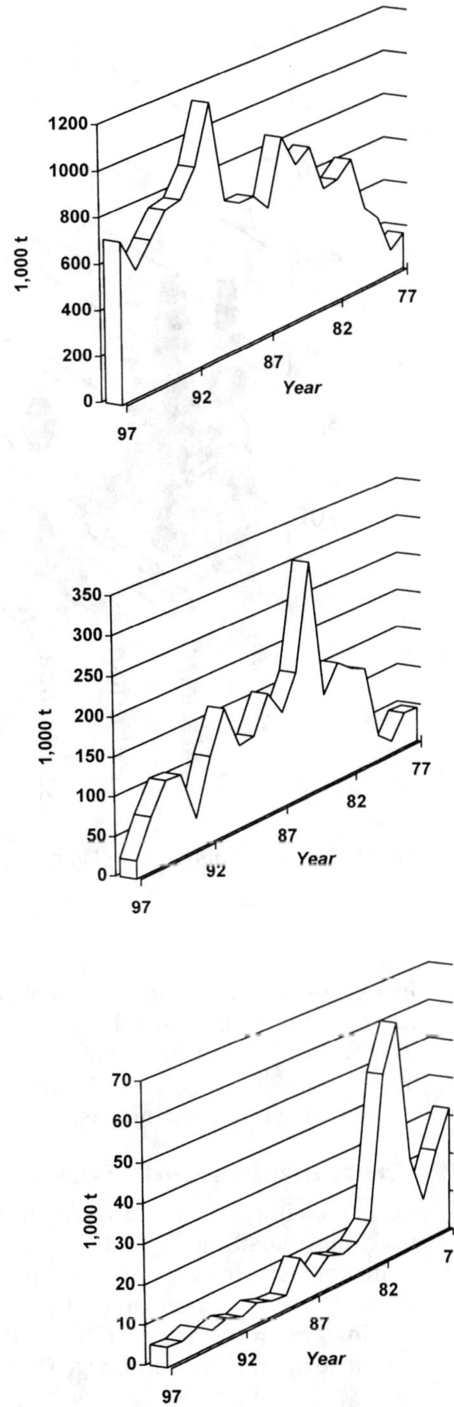

Figure 10. Total consumption by spiny dogfish, bluefish, and goosefish during 1977-1997.

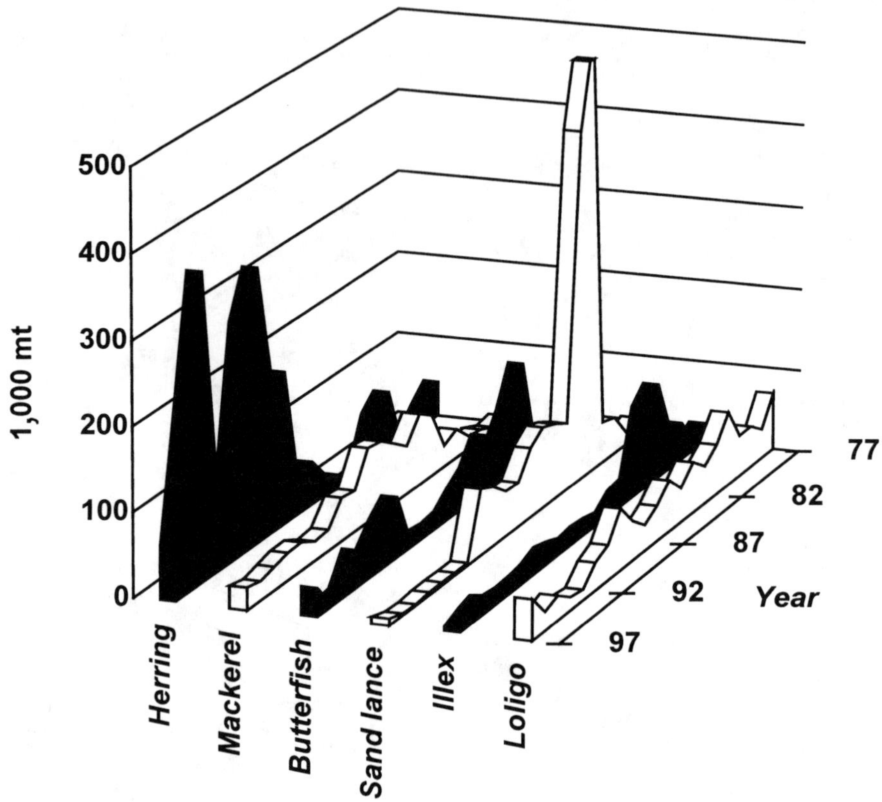

Figure 11. Consumption of pelagic prey by 12 predatory fish during 1977-1997.

increased rapidly as herring biomass began to increase after 1989 (Fig. 13 bottom). Consumption of herring remained high during 1989-1996 with the exception of sharp drop during 1993-1994 (Fig. 13 bottom). Overall consumption of herring by the 12 predators was positively correlated ($r = 0.61$) with herring biomass, increasing substantially in the 1990s (Fig. 13 bottom).

Consumption and Fishing

Recent changes in the Sustainable Fisheries Act (SFA) have meant that maximum sustained yields are again an important biological reference point for fish stocks; MSYs for the pelagic fishes and squids in this analysis were recently reevaluated (NEFMC 1998). The MSY for mackerel and herring are on the order of 300,000 t, while averaging only about 20,000 t for butterfish, *Illex*, and *Loligo* (NEFMC 1998).

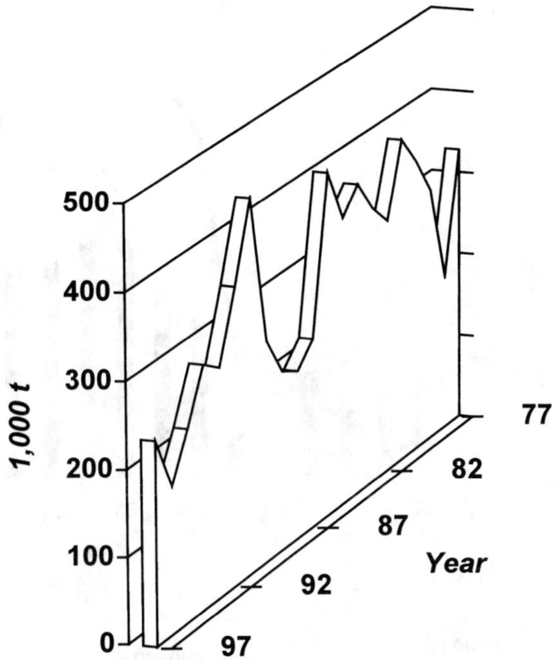

Figure 12. Estimates of the unidentified fish component consumed by 12 species of predatory fish during 1977-1997.

Commercial landings of prey species during 1973-1997 ranged over 2 orders of magnitude (Fig. 14). Squid and butterfish landings were relatively constant and low, except for *Illex* in the mid-1970s. Herring and mackerel landings were very large in the early to mid-1970s and then dropped to much lower values thereafter (Fig. 14). There is no commercial fishery on sand lance.

Consumption of *Loligo* exceeded landings in all years during 1977-1997 (Fig. 15 top). Mackerel consumption and landings were similar during this period and both were well below the MSY for this stock (Fig. 15 middle). Annual herring landings averaged 82,000 t during 1977-1997, ranging from 36,000 to 120,000 t per year (Fig. 15 bottom). Consumption by predators was below 50,000 t during 1977-1987 and then increased dramatically in the 1990s, ranging from 100,000 to 270,000 t (Fig. 15 bottom). Consumption by piscivorous fish in recent years approaches the estimated MSY for the herring complex (Fig. 15 bottom).

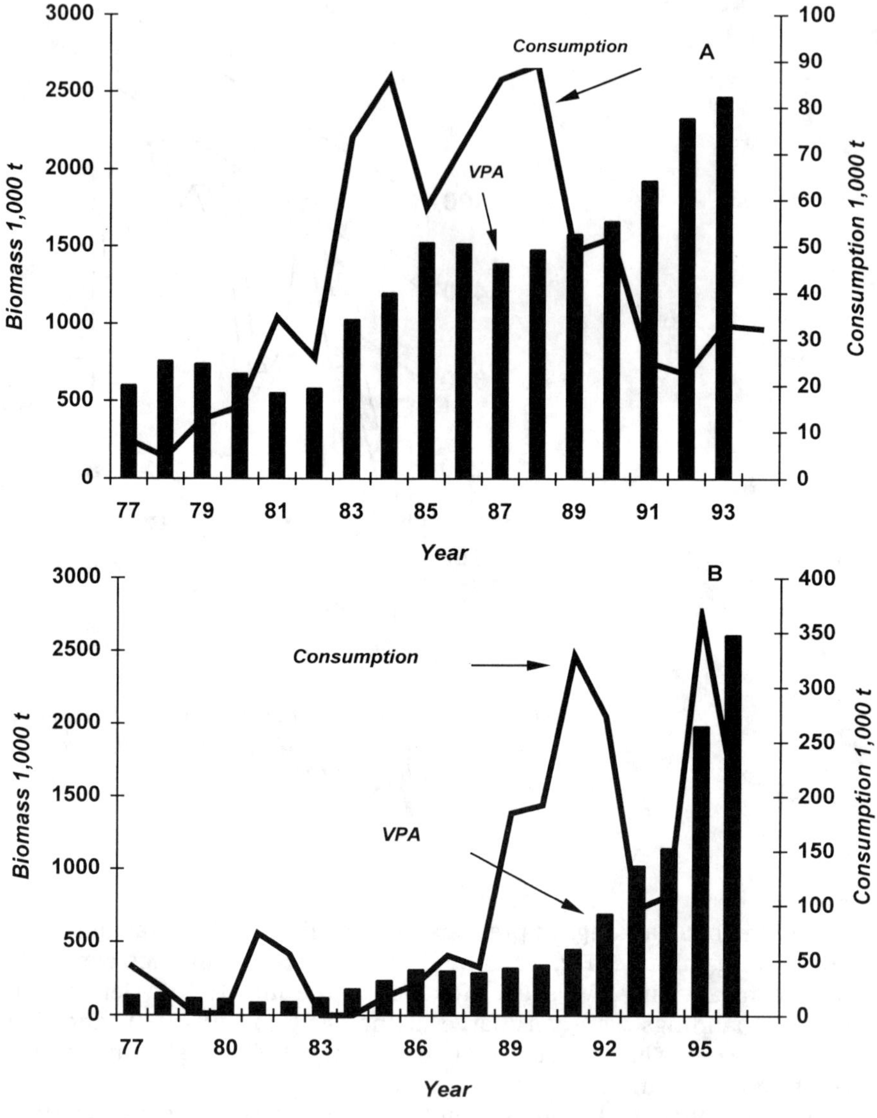

Figure 13. (Top) Consumption of mackerel by 12 species of predatory fish and average mackerel VPA biomass during 1977-1994. (Bottom) Consumption of herring by 12 species of predatory fish and average herring VPA biomass during 1977-1996.

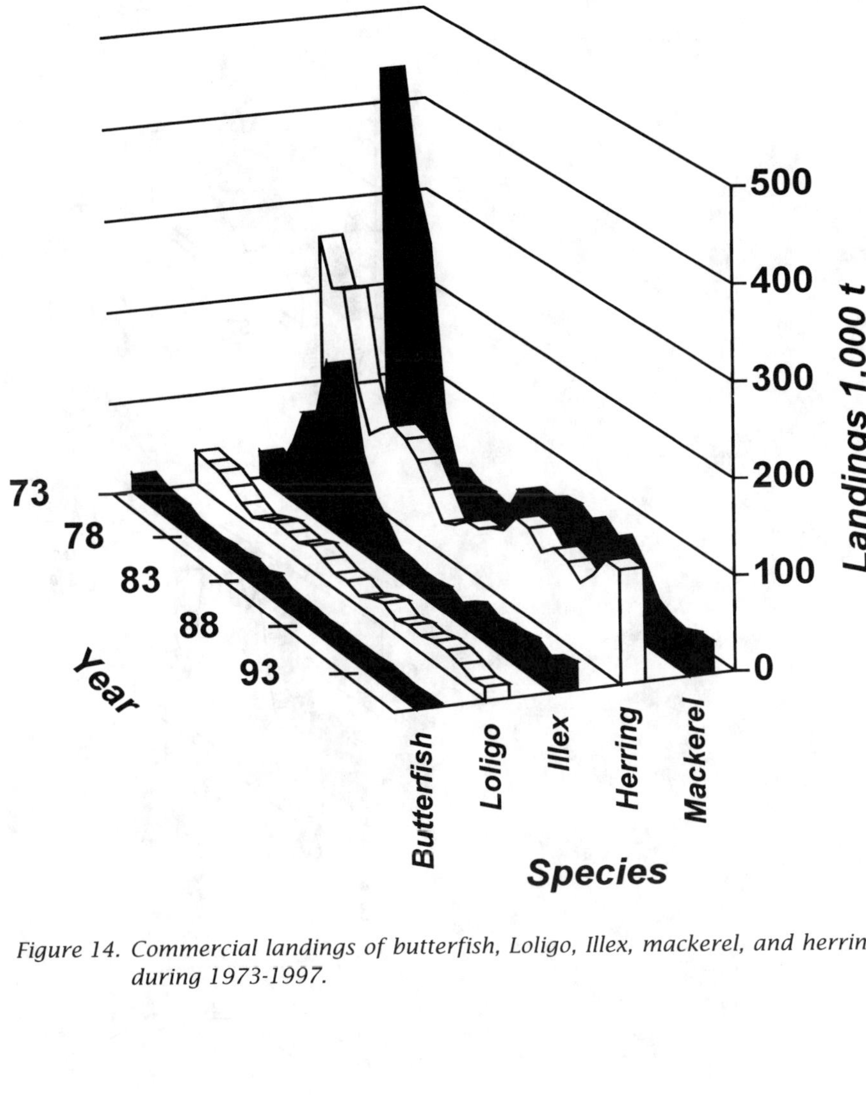

Figure 14. Commercial landings of butterfish, Loligo, Illex, mackerel, and herring during 1973-1997.

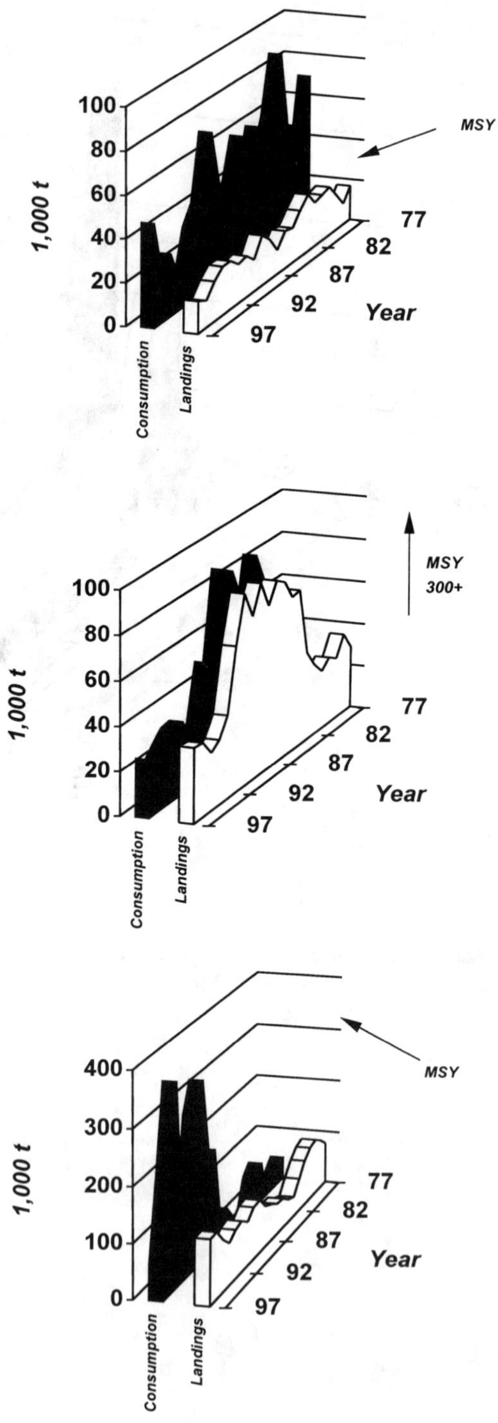

Figure 15. Consumption of (top) Loligo, (middle) mackerel, and (bottom) herring by 12 species of predatory fish and commercial landings during 1977-1997.

Discussion

Results from this analysis suggest that the magnitude of total consumption by these predatory fish is large, ranging from 1.5 million to 3.0 million t during 1977-1997. Predation on pelagic fish and squids is an important and large component of the overall dynamics of this Northeast shelf ecosystem. In many instances that diet composition data collected during bottom trawl surveys follows prey abundance rather closely. For example, during the collapse of the herring and mackerel stocks in the mid-1970s, the consumption of these species dropped and a surge in sand lance abundance and consumption occurred (Fogarty et al. 1991). Similarly, mackerel from the large 1982 year class were present in many predators in 1983 and 1984 as sand lance were declining very rapidly in abundance and importance in predator diets. Stock rebuilding of the herring complex in the 1990s was also apparent in predator diets as percentages increased considerably during this period.

Interestingly, mackerel, although very abundant now, appear to be less important than herring in the diets of piscivorous fish. This may be related to predator preference or availability of these prey fish to predators. Mackerel grow rapidly and are probably only vulnerable for the first several years of their life history, unless affected by mechanisms such as density dependence (Overholtz et al. 1991b). Herring, except at very large sizes (> 30 cm), seldom grow out of the window of predation vulnerability and are subject to predation by fish over most of their life history. Predation may be positively related to the available biomass of mackerel and herring, but consumption of mackerel appears to have declined in conjunction with major increases in herring biomass, at least for this set of predators.

Several issues need to be addressed in the design of future fish stomach sampling regimes to address some of the possible uncertainties encountered in this analysis. The percentage of empty stomachs in the current database may be in excess of actual values encountered in other ecosystems (Daan 1973, Bowman 1976). The problem of unidentified fish in the stomachs is very important because these are likely the same small pelagics of interest to this study. This means that consumption on these species of fish may be underestimated in the current analysis. Daan (1973) was able to closely examine digested fish to provide positive identification of the remains. This can be done by examining hard parts such as otoliths, skeletal structure, musculature, and other forensic approaches. Clearly this is an area for further investigation.

It appears that low-frequency monitoring of predator diet compositions such as the current program can provide an adequate database for drawing conclusions concerning the impact of biological interactions. If predators change their preference for different food items based on several factors, it may be necessary to establish a long-term monitoring program to document patterns, instead of extrapolating results from one or two intensive sample periods. On the other hand, intense seasonal sampling may be necessary to fully elucidate stomach content and diet composition of key predators. Several years of intense seasonal sampling would provide important information for a comparison with the current low-frequency monitoring program. This would improve our knowledge base on the seasonal mean and variance of stomach content data and diet compositions. It would also confirm if extrapolations to half-year intervals based on spring and autumn sampling are valid.

Our results may underestimate consumption since we choose to use relatively conservative parameters in our estimation. The sensitivity analysis we conducted indicates that consumption estimates are very dependent on the choice of parameters. Further work in this important area would allow for more precise statements about the impacts of predation.

Finally, consumption of pelagic fish and squid by predatory fish appears to equal or exceed landings in most years from 1977 to 1997. In several cases, notably for *Loligo*, herring, and butterfish, consumption also exceeds the current value of MSY for these stocks. Length compositions of mackerel in landings and predators do not overlap to any great degree because most of the commercial fishery is centered on age 3+ fish. However, for herring, butterfish, *Loligo*, and *Illex*, there is probably some overlap between the fishery and predators. This means that to an extent the fishery and predators are competing for the same fish.

Acknowledgments

We would like to thank D. Mountain for providing the temperature profiles used in the analysis. We also thank the many seagoing personnel who collected the stomach content data during the 1970s, 1980s, and 1990s.

References

Bowman, R.E. 1986. Effect of regurgitation on stomach content data of marine fishes. Environ. Biol. Fishes 16:171-181.

Bowman, R.E., and W.L. Michaels. 1984. Food of seventeen species of northwest Atlantic fish. NOAA Tech. Memo. NMFS-F/NEC-28. 183 pp.

Clark, S.H., and B.R. Brown. 1977. Changes in biomass of fin fishes and squids from the Gulf of Maine to Cape Hatteras, 1963-1974, as determined from research vessel survey data. Fish. Bull., U.S. 75:1-21.

Cleveland, W.S. 1979. Robust locally weight regression and smoothing scatterplots. Journal of the American Statistical Association 74:829-836.

Cohen, E.B., M.D. Grosslein, M.P. Sissenwine, F. Steimle, and W.R. Wright. 1982. Energy budget of Georges Bank. In: M.C. Mercer (ed.), Multispecies approaches to fisheries management advice. Can. Spec. Publ. Fish. Aquat. Sci., pp. 95-107.

Daan, N. 1973. A quantitative analysis of the food intake of North Sea cod, *Gadus morhua.* Netherlands J. Sea Res. 6:479-517.

Durbin, E.G., A.G. Durbin, R.W. Langton, and R.E. Bowman. 1983. Stomach contents of silver hake, *Merluccius bilinearis,* and Atlantic cod, *Gadus morhua,* and estimation of their daily rations. Fish Bull., U.S. 81:437-454.

Eggers, D.M. 1977. Factors in interpreting data obtained by diel sampling of fish stomachs. J. Fish. Res. Board Can. 34:290-294.

Elliott, J.M., and L. Persson. 1978. The estimation of daily rates of food consumption for fish. J. Anim. Ecol. 47:977-991.

Fogarty, M.J., and S.A. Murawski. 1998. Large scale disturbance and the structure of marine systems: Fishery impacts on Georges Banks. Ecol. Appl. 8:S6-S22.

Fogarty, M.J., E.B. Cohen, W.L. Michaels, and W.W. Morse. 1991. Predation and the regulation of sand lance populations: An exploratory analysis. ICES Symp. Mar. Sci. 193:120-124.

Holzwarth, T., and D. Mountain. 1992. Surface and bottom temperature distributions from the Northeast Fisheries Science Center spring and fall bottom trawl survey program, 1963-1987. Northeast Fisheries Science Center Reference Document 90-03. Woods Hole Laboratory, Woods Hole, Massachusetts 02543. 62 pp.

Jones, R. 1974. The rate of elimination of food from the stomachs of haddock, (*Melanogrammus aeglefinus*), cod (*Gadus morhua*), and whiting, (*Merlangius merlangus*). J. Cons. Cons. Int. Explor. Mer. 35:225-243.

Kiorboe, T. 1978. Laboratory measurements of digestion rate in 0-group flounder (*Platichthys flesus* L.) at two temperatures. Ophelia 17:155-161.

Larkin, P.A. 1996. Concepts and issues in marine ecosystem management. Rev. Fish Biol. Fish. 6:139-164.

Mayo, R.K., M.J. Fogarty, and F.M. Serchuk. 1992. Aggregate fish biomass and production on Georges Bank 1960-1987. J. Northwest Atl. Fish. Sci. 14:59-78.

NEFMC. 1998. Evaluation of existing overfishing definitions and recommendations for new overfishing definitions to comply with the Sustainable Fisheries Act. New England Fishery Management Council. Final Report. June 17, 1998. 179 pp.

NEFSC. 1996. Atlantic mackerel. In: Report of the 20th Northeast Regional Stock Assessment Workshop. Northeast Fisheries Science Center Reference Document 95-18, pp. 60-96.

NEFSC. 1998. Spiny Dogfish. In: Report of the 26th Northeast Regional Stock Assessment Workshop. Northeast Fisheries Science Center Reference Document 98-03, pp. 224-283.

Overholtz, W.J., S.A. Murawski, and K.L Foster. 1991a. Impact of predatory fish, marine mammals, and seabirds on the pelagic fish ecosystem of the northeastern USA. ICES Mar. Sci. Symp. 193:198-208.

Overholtz, W.J., S.A. Murawski, and W.L. Michaels. 1991b. Impact of compensatory responses on assessment advice for the Northwest Atlantic mackerel stock. Fish. Bull., U.S. 89:117-128.

Pennington, M. 1985. Estimating the average food consumption by fish from stomach content data. Dana 5:81-86.

Tyler, A.V. 1970. Rates of gastric emptying in young cod. J. Fish. Res. Board Can. 27:1177-1189.

Multispecies Interactions in the Georges Bank Fish Community

Jeremy S. Collie and Allison K. DeLong
University of Rhode Island, Graduate School of Oceanography, Narragansett, Rhode Island

Abstract

Dramatic shifts in the species composition of the Georges Bank fish community have occurred during the past 30 years. The bottomfish community, once dominated by cod, haddock, and flounder, shifted to lower-value species such as skates and dogfish. Herring and mackerel declined during the 1960s and 1970s, but have since recovered to record levels. These shifts are attributed primarily to high fishing pressure, but predation is the largest source of pre-recruit mortality, and is also considered important in controlling the dynamics of the fish community. Prior studies have identified the important predators as cod, silver hake, and spiny dogfish. Herring, mackerel, silver hake, and yellowtail flounder all experience high predation mortality, especially in the first and second years of life. To construct dynamic production models of the Georges Bank fish community, we aggregated the dominant species into four groups: gadoids, flatfishes, pelagics, and elasmobranchs. Inclusion of species interactions in the dynamic models was based on statistical fits to the biomass data and groundtruthed with known species interactions. The most important interactions are predation of gadoids and elasmobranchs on pelagics and apparent competition between gadoids and elasmobranchs. Equilibrium yields of each group depend on the abundances and hence the exploitation rates of the other species. Harvest strategies are simulated for both the multispecies model and the corresponding single-species equations, to illustrate the sensitivity of medium-term projections to species interactions.

Introduction

The Georges Bank fish community has undergone profound shifts in abundance and species composition during the past 3 decades (Fogarty and Murawski 1998). From 1963 to 1993 the principal groundfish species (cod, haddock, yellowtail flounder) were progressively depleted, first by distant-water fleets and then by domestic fisheries. The principal pelagics (herring and mackerel) initially declined due to high exploitation, and subsequently recovered to record abundance levels as exploitation was reduced. Dogfish and skates, which have experienced low exploitation until recent years, increased to maximal levels in 1990 and have since begun to decline. The end result was a shift in community structure from demersal groundfish to pelagics and elasmobranchs.

Were these shifts in community structure independent responses to fishing? If so, single-species management approaches should be sufficient to prevent overfishing and to rebuild depleted stocks. Or, did trophic interactions contribute to the shift in species composition? Did predation contribute to the decline of the groundfish stocks, and does it delay their recovery from overfishing? What caused the dramatic increase in elasmobranchs and, more recently, the pelagic species? To the extent that multispecies interactions occur, their effects may need to be considered in restoring the Georges Bank fish community to a more desirable and valuable species composition (Sissenwine and Cohen 1991, Fogarty and Murawski 1998).

Several lines of evidence support the importance of species interactions in the Georges Bank fish community. A biomass-dynamics model, fit to the aggregate biomass of fish species, indicated that, although the species composition of the fish community may change, aggregate yield is limited by community-level density dependence (Brown et al. 1976). The Georges Bank energy budget is relatively tight in the sense that most of the production at lower trophic levels can be accounted for by predation at higher trophic levels (Sissenwine et al. 1984). By including a juvenile fish component in the energy budget, Sissenwine et al. (1984) demonstrated that the largest source of mortality of prerecruit fish is predation by other fish. Hence predation can potentially affect the survival of prerecruit fish and/or the growth and production of predator populations.

The feeding habits of the Georges Bank fish species have been analyzed in conjunction with the National Marine Fisheries Service (NMFS) bottom trawl survey (Grosslein et al. 1980). This food habits database has been used to calculate diet overlap by Langton (1982) and as input to age-structured multispecies models of the fish community by Overholtz et al. (1991). These authors found that predation mortality rates were highest on the youngest ages of fish and that predation mortality varied during the 5 years of their analysis. Multispecies virtual population analysis (MSVPA) is a form of age-structured multispecies analysis that relates changes in diet composition to changes in the abundances of prey and predator species (Sparre 1991). A nine-species MSVPA of the Georges Bank fish community from 1978 to 1992 was performed by Tsou and Collie (In press). The important predators were cod, silver hake, dogfish, and skates, whereas the important prey species were herring, mackerel, silver hake, and yellowtail flounder. The MSVPA confirmed the high levels of predation mortality on age-0 and age-1 fish. The biomass of prey species consumed was of the same order as the commercial harvest.

Other types of multispecies analysis have been performed with the NMFS trawl survey data. Fogarty and Brodziak (1994) aggregated the survey data into gadids, flatfishes, skates, other groundfish, and small pelagics to reduce the complexity of the multispecies system. Lagged correlations between the aggregate indices revealed inverse relationships between species groups that are known from food-habits data to interact (e.g., flounder and skates). This correlation structure was used to select the terms of a multivariate time-series model.

In this study we aggregated biomass and landings data from 1963 to 1993 of 10 biologically or commercially important species on Georges Bank into four groups. These data were used to fit single and multispecies biomass-dynamics models. Unlike cohort-based methods, biomass-dynamics models can be applied to species for which age-composition data are lacking. The multispecies model parameter estimates identify and quantify the strongest species interactions. We explore how species interactions cause biological reference points to vary as other species are exploited. Finally, the single and multispecies model fits are used to simulate population trajectories and to evaluate management strategies. The results of this study can help evaluate the ability of single-species models to be used to successfully manage species that interact strongly with one another.

Methods

Ten fish species were selected for this study based on their commercial and/or trophic importance (Table 1). We compiled biomass and landings data for these 10 species for the years 1963-1993. This time period corresponds to the start of the autumn trawl survey in 1963 up until 1994 when catch reporting requirements were changed and consistent landings estimates are not yet available for all species. Sand lance (*Ammodytes* spp.) and squid (*Loligo pealei* and *Illex illecebrosus*) are important as prey of other fishes, but were not included in this study because biomass estimates for these taxa are not available.

Biomass and landings data were compiled from the most recent stock assessments, where available (Table 1). For species with age-structured assessments, we summed ages 2 and older, as age 2 corresponds roughly to the age of recruitment and maturity for many of these species. Data for cod (*Gadus morhua*), haddock (*Melanogrammus aeglefinus*), yellowtail flounder (*Pleuronectes ferrugineus*), herring (*Clupea harengus*), and mackerel (*Scomber scombrus*) were extracted from recent ADAPT runs. Historical virtual population analyses (VPA) were used to fill in the early years for herring and silver hake (*Merluccius bilinearis*), for which there is no recent stock assessment. Catch-survey analysis (CSA) was used to fill in the early years of cod and yellowtail flounder prior to the age-structured assessments.

Table 1. Sources of biomass and landings data.

Method	Species	Years	Reference
Age-structured assessment (ADAPT)	Atlantic cod	1978-1993	NEFSC (1997)
	Haddock	1963-1993	NEFSC (1997)
	Yellowtail flounder	1973-1993	NEFSC (1997)
	Atlantic herring	1967-1993	NEFSC (1996)
	Atlantic mackerel	1963-1993	NEFSC (1995b)
Virtual population analysis (VPA)	Silver hake	1963-1987	Almeida (1987)
	Atlantic herring	1963-1966	Anthony & Waring (1980)
Biomass-dynamics model	Spiny dogfish	1963-1993	NEFSC (1994)
Catch-survey analysis	Atlantic cod	1964-1977	NEFSC (1997)
	Yellowtail flounder	1964-1972	NEFSC (1997)
			Catchability coefficient (q)
Trawl-survey indices	Silver hake	1988-1993	Calibrated to VPA
	Winter flounder	1963-1993	q for yellowtail flounder
	Winter skate	1963-1993	Fall survey area swept
	Little skate	1963-1993	Fall survey area swept

To accomplish this we used the method of Collie and Sissenwine (1983) as implemented on a spreadsheet. The biomass of spiny dogfish (*Squalus acanthias*) has been estimated with a biomass-dynamics model (Northeast Fisheries Science Center [NEFSC] 1994). The biomass estimates of silver hake were updated to 1993 by regressing 1968-1987 VPA estimates on survey weight per tow. For the remaining species we used trawl survey indices scaled by appropriate catchability coefficients (q). The survey q estimated for yellowtail flounder with CSA was used to scale the survey indices for winter flounder (*Pleuronectes americanus*). The Georges Bank trawl survey indices were based on survey strata 15-23. Landings data were compiled from the corresponding assessments or extracted from NMFS catch tables. The statistical areas corresponding to Georges Bank are 522-525, 551, 552, 561, and 562. The tables of biomass and landings estimates are available from the authors.

Several of these fish species migrate (herring, mackerel, and dogfish) and are assessed over a wider geographic area than Georges Bank. For these species, we used the trawl survey data to calculate the proportion of the species total biomass found on Georges Bank each year. We used spring survey data to map the spatial distributions of these species because the spring survey data are considered the best index of their abundance and are therefore used for stock assessment. The estimated proportions on Georges Bank were quite variable because of the high spatial variability in mean catch per tow. We therefore chose to use the mean proportions on Georges Bank across years to scale the total biomasses and catches of herring, mackerel, and dogfish. This proration preserves the temporal trends for these species but scales their abundance to that of species resident on Georges Bank.

The biomasses and landings of the 10 fish species were aggregated into four groups to simplify the model and to reduce the number of interaction terms that would need to be estimated. The groups were primarily taxonomic: gadoids (cod, haddock, silver hake), flatfishes (yellowtail flounder, winter flounder), pelagics (herring, mackerel), and elasmobranchs (spiny dogfish, winter skate, little skate). We expect the species in each group to have similar rates of increase and similar ex-vessel prices. Some of the species also have similar exploitation histories (e.g., haddock and silver hake) and similar feeding habits (e.g., yellowtail and winter flounders). In fitting the biomass-dynamics model we treated each aggregate group as a "species."

Correlation analyses were conducted on the aggregate biomass data to determine at which time lags the potential interactions were the strongest. A 2-year time lag was selected for the biomass-dynamic models because it approximates the age of recruitment and maturity of many of these species. Lags greater than 2 years reduced the apparent correlation between taxonomic groups and would result in fewer observations for model fitting.

The biomass-dynamics model consists of delayed-difference equations of the form:

$$G_{t+1} = G_t + rG_{t-1} - \frac{r}{k}G_{t-1}^2 - Y_t + \text{species interaction terms} \tag{1}$$

This equation applies to gadoids (G, g) and there are three equivalent equations for flatfishes (F, f), pelagics (P, p), and elasmobranchs (E, e). Equation (1) is a discrete form of the Schaefer model with rate of increase r, carrying capacity k and commercial landings Y. The species interaction terms can represent competition or predation. A competition term, with competition coefficient, c_{ge}:

$$-c_{ge}G_{t-1}E_{t-1}\cdot$$

represents the negative effect of elasmobranchs on gadoids. We investigated three types of predator functional response:

Type I $$d_{gp}G_{t-1}P_{t-1},$$

Type II $$\frac{d_{gp}G_{t-1}P_{t-1}}{1 + a_{gf}F_{t-1} + a_{gp}P_{t-1}}, \text{ and}$$

Type III
$$\frac{d_{gp}G_{t-1}P_{t-1}^2}{1+a_{gf}F_{t-1}^2+a_{gp}P_{t-1}^2}.$$

The d_{gp} coefficients represent the positive effect of pelagics on gadoids. The a coefficients combine search time and probability of capture and represent the suitability of each prey type to the predator. Complementary negative terms represent the effects of predators on the prey. By specifying the t-1 lag for the biomass variables, we assume that the important trophic interactions occur during the first year of life and recruitment occurs at age 2.

A hierarchical set of models was fit that differed in complexity and hence number of parameters to estimate. To identify the most important subset of interaction terms, we fitted process-error versions of the single-species (no interactions) and Type-I equations (linear interactions). "Leaps and bounds" regressions were performed to identify which subset of interaction terms minimized Mallow's Cp statistic (Statistical Sciences 1991). Once the important interaction terms had been identified, non-linear estimation was used to fit observation-error versions of the biomass-dynamics models. The biomass values for the first 2 years of each species group were used as starting values and therefore assumed to be measured without error. The remaining biomass values were estimated by minimizing the sum of squares objective function:

$$\sum_{i}^{4}\sum_{t+2}^{T}(\ln B_{i,t}-\ln \hat{B}_{i,t})^2$$

where $B_{i,t}$ is the observed biomass of group$i$ in year t and $\hat{B}_{i,t}$ is the predicted biomass. Due to some missing values in 1963 we had a total (T) of 30 years of data and four species groups. The nonlinear estimation was performed with AD Model Builder (Otter Research 1994).

The Type-I model was fitted with those species interactions that were consistent with food-habits data and gave a low Cp statistic. To determine the important interaction terms in the Type II and III models we started with the single-species fit and added in the terms identified in the Type-I fits one at a time. The term that caused the largest reduction in the sum of squared residuals was added in each step. As terms were included, we tested opposite interactions regardless of whether they were significant in the single-species fits.

The goodness of fit of the alternative model forms was compared with F-tests and the Akaike Information Criterion (AIC). The AIC is defined as:

$$\mathrm{AIC} = -2\ln L + 2p,$$

where ln$L$ is the log likelihood evaluated at the maximum likelihood estimates and p is the number of model parameters (Quinn and Deriso 1999).

Equilibrium analyses were performed to calculate the single-species biological reference points, h_{msy}, the exploitation fraction for maximum sustainable yield. Equilibrium yields were calculated by simulating the system of equations for 600 years. Response surfaces were plotted to investigate the sensitivity of each species group equilibrium yield to the exploitation rate and hence equilibrium biomass of the other groups. These biological reference points and sensitivities were used to guide simulations of the multispecies fish community.

Results

Biomass estimates of the 10 fish species are first compared to check for consistent patterns within taxonomic groups (Fig. 1). Haddock and silver hake were the dominant gadoids in the 1960s, but were fished down by distant-water fleets in the late 1960s (Fogarty and Murawski 1998). Cod was less abundant in the 1960s, reaching its peak biomass in the 1970s. All three gadoids had declined to low levels by the late 1980s. The variability in flounder biomass in the 1960s is due to variability in the survey indices from which these biomass estimates were calculated. Yellowtail flounder biomass declined sharply in the 1970s and collapsed to very low levels in the late 1980s. Winter flounder biomass has less of a trend, but also declined to low levels by 1990.

The biomasses of herring and mackerel were scaled by the proportions found on Georges Bank in the spring trawl survey; the mean proportions were 0.121 for herring and 0.126 for mackerel. Herring biomass declined in the late 1960s, again due to high catches by distant-water fleets. During the 1980s herring was virtually absent from Georges Bank and present at low levels in the rest of the Gulf of Maine (NEFSC 1996). Since then herring biomass has increased to record high levels. Fluctuations in mackerel biomass are less dramatic than for herring, but mackerel has also increased to record high biomass levels. Spiny dogfish biomass was also scaled by the mean proportion found on Georges Bank in the spring survey (0.124). Dogfish abundance increased almost monotonically from 1963 to 1993. The skate data are minimum biomass estimates based on area swept calculations from the autumn trawl survey data. Winter skate biomass increased to a maximum in 1986 and by 1993 had declined to values as low as those seen in the 1970s. Unlike the other two elasmobranchs, there was apparently no concurrent increase in the abundance of little skate.

The aggregate biomass and landings data (Fig. 2) reflect the abundance patterns of the component species. Gadoid abundance declined sharply in the late 1960s and more gradually from 1975 to 1985. The corresponding decreases in flatfish biomass occurred in the early 1970s and the early 1980s, about 5 years later than the gadoids. The exploitation rate of flatfishes (landings per biomass) appears to have been very high in 1975 and in several of the years since 1985. By applying the catchability

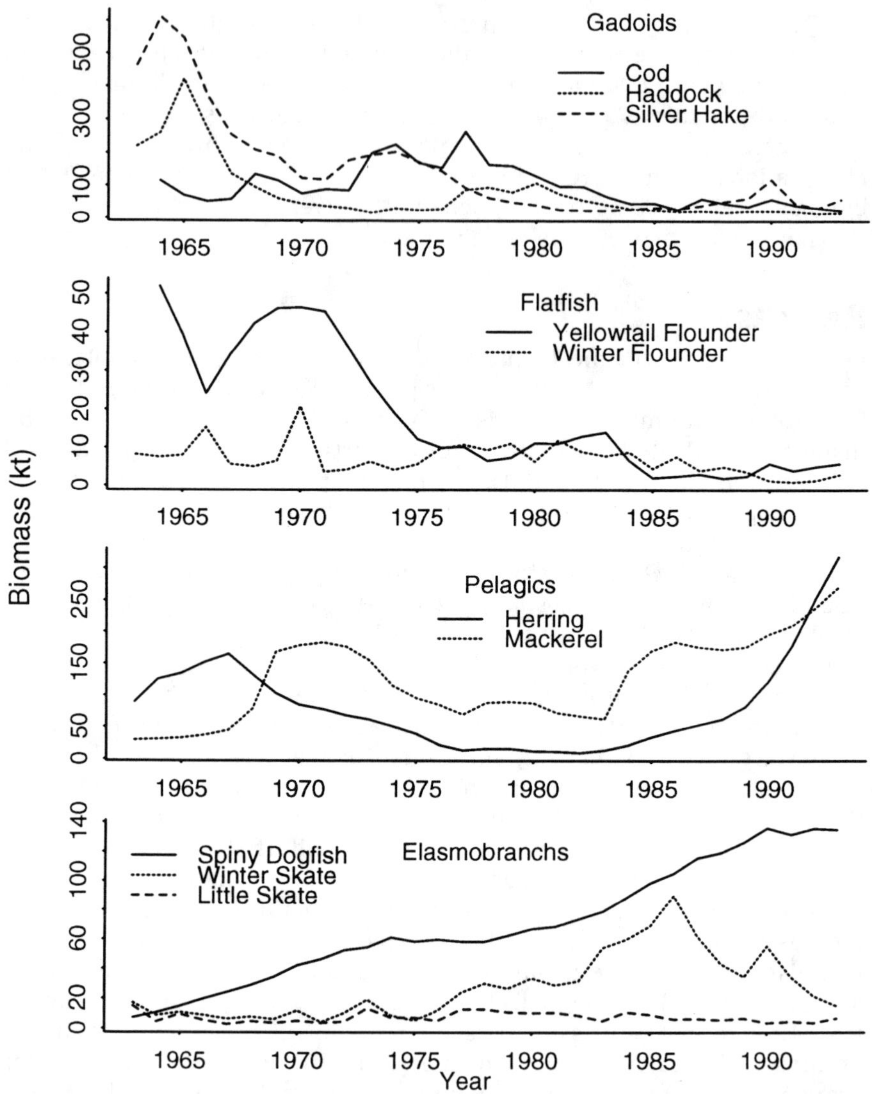

Figure 1. Biomass of 10 fish species on Georges Bank. The biomass units used throughout this paper are thousand metric tons (kt).

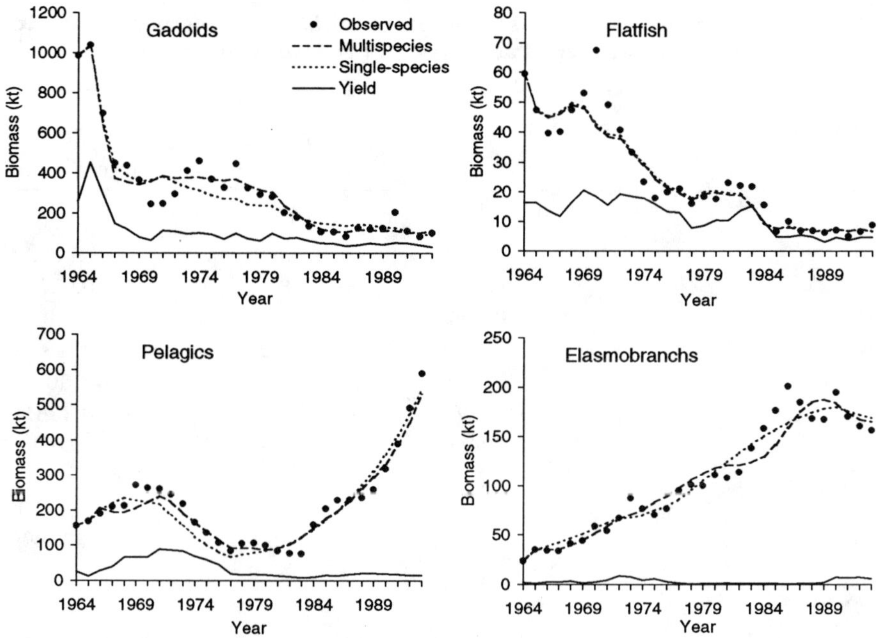

Figure 2. Single and multispecies model fits.

coefficient for yellowtail flounder to winter flounder, we may have underestimated flounder biomass. On the other hand, the Georges Bank yellowtail flounder stock did sustain high levels of fishing mortality before collapsing (NEFSC 1997). The relatively high harvests of pelagics in the early 1970s contributed to their decline. Harvests since then have remained low while the biomass of pelagics increased rapidly to record levels. Elasmobranchs have low fecundities and therefore cannot sustain high exploitation rates. Low catches in the 1980s may have contributed to peak biomass levels, whereas increased harvests in the 1990s may partially explain the downturn in abundance.

The single-species biomass-dynamics models reproduce the general abundance patterns of the four groups (Fig. 2). The pelagic carrying capacity (k_p) cannot be estimated reliably from these data because of the near exponential increase in pelagic biomass in recent years. In all model fits k_p was fixed at 1,000 kt. (The biomass units used throughout this paper are thousand metric tons [kt].) The nonlinear model with Type-I interactions fit significantly better than the single-species models, according to the *F*-test and AIC (Table 2). The interactions identified in this fit were positive effects of flatfishes and pelagics on gadoids and negative effects of elasmobranchs and gadoids on flatfishes. Elasmobranchs had a

Table 2. Parameters of the single- and multispecies biomass-dynamics models with Type-I and Type-III predator functional responses.

Parameter	Single-species	Model Type-I interactions	Type-III interactions
r_g	0.3110	0.1343	0.4106
k_g	1,323	356.7	1,208.2
r_f	0.6125	1.814	0.6371
k_f	97.25	85.22	92.99
r_p	0.3398	0.3627	0.781
k_p	1,000[a]	1,000[a]	1,000[a]
r_e	0.2178	0.6166	0.3231
k_e	194.4	171.4	108.1
d_{gf}		0.001911	
d_{gp}		0.001073	6.361×10^{-5}
c_{ge}			0.002399
c_{fg}		2.99×10^{-4}	
c_{fe}		0.006950	
c_{pg}			9.421×10^{-5}
c_{pe}		3.967×10^{-4}	0.3803
d_{ep}		8.771×10^{-5}	1.437
c_{eg}		5.203×10^{-4}	2.436×10^{-4}
a_{gf}			0.01150
a_{gp}			1.158×10^{-4}
a_{ef}			128.5
a_{ep}			5.660
Number of observations	112	112	112
Number of parameters	8	14	17
Sum of squares	4.838	3.816	3.479
Stepwise *F* statistic	NA	4.374[b]	3.067[c]
Akaike Information Criterion	–17.76	–31.68	–35.56

[a] Fixed value

[b] $p < 0.001$

[c] $p < 0.05$

negative effect on pelagics and pelagics had a small, reciprocal positive effect on elasmobranchs. Gadoids had a negative effect on elasmobranchs which could be interpreted as competition. When this model was simulated with no harvesting, the elasmobranchs and flatfishes both went extinct within 30 years, which suggests that the interaction terms may be statistically significant but not biologically realistic.

The model with Type-II interactions fit slightly better than the Type-I model but the reduction in sum of squares was not significant and the AIC was higher (not shown). The same species interaction terms were present in the Type-II model as in the Type-I model. When this model was simulated with no harvest, the flatfishes and elasmobranchs again were eliminated by the negative interactions. We rejected this model because it lacked biological realism and did not significantly improve the fit to the biomass data.

When fitting the Type-III model, the effect of pelagics on gadoids (d_{gp}) caused the largest reduction in the sum of squared residuals from the single-species fit. The final term added was the effect of elasmobranchs on gadoids (c_{ge}) which did not substantially reduce the sum of squares until it was included with the other interaction terms. The Type-III predation model fit significantly better than either the single-species or the Type-I model. Based on the *F*-tests and the low AIC score we accepted it as the best multispecies model (Table 2). The improvements in the fits over the single-species models are apparent for gadoids and pelagics in the 1970s (Fig. 2). The full equations of the Type-III model are:

$$G_{t+1} = G_t + r_g G_{t-1} - \frac{r_g}{k_g} G_{t-1}^2 - Y_{g,t} - c_{ge} G_{t-1} E_{t-1} + \frac{d_{gp} G_{t-1} P_{t-1}^2}{1 + a_{gf} F_{t-1}^2 + a_{gp} P_{t-1}^2},$$

$$F_{t+1} = F_t + r_f F_{t-1} - \frac{r_f}{k_f} F_{t-1}^2 - Y_{f,t},$$

$$P_{t+1} = P_t + r_p P_{t-1} - \frac{r_p}{k_p} P_{t-1}^2 - Y_{p,t} - \frac{c_{gp} G_{t-1} P_{t-1}^2}{1 + a_{gf} F_{t-1}^2 + a_{gp} P_{t-1}^2} - \frac{c_{ep} E_{t-1} P_{t-1}^2}{1 + a_{ef} F_{t-1}^2 + a_{ep} P_{t-1}^2}$$

and $$E_{t+1} = E_t + r_e E_{t-1} - \frac{r_e}{k_e} E_{t-1}^2 - Y_{e,t} - c_{eg} E_{t-1} G_{t-1} + \frac{d_{ep} E_{t-1} P_{t-1}^2}{1 + a_{ef} F_{t-1}^2 + a_{ep} P_{t-1}^2}.$$

A different set of interaction terms was included in this model than for the Type-I and Type-II models (Fig. 3). The gadoids and elasmobranchs operate as Type-III predators of pelagics, which is consistent with food-habits data. The gadoids and elasmobranchs have reciprocal negative interactions which can be interpreted as competition. Gadoids are not known to

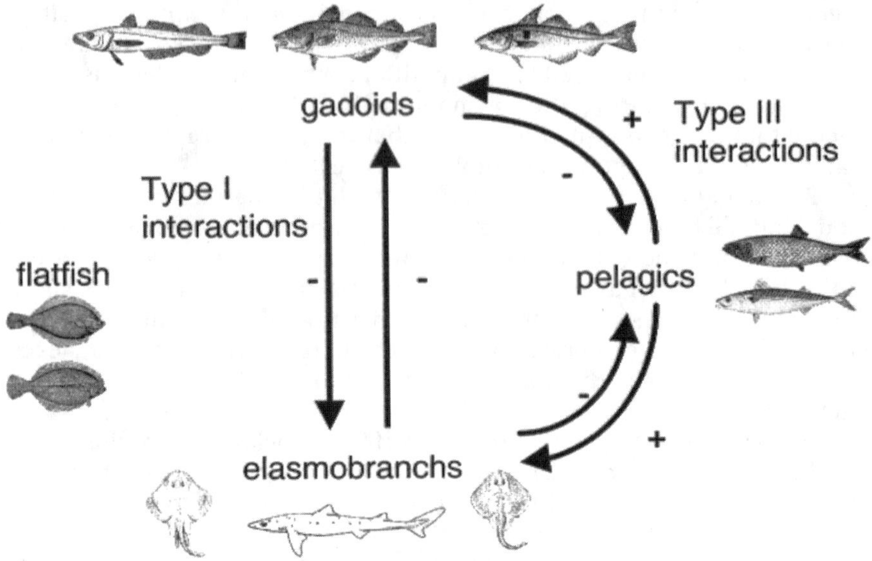

Figure 3. Simplified Georges Bank food web showing the interactions included in the multispecies model with Type-III predation terms.

feed on elasmobranchs, but elasmobranchs do prey on gadoids (Tsou and Collie 1998). The negative effect of elasmobranchs on gadoids could be interpreted as predation; we chose to retain the Type-I interaction so as not to estimate additional parameters. The interactions with flatfishes were not important in this model fit, so the flatfish model is essentially the same as the single-species fit (Fig. 2). However, flatfish biomass still influences the other taxa through the denominators of the predator functional responses. If flatfish biomass increases, the feeding rates on pelagics decline, which benefits the pelagics and negatively effects the gadoids and elasmobranchs.

The interaction terms have units of biomass per year and can therefore be directly compared with the annual yields (Fig. 4). High catches of gadoids account for the sharp decline in biomass in the late 1960s (Fig. 4a). After 1975 the interaction terms were about the same order of magnitude as the yield. From 1975 to 1985 the negative effect of elasmobranchs exceeded the positive effect of pelagics, which helps explain the decline in gadoid biomass during that period. The negative effect of gadoids on pelagics has exceeded the commercial yield since 1975 (Fig. 4b). Consumption by elasmobranchs increased over time, equaling the commercial harvest by the 1990s. With the combination of low harvest rates and low gadoid abundance in the 1980s, the pelagics appear to have escaped

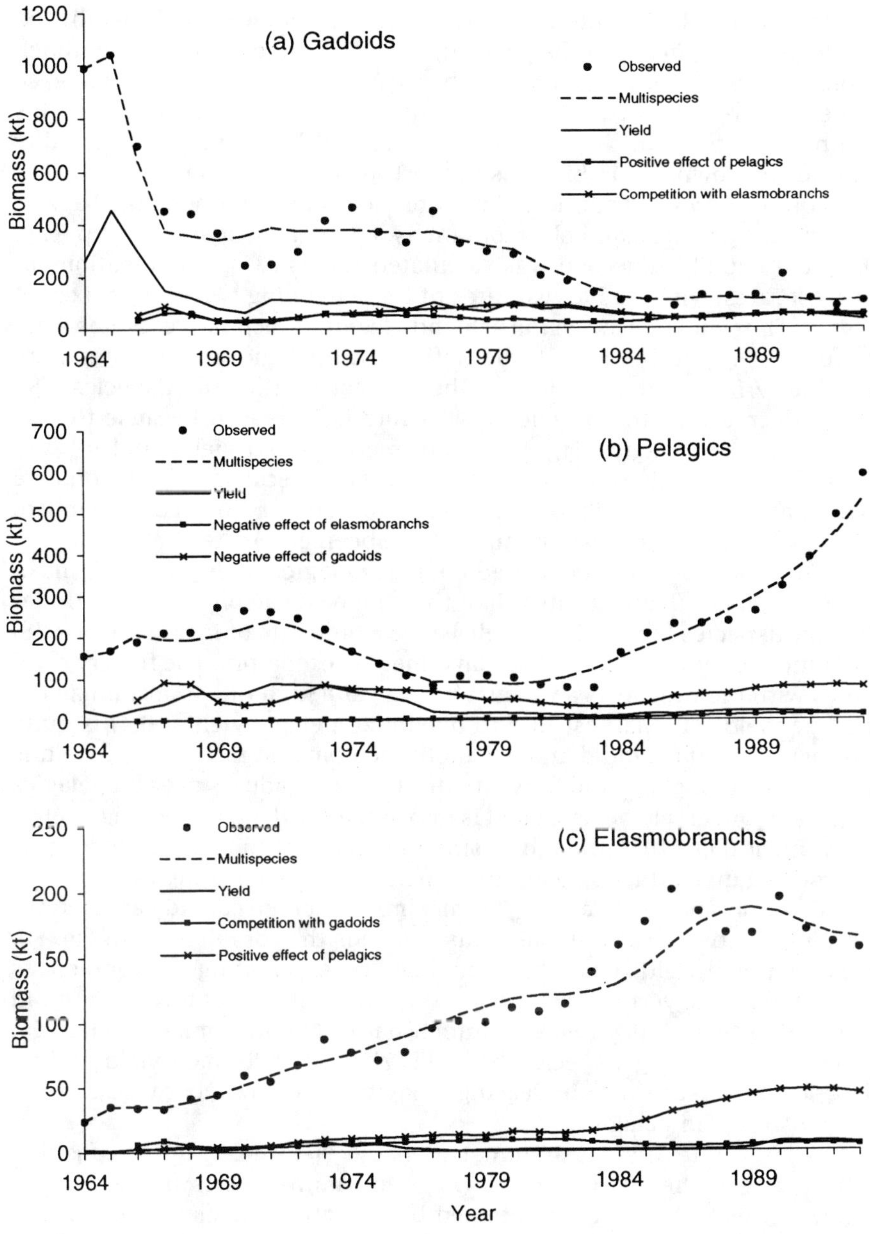

Figure 4. Interaction terms for three species groups as estimated with the Type-III model.

predator control. The interaction terms for elasmobranchs generally exceeded the magnitude of the yield (Fig. 4c). The increase in elasmobranch biomass is partly explained by low harvesting in the 1970s. The increase in pelagic prey coupled with the decrease in gadoid competition explains the peak biomass in the late 1980s. Increased harvesting has probably caused the downturn in biomass in the 1990s.

Equilibrium yield surfaces illustrate the trade-offs among the harvest of gadoids, pelagics, and elasmobranchs. For a given species group (e.g., gadoids) equilibrium yield was calculated for pairwise combinations of harvest rates, with the harvest rates of the remaining two groups fixed at $^1/_2 h_{msy}$. Thus, in Fig. 5a, the flatfish and elasmobranch harvest rates were fixed at $^1/_2 h_{msy}$ and in Fig. 5b the flatfish and pelagic harvest rates were fixed at $^1/_2 h_{msy}$. Gadoid yield and the harvest rate for multispecies MSY (h_{mmy}) decrease as the pelagic harvest rate is increased because the gadoids prey on pelagics (Fig. 5a). Conversely, gadoid yields and h_{mmy} increase with increasing elasmobranch harvest rate because gadoids compete with elasmobranchs (Fig. 5b). The vertical contour lines correspond to regions in which the elasmobranchs have been eliminated.

The multispecies model predicts that pelagic yields would be maximized at higher harvest rates than the single-species h_{msy} (Fig. 6). Under the multispecies model the pelagics have a high rate of increase ($r = 0.78$) but much of this productivity is consumed by predators. The fishery competes with the predators and must harvest at a high rate to maximize the pelagic yield. The harvest rate to maximize pelagic yield (~0.4) is quite insensitive to the gadoid and elasmobranch harvest rate. Pelagic yield is maximized at high gadoid harvest rates because gadoids prey on pelagics (Fig. 6a). Conversely pelagic yield is maximized with no harvesting of elasmobranchs (Fig. 6b). Although elasmobranchs prey on pelagics, they exert a large negative effect on gadoids, which also prey on pelagics.

Elasmobranch yield and h_{mmy} is maximized at high gadoid harvest rates because the two taxa compete. Thus Fig. 7a is the complement of Fig. 5b. Elasmobranch yields could be maximized by eliminating the gadoids, as indicated by the vertical contours in this plot. With lower levels of gadoid harvesting (e.g., $^1/_2 h_{msy}$) elasmobranch harvest is maximized at harvest rates below its single-species h_{msy} (Fig. 7b). Elasmobranch yield and its h_{mmy} level decrease with increasing pelagic harvest rate because elasmobranchs prey on pelagics.

Harvest rates for maximum sustainable yield (MSY) differ considerably between the single-species (h_{msy}) and multispecies models (h_{mmy}). Multispecies MSY would be obtained by depleting the elasmobranchs to remove their consumption of pelagics and competition with gadoids. Flatfishes would be lightly harvested because their biomass indirectly benefits the pelagics. Yield of the pelagics and gadoids would thus be maximized. Interestingly, total yield under the multispecies model with all species harvested at the h_{mmy} levels (290 kt) exceeds the total yield of the single-species models with all species harvested at h_{msy} (213 kt). In the

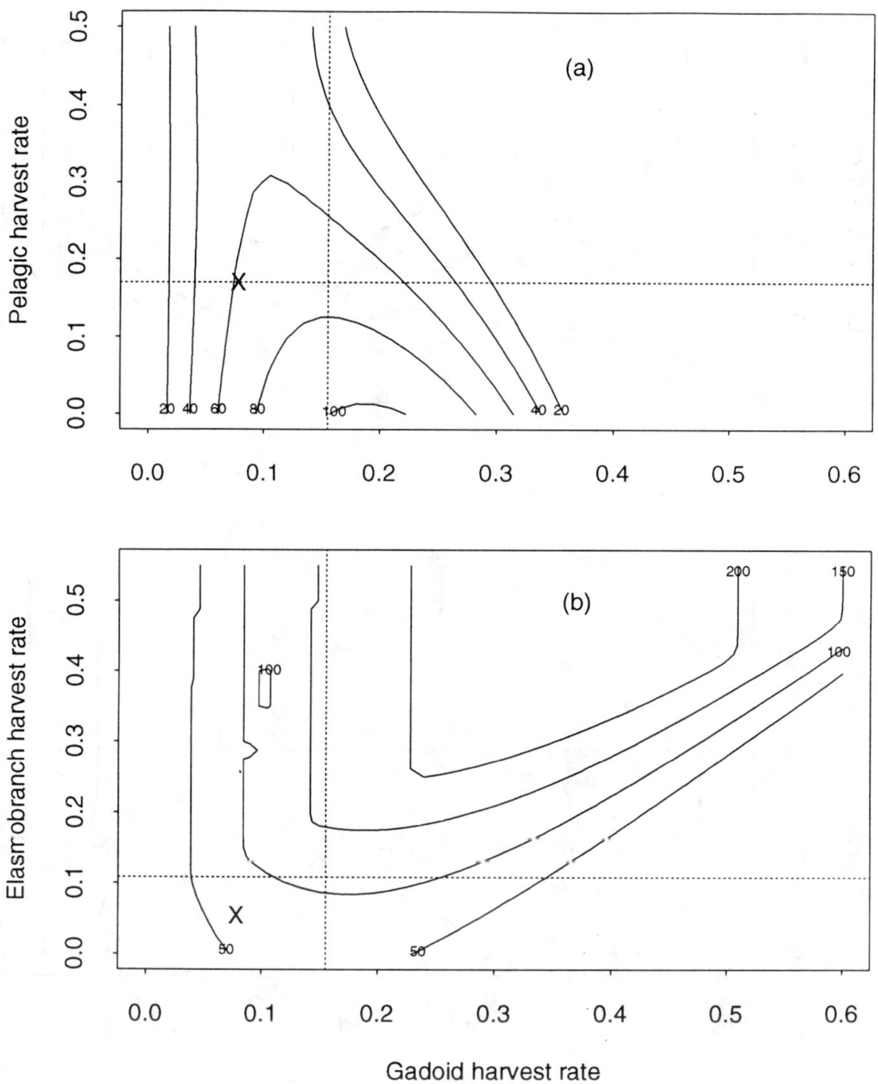

Figure 5. Gadoid equilibrium yield (kt) obtained for pairs of harvest rates (solid contours). Broken lines indicate h_{msy} from single-species model. The combination of harvest rates used in the forward simulation (Fig. 8) is denoted with an X.

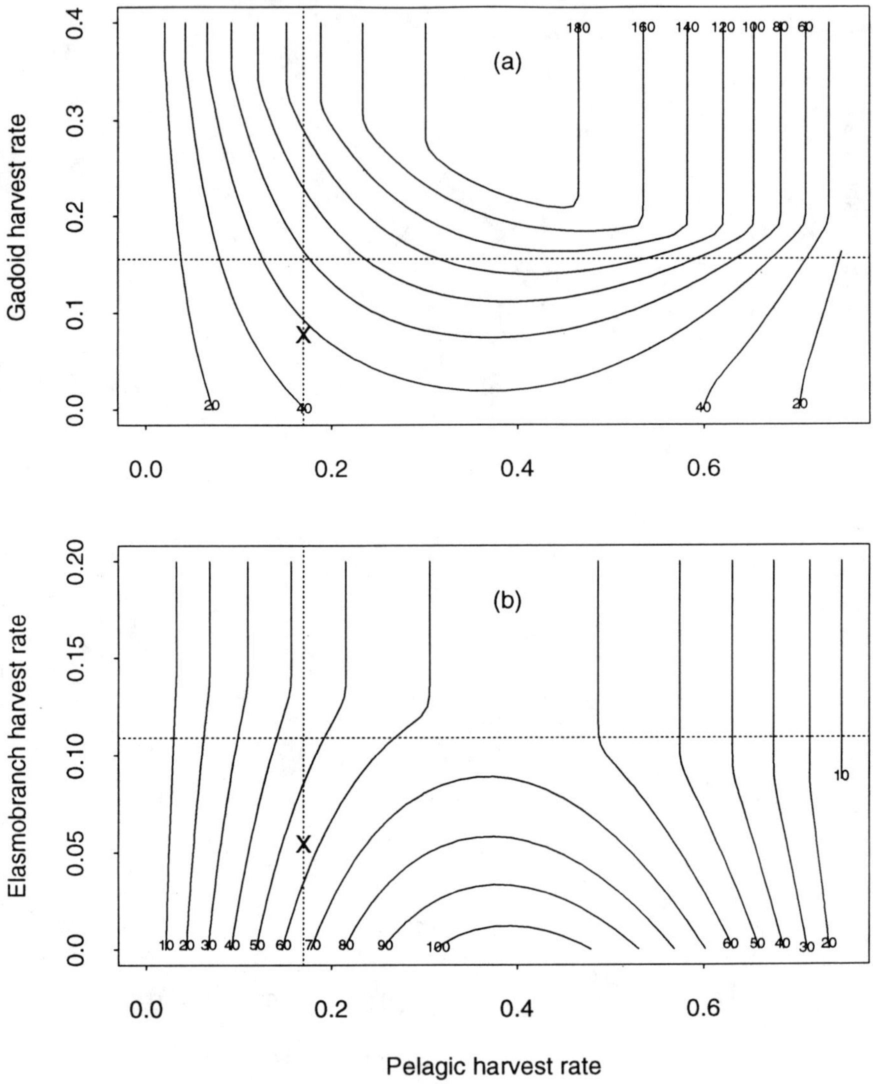

Figure 6. Pelagic equilibrium yield (kt) obtained for pairs of harvest rates (solid contours). Broken lines indicate h_{msy} from single-species model. The combination of harvest rates used in the forward simulation (Fig. 8) is denoted with an X.

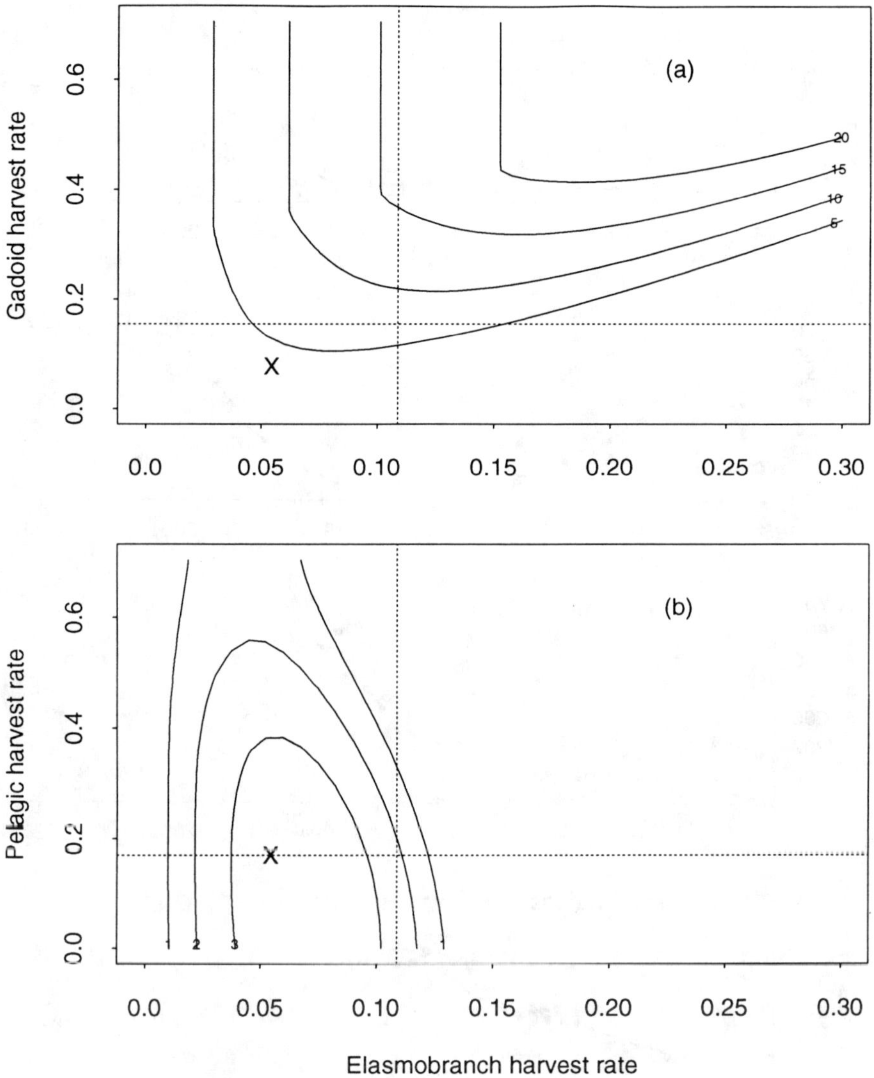

Figure 7. Elasmobranch equilibrium yield (kt) obtained for pairs of harvest rates (solid contours). Broken lines indicate h_{msy} from single-species model. The combination of harvest rates used in the forward simulation (Fig. 8) is denoted with an X.

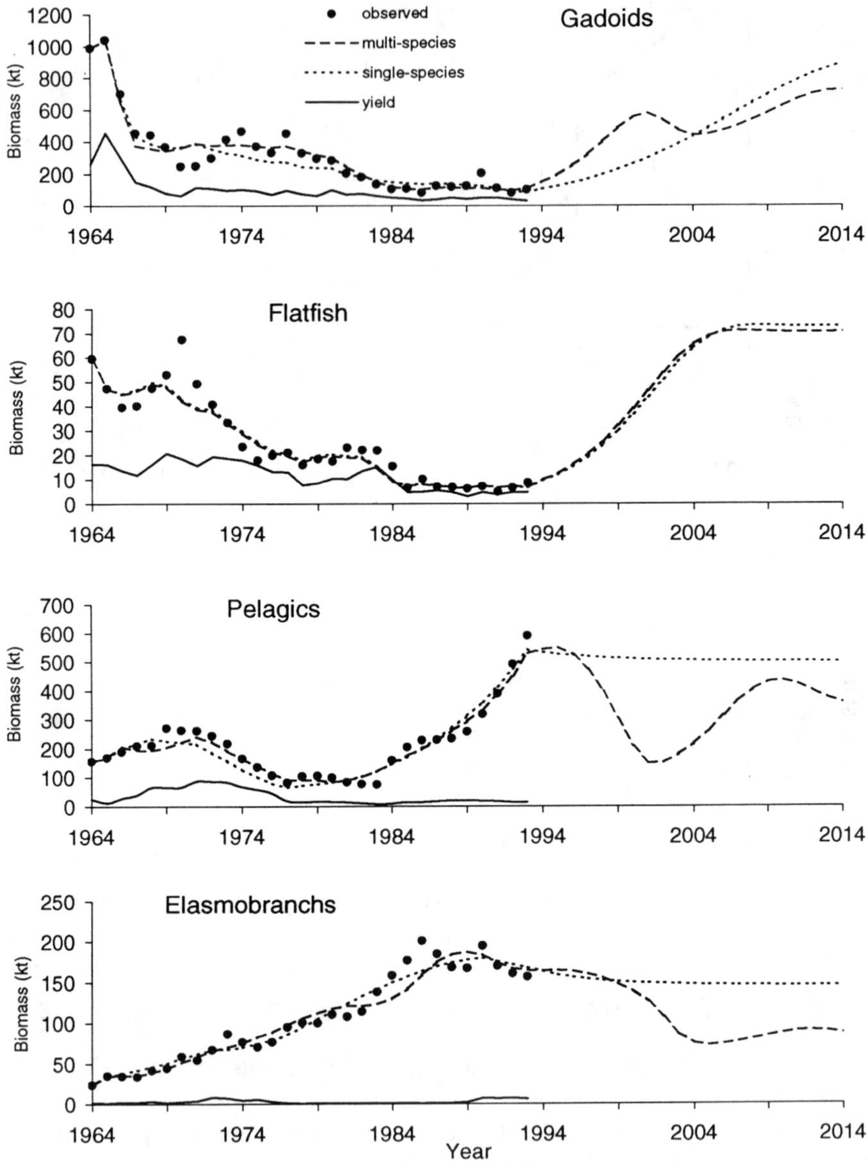

Figure 8. Result of 20-year simulations with single-species and multispecies models. The gadoid, flatfish, and elasmobranch harvest rates were set to $^1/_2 h_{msy}$ and the pelagic harvest rate was h_{msy}.

multispecies case, unproductive biomass (elasmobranchs) would be eliminated and the "surplus" production harvested as pelagics and gadoids.

We also solved for the harvest rates that would maximize the revenue from the multispecies fishery. From 1989 to 1993 the average prices in dollars per kilogram were 1.64 for gadoids, 2.54 for flatfishes, 0.23 for pelagics, and 0.21 for elasmobranchs (NEFSC 1995a). To maximize total revenues, elasmobranchs would be eliminated due to their low value and negative interactions with gadoids and pelagics. Pelagics would not be harvested at all because their production is more valuable when converted to gadoids. Revenues would thus be maximized by enhancing the sustainable yield of valuable gadoids and flatfishes.

Deterministic simulations were run with constant harvest rates to illustrate the differences between single-species and multispecies projections. In this example we simulated the rebuilding of the groundfish by setting the gadoid and flatfish harvest rates to $^1/_2 h_{msy}$. The elasmobranch harvest rate was also set at $^1/_2 h_{msy}$ based on the yield contours in Fig. 7b and concerns that harvesting has caused the recent downturn in biomass. The pelagic harvest rate was set at h_{msy} to take advantage of high abundance levels.

With the single-species models, the gadoids and flatfishes are projected to recover in response to low harvest rates while the pelagics and elasmobranchs equilibrate slightly below their peak biomass levels (Fig. 8). The flatfish single-species and multispecies projections are virtually identical because there are no interaction terms in the flatfish equation. The multispecies model predicts that the other three taxa would oscillate toward different equilibria. Gadoid biomass increases rapidly to a peak in 2001 after which the pelagics become depleted. Elasmobranchs also decline until 2005 due to the increase in gadoids and decrease in pelagics. After 2005 flatfish biomass has increased, and because the flatfish group occurs in the denominator of the predation term, this decreases the predation rate on pelagics. This interaction reverses the decline in pelagic and elasmobranch biomass, allowing the system to converge toward the multispecies equilibrium.

Discussion

Our results provide evidence of trophic interactions between groups of fish species on Georges Bank. Unlike simpler correlative studies (e.g., Fogarty and Brodziak 1994), the multispecies biomass-dynamics model accounts for changes in biomass due to commercial catch. The estimated trophic interactions are the same order of magnitude as the catch and therefore large enough to influence the population dynamics. Before these results can be used in a management context, the assumptions underlying the multispecies model need to be critically examined.

One important assumption was the proration of the biomasses of migrant species, which was done to scale their abundances to those of the species considered to reside on Georges Bank. However, these migrants are also subjected to processes that occur off Georges Bank. For herring, mackerel, and dogfish we must assume that the trophic interactions occurring off Georges Bank are either contained in the single-species terms of the model (r and k) or that they have the same temporal trend as the interaction terms that are included in the model (the c and d terms). In extensions of this work it will be desirable to model the seasonal migration of these species and to include fish stocks residing in southern New England and the Gulf of Maine.

By aggregating the species taxonomically, we pooled some species with different feeding habits. Haddock, for example, feed primarily on benthic invertebrates and would not be expected to prey on herring or mackerel (Grosslein et al. 1980). One reason for aggregating by taxonomic group was that we expected related species to have more similar rates of increase (r) and compensatory capacity (k). Haddock's feeding habits are similar to the flounders' (Grosslein et al. 1980) but the flounders can sustain higher exploitation rates (NEFSC 1997). Pooling the species by aggregate groups was necessary to minimize the number of interaction terms to estimate. Even so, there is a risk of overparameterizing the biomass-dynamics model. The multispecies model appears to fit better than the single-species model (Table 2) but some of the parameters were estimated with low precision. These interaction terms were retained in the model based on known feeding habits and to maintain symmetry in the food web.

Type-III predator functional responses appeared necessary to obtain biologically realistic predator-prey dynamics. With Type-I and Type-II functional responses the flatfishes and elasmobranchs would go extinct even without harvesting. This process has been termed "self simplification" and is common when multispecies models are simulated under exploitation regimes differing from those when the data were collected (Walters et al. 1998). Type-III functional responses are considered appropriate for vertebrate predators (Murdoch and Oaten 1975) but they require estimation of more parameters (a coefficients) than the Type-I response and the squared prey variables make it difficult to solve analytically for equilibrium levels or to conduct a stability analysis. Simulation is the most tractable way of examining the dynamic behavior of these equations. Under all the combinations of harvest rates we considered, the system of equations converged to a multispecies equilibrium. However, with four interacting species and a 2-year time delay, more complex dynamic behavior might be expected (e.g., Basson and Fogarty 1997).

The trophic interactions identified in our multispecies model are consistent with the NMFS food habits data and with previous multispecies analyses of the shelf fish community of the northeastern United States. The magnitudes of the interaction terms are consistent with Sissenwine and Cohen's (1991) finding that the fish community consumes most of its own production. Herring and mackerel are known to be important prey of cod, silver hake, dogfish, and winter skate (Overholtz et al. 1991; Tsou and Collie, In press). MSVPA estimates of the biomass of prey consumed by predator groups (Tsou and Collie, In press) may be analogous to the species interaction terms estimated in this study. MSVPA estimates of yellowtail founder consumption by gadoids and elasmobranchs were quite small (<0.2 kt per year) and may help to explain the lack of significant interaction terms for flatfishes in the Type-III biomass-dynamics model. Furthermore, the negative effect of elasmobranchs on pelagics as estimated by the biomass-dynamics model was quite similar (3-10 kt per year) to MSVPA estimates of elasmobranch consumption. Conversely, the biomass-dynamics model estimates of the negative effects of gadoids on pelagics were considerable larger than the MSVPA predation estimates. This difference may be because interaction terms in the biomass-dynamics model are estimates of the negative effect of predators on the recruited biomass of the prey, rather than direct consumption estimates.

It has been hypothesized that predation may have hastened the collapse of the gadoid stocks or that they may be stuck in a "predator pit" from which escape is possible only with very low exploitation rates (Spencer and Collie 1995). The negative effect of elasmobranchs on gadoids could be interpreted as predation, but MSVPA estimates of gadoid consumption by elasmobranchs were much smaller (1-7 kt per year) than the negative effect estimated with the biomass-dynamics model. Conversely, gadoids are not known to prey on elasmobranchs. We chose to interpret the reciprocal negative interaction between gadoids and elasmobranchs as competition, which is consistent with the diet overlap between these two groups (Grosslein et al. 1980). It is difficult to explain the steady increase in elasmobranch biomass without some concurrent change in their ecosystem. Cod and silver hake are known to be cannibalistic but the consumption of gadoids by gadoids in the MSVPA was low relative to the gadoid biomass (Tsou and Collie, In press). In the biomass-dynamics model, predation among the gadoids should be accounted for by the intraspecific terms (r and k).

Biomass-dynamics models are not commonly used as primary stock-assessment methods (National Research Council 1998) but they are useful for elucidating medium and longer term management strategies and for conducting bioeconomic analyses (Edwards and Murawski 1993). If we accept that trophic interactions occur among species groups, how should these multispecies interactions influence management advice? In our simulation (Fig. 8) the multispecies yield was less than the single-species yields at the same combination of harvest rates because some of the "surplus production" is consumed by predators. However, the management considerations extend beyond simply reducing our expectations of yield from the fish community. There are explicit trade-offs between harvesting the species groups which have different productivities and different economic values and are caught by different segments of the fishing fleet. Total multispecies yield could exceed the sum of the single-species yields if the less productive taxa were depleted (cf. Brander and Mohn 1991). Revenue might be maximized by harvesting only the most valuable species (Clark 1985). We do not advocate a particular harvest strategy but have examined the trade-offs in yield to illustrate that the present pelagic-dominated fish community is less commercially valuable than the groundfish-dominated community of the 1960s and 1970s.

Despite evidence that trophic interactions are strong enough to influence fish population dynamics, our parameter estimates are highly uncertain. The best test of a model is its ability to predict "out of sample data" that were not used to estimate the model parameters. Such a test will be possible when biomass and landings data from after 1993 become available for all 10 species included here. Recent assessments of some of these species suggest that the community has started to shift back toward groundfish. The biomasses of cod, haddock ,and yellowtail flounder have increased somewhat (NEFSC 1997), herring and mackerel biomasses have peaked since 1993, and elasmobranch biomass has declined. This response is consistent with the multispecies model.

Until a more definitive test is possible, the single-species and multispecies models should be considered as alternatives in an adaptive management design (Sainsbury 1991). Harvest strategies (combinations of harvest rates) should be sought that are expected to perform well regardless of which model is correct. An informative harvest policy will lead to different trajectories for the single- and multispecies models, allowing the alternatives to be distinguished more quickly. The harvest rates in Fig. 8 illustrate a policy which would allow the groundfish to recover while providing informative variation between the single- and multispecies population projections.

Acknowledgments

We thank Ralph Mayo and the other stock assessments scientists at the Northeast Fisheries Science Center for helping us to access the biomass and landings data. Michael Fogarty was instrumental in formulating this multispecies model. We thank Jason Link for reviewing an earlier draft. Funding was provided by the Rhode Island Sea Grant College Program, Project No. R/F-9501.

References

Almeida, F.P. 1987. Status of the silver hake resource off the northeast coast of the United States, 1987. NMFS, NEFSC, Woods Hole Lab. Ref. Doc. 87-03.

Anthony, V.C., and G. Waring. 1980. The assessment and management of the Georges Bank herring fishery. Rapp. P.-V. Reun. Cons. Int. Explor. Mer 177:72-111.

Basson, M., and M.J. Fogarty. 1997. Harvesting in discrete-time predator-prey systems. Math. Biosc. 141:41-74.

Brander, K.M., and R.K. Mohn. 1991. Is the whole always less than the sum of the parts? ICES Mar. Sci. Symp. 193:117-119.

Brown, B.E., J.A. Brennen, E.G. Heyrdahl, M.D. Grosslein, and R.C. Hennemuth. 1976. The effects of fishing on the marine finfish biomass in the Northwest Atlantic from the eastern edge of the Gulf of Maine to Cape Hatteras. ICNAF Res. Bull. 12:49-68.

Clark, C.W. 1985. Bioeconomic modeling and fisheries management. Wiley, New York. 291 pp.

Collie, J.S., and M.P. Sissenwine. 1983. Estimating population size from relative abundance data measured with error. Can. J. Fish. Aquat. Sci. 40:1871-1879.

Edwards, S.F., and S.A. Murawski. 1993. Potential economic benefits from efficient harvest of New England groundfish. N. Am. J. Fish. Manage. 13:437-449.

Fogarty, M.J., and J.K.T. Brodziak. 1994. Multivariate time series analysis of fish community dynamics on Georges Bank. ICES C.M. 1994/P:3.

Fogarty, M.J., and S.A. Murawski. 1998. Large-scale disturbance and the structure of marine systems: Fishery impacts on Georges Bank. Ecol. Appl. 8(Suppl. 1):S6-S22.

Grosslein, M.D., R.W. Langton, and M.P. Sissenwine. 1980. Recent fluctuations in pelagic fish stocks of the northwest Atlantic, Georges Bank region, in relation to species interactions. Rapp. P.-V. Reun. Cons. Int. Explor. Mer 177:374-404.

Langton, R.W. 1982. Diet overlap between Atlantic cod, *Gadus morhua*, Silver hake, *Merluccius bilinearis*, and fifteen other northwest Atlantic finfish. Fish. Bull., U.S. 80:745-759.

Murdoch, W.W., and A. Oaten. 1975. Predation and population stability. Adv. Ecol. Res. 9:1-132.

National Research Council. 1998. Improving fish stock assessment. National Academy Press. Washington, D.C. 177 pp.

Northeast Fisheries Science Center. 1994. Report of the 18th Northeast Regional Stock Assessment Workshop. NEFSC Ref. Doc. 94-22.

Northeast Fisheries Science Center. 1995a. Status of the fishery resources off the northeastern United States for 1994. NOAA Tech. Memo. NMFS NE 108. 140 pp.

Northeast Fisheries Science Center. 1995b. Report of the 20th Northeast Regional Stock Assessment Workshop. NEFSC Ref. Doc. 95-20.

Northeast Fisheries Science Center. 1996. Report of the 21st Northeast Regional Stock Assessment Workshop. NEFSC Ref. Doc. 96-05d.

Northeast Fisheries Science Center. 1997. Report of the 24th Northeast Regional Stock Assessment Workshop. NEFSC Ref. Doc. 97-12.

Otter Research. 1994. An introduction to AD model builder: For use in nonlinear estimation and statistics. Otter Research Ltd., Nanaimo, British Columbia.

Overholtz, W.J., S.A. Murawski, and K.L Foster. 1991. Impact of predatory fish, marine mammals, and seabirds on the pelagic fish ecosystem of the northeastern USA. ICES Mar. Sci. Symp. 193:198-208.

Quinn, T.J., and R.B. Deriso. 1999. Quantitative fish dynamics. Oxford University Press. 542 pp.

Sainsbury, K.J. 1991. Application of an experimental approach to management of a tropical multispecies fishery with highly uncertain dynamics. ICES Mar. Sci. Symp. 193:301-320.

Sissenwine, M.P., and E.B. Cohen. 1991. Resource productivity and fisheries management: northeast shelf ecosystem. In: K. Sherman and G. Gold (eds.), Food chains, yields, models and management of large marine ecosystems. AAAS Selected Symposia 99. Westview, Boulder, Colorado, pp. 107-123.

Sissenwine, M.P., E.B. Cohen, and M.D. Grosslein. 1984. Structure of the Georges Bank ecosystem. Rapp. P.-V. Reun. Cons. Int. Explor. Mer 183:243-254.

Sparre, P. 1991. Introduction to multispecies virtual population analysis. ICES Mar. Sci. Symp. 193:12-21.

Spencer, P.D., and J.S. Collie. 1995. A simple predator-prey model of exploited marine fish populations incorporating alternative prey. ICES J. Mar. Sci. 53:615-628.

Statistical Sciences. 1991. S-Plus Reference Manual. Vol. 1. Seattle, Washington.

Tsou, T.-S., and J.S. Collie. In press. Estimating predation mortality in the Georges Bank fish community. Proceedings of the ICES Symposium on Recruitment Dynamics of Exploited Marine Populations: Physical-Biological Interactions.

Walters, C.J., V. Christensen, and D. Pauly. 1998. Structuring dynamic models of exploited ecosystems from trophic mass-balance assessments. Reviews in Fish Biology and Fisheries 7:139-172.

Ecosystem Approaches for Fisheries Management
Alaska Sea Grant College Program • AK-SG-99-01, 1999

Ecosystem Impacts of the KwaZulu-Natal Reef Fishery, South Africa: An Exploratory Model

M. Verónica Toral-Granda
University of Cape Town, Percy FitzPatrick Institute of African Ornithology, Rondebosch, South Africa

Coleen L. Moloney
Marine and Coastal Management, Cape Town, South Africa

Jean M. Harris
KwaZulu-Natal Nature Conservation Service, Durban, South Africa

Bruce Q. Mann
Oceanographic Research Institute, Durban, South Africa

Abstract

The linefishery in the subtropical waters off KwaZulu-Natal, South Africa, targets a variety of species associated with rocky reefs at depths of up to 200 m. Populations of many of these reef fish are severely depressed by past and current fishing. Current management objectives aim to rebuild these stocks and to conserve the ecosystems, but means of implementing these objectives are unclear. This paper presents a mass-balance, ecosystem model of the KwaZulu-Natal reef ecosystem. It divides the reef community into 18 groups, and quantifies biomass, production, and consumption for each group. Flows among groups are calculated using a diet composition matrix that changes according to the relative productivity of food groups. Model results indicate that community biomass is dominated by suspension-feeders. Detritus is assumed to be an important source of food for consumers, but this needs to be substantiated. The biomass and production of selected groups were changed in a series of experiments to investigate the impacts of the fishery. The results were evaluated in terms of ecotrophic efficiencies, and sometimes were unexpected because of factors such as redirected predation from previously abundant groups onto other prey groups. The model was sensitive to changes in biomass of the fish groups, including that of the group that was historically dominant in the fishery. Given the large historical changes, the current reef ecosystems are probably markedly different from the pristine state. However, it is difficult to predict the likely impacts of past and future changes on single species. Future management should aim to maintain unfished areas of representative reef, and future data collection should be directed at comparative studies of protected and unprotected reef communities, to complement the ongoing fisheries data. The model, despite its limitations, is guiding future research so that limited resources can be directed appropriately to achieve management goals.

Introduction

Off KwaZulu-Natal, South Africa, the continental shelf is narrow, reaching a maximum width of 45 km in the central area and narrowing to 3-11 km in the north and south (Garratt 1984; Fig. 1). Scattered rocky reefs occur along the shelf in depths of up to 200 m (Fig. 1). The shelf waters are warm (21-26°C) and the rocky reef ecosystems support diverse, subtropical communities containing more than 1,000 fish species belonging to 150 families (van der Elst 1988). Many of the demersal reef fish species are targeted by a diffuse, mostly small-scale linefishery that operates in the coastal areas (Garratt 1986), and supports recreational, subsistence, and commercial sectors. The mean annual landings of the linefishery over the last decade have been approximately 1,270 t (Mann-Lang et al. 1997).

Fishing effort by all sectors of the KwaZulu-Natal reef fishery increased dramatically from the turn of the century. In early years, the large endemic reef species were important in the catches, including species such as seventy-four (*Polysteganus undulosus*), red steenbras (*Petrus rupestris*), and other sparids that together made up 60-80% of the total catch by weight. With the fishing down of these previously abundant species to very low levels, there have been marked shifts in the species composition of the catch (Garratt 1986, van der Elst 1988, van der Elst and Adkin 1988, van der Elst and de Freitas 1988, Mann-Lang et al. 1997). Small, less favored species are now caught, particularly slinger (*Chrysoblephus puniceus*) and soldier (*Cheimerius nufar*). In addition, demersal migrants have become more important in the catches. Most of the species currently caught by the linefishery are now considered threatened or endangered.

Management of the KwaZulu-Natal linefishery has been carried out primarily by means of species-specific restrictions, including bag limits, size limits, and closed seasons (Mann-Lang et al. 1997). However, in a multispecies, multi-user fishery of this sort, it is difficult to predict the effects of single-species measures on the dynamics of other species in the ecosystem. Furthermore, there are concerns about the impacts that the

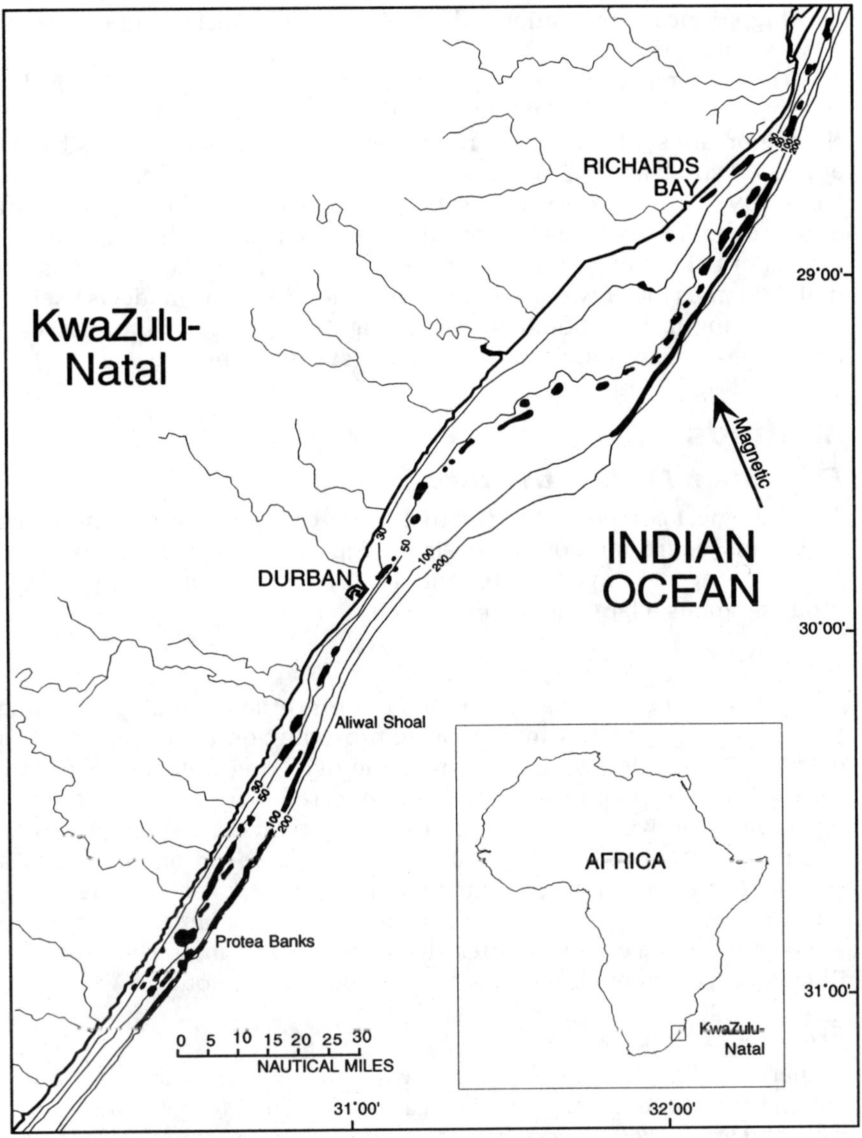

Figure 1. The distribution of rocky reefs on the continental shelf off KwaZulu-Natal, South Africa. After Garratt (1984).

shifting species composition will have on other species and on the reef ecosystem as a whole.

This paper attempts to understand the complex trophic relationships that occur in the rocky reef ecosystem. The limited available data from the region are synthesised in an ecosystem model, using a mass balance approach similar to that used in the Ecopath software (Christensen and Pauly 1992). The model is necessarily speculative and is used primarily as a tool to aid understanding about the possible past and future ecological effects of fishing on the rocky reef ecosystem. The model is also used to highlight gaps in knowledge. Despite its limitations, the model is useful in synthesising information, and it is being used to guide future research and management in the context of an ecosystem approach.

Methods

Ecosystem Model Structure

The biological component of the rocky reef ecosystem is represented by 18 groups, aggregated on the basis of trophic function: 1 detrital, 2 primary producer, 5 invertebrate, and 10 vertebrate groups (Fig. 2). Each group contains a large number of species.

Detritus

In deep reefs, light availability limits photosynthesis, and it is assumed that the energetic basis for the system production comes mainly from detritus. This includes material from the overlying water that has sunk through the water column, as well as imported material, much of which presumably derives from land via the many rivers in the region (Fig. 1). The detrital pool receives contributions from all dead organic material, and the excretions and egestions of all living components of the ecosystem. It consists of fecal material, dead plants and animals, exoskeleton molts, particulate organic matter, dissolved organic matter, and bacteria. Detritus is consumed by many of the groups in the model.

Primary Producers

Primary production is derived from two main sources: macroscopic algae and phytoplankton. Macroscopic algae are attached to the rocky substratum, and are grazed by vertebrates and invertebrates. Phytoplankton provide food for some suspension-feeding, reef-dwelling organisms as well as for pelagic species that are migrant visitors to the reef system.

Invertebrates

All small, invertebrate heterotrophs suspended in the water column are represented by a single zooplankton category (which also contains fish larvae). This aggregation necessarily results in an over-simplification of the food web, because zooplankton communities typically contain herbivores

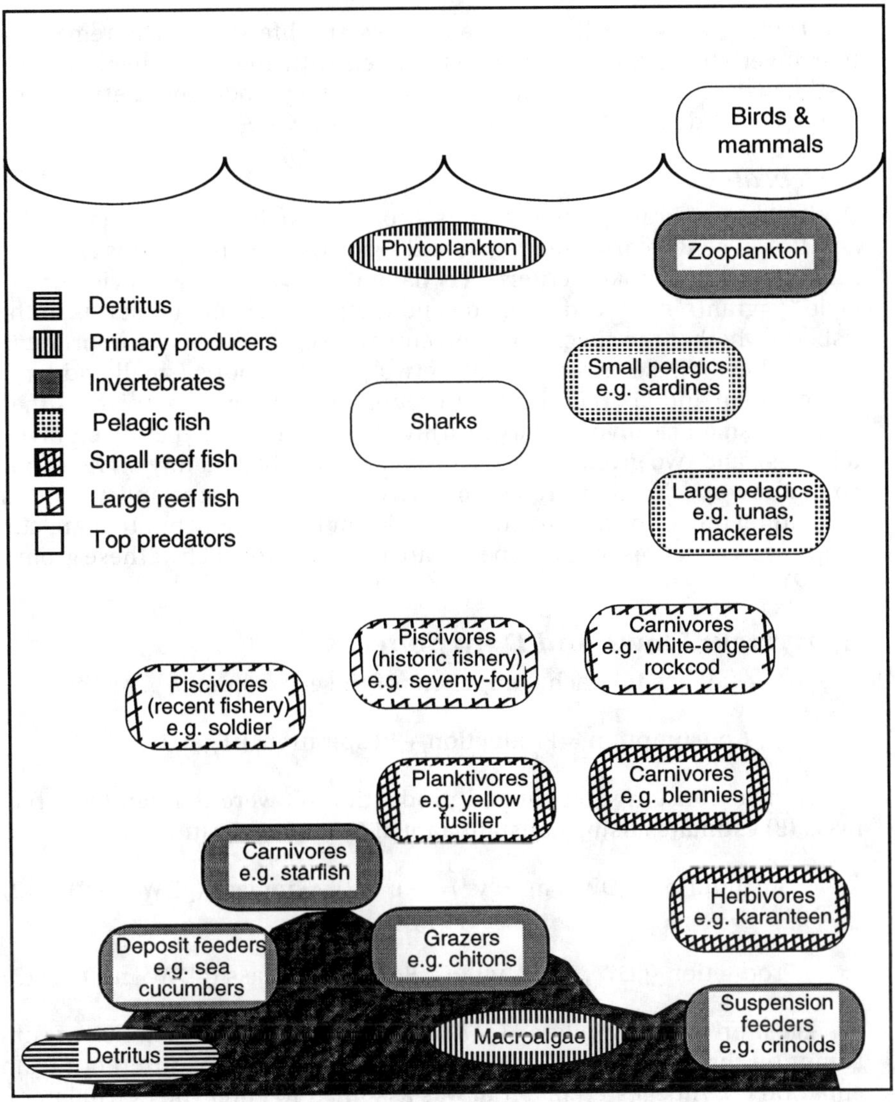

Figure 2. Diagrammatic representation of the major components of the KwaZulu-Natal rocky-reef ecosystem, and their relative positions in the water column. Model groups can be further aggregated into the seven groups shown in the key.

and carnivores with a large range of sizes and life-styles. The remaining four invertebrate groups are all associated with the rocky reef, and are distinguished on the basis of their main feeding mode and diet: suspension-feeders, deposit feeders, grazers, and carnivores.

Vertebrates

Of the 10 vertebrate groups, 8 consist of bony fish, 1 represents sharks, and 1 represents marine mammals and seabirds. The fish groups are separated using four linked criteria: (1) habitat (reef-dwelling species or pelagic "migrants"); (2) feeding mode and diet; (3) size; and (4) status in the fishery (whether the species are currently or were previously important in the catches). These criteria result in two pelagic groups (small and large pelagic fish) and six reef-dwelling groups: planktivores, herbivores (i.e., grazers), small carnivores, large carnivores (i.e., feeding mainly on invertebrates), and two groups of large piscivores (feeding mainly on fish), one containing species that are commonly caught at present and the other containing those species that were important historically but that are currently scarce. Representative species are indicated for each of these groups (Fig. 2).

Ecosystem Flows and Data Sources

The mass balance for each component of the ecosystem is given by:

$$\text{Consumption}_i = \text{Production}_i + \text{Respiration}_i + \text{Egestion}_i \tag{1}$$

Annual consumption (Q) and production (P) were derived from biomass (B) estimates using constant annual Q:B and P:B ratios:

$$\text{Consumption (gDW} \times \text{m}^{-2} \times \text{y}^{-1}\text{)} = \text{Q:B (y}^{-1}\text{)} \times \text{Biomass (gDW} \times \text{m}^{-2}\text{)} \tag{2}$$

$$\text{Production (gDW} \times \text{m}^{-2} \times \text{y}^{-1}\text{)} = \text{P:B (y}^{-1}\text{)} \times \text{Biomass (gDW} \times \text{m}^{-2}\text{)} \tag{3}$$

Egestion (non-assimilated food) was assumed to represent 20% of consumption throughout, and respiration was calculated by difference from equation (1). All egested material was assumed to enter the detrital pool.

The losses to consumers from each category i were calculated as:

$$\text{Losses of } i \text{ to consumers } j = \sum_{j=1}^{18}\left[B_j \times \left(\frac{Q}{B}\right)_j \times DC_{ij}\right] \tag{4}$$

where $DC_{i,j}$ is the diet composition, expressed as the fraction by weight that food item i forms of consumer *j*'s diet. Diet compositions are known for only a few of the species that make up the KwaZulu-Natal rocky reef community. Because one would expect diet compositions to change in response to community changes, a dynamic approach was used to assign the diet proportions. A diet allocation matrix was developed, based on an assumed presence (1) or absence (0) of each group in every other group's diet, with some of the "presence" scores down-weighted from 1 to values of 0.5, 0.1, 0.01, 0.001, or 0.0001, to account for assumed or known preferences (Table 1). For example, many piscivorous and carnivorous fish eat algae, but the proportion in their diets is unlikely to be the same as the relative availability of algae. The diet proportions by mass were calculated using the relative production of each food group, scaled by the "presence" score. The resulting diet matrices (Table 2) are dynamic because they change each time the biomass or production of a food group is changed.

Catches from different species groups were based on information compiled by Mann-Lang et al. (1997), who estimated an approximate mean annual catch of 1,270 t between 1985 and 1995. This catch was converted to appropriate units and allocated among groups as described by Toral-Granda (1998).

The fraction of each component's production that is eaten or harvested is the ecotrophic efficiency (EE), calculated as:

$$EE_i = \frac{\text{Consumption of } i + \text{Harvest of } i}{\text{Production by } i} \tag{5}$$

The EE values were used as an index of intergroup impacts in the ecosystem. Model experiments were carried out in which biomass and production values were changed for single groups, and the resulting effect on the EEs was assessed. All calculations were carried out using a computer spreadsheet.

The data used in the model (Table 3) were obtained from Toral-Granda (1998), who also gives a detailed description of the original data sources. The KwaZulu-Natal rocky reef ecosystem contains complex, diverse communities for which few data were available. Where no data were available for the local system, they were estimated from studies of comparable systems elsewhere; e.g., Opitz's (1996) study of a Puerto Rico–Virgin Islands tropical reef ecosystem in the Caribbean. Some of Toral-Granda's (1998) original values were altered in this study, and the changes and the reasons for them are explained briefly in Table 3.

Results

The pre-exploitation ecosystem structure is difficult to infer. The current (exploited) system is dominated by invertebrate groups (Fig. 3), with the suspension-feeders forming two-thirds of the invertebrate biomass. Second in abundance are the small reef fish groups, which are approximately 6 times as abundant as the large reef fish groups. Pelagic fish make up

Table 1. **Diet composition matrix for the species groups of the KwaZulu-Natal rocky reef ecosystem model, based on assumed presence (1) and absence (0) of items in the diet, downweighted by assumed preferences.**

No.	Prey group	Predator groups														
		18	17	16	15	14	13	12	11	10	9	8	7	6	5	4
1	Detritus									0.01			1		1	0.001
2	Macroalgae			0.001	0.001	0.1			0.1	0.1	1	0.1		1	1	
3	Phytoplankton							0.5		0.5			0.5			1
4	Zooplankton	1		0.001	0.001	0.5	0.1	1	0.1	1			0.5			1
5	Invertebrate deposit feeders	0.1	1	0.1	0.1	1			1			1				
6	Invertebrate grazers	0.1	1	0.1	0.1	1			1			1				
7	Invertebrate suspension feeders	<0.001	0.001	0.001	0.001	1			0.5			1				
8	Invertebrate carnivores	0.1	1	0.1	0.1	1	0.5		0.1			1				
9	Reef fish herbivores		1	1	1	1	0.5									
10	Reef fish planktivores		1	1	1		0.5									
11	Small reef fish carnivores	0.1	1	1	1	1	0.5									
12	Small pelagics	1	1				1									
13	Large pelagics	1	1				1									
14	Large reef fish (carnivores)	1	1	1	1											
15	Large reef fish (pisc. & carn., recent)	1	1	1	1											
16	Large reef fish (pisc. & carn., historic)	1	1	1	1											
17	Sharks	1	0.5	1	1											
18	Mammals & birds		0.1													

Table 2. Diet composition matrix for the species groups of the KwaZulu-Natal rocky reef ecosystem for the "standard case" model. Proportions (by weight) are based on presence/absence values (Table 1) and scaled according to the relative productivity of prey items.

		Predator groups														
No.	Prey group	18	17	16	15	14	13	12	11	10	9	8	7	6	5	4
1	Detritus									0.003			0.234		0.750	<0.001
2	Macroalgae			<0.001	<0.001	0.018			0.047	0.010	1.000	0.025		1.000	0.250	
3	Phytoplankton							0.899		0.888			0.725			0.946
4	Zooplankton	0.860		<0.001	<0.001	0.095	0.124	0.101	0.050	0.100			0.041			0.053
5	Invertebrate deposit feeders	0.004	0.029	0.004	0.004	0.008			0.022			0.011				
6	Invertebrate grazers	0.002	0.013	0.002	0.002	0.004			0.010			0.005				
7	Invertebrate suspension feeders	<0.001	0.002	0.003	0.003	0.657			0.859			0.894				
8	Invertebrate carnivores	0.022	0.166	0.021	0.021	0.048	0.156		0.013			0.065				
9	Reef fish herbivores		0.030	0.037	0.037	0.009	0.028									
10	Reef fish planktivores		0.170	0.210	0.210		0.159									
11	Small reef fish carnivores	0.073	0.560	0.691	0.691	0.161	0.525									
12	Small pelagics	0.003	0.002				0.005									
13	Large pelagics	0.002	0.001				0.002									
14	Large reef fish (carnivores)	<0.001	<0.001	<0.001	<0.001											
15	Large reef fish (pisc. & carn., recent)	0.034	0.026	0.032	0.032											
16	Large reef fish (pisc. & carn., historic)	<0.001	<0.001	<0.001	<0.001											
17	Sharks	<0.001	<0.001	<0.001	<0.001											
18	Mammals & birds		<0.001													

Table 3. Initial and final parameter sets applied to the ecosystem model of the KwaZulu-Natal rocky reef ecosystem.

	B(gDW/m²)		Q:B(/yr)		P:B(/yr)		Catch	
	initial	final	initial	final	initial	final	(gDW/m²)	Explanation for changes
Detritus	–	–	–	–	–	–	–	
Macroalgae	133.53[a]	5[d]	–	–	13.25[a]	–	–	Biomass likely to be much smaller than for coral reefs, because water is murky and some rocky reefs are very deep.
Phytoplankton	150[b]	3.402[c]	–	–	154.4[b]	365[d]	–	Final biomass based on local study (see Toral-Granda 1998). P:B ratio of 1/d is reasonable for warm temperatures. Reverted to values from Opitz (1996).
Zooplankton	3.5[b]	1.75	165[a]	–	80[b]	40[a]	–	
Invertebrate filter feeders	3107[c]	55[d]	14.09[a]	20[d]	4.39[c]	–	–	Initial biomass based on studies from shallow reefs, where biomass is dominated by red bait. Deep reefs have a much smaller biomass of suspension-feeders. Q:B was adjusted to ensure that NGE was realistic.
Invertebrate deposit feeders	10.46[c]	–	3.36[a]	–	0.6[b]	0.29[a]	–	Reverted to values from Opitz (1996).
Invertebrate grazers	4.168[c]	–	10.8[a]	2[d]	0.33[a]	–	–	Q:B was adjusted to ensure that NGE was realistic.
Invertebrate carnivores	10.112[b]	–	13[a]	–	1.74[c]	–	–	
Reef fish planktivores	5.073[a]	–	18.85[a]	–	3.54[a]	–	0.043[b]	
Reef fish herbivores	2.22[c]	–	3.34[b]	6[d]	1.41[c]	–	0.060[b]	Q:B was adjusted to ensure that NGE was realistic.
Small reef fish carnivores	11.315[c]	22[d]	7.48[b]	12[d]	2.69[c]	–	0.043[b]	Biomass was estimated from just two species, so estimate was doubled to allow for other species. Q:B was adjusted to ensure that NGE was realistic.
Small pelagics	0.24[c]	–	11.3[c]	–	1.1[c]	–	0.014[b]	
Large pelagics	0.207[b]	–	4.6[c]	–	0.608[c]	–	0.200[d]	Catch adjusted from Toral-Granda (1998), to allow for new species groupings.
Large reef fish (carnivores)	0.001[c]	–	3.92[b]	–	0.37[c]	–	–	
Large reef fish (pisc & carn, recent)	4.64[a]	–	4.68[b]	–	0.6[c]	–	1.085[d]	Catch adjusted from Toral-Granda (1998), to allow for new species groupings.
Large reef fish (pisc & carn, historic)	0.073[c]	–	3.39[b]	–	0.478[c]	–	0.011[d]	Catch adjusted from Toral-Granda (1998), to allow for new species groupings.
Sharks	0.001[b]	0.01[c]	4.9[b]	–	0.53[c]	–	0.025[b]	Final biomass based on local study (see Toral-Granda 1998).
Mammals & birds	0.001[c]	–	10[c]	–	0.6[c]	–	–	

a. Values derived directly from a model of a coral-reef ecosystem (Opitz 1996). b. Values assumed or obtained by tuning (Toral-Granda 1998). c. Values based on studies from the region, with references given by Toral-Granda (1998). d. Values assumed for this study (with explanations).

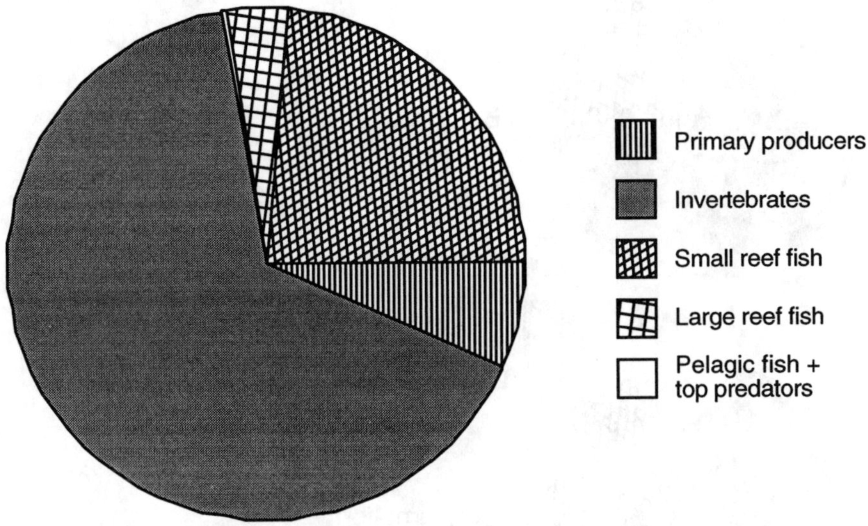

Figure 3. The relative biomasses of aggregated model groups.

only a small part of the ecosystem, because of their migratory habits, and the primary producers form only a small part of overall system biomass.

The estimated primary production in the system is 1,310 gDW per m^2 per year, of which 95% is attributable to phytoplankton and 5% to macroalgae. The patterns for heterotrophic production and consumption are similar, with invertebrates the dominant consumers and producers (Figs. 4a and b), followed in importance by small reef fish groups.

The most important food sources in the ecosystem derive from plant material (Fig. 5), although detritus contributes a substantial proportion (approximately 20%) of consumers' diets. Invertebrates also are important as food items, as would be expected from their large biomass and production.

The ecotrophic efficiencies resulting from parameters in Table 3 ranged from values less than 1 for plants, small reef fish groups, and large reef fish groups, to values greater than 1 for some invertebrates, large pelagics, and sharks (Fig. 6a). The two groups containing mammals/birds and large carnivorous reef fish were a very small component of the ecosystem, and their EEs were extremely sensitive to values used for the other groups; they will not be discussed further.

When the values for biomass and production of some of the major groups were changed, this automatically caused changes in diet compositions, and affected the EEs. When the production of phytoplankton was reduced by a factor of 5 (Fig. 6b), there was an increased demand for detritus and zooplankton, but insufficient quantities of all three groups

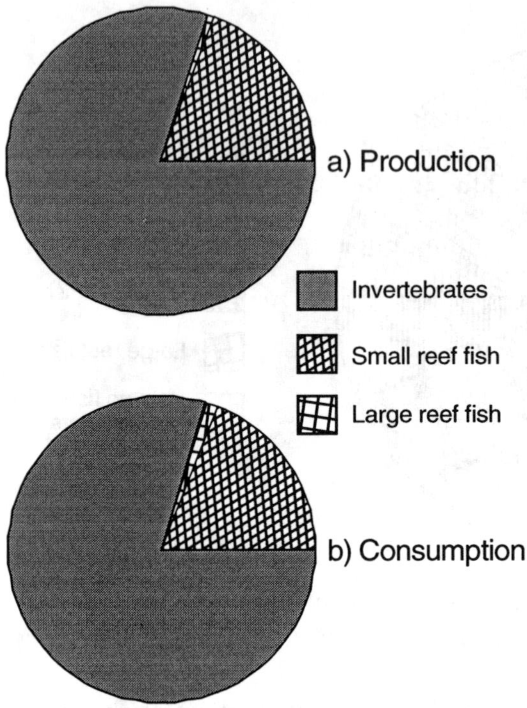

Figure 4. The relative (a) production and (b) consumption of aggregated consumer groups.

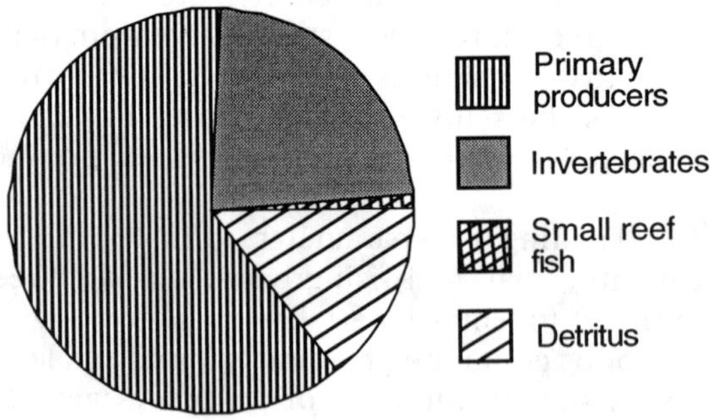

Figure 5. The relative contributions of aggregated model groups as food items for consumers.

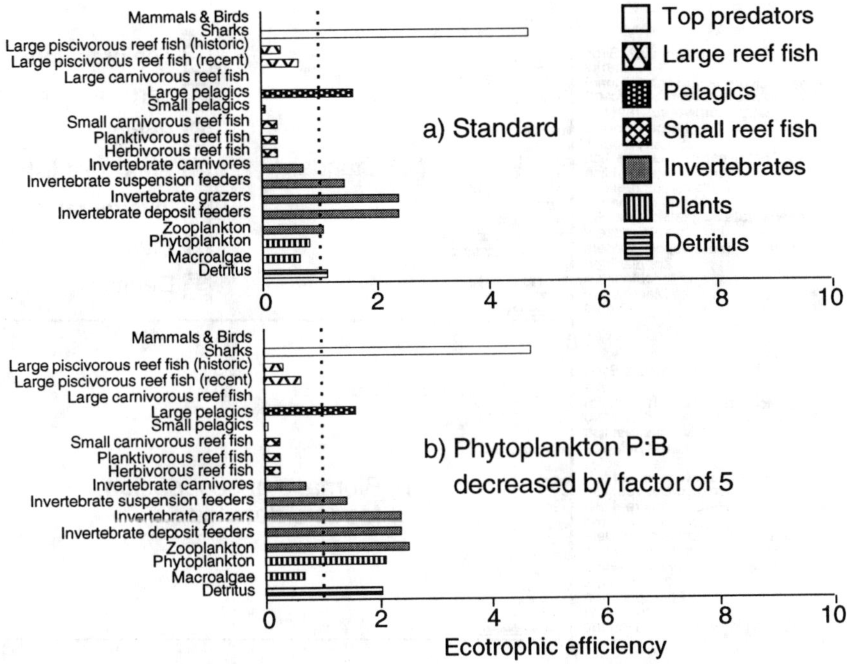

Figure 6. Results of the model calculations showing ecotrophic efficiencies. (a) Standard case. (b) Results when phytoplankton production reduced by a factor of 5.

(phytoplankton, zooplankton and detritus) to meet consumers' requirements.

The dominant consumer group in the ecosystem was the invertebrate suspension-feeders. When their biomass was doubled (Fig. 7b), the EEs increased for their main food items (phytoplankton and detritus), but decreased for themselves, because their production increased proportionately more than did predation on them. Similarly, the EEs for other invertebrate groups (grazers and deposit feeders) also decreased, because predation was deflected from these groups to the more abundant suspension feeders. The converse was true when suspension-feeder biomass was substantially reduced (Fig. 7c).

Three scenarios were tested for the reef-fish groups. In the first scenario, the biomass of the group of large piscivorous reef fish that was important in catches historically was increased by a factor of 1,000 (Fig. 8b). This emulated "pristine" levels when groups such as the rockcods and seventy-four were abundant in the ecosystem. Increasing the biomass resulted in increases in the EEs of small- and large-reef-fish groups, because

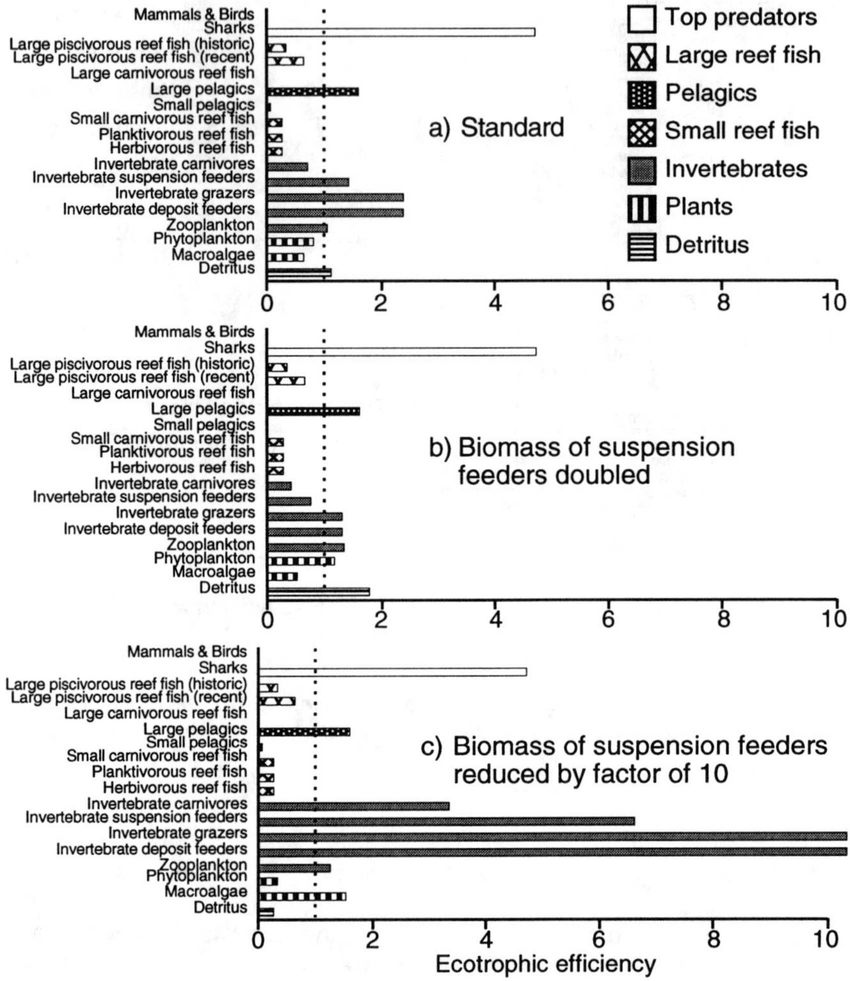

Figure 7. Results of the model calculations showing ecotrophic efficiencies. (a) Standard case. (b) Results when biomass of suspension-feeders doubled. (c) Results when biomass of suspension-feeders decreased by a factor of 10.

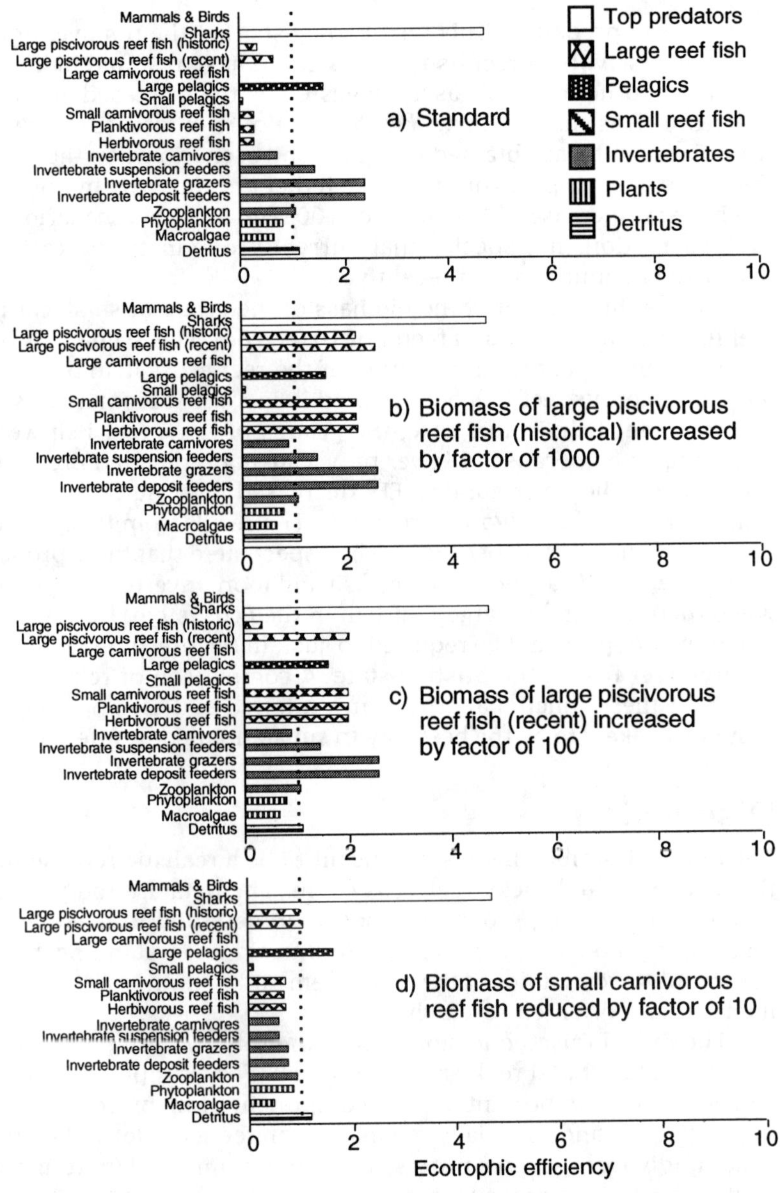

Figure 8. *Results of the model calculations showing ecotrophic efficiencies. (a) Standard case. (b) Results when biomass of large reef fish (piscivorous, historical) increased by a factor of 1,000. (c) Results when biomass of large reef fish (piscivorous, recent) increased by a factor of 100. (d) Results when biomass of small carnivorous reef fish decreased by a factor of 10.*

of increased predation. This result suggests that the removal of many of the large piscivorous reef fish species could have resulted in increases in biomass of other reef fish, as is suggested by the increased importance of species such as slinger (*Chrysoblephus puniceus*) in catches in recent years. A similar result was obtained in the second scenario when the biomass of the group of large piscivorous reef fish that are currently important in the catches was increased by a factor of 100 (Fig. 8c). This scenario supports the observation that species that currently dominate the catches have decreased in abundance in recent years.

For the third scenario, the biomass of the group of small carnivorous reef fish was decreased by a factor of 10 (Fig. 8d). This scenario is based on the assumption that transformation of the reef ecosystem by overfishing large piscivorous reef fish has resulted in increased biomass of small carnivorous reef fish. In this case, the groups of large reef fish were negatively impacted because of fewer prey, causing redirected predation onto themselves. The invertebrates' EEs decreased because of reduced predation by the small carnivorous reef fish. The result from this scenario was somewhat unexpected, because we had speculated that there probably had been fewer small carnivorous reef fish and more invertebrates in the past, when there were more large reef fish in the ecosystem. However, alternative prey groups would be required to sustain the presumed large biomass of large reef fish in the pristine state. A comparison of relative standing stocks of these functional groups in exploited and unexploited reef ecosystems is likely to be the best way to substantiate or refute these results.

Discussion

The model described here is not meant to be a realistic representation of the KwaZulu-Natal rocky reef ecosystem. The data are too poor and the relationships among groups too poorly understood to make that claim. Instead, the model is proposed as a tool to aid understanding, both in the context of the dynamics of the ecosystem and in terms of directing future management-oriented research.

The model raised questions about the roles of different components of the KwaZulu-Natal rocky reef ecosystem. For example, detritus was assumed to be an important dietary component of many consumers in the model (Fig. 5), and especially of suspension feeders. Detrital material can sink rapidly from surface waters, and often is enriched (in terms of nitrogen) by bacteria to make it more nutritious for animals (Smetacek 1984). To what extent is the production of the rocky reef ecosystem dependent on detritus? In coral reef ecosystems, much of the primary production is by epilithic algae, such as small filamentous and crustose coralline algae (Klumpp and Polunin 1990). However, coral reefs typically occur in clear, well-lit waters, whereas many of the KwaZulu-Natal rocky reefs lie at great depths and often in poor light environments because of turbidity caused by riverine inputs. Approximately 30 million t of suspended material enters the sea annually via the rivers that occur along the KwaZulu-Natal coastline (Mann-Lang et al. 1997). Autochthonous production (mainly by macroalgae) in the rocky reef ecosystem was assumed to be small. Allochthonous production can come from phytoplankton production in the surface waters overlying the reefs, from riverine inputs, and/or from algal production on shallow coastal reefs. The relative importance of these different production sources is not known, and should be investigated.

The rocky reef community was dominated by invertebrate suspension-feeders, with an assumed biomass of 55 gDW per m^2. This value is 2 orders of magnitude smaller than that found for shallow-reef communities in the region (Lambert and Steinke 1986, Fielding et al. 1994), which have dense populations of mussels and red bait (*Pyura stolonifera*). Community structure on the deep reef systems is poorly known because of the difficulties of sampling benthic communities of the continental shelf. This makes it difficult to assess whether any marked community changes have occurred over the past century of fishing. In the model, the invertebrate groups were linked to the commercially important reef-fish species primarily via the small reef-fish species. Fishing has resulted in a substantial reduction in large reef fish, but the extent to which this may have disrupted food webs and transformed benthic community structure is unknown. One possible consequence of the reduction in large reef fish might be an increase in the biomass of small reef fish (Figs. 7b, c), which in turn could negatively impact invertebrates (Fig. 8d). Such possible cascading effects require an understanding of the relative significance of top-down control in the food web in contrast to bottom-up processes. This requires quantitative information about primary production and detrital inputs as well as invertebrate standing stocks and dynamics, to assess whether primary consumers might be limited by available production.

The model was used to investigate the likely changes that would occur if biomass and production of different species groups were altered. This was done by examining the ecotrophic efficiencies, which indicate whether groups are under- or over-utilized in the model ecosystem. In the standard case, many of the EEs were much less than or greater than unity (Fig. 6a), indicating that the model ecosystem was not in an equilibrium situation. This is not surprising, given both the poor quality of the data and the disturbed nature of the ecosystem, and for this reason no attempt was made to try and balance the flows. Instead, the sensitivity of the model results was tested in relation to changes in biomass and production of some of the major groups, and the resultant model output (Figs. 6-8) was not easy to interpret. The biological processes are complex, even though the model equations are linear. Thus the effect of competition could not be assessed directly in this model, because the consumption of each group was dependent only on its own biomass, and not on that of other groups. However, it was possible to investigate predatory interactions, because these affected the EEs. A biological interaction that caused some unexpected results was that of redirected predation. This interaction manifested when the biomass of a prey group was reduced or increased, causing other prey groups to be more heavily or lightly predated. For example, the grazers and deposit feeders of Fig. 7b benefited from reduced predation when the biomass of suspension feeders increased (Fig. 7b), and the large reef fish were adversely affected when the biomass of small reef fish decreased (Fig. 8d). These effects are indirect, and would probably be difficult to interpret if they were operating in tandem with competition and predation, as would occur in simulation models or in reality.

Ecosystem Management

Current management objectives for the KwaZulu-Natal reef-fishery include the rebuilding of depleted stocks and conservation of ecosystem structure and function. These objectives need to be reached in a climate of increasing fishing and political pressures from many user-groups. Because of the complexity of the ecosystem and the large number of species involved, impacts on single species are difficult to predict. Natural mortality through predation is dependent on the relative population sizes of different species. It is likely that any reference points for management that are applied in a single species context and based on single species attributes would differ from reference points that take into account changing mortality caused by community changes in the reefs. We suggest that the effects of interspecies interactions on the accepted values of key biological reference points warrant further investigation. For example, spawner biomass per recruit reference values for one species may have to be adjusted to compensate for the effects of overfishing on species with which it interacts.

Future management should aim to maintain unfished areas of representative reef in marine protected areas (MPAs) (Attwood et al. 1998). Although we may not fully understand the impacts of the reef fishery, it is likely to have transformed the community structure. The MPAs will provide opportunities for comparative studies between exploited and unexploited reef ecosystems, giving an empirical basis for ecosystem-level management advice.

In addition to ongoing monitoring of the fishery and the collection of fisheries data, future data collection should be conducted within the framework presented here. Studies of the reef ecosystem should be holistic, using functional groups to aggregate the data and attempting to fill in the data gaps in Tables 1, 2, and 3. As standard practice, samples of individual diet items should be reconstituted to give bulk estimates of biomass or volume in the diets, rather than numbers or frequency of occurrence. Future studies should attempt to estimate both the standing stocks and also the associated dynamics encapsulated by the various ecological rates of Table 3. The sources and magnitude of primary production in the reef ecosystem should be resolved. In particular, it is important to know to what extent production is being imported to the system (through advection of detritus and sinking of plankton), and to what extent it is produced locally by macroalgae.

Effective ecosystem management advice requires understanding of species and community interactions. Ultimately, incorporation of ecosystem considerations into management of the KwaZulu-Natal reef fishery requires that field experiments be carried out to test hypotheses about likely effects. This model is a useful tool for generating hypotheses and exploring linkages in the ecosystem, along with the possible consequences of disrupting these linkages. The results presented in this paper sometimes yielded unexpected results, and this should serve as a caution against invoking speculative arguments without careful scrutiny of ecosystem-level models. Ecosystem models allow both scientists and managers to assess what is and what is not possible in the context of ecosystem management, before expending time and effort in expensive surveys and data-collection exercises.

Acknowledgments

We thank the Oceanographic Research Institute (ORI), Durban, for the use of facilities and access to unpublished data that formed that basis of much of this study. Staff from ORI were generous in giving their time and expertise to advise on aspects of this study, and in particular we thank Sean Fennessey, Peter Fielding, Anesh Govender, Judy Mann-Lang, Wendy Robertson, Michael Schleyer, Bruce Tomalin, and Rudy van der Elst. Financial and logistical support was supplied by the KwaZulu Natal Nature Conservation Service, the Chief Directorate: Marine and Coastal Management of the Department of Environmental Affairs and Tourism, the University of Cape Town, and ORI.

References

Attwood, C.G., B.Q. Mann, J. Beaumont, and J.M. Harris. 1998. Review of the state of marine protected areas in South Africa. S. Afr. J. Mar. Sci. 18:341-367.

Christensen, V., and D. Pauly. 1992. Ecopath II: A software for balancing steady-state ecosystem models and calculating network characteristics. Ecol. Modell. 61:169-185.

Fielding, P.J., K.A. Weerts, and A.T. Forbes. 1994. Macroinvertebrate communities associated with intertidal and subtidal reefs of *Pyura stolonifera* (Heller) (Tunicata: Ascidiacea) on the Natal coast. S. Afr. J. Zool. 29:46-53.

Garratt, P.A. 1984. The biology and fishery of *Chrysoblephus puniceus* (Gilchrist and Thompson, 1917) and *Cheimerius nufar* (Ehrenberg, 1830), two offshore sparids in Natal waters. M.S. thesis, University of Natal, Durban. 139 pp.

Garratt, P.A. 1986. The offshore linefishery of Natal. I. Exploited population structures of the sparids *Chrysoblephus puniceus* and *Cheimerius nufar*. Oceanographic Research Institute Investigational Report 62:1-17.

Klumpp, D.W., and N.V.C. Polunin. 1990. Algal production, grazers and habitat partitioning on a coral reef: Positive correlation between grazing rate and food availability. In: M. Barnes and R.N. Gibson (eds.), Trophic relationships in the marine environment. Proc. 24th Europ. Mar. Biol. Symp., pp. 372-388.

Lambert, G., and T.D. Steinke. 1986. Effects of destroying juxtaposed mussel-dominated and coralline algal communities at Umdoni Park, Natal coast, South Africa. S. Afr. J. Mar. Sci. 4:203-217.

Mann-Lang, J.B., R.P. van der Elst, and A. Penney. 1997. An analysis of the KwaZulu-Natal commercial and recreational reef fisheries. Oceanographic Research Institute Unpublished Report No. 137, available from Oceanographic Research Institute, P.O. Box 10712, Marine Parade 4056, South Africa. 43+6 pp.

Opitz, S. 1996. Trophic interactions in Caribbean coral reefs. ICLARM Report. 431 pp.

Smetacek, V. 1984. The supply of food to the benthos. In: M.J.R. Fasham (ed.), Flows of energy and materials in marine ecosystems: Theory and practice. Plenum Press, New York, pp. 517-547.

Toral-Granda, M.V. 1998. Understanding trophic interactions and fishery impacts in a rocky reef ecosystem in KwaZulu-Natal, South Africa. M.S. thesis, University of Cape Town. 52+18 pp.

van der Elst, R.P. 1988. The common sea fishes of southern Africa. Struik, Cape Town. 398 pp.

van der Elst, R.P., and F. Adkin. 1988. National Marine Linefish System: Recreational data. In: I.A.W. Macdonald and R.J.M. Crawford (eds.), Long-term data series relating to South Africa's natural renewable resources. South African National Scientific Programmes Report No. 157, pp. 72-75.

van der Elst, R.P., and A.J. de Freitas. 1988. Long-term trends in Natal marine fisheries. In: I.A.W. Macdonald and R.J.M. Crawford (eds.), Long-term data series relating to South Africa's natural renewable resources. South African National Scientific Programmes Report No. 157, pp. 76-84.

Ecosystem Approaches for Fisheries Management
Alaska Sea Grant College Program • AK-SG-99-01, 1999

Trophic Relationships of Commercial Fishes in the Pacific Waters off Southeastern Kamchatka and the Northern Kuril Islands

Alexei M. Orlov
Russian Federal Research Institute of Fisheries and Oceanography (VNIRO), Moscow, Russia

Abstract

Of the fish species in Pacific waters off the northern Kuril Islands and southeastern Kamchatka, 19 are commercially important and most abundant. Among them are three groups with different feeding habits: (1) planktophages (walleye pollock, *Theragra chalcogramma*; Atka mackerel, *Pleurogrammus monopterygius*; Pacific ocean perch, *Sebastes alutus*); (2) benthophages (rock greenling, *Hexagrammos lagocephalus*; rock sole, *Pleuronectes bilineatus*; Bering skate, *Bathyraja interrupta*; Okhotsk skate, *Bathyraja violacea*); and (3) predators (sablefish, *Anoplopoma fimbria*; Pacific cod, *Gadus macrocephalus*; Greenland turbot, *Reinhardtius hippoglossoides*; Kamchatka flounder, *Atheresthes evermanni*; Pacific halibut, *Hippoglossus stenolepis*; Alaska skate, *Bathyraja parmifera*; shortraker rockfish, *Sebastes borealis*; broadbanded thornyhead, *Sebastolobus macrochir*; shortspine thornyhead, *Sebastolobus alascanus*; Aleutian skate, *Bathyraja aleutica*; whiteblotched skate, *Bathyraja maculata;* and Matsubara skate, *Bathyraja matsubarai*). The biology of the 19 species in these waters has not been studied in detail and little investigation has been done on their trophic relationships.

Based on shipboard scans of stomach contents in 1992-1996 (15,763 stomachs were analyzed, of which 9,429 contained food), dietary composition was investigated and some aspects of trophic relationships were analyzed. For all species of each group, Schoener's indices of dietary overlap were calculated. Level of feeding competition was determined based on intra-annual dynamics of feeding intensity and features of spatial (vertical and horizontal) distribution of fish concentrations.

Among planktophages the maximum index value (60.0%) was characteristic of diets of Pacific ocean perch and pollock. The maximum feeding intensity for both species appeared during similar periods (May-July). At the same time, the characteristics of spatial distribution of concentrations of pollock and Pacific ocean perch differed widely indicating a low level of feeding competition between them.

Among benthophages the greatest dietary overlap was noted for rock sole and Bering skate (73.3%) and Bering and Okhotsk skates (65.5%). The dissimilarity of characteristics of feeding intensity dynamics and spatial distribution showed a low level of feeding competition between these species.

Among predatory fishes the maximum index values of dietary overlap were characteristic of shortraker rockfish and broadbanded thornyhead (82.7%), whiteblotched skate and sablefish (62.2%), shortraker rockfish and shortspine thornyhead (61.4%), Greenland turbot and Kamchatka flounder (60.1%), and whiteblotched and Aleutian skates (60.0%). For all of those pairs (except for sablefish and whiteblotched skate) the high similarity of intra-annual dynamics of feeding intensity was noted. However, similar characteristics of spatial distribution of fish assemblages were noted only for shortraker rockfish and broadbanded thornyhead, which indicates a high level of feeding competition between these species.

Introduction

Among planktophage fishes of the Pacific waters off the northern Kuril Islands and southeastern Kamchatka, Pacific ocean perch, *Sebastes alutus*; walleye pollock, *Theragra chalcogramma*; and Atka mackerel, *Pleurogrammus monopterygius*, are most abundant and commercially important. Data on feeding of Pacific ocean perch in this area are limited to only one paper (Novikov 1974). Feeding habits of Atka mackerel in this area are well researched only in coastal waters (Zolotov and Medveditsyna 1978, Zolotov and Tokranov 1991). Published data on the diet of Atka mackerel in offshore waters (Onishchik 1997, Orlov 1997) are limited to one summer. Feeding habits of walleye pollock in this area are the least studied, with diets described only for some size groups (Volkov et al. 1990, Shuntov et al. 1993).

Among commercially important benthophage species the most abundant are rock sole, *Pleuronectes bilineatus*, and rock greenling, *Hexagrammos lagocephalus*. Bering skate, *Bathyraja interrupta*, and Okhotsk skate, *Bathyraja violacea*, are prospective targets for a fishery. The feeding habits of rock sole and benthophagous skates in this area have not been investigated, and the feeding habits of rock greenling were studied only in coastal waters (Klyashtorin 1962, Zolotov 1993).

The most abundant and commercially important predatory fishes of the Pacific waters off the northern Kuril Islands and southeastern Kamchatka are Greenland turbot, *Reinhardtius hippoglossoides*; Kamchatka flounder, *Atheresthes evermanni*; Pacific halibut, *Hippoglossus stenolepis*; sablefish, *Anoplopoma fimbria*; Pacific cod, *Gadus macrocephalus*; Aleutian skate, *Bathyraja aleutica*; whiteblotched skate, *Bathyraja maculata*; Matsubara skate, *Bathyraja matsubarai*; Alaska skate, *Bathyraja parmifera*; shortraker rockfish, *Sebastes borealis*; shortspine thornyhead, *Sebastolobus alascanus*; giant grenadier, *Albatrossia pectoralis*; and large sculpins of the genus *Myoxocephalus*. According to their feeding behavior broadbanded thornyhead, *Sebastolobus macrochir*, is similar to the predatory fishes. The biology of most of the above species in these waters is little studied, and least of all their feeding habits. Food habits of Pacific cod, giant grenadier, and large sculpins are well investigated compared to other Pacific predators off the northern Kurils and southeastern Kamchatka (Novikov 1970; Tokranov 1986a,b; Tuponogov and Kurennoi 1986; Tokranov and Vinnikov 1991). The diets of halibuts and shortraker rockfish were briefly described by Novikov (1974). Trophic relations among fishes in these waters have not been analyzed. The main purpose of this paper is to compare their diets and consider some aspects of their trophic relationships.

Material and Methods

The results in this paper are based on data obtained by shipboard scans of stomach contents of commercially important fishes from the Pacific waters off the northern Kuril Islands and southeastern Kamchatka aboard Japanese trawlers during 1992-1996. Quantitative stomach content analysis, which measures stomach content weight and the weight of prey items, was not performed. For diet descriptions, the normalized frequencies of occurrence of prey items were used. Dietary overlap indices, which are commonly used to compare fish diets (Livingston et al. 1986; Yang and Livingston 1986; Brodeur and Livingston 1988; Lang et al. 1991; Yang 1993, 1996; Lang and Livingston 1996), were calculated according to the method of Schoener (1970). The same method (Schoener's dietary overlap index) was used to estimate interspecific bathymetric overlap based on the relative abundance of fish. The relative abundance of one species at each depth range was calculated as the percentage of total number of all species caught at this depth range during 1-hour trawls divided by the total number of fish of species caught during 1-hour trawls (Shuntov 1965, Borets 1985) proportionally to the number of hauls within each depth range (Orlov 1998). The stomach fullness index was determined on a 5-step scale (0 = empty stomach, 4 = full stomach). Maps of spatial distribution were drawn using data from 1,083 trawls made during the summer-autumn of 1992-1996, when feeding intensity of most of these species was considered to be highest.

Results and Discussion

Planktophages

All three species under consideration are typically planktivorous fishes. Their diets were rather wide (from coelenterates to fish) and consisted mainly of planktonic organisms. Benthic prey items did not play an essential role in their diets (Table 1). The majority of their food consisted of copepods and euphausiids.

The frequency of occurrence (FO) of euphausiids in walleye pollock stomachs was highest (61.5%). Second highest was copepods (13.6%). The FO of fishes was 8.1% which were represented mostly by mesopelagic species and by juvenile walleye pollock, rockfishes (*Sebastes* spp.), snailfishes (*Careproctus* spp.), and darkfin sculpin (*Malacocottus zonurus*), and also by adult smelt (*Mallotus villosus*), which is characteristic of other continental slope regions (Lang and Livingston 1996). Typically benthic prey items did not play an essential role in the diet of pollock and were consumed mainly by larger fishes observed in other areas (Shuntov et al. 1993; Yang 1993, 1996; Lang and Livingston 1996).

The most important prey in the diet of Atka mackerel were euphausiids and copepods (36.7% and 31.7% FO, respectively). According to frequency of occurrence the chaetognaths were the third highest (10.8%). Benthic organisms were consumed mainly during spawning (July-September), when Atka mackerel inhabited the near-bottom layers. Our data on Atka mackerel diets are similar to data obtained off the Aleutians (Yang 1996) and differed from that of coastal waters off the northern Kurils and eastern Kamchatka (Zolotov and Medveditsyna 1978, Zolotov and Tokranov 1991).

The ephausiids were the basic prey item of Pacific ocean perch (FO 25.2%). Other significant portions of their diet were copepods (15.1%) and amphipods (12.6%). Fishes (mostly mesopelagic myctophids and bathylagids) also were significant in their diet. The data obtained on Pacific ocean perch diets were also similar to data from the Aleutians (Yang 1996).

Dietary overlaps of all three species were moderate. Calculated values of Schoener's dietary overlap index were 53.2% for walleye pollock and Atka mackerel, 60.0% for walleye pollock and Pacific ocean perch, and 53.0% for Atka mackerel and Pacific ocean perch (Table 2).

Feeding competition between species is defined not only by degree of interspecific dietary overlap, but also by features of intra-annual feeding intensity and the vertical and spatial distribution patterns of these species (Shorygin 1946, Ivlev 1977).

Stomach fullness analyses showed that pollock fed most intensively from May to November (Fig. 1); i.e., in postspawning, feeding, and prespawning periods. Minimum values of stomach fullness occurred from December to April (spawning time).

Our data on feeding intensity of Atka mackerel are limited to May through December. In coastal waters, a decrease of Atka mackerel feeding intensity was noted during July-September that corresponded with their spawning there (Zolotov 1984). Outside of the spawning areas, where Atka mackerel are represented mainly by immature and prespawning individuals, and by females spawning and migrating between coastal and offshore waters, the most intensive feeding occurred from May to July (Fig. 1) followed by a sharp decrease in August that probably related to the peak of spawning in coastal waters. In the winter, according to Zolotov and Medveditsyna (1978), the feeding intensity of Atka mackerel is very low.

The highest feeding intensity of Pacific ocean perch occurred in June-July and in November (Fig. 1). The summer peak of feeding intensity is related to the end of spawning that is characteristic of other regions (Novikov 1974). Although Novikov (1974) noted that Pacific ocean perch did not feed during November-March, we found a rather high feeding intensity in November that may correspond with the end of copulation, which occurred in various regions from September to October.

Thus, all three species fed most intensively at almost the same time, which would increase the potential for feeding competition.

Walleye pollock occurred in bottom trawls in the depth range of 117-630 m, while the maximum values of their biomass (92.6%) were noted at depths of 250-400 m. The vertical range of occurrence of Atka mackerel in bottom trawls was 85-504 m, while only 6.4% of their biomass was concentrated in the upper bathyal (200-500 m). Pacific ocean perch were caught at a depth of 85-630 m, and 73.5% of their biomass occurred at depths of 200-350 m. Analysis of relative abundance with depth showed that bathymetric distribution of pollock and Pacific ocean perch was the most similar (Table 3). At the same time, the different bathymetric patterns of Atka mackerel and Pacific ocean perch almost eliminate feeding competition between these species.

Comparison of spatial distribution maps (Fig. 2) showed that there were some similarities in the locations of most dense concentrations between walleye pollock and Atka mackerel, and Atka mackerel and Pacific ocean perch. The similarity levels for both pairs, however, were moderate.

Summarizing the factors above, we conclude that in spite of similar diets and the same traits of feeding intensity the different patterns of vertical and spatial distribution make feeding competition between walleye pollock, Atka mackerel, and Pacific ocean perch insignificant.

Benthophages

All four species are typically benthophagous fishes, whose diets consist mainly of worms and crustaceans (Table 1). It is accepted that rock greenling are benthophage-polyphage fish (Klyashtorin 1962). In coastal waters they feed mainly on the eggs of Atka mackerel (Zolotov 1993). Our study showed that in offshore waters the most important prey of rock greenling

Table 1. Prey items (expressed as normalized frequency of occurrence, %) of commercially important fishes collected in the Pacific waters off the northern Kuril Islands and southeastern Kamchatka in 1992–1996. (+ = <0.1%)

	Planktophages			Benthophages				Predators												
	Walleye pollock	Atka mackerel	Pacific ocean perch	Rock greenling	Rock sole	Bering skate	Okhotsk skate	Greenland turbot	Kamchatka flounder	Pacific halibut	Sablefish	Pacific cod	Aleutian skate	Whiteblotched skate	Matsubara skate	Alaska skate	Shortraker rockfish	Shortspine thornyhead	Broadbanded thornyhead
Algae	–	–	–	0.4	0.1	–	–	–	–	–	–	–	–	–	–	–	–	–	0.5
Porifera	–	–	–	–	–	–	–	–	0.8	–	–	–	–	–	–	–	–	–	–
Hydrozoa	–	0.1	–	0.4	–	–	–	–	–	0.2	–	–	–	–	–	–	–	–	–
Scyphozoa	0.4	–	–	–	–	–	–	–	–	–	–	0.1	–	–	–	–	–	–	–
Anthozoa	–	–	–	1.2	–	–	–	–	–	–	–	–	–	–	–	–	–	–	–
Actiniaria	–	–	–	–	–	–	–	–	0.3	–	0.3	–	–	–	–	–	–	–	–
Ctenophora	–	1.9	1.9	–	–	–	–	–	–	–	0.3	0.1	–	–	–	–	–	–	–
Nematoda	–	–	–	–	9.0	2.3	–	–	–	–	–	–	0.5	0.4	–	–	–	–	–
Polychaeta	–	+	–	6.8	11.0	1.7	0.8	0.1	0.5	0.2	–	1.1	0.5	2.9	1.0	–	–	–	0.9
Oligochaeta	–	+	–	1.6	8.8	15.1	8.3	–	–	0.2	–	1.0	2.2	1.6	1.3	–	–	–	–
Echiurida	–	–	–	–	1.4	0.3	–	–	–	0.9	–	0.9	–	–	–	–	–	–	–
Sipunculida	–	–	–	–	0.2	–	–												
Other worms	–	–	–	–	–	–	–	–	–	–	–	0.3	–	–	–	–	–	–	–
Copepoda	13.6	36.7	15.1	–	–	–	–	–	–	–	–	–	–	–	–	–	–	–	–
Mysidacea	0.3	0.1	1.9	–	–	–	–	–	0.3	–	–	–	–	0.5	–	–	–	–	2.3
Euphausiacea	61.5	31.7	25.2	0.4	0.1	–	–	0.8	1.0	–	0.6	0.3	–	–	–	–	0.7	1.2	0.9
Isopoda	–	–	–	2.8	1.1	1.7	1.5	–	–	1.9	–	0.3	–	0.3	–	–	–	–	–
Amphipoda	1.9	0.6	12.6	22.3	55.1	65.7	43.6	1.2	0.5	1.67	22.2	8.5	6.7	23.8	0.7	16.0	21.3	12.2	17.7
Cirripedia	–	1.2	–	–	–	–	–	–	–	–	–	–	–	–	–	–	–	–	–

Table 1. (Continued.)

	Walleye pollock	Atka mackerel	Pacific ocean perch	Rock greenling	Rock sole	Bering skate	Okhotsk skate	Greenland turbot	Kamchatka flounder	Pacific halibut	Sablefish	Pacific cod	Aleutian skate	Whiteblotched skate	Matsubara skate	Alaska skate	Shortraker rockfish	Shortspine thornyhead	Broadbanded thornyhead
	Planktophages			Benthophages				Predators											
Decapoda																			
Pandalidae	2.3	0.1	6.3	0.4	0.2	4.7	5.3	0.3	14.3	2.8	2.5	7.4	23.4	10.2	5.7	–	11.3	28.1	11.2
Crangonidae	2.4	–	1.9	–	0.8	2.0	12.0	–	1.6	0.7	–	8.6	0.5	0.8	0.3	4.0	10.0	7.3	7.9
Paguridae	–	–	–	3.6	–	0.3	2.3	–	0.3	7.0	–	2.2	0.5	3.3	3.7	4.0	–	–	–
Lithodidae																			
Lithodes aequispina	–	–	–	–	–	–	–	0.1	–	2.4	3.8	11.5	–	0.1	2.3	–	–	3.7	–
Majidae																			
Chionoecetes opilio	–	–	–	–	–	–	–	–	–	7.2	3.8	13.2	9.0	4.5	31.8	8.0	–	1.2	–
Chionoecetes bairdi	–	–	–	–	–	–	–	–	–	–	0.3	–	–	–	–	–	–	1.2	–
Chionoecetes sp.	–	–	–	–	–	–	18.8	–	–	3.3	–	–	–	–	–	–	–	–	0.5
Hyas coarctatus	–	–	–	–	–	–	–	–	–	0.2	–	–	–	–	–	–	–	–	–
Other crustaceans	0.1	–	–	0.8	1.0	–	–	0.4	1.0	–	–	0.5	–	–	–	–	4.0	2.4	0.9
Gastropoda																			
Buccinidae	–	–	–	–	–	–	–	–	–	–	–	1.1	–	–	–	–	–	–	–
Buccinidae (eggs)	–	–	–	–	–	–	–	–	–	–	–	0.1	–	–	–	–	–	–	–
Other snails	–	+	–	4.0	–	–	–	0.1	–	0.9	–	0.1	–	–	0.3	–	–	–	0.5

Table 1. (Continued.) Prey items (expressed as normalized frequency of occurrence, %) of commercially important fishes collected in the Pacific waters off the northern Kuril Islands and southeastern Kamchatka in 1992–1996. (+ = <0.1%)

	Walleye pollock	Atka mackerel	Pacific ocean perch	Rock greenling	Rock sole	Bering skate	Okhotsk skate	Greenland turbot	Kamchatka flounder	Pacific halibut	Sablefish	Pacific cod	Aleutian skate	Whiteblotched skate	Matsubara skate	Alaska skate	Shortraker rockfish	Shortspine thornyhead	Broadbanded thornyhead
	Planktophages			Benthophages				Predators											
Bivalvia																			
Chlamys sp.	–	–	–	–	–	–	–	–	–	–	–	0.1	–	–	–	–	–	–	–
Spisula sakhalinensis	–	–	–	–	–	–	–	–	–	–	–	0.1	–	–	–	–	–	–	–
Other bivalves	–	–	–	1.6	2.5	–	–	–	–	–	–	0.5	–	0.1	–	–	–	–	–
Nudibranchia	–	+	–	–	–	–	–	–	0.3	–	–	–	–	–	–	–	–	–	–
Teuthida																			
Belonella borealis	–	–	–	–	–	–	–	–	–	–	–	0.2	–	0.5	–	–	–	–	–
Galiteuthis phyllura	–	–	–	–	–	–	–	–	–	–	–	–	–	0.1	–	–	–	–	–
Gonatopsis borealis	–	–	–	–	–	–	–	–	–	–	–	0.1	–	0.4	–	–	–	–	–
Berryteuthis magister	–	–	–	–	–	–	–	46.1	15.5	13.9	15.9	1.1	13.9	22.8	16.1	8.0	0.7	1.2	1.4
Other squids	4.0	2.7	5.7	0.4	0.7	3.2	3.8	0.1	–	0.2	–	8.5	0.5	1.3	0.7	4.0	20.0	1.2	29.8
Octopoda	–	–	–	18.6	–	–	–	1.3	0.8	14.8	7.9	3.1	4.9	3.2	2.7	12.0	–	–	–
Octopoda (eggs)	–	0.2	–	4.4	–	–	–	0.1	–	–	–	–	–	–	–	–	–	–	–
Brachiopoda	–	–	–	0.8	–	–	–	–	–	–	–	–	–	–	–	–	–	–	–
Other mollusks	–	–	–	2.4	–	–	–	–	0.5	0.4	–	0.1	–	–	–	–	–	–	–
Chaetognatha	1.5	10.8	–	–	–	–	–	–	–	–	–	–	–	–	–	–	–	–	–

Table 1. (Continued.)

	Planktophages			Benthophages				Predators												
	Walleye pollock	Atka mackerel	Pacific ocean perch	Rock greenling	Rock sole	Bering skate	Okhotsk skate	Greenland turbot	Kamchatka flounder	Pacific halibut	Sablefish	Pacific cod	Aleutian skate	Whiteblotched skate	Matsubara skate	Alaska skate	Shortraker rockfish	Shortspine thornyhead	Broadbanded thornyhead
Ophiuroidea	–	+	–	0.4	0.2	0.6	–	–	–	–	–	0.2	–	–	–	–	–	1.2	0.5
Holoturoidea	–	–	–	0.4	–	–	–	–	–	–	–	0.2	–	–	–	–	–	–	–
Asteroidea	–	–	–	–	0.2	–	–	–	–	–	–	–	–	–	–	–	–	–	–
Ehinoidea	–	–	–	0.4	–	–	–	–	–	–	–	–	–	–	–	–	–	–	–
Rajidae																			
Bathyraja interrupta	–	–	–	–	–	–	–	–	–	–	–	–	0.5	–	–	–	–	–	–
Salmonidae																			
Oncorhynchus sp.	–	–	–	–	–	–	–	–	–	0.4	1.9	–	0.5	–	–	–	–	–	–
Osmeridae																			
Mallotus villosus	0.1	–	–	–	–	–	–	–	–	0.2	–	–	0.1	–	–	–	–	–	–
Bathylagidae																			
Leuroglossus schmidti	1.1	0.4	2.5	–	–	–	–	2.4	0.3	–	–	–	–	0.9	–	4.0	–	–	1.4
Bathylagus pacificus	–	–	–	–	–	–	–	0.1	–	–	–	–	–	–	–	–	–	–	–
Lipolagus ochotensis	–	0.2	–	–	–	–	–	–	–	–	–	–	–	–	–	–	–	–	–
Bathylagidae gen. sp.	–	–	–	–	–	–	–	0.8	–	–	–	–	–	–	–	–	–	–	–
Gonostomatidae																			
Gonostoma gracile	–	0.1	–	–	–	–	–	–	–	–	–	–	–	–	–	–	–	–	–

Table 1. (Continued.) Prey items (expressed as normalized frequency of occurrence, %) of commercially important fishes collected in the Pacific waters off the northern Kuril Islands and southeastern Kamchatka in 1992–1996. (+ = <0.1%)

	Planktophages			Benthophages				Predators											
	Walleye pollock	Atka mackerel	Pacific ocean perch	Rock greenling	Rock sole	Bering skate	Okhotsk skate	Greenland turbot	Kamchatka flounder	Pacific halibut	Sablefish	Pacific cod	Aleutian skate	Whiteblotched skate	Matsubara skate	Alaska skate	Shortraker rockfish	Shortspine thornyhead	Broadbanded thornyhead
Chauliodontidae																			
Chauliodus macouni	–	–	–	–	–	–	–	–	–	–	–	–	–	–	0.3	–	–	–	–
Paralepididae																			
Arctozenus rissoi	–	–	–	–	–	–	–	–	–	–	–	–	–	0.1	–	–	–	–	–
Myctophidae																			
Protomyctophum thompsoni	0.1	2.7	1.9	–	–	–	–	–	–	–	–	–	–	0.3	0.3	–	–	–	–
Stenobrachius leucopsarus	–	0.5	–	–	–	–	–	0.1	–	–	–	–	–	0.8	1.7	–	–	–	–
Stenobrachius nannochir	0.1	0.1	–	–	–	–	–	–	0.3	–	–	–	–	0.1	0.3	–	–	–	–
Lampanyctus jordani	–	–	–	–	–	–	–	0.1	–	–	–	–	–	–	0.3	–	–	–	–
Myctophidae gen. sp.	–	1.0	–	–	–	–	–	2.4	4.4	0.7	–	–	–	–	0.7	–	–	–	–
Notosudidae																			
Scopelosaurus harryi	–	–	–	–	–	–	–	–	–	–	–	–	0.5	–	–	–	–	–	–

Table 1. (Continued.)

	Walleye pollock	Atka mackerel	Pacific ocean perch	Rock greenling	Rock sole	Bering skate	Okhotsk skate	Greenland turbot	Kamchatka flounder	Pacific halibut	Sablefish	Pacific cod	Aleutian skate	Whiteblotched skate	Matsubara skate	Alaska skate	Shortraker rockfish	Shortspine thornyhead	Broadbanded thornyhead
	Planktophages			Benthophages				Predators											
Gadidae																			
Theragra chalcogramma	1.1	–	–	–	–	–	–	0.5	1.8	7.0	–	2.0	3.1	1.1	–	4.0	–	–	–
Gadus macrocephalus	–	–	–	–	–	–	–	–	–	0.6	–	–	–	–	–	–	–	–	–
Macrouridae																			
Albatrossia pectoralis	–	–	–	–	–	–	–	–	0.5	–	3.8	–	–	–	–	–	–	–	–
Coryphaenoides cinereus	–	–	–	–	–	–	–	0.4	0.8	–	1.3	–	–	0.1	–	–	–	–	–
Macrouridae gen. sp.	–	–	–	–	–	–	–	0.9	0.5	–	–	–	–	–	–	–	–	–	–
Gasterosteidae																			
Gasterosteus aculeatus	–	–	–	–	–	–	–	–	–	–	–	–	0.5	–	0.3	–	–	–	–
Stichaeidae																			
Bryozoichthys lyzimus	–	–	–	–	–	–	–	–	–	0.6	–	–	–	–	–	–	–	–	–
Zoarcidae																			
Lycodes sp.	–	–	–	–	–	–	–	–	–	–	–	0.1	0.5	–	–	–	–	–	–
Lycenchelys sp.	–	–	–	–	–	–	–	–	0.8	–	–	0.1	0.5	–	–	–	–	–	–

Table 1. **(Continued.) Prey items (expressed as normalized frequency of occurrence, %) of commercially important fishes collected in the Pacific waters off the northern Kuril Islands and southeastern Kamchatka in 1992–1996. (+ = <0.1%)**

	Planktophages			Benthophages				Predators												
	Walleye pollock	Atka mackerel	Pacific ocean perch	Rock greenling	Rock sole	Bering skate	Okhotsk skate	Greenland turbot	Kamchatka flounder	Pacific halibut	Sablefish	Pacific cod	Aleutian skate	Whiteblotched skate	Matsubara skate	Alaska skate	Shortraker rockfish	Shortspine thornyhead	Broadbanded thornyhead
Zaproridae																			
Zaprora silena	–	–	–	–	–	–	–	–	–	–	–	–	–	0.1	–	–	–	–	–
Scorpaenidae																			
Sebastolobus macrochir	–	–	–	–	–	–	–	0.1	–	–	3.5	–	–	–	–	–	–	–	–
Sebastes borealis	–	–	–	–	–	–	–	–	–	–	–	–	–	0.3	–	–	–	–	–
Sebastes alutus	–	–	–	–	–	–	–	–	–	–	–	0.1	–	–	–	–	–	–	–
Sebastes sp.	0.1	–	–	–	–	–	–	0.1	0.3	–	–	–	–	–	–	–	–	–	–
Hexagrammidae																			
Pleurogrammus monopterygius	–	–	–	–	–	–	–	–	0.5	0.6	0.3	4.2	4.5	3.8	0.7	12.0	–	–	–
Cottidae																			
Triglops forficatus	–	–	–	–	–	–	–	–	–	0.2	–	0.2	0.5	–	–	–	1.3	–	–
Triglops scepticus	–	–	–	–	–	–	–	–	–	–	–	0.3	–	1.2	–	8.0	–	–	–
Triglops sp.	–	–	–	–	–	–	–	0.1	2.1	1.9	–	–	–	–	–	–	–	–	–
Icelus canaliculatus	–	–	–	–	–	–	–	–	–	–	0.3	–	0.9	–	0.3	–	1.3	–	0.9
Icelus sp.	–	–	–	–	–	–	–	0.1	1.0	–	–	–	1.8	–	–	–	–	–	–
Hemilepidotus gilberti	–	–	–	–	–	–	–	–	–	–	–	0.1	1.3	0.1	–	4.0	–	–	–

Table 1. (Continued.)

	Walleye pollock	Atka mackerel	Pacific ocean perch	Rock greenling	Rock sole	Bering skate	Okhotsk skate	Greenland turbot	Kamchatka flounder	Pacific halibut	Sablefish	Pacific cod	Aleutian skate	Whiteblotched skate	Matsubara skate	Alaska skate	Shortraker rockfish	Shortspine thornyhead	Broadbanded thornyhead
	Planktophages			Benthophages				Predators											
Hemilepidotus hemilepidotus	–	–	–	–	–	–	–	–	–	–	–	0.1	–	–	–	–	–	–	–
Artediellichthys nigripinnis	–	–	–	–	–	–	–	–	0.5	–	–	–	–	–	–	–	–	–	–
Psychrolutidae																			
Malacocottus zonurus	0.1	–	–	–	–	–	–	0.3	2.1	0.6	1.9	0.2	4.9	0.4	2.7	–	2.0	–	0.9
Agonidae																			
Aspidophoroides bartoni	–	–	–	–	–	–	–	–	–	0.2	–	0.3	–	–	–	–	–	–	–
Bathyagonus nigripinnis	–	–	–	–	–	–	–	–	–	–	0.3	–	1.3	–	–	–	–	–	–
Sarritor frenatus	–	–	–	–	–	–	–	–	–	0.6	–	0.5	0.5	–	–	–	–	–	–
Sarritor leptorhynchus	–	–	–	–	–	–	–	–	–	–	–	0.1	–	–	–	–	–	–	–
Agonidae gen. sp.	–	–	–	–	–	–	–	0.4	–	–	–	–	1.3	–	–	–	–	–	–
Cyclopteridae																			
Aptocyclus ventricosus	–	–	–	–	–	–	–	–	–	–	0.3	–	–	–	–	–	–	–	–
Liparidae																			
Allocaraproctus jordani	–	–	–	–	–	–	–	–	–	0.2	–	–	–	–	–	4.0	–	–	–
Careproctus cypselurus	–	–	–	–	–	–	–	0.1	–	–	–	–	–	–	–	–	–	–	–

Table 1. (Continued.) Prey items (expressed as normalized frequency of occurrence, %) of commercially important fishes collected in the Pacific waters off the northern Kuril Islands and southeastern Kamchatka in 1992–1996. (+ = <0.1%)

	Planktophages			Benthophages				Predators											
	Walleye pollock	Atka mackerel	Pacific ocean perch	Rock greenling	Rock sole	Bering skate	Okhotsk skate	Greenland turbot	Kamchatka flounder	Pacific halibut	Sablefish	Pacific cod	Aleutian skate	Whiteblotched skate	Matsubara skate	Alaska skate	Shortraker rockfish	Shortspine thornyhead	Broadbanded thornyhead
Careproctus rastrinus	–	–	–	–	–	–	–	–	–	–	–	0.1	–	–	–	–	–	–	–
Careproctus zachirus	–	–	–	–	–	–	–	–	–	–	–	–	–	–	0.3	–	–	–	–
Careproctus sp.	0.1	–	–	–	–	–	–	–	0.3	–	1.3	–	–	–	–	–	–	–	–
Elassodiscus tremebundus	–	–	–	–	–	–	–	0.4	0.8	–	1.9	–	0.5	–	1.7	–	–	–	–
Elassodiscus obscurus	–	–	–	–	–	–	–	0.1	–	–	–	–	–	–	–	–	–	–	–
Elassodiscus sp.	–	–	–	–	–	–	–	0.5	0.5	0.2	–	–	–	–	–	–	–	1.2	–
Paraliparis grandis	–	–	–	–	–	–	–	0.3	–	–	0.6	–	–	–	–	–	–	–	–
Paraliparis sp.	–	–	–	–	–	–	–	0.1	–	–	–	–	–	–	–	–	–	–	–
Liparidae gen. sp.	–	–	–	–	–	–	–	3.3	2.9	0.7	2.2	0.1	1.5	–	3.0	–	–	3.7	–
Pleuronectidae																			
Hippoglossoides elassodon	–	–	–	–	–	–	–	–	–	–	–	0.1	–	–	–	–	–	–	–
Pleuronectes bilineatus	–	–	–	–	–	–	–	–	–	0.4	–	0.3	–	–	–	–	–	–	–

Table 1. (Continued.)

	Walleye pollock	Atka mackerel	Pacific ocean perch	Rock greenling	Rock sole	Bering skate	Okhotsk skate	Greenland turbot	Kamchatka flounder	Pacific halibut	Sablefish	Pacific cod	Aleutian skate	Whiteblotched skate	Matsubara skate	Alaska skate	Shortraker rockfish	Shortspine thornyhead	Broadbanded thornyhead
	Planktophages			Benthophages				Predators											
Fish unidentified	5.5	1.2	5.7	0.4	0.9	0.3	3.0	18.2	16.1	10.6	12.7	15.9	9.9	9.7	13.0	8.0	6.0	4.9	6.1
Fish eggs	–	3.7	–	3.2	–	0.3	–	–	–	–	1.0	0.1	–	1.1	0.3	–	0.7	–	–
Fish larvae	–	+	–	–	–	–	–	–	–	–	–	–	–	–	–	–	–	–	–
Unidentified organic material	3.8	3.9	19.5	17.4	6.5	1.7	0.8	17.1	26.2	15.7	8.9	4.0	2.7	3.0	7.4	–	20.7	29.3	15.8
Total stomachs with food	1,612	2,008	118	177	791	253	83	411	752	468	277	1,113	141	542	239	18	141	78	207
Empty stomachs	791	1,214	621	9	389	14	8	872	689	132	254	71	26	70	69	1	500	115	489

Table 2. **Diet overlap of commercially important fishes during 1992-1996 in the Pacific waters off the northern Kuril Islands and southeastern Kamchatka calculated using Schoener's Index.**

	Walleye pollock	Atka mackerel	Pacific ocean perch	Rock greenling	Rock sole	Bering skate	Okhotsk skate	Greenland turbot	Kamchatka flounder	Pacific halibut	Sablefish	Pacific cod	Aleutian skate	Whiteblotched skate	Matsubara skate	Alaska skate	Shortraker rockfish	Shortspine thornyhead	Broadbanded thornyhead
	Planktophages			Benthophages				Predators											
Walleye pollock		53.2	60.0																
Atka mackerel	M		53.0																
Pacific Ocean perch	M	M																	
Rock greenling					41.6	31.1	30.8												
Rock sole				M		73.3	57.5												
Bering skate				L	H		65.5												
Okhotsk skate				L	M	M													
Greenland turbot									60.1	46.0	44.7	25.8	32.5	40.6	43.9	21.6	26.7	30.0	28.0
Kamchatka flounder								M		52.2	48.6	35.5	50.8	44.0	51.2	21.7	44.4	54.0	40.3
Pacific halibut								M	M		21.1	41.7	50.5	45.6	54.9	46.5	28.5	31.7	30.2
Sablefish								M	M	L		40.6	49.5	62.2	54.8	44.1	43.0	44.1	38.4
Pacific cod								L	M	M	M		50.0	50.0	49.1	45.4	44.9	40.4	45.7
Aleutian skate								L	M	M	M	M		60.0	53.2	45.9	31.6	42.7	31.1
Whiteblotched skate								M	M	M	M	M	M		51.6	52.3	44.3	34.8	43.2
Matsubara skate								M	M	M	M	M	M	M		32.7	24.1	27.4	24.7
Alaska skate								L	L	M	M	M	M	M	L		30.7	24.7	32.8
Shortraker rockfish								L	M	L	M	M	L	M	L	L		61.4	82.7
Shortspine thornyhead								L	M	L	M	M	M	M	L	L	M		56.1
Broadbanded thornyhead								L	M	L	M	M	L	M	L	L	H	M	

Numbers represent the species dietary overlap value and letters correspond to the degree of overlap (L = low, ≤33%; M = moderate, 34-66%; H = high, ≥67%).

Table 3. Bathymetric overlap of commercially important fishes during 1992-1996 in the Pacific waters off the northern Kuril Islands and southeastern Kamchatka calculated using Schoener's Index.

	Planktophages			Benthophages				Predators											
	Walleye pollock	Atka mackerel	Pacific ocean perch	Rock greenling	Rock sole	Bering skate	Okhotsk skate	Greenland turbot	Kamchatka flounder	Pacific halibut	Sablefish	Pacific cod	Aleutian skate	Whiteblotched skate	Matsubara skate	Alaska skate	Shortraker rockfish	Shortspine thornyhead	Broadbanded thornyhead
Walleye pollock		29.7	63.8																
Atka mackerel	L		6.5																
Pacific ocean perch	M	L																	
Rock greenling					41.2	29.1	29.1												
Rock sole				M		57.6	27.3												
Bering skate				L	M		35.0												
Okhotsk skate				L	L	M													
Greenland turbot									59.8	23.5	57.1	9.0	66.7	53.8	71.3	32.0	56.5	74.2	73.3
Kamchatka flounder								M		50.4	36.0	35.2	84.9	82.9	50.7	47.1	68.4	43.1	61.0
Pacific halibut								L	M		6.5	78.4	46.8	52.8	27.5	56.8	23.6	15.2	22.8
Sablefish								M	M	L		1.0	41.9	26.2	53.9	17.2	39.4	70.7	63.0
Pacific cod								L	M	H	L		30.6	37.8	13.0	60.1	9.1	2.5	8.3
Aleutian skate								M	H	M	M	L		81.5	61.0	43.4	63.3	52.0	69.0
Whiteblotched skate								M	H	M	L	M	H		52.2	49.3	61.8	37.6	52.3
Matsubara skate								H	M	L	M	L	M	M		34.2	53.7	62.8	66.2
Alaska skate								L	M	M	L	M	M	M	M		29.0	25.2	30.7
Shortraker rockfish								M	H	L	M	L	M	M	M	L		44.9	67.3
Shortspine thornyhead								H	M	L	H	L	M	M	M	L	M		63.1
Broadbanded thornyhead								H	M	L	M	L	H	M	M	L	H	M	

Numbers represent the species bathymetric overlap value and letters correspond to the degree of overlap (L = low, ≤33%; M = moderate, 34-66%; H = high, ≥67%).

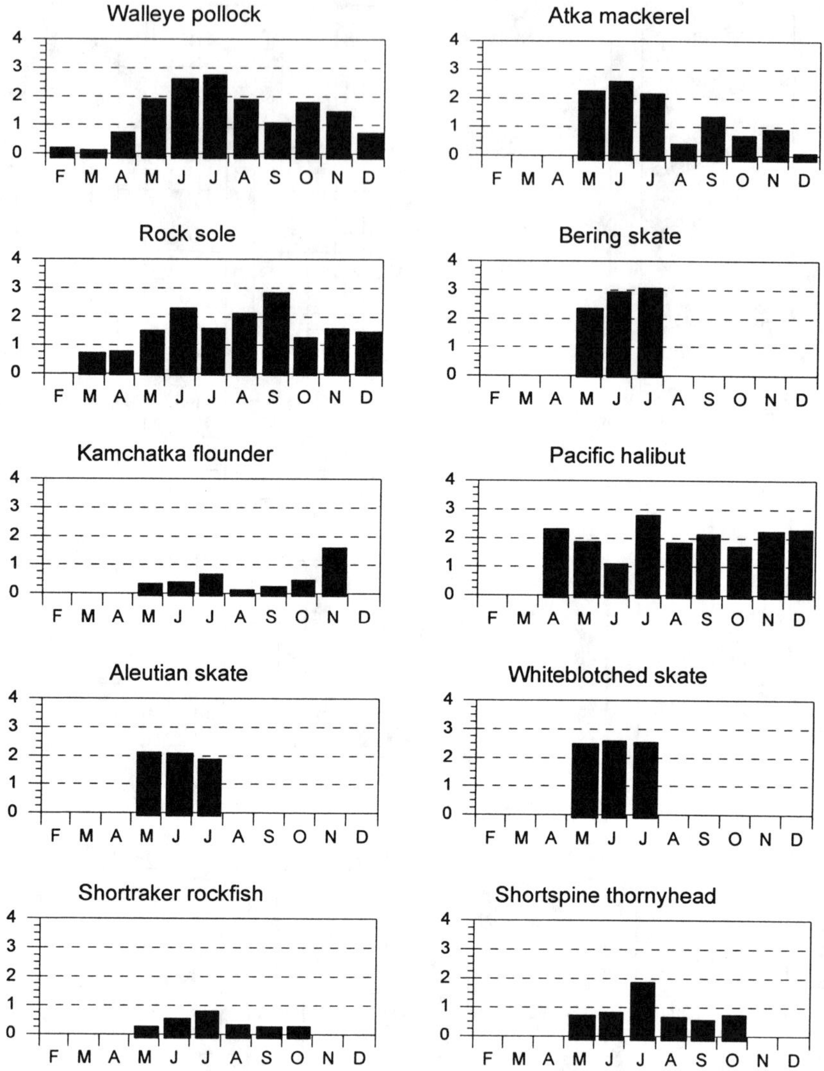

Figure 1. (This and facing page.) Average food consumption by month of commercially important fishes in the Pacific waters off the northern Kuril Islands and southeastern Kamchatka in 1992-1996 (X axis is month of observation, and Y axis is average stomach fullness index).

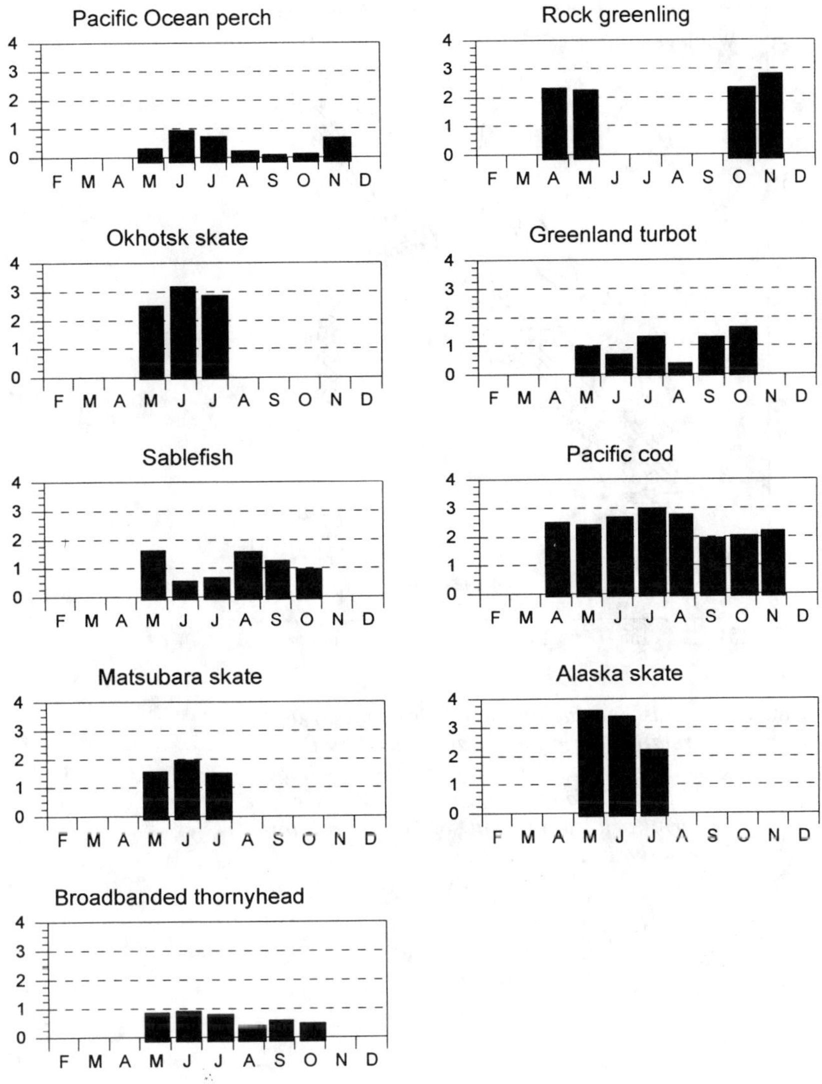

Figure 1. (Continued.)

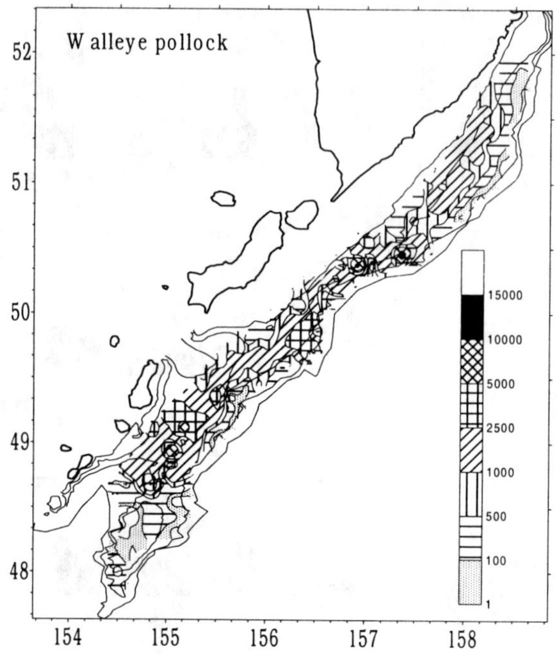

Figure 2. (This and facing page.) Distribution and relative abundance categorized by CPUE (kg per hour trawling) of planktophage fishes in the Pacific waters off the northern Kuril Islands and southeastern Kamchatka in 1992-1996.

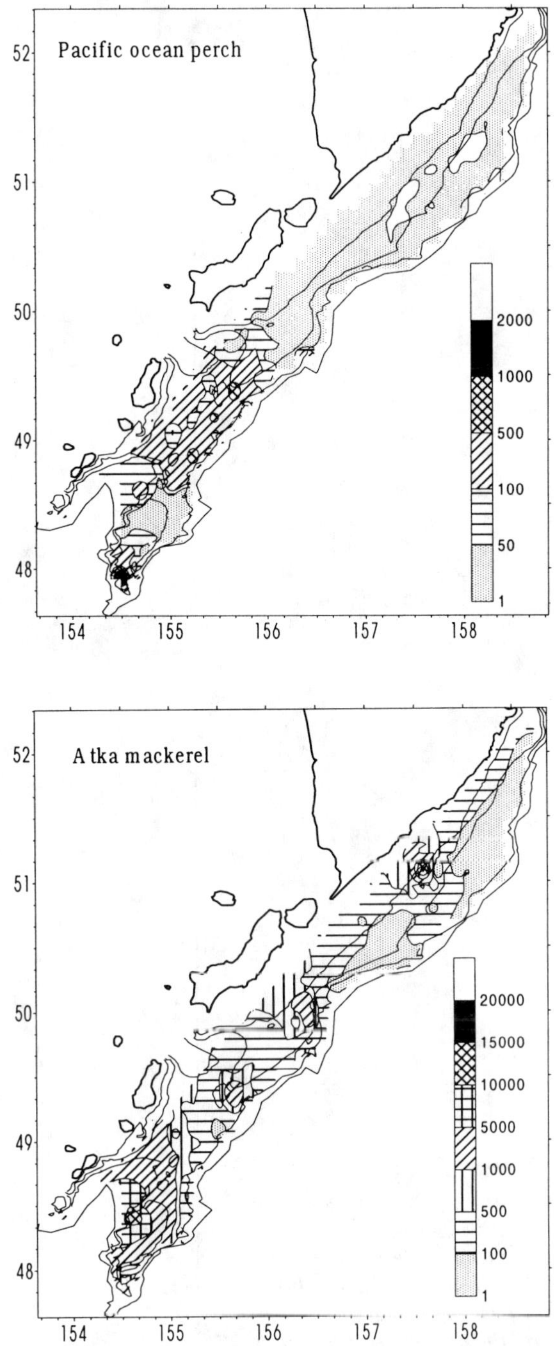

Figure 2. (Continued.)

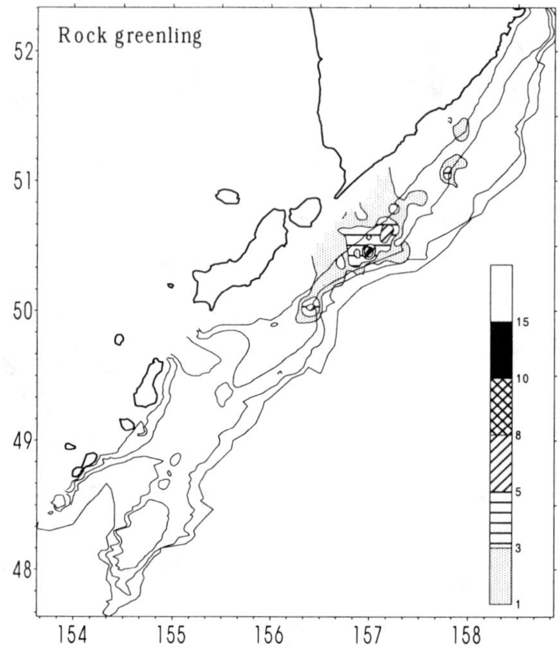

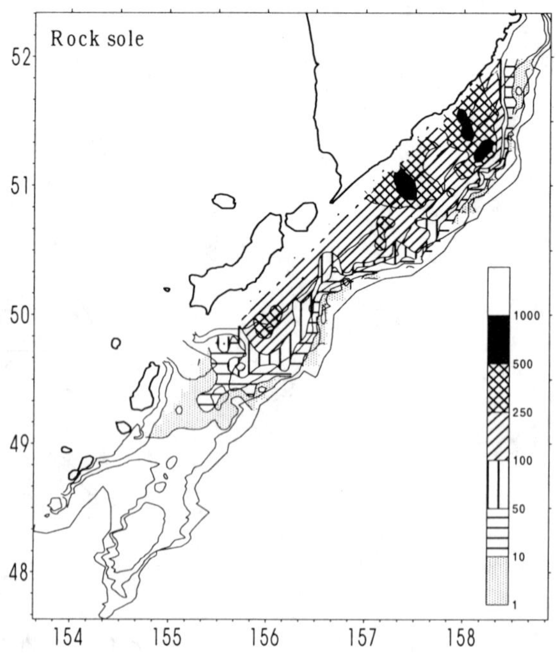

Figure 3. Distribution and relative abundance categorized by CPUE (kg per hour trawling) of benthophage fishes in the Pacific waters off the northern Kuril Islands and southeastern Kamchatka in 1992-1996.

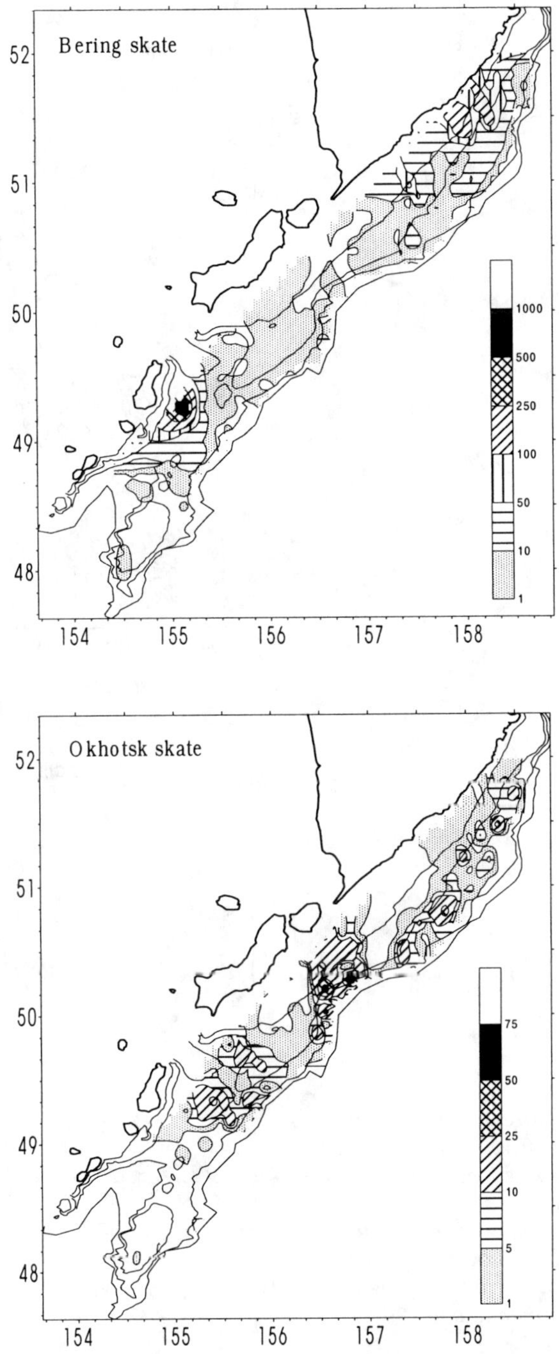

Figure 3. (Continued.)

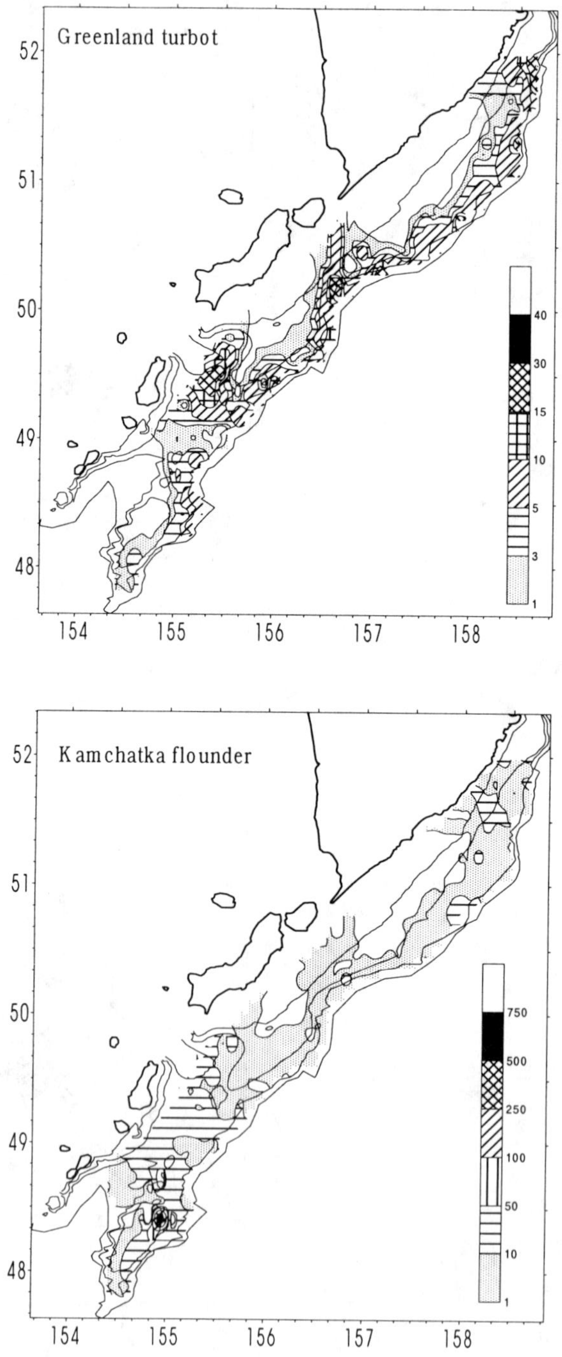

Figure 4. Distribution and relative abundance categorized by CPUE (kg per hour trawling) of predatory fishes in the Pacific waters off the northern Kuril Islands and southeastern Kamchatka in 1992-1996.

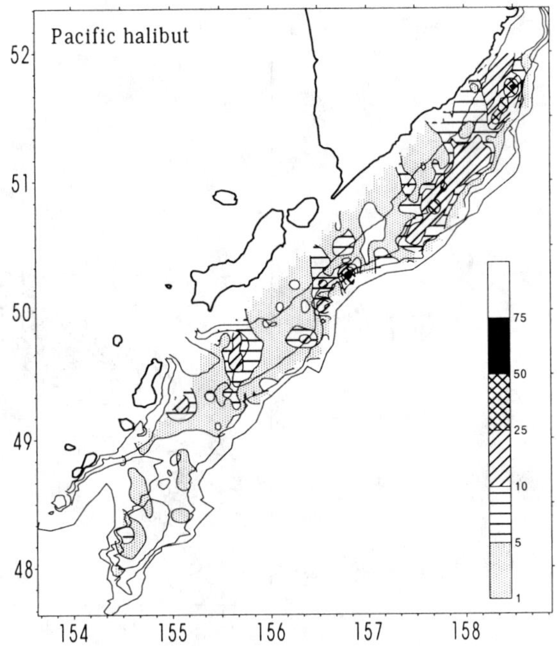

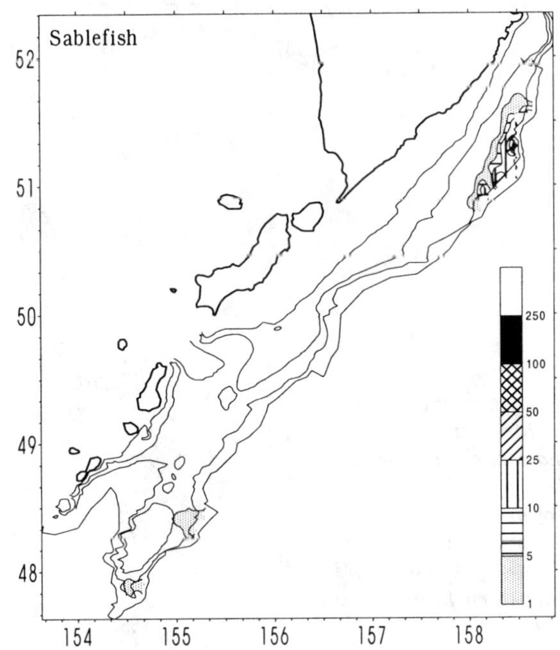

Figure 4. (Continued.)

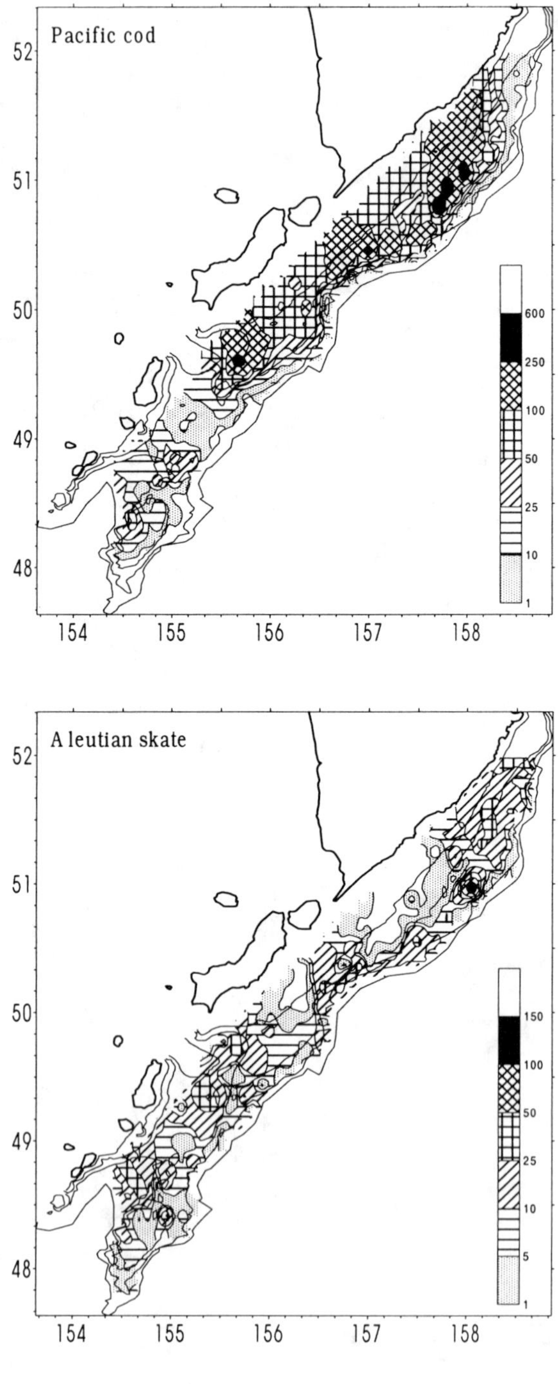

Figure 4. (Continued.) Distribution and relative abundance categorized by CPUE (kg per hour trawling) of predatory fishes in the Pacific waters off the northern Kuril Islands and southeastern Kamchatka in 1992-1996.

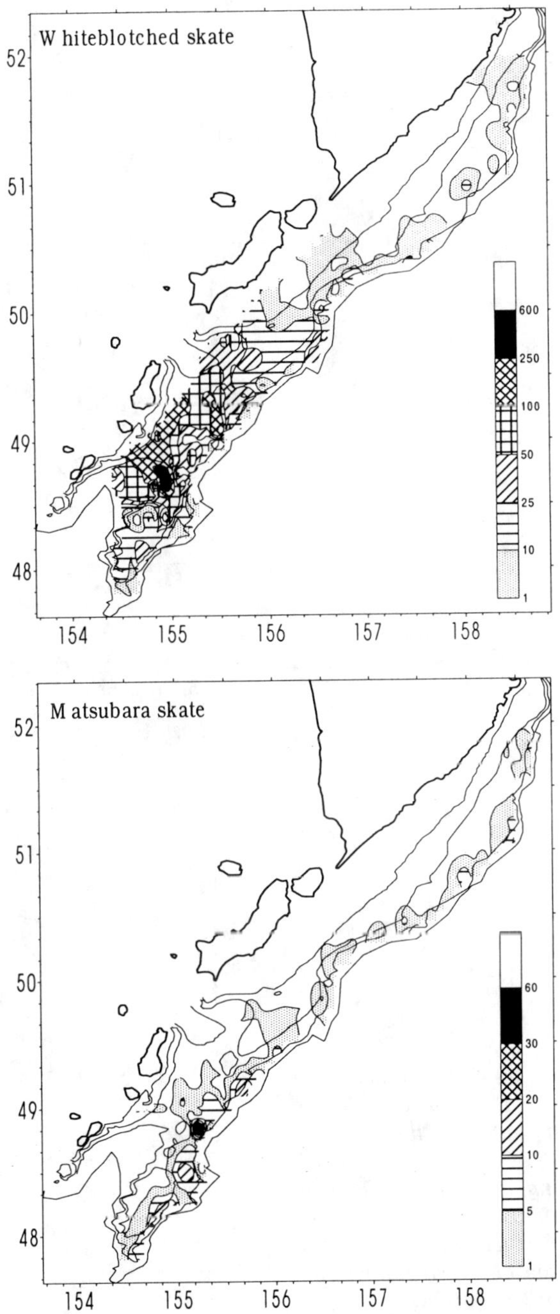

Figure 4. (Continued.)

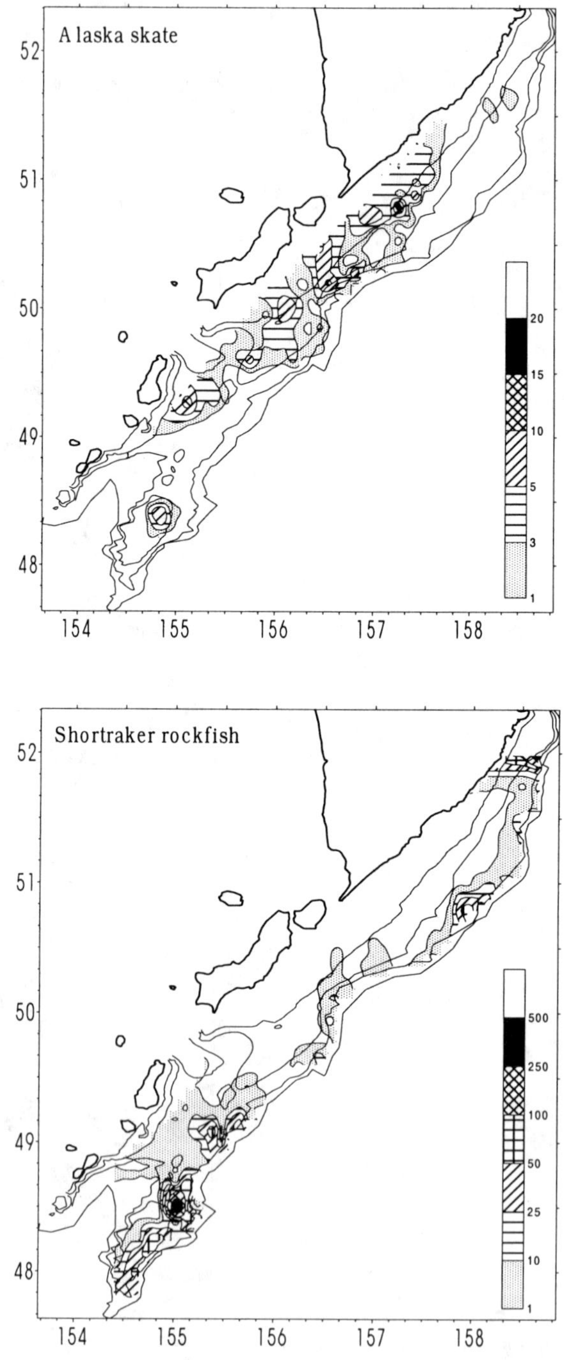

Figure 4. (Continued.) Distribution and relative abundance categorized by CPUE (kg per hour trawling) of predatory fishes in the Pacific waters off the northern Kuril Islands and southeastern Kamchatka in 1992-1996.

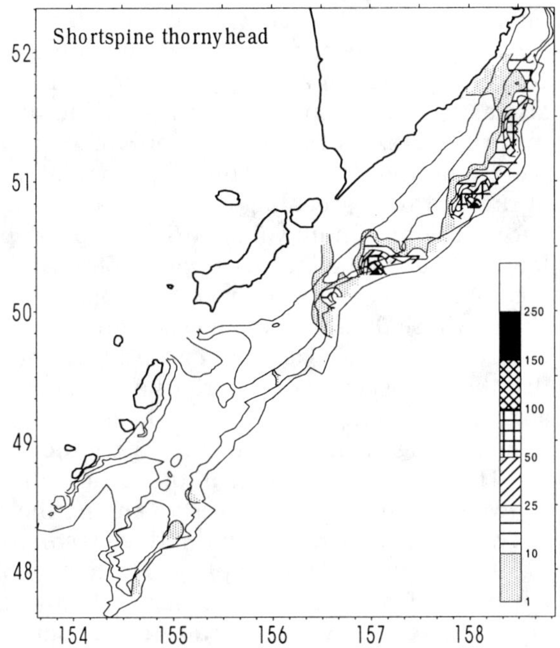

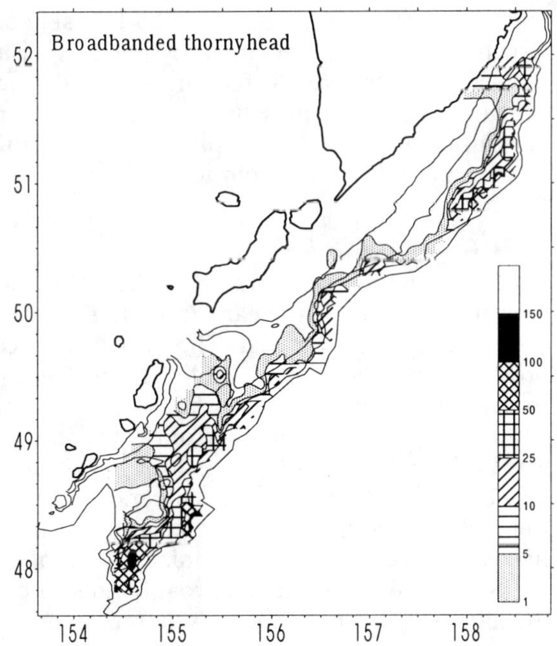

Figure 4. (Continued.)

were mollusks (mainly octopuses). In their stomachs we found eggs, cupules, and fragments of pulpi of North Pacific giant octopus, *Octopus dofleini.* Some fishes had wounds and traces of octopus bites. Female octopus protect the eggs for a long time (Roper et al. 1984). Some authors consider predation among fishes not only as consumption of fish, but also as feeding on mobile animals, including actively finding and pursuing prey (Berezina 1973). Therefore we examined the similarity in feeding pattern between rock greenling and predators. However, the comparison of diets of rock greenling and predatory fishes showed essential differences. The maximum value of dietary overlap calculated for rock greenling and Pacific halibut was only 43.8%. All benthophages had a low or medium degree of dietary overlap (Table 2), except for rock sole and Bering skate, whose dietary overlap was high (73.3%).

Comparison of diagrams of intra-annual feeding intensity (Fig. 1) showed that the highest overlap of periods of maximum stomach fullness was characteristic of Bering and Okhotsk skates. The similarity of periods of maximum feeding intensity among other benthophages was insignificant.

The analysis of patterns of vertical distribution of benthophages showed that bathymetric overlap among them was low and moderate in degree (Table 3), which would reduce feeding competition.

The majority of benthophages formed dense concentrations in different parts of the area studied (Fig. 3). Only rock greenling and Okhotsk skate had relatively similar patterns of spatial distribution, which degree may be characterized as moderate.

Summarizing all factors, it may be concluded that among benthophages only rock sole and Bering skate had closely similar diets. These species also had the most similar bathymetric overlap. At the same time their patterns of intra-annual feeding intensity and spatial distribution were considerably different, which makes feeding competition between them insignificant. Other benthophages with significant differences in these factors did not notably compete.

Predators

The basic diet of predators was fishes. Mollusks (mainly squids and octopuses) and large crustaceans (mostly shrimps and crabs) also played an essential role in their diet. The variation in consumption of three basic prey groups is shown in Table 1. The feeding pattern of broadbanded thornyhead was described by Novikov (1974) as mixed (predatory-benthophage). Observations from a submersible (Orlov 1993) showed that this fish used natural cover (sponges, stones, bottom hollows) for hunting. The total frequency of occurrence of fish in stomachs of this species was 9.8%. Consequently, we consider the broadbanded thornyhead as a fish with a feeding pattern similar to that of predators.

Shortraker rockfish and broadbanded thornyhead were found to have the highest value of Schoener's dietary overlap index among all predators considered (82.7%) (Table 2). Indices for the rest of the species varied from 9.9% to 62.2% (low and moderate degrees of dietary overlap).

Comparison of diagrams of intra-annual feeding intensity (Fig. 1) showed that the highest overlap of periods of maximum feeding intensity was characteristic of Greenland turbot and Kamchatka flounder, Aleutian skate and whiteblotched skate, whiteblotched skate and Alaska skate, shortraker rockfish and shortspine thornyhead, and shortraker rockfish and broadbanded thornyhead.

The analysis of similarity of patterns of vertical distribution among species considered showed that only seven of them had a high degree of bathymetric overlap (Table 3). They are Kamchatka flounder and Aleutian skate (84.9%), Aleutian skate and whiteblotched skate (81.5%), whiteblotched skate and Kamchatka flounder (82.9%), Matsubara skate and Greenland turbot (71.3%), shortspine thornyhead and Greenland turbot (74.2%), broadbanded thornyhead and Greenland turbot (73.3%), and shortraker rockfish and broadbanded thornyhead (67.3%). The patterns of vertical distribution of other species differed considerably.

Spatial distribution maps for all predators were drawn (Fig. 4). These maps showed that only three pairs had similar patterns of spatial distribution (locations of most dense concentrations): Pacific cod and Aleutian skate, shortraker rockfish and broadbanded thornyhead, and Kamchatka flounder and shortraker rockfish. The degree of this similarity may be characterized as moderate.

Thus, among the 12 predators examined according to four factors the maximum feeding competition was noted for shortraker rockfish and broadbanded thornyhead, the degree of which may be characterized as moderate. Other species, due to essential differences in their diets, features of intra-annual feeding intensity, and patterns of vertical and spatial distribution, did not compete significantly.

References

Berezina, M.A. 1973. Hydrobiology. Pishchevaya Promyshlennost', Moscow. 496 pp. (In Russian.)

Borets, L.A. 1985. Composition and present status of benthic fish community of Karaginsky Bay. Izv. TINRO 110:20-28. (In Russian.)

Brodeur, R.D., and P.A. Livingston. 1988. Food habits and diet overlap of various Eastern Bering Sea fishes. NOAA Tech. Memo. NMFS F/NWC-127. 76 pp.

Ivlev, V.S. 1977. Experimental feeding ecology of fishes. Naukova Dumka, Kiev. 272 pp. (In Russian.)

Klyashtorin, L.B. 1962. Studies on greenlings (Hexagrammidae, Pisces) inhabiting waters of Kuril Islands. Tr. Inst. Okeanol. Akad. Nauk SSSR 59:104-109. (In Russian.)

Lang, G.M., and P.A. Livingston. 1996. Food habits of key groundfish species in the eastern Bering Sea slope region. NOAA Tech. Memo. NMFS-AFSC-67. 111 pp.

Lang, G.M., P.A. Livingston, R. Pacunski, J. Parkhurst, and M.-S. Yang. 1991. Groundfish food habits and predation on commercially important prey species in the eastern Bering Sea from 1984 to 1986. NOAA Tech. Memo. NMFS F/NWC-207. 240 pp.

Livingston, P.A., D.A. Dwyer, D.L. Wencker, M.-S. Yang, and G.M. Lang. 1986. Trophic interactions of key fish species in the eastern Bering Sea. Bull. Int. North Pac. Fish. Comm. 47:49-65.

Novikov, N.P. 1970. Biology of giant grenadier (*Nematonurus pectoralis*) in the North Pacific Ocean. Tr. VNIRO 70:300-326. (In Russian.)

Novikov, N.P. 1974. Commercially important fishes of the continental slope of the North Pacific Ocean. Pishchevaya Promyshlennost', Moscow. 308 pp. (In Russian.)

Onishchik, N.A. 1997. Feeding of Atka mackerel, *Pleurogrammus monopterygius* (Hexagrammidae) in the area off the underwater Vitiaz' Chain. Vopr. Ikhtiol. 37(5):647-652. (In Russian.)

Orlov, A.M. 1993. Behavior features of some groundfishes of the shelf of the northern Okhotsk Sea. In: Biology and rational use of hydrobionts, their role in ecosystems. Tez. Dokl. Konf. Molod. Uchyon. TINRO, Vladivostok, pp. 26-27. (In Russian.)

Orlov, A.M. 1997. Mesopelagic fishes as prey of Atka mackerel (*Pleurogrammus monopterygius,* Hexagrammidae, Scorpaeniformes) off the northern Kuril Islands. In: Forage fishes in marine ecosystems. Proceedings of the International Symposium on the Role of Forage Fishes in Marine Ecosystems. University of Alaska Sea Grant AK-SG-97-01, Fairbanks, pp. 323-335.

Orlov, A.M. 1998. Demersal ichthyofauna of the Pacific waters off the northern Kuril Islands and southeastern Kamchatka. Biol. Morya 24(3):146-160 (In Russian.)

Roper, C.F.E., M.J. Sweeney, and C.E. Nauen 1984. Cephalopods of the world. FAO Fisheries Synopsis 125(3). FAO, Rome. 277 pp.

Schoener, T.W. 1970. Nonsynchronous spatial overlap of lizards in patchy habitats. Ecol. 51(3):408-418.

Shorygin, A.A. 1946. Quantitative method of the study of feeding competition of fishes. Zool. Zh. 25(1):45-60. (In Russian.)

Shuntov, V.P. 1965. Vertical zonality in fish distribution in the upper bathyal of the Sea of Okhotsk. Zool. Zh. 44(11):1678-1689. (In Russian.)

Shuntov, V.P., A.F. Volkov, O.S. Temnykh, and E.P. Dulepova. 1993. Walleye pollock in the ecosystems of the Far East Seas. TINRO, Vladivostok. 426 pp. (In Russian.)

Tokranov, A.M. 1986a. Food habits of great sculpin, *Myoxocephalus polyacanthocephalus* Pallas and plain sculpin, *Myoxocephalus jaok* (Cuvier) (Cottidae) in coastal waters of Kamchatka. Vopr. Ikhtiol. 26(6):980-989. (In Russian.)

Tokranov, A.M. 1986b. Food habits of Pacific cod off the eastern Kamchatka. In: Codfishes of the Far East seas. TINRO, Vladivostok, pp. 102-111. (In Russian.)

Tokranov, A.M., and A.B. Vinnikov. 1991. Food habits of Pacific cod, *Gadus morhua macrocephalus*, and their position in the trophic system of coastal Kamchatka waters. Vopr. Ikhtiol. 31(2):253-265. (In Russian.)

Tuponogov, V.N., and A.A. Kurennoi. 1986. Giant grenadier. In: Biological resources of the Pacific ocean. Nauka, Moscow, pp. 233-241. (In Russian.)

Volkov, A.F., K.M. Gorbatenko, and A.Ya. Efimkin. 1990. Strategies of walleye pollock feeding. Izv. TINRO 111:123-132. (In Russian.)

Yang, M-S. 1993. Food habits of the commercially important groundfishes in the Gulf of Alaska in 1990. NOAA Tech. Memo. NMFS-AFSC-22, 150 pp.

Yang, M-S. 1996. Diets of the important groundfishes in the Aleutian Islands in summer 1991. NOAA Tech. Memo. NMFS-AFSC-60. 105 pp.

Yang, M-S., and P.A. Livingston. 1986. Food habits and diet overlap of two congeneric species, *Atherestes stomias* and *Atherestes evermanni*, in the eastern Bering Sea. U.S. Fish. Bull. 82(3):615-623.

Zolotov, O.G. 1984. Biology of Atka mackerel *Pleurogrammus monopterygius* (Pallas) in the waters off the Kamchatka and Kuril Islands. Ph.D. thesis. VNIRO, Moscow. 22 pp. (In Russian.)

Zolotov, O.G. 1993. Some ecological features of rock greenling, *Hexagrammos lagocephalus* (Pallas), in coastal waters off Kamchatka and northern Kuril Islands. In: Investigation of biology and abundance dynamics of commercially important fishes of Kamchatka shelf, Vol. 2. KamchatNIRO, Petropavlovsk-Kamchatsky, pp. 190-201. (In Russian.)

Zolotov, O.G., and A.V. Medveditsyna. 1978. Feeding of Atka mackerel in coastal waters off the northern Kuril Islands. Biol. Morya 4:84-86. (In Russian.)

Zolotov, O.G., and A.M. Tokranov. 1991. Feeding habits of greenlings and Irish lords in the upper sublittoral of the eastern Kamchatka in spawning period. Vopr. Ikhtiol. 31(1):130-137. (In Russian.)

Comparative Studies on Effects of Predation by Fish on *Daphnia* in Lake Donghu (China) and Loch Leven (Scotland)

Yufeng Yang
Shantou University, Science Center, Shantou, China

Linda May and Iain Gunn
Institute of Freshwater Ecology, Edinburgh Laboratory, Edinburgh, United Kingdom

Xiangfei Huang and Jiankang Liu
Chinese Academy of Sciences, Institute of Hydrobiology, Wuhan, China

Abstract

From the data of *Daphnia* surveys made in 1979-1996 in Lake Donghu (China) and in 1978-1982 and 1992-1995 in Loch Leven (Scotland), two shallow eutrophic lakes near Wuhan and Edinburgh, respectively, the authors analyzed the long-term changes in densities and size frequency of *Daphnia*. The results showed annual average densities of *Daphnia* (station I + station II) in Lake Donghu were negatively correlated with fish yield. *Daphnia* densities tended to decrease after the introduction of rainbow trout to Loch Leven from 1993 on. The size-frequency distributions of *Daphnia* showed that small-sized *Daphnia* were dominant in Lake Donghu, but the distributions in Loch Leven changed little.

The comparative studies showed that the *Daphnia* population was determined largely by fish in Lake Donghu, with a high stocking rate of planktivorous fish, and also by predatory copepods. In Loch Leven, food resources, water temperature, etc., were key factors causing a stable *Daphnia* population, the introduction of rainbow trout having little effect on *Daphnia*, but analysis of gut contents of rainbow trout demonstrated an apparent size-selective predation for *Daphnia* larger than 1.4 mm in length. A low fish stocking rate may be one of the important reasons for the stable *Daphnia* population in Loch Leven.

Introduction

Lake Donghu (30°33′N, 114°23′E), Wuhan, and Loch Leven (56°10′N, 3°30′W), Scotland, are two of very few lakes in China and Great Britain for which hydrobiolological long-term monitoring data are available. Lake Donghu is a lake of multiple uses: water supply, aquatic sport, commercial fishery, and sightseeing (Liu 1984). Loch Leven has both national and international importance for four reasons: (1) it has great and renowned historical importance, (2) it has scenic importance, (3) it has environmental importance as a wetland site for plants and wildlife, and (4) it is important for its angling fishery, providing sport for 50,000 anglers per year (Montgomery 1994). So, for many years, Loch Leven has been regarded as Scotland's "crown jewel" as far as brown trout (*Salmo trutta* L.) fishing is concerned.

A warm, humid, northern subtropical climate predominates in Lake Donghu catchment. The maximum monthly temperature, usually in July, varied from 28.8°C to 31.4°C. The minimum monthly temperature, usually in January, varied from 2.6°C to 4.6°C during 1960-1978. The annual average temperature of the lake water was 19.25°C, varying from 5°C to 32°C in 1995. In most years there has been no freezing at all in the lake (Liu 1984).

The weather of Britain, situated on the eastern edge of the Atlantic Ocean, is dominanted by maritime air masses. It is characterized by its lack of extremes: small range in temperature, rain at any time of the year, and liability to wind at any time of year. Loch Leven is situated in an area with a temperate climate and its water mass is well mixed and unstratified (Smith 1974). Mean monthly water temperature is relatively high during June-August (13.2-18.5°C).

There are some common characteristics in Lake Donghu and Loch Leven: (1) they are shallow eutrophic lakes (mean depth less than 4 m), with severe algae bloom in the mid-1980s and the early 1990s, respectively; (2) long-term ecological data was available, with detailed monitoring of the zooplankton in the two lakes beginning in the 1960s; (3) they are similar in lake water volume and in the size of catchment; and (4) increasing fish yield is the common aim to both lakes, one for the commercial fishery, the other for the sport fishery. However, the stocked fish species and the stocking rate were quite different in the two lakes. Planktivorous silver carp (*Hypophthalmichthys molitrix*) and bighead carp (*Aristichthys nobilis*) are the main species stocked in Lake Donghu; predatory brown trout and rainbow trout (*Oncorhynchus mykiss*) are the main species in Loch Leven. Fish yield of Lake Donghu increased from 39 kg/ha in the 1950s to over 1,000 kg/ha in the 1990s (Liu 1984, Yang et al. 1994), but the yield of Loch Leven was less than 30 kg/ha during 1872-1993 (Duncan 1994).

The aim of the present study is to compare the ecological changes of *Daphnia* in densities and size-frequency distribution in the two lakes, and to evaluate the effects of the fisheries on the *Daphnia* population.

Material and Methods

Lake Donghu

Field Sampling

Daphnia samples were taken at two permanent stations (stations I and II, Fig. 1) at more or less monthly intervals. Station I is situated in the middle of the Shuiguohu Bay, near the west end of the lake. Station II is in the central part of the Guozhenghu area, about 1,500 m off the southern shore. Trophic levels have been different at the two sampling stations. Main limnological parameters at stations I and II are listed in Table 1. At station I, 15-30 liters of lake water were collected at 4 depths (0, 1, 2, 2.5 m). At station II, 25-50 liters of lake water were collected at 5 depths (0, 1, 2, 3, 4 m). *Daphnia* samples were obtained by filtering the lake water through a plankton net (mesh size 64 μm) and preserved with 5% formalin after collection. In the laboratory, *Daphnia* samples were counted in total numbers. In general, average body length of *Daphnia* was calculated by measurements of 50 animals.

Aquarium Experiment

An aquarium experiment was carried out in 1994 so as to demonstrate the feeding effects of silver carp and bighead carp on *Daphnia*. On April 25, 1994, about 0.5 m^3 tap water was pumped into the aquarium (diameter 0.9 m, height 0.7 m). After 2 days, about 20 kg sediment with *Daphnia* resting eggs (sediment dried for a week before experiment) from station II was introduced into the aquarium. Thickness of the sediment was 10 cm or so. *Daphnia* samples were taken with plankton net (mesh size 64 μm) every 10 days. On June 8, three silver carp (125-135 g/fish, average weight 130 g/fish) and five bighead carp (95-200 g/fish, average weight 118 g) were stocked into the aquarium. On June 13, all fish were caught. The experiment ended on July 2.

A similar experiment was carried out in 1996. The experiment began on September 26, and ended on December 25. Only three bighead carp (236-257 g/fish) were stocked into the aquarium on November 5, and then caught on November 9. Sediment was collected from station I.

Loch Leven

Samples of *Daphnia* were taken at more or less weekly intervals at a site near Reed Bower in Loch Leven during May-September of 1982, 1992, 1993, and 1995 (Fig. 2). The samples for 1982 were collected either by vertical net haul (mesh size 100 μm) from a depth of 4 m to the surface, or with a tube sampler incorporating a filter (mesh size 125 μm) which was lowered to a depth of 4 m. From 1992 to 1995, samples were collected and concentrated by vertical net haul (mesh size 118 μm). All samples were preserved with 4% formaldehyde. The preserved *Daphnia* samples were placed in a glass vessel and made up to the final volume of 250 ml with distilled

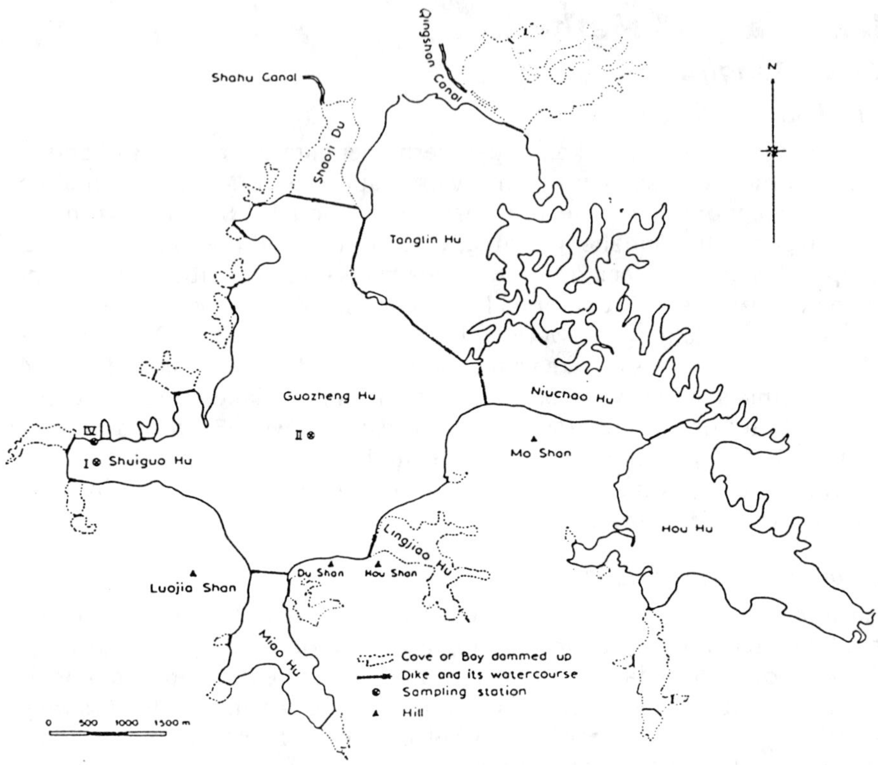

Figure 1. Map of Lake Donghu, Wuhan (modified from Gong et al. 1965).

Table 1. Main limnological characteristics at stations I and II in Lake Donghu.

	Station I	Station II
Water depth (1988-1996)	2.8-3.2 m	4.0-5.0 m
Sediment depth (1994-1996)	15-35 cm	>35 cm
pH (1995)	8.11	8.31
TN (1995)	7.04 mg/L	3.07 mg/L
TP (1995)	0.794 mg/L	0.402 mg/L
Chlorophyll *a* (1995-1996)	55.92 μg/L	34.59 μg/L

TN = total nitrogen. TP = total phosphorus.

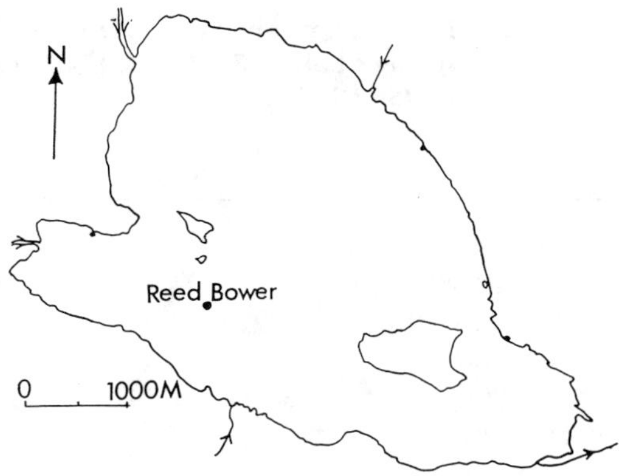

Figure 2. Map of Loch Leven, Scotland.

water. Each sample was thoroughly mixed to ensure a random distribution of the animals, and then sub-sampled with a Stempel pipette (volume 5 ml) to estimate the density of *Daphnia. Daphnia* were taken at random from each sample for body-size measurement. About 100 individuals were measured for each sample. The samples of gut contents of rainbow trout were provided by the Institute of Fisheries Management, Pitlochry, U.K.

Results

Morphometric Data for Lake Donghu and Loch Leven

Some morphometric features for Lake Donghu and Loch Leven are given in Table 2. The two lakes are shallow, with mean depths less than 4 m. However, they are quite different in maximum depth, surface area, lake length, and length of shoreline.

Changes in Densities of Daphnia

In the 1960s, *Daphnia hyalina* and *D. carinata* ssp. were the dominant cladocerans in Lake Donghu. The maxmum density (60 individuals per liter) was recorded in June. Their population densities were of paramount importance to the population size of the cladocerans as a whole. The two species prevailed alternately. In general, the population density of *D. hyalina* displayed a maximum in spring and winter, but the occurrence of a summer or autumn maximum was recorded for *D. carinata* ssp. The shift from one to the other took place first in late spring and then at the end of autumn; in either case the water temperature was in the range from 16°C to 19°C. The seasonal fluctuation in the abundance of *Daphnia* seems be governed by water temperature and food supply in the main (Liu 1984).

Table 2. Main morphometric data for Lake Donghu (Gong et al. 1965, Liu 1984) and Loch Leven (Smith 1974).

	Lake Donghu	Loch Leven
Mean depth (m)	2.21	3.9
Maximum depth (m)	4.75	25.5
Surface area (km^2)	32	13.3
Volume (m^3)	62×10^6	52.4×10^6
Length (km)	11.24	5.9
Width (km)	5.86	2.3
Length of shoreline (km)	92	18.5
Catchment (km^2)	187	170

The most striking changes in densities of *Daphnia* took place during 1986-1987, annual average densities of *Daphnia* at the two sampling stations decreasing markedly from 1987 on. At station I, annual average densities of *Daphnia* decreased from 47.36 individuals per liter in 1979 to 14.22 in 1986, to 1.91 in 1987, to 0.61 in 1995; at station II, densities of *Daphnia* decreased from 37.32 individuals per liter in 1979 to 31.62 in 1986, to 3.19 in 1987, to 0.80 in 1995. Densities of *Daphnia* tended to diminish to nothing, while fish yield tended to increase with year (Fig. 3). Annual average densities (*D*, individuals per liter) of *Daphnia* (station I + station II) were negatively correlated with fish yield (*Y*, kilograms per hectare). The regression equation is:

$$D = 62.9583 - 0.0619\ Y \text{(correlation coefficient} - 0.8341)$$

Research results of enclosure experiments in 1990 showed silver carp and bighead kept high feeding pressure on *Daphnia*. Average densities of *Daphnia* were 11 and 14 per liter in the two enclosures with fish, and 307 and 237 per liter in the two enclosures without fish (Yang and Huang 1992).

Daphnia has been the dominant crustacean zooplankton in Loch Leven. Long-term monitoring showed that the peaks of *Daphnia* occurred in May-June during 1982 (54 per liter), 1992 (66 per liter) and 1993 (55 per liter), but the peak of *Daphnia* fell in early September (113 per liter) in 1995. The peak of 1995 was higher than anything recorded in the lake during 1978-1995; low densities (<10 per liter) of *Daphnia* were recorded on most sampling dates in 1995 (Fig. 4). The densities of *Daphnia* tended to decrease after introduction of rainbow trout into the lake (excluding 1995). Such extreme fluctuation suggested that the *Daphnia* population had become less stable in Loch Leven.

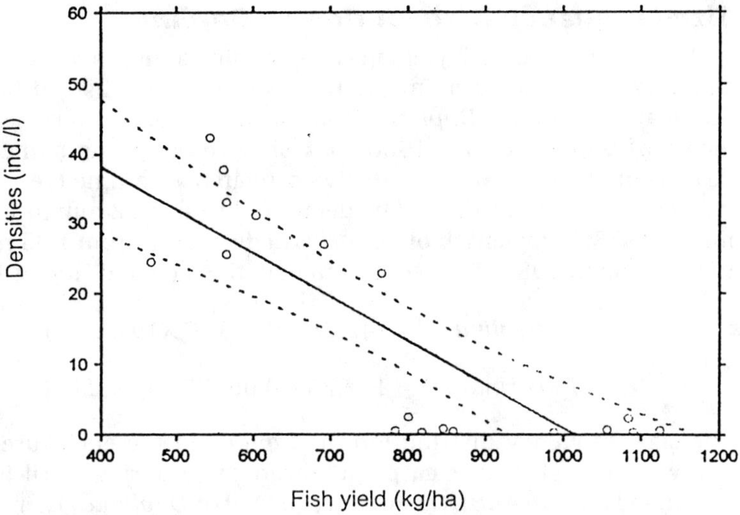

Figure 3. *The relationship between annual average densities of Daphnia (stations I + II) and fish yield (kilograms per hectare) in Lake Donghu.*

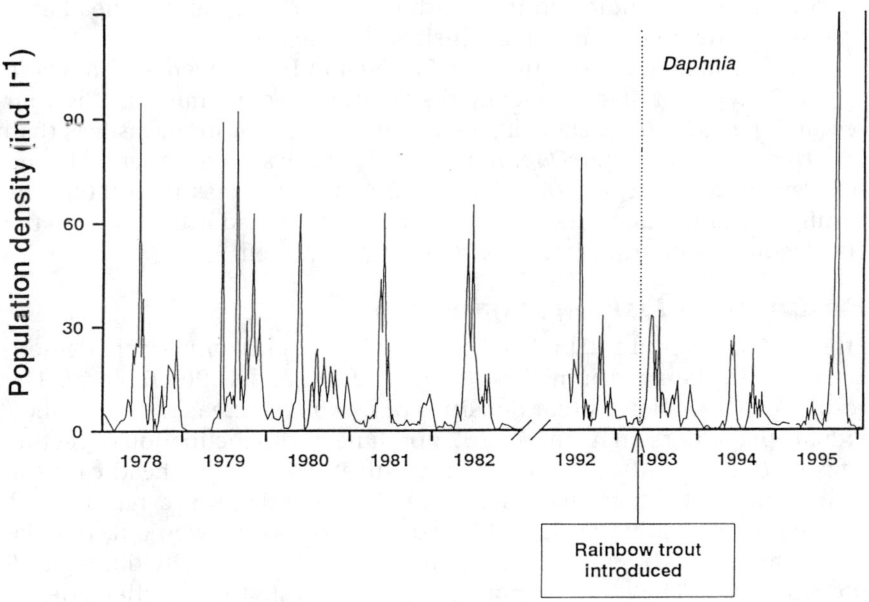

Figure 4. *Changes in population densities (individuals per liter) of Daphnia in Loch Leven during 1978-1995 (from Gunn and May 1996).*

Size-Frequency Distribution of Daphnia

With the increase of fish yield in Lake Donghu, annual average densities of *Daphnia* decreased with time; in the meantime, small-sized *Daphnia* predominated over large *Daphnia*. Monitoring data of measurements of body length of *Daphnia* during 1980-1988 showed body length (mean value of year, *Y*) of *Daphnia* was negatively correlated with time (year, *X*). Annual mean body length of *D. hyalina* decreased from 1.22 mm in 1980 to 0.65 mm in 1988. Body length of *D. carinata* decreased from 1.33 mm in 1980 to 0.78 mm in 1988. The regression equations (Li and Huang 1992) are:

$$D.\ hyalina:\ Y = 142.28 - 0.07117\ X\ (p < 0.05)$$

$$D.\ carinata:\ Y = 122.23 - 0.06107\ X\ (p < 0.05)$$

The size-frequency distribution of *Daphnia* in the enclosures with fish and without fish (silver carp + bighead carp) during April-June1990 is shown in Fig. 5. In April, the total frequency of *Daphnia* >1.5 mm was 0.37 in the enclosures without fish and 0.16 in the enclosure with fish; the frequency >2.0 mm was 0.07 and 0.004, respectively. In May, the total frequency >1.5 mm was 0.38 in the enclosures without fish and 0.06 in the enclosure with fish; the total frequency >2.0 mm was 0.02 and 0, respectively. In June, the total frequency of *Daphnia* >1.5 mm was 0.30 in the enclosures without fish and 0 in the enclosures with fish (Yang et al. 1995). The results indicated that feeding of silver carp and bighead carp resulted in the dominance of small-sized *Daphnia*.

Size-frequency distributions of *Daphnia* in Loch Leven are shown in Fig. 6. Patterns of size-frequency distributions were similar in the four research periods: frequency of small *Daphnia* (0.6-1.4 mm) was less than 0.1, frequency of larger *Daphnia* (1.8-2.2 mm) was more than 0.1; and frequency of the largest *Daphnia* (2.6-2.8 mm) was less than 0.05. The results indicated the introduction of rainbow trout had little effect on the size-frequency distribution of *Daphnia* in Loch leven.

Predation of Fish on Daphnia

Effects of silver carp and bighead carp predation on *Daphnia* were demonstrated with the aquarium experiment of 1994 and 1996 (Table 3). The results of 1994 showed that densities of *Daphnia* increased from 0 individuals per 5 liters on April 27 to 373 on June 8 with incubation of resting eggs of *Daphnia*. After introduction of silver carp and bighead carp (on June 8) into the aquarium, densities of *Daphnia* decreased rapidly to 1 individual per 5 liters on June 13. After the fish were taken out, population densities of *Daphnia* began to increase, up to 60 individuals per 5 liters on July 2. The experimental results of 1996 showed that densities of *Daphnia* increased from 0 individuals per 5 liters on September 29 to 845 on November 5. After introduction of fish into the aquarium on October 6, *Daphnia* densities decreased from 40 to 0 individuals per 5 liters. After

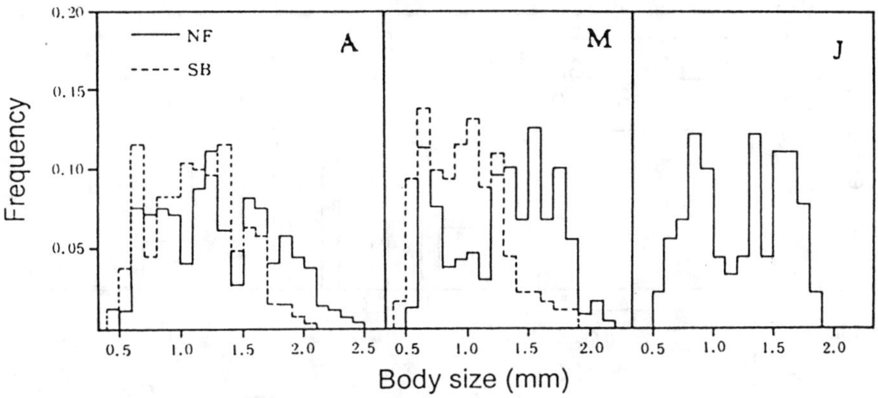

Figure 5. Size-frequency distributions of Daphnia in enclosures with silver carp and bighead carp (SB) and without fish (NF) during April-June 1990 (from Yang et al. 1995).

Table 3. *Daphnia* population fluctuation (individuals per 5 liters) in aquarium in 1994 and 1996.

1994	Apr. 27	Jun. 8*	Jun. 13	Jul. 2
Daphnia	0	373	1	60
1996	Sep. 29	Nov. 5*	Nov. 6	Nov. 8
Daphnia	0	845	40	0

*Introduction of fish after sampling in aquarium.

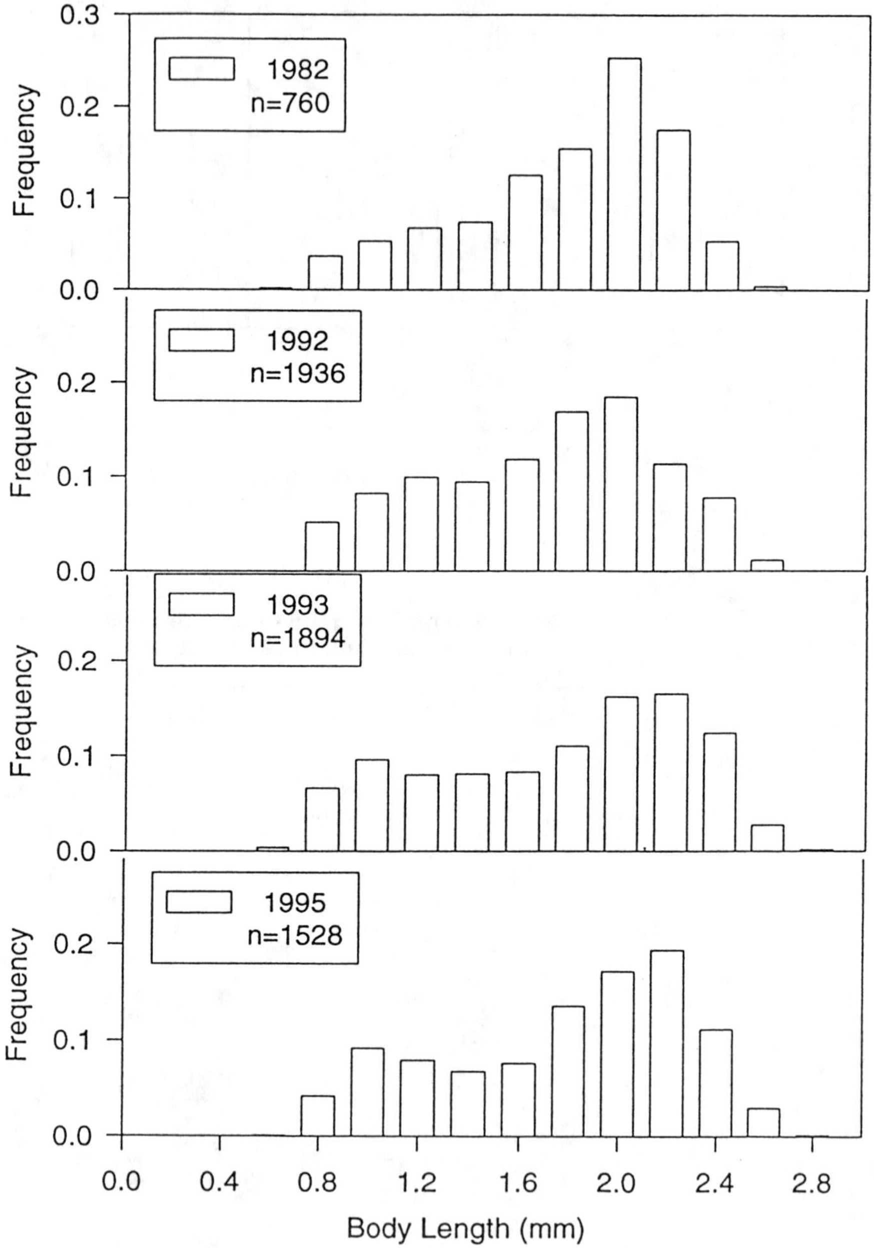

Figure 6. Size-frequency distributions of Daphnia in Loch Leven.

Table 4. Feeding of two bighead carp on *Daphnia* (1.5-2.0 mm).

Date	Water temp. (°C)	Bighead carp (weight)	*Daphnia* eaten / *Daphnia* provided
Oct. 30	17	409 g, 375 g	67/100
Oct. 31	15	393 g, 305 g	83/100
Nov. 5	15	365 g, 330 g	69/100
Nov. 6	15	350 g, 315 g	66/100

the fish were taken out, the densities of *Daphnia* were still quite low. The aquarium experimental results demonstrated that predation by silver carp and bighead carp resulted in decreased *Daphnia* densities.

Analyses of gut contents of silver carp and bighead carp from Lake Donghu and at the termination of the aquarium experiment showed that remains of *Daphnia* were often observed in gut contents of the two fishes. These indicated *Daphnia* was a favorite prey for them. The first author of the present paper fed 100 *Daphnia* (1.5-2.0 mm) to two bighead carp in a 20-liter aquarium, each experiment lasting for 3 hours (11:00 am-2:00 pm), in 1996. More than 65 *Daphnia* were eaten by the bighead carp (Table 4). The result suggested that the scarcity of *Daphnia* in Lake Donghu is related to the feeding activity of fish in the lake.

Daphnia from the gut contents of 11 rainbow trout from Loch Leven were examined during March, June, and August of 1995. Remains of *Daphnia* such as head with compound eye and post-abdomen with claw are common in stomachs of rainbow trout. Body length (BL), eye diameter (ED) and claw size (CS) of *Daphnia* (over 100 individuals during March-Octber 1995) were measured. The authors found that the body length of *Daphnia* is correlated with eye diameter and claw size. The regression equations are:

$$BL = 300.2233 + 12.7050\ ED\ (\mu m,\ r^2 = 0.8)$$

$$BL = 110.3307 + 12.9838\ CS\ (\mu m,\ r^2 = 0.8)$$

These equations should allow total lengths to be estimated from measurements of eye-diameter and claw-size of *Daphnia* in the stomachs of fish. In gut contents of rainbow trout, only large-sized (1.6-2.8 mm) *Daphnia* were recovered (Fig. 7). The size-frequency distributions of *Daphnia* were quite different in the lake and in fish gut contents. The body length of *Daphnia* ranged from 0.6 mm to 2.8 mm in the lake (Fig. 6). *Daphnia* lengths in trout stomachs were greater than those observed in the lake. In stomachs of rainbow trout, the percentage of *Daphnia* (>2 mm) was up to 38.84%. The results showed that rainbow trout only chose *Daphnia* over 1.4 mm long and ignored the often more numerous, but smaller individuals. These support the idea that rainbow trout select the larger *Daphnia*.

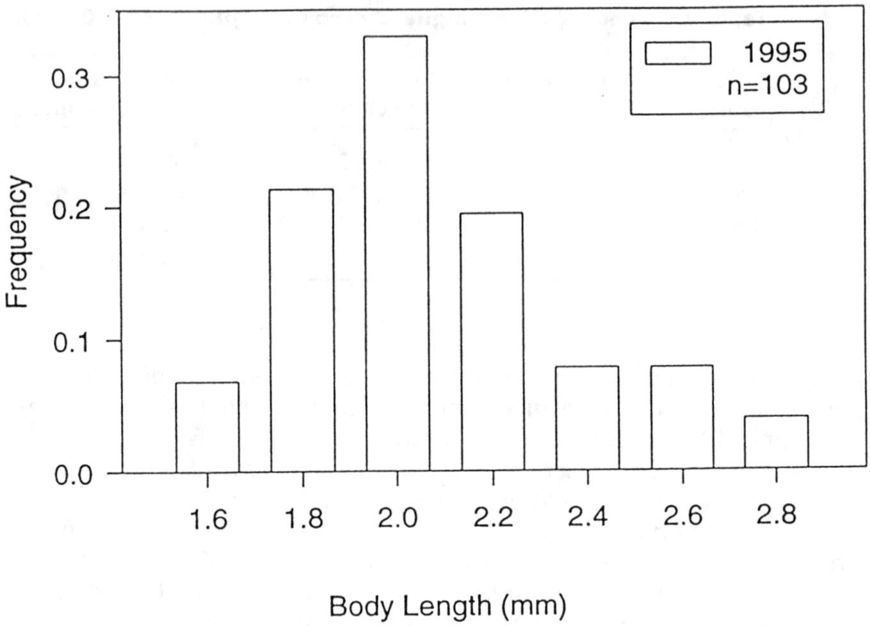

Figure 7. Size-frequency distributions of Daphnia in rainbow trout in Loch Leven (March, June, and August of 1995).

Discussion

Lake Donghu

From the 1960s to the 1990s, great changes have taken place in the Lake Donghu ecosystem: the ever-increasing human population in the watershed (now estimated at 20,000 by the present authors) and the rapid development of industry, agriculture, and animal husbandry in the drainage basin pour hundreds of tons of nitrogen and phosphorus into the lake each year in the form of semi-treated effluents (Jao and Zhang 1980). The commercial fishery, but not the sport fishery, plays an important part in the integrated utilization of Lake Donghu. With a warm climate, abundant rainfall, and a rich supply of food resources in the catchment, Lake Donghu is most suitable for fish stocking. According to the production data from Donghu Fish Farm, planktivorous silver carp and bighead carp are two main species for stocking in the lake. The two fishes constituted 83-98% of the total fish yield. During the 1950s and 1960s, the fish yield was low (less than 100 kg/ha in most years). With the improvement in fishery management, annual fish yield increased steadily from 124.5 kg/ha in 1971, to 466.9 kg/ha in 1981, and to 858.9 kg/ha in 1991. From 1993 on, fish yield was over 1,000 kg/ha (1,056.6-1,124.7 kg/ha). Heavy stocking of grass carp (*Ctenopharyngodon idellus*) during the latter half of the 1960s and the 1970s resulted in the disappearance of one-third of the submerged and floating leaf macrophytes from the lake; vegetation occurred in 1-28% of the surface area of various parts of the lake, with the largest vegetation area in Tanglin Hu, and the smallest in the Guozheng Hu and Shuiguo Hu area (Ni 1995).

Plankton-eating fish tend to remove the larger species; invertebrate predators tend to remove the smaller species (Brooks and Dodson 1965, Lair 1990). As plankton feeders, silver carp and bighead carp exert a strong predation pressure on *Daphnia*. Yang and Huang (1992) demonstrated feeding effects of the two fish on *Daphnia*. Enclosure experiments showed that the densities of *Daphnia* were lower in the enclosures with fish than in the enclosures without fish. Long-term monitoring data (at station II) indicated that annual average densities of *Daphnia* decreased from 44.16 individuals per liter in 1962-1963 to 42.57 in 1980, and then to 1.21 in 1990, whereas fish yield increased during 1962-1990 (Yang et al 1994). Changes in submerged macrophytes and fish density were accompanied by major changes in the zooplankton community.

The refuge capacity of macrophytes decreased markedly with increasing fish density (Schriver et al. 1995). In Lake Donghu, macrophytes have been disappearing and fish yield increasing in the Shuiguo Hu and Guozheng Hu area of the lake since the 1970s. The long-term high fish predation resulted in the reduction of *Daphnia* density in Lake Donghu. Moreover, *Cyclops vicinus*, being the dominant copepod in the lake, can feed effectively on young and small-sized *Daphnia* in indoor and field experiments (Yang et al. 1998). These results lead us to conclude that feeding by fish and predatory copepods is the primary cause of the reduction in the density of *Daphnia* in the lake in recent years.

Some results showed that predation by fish resulted in dominance of small-sized zooplankton. Arcifa et al. (1986) demonstrated the influence of zooplanktivorous fish on *Daphnia*: large-sized *Daphnia* were present in the fish-free enclosures, but not in the enclosures with fish. Field investigations from Americana Reservoir showed fish (*Astyanax*) preyed heavily on daphnids. Brooks and Dodson (1965) indicated that intense predation from fish will eliminate the large species, and the relatively immune small species will predominate. In Lake Donghu, quite high feeding pressure of silver carp and bighead carp (heavily stocked) resulted in a decrease in densities of *Daphnia* and in the dominance of small-sized *Daphnia* (Li and Huang 1992, Yang et al. 1995).

Loch Leven

The structure of a zooplankton community is generally thought to be determined by two major factors: predation (by fish and invertebrates) and food availability (Harper 1986). Cladocerans, especially the genus *Daphnia*, whose members belong to the larger zooplankters, are considerably exposed to fish predation (Brooks and Dodson 1965, Li and Huang 1992).

The most significant change recorded in the crustacean zooplankton community in Loch Leven in the latter half of the century is probably the disappearance of *Daphnia* sometime between 1954 and 1960, and their reappearance 16 years later, in 1970 (Johnson and Walker 1974). However, this temporary loss of *Daphnia* has been attributed to the organochlorine insecticide dieldrin used as a moth-proofing agent in a local woolen mill during the late 1950s and early 1960s (Morgan 1970). *Daphnia* is a filter-feeder and has been shown to be the main grazer of phytoplankton, particularly the small species, in Loch Leven (Bailey-Watts 1986). For this reason, any decline in the *Daphnia* population would be expected to result in a corresponding reduction in grazing pressure, especially on the population of small algae, allowing the latter to proliferate. It is of interest that a significant bloom of small phytoplankton, *Oscillatoria,* occurred in Loch Leven during 1968-1971, when *Daphnia* were absent or scarce (Holden and Craines 1974).

Brown trout is a native fish in Loch Leven. Angling records, which have been maintained since 1872, show large annual variations in catch statistics, varying from 2,137 to 85,883 fish. The general trend in these statistics is an increase in catch up to the 1950s and a decline thereafter. Between 1872 and 1969, the annual mean weight of fish caught was 0.4 kg but this increased to 0.61 kg during 1970-1993 (Duncan 1994).

Loch Leven's problems arose in the spring of 1992 when a severe algal bloom resulted in the closure of the fishery and concerns about the risk to public health (Bailey-Watts et al. 1994). In March of 1993, approximately 40,000 rainbow trout with an average weight of approximately 0.25 kg were introduced to improve the angling catches and, to this end, it has been a success. This was done by the lake owners in an attempt to bolster the fishery which had been declining over at least the last 30 years, in spite of stocking each spring over the last decade with some 140,000 home-bred brown trout averaging 12-18 cm in length. It is a policy to endeavor to restore Loch Leven, giving great sport to the many anglers who visit each year. This stocking exercise was repeated in 1994 and 1995 with approximately 30,000 rainbow trout of a mean individual weight of about 0.23 kg (Montgomery 1994, Gunn and May 1996).

In 1993, 13,150 brown trout and 10,456 rainbow trout were caught by anglers (Duncan 1994). The rainbow trout is native to the western seaboard of North America and is believed to provide better angling than brown trout during algal blooms, although when the blooms are severe their catchability also declines (Cragg-Hine 1975).

Introduction of rainbow trout into the lake started in March 1993. Initially, the crustacean zooplankton community seemed unaffected by this activity and continued to be dominated by *Daphnia galeata* However, after a few months, some changes began to occur. A decrease in abundance of *Daphnia* was recorded during the first 2 years after the introduction of rainbow trout (1993-1994). But in September 1995, the *Daphnia* population had rapidly increased to record levels, suggesting a reduction in predation pressure (owing to the removal of trout by anglers). This period was also noteworthy for the very high abundance of the large blue-green algae *Anabaena spiroides* and high temperature. In 1995, spells of extraordinarily warm weather resulted in a water temperature of 21°C in mid-June and from late July to mid-August exceeded values not recorded for some 400 years (Bailey-Watts et al. 1996). This could result in a drastic increase in the *Daphnia* population because of plentiful food supply and favorable water temperature.

The rainbow trout only chose *Daphnia* over 1.4 mm long and ignored the often more numerous, but smaller individuals as evidenced from analysis of fish gut contents. But the size-frequency distribution of *Daphnia* in Loch Leven showed that the large-size individuals changed little in the 4 years. This suggests that the effects of introduction of rainbow trout on *Daphnia* is not apparent. The low stocking rate of fish is one of the important reasons.

Initial evidence from dietary analyses of brown trout gut contents, taken during 1993, indicated that zooplankton (including *Daphnia*) also formed a major dietary component for this species (Duncan 1994). *Daphnia* comprised over 90% of the rainbow trout diet for most age classes (Knapp and Sotero 1983). In the gut contents of this fish, a lot of remains of *Daphnia* were observed. This suggests *Daphnia* is a preferred pey for rainbow trout.

Conclusions

Lake Donghu, Wuhan, China, and Loch Leven, Scotland, U.K. differ in geographical position, climate, lake morphometry, and fishery. The ecology of *Daphnia* in both water bodies was studied in long-term field investigations and also by way of experimentation. The comparative research results showed that *Daphnia* population size was determined largely by fish in Lake Donghu, with high stocking rate of planktivorous fish, and also by predatory copepods. Feeding by silver carp and bighead carp was one of the most important causes of density reduction in large-sized *Daphnia*, followed by the dominance of small-sized *Daphnia*. In Loch Leven, food resources, water temperature, etc., were key factors causing a stable *Daphnia* population. The introduction of rainbow trout had little effect on the population size of *Daphnia*, but analysis of gut contents demonstrated an apparent size-selective predation on *Daphnia* 1.4 mm in length or larger. The low stocking rate of fish may be one of the important reasons for the stable population of *Daphnia* in Loch Leven.

Acknowledgments

This work was supported by Donghu Experimental Station of Lake Ecosystems, which is affiliated with the Chinese Ecological Research Network, Chinese Academy of Sciences, and by Edinburgh Laboratory, Institute of

Freshwater Ecology (Great Britain). Some relevant data were provided by Donghu Experimental Station and Edinburgh Laboratory. We express our thanks to Prof. Xie Ping and all members of Donghu Experimental Station for their assistance in carrying out the field work. Many thanks to Dr. A.E. Bailey-Watts, Mr. A. Kirika, and A. Lyle for their help. This study was also supported partly by China's National Science Foundation Grant 39630090 and 39625008.

References

Arcifa, M.S., T.G. Northcote, and O. Froehlich 1986. Fish-zooplankton interactions and their effects on water quality of a tropical Brazilian reservoir. Hydrobiologia 139:49-58.

Bailey-Watts, A.E. 1986. Seasonal variation in size spectra of phytolankton assemblage in Loch Leven, Scotland. Hydrobiologia 138:25-42.

Bailey-Watts, A.E., N.J. Wiltshire, and A. Kirika. 1996. The ecology of the phytoplankton of Loch Leven during 1995. Report to Forth River Purification Board, Edinburgh.

Bailey-Watts, A.E., A. Kirika, I.D.M. Gunn, C.L. Bryant, and N.J. Wiltshire. 1994. The environment of fish: Physics, phosphorus, phytoplankton and fleas. In: The Loch Leven trout fishery: Its future. Institute of Fisheries Management, Pitlochry, U.K., pp. 7-15.

Brooks, J.L., and S.I. Dodson. 1965. Predation, body-size and composition of plankton. Science 150:28-35.

Cragg-Hine, D. 1975. Studies on overwinter mortality of autumn stocked rainbow trout in some lakes in Northern Ireland. J. Inst. Fish. Manage. 6(1):1-7.

Duncan, W. 1994. The fish community in Loch Leven. In: The Loch Leven trout fishery: Its future. Institute of Fisheries Management, Pitlochry, U.K., pp. 29-38.

Gong, L.J., Z.H. Guan, Y.T. Hwang, and R.H. Hu. 1965. On the types and distribution of sediments in Lake Tunghu. Oceanol. Limnol. Sin. 7:181-194. (In Chinese with English abstract.)

Gunn, I.D.M., and L. May. 1996. Analysis of 1995 zooplankton samples: Loch Leven NNR. Report to Scottish Natural Heritage, Institute of Freshwater Ecology, Edinburgh.

Harper, D.M. 1986. The effects of artificial enrichment upon the planktonic and benthic communities in a mesotrophic to hypertrophic loch series in lowland Scotland. Hydrobiologia 137:9-19.

Holden, A.V., and L.A. Craines. 1974. Nutrient chemistry of Loch Leven, Kinross. Proc. R. Soc. Edinb. Sect. B (Biol.) 74:101-121.

Jao, C.C., and Z.S. Zhang. 1980. Ecological changes of phytoplankton in Lake Donghu, Wuhan during 1956-1975 and the eutrophication problem. Acta Hydrobiol. Sin. 7(1):1-17. (In Chinese with English abstract.)

Johnson, D., and A.E. Walker, 1974. The zooplankton of Loch Leven, Kinross. Proc. R. Soc. Edinb. Sect. B (Biol.) 74:285-294.

Knapp, S.M., and R.A. Soltero. 1983. Trout-zooplankton relationships in Medical Lake, WA following restoration by aluminum sulfate treatment. J. Freshwater Ecol. 2:1-12.

Lair, N. 1990. Effects of invertebrate predation on the seasonal succession of a zooplankton community: A two year study in Lake Aydat, France. Hydrobiologia 198:1-12.

Li, C.H., and X.F. Huang. 1992. Species succession of Cladocera in Lake Donghu in relation to various ecological factors. Acta Hydrobiol. Sin. 16(2):101-112.

Liu, J.K. 1984. Lakes of the middle and lower basins of the Chang Jiang (China). In: F.B. Taub (ed.), Lakes and reservoirs. Elsevier Science Publishers, Amsterdam, pp. 331-355.

Montgomery, D. 1994. Management of the Loch Leven trout fishery. In: The Loch Leven trout fishery: Its future. Institute of Fisheries Management, Pitlochry, U.K., pp. 26-28.

Morgan, N.C. 1970. Changes in the fauna and flora of a nutrient enriched lake. Hydrobiologia 35:545-553.

Ni, L.Y. 1995. Study on structure and biomass of aquatic vegetation and its long-term changes in the Donghu Lake, Wuhan. In: J.K. Liu (ed.), Studies on ecology of Lake Donghu (II). Science Press, Beijing, pp. 287-301. (In Chinese with English abstract.)

Schriver, P., J. Bogestrand, E. Jeppesen, and M. Sondergaard. 1995. Impact of submerged macrophytes on fish-zooplankton-phytoplankton interaction: Large-scale enclosure experiments in a shallow eutrophic lake. Freshwater Biol. 33:255-270.

Smith, I.R. 1974. The structure and physical environment of Loch Leven, Scotland. Proc. R. Soc. Edinb. Sect. B (Biol.):81-100.

Yang, Y.F., and X.F. Huang. 1992. The influence of silver carp and bighead carp on zooplankton community structure. J. Lake Science 4(3):78-86. (In Chinese with English abstract.)

Yang, Y.F., X.F. Huang. and J.K. Liu. 1994. Long-term changes in zooplankton of subtropical Lake Donghu (Wuhan, China) under conditions of increased fish predation. Russian J. Aquat. Ecol. 3(1):23-30.

Yang, Y.F., X.F. Huang, and P. Xie. 1995. Effects of silver carp and bighead carp on *Daphnia* population in the Donghu Lake, Wuhan. In: J.K. Liu (ed.), Ecological studies of Lake Donghu (II). Science Press, Beijing, pp. 415-421.

Yang, Y.F., J.X. Yang, and X.F. Huang. 1998. Feeding of *Cyclops vicinus* on zooplankton. Acta Hydrobiol. Sin. 22(1):71-78. (In Chinese with English abstract.)

Ecosystem Approaches for Fisheries Management
Alaska Sea Grant College Program • AK-SG-99-01, 1999

Managing the Ecosystem by Leverage Points: A Model for a Multispecies Fishery

Nicholas Bax, Alan Williams, Stevie Davenport, and Cathy Bulman
CSIRO (Commonwealth Scientific and Industrial Research Organisation) Marine Research, Hobart, Tasmania, Australia

Abstract

Ecosystem management has been popularly adopted as a goal of fisheries management, but what does it really mean? In stressed ecosystems, ecosystem functions may remain unchanged while changes in species composition—particularly dominance—and the health of individuals can change dramatically (Schindler et al. 1985). The species are more sensitive indicators of stress than the system. We suggest, therefore, that there is little value in trying to manage a marine ecosystem as a whole. Instead one should manage people's interaction with the particular ecosystem components that influence the quantity and quality of valued products. We call these particular interactions "leverage points" and use a 5-year study of the southeastern Australian continental shelf fishery ecosystem to show how leverage points can be found. Neither nearshore production nor predation on commercial fish species has a major influence on fisheries production; they therefore have little leverage potential. However, the interaction of benthic habitat with fish and fishers is a potential leverage point. Benthic habitat directly influences the fish community, and specific benthic habitats are vulnerable to fishing. Their vulnerability is increasing because of technological advances in accurate positioning (Global Positioning System) and position recording (trackplotters), fishing vessel power, and fishing gear. Accurate positioning can also be used to manage the location of fishing effort. For this leverage point to be successful, the spatial management of fishing effort would need to be developed in collaboration with the fishing industry and to be targeted specifically at the vulnerable habitats. We suggest that new management instruments such as the transferable ecological stock rights are needed to link fisheries management directly to the ecosystem.

Introduction

Ecosystem management is a commonly stated goal of marine fisheries management in Australia and overseas, but what does it mean? Do ecosystems even exist? In 1953, Odum defined "any entity or natural unit that includes living and non-living parts interacting to produce a stable system in which the exchange of materials between the living and nonliving parts follows circular paths is an ecological system or ecosystem." Forty years later, Golley (1993) reviewing the history of the ecosystem concept in ecology, questions whether the ecosystem concept is reasonable—philosophers require that wholes have genuine properties, where a genuine property is one that is unique to the whole and not reducible to the properties of its components. As a whole, ecosystem properties are not studied in relation to marine fisheries (though see Caddy 1993). Extension of traditional marine fisheries management has frequently been through amalgamating single-species models linked by predation into multispecies models, rather than looking at system properties (e.g., Laevastu and Larkins 1981, Gislason and Helgason 1985). Even in these multispecies models, few studies considered levels higher than individual species (though see May et al. 1979). Given the high public and political profile of ecosystem management, why is this so?

The Experimental Lakes project in Canada (Schindler et al. 1985) provides one answer, and answers Golley's (1993) question. Systematic acidification of the lakes changed the abundance, composition, and dominance of species, but did not affect genuine ecosystem properties such as productivity and nutrient cycling. Rare species became common and common species of interest to fishery managers, such as trout, became rare. Biological redundancy in the lakes meant that the abundance and activities of individual species changed in response to changing environmental conditions, but the higher-level system properties of the lake, which are a function of the watershed and the atmosphere as well as the biota, were more robust and varied much less in response to environmental change. The authors concluded that higher-level system properties (including species diversity, numbers, and abundance) were not sensitive indicators of ecosystem stress. The more sensitive indicators are species dominance, and species' physical condition (including disease prevalence). A similar conclusion can be drawn from changes in the continental shelf ecosystem of the U.S. Northeast, where large-scale shifts in species composition resulted from fishing (Sissenwine 1986), while similar changes in system-level processes have not been reported.

Until recently, ecosystem approaches in marine fisheries have concentrated on species interactions. This has happened (1) because the multispecies models were a logical adaptation of familiar single-species models; (2) because the models have an extensive theoretical background; and (3) because they use the sorts of data (species, abundance, diets, growth, and natural mortality rates) that are the fodder of fisheries science (they can be collected from fishing vessels or fish markets). These models have been used to correct misconceptions of processes at the single-species level and to provide advice on managing multispecies communities (e.g., Gislason and Helgason 1985).

But species interactions, and associated energy flows, are only one facet of ecosystem functioning. Recent technological advances in remote sensing and geographic positioning systems are changing the ways in which we can study the marine environment and therefore the ways in which we can monitor and manage ecosystem processes. We are now no longer absolutely limited by technology in our choice of which aspect of the marine ecosystem to study, but can now choose aspects of marine ecosystems that are likely to benefit most from management intervention.

CSIRO Marine Research is concluding a 5-year study of the southeast Australian continental shelf ecosystem to determine which new management measures would usefully supplement the current single-species management of the South East Fishery. The South East Fishery is Australia's largest fishery for domestic scalefish markets. Trawling started on the continental shelf in the early 1900s and expanded to the slope in the 1970s and to deepwater pinnacles in the late 1980s. A non-trawl sector uses dropline, demersal longline/setline, gill nets, and traps to target many of the same species as the trawl fishery. Until recently, management of this multispecies fishery (there are about 90 retained species) centered on minimum lengths for some species, minimum cod-end mesh sizes, and protection of estuarine and nearshore waters. In 1985, management tried, but failed, to cap effort through restricting vessel numbers and size. In 1989, total allowable catches and individual transferable quotas were introduced for the 16 main targeted species. However, effort is still increasing—reported trawling in the study area increased 25% in the last 10 years (Unpublished data, N.J. Bax)—and effort is increasingly targeted on specific bottom features. Our study of this system was based on the premise that management interventions aimed at units greater than single species could improve sustainability of this fishery and ecosystem. We present an overview of our results here.

Methods

One approach to managing complex systems is to begin by determining where the "leverage" is greatest (Senge 1990). Leverage is based on the notion that small, well-focused actions can produce enduring improvements if they are directed at sensitive system components. We used the notion of leverage to direct our research.

At the outset of the study, a conceptual model of the factors that could affect productivity of the fish community was developed (Fig. 1a) and refined (Fig. 1b). The sampling program focused on key factors and their potential as leverage points for management of fish resources. Four potential leverage points were identified:

1. Primary production from coastal seagrass beds.
2. Predation on commercial fish species.
3. Effects of fishing on commercial fish species.
4. Effects of fishing on benthic habitat.

Results that address leverage points 1, 2, and 4 are presented here.

There were four surveys: July 1993, August 1994, April 1996, and November 1996. Each survey consisted of a broad-scale survey examining the seasonal distribution of biota and physical oceanography of the region, followed by intensive sampling of selected habitats. Collected samples were used to determine the relationships between biological species and the habitat they occupy by analyzing production sources, trophic position, diet, and morphological adaptation.

Study Area

The study area was the Australian continental shelf between the latitudes of 36° and 39°S—the southeastern point of the continental margin where east and south coasts meet (Fig. 2). The shelf, which is defined as the area from the coast out to water depths of ~170-200, is narrow on the east coast (~25 km) and wide on the south coast (~175 km). Several small rivers flow onto the shelf, but their discharge is low. The area has a complex oceanography that is variable between the south and east coasts. Eddy fields from the seasonally variable southward-flowing East Australian Current bring intrusions of continental slope water onto the shelf, particularly in spring and summer (Church and Craig 1998). An underlying northward countercurrent at the shelf break also transports cool slope water onto the shelf (Cresswell 1994). Northerly winds sometimes enhance these intrusions by bringing nutrient-rich water to the surface (Cresswell 1994). Associated with northeasterly winds, intermittent upwellings off the east coast bring cool, nutrient-rich water to the surface (Edwards 1990). An eastward outflow of water from Bass Strait in winter, driven by strong prevailing westerly winds, cascades down the slope to the east of Bass Strait and can be detected over 11,000 km northward along the slope (Tomczak 1985). Average waves in the area are 1-3 m in height with periods of 5-6 seconds, and penetrate to depths 60 m or more (Morrow and Jones 1988). The southeast Australian continental shelf is, therefore, a moderate- to high-energy, wave-dominated environment.

The sediments on the shelf are autochthonous. Continental basement rocks that underlie the sediments outcrop on the inner shelf near the coast. Fossiliferous limestone reefs, consisting largely of bivalve and bryozoan clasts, occur across the shelf. The coarse-grained quartzite sandstones that occur on the inner shelf of the south coast could have originated as elongate sand bodies formed in a high-energy coastal plain environment parallel to palaeo-shorelines.

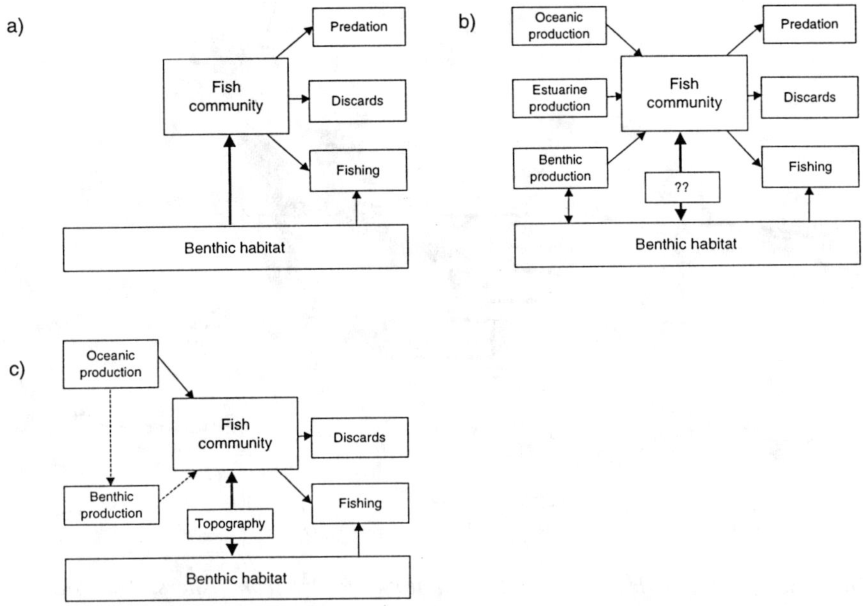

Figure 1. *Conceptual models of factors influencing the southeast Australia continental shelf fish community (a) before sampling began, (b) after the first survey, and (c) at the end of the study.*

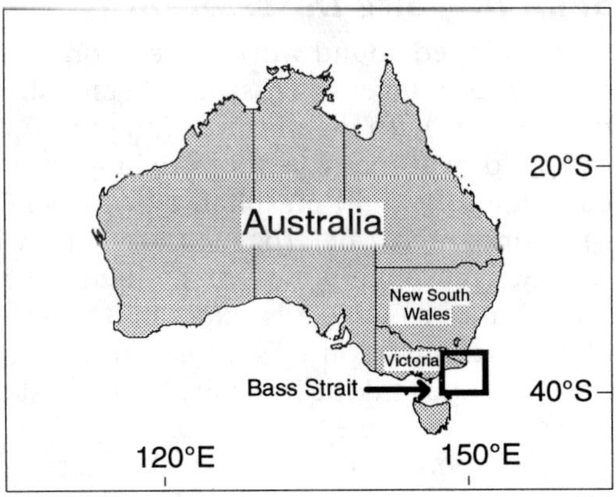

Figure 2. *Location of the study area in southeastern Australia.*

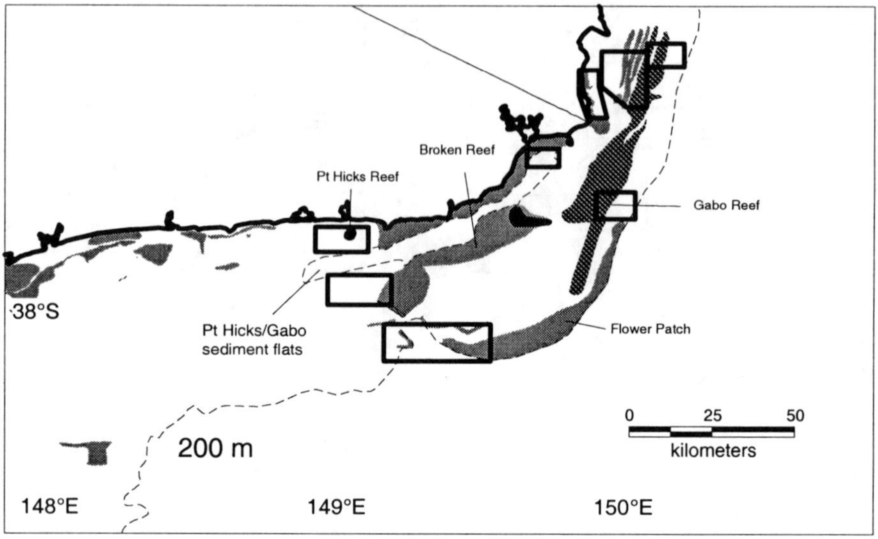

Figure 3. Coarse-scale map of benthic habitat compiled from fishers' observations, interpreted after biological and geological sampling, with the perimeters of eight study sites (mesohabitats) shown. Substrate types: sediment flats (unshaded), granite outcrops (black), low-relief limestones/consolidated carbonates (gray), high-relief limestones (gray, patterned).

Sampling-Site Description

We developed sound working relationships with key fishers of our study area. Once a level of trust had been established, the fishers generously provided us with details on fishing grounds and fish-habitat associations collected over many years of fishing. A map of this information enabled us to identify eight mesohabitats (sensu Greene et al. 1994) with significant topographic heterogeneity to survey (Fig. 3).

Each mesohabitat was divided into two or three macrohabitats—"soft," "hard," and "rough"—discriminated from echo returns at 12, 38, and 120 kHz. Discrimination was based on the length and intensity of the tail of the first echo and the intensity of the first and second echoes (Bax et al., In press).

The geomorphology of each macrohabitat was described from sediment and rock samples, and from video and still photographs collected from a towed camera system developed for the study (Barker et al., In press). Current flows around selected topographic features were interpreted from CTD and hydrographic profiles, supplemented in one instance with moored current arrays.

Sources of Production and Trophic Position

A broad collection of biological material was analyzed for stable isotopes to describe production sources and overall system structure. Biological material included hand-collected estuarine seagrasses and marine algae; particulate organic matter and phytoplankton collected on fiberglass filters from water sampled in niskin bottles at the surface and below; zooplankton collected with oblique bongo net (500 μm mesh) tows and a 56-cm drop net (100 μm mesh); sediments and macroalgae collected with benthic grabs and sleds; invertebrates collected with an infauna/epifauna benthic sled; fish collected with demersal and midwater trawls (40 mm cod-end liner), traps, and variable-mesh gillnet; and marine mammals and seabirds collected adventitiously from animals that died of natural causes.

Ratios of $^{13}C/^{12}C$ and $^{15}N/^{14}N$ were expressed as the relative per mil (‰) difference between the sample and conventional standards (Pee Dee Belemnite carbonate and N_2 in air). The formula used to express these values is:

$$\delta X = \{[R\ (\text{sample})]\ /\ [R\ (\text{standard})] - 1\}1{,}000\ ‰$$

where

$X = {}^{13}C$ or ${}^{15}N$; and

$R = {}^{13}C/{}^{12}C$ or ${}^{15}N/{}^{14}N$.

Diet

Fish stomachs were collected throughout each cruise to obtain samples of (ideally) each species caught at a range of depths, times, geographical locations, and sizes of fish. Up to 10 stomachs per species per tow and 50 stomachs per species per cruise were taken. Large stomachs were frozen at –20°C and small preserved in 10% formalin. The length, weight, and sex of donor fish were recorded. Stomachs were assessed for fullness in the laboratory and then dissected. Prey items were identified to the lowest possible taxon, counted, blotted on absorbent paper to remove excess moisture, and weighed (to 0.001 g in the case of very small items). Fish digested beyond recognition were identified from an otolith guide created for this study. Diets were described in terms of wet weight of prey. For the purposes of the present paper all prey items were classified into broad trophic categories.

Fish Communities and Habitat

Following Hudson et al. (1992), we define habitat as "simply the place where an organism lives." The 19 macrohabitat sites were grouped into habitat types by analyzing fish community structure. Three fishing gears—trawl, gillnet, and trap—were needed to effectively sample the range of bottom types (Bax et al., In press).

Fish caught by all gears were sorted to species, weighed, and counted; all species represented by more than five individuals were measured, and a broad selection of species was retained for morphometric analysis in the laboratory. Abundance data were standardized and communities determined from non-metric multi-dimensional scaling (MDS) and hierarchical agglomerative clustering based on between-sample, Bray-Curtis similarities (Clarke 1993). Data were double-square-root transformed and analyzed with the PRIMER statistical package. A similarities analysis (SIMPER) identified species making the greatest contributions to grouping of macrohabitats by community. "Typifying" species were those contributing most to within-community similarity, while "discriminating" species were those contributing most to between-community dissimilarity. In the figures in this paper we show the species making up 80% biomass (ranked abundance) caught by all gears ("abundant"), less abundant species with typifying or discriminating power ("distinctive"), and species restricted to one habitat type ("indicators").

Results

Sampling-Site Description

The sea floor may be visualized as a series of extensive sediment flats ("soft grounds") with interspersed outcrops of consolidated material ("hard grounds"). Physical and photographic sampling showed soft grounds are composed of particulate material, primarily sands, with areas of mud and gravel, whereas hard grounds include cemented sediments, reefs, and bedrock. These geological features are primary attributes of seafloor habitat in a demersal fishery and determine the distribution and abundance of fishes, and therefore fishing effort.

Geological properties also determine the vulnerability of hard grounds to modification or permanent damage by fishing gear. The key attributes of vulnerability are hardness, relief, and patch size. Fossiliferous limestones, probably often in conjunction with sandstones, comprise most of the hard grounds in the study area, including most large tracts of reef; numerous scattered small outcrops throughout the study area; components of the elongate inner shelf south coast reefs; and patchy, low-relief hard grounds in the deep-shelf "Flower Patch" (Fig. 3). Devonian granite bedrock (probably lateral submarine extensions of adjacent rocky headlands) also outcrops from soft sediments on the inner-shelf south coast.

Composition, relief, and spatial extent of hard grounds are among the factors determining their vulnerability to damage by fishing gears. The extensive, high-relief and heavily cemented mid- to deep-shelf limestone reefs such as "Gabo Reef" are relatively resilient to trawling, as are the localized, high-relief granite outcrops such as "Point Hicks Reef" (Fig. 3). Patchy, low-relief, inner- to mid-shelf limestone reefs that were weathered during previous sea-level regressions, such as "Broken Reef," are relatively vulnerable to progressive erosion or removal by trawling. Fishers report that "Broken Reef" is gradually being opened up to trawling as prominent features are mapped and obstructions removed. The patchy hard grounds include relatively small "bryozoan" reefs in mobile substrates towards the shelf edge that support stands of stalked crinoids as in the "Flower Patch" (Fig. 3). Fishers report that some hard ground areas that supported local aggregations of important commercial species have now disappeared, together with the fish.

Sources of Production

Stable nitrogen and stable carbon isotopes were analyzed in 1,214 fish (teleost and elasmobranch) samples representing 87 species; 153 invertebrate (benthic and pelagic) samples from eight phyla; 10 species of marine mammal; 1 seabird; 9 species of algae; 91 samples of water column particulates from four surveys and 103 samples of sediment from three surveys (Fig. 4).

Stable carbon ratios for seagrasses are higher ($\delta^{13}C$ less negative) than those of the benthos, fish and higher predators, and therefore cannot have contributed significantly to their trophic pathway. Stable carbon ratios for phytoplankton and zooplankton (and their derivatives—particulate organic matter [POM] and sediments) are consistent with marine phytoplankton being the primary source of production. Therefore the foundations of the ecosystem in the study region are marine phytoplankton. Benthic macroalgae (red and brown) may also contribute to a lesser degree, but pigment analysis of POM and sediments (Unpublished data, S. Davenport) indicate their input to be limited to water depths of 25 m or less. No primary production of terrestrial origins was detected by isotope or pigment analysis (Unpublished data, S. Davenport).

Predation

Quota species and 16 other species were examined because of their commercial value or potential importance as predators. Their diets were analyzed and summarized by major prey group: 10 were markedly piscivorous (Fig. 5a). Piscivorous fish occurred in all habitats: for example tiger flathead (*Neoplatycephalus richardsoni*) is a dorsoventrally flattened fish associated with sediment, while striped trumpeter (*Latris lineata*) is an obligate reef dweller. Of the 28 species, the velvet leatherjacket (*Parika scaber*) had no fish in its stomach, while spotted warehou (*Seriolella punctata*), blue warehou (*S. brama*), butterfly perch (*Caesioperca lepidoptera*), sparsely-spotted stingaree (*Urolophus paucimaculatus*), stinkfish (*Synchiropus calauropomus*), and blue morwong (*Nemadactylus douglasii*) had unidentifiable fish in their stomachs. In the remaining 21 piscivorous fishes, we subdivided the fish component of the diet into commercial and non-commercial species. The proportions of commercial fish were small (Fig. 5b). The species with the largest proportion—the striped trumpeter—

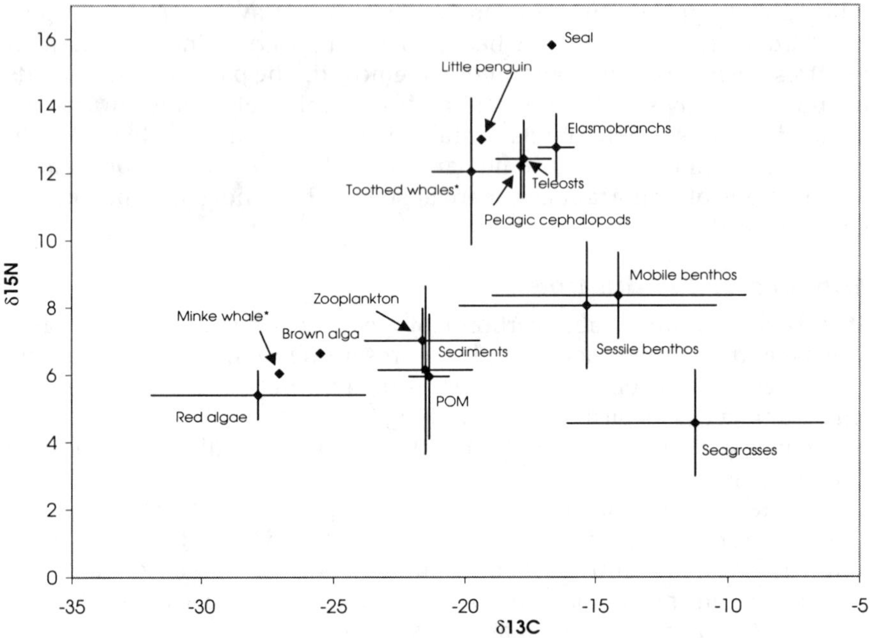

Figure 4. Mean stable isotope ($\delta^{13}C$ and $\delta^{15}N$) values (± 1SD) of primary producers to apex predators. Cetacean samples integrate isotope values from several ecosystems. Mobile benthos includes: asteroids, bivalves, crustaceans, echinoids, gastropods, octopus, and ophiuroids. Sessile benthos includes: anemones, ascidians, bryozoans, crinoids, soft corals, and sponges.

is a highly valued fish, but is seldom caught in this area so can be considered to be of low abundance. No major predator of commercial fish species was found.

Piscivores other than fish in this area include seals (*Arctocephalus pusillus doriferus*), dolphins (*Delphinus delphis, Tursiops truncatus*), shearwaters (*Puffinus tenuirostris*), and penguins (*Eudyptula minor*). These animals have a high trophic position (Fig. 4), and are locally (some seasonally) abundant. They eat mainly small pelagic fish (e.g. anchovies, pilchards, mackerel), and cephalopods (e.g., arrow squid), while shearwaters also eat crustaceans. Juveniles of important commercial species in this area are not a large part of their diet.

Fish Communities and Habitat

Seven habitat types were defined by the groups of macrohabitats formed from the analysis of fish abundance data (Unpublished data, A. Williams). Habitats were also clearly differentiated by physical features—depth,

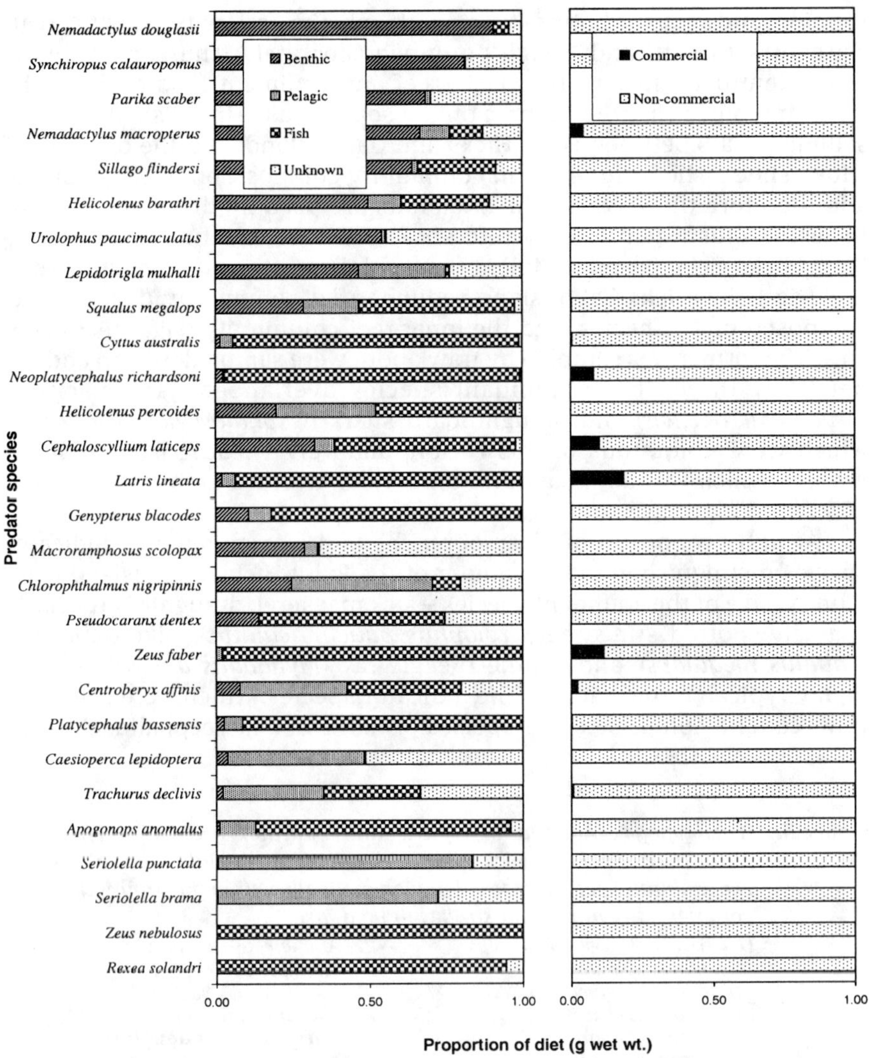

Figure 5. Dietary analysis of 28 key South East Fishery fishes: (a) proportion of diet due to pelagic invertebrates, fish, and benthos; and (b) proportion of commercial species in diet.

location, and acoustic bottom type—with only the bottom types with greatest contrast ("soft" and "rough") grouped separately. Thus four habitats were identified on the east coast—reef and sediment flats on both the inner and outer shelf—and three on the south coast—inner shelf reef and sediment flats between Point Hicks and Gabo Island, and the outer shelf "Horseshoe." The differences in community composition of the habitats, and therefore the ways in which habitats are used by different suites of fishes, is illustrated here by reference to two contrasting habitats: the Point Hicks-Gabo Island sediment flats and the east coast inner-shelf reef.

The Point Hicks-Gabo Island sediment-flat community (Fig. 6a) was composed of 76 species, and the inner-reef community (Fig. 6b) 55 species. The former was sampled by trawl; both were sampled by trap and gill net. Only three of the abundant species overlapped: jack mackerel (*Trachurus declivis*) and draughtboard shark (*Cephaloscyllium laticeps*), which were ubiquitous in this region; and velvet leatherjacket (*Parika scaber*), which was abundant and broadly distributed across many habitats.

A high proportion of elasmobranchs and a large suite of indicator species distinguished the remainder of the Point Hicks-Gabo Island community. Five of the abundant species—jack mackerel, draughtboard shark, sparsely-spotted stingaree (*Urolophus paucimaculatus*), piked spurdog (*Squalus megalops*), and southern eagle ray (*Myliobatis australis*)—were highly typical of this habitat. The indicator species, which were primarily trawl-caught, comprised four species of flathead (Platycephalidae), three

Figure 6. (Facing page.) "Abundant," "typical," and "indicator" groups from fish communities at (a) Point Hicks–Gabo Island sediment flats, and (b) inshore reef habitat. Species, ordered phylogenetically within the groups and with commercially important species in bold, are:

(1,23) Cephaloscyllium laticeps
(2) Squalus megalops
(3) Urolophus paucimaculatus
(4) Urolophus cruciatus
(5) Narcine tasmaniensis
(6) ***Cyttus australis***
(7,27) Trachurus declivis
(8) Synchiropus calauropomus
(9,30) Parika scaber
(10,33) Thyrsites atun
(11) Diodon nicthemerus
(12) ***Callorhynchus milii***
(13) Myliobatis australis
(14) ***Galeorhinus galeus***
(15) ***Squatina australis***
(16) Trygonorrhina fasciata
(17) ***Neoplatycephalus richardsoni***
(18) Platycephalus arenarius
(19) Neoplatycephalus aurimaculatus
(20) Platycephalus longispinis
(21) Kathetostoma laeve
(22) Ammotretis rostratus
(24) ***Helicolenus percoides***
(25) Pseudophycis bachus
(26) Latridopsis forsteri
(28) ***Nemadactylus douglasii***
(29) ***Nemadactylus macropterus***
(31) ***Seriolella brama***
(32) Caesioperca lepidoptera
(34) Hypoplectrodes annulatus
(35) Hypoplectrodes maccullochi
(36) Callanthias australis
(37) Pempheris multiradiata
(38) Scorpis lineolatus
(39) Atypichthys strigatus
(40) ***Seriola lalandi***
(41) ***Pagrus auratus***
(42) Cheilodactylus spectabilis
(43) Ophthalmolepis lineolatus
(44) Notolabrus tetricus
(45) Bodianus unimaculatus

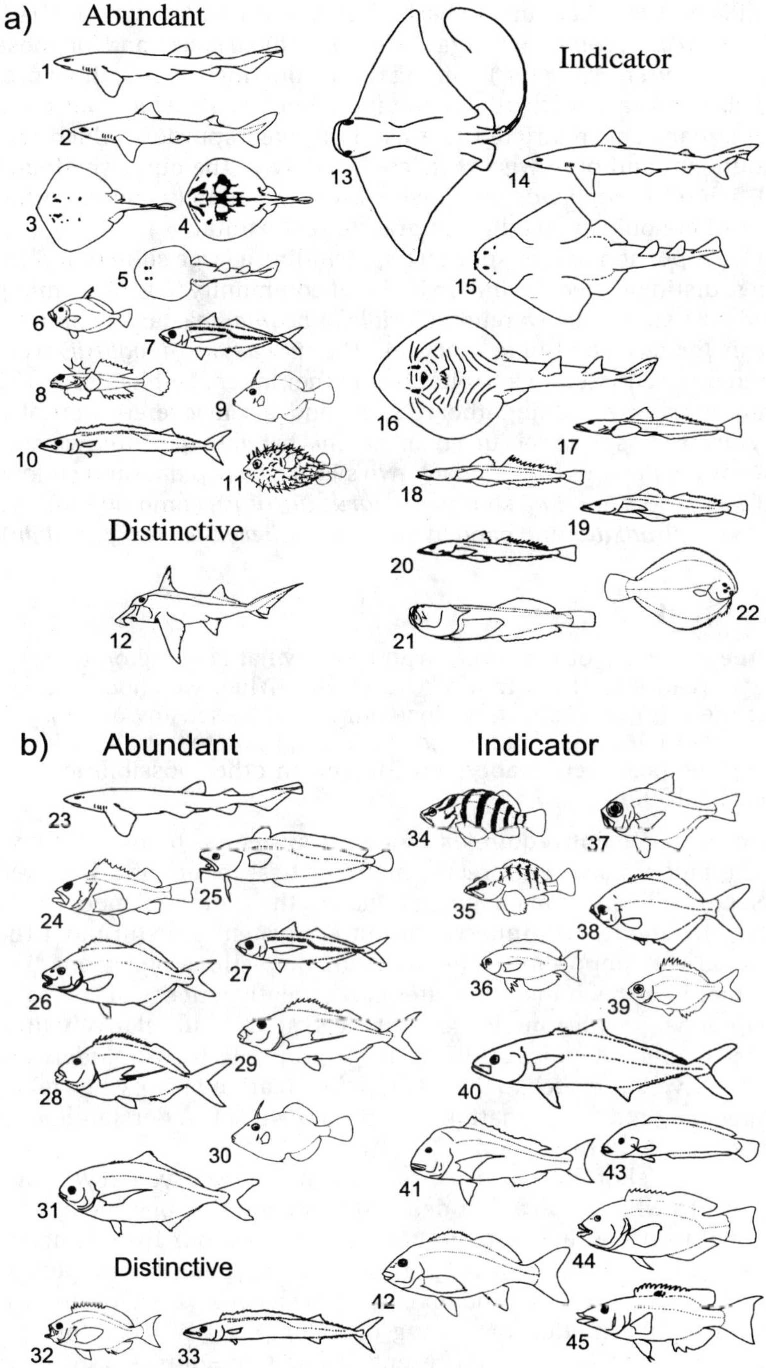
a) Abundant
Indicator
Distinctive
1
2
3
4
5
6
7
8
9
10
11
12
13
14
15
16
17
18
19
20
21
22
b) Abundant
Indicator
Distinctive
23
24
25
26
27
28
29
30
31
32
33
34
35
36
37
38
39
40
41
42
43
44
45

rays (Rajiidae), Australian angel shark (*Squatina australis*), school shark (*Galeorhinus galeus*), common stargazer (*Kathetostoma laeve*), and longnose flounder (*Ammotretis rostratus*). Among discriminating species, three provided reliable contrasts with other habitats: school shark, which was moderately abundant and restricted to this site; piked spurdog, which was highly abundant and otherwise restricted mainly to the outer shelf; and elephant fish (*Callorhynchus milii*), which was moderately abundant but found elsewhere only at an adjacent granite reef habitat.

A high proportion of reef specialists, including a large suite of indicator species, distinguished the inner shelf reef community (Fig. 6b). Among the abundant fishes, blue warehou (*Seriolella brama*), bastard trumpeter (*Latridopsis forsteri*), and blue morwong (*Nemadactylus douglasii*) were typical and differentiated this habitat from other inner shelf habitats. The two ubiquitous species—draughtboard shark and jack mackerel—were also typical. A further 12 species occurred only in this habitat type: three wrasses (Labridae), three perches (Serranidae), two sweeps (Scorpidae) and yellowtail kingfish (*Seriola lalandi*), snapper (*Pagrus auratus*), common bullseye (*Pempheris multiradiata*), and banded morwong (*Cheilodactylus spectabilis*).

Discussion

> It is the existence of observers who notice what is going on that imparts reality to the origin of everything. When we choose to experiment for one aspect, we lose our ability to see any others. Every act of measurement loses more information that it obtains, closing the box irretrievably and forever on other possibilities. (Wheatley 1992)

Ecosystem management requires a model of system structure and processes. The model may be a precise mathematical model or a less well defined set of beliefs, but in choosing it many other potential models will be omitted. No model will fully represent ecosystem structure and the processes, so it is important to be aware of the serious omissions. For example, there has been much debate on the relative merits of "bottom-up" or "top-down" trophic models of aquatic systems, but relatively little on whether trophodynamics itself is the appropriate focus. Models are not only a "marvelous crutch to the imagination" (Larkin 1978); when taken too literally they limit imagination and the growth of understanding. A good model is a disposable model.

Our first model of the southeast Australian shelf ecosystem was that the demersal trawl fishery caught demersal fish and that benthic habitat was essential to these fish communities (Fig. 1a). On our first demersal trawl survey we caught a high proportion of pelagic and benthopelagic fish—for example, the very abundant carangid, jack mackerel. It was clear that our conceptual model was wrong or incomplete. We therefore extended the model to coarsely represent production sources as well as extractive processes (Fig. 1b), but we left the link between benthic habitat and fish communities unspecified. Given the broadened scope, it was clear that we did not have sufficient resources to study all aspects of system structure, so we concentrated on those we thought had leverage potential. For our purposes we defined potential leverage points as system structures or processes to which our chosen output measure (fisheries production) was sensitive and, as importantly, structures or processes that were amenable to management intervention.

The first potential leverage point that we identified was the input of primary production from seagrass. Estuarine and terrestrial sources of primary production, including seagrasses, have been identified as contributing to production over the entire continental shelf for 110 km off northeast Australia (Risk et al. 1994), and seagrass is important in the trophic ecology of juvenile blue grenadier off southeast Australia (Thresher et al. 1992). Thus it seemed plausible that seagrass production was an important source of primary production for the southeast shelf. Seagrass conservation also provided an attractive management option, because seagrass acreage in Australia has been greatly reduced (Poiner and Peterken 1995), seagrass coverage is easily monitored, and seagrass conservation could involve fishers in ecosystem management without affecting their own livelihoods. However, stable isotope analyses (backed up by analysis of photoreactive pigments; Unpublished data, S. Davenport) could detect no contribution of seagrass or terrestrial production to the continental shelf food webs. Shallow water red and brown algae may contribute to local primary production, but sources are local and not amenable to management intervention. The primary source of production for the shelf ecosystem is pelagic phytoplankton in the open ocean. This production source is also not amenable to management intervention at the local scale.

Our second potential leverage point was predation on fish, a well-studied aspect of ecosystem interactions (e.g., Bax, In press). It has been suggested that the abundance of desirable fish species could be increased by removing their predators (Gulland 1982, Harwood and Greenwood 1985). Marine mammals and birds in the area are strongly piscivorous, as indicated by their enriched $\delta^{15}N$ ratios (Fig. 4). Diet studies, however, show that they eat mainly surface and midwater pelagic species, including jack mackerel and Australian pilchard (*Sardinops neopilchardus*). These species are part of the midwater prey community, sustained by euphausiids and lanternfish, and exploited by many taxa including tuna and pelagic sharks (e.g., Young et al. 1997). Some of the fish species caught with demersal nets were piscivorous, but ate few commercial species. If the more abundant piscivorous species, such as jack mackerel, eat commercial fish (even occasionally), they could have a marked impact. However, the abundant piscivores had essentially no commercial species in their diets, although unidentified fish in stomach contents could have hidden predation on the larvae of commercial species. Many taxa feed on the midwater food and there may be competitive interactions among them, but in practice it would be very difficult to demonstrate that resources were limiting to the extent that competition was occurring. Monitoring and managing competitive interactions would prove even harder.

The third potential leverage point was the direct impacts of fishing on fish populations; indirect impacts, for example, fish feeding on discards, has yet to be addressed for this system. Direct impacts are well covered in annual assessment reports (summarized in Caton et al. 1997) and in focused discarding studies (Liggins 1996). Discarding of commercial species can be high, either because they are too small for the market or because market prices are temporarily too low to cover transport costs. Discarding of juvenile redfish (*Centroberyx affinis*), for example, can exceed 90% of catch in some ports (Liggins 1996). There is an ontogenetic change in habitat with movement to greater depth for many commercial species on the southeast Australian continental shelf (Unpublished data, A. Williams). Therefore, most discards of many commercial species are caught in shallow waters, typically either when sea conditions prevent vessels from fishing offshore, or when they are targeting marketable commercial species whose adults occur in shallow waters. This is an obvious leverage point. Modifications to gear and fishing practices have the potential to reduce discarding (Bax 1997) and thereby affect fish populations, but the implications for fishers' activities and financial return have not been determined.

The link between the fish community and habitat was one potential leverage point we identified. The impacts of demersal trawling on benthic organisms, habitat, and fish communities have been well documented (e.g., Jones 1992, Schwinghammer et al. 1996, Sainsbury et al. 1997). Comparisons of the diets of fish species caught in different habitats did not indicate any particular trophic link with habitat (Unpublished data, C. Bulman). However, multispecies abundances clearly delineated fish communities associated with distinct habitats. Individual species were mostly either obligate or facultative users of particular habitat types, and rarely ubiquitous. Analysis of the shape and morphology of obligate and facultative habitat users suggested that the relationship between habitat and fish might be mediated through fish seeking refuge from prevailing currents. Fish found in current-swept sediment flat habitats were frequently dorsoventrally flattened for low drag, or were burrowers or sustained swimmers (Fig. 6a). Fishes found in topographically complex reef areas were mostly deep-bodied, with specializations such as fin shape and positioning that would confer good maneuverability (Fig. 6b). Although we cannot determine the full scope of relationship between benthic habitat and fish community, the distribution of morphotypes together with measurements of water chemistry and currents around reefs, indicates that habitat topography has a role through changing current flow. It may not be necessary to define the link between benthic habitat and fish populations precisely because the association of many taxa with structural habitat implies an increase in individual fitness that would be lost if the structural features were lost (Auster and Malatesta 1995). In addition, even if particular benthic habitat conferred no increase in individual fitness, the role of particular habitat types in aggregating particular species would increase fishers' effectiveness. Because major commercial species in the South East Fishery are managed by quota, measures that increase fishers' effectiveness without increasing habitat impacts may aid conservation by reducing effort.

Fishers target very specific habitats on the southeast Australian shelf. They report that some key habitats have been, and are being, impacted by fishing. For example, low-relief limestone reefs that traditionally yielded good catches of high-value fishes such as snapper are being eroded or removed. Patchy mosaics of low-relief reef are particularly vulnerable to being "opened-up" as vessels become more powerful and use thicker warps and heavier bottom gear on trawls. The gear development that has made precise targeting possible is the combination of GPS and electronic trackplotters, which enable skippers to plot obstacles precisely and to either avoid or remove them.

The links between fish communities and benthic habitat suggest that habitat preservation could be a strong leverage point. Some fishers have spoken out on the need to preserve habitat, but may be reluctant to diminish their own catching efficiency unless other fishers also avoid—and are seen to avoid—the sensitive habitat. For fishers to agree to limitations on their fishing practices they must see the potential benefits clearly and also accept that any restrictions are not excessive. For example, although some topographically complex habitats are vulnerable to fishing impacts, other complex habitats (for example, those based on granite or large contiguous areas of fossiliferous limestone) are less vulnerable. At the moment they are considered untrawlable. However, trawlable areas close to these complex habitats are prime fishing grounds. Restricting fishing on all complex habitat, regardless of its vulnerability to fishing, would unnecessarily impede fishing on these prime grounds. Other topographically complex habitats are vulnerable to fishing and it is these that should be targeted by habitat-based management. Habitat-based management need not require that habitats be closed to all fishing, so long as management objectives are clearly specified and outcomes monitored. Satellite-linked vessel monitoring systems, as used to manage effort in the Australian orange roughy fishery, provide one means of monitoring.

An alternative to closing particular habitats is to limit their use through economic means. Fishers in the South East Fishery pay an annual levy for fishery management based on the estimated market value of their individual transferable quota (ITQ) holdings. No account is taken of the biological or environmental impacts of their fishing practices, although managing biological impacts is the goal of single-species management, and managing broader environmental impacts is one goal of ecologically sustainable development (ESD)—a legislative requirement for the Australian Fisheries Management Authority. As Alain Laurec of the European Union said in reference to sustainable fisheries: "Limiting catches is a symptom of the disease rather than the cure" (Senior 1996). One proposed alternative to ITQs is transferable dynamic stock rights based on a fraction of a year class rather than a set tonnage, enabling a fisher to profit from catching his/her fraction of the year class at an appropriate biological (or economic) age (Townsend 1995). Future stock rights could also be dependent on the opportunity a fisher's year-class fraction has had to contribute to future generations before being caught.

Transferable dynamic stock rights have attractions, but because they require monitoring of catch and discarded catch to be effective, they would be cumbersome to monitor and enforce in most fisheries. We propose a modification of these rights: transferable ecological stock rights. In this instance a fisher would be given the right to harvest a certain fraction of a year class subject to the perceived ecological damage associated with harvesting. Monitoring (satellite-derived positions for fishing vessels) and enforcement would be based on the distribution of fishing effort in relation to habitat as a proxy for the likelihood of catching (and discarding) immature fish or causing ecological damage. If fishing in shallow waters where smaller fish reside would be expected to lead to higher discarding, then landed catch would count more against stock rights than a similar tonnage landed in deeper waters. In a similar fashion, fish caught from fishing in sensitive areas or with gear that damages benthic habitat would attract a higher deduction from that year's stock rights. Transferable ecological stock rights would provide managers an instrument more clearly linked with the goals of ecosystem management and ESD than ITQs are—and would treat the problem, not the symptoms.

Improved remote sensing and satellite-tracking technology has enabled scientists to cost-effectively research new features of marine ecosystems. The same technology has enabled fishers to target particular habitats more precisely, increasing their impact on particular productive habitats. Limiting landed catch no longer meets the requirements of managers attempting to satisfy goals of ecosystem management and ecologically sustainable development. Management of marine ecosystems requires more than management of landed catches. "Fisheries management is environmental management" (Martin Cabot, head, Newfoundland Inshore Fishermen's Association, quoted in Griffin 1993). If fisheries managers are to become environmental managers, then fisheries (environmental?) scientists must provide them with the appropriate concepts, tools, and information. In a complex system it will be essential to understand where the leverage points are. We have identified one such point for the continental shelf off southeast Australia, but it remains for managers and fishers, supported by scientists to determine how this particular leverage point can be used profitably.

Acknowledgments

We gratefully acknowledge our many colleagues at CSIRO who have contributed to this project. Of special note are the biological technicians who have processed and analyzed data, the crew of the *Southern Surveyor* who helped collect the data, and the technical and workshop staff who designed and fabricated items of equipment for us. We also thank the commercial fishers who gave us the benefit of their accumulated knowledge of the area, which was critical to planning and interpreting our comparatively brief scientific surveys. Drs. Tony Smith and Xi He and our editor Vivienne Mawson improved the manuscript with their reviews. This study was supported by research grant 94/040 from the Fisheries Research and Development Corporation.

References

Auster, P.J., and R.J. Malatesta. 1995. Assessing the role of non-extractive reserves for enhancing harvested populations in temperate and boreal marine systems. In: N.L. Shackell and J.H.M. Wilson (eds.), Marine protected areas and sustainable fisheries. Science and Management of Protected Areas Association, Wolfville, Novia Scotia, Canada, pp. 82-89.

Barker, B., I. Helmond, N. Bax, A. Williams, S. Davenport, and V. Wadley. In press. A vessel-towed camera platform for sea-floor surveys of the continental shelf. Continental Shelf Res.

Bax, N.J. 1997. The potential for spatial management and mesh size restriction for reducing bycatch of undersized tiger flathead off Southeast Australia. In: Fisheries bycatch. University of Alaska Sea Grant, AK-SG-97-02, Fairbanks, pp. 87-90.

Bax, N.J. 1998. The significance and prediction of predation in marine fishery ecosystems. ICES J. Mar. Sci. 55:997-1030.

Bax, N.J., R. Kloser, A. Williams, K. Gowlett-Holmes, and T. Ryan. In press. Seafloor habitat definition for spatial management in fisheries: A case study on the continental shelf of southeast Australia using acoustics and biotic assemblages. Oceanol. Acta.

Caddy, J.F. 1993. Toward a comparative evaluation of human impacts on fishery ecosystems of enclosed and semi-enclosed seas. Rev. Fish. Sci. 1(1):57-95.

Caton, A., K. McLoughlin, and D. Staples (eds.). 1997. Fishery status reports 1997: Resources assessments of Australian Commonwealth fisheries. Bureau of Resource Sciences, Canberra, Australia. 139 pp.

Church, J.A., and P.D. Craig. 1998. Australia's shelf seas: Diversity and complexity. In: A.R. Robinson and K.H. Brink (eds.), The sea, vol. 11. John Wiley & Sons, pp. 933-964.

Clarke, K.R. 1993. Non-parametric multivariate analyses of changes in community structure. Aust. J. Ecol. 18:117-143.

Cresswell, G. 1994. Nutrient enrichment of the Sydney continental shelf. Aust. J. Mar. Freshwater Res. 45:677-91.

Edwards, R.J. 1990. Upwelling could hold clues to fish patterns. Aust. Fish. 49:18-20.

Gislason, H., and T. Helgason. 1985. Species interaction in assessment of fish stocks with special application to the North Sea. Dana 5:1-44.

Golley, F.B. 1993. A history of the ecosystem concept in ecology: More than the sum of the parts. Yale University Press, New Haven.

Greene, H.G., M.M. Yoklavich, D. Sullivan, and G.M. Cailliet. 1994. A geophysical approach to classifying marine benthic habitats: Monterey Bay as a model. In: Applications of side-scan sonar and laser-line systems in fisheries research. Alaska Department Fish and Game, Juneau, Alaska, pp. 15-30.

Griffin, N. 1993. Newfoundland's plight offers many lessons. National Fisherman 74(8):16.

Gulland, J.A. 1982. Long-term potential effects from management of the fish resources of the North Atlantic. J. Cons. Int. Explor. Mer 40(1):8-16.

Harwood, J., and J.J.D. Greenwood. 1985. Competition between British grey seals and fisheries. In: J.R. Beddington, R.J.H. Beverton, and D.M. Lavigne (eds.), Marine mammals and fisheries. George Allen & Unwin, London, pp. 153-169.

Hudson, P.L., R.W. Griffiths, and T.J. Wheaton. 1992. Review of habitat classification schemes appropriate to streams, rivers, and connecting channels in the Great Lakes drainage basin. In: W.D.N. Busch and P.G. Sly (eds.), The development of an aquatic habitat classification system for lakes. CRC Press, Ann Arbor, Michigan, pp. 73-107.

Jones, J.B. 1992. Environmental impact of trawling on the seabed—a review. N. Z. J. Mar. Freshwater Res. 26:59-67.

Laevastu, T., and H.A. Larkins. 1981. Marine fisheries ecosystem: Its quantitative evaluation and management. Fishing News Books Ltd., Surrey, U.K.

Larkin, P.A. 1978. Fisheries management—an essay for ecologists. Ann. Rev. Ecol. Syst. 9:57-73.

Liggins, G. 1996. The interaction between fish trawling (in NSW) and other commercial and recreational fisheries. Final Report to the Fisheries Research and Development Corporation 92/79. Canberra, Australia.

May, R.M., J.R. Beddington, C.W. Clark, S.J. Holt, and R.W. Laws. 1979. Management of multispecies fisheries. Science 205:267-277.

Morrow, R.A., and I.S.F. Jones. 1988. Sea floor drag coefficients—Eastern Bass Strait. Report No. 29, University of Sydney, Ocean Sciences Institute.

Odum, E.P. 1953. Fundamentals of ecology. W.B. Saunders, Philadelphia.

Poiner, I.R., and C. Peterken. 1995. Seagrasses. In: L.P. Zann and P. Kailola (eds.), The state of the marine environment: Report for Australia. Technical annex 1: The marine environment. Environment Australia, DEST, Canberra, Australia, pp.107-117.

Risk, M.J. P.W. Sammarco, and H.P. Schwarz. 1994. Cross-continental shelf trends in $\delta^{13}C$ in coral on the Great Barrier Reef. Mar. Ecol. Prog. Ser. 106:121-130.

Sainsbury, K.J., R.A. Campbell, R. Lindholm, and A.W. Whitelaw. 1997. Experimental management of an Australian multispecies fishery: Examining the possibility of trawl-induced habitat modification. In: E.K. Pikitch, D.D. Huppert, and M.P. Sissenwine (eds.), Global trends in fisheries management. American Fisheries Society, Bethesda, Maryland, pp. 107-112.

Schindler, D.W., K.H. Mills, D.F. Malley, D.L. Findlay, J.A. Sheaver, I.J. Davies, M.A. Turner, G.A. Lindsey, and D.R. Cruikshank. 1985. Long-term ecosystem stress: The effects of years of experimental acidification on a small lake. Science 228:1395-1401.

Schwinghammer, P., J.-Y. Guigne, and W.C. Siu. 1996. Quantifying the impact of trawling on benthic habitat using high-resolution acoustics and chaos theory. Can. J. Fish. Aquat. Sci. 53:288-296.

Senge, P. 1990. The fifth discipline: The art and practice of the learning organization. Doubleday/Currency, New York.

Senior, G. 1996. Fisheries in the future: Sustainability or extinction? Marine Policy 20:91-97.

Sissenwine, M.P. 1986. Perturbation of a predator-controlled continental shelf ecosystem. In: K. Sherman and L.M. Alexander (eds.), Variability and management of large marine ecosystems. AAAS Selected Symposium 99. American Fisheries Society, Bethesda, Maryland.

Thresher, R.E., P.D. Nichols, J.S. Gunn, B.D. Bruce, and D.M. Furlani. 1992. Seagrass detritus as the basis of a coastal planktonic food chain. Limnol. Oceanogr. 37:1754-1758.

Tomczak, M. Jr. 1985. The Bass Strait water cascade during winter 1981. Continental Shelf Res. 4:255-278.

Townsend, R.E. 1995. Transferable dynamic stock rights. Marine Policy 19:153-158.

Wheatley. M.J. 1992. Leadership and the new science: Learning about organization from an orderly universe. Berrett-Koehler Publishers Inc., San Francisco.

Young, J.W., T.D. Lamb, D. Le, R.W. Bradford, and A.W.Whitelaw. 1997. Feeding ecology and interannual variations in diet of southern bluefin tuna, *Thunnus maccoyii*, in relation to coastal and oceanic waters off eastern Tasmania, Australia. Environ. Biol. Fishes 50:275-291.

Ecosystem Approaches for Fisheries Management
Alaska Sea Grant College Program • AK-SG-99-01, 1999

Sustainability: Empirical Examples and Management Implications

Charles W. Fowler, Jason D. Baker, Kim E.W. Shelden, Paul R. Wade, Douglas P. DeMaster, and Roderick C. Hobbs
National Marine Fisheries Service, National Marine Mammal Laboratory, Seattle, Washington

Abstract

Management in regard to ecosystems must meet a number of criteria. As indicated in Table 1, such management must: (1) be consistent with management in respect to other biological systems; (2) account for reality, including uncertainty; (3) result in components of each level of biological organization falling within normal ranges of natural variation; (4) exercise precaution and consider risk in achieving sustainability; (5) be information-based and interdisciplinary; (6) include monitoring and assessment; (7) have clear objectives; (8) recognize that our ability to control is limited to managing human activity; and (9) consider humans as part of ecosystems.

Management to meet these requirements can be guided by empirical examples of sustainability demonstrated by other species. An example is consumption of resources within ecosystems by predators. Limits to sustainability are demonstrated by frequency distributions of consumption rates observed among predators. These empirical examples are the result of a variety of influences and constraints, including processes of natural selection resulting from exposure to the suite of factors we wish to account for in management. In this paper, the consumption by marine vertebrates is presented to empirically exemplify sustainable resource consumption from both ecosystems and individual species.

The implications of empirical examples of sustainability for management are emphasized by the central tendencies in frequency distributions among species. The lack of consumption rates beyond the tails of observed distributions indicates that consumption rates in these two extremes are not sustainable over long time scales.

Introduction

Much has been published on the topic of management involving ecosystems (e.g., Christensen et al. 1996; Mangel et al. 1996; Fowler 1999, In press). The requisites for successful management are numerous but can be condensed to the nine essential elements or criteria presented in Table 1, all of which need to be adhered to simultaneously (Fowler 1999). While conceptually rich, the available literature merely describes reasonable management methods and serves well to eliminate alternatives but does not specify an acceptable approach.

Existing forms of management routinely fail to conform to one or more of the criteria presented in Table 1, exposing the difficulty of finding a practical form of management that meets all nine. In this paper, we suggest a way forward in the form of management that follows the example of other species, treating them as empirical models of sustainability (Fowler 1999; Fowler, In press). We exemplify this approach by considering management of the harvest (take, utilization, or consumption) of biomass from both ecosystems and individual resource species as two levels of biological organization.

Management at the Ecosystem Level

What level of biomass consumption in an ecosystem achieves the greatest sustainabilty? To answer this question, we begin by examining relevant empirical information provided by other species, then turn to management based on this information.

Figure 1 shows frequency distributions for estimated rates of consumption among 13 species of marine mammals found in the Bering Sea. Information for each species is represented through its contribution to the height of the bar corresponding to the biomass it is estimated to consume ($\log_{10}$ metric tons; Sobolevsky and Mathisen 1996). Figure 2 shows a similar distribution representing biomass consumption by 24 species of marine mammals and seabirds from the Georges Bank ecosystem off the east coast of the United States. These nonhuman species of the Bering Sea and Georges Bank are empirical examples of varying degrees of sustainability within their ecosystems having, in part, survived the risks of extinction to which they are exposed. Thus, Figs. 1 and 2 illustrate the types of information that can be used to answer the question raised at the outset of this section.

How can we base management on such information? Generally, management would restrict consumption by humans to within the bounds of the normal range of natural variation in the sustainability exhibited by other species. Specifically, in this case, catches by commercial fisheries would be confined to within the range shown in Figs. 1 and 2 as one step toward assuring that humans consume resources sustainably.

Table 1. A list of criteria that must be met, or principles to be adhered to, by any form of management, particularly any that applies to the human use of natural resources (Christensen et al. 1996, Mangel et al. 1996, Fowler 1999).

1. Management of the harvest of biomass from individual resource species cannot be in conflict with management of the harvest of biomass from the ecosystems in which the harvested species occur. Similarly, biomass consumption by humans from the biosphere must be guided by principles that are not in conflict with those guiding the harvest of biomass from either an individual resource species or any particular ecosystem. Any form of management must apply simultaneously at the various levels of biological organization and it must do so consistently.
2. Management action must be based on an approach that accounts for reality in its complexity over the various scales of time, space, and biological organization. The context of environmental factors (e.g., ecological complexity) must be accounted for along with the elements of stochasticity and the diversity of processes, mechanics, and dynamics. The complexes of chemical and physical substances and processes as well as energetic dynamics must be taken into account. Furthermore, we must be able to deal with uncertainty, including what we cannot know, or may never know.
3. A core principle of management is that of undertaking actions that ensure that individuals, species, and ecosystems are within (or will return to) their respective normal ranges of natural variation as components of the more aggregated levels of biological organization (Rapport et al. 1981, 1985; Christensen et al. 1996; Holling and Meffe 1996; Mangel et al. 1996). Any form of management must apply this principle.
4. Management must be risk-averse and exercise precaution in achieving sustainability. Sustainability is, by definition, not achieved by any form of management that generates risk rather than minimizing it.
5. Guidance must be available to management in the form of useful information that enables managers to develop meaningful, measurable, and reasonable goals and objectives. This information must be based on interdisciplinary approaches to adhere to the principle behind criterion 2.
6. Management must include science (scientific methods and principles) in research, monitoring and assessment, not only to produce the information that is used for guidance (criterion 5), but also for evaluation of progress in achieving established goals and objectives (criterion 7).
7. There must be clearly defined goals and objectives that are measurable (quantifiable) to provide quantitative evaluation of problems to be solved and gauge progress in solving them.
8. It must be recognized that control over other species and ecosystems is impossible. The only option for control is the control of human action (Christensen et al. 1996, Holling and Meffe 1996, Mangel et al. 1996). For example, we can control fishing effort but not the resource population or ecosystem in which it occurs. We can influence the resource population and its ecosystem, but we cannot control them to avoid indirect changes, side effects, or secondary reactions brought about by our influence. The guidance (criterion 7) we need for management is guidance regarding the level of influence (e.g., harvest rate) that meets the other criteria of this list.
9. Humans must have the option of being components of at least some ecosystems to avoid the unrealistic option of precluding human existence.

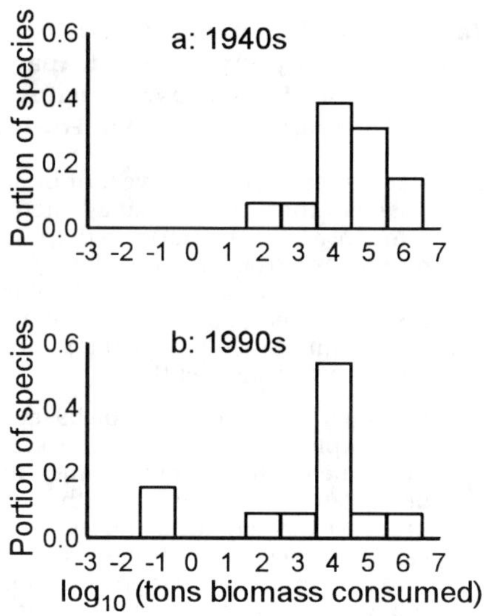

Figure 1. Two frequency distributions representing consumption rates ($\log_{10}$ metric tons consumed annually) by 13 cetacean (whale) species from the Bering Sea showing variability and its limits among these species. This figure compares distributions for the mid-1900s (a) and the late 1980s and early 1990s (b). From Sobolevsky and Mathisen (1996).

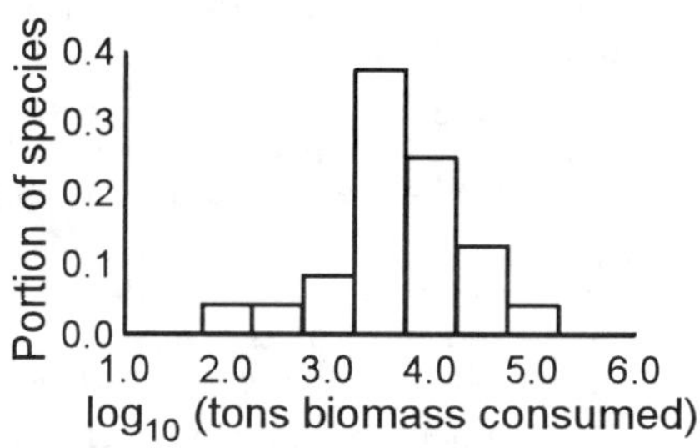

Figure 2. A frequency distribution representing the Georges Bank ecosystem showing variability among 24 species of marine mammals and birds, as distributed according to estimated annual biomass consumption for each species ($\log_{10}$ metric tons) within this region. From Backus and Bourne (1986).

The concept of using other species as empirical examples of sustainability may seem simple on the surface. In practice, however, implementing this form of management is extremely difficult. This is true at the ecosystem level as introduced above as well as at the individual species level discussed in our next example.

Management at the Single-Species Level

Criterion 1 (Table 1) requires that management be consistent in its application at both the single-species level and the ecosystem level. One of the many questions facing managers regarding individual species is: "What level of harvest (e.g., biomass consumption) is the most sustainable?" To address such questions we must consider more than the population dynamics of the resource species. It is important to account for evolution and coevolution, including anthropogenic impacts (i.e., the genetic influence of harvesting; Law et al. 1993). This helps account for systemic complexity (criterion 2, Table 1).

Frequency distributions of estimated rates of consumption among fishes, birds, and marine mammals as predators on specific prey species are shown in Figs. 3 and 4. Similar to the ecosystem approach outlined above, a sustainable level of harvest from each individual species would be achieved by management to restrict commercial harvests to levels within the normal range of natural variation exhibited by the respective set of consuming species exemplified in Figs. 3 and 4. Again, implementation of such a strategy is complicated. Yet if such an approach were employed, the risks that preclude species from occurring beyond the range of such distributions would be avoided. These risks include those posed by long-term ecosystem change stimulated by the influence of species that even briefly occur in the upper extreme of the normal range of natural variation. This approach leads to what can be called ecologically sustainable yield rates (ESY for the yield, ESYR for the rate; Fowler 1999) which correspond to the central tendencies (e.g., mean) of frequency distributions similar to those illustrated in Figs. 3 (yields) and 4 (rates).

Meeting Management Standards

How does the management described above meet the criteria presented in Table 1? Fishing practices in our examples would be constrained based on information like that exhibited in Figs. 1-4, but not prohibited. Thus, humans would be constrained (criterion 8) to fall within the normal range of natural variation (criterion 3), rather than being excluded (criterion 9). The approach is risk-aversive and precautionary (criterion 4) because commercial harvests confined to levels near the central tendencies of such distributions avoid the collective risks and constraints that have prevented the accumulation of species in the tails of such distributions. Thus, sustainability (criterion 4) is more likely in the central regions of species

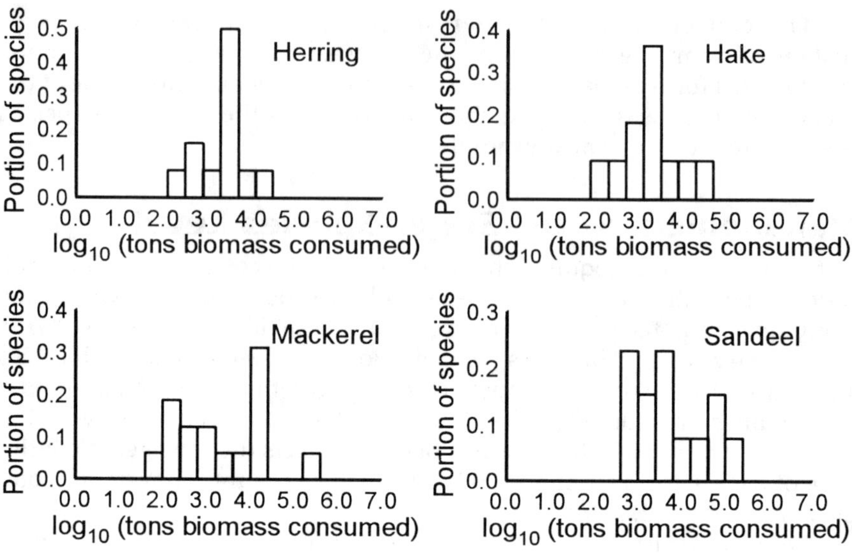

Figure 3. *The frequency distribution of nonhuman vertebrate species that consume hake (Merluccius bilinearis; N = 11), herring (Clupea harengus; N = 12), mackerel (Scomber scombrus; N = 16), and sand eel (Ammodytes americanus; N = 13) measured as the $\log_{10}$ (metric tons) of biomass consumed in an area (ecosystem) of the northwest Atlantic Ocean (Overholtz et al. 1991; pers. comm., S.A. Murawski and W.J. Overholtz, National Marine Fisheries Service, Northeast Fishery Science Center, Woods Hole, MA 02543).*

Figure 4. *The frequency distribution of 21 nonhuman vertebrate species that consume walleye pollock (Theragra chalcogramma) in the Bering Sea and North Pacific, according to the $\log_{10}$ of the fraction of the standing stock biomass of pollock they consume (Livingston 1993 and pers. comm. 1994).*

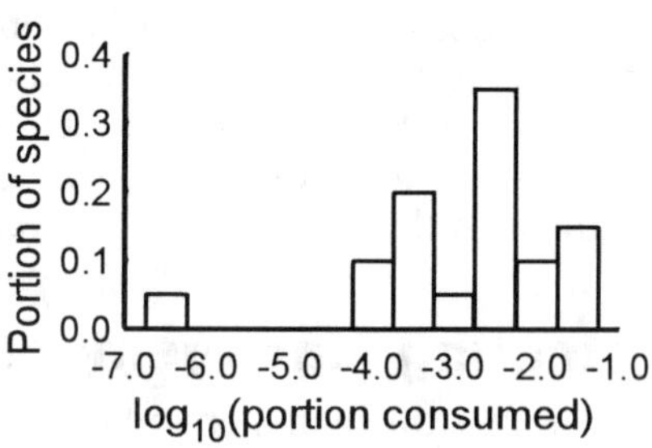

frequency distributions than it is in the tails. As such, the central tendencies of frequency distributions provide specific measures that can be used to define goals or objectives for management (criteria 5 and 7).

The various temporal and spatial scales and elements of complexity are accounted for in this approach (criterion 2), although the available data are limited in this regard. Long time scales and broad spatial scales are accounted for, in part, by the evolutionary dynamics that influence where species fall in the range represented by species frequency distributions. Natural selection affects these distributions both through classical individual selection and the dynamics of selective extinction and speciation (Lewontin 1970; Slatkin 1981; Arnold and Fristrup 1982; Fowler and MacMahon 1982; Levinton 1988; Eldredge 1989; Williams 1992; Fowler, In press). Extinction is one of the risks that prevent the accumulation of species in the tails of such distributions. All forms of selection are influenced by the variety of factors making up the reality (e.g., ecological conditions, criterion 2) to which species are exposed (and simultaneously a part).

Applied at both the single-species and ecosystem levels (as well as other levels of biological organization, criterion 1 of Table 1), this approach meets all of the requirements of management presented in Table 1. Criterion 6 is met when science is used to produce data for species frequency distributions and to monitor change.

Limitations to Existing Data

Existing data are clearly imperfect. The information in species frequency distributions would be more reliable if integrated (e.g., averaged) over longer periods of time. Patterns (e.g., correlations) across environmental factors such as mean temperature and annual variance in temperature should be taken into account. The same holds for spatial patterns. Scientists are faced with an immense challenge in providing such information.

It is important to remove the effects of factors that may introduce bias. The species frequency distributions in Figs. 1 and 2 are probably biased. Ideally, we want data that represent ecosystems for which human influence is either absent or historically within the normal ranges of natural variation. The information in Fig. 1a cannot be considered free of such influence, but may better reflect the natural state of the ecosystem than that presented in Fig. 1b. Different levels of human influence may explain part of the differences between Figs. 1a and 1b. These include the cumulative effects of considerable sealing, whaling, and fishing, plus a number of other factors such as atmospheric transport of pesticides and their degradation products (National Research Council 1996). Interdisciplinary efforts will be necessary to produce adequate data by taking such factors into account.

Similarly, the data summarized in Fig. 2 may be influenced by commercial fishing, especially as harvest rates have often been outside the normal range of natural variation for the Georges Bank ecosystem. Changes introduced to such distributions by human influence could include increased variation, biased mean, or modified shape.

Temporal dynamics and measurement bias also likely affect species frequency distributions. Figures 1 and 2 represent little more than a snapshot of a dynamic system. Beyond process-related variance, there is error and bias attributable to measurement and estimation procedures that influence observed distributions.

Thus, the quality of existing data is one problem. Our choice and interpretation of data can result in others. It is important to avoid being misled. It is important to account for factors that are involved in correlative relationships as would be the case in our examples where trophic level and total consumption are presumably interrelated. Tiny invertebrates might be inferior as examples of sustainable biomass consumption compared to small cetaceans that consume at the same trophic level as humans. Care, and often much more research, are needed to find the subsets of data that provide adequate and relevant information regarding the normal ranges of natural variation to guide decision-making.

Discussion

In this paper, we have formulated management by using nonhuman species as natural empirical examples of sustainability. Humans cannot claim to be operating sustainably when we are found at the extremes of the range of natural variation. Species in the tails of frequency distributions like Figs. 1-4 do not serve as good examples, especially if they have been there only briefly.

We must now deal with the burden of proof (Mangel et al. 1996, Dayton 1998). Consumption rates by nonhuman predators empirically exemplify sustainability and are clearly more proven alternatives for human fisheries management than harvest rates based on population models (e.g., maximum sustainable yield [MSY] rates). For a variety of reasons it would seem on the surface that sustainability can be maximized for harvest levels in or beyond the upper tails of unbiased frequency distributions (e.g., through consideration of population dynamics and resulting estimates of MSY). There certainly is no lack of desire for greater harvest rates (e.g., considering economic, historical, or cultural factors). But now we must prove that harvest rates larger than the central tendencies of unbiased frequency distributions achieve greater sustainability if we are going to use them. To do otherwise is not precautionary.

We recommend consideration of the approach described herein by using the preliminary information in cases represented by frequency distributions such as those of Figs. 1-4 (Fowler 1999; Fowler, In press), in spite of their limitations. Management based on such information satisfies all of the nine criteria laid out in Table 1 and therefore satisfies the need for management that applies at the ecosystem level. Doing so leads to application at other levels of biological organization and to other ways of measuring species, thereby further adhering to criteria 1 and 2 of Table 1.

However, the acceptance and implementation of the management approach that we propose are extremely difficult. This difficulty emphasizes the complexity with which we are dealing. Existing fisheries harvest at rates that are often more than ten-fold greater than the mean consumption rates for nonhuman predators. Making the changes required to implement management as we describe it here is even more of a challenge than the research needed for more reliable information. Involved are institutional, social, economic, political, and behavioral changes, many of which are beyond comprehension. The magnitude of these challenges, however, is more a measure of the size of the problems to solve than justification for avoiding the work required.

Acknowledgments

We thank Pat Livingston for help and information necessary for the construction of Fig. 4 and Steve Murawski and William Overholtz for their help in providing the information necessary to construct Fig. 3. We also thank Gary Duker, John Jansen, James Lee, Michael Perez, and Bruce Robson who kindly provided very useful reviews of earlier drafts of this paper.

References

Arnold, A.J., and K. Fristrup. 1982. The theory of evolution by natural selection: A hierarchical expansion. Paleobiology 8:113-129.

Backus, R.H., and D.W. Bourne. 1986. Georges Bank. MIT Press, Boston.

Christensen, N.L., A.M. Bartuska, J.H. Brown, S.R. Carpenter, C. D'Antonio, R. Francis, J.F. Franklin, J.A. MacMahon, R.F. Noss, D.J. Parsons, C.H. Peterson, M.G. Turner, and R.G. Woodmansee. 1996. The report of the Ecological Society of America Committee on the Scientific Basis for Ecosystem Management. Ecol. Appl. 6:665-691.

Dayton, P. 1998. Reversal of the burden of proof in fisheries management. Science 279:821-822.

Eldredge, N. 1989. Macroevolutionary dynamics: Species, niches, and adaptive peaks. McGraw-Hill Publishing Company, New York.

Fowler, C.W. 1999. Nature's Monte Carlo experiments in sustainability. In: V.R. Restrepo (ed.), Proceedings of the Fifth NMFS Stock Assessment Workshop: Providing scientific advice to implement the precautionary approach under the Magnuson-Stevens Fishery Conservation and Management Act. NOAA Tech. Memo. NMFS-F/SPO, pp. 25-32.

Fowler, C.W. In press. A niche for humans. U.S. Government Printing Office.

Fowler, C.W., and J.A. MacMahon. 1982. Selective extinction and speciation: Their influence on the structure and functioning of communities and ecosystems. Am. Nat. 119:480-498.

Holling, C.S., and G.K. Meffe. 1996. Command and control and the pathology of natural resource management. Conserv. Biol. 10:328-337.

Law, R., J.M. McGlade, and T.K. Stokes (eds.). 1993. The exploitation of evolving resources: Proceedings of an international conference held at Julich, Germany, Sept. 3-5, 1991. (Lecture notes in biomathematics, 99). Springer-Verlag, Berlin.

Levinton, J.S. 1988. Genetics, paleontology, and macroevolution. Cambridge University Press, New York.

Lewontin, R.C. 1970. The units of selection. Ann. Rev. Ecol. Syst. 1:1-18.

Livingston, P. A. 1993. Importance of predation by groundfish, marine mammals and birds on walleye pollock *Theragra chalcogramma* and Pacific herring *Clupea pallasi* in the eastern Bering Sea. Mar. Ecol. Prog. Ser. 102:205-215.

Mangel, M., L.M. Talbot, G.K. Meffe, M.T. Agardy, D.L. Alverson, J. Barlow, D.B. Botkin, G. Budowski, T. Clark, J. Cooke, R.H. Crozier, P.K. Dayton, D.L. Elder, C.W. Fowler, S. Funtowicz, J. Giske, R.J. Hofman, S.J. Holt, S.R. Kellert, L.A. Kimball, D. Ludwig, K. Magnusson, B.S. Malayang, C. Mann, E.A. Norse, S.P. Northridge, W.F. Perrin, C. Perrings, R.M. Peterman, G.B. Rabb, H.A. Regier, J.E. Reynolds III, K. Sherman, M.P. Sissenwine, T.D. Smith, A. Starfield, R.J. Taylor, M.F. Tillman, C. Toft, J.R. Twiss, Jr., J. Wilen, and T.P. Young. 1996. Principles for the conservation of wild living resources. Ecol. Appl. 6:338-362.

National Research Council. 1996. The Bering Sea ecosystem. National Academy Press, Washington, DC. 307 pp.

Overholtz, W.J., S.A. Murawski, and K.L. Foster. 1991. Impact of predatory fish, marine mammals, and seabirds on the pelagic fish ecosystem of the northeastern USA. ICES Marine Science Symposia 193:198-208.

Rapport, D.J., H.A. Regier, and T.C. Hutchinson. 1985. Ecosystem behavior under stress. Am. Nat. 125:617-640.

Rapport, D.J., H.A. Regier, and C. Thorpe. 1981. Diagnosis, prognosis, and treatment of ecosystems under stress. In: G.W. Barrett and R. Rosenberg (eds.), Stress effects on natural ecosystems. John Wiley and Sons, New York, pp. 269-280.

Slatkin, M. 1981. A diffusion model of species selection. Paleobiology 7:421-425.

Sobolevsky, Y.I., and O.A. Mathisen. 1996. Distribution, abundance, and trophic relationships of Bering Sea cetaceans. In: O.A. Mathisen and K.O. Coyle (eds.), Ecology of the Bering Sea: A review of Russian literature. University of Alaska Sea Grant, AK-SG-96-01, Fairbanks, pp. 265-275.

Williams, G.C. 1992. Natural selection: Domains, levels, and challenges. Oxford University Press, New York.

Incorporating Ecosystem Considerations into Management of Bering Sea Groundfish Fisheries

David Witherell
North Pacific Fishery Management Council, Anchorage, Alaska

Abstract

The North Pacific Fishery Management Council has made significant progress toward ecosystem-based management of the Bering Sea and Aleutian Islands area. First and foremost, the council has taken a precautionary approach to extraction of fish resources. As a result, all groundfish stocks covered under the fishery management plan are considered healthy and not overfished. Potential impacts of fishing on other ecosystem components such as marine mammals and seabirds are also considered. The council's approach involves public participation, reliance on scientific research and advice, conservative catch quotas, comprehensive monitoring and enforcement, bycatch controls, habitat conservation areas, and other biological and socioeconomic considerations.

Introduction

Ecosystem-based management strategies are being adopted throughout the United States in response to biodiversity concerns. As defined by the Ecological Society of America, ecosystem management is "management driven by explicit goals, executed by policies, protocols, and practices, and made adaptable by monitoring and research based on our best understanding of the ecological interactions and processes necessary to sustain ecosystem composition, structure, and function" (Christensen et al. 1996). In more general terms, ecosystem-based management is shorthand for a more holistic approach, which focuses more on maintaining system integrity than on maximizing extraction of certain resources.

The precautionary principle has developed over the past 10 years as a policy measure to address sustainability of natural resources in the face of uncertainty. The principle is becoming widely adopted throughout the world in both national and international environmental policies. The central element of the precautionary principle is that precise impacts caused by human activity cannot be known with certainty, so that a more cautious approach is required (Dovers and Handmer 1995). This is particularly true when there is a high level of uncertainty and there are large (potentially irreversible) costs if a mistake is made (Garcia 1995). Fisheries management around the world had traditionally been based on a preventative and trial-and-error approach, yet the collapse of some fisheries indicates that a more precautionary approach should have been applied.

Although the precautionary principle and ecosystem-based management are relatively new concepts being debated in the scientific literature, the North Pacific Fishery Management Council has used these approaches for management of North Pacific groundfish fisheries since 1976. The council is a regional organization established by the Magnuson-Stevens Act in 1976 when the United States extended its fisheries jurisdiction out to 200 miles. The council, together with the National Marine Fisheries Service, has primary responsibility for groundfish management in the Gulf of Alaska, Bering Sea, and Aleutian Islands area, encompassing about 900,000 square nautical miles. Conservative management policies were implemented right from the start, with adoption of the first fishery management plans (FMPs). Since 1984, the council's number one comprehensive management goal has been to conserve fishery resources, maintain habitats, and give full consideration for interactions with other elements of the ecosystem. The council has also had a comprehensive policy on habitat since 1988. The objectives of this policy are to maintain the current quantity and productive capacity of habitats, and restore and rehabilitate any habitats previously degraded.

The primary goal of the Bering Sea and Aleutian Islands (BSAI) groundfish FMP is to promote conservation while providing optimum yield from marine resources. To accomplish this goal, the FMP specifies six objectives for conservative management (Table 1), all of which are consistent with a precautionary approach and ecosystem-based management. Clearly, these objectives have been met successfully, as all BSAI groundfish stocks are considered healthy after 20 years of sustained annual harvests of nearly 2 million t. No fish stocks were deemed overfished in a recent evaluation of the status of U.S. fisheries (National Marine Fisheries Service 1997). When revised overfishing definitions are implemented in 1999, only one fish stock in the region (Bering Sea *Chionoecetes bairdi* crab) is expected to be below its minimum stock size threshold and declared overfished.

Concerns about the impacts of fish removals on other components of the ecosystem have motivated the council to continue development of a more ecosystem-based management strategy for the Bering Sea. This development has progressed at all levels, from science to policy making. Since 1995, the Groundfish Plan Teams have prepared an Ecosystem Considerations section to supplement the annual Stock Assessment and Fish-

Table 1. Objectives of the Bering Sea/Aleutian Islands area groundfish fishery management plan.

1. Conservation and management measures take into account the unpredictable characteristics of future resource availability.
2. Individual stocks of fish should be managed as a unit throughout their range.
3. Stocks shall be rebuilt if they have declined to a level below that capable of producing MSY.
4. Management should avoid disruption of social and economic structures where practicable.
5. In the face of uncertainty, allowable harvest levels should contain a margin of safety.
6. Minimize impacts of fisheries on other fish and the environment.

ery Evaluation report (North Pacific Fishery Management Council 1997). This chapter provides an annual assessment of the ecosystem, a review of recent ecosystem-based management literature, updates of ongoing ecosystem research, and new information on the status of seabirds, marine mammals, habitat, and other components of the North Pacific ecosystem.

In 1996, the council established an Ecosystem Committee to discuss possible approaches to incorporating ecosystem concerns into the fishery management process. The committee has held workshops on ecosystem research, held several meetings to discuss essential fish habitat, and has hosted numerous informal discussions on ecosystem-based management and habitat concerns. A major role of this committee has been to provide the council and stakeholders information on ecosystem-based management in the North Pacific. The committee identified primary principles and elements of ecosystem management from scientific literature (e.g., Grumbine 1994, Mangel et al. 1996, Christensen et al. 1996). The concept of ecosystem-based management includes the elements of sustainability, goals, ecological models and understanding, complexity, dynamic character, context and scale, adaptability, and humans as ecosystem components. The committee's draft policy for ecosystem-based management of North Pacific fisheries is shown in Table 2.

Fisheries can impact ecosystems in numerous ways, including the selectivity, magnitude, timing, location, and methods of fish removals. Fisheries can also impact ecosystems by gear interactions, vessel disturbance, nutrient cycling, introduction of exotic species, pollution, and habitat alteration. Managers attempt to minimize potential impacts while at the same time allowing the extraction of fish resources at sustainable levels. A review of ecosystem-based management actions taken by the council is provided below.

Table 2. Draft ecosystem-based management policy of the North Pacific Fishery Management Council.

Definition	Ecosystem-based management, as defined by the NPFMC, is a strategy to regulate human activity towards maintaining long-term system sustainability (within the range of natural variability as we understand it) of the North Pacific, covering the Gulf of Alaska, the Eastern and Western Bering Sea, and the Aleutian Islands region.
Objective	Provide future generations the opportunities and resources we enjoy today.
Goals	1. Maintain biodiversity consistent with natural evolutionary and ecological processes, including dynamic change and variability. 2. Maintain and restore habitats essential for fish and their prey. 3. Maintain system sustainability and sustainable yields of resources for human consumption and non-extractive uses. 4. Maintain the concept that humans are components of the ecosystem.
Guidelines	1. Integrate ecosystem-based management through interactive partnerships with other agencies, stakeholders, and public. 2. Utilize sound ecological models as an aid in understanding the structure, function, and dynamics of the ecosystem. 3. Utilize research and monitoring to test ecosystem approaches. 4. Use precaution when faced with uncertainties to minimize risk; management decisions should err on the side of resource conservation.
Understanding	1. Uncontrolled human population growth and consequent demand for resources is inconsistent with resource sustainability. 2. Ecosystem-based management requires time scales that transcend human lifetimes. 3. Ecosystems are open, interconnected, complex, and dynamic; they transcend management boundaries.

Management Actions

Conservative Catch Limits

Total removals of groundfish are controlled by annual catch limits established for each stock. For target species, three harvest levels are set, corresponding to the overfishing level (OFL), the acceptable biological catch (ABC), and total allowable catch (TAC). TACs are essentially annual catch limits for the fishery, and are established at or below the ABC. ABCs generally define acceptable harvest levels from a biological perspective, and OFL defines the unacceptable harvest level. Specification of harvest limits is done in a precautionary manner, due to a number of reasons as explained below.

Harvest rates specifications are more conservative when less information is available (Thompson 1997). The maximum allowable rates are prescribed through a set of six tiers which are listed below in descending order of preference, corresponding to descending order of information availability (Table 3). For most tiers, ABC is based on $F_{40\%}$, which is the fishing mortality rate associated with an equilibrium level of spawning per recruit (SPR) equal to 40% of the equilibrium level of spawning per recruit in the absence of any fishing. The $F_{40\%}$ rate is considered to be a very conservative harvest rate for most fish stocks (Clark 1993, Rosenberg and Restrepo 1995). To further minimize the possibility of catches jeopardizing a stock's long term productivity, there is a buffer established between ABC and OFL. The OFL definition was recently increased from $F_{30\%}$ to $F_{35\%}$ for stocks having tiers 2-4 information.

Harvest rates used to establish ABCs are reduced at low stock size levels, thereby allowing rebuilding of depleted stocks. If the biomass of any stock falls below B_{msy} or $B_{40\%}$ (the long-term average biomass that would be expected under average recruitment and $F = F_{40\%}$), the fishing mortality is reduced relative to stock status. This serves as an implicit rebuilding plan should a stock fall below a reasonable abundance level. The council has a record of rebuilding groundfish stocks that were depleted prior to implementation of the Magnuson-Stevens Act. For example, conservative harvest policies adopted in the 1980s had the effect of restoring depleted stocks such as yellowfin sole, *Pleuronectes asper*, and sablefish, *Anoplopoma fimbria* (Megrey and Wespestad 1990).

As a result of these definitions, specified harvest rates for groundfish stocks are very low. Actual harvest rates are significantly lower for many species, as the TAC may be set lower than ABC, and harvests may be less than TAC due to regulatory closures. All fish caught in any fishery (including bycatch), whether landed or discarded, accrue towards the TAC for that stock. Although 100% mortality for all discards is assumed, some fish likely survive, so actual removals are lower than published catch numbers indicate.

Additional precaution is incorporated into the BSAI groundfish catch specification. Since 1981, the total annual allowable catch for BSAI ground-

Table 3. Tiers used to determine catch specifications for BSAI groundfish stocks as approved under Amendment 56.

1. Information available: Reliable point estimates of B and B_{MSY} and reliable pdf of F_{MSY}.
   - 1(a) Stock status: $B/B_{MSY} > 1$
     - $F_{OFL} = m_A$, the arithmetic mean of the pdf
     - $F_{ABC} \leq m_H$, the harmonic mean of the pdf
   - 1(b) Stock status: $a < B/B_{MSY} \leq 1$
     - $F_{OFL} = m_A \times (B/B_{MSY} - a)/(1 - a)$
     - $F_{ABC} \leq m_H \times (B/B_{MSY} - a)/(1 - a)$
   - 1(c) Stock status: $B/B_{MSY} \leq a$
     - $F_{OFL} = 0$
     - $F_{ABC} = 0$
2. Information available: Reliable point estimates of B, B_{MSY}, F_{MSY}, $F_{30\%}$, and $F_{40\%}$.
   - 2(a) Stock status: $B/B_{MSY} > 1$
     - $F_{OFL} = F_{MSY}$
     - $F_{ABC} \leq F_{MSY} \times (F_{40\%}/F_{35\%})$
   - 2(b) Stock status: $a < B/B_{MSY} \leq 1$
     - $F_{OFL} = F_{MSY} \times (B/B_{MSY} - a)/(1 - a)$
     - $F_{ABC} \leq F_{MSY} \times (F_{40\%}/F_{35\%}) \times (B/B_{MSY} - a)/(1 - a)$
   - 2(c) Stock status: $B/B_{MSY} \leq a$
     - $F_{OFL} = 0$
     - $F_{ABC} = 0$
3. Information available: Reliable point estimates of B, $B_{40\%}$, $F_{30\%}$, and $F_{40\%}$.
   - 3(a) Stock status: $B/B_{40\%} > 1$
     - $F_{OFL} = F_{35\%}$
     - $F_{ABC} \leq F_{40\%}$
   - 3(b) Stock status: $a < B/B_{40\%} \leq 1$
     - $F_{OFL} = F_{35\%} \times (B/B_{40\%} - a)/(1 - a)$
     - $F_{ABC} \leq F_{40\%} \times (B/B_{40\%} - a)/(1 - a)$
   - 3(c) Stock status: $B/B_{40\%} \leq a$
     - $F_{OFL} = 0$
     - $F_{ABC} = 0$
4. Information available: Reliable point estimates of B, $F_{30\%}$, and $F_{40\%}$.
   - $F_{OFL} = F_{35\%}$
   - $F_{ABC} \leq F_{40\%}$
5. Information available: Reliable point estimates of B and natural mortality rate M.
   - $F_{OFL} = M$
   - $F_{ABC} \leq 0.75 \times M$
6. Information available: Reliable catch history from 1978 through 1995.
   - OFL = the average catch from 1978 through 1995, unless an alternative value is established by the SSC on the basis of the best available scientific information
   - ABC $\leq$ 0.75 $\times$ OFL

ABC is acceptable biological catch, *OFL* is the overfishing level, *F* is instantaneous fishing mortality rate, *B* is exploitable biomass, "pdf" is probability density function, and *MSY* is maximum sustainable yield.

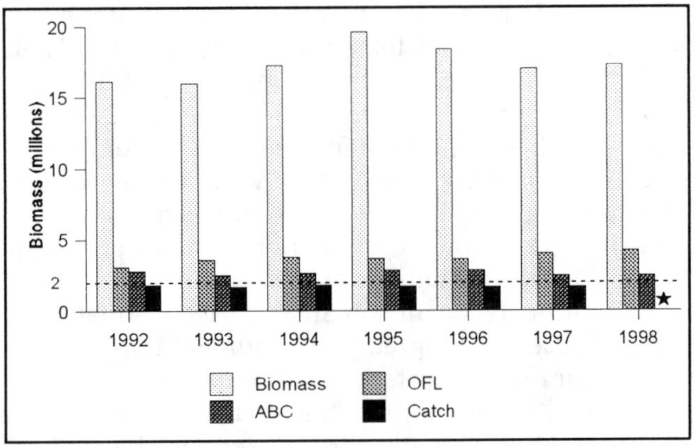

Figure 1. Exploitable biomass and catch specifications for Bering Sea and Aleutian Islands groundfish complex, 1992-98. The dashed line shows the 2 million t optimum yield limit. ABC is acceptable biological catch, and OFL is the overfishing level.

fish complex must fall within an optimum yield range of 1.4-2.0 million t. This has limited the sum of TACs for all species to 2 million t per year, which has been considerably less than the sum of all ABCs (averaging about 2.8 million t per year). As a result, most groundfish stocks, particularly flatfish stocks, are being exploited well below sustainable levels (Witherell 1995). Based on observer data and reports provided by the fleet, the National Marine Fisheries Service closes directed fisheries for each species or complex prior to when the TAC is taken. As such, inseason managers have been effective at maintaining catches of groundfish within biologically acceptable levels (Fig. 1).

Bycatch Controls

Bycatch management measures implemented for groundfish fisheries of the eastern Bering Sea have focused on reducing the incidental capture and injury of species traditionally harvested by other fisheries. These species include king crab, *Paralithodes* and *Lithodes* spp.; Tanner crab, *Chionoecetes* spp.; Pacific herring, *Clupea pallasi*; Pacific halibut, *Hippoglossus stenolepis*; and Pacific salmon and steelhead trout, *Oncorhynchus* spp. Collectively, these species are called "prohibited species," as they cannot be retained as bycatch in groundfish fisheries and must be discarded with a minimum of injury.

Bycatch controls were instituted on foreign groundfish fisheries prior to passage of the Magnuson-Stevens Act in 1976 and have become more restrictive in recent years (Witherell and Pautzke 1998). Bycatch limits for 1998 BSAI groundfish fisheries included 3,775 t of halibut mortality, 1,697 t of herring, 100,000 red king crabs, 2,850,000 *C. bairdi* crab, 4,654,000 *C. opilio* crab, 48,000 chinook salmon, and 42,000 other salmon. Bycatch limits for herring and crab are based on biomass of those stocks, and therefore fluctuate from year to year. The bycatch limits are apportioned to specific groundfish target fisheries and may also be seasonally apportioned. Attainment of any apportionment closes that groundfish target fishery for the remainder of the season.

To address Magnuson-Stevens Act mandates to reduce bycatch, the council recently adopted an amendment to prohibit the use of non-pelagic trawl gear for vessels targeting pollock in the BSAI. Only pelagic trawl gear as defined in regulations (together with the performance-based bycatch standard of 20 crabs) will be allowed in the directed pollock fishery. Although this action could have been taken annually as part of the BSAI TAC specification process, the plan amendment will make this prohibition a permanent regulation. Total bycatch limits of prohibited species will be reduced to reflect this gear prohibition. Prohibited species bycatch will be reduced by 100 t of halibut mortality, 3,000 red king crab, 50,000 *C. bairdi* crab, and 150,000 *C. opilio* crab.

Limits on Discards and Waste

The issue of discarding and waste of fish resources stems from social, economic, and conservation concerns. Fish are discarded for two reasons: either they are required to be thrown back due to regulations, or they are unwanted by that fishing vessel. Discards of unwanted fish (so-called economic discards) result when fishermen do not have markets, sufficient equipment, time, or return to retain and process the catch. In the 1997 BSAI fisheries, a total of 258,000 t of groundfish were discarded, equating to about 15% of the total groundfish catch. Although this discard rate is much lower than most of the world's fisheries (Alverson et al. 1994), the sheer volume of the discards is troublesome to many people who consider economic discards as waste of food and an impact to the ecosystem.

To reduce discarding, the council adopted an improved retention and utilization program for all Bering Sea and Aleutian Islands groundfish fisheries. Beginning in 1998, 100% retention of pollock *(Theragra chalcogramma)* and Pacific cod *(Gadus macrocephalus)* was required. Rock sole *(Lepidopsetta bilineata)* and yellowfin sole *(Limanda aspera)* retention will be required beginning in 2003, with the delay time necessary to allow for development of markets and gear technological responses by the vessels engaged in these fisheries. These retention requirements are expected to reduce overall discard rates from 15% to less than 5%. The council addressed the utilization side of the program not by mandating specific product forms, but instead by requiring a minimum required product recovery rate of 15%.

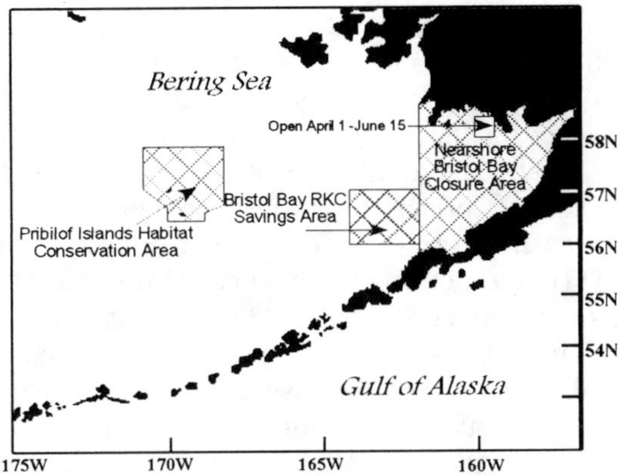

Figure 2. Location of marine protection areas in the Berinq Sea.

Marine Protected Areas

The council and the Alaska Board of Fisheries have established several marine protected areas to protect habitat for fish, crabs, and marine mammals. Three large areas of the Bering Sea have been closed to groundfish trawling and scallop dredging to reduce potential adverse impacts on the habitat for crab and other resources (Fig. 2). The Pribilof Islands Conservation Area was closed to protect blue king crab habitat (primarily shell hash). The Red King Crab Savings Area was established to protect adult red king crab and their habitat, and as a precautionary measure to reduce potential unobserved mortality. To protect juvenile red king crab and critical rearing habitat (stalked ascidians and other living substrate), Bristol Bay was declared another marine protected area. Specifically, the area east of 162°W longitude is closed to trawling and dredging, with the exception of a small area that opens for a brief period each spring to accommodate a yellowfin sole trawl fishery. A limited amount of longlining for Pacific cod and halibut, as well as pot fishing for Pacific cod and crabs occurs within all three of these marine protected areas.

These marine protected areas comprise a relatively large portion of the continental shelf, and in many respects, serve as marine reserves. In total, these three area closures encompass about 30,000 square nautical miles. To put this in perspective, this is an area more than twice the size of Georges Bank off the New England coast. Lauck et al. (1998) recently suggested that marine reserves should be at least 20% of available habitat in order to be effective. The Bering Sea marine protection areas exceed this threshold by encompassing about 25% of the shelf where commercial quantities of groundfish can be taken with bottom trawl gear, based on interpolation of data from Fritz et al. (1998). Marine protection areas of the Bering Sea include essential fish habitat (EFH) for crab species and fish species such as walleye pollock, Pacific cod, yellowfin sole, rock sole, and other flatfish species (North Pacific Fishery Management Council 1998).

The Sustainable Fisheries Act of 1996 amended the Magnuson-Stevens Fishery Conservation and Management Act to require the description and identification of essential fish habitat (EFH) in FMPs, adverse impacts on EFH, and actions to conserve and enhance EFH. EFH assessments for fish species covered under Alaska's five FMPs (BSAI groundfish, GOA groundfish, BSAI crab, Alaska scallops, Alaska salmon) were adopted by the council in June 1998. EFH assessment reports will form the basis for management actions taken to conserve and enhance essential fish habitat in the Alaska region. Additional marine protected areas may be required to minimize fishing impacts on EFH, particularly in sensitive, rare, and vulnerable habitats.

Marine Mammal and Seabird Considerations

In addition to setting maximum harvest levels, fisheries have been both seasonally and spatially allocated to reduce potential impacts of localized depletion of prey. For example, the Bering Sea pollock TAC is split among a winter fishery (A-season) and a late summer fishery (B-season). Seasonal and regional apportionment is also done for Atka mackerel *(Pleurogrammus monopterygius)* in the Aleutian Islands. The council recently adopted a regulation to reduce fishing for Atka mackerel near rookeries to reduce potential for localized depletion of Atka mackerel and competition with Steller sea lions *(Eumetopias jubatus)*, an endangered species. Because Atka mackerel and pollock are important prey for higher trophic levels, these measures reduce the impacts of harvesting on the ecosystem.

Area closures have also been implemented to prevent disrupting marine mammals at rookeries and haulouts, and to reduce competition from fisheries. To protect Pacific walrus *(Odobenus rosmarus)*, fishing vessels are prohibited in that part of the Bering Sea within 12 miles of Round Island, the Twins and Cape Peirce in northern Bristol Bay during the period April 1 through September 30. To protect Steller sea lions, no trawling is allowed year round in the BSAI within 10 nautical miles of 27 Steller sea lion rookeries (Fig. 3). In addition, six of these rookeries have 20 nautical mile trawl closures during the winter pollock season. There are additional rookery closures in the Gulf of Alaska as well.

In 1997, the council adopted an amendment that prohibits directed fishing for forage fish, which are prey for groundfish, seabirds, and marine mammals. Under this amendment, protection is provided for forage fish species such as capelin *(Mallotus villosus)* and a host of other forage species such as euphausiids (krill). The council took this proactive approach by preventing fisheries for these important species from expanding or developing.

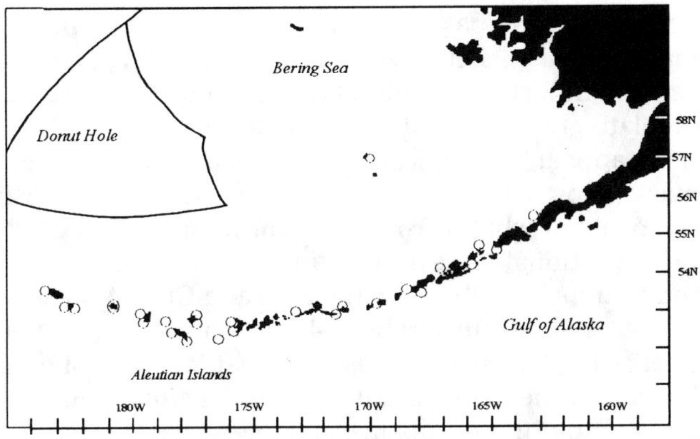

Figure 3. Location of the no-trawl zones around Steller sea lion rookeries in the Bering Sea and Aleutian Islands area.

Concern for the incidental bycatch of seabirds (the endangered short-tailed albatross, Diomedea albatrus, in particular) led to regulations requiring deterrent devices to be employed on groundfish longline vessels beginning in 1997. Approximately 9,600 seabirds (including about 1 albatross) are incidentally killed in Alaska groundfish fisheries each year (Wohl et al. 1995). It is hoped that these deterrent devices, which are actively being developed and improved upon by fishermen, will significantly reduce incidental mortality.

Discussion

Fisheries management of Bering Sea groundfish provides an excellent example of sustainable fisheries using a precautionary approach. Bering Sea fisheries have produced annual landings of nearly 2 million t of groundfish over the past 20 years, yet groundfish stocks in the Bering Sea have remained healthy to this day. The council and the National Marine Fisheries Service have used a precautionary approach by relying on scientific research and advice, conservative catch quotas, comprehensive monitoring and enforcement, bycatch controls, habitat conservation areas, and additional ecosystem considerations. By incorporating ecosystem considerations into management of groundfish fisheries, human impacts can be lessened, while at the same time providing sustained yields of fishery resources.

It is widely recognized that the Bering Sea ecosystem is driven by climate-induced variability and is subject to regime shifts, which have occurred in the past (McGowan et al. 1998). Populations of some marine mammals (e.g., Steller sea lions), seabirds (e.g., red legged kittiwakes at St. Paul Island), crab (e.g, Tanner crab), and prey species (such as capelin) are in relatively low abundance compared to historical records. Other populations appear to be in relatively high abundance (e.g., orcas, northern fulmars, snow crab, rock sole). It has been hypothesized that large-scale removals of whales from 1848 through 1976 caused major cascading effects (National Research Council 1996). From 1950 to 1976, estimated total catches of whales in the BSAI and Gulf of Alaska exceeded 5,700 blue whales, 26,000 fin whales, 74,000 sei whales, 30,000 humpback whales, and 210,000 sperm whales (National Marine Fisheries Service 1991). It is likely that the Bering Sea ecosystem, including groundfish populations, is still responding to this massive perturbation. Future changes to the ecosystem should be expected.

References

Alverson, D.L., M.H. Freeberg, S.A. Murawski, and J.G. Pope. 1994. A global assessment of fisheries bycatch and discard. FAO Fish. Tech. Pap. 339.

Christensen, N.L., and 12 others. 1996. The report of the Ecological Society of America committee on the scientific basis for ecosystem management. Ecol. Appl. 6(3):665-691.

Clark, W.G. 1993. The effects of recruitment variability on the choice of a target level of spawning biomass pre recruit. In: G. Kruse, D.M. Eggers, R.J. Marasco, C. Pautzke, and T. Quinn II (eds.), Proceedings of the International Symposium on Management Strategies for Exploited Fish Populations. University of Alaska Sea Grant, AK-SG-93-02, Fairbanks, pp. 233-246.

Dovers, S.R., and J.W. Handmer. 1995. Ignorance, the precautionary principle, and sustainability. Ambio 24(2):92-97.

Fritz, L.W., A. Greig, and R.F. Reuter. 1998. Catch-per-unit-effort, length, and depth distributions of major groundfish and bycatch species in the Bering Sea, Aleutian Islands, and Gulf of Alaska regions based on groundfish fishery observer data. NOAA Tech. Memo. NMFS-AFSC-88.

Garcia, S.M. 1995. The precautionary approach to fisheries and its implications for fishery research, technology and management: An updated review. Precautionary approach to fisheries. Part 2: Scientific papers. FAO Fish. Tech. Pap. 350(2):1-76.

Grumbine, R.E. 1994. What is ecosystem management? Conserv. Biol. 8(1):27-38.

Lauck, T., C.W. Clark, M. Mangel, and G.R. Munro. 1998. Implementing the precautionary principle in fisheries management through marine reserves. Ecol. Appl. 8(1):S72-S78.

Mangel, M., and 41 others. 1996. Principles for the conservation of wild living resources. Ecol. Appl. 6(2):338-362.

McGowan, J.A., D.R. Cayan, and L.M. Dorman. 1998. Climate-Ocean variability and ecosystem response in the northeast Pacific. Science 281:210-217.

Megrey, B.A., and V.G. Wespestad. 1990. Alaskan groundfish resources: 10 years of management under the Magnuson Fishery Conservation and Management Act. N. Am. J. Fish. Manage. 10(2):125-143.

National Marine Fisheries Service. 1991. Endangered whales: Status update. National Marine Fisheries Service, Silver Spring, Maryland. 52 pp.

National Marine Fisheries Service. 1997. Report to Congress on the status of fisheries of the United States, 1997. National Marine Fisheries Service, Silver Spring, Maryland.

National Research Council. 1996. The Bering Sea ecosystem. National Academy Press, Washington, DC. 307 pp.

North Pacific Fishery Management Council. 1997. Ecosystem considerations for 1998. North Pacific Fishery Management Council, Anchorage, Alaska. 56 pp.

North Pacific Fishery Management Council. 1998. Draft environmental assessment/ regulatory impact review for essential fish habitat. North Pacific Fishery Management Council. May 12, 1998. 362 pp.

Rosenberg, A.A., and V.R. Restrepo. 1995. Precautionary management reference points and management strategies. Precautionary approach to fisheries. Part 2: Scientific papers. FAO Fish. Tech. Pap. 350(2):129-140.

Thompson, G. 1997. The precautionary principle in North Pacific groundfish management. National Marine Fisheries Service, Alaska Fisheries Science Center, Seattle, Washington.

Witherell, D. 1995. Management of flatfish fisheries in the North Pacific. In: Proceedings of the International Symposium on North Pacific Flatfish. University of Alaska Sea Grant, AK-SG-95-04, Fairbanks, pp. 573-589.

Witherell, D., and C. Pautzke. 1998. A brief history of bycatch management measures for eastern Bering Sea groundfish fisheries. Mar. Fish. Rev. 59(4):15-22.

Wohl, K.D., P.J. Gould, and S.M. Fitzgerald. 1995. Incidental mortality of seabirds in selected commercial fisheries in Alaska. Unpublished report. U.S. Fish and Wildlife Service, Migratory Bird Management, Anchorage, Alaska.

Integrated Operational Rule Curves for Montana Reservoirs and Application for Other Columbia River Storage Projects

Brian L. Marotz
Montana Fish, Wildlife and Parks, Kalispell, Montana

Daniel Gustafson
Montana State University, Department of Biology, Bozeman, Montana

Craig Althen and Bill Lonon
Montana Fish, Wildlife and Parks, Kalispell, Montana

Abstract

Reservoir operation guidelines were developed to balance resident fisheries concerns with anadromous species recovery actions in the lower Columbia River. Fisheries requirements were integrated with power production and flood control to reduce the economic impact of basin-wide fisheries recovery actions. These Integrated Rule Curves (IRCs) were developed simultaneously in the Columbia Basin System Operation Review (SOR 1995), the fish and wildlife program of the Northwest Power Planning Council (NPPC), and recovery actions for endangered fish species (Marotz et.al. 1996). IRCs were adopted into the council's fish and wildlife program (NPPC 1994). However, the IRC operations were supplanted by the Biological Opinion (BiOp) of the National Marine Fisheries Service (NMFS 1995). BiOp operations are designed to enhance the downstream migrational success of populations of juvenile chinook and sockeye salmon from the Snake River listed under the Endangered Species Act. The BiOp calls for summer releases of water from storage projects, including Hungry Horse and Libby reservoirs. BiOp operations cause drawdown of these reservoirs during the summer and unnatural flow fluctuations downstream causing impacts to the aquatic ecosystem. The decision to release water from headwater dams to augment summer flows downstream pivots on the potential benefits to anadromous fish relative to the potential impacts to resident fish.

Introduction

The construction and operation of dams in the Columbia River basin in western North America negatively impacted aquatic and riparian environments. Dams alter stream hydrology, isolate fish spawning migrations, and directly kill fish. Snake River chinook salmon (*Oncorhynchus tshawytscha*), sockeye salmon (*O. nerka*), several steelhead stocks (*O. mykiss*), Kootenai River white sturgeon (*Acipenser transmontanus*), and bull trout (*Salvelinus confluentus*) have been listed as threatened or endangered under the Endangered Species Act. Recently, dam operations at many federally operated dams were altered to restore these dwindling fish populations (USFWS 1994; NMFS 1995, 1998). As restoration actions are taken, system operations should be designed to benefit all native species in the watershed.

Prior to dam construction, the hydraulic cycle in Columbia River tributaries included a high flow event during spring snowmelt (late May through early June) and a stabilized low flow period throughout the remainder of the year (Parret and Hull 1985). Hydropower operations reversed this discharge pattern by storing water during the runoff and releasing water during the fall and winter, when flows were naturally at their lowest level. Loss of the spring freshet is believed to be a primary factor in the decline of anadromous and resident fish populations in the Columbia River basin (ISG 1996).

Montana has resident fish at risk in the Flathead and Kootenai river systems (Fig. 1), including the westslope cutthroat trout (*O. clarki lewisi*), interior redband trout (*O. mykiss*), the bull trout, and the Kootenai river white sturgeon. The Kootenai River white sturgeon is listed as endangered under the Endangered Species Act (59 Fed. Reg. 45989; 1994), the bull trout as threatened (63 Fed. Reg. 31647; 1998), and the westslope cutthroat trout and interior redband trout are candidate species of special concern (Williams et al. 1989). These fish were adversely affected by the Montana dams.

The Kootenai River white sturgeon is a genetically distinct population that occurs from Kootenai Falls below Libby Dam downstream to Kootenay Lake in British Columbia. There has been no significant recruitment of juveniles into this population for 25 years since the construction of Libby Dam. Less than 2,000 individuals remain. The aging fish in the population may not recruit young in sufficient numbers to avoid extinction. Conservation stocking has been initiated to avoid extinction while actions are taken to enhance natural reproduction. Evidence suggests that flow and temperature are related to reproduction and survival (Apperson and Anders 1991, Apperson 1992). The operation of Libby Dam is critical to white sturgeon recovery.

The bull trout is the largest native trout in Montana. Bull trout numbers have been reduced by habitat degradation and negative interactions with non-native species (Montana Bull Trout Scientific Group 1998). Some of the strongest meta-populations exist in the Canadian headwaters of

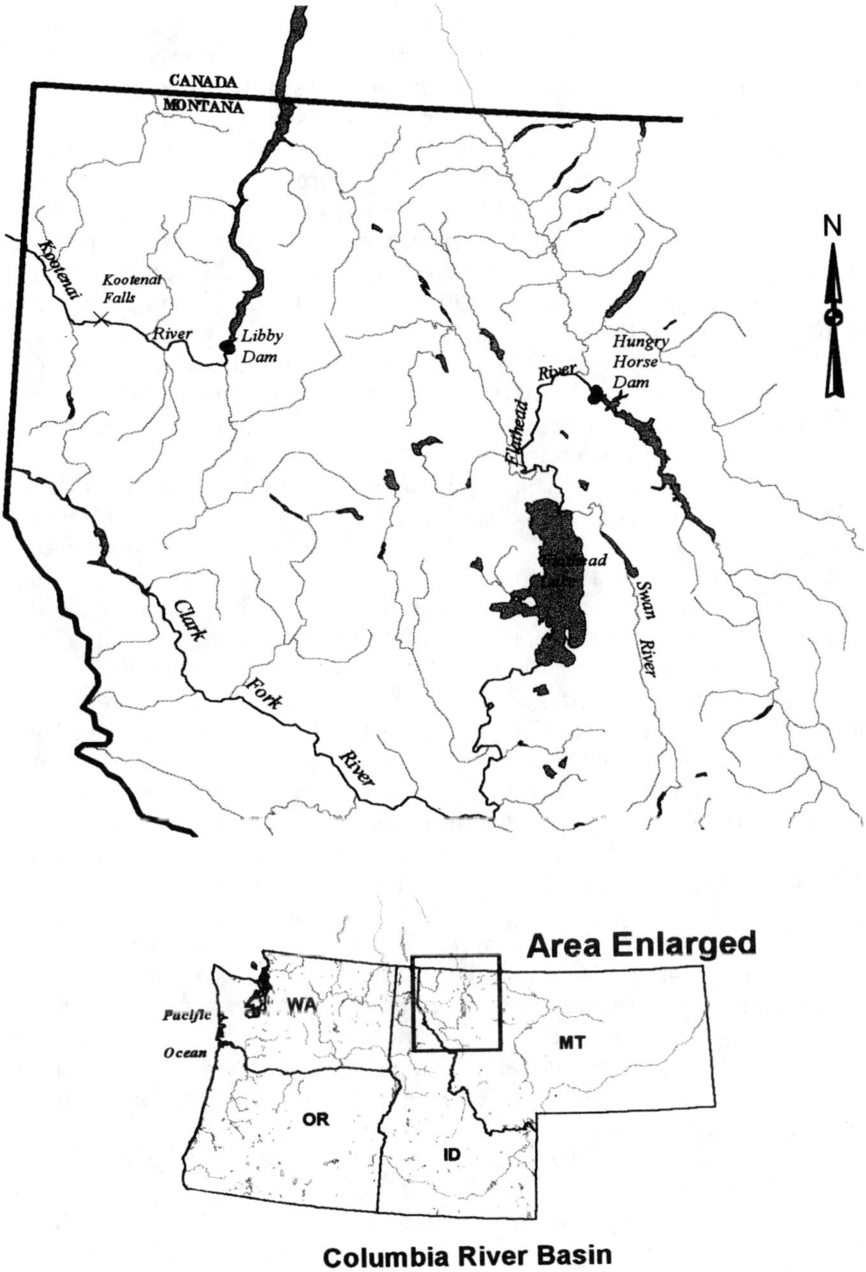

Figure 1. Flathead and Kootenai systems of Montana.

Libby Reservoir and in the South Fork Flathead River upstream of Hungry Horse Dam. Fortunately, Hungry Horse contains a native species assemblage and much of their habitat is in wilderness and national forest that retains many natural attributes. The headwaters of Libby Reservoir also contain sufficient habitat suitable for the persistence of bull trout. Dam operation directly affects reservoir and riverine habitat, insect production for juveniles, and the availability of fish prey for adults.

The westslope cutthroat trout has been reduced to less than 10% (estimated approximately 5%) of the species' historic range. Major threats to this species are habitat loss, genetic introgression, and competition with introduced species (Van Eimeren 1996). The most secure meta-population of the westslope cutthroat trout exists in Hungry Horse Reservoir and the South Fork Flathead River upstream. Libby Reservoir headwaters also contain genetically pure stocks. After westslope cutthroat emigrate from their natal tributaries to the reservoir or river, the availability of insects, zooplankton, and habitat is controlled by dam operation.

The interior redband trout is the only rainbow trout native to Montana (Allendorf et al. 1980). Redband trout have been reduced to a fraction of their historical range (Williams et al. 1989, Behnke 1992). In Montana, redband occur only in five Kootenai River tributaries. Only the Callahan Creek population is known to migrate to the Kootenai River where the availability of food and habitat is directly related to dam operation. Redband trout were also documented in the Yaak River, a Kootenai River tributary. Interior redband are a category 2 subspecies; listing may be warranted but precluded due to lack of biological information in their present range.

Recent changes in dam operation to aid in the recovery of salmon and steelhead in the lower Columbia River basin have caused undesirable reservoir and river operations in headwater storage projects. The operating plan described herein reduces the impacts to native fish species in the Columbia River headwaters while maintaining features essential to the National Marine Fisheries Service's operational strategy (NMFS 1995, 1998). Our Integrated Rule Curves (IRCs) for dam operation are a series of drawdown and refill targets for operating the dams that incorporate incremental adjustments for uncertainties in water availability (Marotz et al. 1996). The IRCs limit the frequency of deep drawdowns and reservoir refill failure and produce a more natural discharge hydrograph (Fig. 2). Actual operations will vary due to inflow forecasting error. This flexibility allows operators to market power.

For Montana reservoirs, the intent is to reduce reservoir drawdown and increase refill probability to protect the most biologically productive period from mid to late summer through fall (May et al. 1988, Chisholm et al. 1989). At full pool, the reservoirs contain the maximum volume of optimal-temperature water for fish growth, and a large surface area for the deposition of terrestrial insects from the surrounding landscape. Reduced drawdown protects aquatic food production in the reservoirs, assuring an ample springtime food supply for fish.

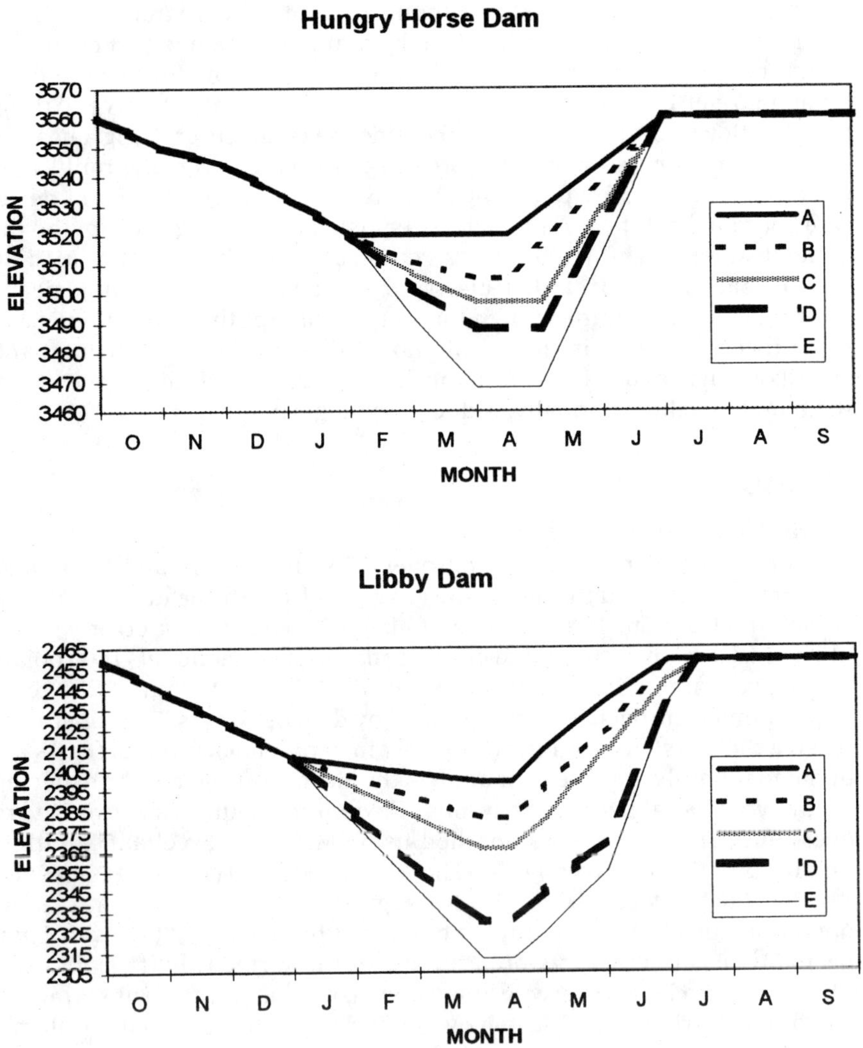

Figure 2. Integrated Rule Curves for Hungry Horse (top) and Libby Dam (bottom) operation. Reservoir elevations are selected on monthly inflow forecasts beginning in January. Elevational targets are interpolated when the inflow forecast is intermediate between volumes described by the curves shown above.

The IRCs create a naturalized annual discharge hydrograph. In a natural river environment, the nearshore habitat is productive and critical to fish. Riparian vegetation reestablishes seasonally, providing secure habitat along river margins and reducing erosion of silt into the river. Spring flushing flows sort river gravel and define the channels creating a healthy environment for fish and their prey organisms. Conversely, rapid flow fluctuations caused by power operations intermittently flood and desiccate shoreline habitats. Aquatic insects, fish, and fish eggs occupying the zone of water level fluctuation, or *varial zone*, may become stranded on the dry banks (Hauer et al. 1994, 1997). The varial zone is biologically unproductive. Intermittent high discharges disrupt the natural revegetation process. Vegetation that would normally provide secure habitat and stabilize soils cannot fully reestablish. The IRCs cause discharges from the dam to decline gradually, reducing biological impacts.

Methods

Hydraulic Modeling

This operating plan was based on over 16 years of field and laboratory research. We modeled the hydropower system from the headwaters downstream (Fraley et al. 1989, Marotz et al. 1996). Forecasting error was applied to each dam to minimize error propagation. The models interpolate elevational targets in the reservoir based on inflow volume. This technique mimics the decision process used by dam operators. Dam operators receive their first annual inflow forecast in early January. As forecasts are updated monthly, the operator adjusts the elevation target to the expected inflow volume, allowing operational flexibility as runoff forecasts vary over time. The IRCs were smoothed using Microsoft Excel and multiple iterations of the reservoir models HRMOD and LRMOD (Marotz et al. 1996).

The IRCs incorporate a new strategy for system flood control. The operation currently being implemented by the U.S. Army Corps of Engineers (ACOE) attempts to store as much of the spring runoff as possible. This requires a large reservoir drawdown to evacuate sufficient storage to contain the spring runoff. As the reservoir refills with snowmelt, discharge is typically held to the minimum allowable flow. Conversely, the variable discharge strategy (VARQ) embodied in the IRCs releases a naturally shaped spring freshet and stores only the amount of water that would exceed flood stage downstream (ACOE 1997). Flood control criteria at downstream locations (e.g., the Flathead River at Columbia Falls; and Kootenai River at Bonners Ferry, Idaho, and Kootenay Lake, British Columbia (IJC 1938)), further limit the maximum allowable flow.

During 1996 and 1997, the ACOE Hydraulics Branch reevaluated flood control requirements in the Columbia River basin. The study was conducted in accordance with NMFS (1995) and USFWS (1994), and the fish and wildlife program of the Northwest Power Planning Council (NPPC 1994). The ACOE critically compared the IRCs and VARQ and determined that the strategies were similar (ACOE 1997). However, in years of less-than-average water availability, VARQ required less drafting for flood control and reservoir elevations were higher than the IRCs. Operational flexibility above the IRCs can be used to save more water during dry years to augment spring flows. It should be noted, however, that in this analysis the model was configured to select the *lower* of the two targets (IRC as opposed to VARQ). Therefore, possible additional benefits to anadromous recovery by implementing VARQ during years of less-than-average water were not presented. For years of average or high water, VARQ and IRCs were modeled identically.

Thermal Modeling

Thermal effects downstream of Hungry Horse and Libby dams are controllable by regulating the depth of water withdrawal at the dams (Christenson et al. 1996, Marotz et al. 1996). For modeling comparisons, withdrawal depths were standardized across operational alternatives using an automated selective withdrawal (thermal control) model component. This resulted in identical discharge temperature under differing operational strategies.

Comparison with the NMFS 1995 Biological Opinion Operations

Comparisons of the IRC with the NMFS operation plan were conducted using the reservoir models (LRMOD and HRMOD). The two operational plans differed from each other in nearly parallel ways at Hungry Horse and Libby dams. For sake of brevity, only information for Libby Dam is presented. The operations specified by the NMFS 1995 Biological Opinion (BIOp) were provided by the Dittmar Control Center of the Bonneville Power Administration (BPA), Study 98C-01.OPERB (NMFS 1995; Pers. comm., Roger Schiewe, BPA, and Michael Newsom, NMFS, Portland, OR).

Plots of reservoir elevations and dam discharge schedules were overlaid for visual comparison of the two alternatives during low (inflow of 6.068 million acre feet [MAF]), average (8.088 MAF), and high (> 10.110 MAF) water availability. A representative NMFS operation for years of low, average, and high water availability was constructed by selecting five or more years with inflows approximately equal to the specified annual inflow volumes (±0.5 SD). We then calculated the mean elevation for each of the 14 periods. This was necessary to mask the effect of differences in water availability in the main-stem Columbia, relative to the Kootenai subbasin (i.e., water availability in the Kootenai system varies somewhat independently from water availability in the lower Columbia River). Water years included in the composite operations are shown in Table 1.

The NMFS operation assumed that storage reservoirs would be drafted to 20 feet below full pool in August only if the seasonal target (July 1-August 31) of 200 thousand cubic feet per second (kcfs) at McNary Dam was not met. This resulted in varying degrees of reservoir drafting (0-20

Table 1. Water years included in the NMFS composite operation.

Water availability	Water year[a]	Annual inflow (MAF)
High	1956	10.863
	1934	10.658
	1959	10.496
	1969	10.068
	1976	9.785
Medium	1963	8.101
	1953	8.088
	1935	8.046
	1932	8.017
Low	1929	6.259
	1970	6.179
	1940	6.014
	1936	5.974
	1945	5.904

[a] A water year extends from October 1 through September 30.

feet from full pool during August) throughout the 50-year record, and caused the composite data to underestimate the effect of summer flow augmentation.

Biological Modeling

Comparisons of trophic responses resulting from the alternative dam operational strategies were examined using the empirically calibrated biological reservoir models HRMOD and LRMOD (Fraley et al. 1989, Marotz et al. 1996). Model simulations were configured for annual, as opposed to continuous runs.

Fish Entrainment

Fish entrainment through Libby Dam was qualitatively assessed using the empirically calibrated entrainment model developed by Skaar et al. (1996). Since the two operational alternatives compared herein are hypothetical, comparative field data were not possible. Nonetheless, trends in fish density and vertical distribution can be extrapolated from sampling conducted by Skaar et al. (1996). Discharges during spring and summer can be accurately estimated through computer modeling. We assumed that seasonal trends in vertical fish distributions were constant to qualitatively assess entrainment under the operational alternatives as correlated to discharge volume.

Results and Discussion

Libby Reservoir Conditions

Alternative 1: IRC/VARQ

The reduced summer drawdown and improved refill probability resulting from the IRC/VARQ operation (Figs. 3, 4, and 5) protects aquatic and benthic food production in the reservoirs during the warm months, late May through early September. Overall, this operation sustains roughly 70% of the optimum reservoir productivity (Table 2).

Entrainment of fish through Libby Dam is proportional to discharge volume. During spring, fish are concentrated near the surface associated with warmer water in the top 20 m (Skaar et al. 1996). Fish densities are highest during spring. As a result of the tiered flow approach, highest entrainment rates would occur in years of above-average water availability. Lowest entrainment rates would occur in years of below-average water, proportional to lower discharge volumes. Fish entrainment during spring would be similar to the NMFS 95 BiOp Alternative 2 in wetter years and less than the NMFS 95 BiOp during average or drier years. During summer, areal fish densities are lower than in May and June, although densities are typically higher in August than in late fall and winter. Entrainment during August resulting from the IRC/VARQ alternative would be the less as compared to the Alternative 2.

Alternative 2: NMFS Biological Opinion

Computer simulations performed at BPA Dittmar Control Center show that the NMFS BiOp, in attempting to meet an August flow target of 200 kcfs at McNary Dam in the lower Columbia River, reduces reservoir refill probability (Wright et al. 1996). In some years, Libby Reservoir fails to refill by 20 feet or more, thus affecting the sustainability of the operation.

Failure to refill the reservoirs by July and draw down during the summer reduce biological production in the reservoirs (Table 2). The food web supporting fish is most productive in the shallower and warmer littoral or nearshore zones during the summer months. Frequent dewatering reduces the biomass of larval insects that are killed as water recedes. A brief deep drawdown which exposes a large percentage of the reservoir bottom requires at least 2 years for aquatic insect populations to rebound. Also, the contribution of terrestrial insects as a food source for fish is reduced as the reservoir surface recedes from shoreline vegetation. Terrestrial insects are most abundant near the shore from June through September when the NMFS operation is lower than the IRCs. Zooplankton, an important food for juvenile trout and adults during winter, is washed out of the reservoirs through dam turbines as the reservoirs shrink. Thus excessive reservoir drawdown and refill failure negatively impact fish food availability and, therefore, fish growth.

The effects of the 1995 BiOp were recently assessed by the Independent Scientific Advisory Board (ISAB 1997a). The NMFS asked whether the

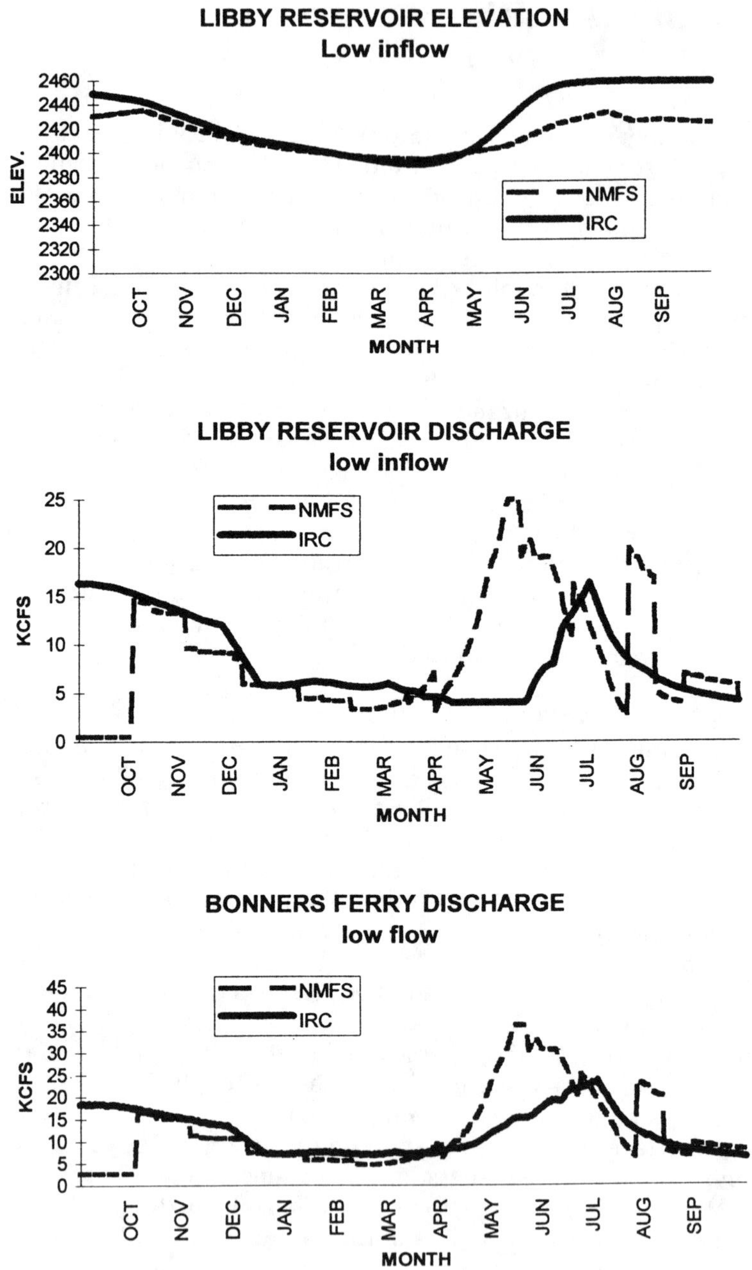

Figure 3. Comparison of Libby Reservoir elevations (top) and dam discharge (middle) and Kootenai River discharge at Bonners Ferry (bottom) under low water conditions.

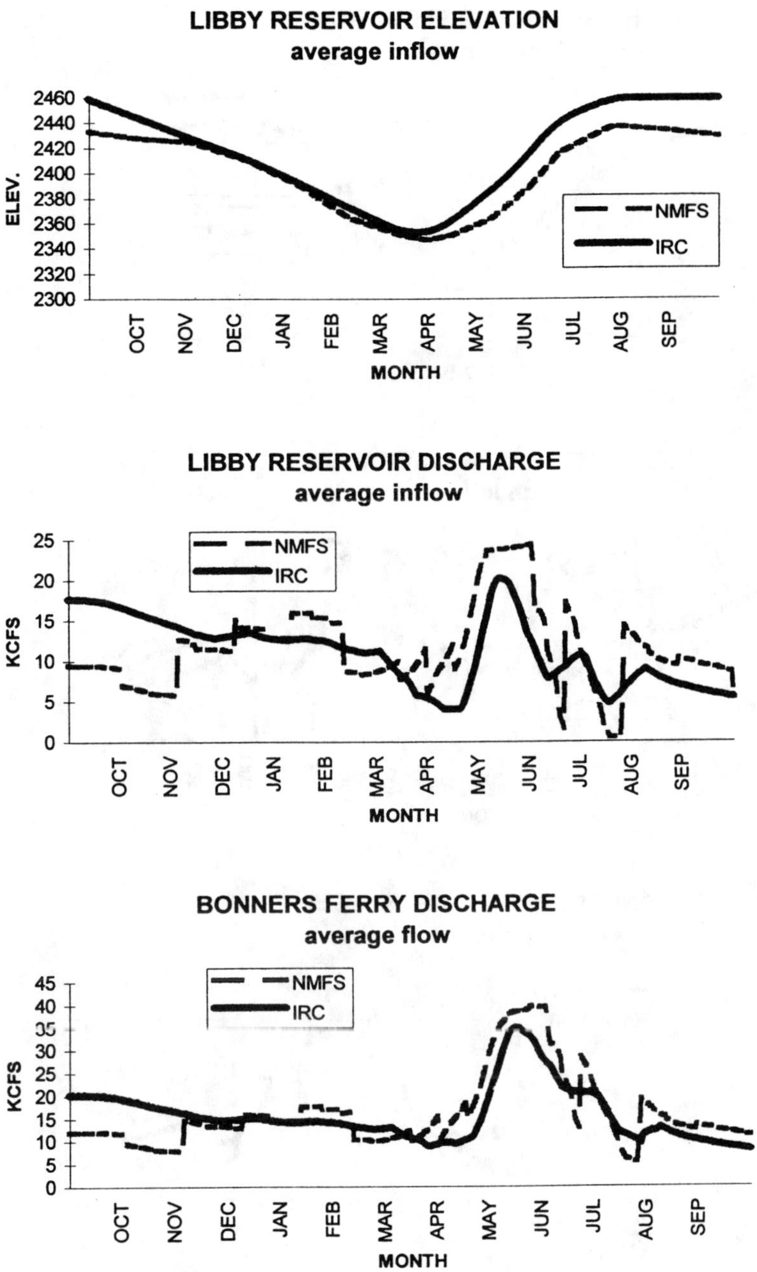

Figure 4. Comparison of Libby Reservoir elevations (top) and dam discharge (middle) and Kootenai River discharge at Bonners Ferry (bottom) under average water conditions.

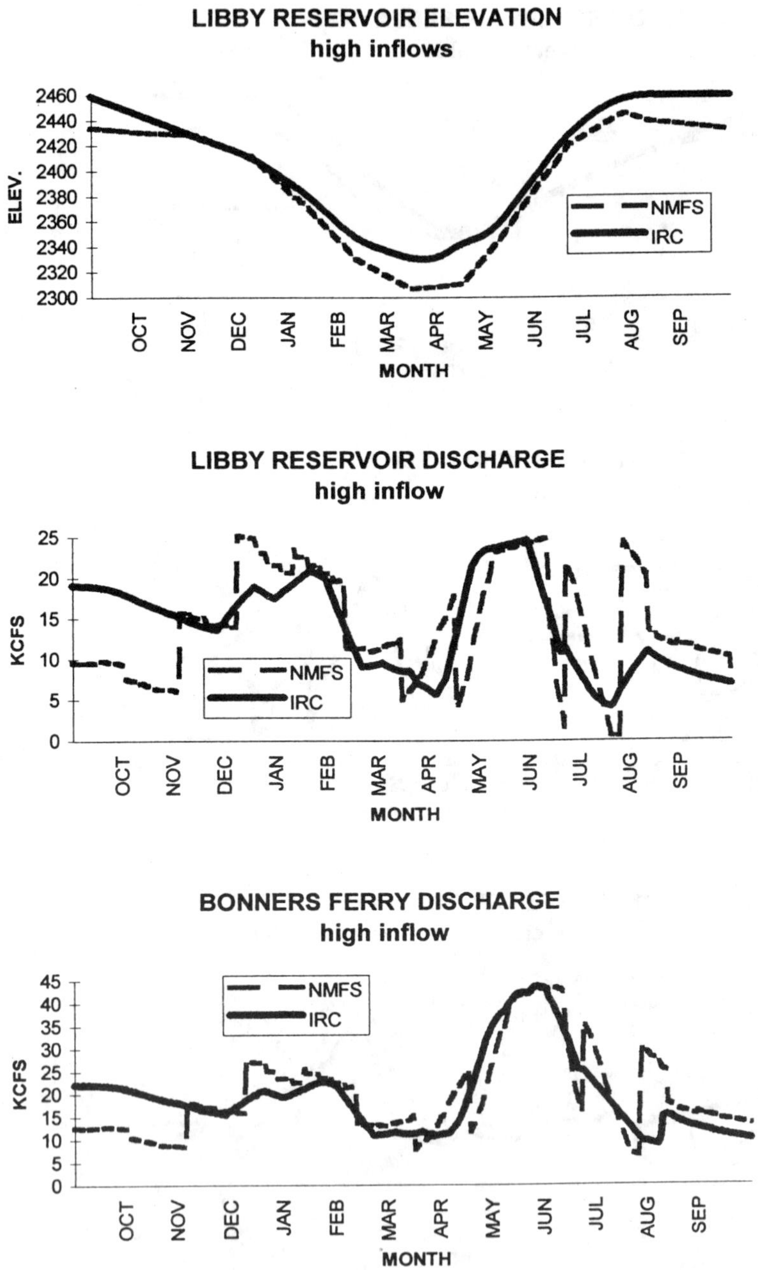

Figure 5. Comparison of Libby Reservoir elevations (top) and dam discharge (middle) and Kootenai River discharge at Bonners Ferry (bottom) under high water conditions.

Table 2. Trophic responses in Libby Reservoir calculated using the reservoir model LRMOD (Marotz et al. 1996).[a]

		Primary production (t)		Secondary production (t)		Terrestrial insect deposition by insect order (% maximum)				Fish growth, kokanee			
										TL (mm)		Weight (g)	
Water avail.	Name	Carbon fixed	Wash-out	Zoop prod	Bent[b]	Col	Hem	Hom	Hym	Age I+	Age II+	Age I+	Age II+
Low	NMFS	11,836	37	1,354	382.1	74.1	83.3	85.0	85.7	285	386	219	576
	KIRC	13,003	30	1,489	367.5	79.0	94.7	97.5	99.9	298	412	252	706
Avg.	NMFS	11,063	39	1,265	337.3	62.0	80.1	83.8	88.7	279	374	205	521
	KIRC	12,178	35	1,393	303.2	68.8	90.6	94.8	99.9	291	397	233	630
High	NMFS	10,820	46	1,236	229.7	56.6	80.3	85.0	90.2	278	372	202	510
	KIRC[c]	11,680	45	1,335	301.5	62.7	87.9	92.9	99.9	287	389	223	589

[a] Results represent phytoplankton production (metric tons of carbon fixed) calibrated by C^{14} liquid scintillation. Phytoplankton washout through the dam (metric tons) calibrated by chlor *a* vertical distribution and entrainment sampling. Total zooplankton production (metric tons) calibrated on phytoplankton production and seasonal measures of carbon transfer efficiencies. Benthic production (metric tons of emergent insects) calibrated on depth distribution of insect larvae and emergence captures. Terrestrial insect deposition (percentage of maximum) by insect order (Col = Coleoptera, Hem = Hemiptera, Hom = Homoptera, and Hym = Hymenoptera), calibrated on nearshore (< 100 m) and offshore surface insect tows. Fish growth (end-of-year kokanee size) in total length (TL) and weight (grams) calculated through multivariate analysis on water temperature structure and food availability.

[b] Benthic insect production is artificially enhanced by reservoir refill failure. This single year event is caused when the warm epilimnetic water settles over substrate containing high larval densities (in the infrequently dewatered zone), thus enhancing larval production and emergence. A single deep drawdown event or reservoir refill failure can impact benthic insect production for 2 years or longer

[c] KIRC = Kootenai Integrated Rule Curve. The KIRC was modified by the Kootenai River white sturgeon recovery team to balance the needs of the endangered white sturgeon with the reservoir fishery and recovery actions for anadromous species in the lower Columbia River.

resident fish populations are at risk of extinction in Libby and Hungry Horse reservoirs due to flow augmentation strategies. The ISAB (1997a) found that the biological effects of summer drafting will not likely drive resident fish populations to extirpation in Montana, however, late summer reservoir drawdown adversely effects resident fishes in the reservoirs and downstream through flow fluctuation in the streams and lakes below the two reservoirs.

Under the NMFS 95 BiOp alternative, fish entrainment through Libby Dam (proportional to dam discharge) would be higher during spring in average and dry years, and higher during August, compared to the IRC/VARQ alternative.

Kootenai River Conditions Downstream of Libby Dam

Alternative 1: IRC/VARQ

A comparison of Columbia River flows during spring performed by Wright et al. (1996) revealed that flows resulting from the IRCs were nearly the same as the NMFS BiOp. Spring discharges in the Kootenai River, resulting from the IRC/VARQ operation are nearly consistent with the NMFS BiOp during years of average to high water. During years of low water, the tiered flow approach incorporated into the IRC/VARQs releases less water than called for by the NMFS BiOp (Fig. 3).

The VARQ flood control strategy, if used, would allow greater operational flexibility during drier years than average years. Reservoir elevations higher than the IRCs could be achieved. Thus, operators could store more water that can be released to augment stream flows for listed resident and anadromous stocks. A naturalized spring freshet (greater than provided by the IRC strategy) could be created, even in dry years, without compromising reservoir refill. In average or wetter years, VARQ and Integrated Rule Curves were identical. This improves on historical operations because reservoir elevations remain higher prior to the spring runoff, so that a larger percentage of the runoff volume can be shaped to create a normalized spring freshet.

The IRC alternative was designed to gradually ramp down from the spring peak to reduce flow fluctuations. During dry years, the maximum drawdown of the reservoir was reduced consistent with the NMFS 95 BiOp and VARQ to increase the volume of pass-through flows during spring runoff. In wetter years, the discharge was smoothed to further extend the descending limb of the hydrograph.

Maximum flows regulated by storage dams are less than would occur prior to dam construction. Maximum allowable dam discharge is dictated by the physical capacity of the turbines and acceptable spill levels. Libby Dam presently contains five turbines that can release a maximum of 27 kcfs, collectively. The spillway entrains atmospheric gas during operation, so only a small percentage of the total flow can be spilled before

Montana water quality laws pertaining to dissolved gas are violated (e.g., not to exceed 110% gas saturation). The maximum flow is typically limited by turbine capacity plus additional flows from unregulated tributaries that enter downstream from the dam. Flood control criteria at Bonners Ferry, Idaho, and Kootenay Lake, British Columbia (IJC 1938), further limit the maximum allowable flow.

The structure of the lotic community is defined by streamflow characteristics (Radar and Ward 1988, Poff and Ward 1989). The naturalized spring freshet re-sorts and cleans river sediments and restores nutrient cycles and floodplain function. The freshet resuspends streambed sediments, redefining the stream channel, backwaters and banks (Wesche and Richard 1980). Clean, unburied cobble provides interstitial spaces and ample surface area suitable for benthic algae, aquatic insects, and young fish (Perry et al. 1986, Hauer et al. 1997).

Clean gravels with subsurface and groundwater inflow are sought by nest-building salmonids and broadcast spawners (Peters 1962). Gravels consisting of less than 30% fines (< 0.65 cm) provide suitable oxygenation for egg incubation and sack fry, enhancing early-life survival (Weaver and Fraley 1993). Preferred spawning substrate for white sturgeon, which are broadcast spawners, consists of gravel, cobbles, and boulders (Parsley et al. 1993, Hildebrand and McKenzie 1994).

Fine clays, silts, sands, and organic materials deposited in low-velocity areas (e.g., high on the streambanks) become dry as spring flows gradually recede. If stream flows stabilize, this rich soil becomes bound by rooted terrestrial vegetation. Erosion and subsequent siltation of the streambed is reduced. Fine materials in the stream support aquatic plants which provide habitat for aquatic and terrestrial organisms. The IRCs gradually ramp down from the spring runoff peak and restore these favorable biological conditions.

The benefit to anadromous salmon would be greater flows in the spring, a gradual rampdown from the spring freshet that enhances summer flows, and improved reservoir levels during drought cycles so that the operation is sustainable from year to year.

Alternative 2: NMFS Biological Opinion

The August releases called for by the NMFS BiOp are designed to aid the migration of juvenile Snake River salmon in the lower Columbia River. The NMFS BiOp calls for maximum dam discharge during August until reservoirs are drafted to 20 feet from full pool. Water from headwater storage projects is released to augment the natural flows in the Columbia River to meet a summer flow target of 200,000 cfs at McNary Dam. The goal is to increase water velocities in the pools upstream from dams in the lower Columbia to speed the juvenile salmon migration toward the ocean.

The NMFS BiOp creates an augmented spring freshet followed by a trough, then a second flow peak in August (Figs. 3, 4, and 5). The rapid flow reduction between the peaks dewaters the varial zone, stranding insects, zooplankton, fish, and fish eggs (Hauer and Stanford 1982, Armitage 1984, Hauer 1997). Unnatural flow fluctuation is not consistent with the normative river concept described by the Independent Scientific Group (ISG 1996).

Fluctuating or abnormally high discharges also disrupt the natural revegetation, insect, and larval fish recolonization process. Aquatic and terrestrial vegetation that would normally provide secure habitat and stabilize soils cannot fully reestablish, so fine materials are more easily eroded into the channel. These discharge fluctuations could be moderated by delaying the date of reservoir refill or by extending the period of flow augmentation.

Hyporheic interactions are also altered by intermittent, abnormally high flows. Augmented summer flows may increase the river stage by up to 4 feet. Fluctuating flows alternately saturate and dewater the streambanks, which can weaken the riverbanks and cause bank failure and increased sedimentation.

Kootenay Lake Conditions (British Columbia)

Releases from Libby Dam affect water retention time, and thus biological productivity in Kootenay Lake, British Columbia. The warm, sunlit epilimnion contains the highest density of phytoplankton and zooplankton. As inflow to the lake increases, more water must flow through the outlet or be stored in the pool. If the pool elevation is stable or declining, inflowing waters displace a commensurate volume through Corra Linn Dam, British Columbia. The physical configuration of Kootenay Lake results in an epilimnetic release of water from the lake, which results in greater downstream loss (entrainment) of organisms through the turbines. High summer discharges from Libby Dam exacerbate this effect during the summer when thermal stratification in Kootenay Lake is well established. Concerns over nutrient levels in the lake are evident by past investigations of nutrient loading (Daley et al. 1981) and ongoing lake fertilization experiments (Ashley and Thompson 1996).

Alternative 1: IRC/VARQ

Dam releases under this alternative were designed to create a gradual rampdown from the spring runoff toward basal flows. Water retention time in the epilimnion of Kootenay Lake would therefore be greater than Alternative 2 during the warm summer months because Libby Dam discharge is less.

Alternative 2: NMFS Biological Opinion

The late summer water releases from Libby Dam would cause a higher rate of water exchange in Kootenay Lake's epilimnion. Downstream loss of the most productive surface layer of Kootenay Lake would reduce food availability for lake-dwelling species, including white sturgeon. This would likely affect survival of fish and may jeopardize the success of the fertilization program (Ashley and Thompson 1996).

The Kootenay River downstream of Kootenay Lake passes through numerous small (and old) hydro dams. This water must be passed relatively quickly and will likely result in increased levels of dissolved gas supersaturation (Pers. comm., Jay Hammond, B.C. Ministry of Environment).

Effects on White Sturgeon

Alternative 1: IRC/VARQ

Maximum flows are regulated by maximum allowable flood stage (approximately 60,000 cfs) at Bonners Ferry, which eliminates the high flows necessary to completely resort the river substrate. The tiered flow approach in the IRC/VARQ alternative reestablishes a more natural spring runoff period (Fig. 6). Model simulations estimate that combined flows in excess of 50,000 cfs can be achieved at Bonners Ferry in approximately 4 out of every 10 years (Marotz et al. 1996). Approximating the bankfull flow on this frequency is expected to reduce imbeddedness and clean interstitial spaces in riffle areas. Flows during dry years are less under the tiered flow approach than those specified by the NMFS 95 BiOp.

Historically, white sturgeon incubation, hatching, and early fry stage coincided with gradually declining flows, immediately after the spring runoff. The gradual flow reduction after the spring peak may reduce predation mortality in larval sturgeon by increasing the area of submerged riverbed, thus increasing security habitat (Pers. comm., Carl Walters, University of British Columbia, Vancouver, B.C.).

Alternative 2: NMFS Biological Opinion

The spring release called for by the NMFS 1995 BiOp is similar to Alternative 1 in that it would mimic the natural spring runoff. The 1995 BiOp differs in dry years, when dam discharge would be greater than the sturgeon tiered flow strategy. Bankfull flows could be achieved on the same frequency as the IRC/VARQ.

However, the August release is inconsistent with the restorative flows recommended in IRCs. The varial zone is flooded, then dewatered twice during the period crucial to sub-yearling sturgeon development. White sturgeon can be directly affected (through stranding of juveniles) or indirectly affected (through food web dynamics) by summertime flow augmentation (Stanford et al. 1996, Hauer et al. 1997). Unseasonably high water velocities during August could displace juvenile sturgeon that evolved under stable low flows during the critical early life stage. Summer releases dictated by the NMFS BiOp, therefore, likely impact postlarval survival.

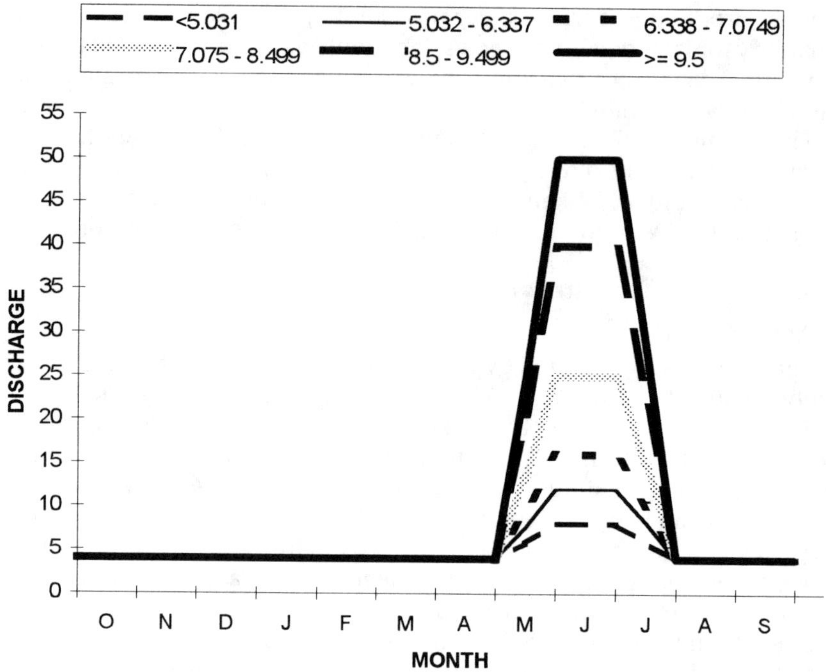

Figure 6. Minimum flow objective for Kootenai River white sturgeon at Bonners Ferry, Idaho, based on water availability (millions of acre feet) at Koocanusa Reservoir (April and May forecasts). Volumes and shapes of the flow from April 1 through July 31 will be adjusted during the runoff period to achieve the optimal mix of flow and water temperature.

Flood Control

A new strategy for system flood control (VARQ) is required to balance the needs of resident and anadromous species in the Columbia system. The IRC/VARQ strategy releases a naturally shaped spring freshet and stores only the amount of water that would exceed flood stage downstream. Less reservoir drafting is required, which benefits reservoir biology and improves reservoir refill probability. Thus, a naturalized spring freshet can be created, even in dry years, without compromising reservoir refill.

ACOE modelers established that the IRCs were nearly identical to a new flood control strategy being developed by the ACOE (1997). The IRCs are consistent with VARQ in years of average and higher water. Differences between VARQ and IRCs during years of lower water are a result of integrating power constraints. A combination of IRCs and VARQ is being explored for Hungry Horse and Libby operation (ACOE 1997). VARQ is a critical tool to simultaneously balance the needs of resident and anadromous fish recovery by providing greater operational flexibility in dry years to help salmon and steelhead without harming native resident fish species.

Economic Effects

Wright et al. (1996) reported that the enhanced reservoir operation (IRC concept) was the least expensive of the alternatives analyzed, saving the power system an annual incremental average of $27 million as compared to the NMFS BiOp. Furthermore, Wright et al. (1996) stated, "the mathematical decision process for establishing reservoir elevations and flow targets, based on updated inflow forecasts, is amenable to power and flood control planning."

Applicability to Other Storage Projects

Integrated Rule Curves are a tool for examining trade-offs among the environmental needs of anadromous fish downstream and resident fish in the Columbia River headwaters. By implementing IRCs at other storage projects, sub-basins experiencing wet conditions can supply spring and summer flow augmentation for anadromous fish whereas dry sub-basins provide less flow, protecting important reservoir and riverine stocks. Preliminary IRCs for other projects can be based on the hydraulic balance in the watershed and physical capabilities of the dam, then modified to site-specific biological concerns as data become available. As flows descend through the system, water can be temporarily delayed en route, then released as flows decline to protract the pulse of water. The result is a sustainable operation that can function within the vagaries of annual water availability.

A number of advisory groups have provided scientific and policy review of the Columbia River salmon recovery efforts and the IRCs. The Independent Scientific Group (ISG 1996) endorsed the IRC concept for other reservoirs in the Federal Columbia River Power System. The Independent Scientific Advisory Board recognized that the IRCs were designed to minimize the impacts of drawdowns on reservoir food webs (ISAB 1997a) and that the Hungry Horse and Libby IRCs "provide seasonality of flow in downstream reaches as called for under our normative river concept."(ISG 1996, ISAB 1997b).

Columbia River Flow Augmentation

Flow augmentation for anadromous species in the lower Columbia is based on a relationship between water velocity and smolt (juvenile salmon or stealhead) migration speed. Evidence suggests that fish generally travel faster at higher water velocities and that more rapid travel results in better survival. The release of water from Montana reservoirs during August is intended to speed, thereby improving the survival of smolts. Water velocity profiles measured by the U.S. Geological Survey (USGS) in McNary reservoir were used to describe the influence of Montana dams on water velocity (Fig. 7). Since most smolts travel in the top 10 m of the reservoirs

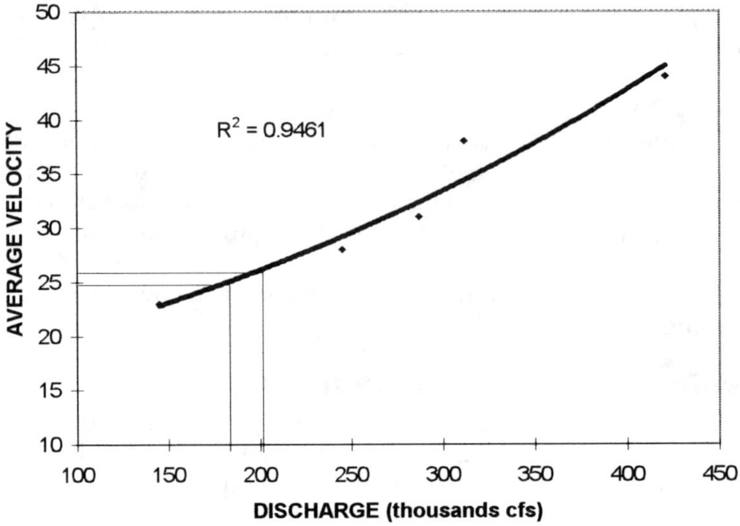

Figure 7. Relation between river discharge at McNary Reservoir and average water velocity in the offshore area less than 10 m depth. The difference in water velocity at 180 kcfs and 200 kcfs is highlighted; approximately 1.6 cm per second. Data source: U.S. Geological Survey.

in the lower Columbia, we can focus on the average velocity in that zone. (Note: USGS data do not include the low-velocity nearshore portion of the channel cross-section.) An exponential regression of average water velocity under varying discharge volumes shows that the average velocity increases 1.6 cm per second when the discharge is augmented by 20 kcfs (roughly the equivalent of the total release from Montana), increasing the flow at McNary from 180 kcfs to 200 kcfs (the August flow target). Thus, if no Montana water was released, the velocity would be reduced by an average of 1.6 cm per second. Since John Day Reservoir, immediately downstream, has a larger cross-section and volume than McNary, the change in water velocity and smolt travel time would be even less.

Conclusion

It our conclusion that the Endangered Species Act and the action agencies operating the dams in the Columbia River basin must consider effects on the ecosystem, including listed and nonlisted stocks of native species to avoid additional listings. The current operational strategy results in unnatural flow fluctuations below Libby and Hungry Horse dams that are harmful to the riverine ecosystem that supports the endangered Kootenai white sturgeon, threatened bull trout, westslope cutthroat trout, and interior redband trout.

Water released for anadromous fish recovery as called for by NMFS disrupts the desired balance between resident fish needs, storage reservoir operation, and river flows. Reservoir refill failures during dry years are expected under the IRC/VARQ operation, but less frequently than would occur by implementing the NMFS operation. Extreme reservoir refill failure (more than 20 feet) negatively affects biological production in the reservoirs, entrains more fish through Libby Dam, and impacts fishing and recreation in the Flathead and Kootenai rivers downstream. Reservoir refill failure also compromises the system's ability to store water for release during the following spring. The best conditions for Columbia Basin fish resources can be achieved by implementing the IRC/VARQ in Montana and other storage projects (e.g., Mica, Arrow, Dworshak). Sub-basins experiencing wet conditions can supply salmon flow augmentation, whereas dry sub-basins would provide less flow, protecting important resident fish stocks. Combined flows from the headwater sub-basins could then be shaped to achieve the greatest benefit for salmon and other anadromous stocks while protecting fish populations in the dry sub-basins. A gradual rampdown from the spring runoff in the sub-basins can be used to normalize the river hydrograph below headwater projects.

References

ACOE (U.S. Army Corps of Engineers). 1997. Columbia River Basin system flood control review preliminary analysis report. U.S. Army Corps of Engineers, North Pacific Division. 98 pp. plus appendices.

Allendorf, F.W., D.M. Esperlund, and D.T. Scow. 1980. Coexistence of native and introduced rainbow trout in the Kootenai River drainage. Proc. Montana Acad. Sci. 39:28-36.

Apperson, K.A. 1992. Kootenai River white sturgeon investigations and experimental culture. Annual Progress Report FY 1991. Idaho Department of Fish and Game and Bonneville Power Administration. Contract No. DE-AI79-88BP93497. Project No. 88-65. Portland, Oregon.

Apperson, K.A., and P.J. Anders. 1991. Kootenai River white sturgeon investigations and experimental culture. Annual Progress Report FY 1990. Idaho Department of Fish and Game and Bonneville Power Administration. Contract No. DE-AI79-88BP93497. Project No. 88-65. Portland, Oregon. 67 pp.

Armitage, P.D. 1984. Environmental changes induced by stream regulation and their effect on lotic macroinvertebrate communities. In: A. Lillehammer and S.J. Saltveit (eds.), Regulated rivers. University Press, Oslo, Norway, pp. 139-165.

Ashley, K., and L. Thompson. 1996. Kootenay Lake Fertilization experiment: Year 3 (1994/1995) report. Fisheries Project Report No. RD 49. Fisheries Branch, Ministry of Environment, Lands and Parks, University of British Columbia.

Behnke, R.J. 1992. Native trout of western North America. American Fisheries Society, Monograph 6. Bethesda, Maryland. 275 pp.

Chisholm, I.M., M.E. Hensler, B. Hansen, and D. Skaar. 1989. Quantification of Libby Reservoir levels needed to maintain or enhance reservoir fisheries. Methods and Data Summary 1983-1987. Prepared for Bonneville Power Administration by Montana Department of Fish, Wildlife and Parks. Kalispell, Montana. Project No. 83-467.

Christenson, D.J., R.L. Sund, and B.L. Marotz. 1996. Hungry Horse Dam's successful selective withrawal system. Hydro Review 15(3):10-15.

Daley, R.J., E.C. Carmack, C.B.J. Gray, C.H. Pharo, S. Jasper, and R.C. Weigand. 1981. The effects of upstream impoundments on the limnology of Kootenay Lake, B.C. Scientific Series No. 117, National Water Research Institute, British Columbia.

Fraley, J.J., B. Marotz, J. Decker-Hess, W. Beattie, and R. Zubik. 1989. Mitigation, compensation, and future protection for fish populations affected by hydropower development in the Upper Columbia system, Montana, U.S.A. Regulated Rivers: Research and Management (3):3-18.

Hauer, F.R., and J.A. Stanford. 1982. Ecological responses of hydropsychid caddisflies to stream regulation. Can. J. Fish. Aquat. Sci. 39(9):1235-1242.

Hauer, F.R., J.T. Gangemi, and J.A. Stanford. 1994. Long-term influence of Hungry Horse Dam operation on the ecology of macrozoobenthos of the Flathead River. Prepared for Montana Fish, Wildlife and Parks, Special Projects Bureau, Kalispell, Montana.

Hauer, F.R., J.B. Biby, J.T. Gangemi, and J.A. Stanford. 1997. Long-term influence of Libby Dam operation on the ecology of macrozoobenthos of the Kootenai River. Report to Montana Fish, Wildlife and Parks. University of Montana, Yellow Bay Biological Station, Polson, Montana.

Hildebrand, L., and S. McKenzie. 1994. Status of white sturgeon in the Columbia River, B.C. Final Report for B.C. Hydro. R.L.&L. Environmental Services, Limited.

International Joint Commission. 1938. Application to construct and operate Corra Linn Dam on Kootenay Lake, B.C. West Kootenay Power and Light Company. Order of Approval.

ISAB (Independent Scientific Advisory Board). 1997a. Ecological impacts of the flow provisions of the Biological Opinion for endangered Snake River salmon on resident fishes in the Hungry Horse, and Libby systems in Montana, Idaho, and British Columbia. Independent Scientific Advisory Board. Report 97-3 for the Northwest Power Planning Council and National Marine Fisheries Service. Portland, Oregon.

ISAB. 1997b. The normative river. Independent Scientific Advisory Board report to the Northwest Power Planning Council and National Marine Fisheries Service. Portland, Oregon.

ISG (Independent Scientific Group). 1996. Return to the river: Restoration of salmonid fishes in the Columbia River ecosystem. Development of an alternative conceptual foundation and review and synthesis of science underlying the Columbia River Basin Fish and Wildlife Program of the Northwest Power Planning Council Document 96-6. Prepublication copy.

Marotz, B., D. Gustafson, C. Althen, and B. Lonon. 1996. Model development to establish Integrated Operation Rule Curves for Hungry Horse and Libby reservoirs, Montana. Montana Fish, Wildlife and Parks report to Bonneville Power Administration. DOE/BP-92452-1. Portland, Oregon. 114 pp.

May, B., S. Glutting, T. Weaver, G. Michael, B. Marotz, P. Suek, J. Wachsmuth, and C. Weichler. 1988. Quantification of Hungry Horse Reservoir water levels to maintain or enhance reservoir fisheries. Montana Department of Fish, Wildlife, and Parks, Kalispell, Montana. Annual report prepared for Bonneville Power Administration. 68 pp.

Montana Bull Trout Scientific Group. 1998. The relationship between land management activities and habitat requirements of bull trout. Prepared for the Montana Bull Trout Restoration Team by the Montana Bull Trout Scientific Group. Montana Fish, Wildlife and Parks, Helena, Montana. 77 pp.

NMFS (National Marine Fisheries Service). 1995. Endangered Species Act—Section 7 Consultation, Biological Opinion: Reinitiation of consultation on 1994-1998 operation of the Federal Columbia River Power System and juvenile transportation program in 1995 and future years. Consultation by National Marine Fisheries Service, Northwest Region, U.S. Army Corps of Engineers, Bonneville Power Administration, Bureau of Reclamation.

NMFS. 1998. Endangered Species Act—Section 7 Consultation, Supplemental Biological Opinion: Operation of the Federal Columbia River Power System including the smolt monitoring program and the juvenile fish transportation program: A supplement to the biological opinion signed on March 2, 1995, for the same projects. Consultation by National Marine Fisheries Service, U.S. Army Corps of Engineers, Bonneville Power Administration, Bureau of Reclamation.

NPPC. 1994. Northwest Power Planning Council Fish and Wildlife Program. NPPC Document 94-55.

Parret, C., and J.A. Hull. 1985. Streamflow characteristics of mountain streams in western Montana. U.S. Geological Survey Water-Supply Paper 2260.

Parsley, M.J., L.G. Beckman, and G.T. McCabe Jr. 1993. Habitat use by spawning and rearing white sturgeon in the Columbia River downstream from McNary Dam. In: Status and habitat requirements of the white sturgeon populations in the Columbia River downstream from McNary Dam. Final Report by Oregon Department of Fish and Wildlife, National Marine Fisheries Service, U.S. Fish and Wildlife Service, and Washington Department of Fisheries to Bonneville Power Administration. Portland, Oregon.

Perry, S.A., W.B. Perry, and J.A. Stanford. 1986. Effects of stream regulation on density, growth, and emergence of two mayflies (Ephemeroptera: Ephemerellidae) and a caddisfly (Trichoptera: Hydropsychidae) in two Rocky Mountain rivers (U.S.A.). Can. J. Zool. 64:656-666.

Peters, J.C. 1962. The effects of stream sedimentation on trout embryo survival. In: Biological Problems in Water Pollution, Third Seminar, 1962. Public Health Service Publication 999-WP-25, pp. 275-279.

Poff, N.L., and J.V. Ward. 1989. Implications of streamflow variability and predictability for lotic community structure: A regional analysis of streamflow patterns. Can. J. Fish. Aquat. Sci. 46(10):1805-1818.

Radar, R.R., and J.V. Ward. 1988. Influence of regulation and environmental conditions and the macroinvertebrate community in the upper Colorado River. Regulated Rivers 2:597-618.

Skaar, D., J. DeShazer, L. Garrow, T. Ostrowski, and B. Thornberg. 1996. Quantification of Libby Reservoir levels needed to maintain or enhance reservoir fisheries: Investigations of fish entrainment through Libby Dam, 1990-1994. Montana Fish, Wildlife and Parks and Bonneville Power Administration. Contract No. DE-AI79-83BP12660. Project No. 83-467. Portland, Oregon. 110 pp.

SOR. 1995. Columbia River System Operation Review. Environmental Impact Statement. Main report and Appendix K, resident fish. Bonneville Power Administration, U.S. Army Corps of Engineers, U.S. Bureau of Reclamation. DOE/EIS-0170.

Stanford, J.A., J.V. Ward, W.J. Liss, C.A. Frissell, R.N. Williams, J.A. Lichatowich, and C.C. Coutant. 1996. A general protocol for restoration of regulated rivers. Regulated Rivers: Research and Management 12:391-413.

USFWS (U.S. Fish and Wildlife Service). 1994. Endangered and threatened wildlife and plants; determination of endangered status for the Kootenai River population of white sturgeon final rule. Federal Register 59(171) (September 6, 1994):45989-46002.

Van Eimeren, P.V. 1996. Westslope cutthroat trout. In: D.A. Duff (ed.), Conservation assessment for inland cutthroat trout status and distribution. U.S. Department of Agriculture, Forest Service, Intermountain Region, Odgen, Utah.

Walters, C. 1997. Personal communication and minutes from the February 18, 1997, Kootenai River Modeling Workshop. Dr. Carl Walters, University of British Columbia, Victoria, B.C.

Weaver, T.M., and J.J. Fraley. 1993. A method to measure emergence success of westslope cutthroat trout fry from varying substrate compositions in a natural stream channel. N. Am. J. Fish. Manage. 13:817-822.

Wesche, T.A., and P.A. Richard. 1980. A summary of instream flow methods for fisheries and related research needs. Eisenhower Consortium Bulletin 9.

Williams, J.E., J.E. Johnson, D.A. Hendrickson, S. Contreras-Balderas, J.D. Williams, M. Avarro-Mendosa, D.E. McAllister, and J.E. Deacon. 1989. Fishes of North America endangered, threatened, or of special concern: 1989. Fisheries 14(6):2-20.

Wright, A., et al. 1996. Review of Columbia River operating criteria. Report to National Marine Fisheries Service and Bonneville Power Administration by Al Wright Consulting, Portland, Oregon.

Utilizing Ecosystem Concepts in Fisheries Management Strategies

Kristine D. Lynch and William W. Taylor
Michigan State University, Department of Fisheries and Wildlife, East Lansing, Michigan

John M. Robertson
Michigan Department of Natural Resources, Forest Management Division, Lansing, Michigan

Kelley D. Smith
Michigan Department of Natural Resources, Fisheries Division, Lansing, Michigan

Abstract

Fisheries ecosystem management programs and policies must incorporate holistic, integrative, and multidisciplinary collaborative strategies. Those who manage fisheries are often administratively, procedurally, and disciplinarily separated from those who manage other related ecosystem elements, such as forests, wildlife, water quality, and land use. This organizational and philosophical division within agencies can render ecosystem management programs incomplete, inefficient, and ineffective. Since management processes influence how managers conceptualize and respond to resource issues, it is important that management systems be congruent with the dynamics of ecosystems. The Michigan Department of Natural Resources (MDNR) is redesigning their management processes to enable more efficient and effective ecosystem management. Boundaries for management units within all divisions (including Fisheries, Wildlife, Forest Management, and Parks and Recreation) are being redrawn along watershed and eco-region lines. Instead of each division developing their management plans in isolation, the design and implementation of all management plans will require input and contributions from managers in all relevant divisions. Furthermore, each management plan will be based on and measured by locally specific criteria and indicators that are most diagnostic for each ecosystem. New management plans will thereby be more comprehensive and ecosystem-specific, although they will require a great deal of coordination, collaboration, and communication among agency personnel. By redesigning their management planning and implementation processes to incorporate ecosystem priorities and activities, the MDNR should experience long-term improvements in the sustainability of fisheries (and other) ecosystems.

Introduction

Regional landscape inputs are integrated in freshwater ecosystems, holistically reflecting complex physical, chemical, biological, and social processes, interrelationships, and dynamics. Therefore, fisheries ecosystem management programs and policies must also incorporate holistic, integrative, and multidisciplinary collaborative strategies. For many natural resource management agencies, these ecosystem management requirements pose a significant challenge to conventional methods of planning and implementing management activities. By reviewing ecosystem management concepts and examining one agency's strategy for ecosystem management, we can better understand the potential successes and pitfalls surrounding the implementation of agency-based ecosystem management. Through learning about one agency's experiences, other agencies can better anticipate and possibly avoid many of the potential difficulties in order to reap the benefits of ecosystem management.

Ecosystem Concepts

Ecosystem scientists have long demonstrated that aquatic ecosystems function as integrators of regional watershed and airshed dynamics (Hynes 1975). The quality and quantity of water in aquatic ecosystems, the basis for fish production, is therefore directly affected by physical, chemical, and biological inputs which are a result of the watershed-specific environmental and social processes (Taylor et al., In press). These include inputs from processes such as erosion and sedimentation, landscape runoff, point and nonpoint source pollution, and atmospheric deposition. Diverse human activities, such as urbanization, deforestation, industrial manufacturing, energy production and consumption, and intensive agriculture, drive these processes and therefore affect aquatic ecosystem productivity and fisheries sustainability. Such land use and development activities are a product of diverse and fragmented planning and policy-making bodies, which allow development to proceed with little or no coordination or regard for the structure and function of ecological systems (Caldwell 1994, Ferreri et al. 1998).

For aquatic ecosystems and their resources to be effectively managed, they must be understood in context of their function as holistic integrators of regional social and environmental processes (Knight and Meffe 1997). This integration is a guiding principle of ecosystem management, a management philosophy that emphasizes: (1) the integration of the physical, chemical, biological, and social components of the resource, (2) the interactions of these four components, and (3) how these components relate to ecosystem productivity and resource sustainability (Ferreri et al. 1998). Management based on these ecosystem principles will help maintain and improve the sustainability of these systems.

While the concept of ecosystem management is widely accepted by managers, scientists, and stakeholders as an appropriate management paradigm, it has undergone intense scrutiny and refinement throughout its evolution (Yaffee et al. 1996). The definitions of ecosystem management are as varied as ecosystems. Recognizing that this kind of management needs a strong conceptual foundation to be useful, the Ecological Society of America (ESA) assembled an ad hoc Committee on Ecosystem Management to address the concept of ecosystem management and its scientific foundation. Their 1995 report established that every ecosystem management effort, regardless of its specific definition, should include eight principles: "(1) long-term sustainability as fundamental value; (2) clear, operational goals; (3) sound ecological models and understanding; (4) understanding complexity and interconnectedness; (5) recognition of the dynamic character of ecosystems; (6) attention to context and scale; (7) acknowledgment of humans as ecosystem components; (8) commitment to adaptability and accountability" (ESA 1995:1).

Adapting Agency Management to Ecosystem Concepts

Due to the diverse and dynamic nature of regional social and environmental processes, holistic management of aquatic ecosystems requires a high level of coordination and collaboration among those in the watershed whose activities affect water quality and quantity (Ferreri et al. 1998). Those who manage fisheries are very often administratively, procedurally, and disciplinarily separated from those who manage other related ecosystem elements, such as forests, wildlife, water quality, and land use (Knight and Meffe 1997). These organizational and philosophical divisions within and between management agencies can render ecosystem management programs incomplete, inefficient, and ineffective (Caldwell 1994). Agency management activities are also usually separated from other watershed activities that influence aquatic ecosystems and fishery production; for example, activities by private landowners, non-governmental organizations, organized citizen groups, and commercial industries. These groups engage in a wide range of water- and land-based activities that directly and indirectly affect various components of aquatic ecosystems, such as nutrient cycling, temperature, and flow rates (Decker and Krueger 1993).

Agency-based management of fisheries and aquatic ecosystems must often be transformed from a "closed system" of short term fish production and habitat manipulation to a more open, dynamic, and integrative system based on long-term collaboration with other ecosystem scientists, managers, and stakeholders (Ferreri et al. 1998). Implementation of ecosystem management will require natural resource agencies to reevaluate their management philosophy, which will enable them to change their management structures and in ways that are consistent with ecosystem concepts (Knight and Meffe 1997). Since management structures and processes influence how managers conceptualize and respond to resource issues, it is essential that agency management systems be congruent with the dynamics of ecosystems.

Case Study: The Michigan Department of Natural Resources

Some natural resource management agencies are redesigning their management structures and processes in order to enable more efficient and effective ecosystem management (Yaffee et al. 1996). The Michigan Department of Natural Resources (MDNR) is one such agency. Recognizing that their traditional management system has been incompatible with many ecosystem concepts, the MDNR is undergoing a comprehensive transformation to make their management activities more congruent with ecosystem structure and function. Following this transformation, the MDNR intends to produce ecosystem management plans that are more comprehensive and ecosystem-specific, thereby producing more effective and efficient management tools. A review of the overall MDNR ecosystem management strategy and a closer look at one division's role in the transformation reveal that the MDNR approach to ecosystem management has potential for success, but the MDNR must overcome many challenges.

Michigan Department of Natural Resources Joint Venture Strategy

For decades, the MDNR has been managing natural resources through the fragmented efforts of narrowly focused divisions, such as Fisheries, Wildlife, Forest Management, and Parks and Recreation. As a result, managers from different divisions often had little involvement with each other, whether they were working for common or conflicting ecosystem objectives. For example, the Fisheries Division's stream habitat management activities are strongly influenced by upstream logging activities, but the Forest Management Division may not be aware of the Fisheries Division fish habitat objectives in that stream. The divisions' planning and management activities were therefore often inefficient and ineffective.

According to its mission statement, the MDNR is committed to the conservation, protection, management, use, and enjoyment of the state's natural resources for current and future generations. In order to satisfy the current multiple values of natural resources and manage on an intergenerational time scale, the MDNR recognizes that the principles of ecosystem management are highly appropriate. The MDNR conceptualizes ecosystem planning and management as a process that *integrates* physical, chemical, biological, and ecological principles, along with economic and social factors, into a *comprehensive* strategy aimed at protecting and enhancing sustainability, diversity, and productivity of a system. This definition of ecosystem management emphasizes the need for an integrative and comprehensive perspective, which is necessary to maintain the long-term multiple values of Michigan's natural resources.

The MDNR's traditional management structure and processes were neither very integrative nor comprehensive. Each division's management activities have traditionally been administratively and scientifically separated from the management activities of other divisions, despite the fact that the resource management divisions both influence and rely upon various aspects of common ecosystems. For example, landscape activities such as canopy removal through logging may increase water temperature in rivers, thereby altering aquatic habitat. Other changes to aquatic habitats can shift the distribution of wildlife, such as waterfowl, thereby altering ecosystem structure and the availability of recreational opportunities. Realizing the close interdependence between their management objectives and activities, each division has been motivated to seek a more collaborative and ecologically appropriate management style. Creating a more ecologically compatible management agency forced the MDNR to critically examine how management procedures and administration should be organized. Through the extensive joint efforts of division leaders and numerous subcommittees charged with designing ecosystem management processes, a new joint venture (JV) strategy for ecosystem management has emerged.

Each division, including Fisheries, Wildlife, Forest Management, and Parks and Recreation, is undergoing a structural and procedural transformation as part of the JV initiative. This JV strategy preserves the MDNR's divisions, although the organization of each division's management units and processes will be realigned according to appropriate ecosystem concepts. For example, since the State of Michigan contains four distinct eco-regions and four distinct Great Lakes basins, these areas will become the fundamental "eco-units" for coordinating regional management activities. While divisions are committed to the eco-unit concept, divisions will not all adopt common boundaries for their eco-units. The ecosystem-based boundaries that are appropriate for fisheries (i.e., watersheds and basins) are not necessarily the most appropriate unit for planning forest or wildlife management, where management units based on eco-regions or habitat type are more appropriate. Management activities will be implemented at the management area level, a new geographic subunit of the eco-units. Furthermore, each division will establish eco-unit teams, consisting of regional division staff and an eco-unit coordinator, to coordinate and facilitate the management activities for their respective eco-units. Coordinators will also serve on a department-wide ecosystem management team, responsible for overseeing, linking, and coordinating statewide ecosystem planning and assessment based on appropriate criteria and indicators. The concept for the revised MDNR structure is illustrated in Fig. 1.

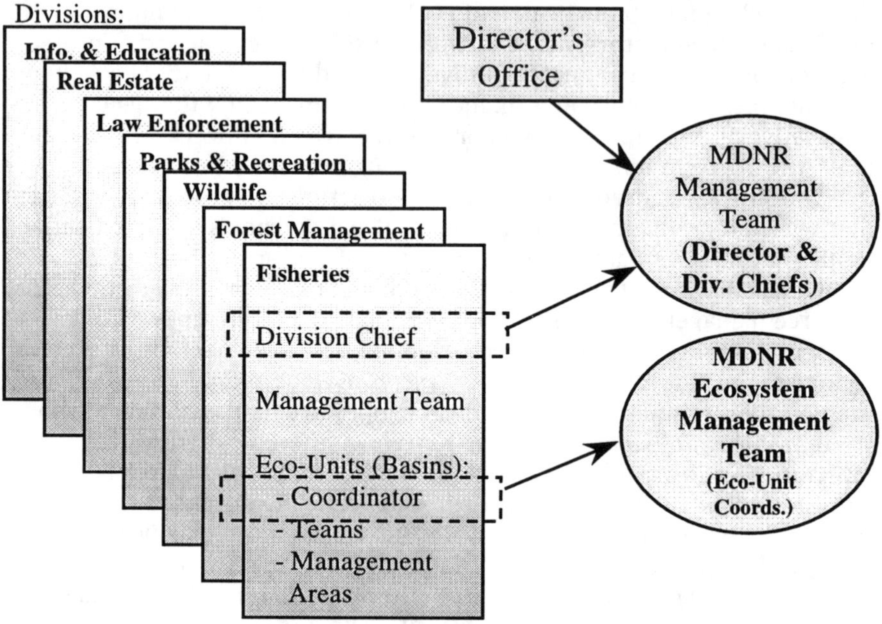

Figure 1. Proposed concept for organization of the Michigan Department of Natural Resources.

Ecosystem management requires more than structural revisions. Divisions are also re-designing their management *processes* to allow for more comprehensive integration of all available information about ecosystems. Instead of independently developing fragmented management plans in isolation, each division will be required to develop management plans at their eco-unit level in cooperation with resource managers from other divisions, agencies, and public groups. This broader base of input and contributions will allow each division to plan on a more holistic eco-regional perspective, and it will enable more appropriate allocation of budget and staffing resources in the implementation of ecosystem projects. Division-based eco-unit teams will be responsible for adapting their management planning processes to incorporate the relevant groups. They will also be fundamentally held accountable for their eco-unit's management objectives and accomplishments. Furthermore, as eco-unit teams develop management plans, they will also cooperatively develop the criteria and indicators, customized for each eco-unit's unique biological, physical, and social conditions, which will be used to gauge management success.

Fisheries Division Joint Venture Strategy

The MDNR Fisheries Division, responsible for managing the fisheries and other aquatic resources in Michigan's waters of the Great Lakes as well as inland aquatic ecosystems, is one of the divisions undergoing a structural and procedural transformation as part of the JV initiative. In the fall of 1997, the Fisheries Division began planning and designing new administrative structures and management processes. This work was undertaken by a special committee, consisting of representatives from Fisheries Division research, management, fish production, and program services sections, charged with identifying a strategy for managing aquatic resources based on ecologically based management units and processes.

The committee began by changing the division's management planning and implementation units within Michigan. Traditional management units were based on county lines for convenience in statistical reporting, not according to watershed or other eco-region boundaries. Recognizing that the structure and function of aquatic ecosystems are influenced by watershed-level processes, the committee recommended establishing watershed-based boundaries for both management areas (smaller local watersheds) and eco-units (regional groupings of adjacent watersheds). Eco-units will be based on the watersheds for the four Great Lakes that border Michigan: Lakes Superior, Michigan, Huron, and Erie (Fig. 2); within the Fisheries Division these are termed "basin units." These four basin units will encompass one (Lakes Erie and Superior), two (Lake Huron), or three (Lake Michigan) management areas.

The fisheries management roles, responsibilities, and processes for basin units and management areas will be distinct, yet interdependent. Fisheries in each of the four basin units are to be primarily managed by a basin team (comparable to other divisions' eco-unit teams), responsible for both Great Lakes and inland waters. Basin team responsibilities will include both long- and short-term management planning. Long-term planning includes goal setting; determining staffing, research, and fish production needs; and providing input for division strategic planning. Basin teams will also use their collaborative, basin-wide perspective to oversee the short-term planning of management area activities, such as stocking fish and setting work priorities. To plan these management activities, the basin teams will consult with other divisions in the MDNR and the Department of Environmental Quality, as well as with other state and federal agencies, universities, consultants, and public groups. Furthermore, basin teams will be responsible for allocating budgets to management areas according to basin-wide work priorities, and they will be responsible for monitoring budget accountability work performance. Management areas, in contrast, will implement management plans approved by the basin team and report on the status of management activities. Due to the collaborative and holistic requirements of ecosystem management, these work plans

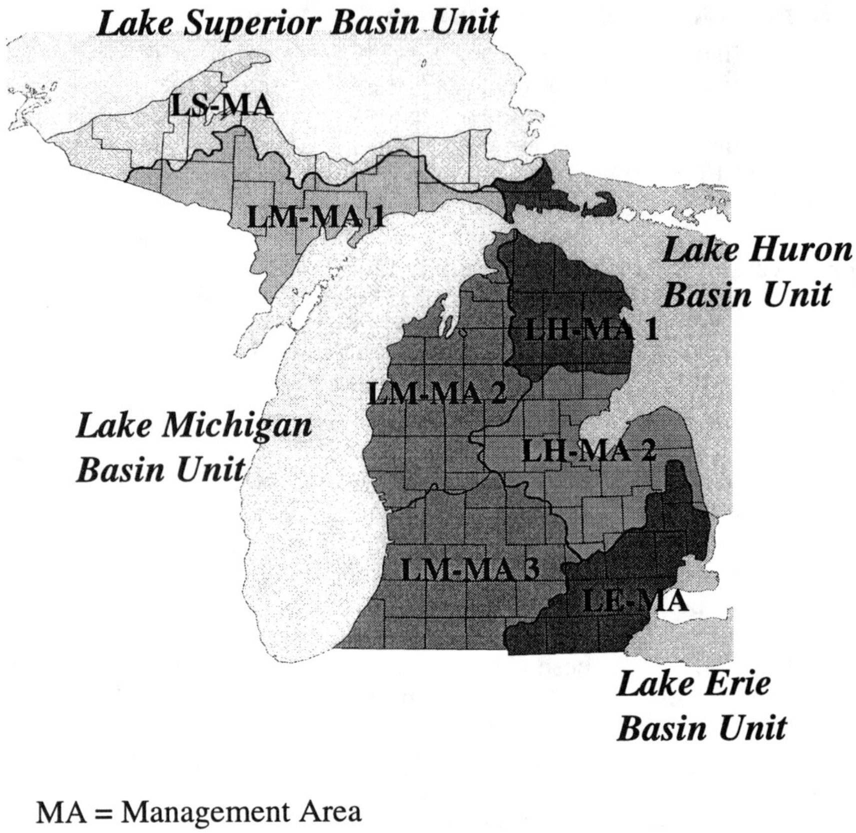

Figure 2. Proposed MDNR Fisheries Division basin units and management areas.

may involve contributing to the management activities of other divisions, such as Forest Management or Wildlife, developed in cooperation with Fisheries Division personnel.

Basin-wide management activities will be coordinated by basin teams, since the team approach enables broad-scale collaboration and encourages the exchange of perspectives and knowledge across program areas. Members of the interdisciplinary basin teams will come from all management areas in each respective basin and will consist of area managers, fisheries management biologists, technician supervisors, inland researchers, Great Lakes researchers, fish production specialists, or other staff specialists as needed. Specific basin team composition and the decision-making structure (e.g., voting, consensus) will vary at different times, depending on the specific issues or problems that need to be addressed.

The activities of each basin unit will be facilitated and represented by a basin coordinator. Each basin coordinator will be responsible for: (1) determining the statewide needs for the basin and incorporating these into area work planning; (2) providing the Fisheries Division management team with annual basin priorities set by the basin team; (3) working with management areas on budget allocation issues; and (4) drafting an annual report on the basin unit's accomplishments. The basin coordinators will also represent their basin units on the department-wide ecosystem management team, with the public, and on other ad hoc ecosystem management initiatives. The four basin coordinators will hold this position on a full-time basis.

At the upper end of the management hierarchy, the Fisheries Division will maintain a single division-wide management team. This team is responsible for long-term oversight, program development, and program review. It will consist of the division chief, section leaders (e.g., fish production, research, program services), basin coordinators, and other division specialists as needed. The management team will ultimately be responsible for setting division priorities and allocating budgets among basins. They will review the accomplishments of the four basin teams to ensure adherence to priorities and implementation of projects. The concept for the proposed Fisheries Division organizational structure is illustrated in Fig. 3.

Evaluation of the Joint Venture Initiative

By developing the JV strategy, the MDNR believes it is adopting and internalizing the concept of ecosystem management. By comparing features of the JV strategy to the eight guiding principles developed by the ESA (1995), it may be possible to determine the extent to which the JV strategy has embraced ecosystem management standards. This may also indicate potential areas of JV success or problems.

(1) *Long-term sustainability as fundamental value*: The MDNR definition of ecosystem management emphasizes the need to protect and enhance the sustainability of natural resource systems. Instead of promoting multiple *uses* of ecosystems, the MDNR acknowledges that sustainability is one of the multiple *values* that must be maintained.

(2) *Clear, operational goals*: At this point, the goals of the MDNR's JV initiative are largely at the conceptual stage; for example, JV goals include: (a) efficient and effective management and supervision, (b) holistic resource management, (c) communication and education, (d) organizational culture, (e) employee development and training, and (f) strategic management. While these goal areas are explained in the JV document, they do not yet consist of measurable goals that specify the outcomes of ecosystem management activities. These will be determined as ecosystem management activities are designed and implemented.

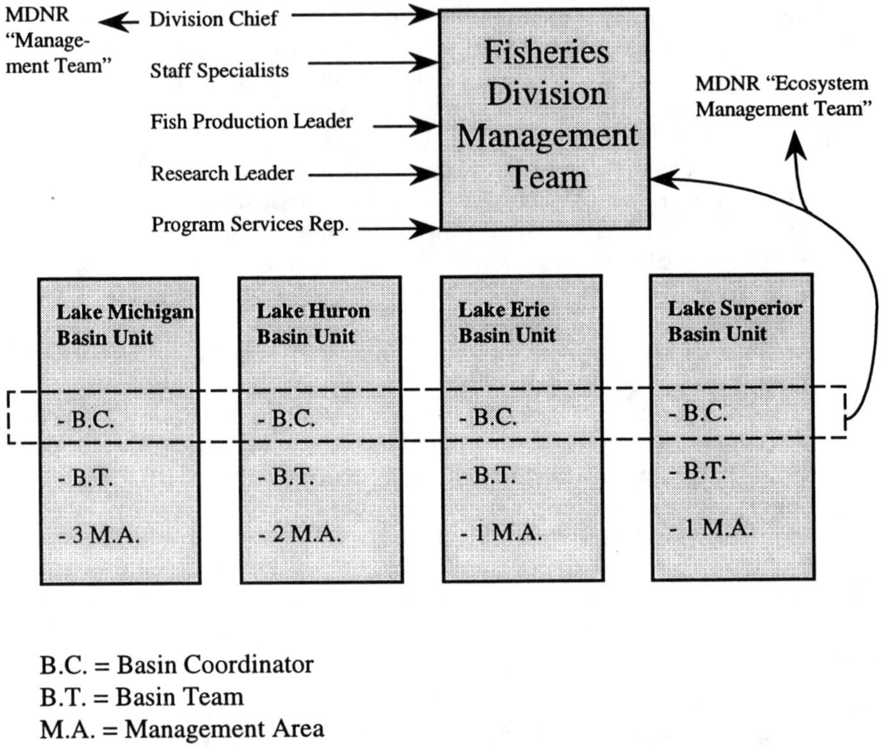

Figure 3. Proposed concept for organization of the MDNR Fisheries Division.

(3) *Sound ecological models and understanding*: The MDNR will continue to conduct research and monitoring at multiple levels of ecological organization. Divisions will continue research on various ecosystem components, and a common GIS (geographic information system) database will be available for use among all divisions. Their ability to develop ecological models and understanding will rely upon their ability to integrate their research activities into a common ecological framework.

(4) *Understanding complexity and interconnectedness*: By requiring cross-divisional collaboration in the management planning and implementation processes, the MDNR understands that ecosystems are far more complex and interconnected than any one division can understand. While the JV initiative requires interdepartmental cooperation and collaboration with other non-MDNR partners in the watersheds and landscapes, the MDNR must still demonstrate how these collaboration requirements will be for-

mally integrated into management processes. Like principle (2) above, these will be determined in the future.

(5) *Recognition of the dynamic character of ecosystems*: Instead of managing for only one set of ecosystem conditions, the MDNR recognizes that ecosystems are dynamic and require adaptive management approaches so that managers can learn and adjust their management plans accordingly. The departmental ecosystem management team and the division eco-unit teams will develop appropriate criteria and indicators that are most diagnostic for different ecosystem conditions and goals; these will be used to gauge the direction of ecosystem trends, not static conditions, for determining management success. Like principles (2) and (4), these criteria and indicators will be developed as ecosystem management evolves.

(6) *Attention to context and scale*: Divisions recognize that there are multiple scales at which ecosystem components function; therefore, each division is independently designing their most appropriate management scales. They recognize that the "ecosystem" for a fish or a tree may encompass different social and environmental processes, and they have established appropriate spatial scales for management. While they also acknowledge the need for a long-term temporal scale, this too will likely be customized for each ecosystem component and demonstrated as the management planning process evolves.

(7) *Acknowledgment of humans as ecosystem components*: The MDNR definition of ecosystem management acknowledges that economic and social factors must be integrated with other ecosystem processes. Divisions know that their management activities must meet a wide range of human needs and values, and they know that various stakeholder groups must be incorporated into management processes. Again, the eco-unit teams must establish appropriate mechanisms for ensuring effective involvement of multiple stakeholders.

(8) *Commitment to adaptability and accountability*: The JV documentation repeatedly stresses the importance of adaptive management that allows for evaluation, learning, and management modification as ecosystems change. The MDNR also intends to hold eco-units accountable for developing and implementing appropriate management activities. Once ecosystem management process are implemented, the extent of the eco-units' adaptability and accountability must be determined.

According to the standards developed by the ESA (1995), the MDNR JV strategy appears to be on the right path for establishing an ecosystem-based approach to management. The extent to which these standards are developed and implemented, however, has yet to be determined. Many of the above standards must be manifested in the specific ecosystem management plans developed by eco-unit teams, which should be developed by mid-1999. At that time, and periodically thereafter, these eight standards should again be reviewed by those on eco-unit teams.

Other Challenges to the Joint Venture Initiative

While the JV initiative demonstrates a solid commitment to ecosystem management, the divisions are experiencing certain problems that may impede their ability to establish successful management. Many of these problems are related to the MDNR's administrative structure and communication strategy, and they may experience other potential problems related to team performance.

As the divisions proceed with developing ecosystem management processes, they will have to reconcile division-specific eco-unit structure with the need for common administrative structures regarding budgets, work locations, and staffing issues. In other words, although the eco-units will be based on different boundaries in different divisions, there is still a need to share facilities, office space, and support staff for managers in different eco-units; this will require coordination of administrative structure across different divisions and management boundaries. This kind of coordination will likely present challenges as their ecosystem management strategy is implemented.

The MDNR will also experience challenges as they communicate the JV strategy to their personnel and other stakeholders. The concept of ecosystem management has been embraced and pursued by MDNR leadership, although the majority of MDNR personnel are not involved in the process of adapting the agency's structure and processes to ecosystem requirements. As a result, they often lack complete understanding of this process; yet this misunderstanding is expected since the process of adopting ecosystem management has gone through several revisions since its initiation. Leaders in the MDNR are aware they must develop a strategy for communicating the new structures and processes to their personnel to ensure staff support, which may require extensive continuing education and training programs on ecosystem management philosophy as well as the new administrative procedures. The MDNR is also aware they must develop a strategy for informing other stakeholders about their new structures and processes, since many stakeholders regularly deal with MDNR staff through both formal partnerships and informal inquiries. They need to understand that the services provided by the MDNR (resource information, visitor centers, etc.) may be provided differently in the future.

The MDNR should anticipate other potential problems surrounding the roles and responsibilities of various teams and personnel in the JV initiative; this may help them avoid any shortcomings in team or job performance. The key to successful ecosystem management planning rests on the eco-unit teams' ability to identify and incorporate the appropriate ecosystem groups and information. Will they comprehensively incorporate other ecosystem stakeholders in their planning *and* become involved in other divisions' planning, potentially at a cost of their own division's budget, time, and personnel? Successful ecosystem management requires successful coordination and collaboration, which, in turn, requires active participation from eco-units, inclusion of other partners, *and* diligent oversight from the ecosystem management team. If these teams fail to function appropriately, the ecosystem management process will break down. The JV initiative must develop safeguards to ensure certain standards of team and personnel performance so that managers do not resort to traditional fragmented management systems.

Conclusions about Implementing Ecosystem Management

In embracing the concept of ecosystem management, the MDNR has undertaken a significant challenge, yet the JV approach appears to have a solid foundation in ecosystem management principles. Their challenge is far from over, however, as there are many areas in which the initiative could break down, and there are many important ecosystem principles that need to be further addressed through specific management plans. While it is too early to predict the outcomes of this ecosystem management initiative, the MDNR has the potential to experience success through appropriate development and implementation of management plans. By redesigning their administrative structure and management processes to incorporate holistic ecosystem priorities and activities, the MDNR may experience long-term improvements in the health and sustainability of fisheries (and other) ecosystems.

The MDNR's experience with ecosystem management, while far from being fully implemented, reveals many issues and challenges that natural resource agencies must address. The MDNR experience should be compared with the experiences of other natural resource agencies to better establish a framework for designing and implementing agency-based ecosystem management. All natural resource managers, scientists, and stakeholders have a great deal to learn about this approach to management, as the future of natural resource use and enjoyment depends on our ability to successfully manage ecosystems and their resources. Successful ecosystem management will provide Michigan, and the Great Lakes region, with long-term benefits by sustaining multiple uses and values of natural resources.

References

Caldwell, L.K. 1994. Disharmony in the Great Lakes Basin: Institutional jurisdictions frustrate the ecosystem approach. Alternatives 20(3):26-32.

Decker, D.J., and C.C. Krueger. 1993. Communication: Catalyst for effective fisheries management. In: C.C. Kohler and W.A. Hubert (eds.), Inland fisheries management in North America. American Fisheries Society, Bethesda, Maryland.

Ecological Society of America (ESA). 1995. The scientific bases for ecosystem management: An assessment by the Ecological Society of America. Washington, DC.

Ferreri, C.P., W.W. Taylor, and J.M. Robertson. 1998. Great Lakes fisheries futures: Balancing the demands of a multi-jurisdictional resource. In: W.W. Taylor and C.P. Ferreri (eds.), Great Lakes fisheries policy and management: A binational perspective. Michigan State University Press, East Lansing, Michigan.

Hynes, H.B.N. 1975. The stream and its valley. Verh. Internat. Verein. Limnol. 19:1-15.

Knight, R.L., and G.K. Meffe. 1997. Ecosystem management: Agency liberation from command and control. Wildl. Soc. Bull. 25(3):676-678.

Taylor, W.W., K.D. Lynch, and K.W. Taylor. In press. Integrating fisheries management with watershed processes. Verh. Internat. Verein. Limnol. 27.

Yaffee, S.L., A.E. Phillips, I.C. Frentz, P.W. Hardy, S.M. Maleki, and B.E. Thorpe. 1996. Ecosystem management in the United States: An assessment of current experience. Island Press, Washington, DC.

A Solution to the Conflict Between Maximizing Groundfish Yield and Maintaining Biodiversity

Albert V. Tyler
University of Alaska Fairbanks, School of Fisheries and Ocean Sciences, Fairbanks, Alaska

Abstract

There is a practical conflict between the goals of biodiversity maintenance and maximization of long-term fishery yields. It is not possible to maximize both simultaneously. On one hand, how to achieve long-term maximum yields while accounting for species interactions cannot be formulated because of variable relationships among species. On the other hand, it is known that catch rates tuned to the most productive species will bring about decreases in biomasses of less productive species taken as bycatch. This means diversity will decrease. Yet the long-term yields of economically valuable assemblages are likely dependent on the very diversity that is threatened by maximizing yields. A comparison of groundfish assemblages of continental shelves from the North Atlantic Ocean demonstrates that whole assemblages have declined—not just a few economically valuable species. An assemblage maintenance approach may be the only method of achieving multispecies persistence. A program of assemblage maintenance is suggested that reduces destructive exploitation risk while achieving economically viable levels of fisheries yields. The proposal involves setting up areas of contrasting fishing effort, with areas based on species stock structure and assemblage maps. It might be possible to hold part of a region with present effort levels, with increases encouraged in another part. The establishment of management areas would be followed by monitoring commercial catches of selected species, and by a program of research surveys that included assemblage analysis. Criteria are presented for adjustments of catch rates in response to assemblage changes related to fishing effort.

Introduction

My contention is that it is impossible to maximize both the fishery yield of major species and to maintain the biodiversity (Pielou 1975) of an exploited system. High rates of exploitation inevitably lead to changes in species composition and decreases in diversity that in turn feed back to the productivity of the economically important stocks and interfere with that productivity. I have come to conclude that without the original, natural diversity of the production system, the high productivity of the original target species cannot be maintained.

In this paper I will review some concerns about existing changes in a number of major groundfish production systems in regions of North America. I will then show reasons why the present paradigm of maximizing the yields of the main economic species in these systems is not only inadequate for maintaining total yield, but will lead to unlooked-for changes in the total production system that will decrease the value of the system to fisheries. Some of these changes will likely be irreversible in terms of the composition of fish and invertebrate fauna. Yet there is an approach to stock assessment that could be used as an adjunct to the yield maximization paradigm that would likely prevent these damaging changes. I will describe this alternative, empirically based suggestion that is similar to an earlier, but less complete, attempt (Tyler et al. 1982).

Catch Rate and the Mixed-Stock Management Problem

The mixed stock management problem is well known to biologists in the North Pacific as it applies to salmon. Paulik et al. (1967) showed that stocks of a single species of salmon often have differing productivity rates, and that when these stocks are caught together in an offshore fishery at a high catch rate, the least productive of the stocks will decline in spawning biomass. He showed how to calculate the rates of decline in relation to offshore catch rate and how to arrive at the rate that would sustain all of the stocks. He pointed out that maintaining the least productive stock means giving up catch potential of the most productive stock. The catch rate that maximizes the take for the most productive stock will cause the least productive stock to disappear.

When hatchery programs and ocean ranching became big along the Pacific coast, the mixed stock problem took on new dimensions. Hatcheries and ocean ranching operations could put so many young fish in the ocean that the catch rates for fisheries on returning runs could be very high, leading to increased catches of the hatched species. This catch is often taken in areas just offshore from the hatchery along with fish from wild stocks. The result is that the nearby wild stocks decline because they do not have the rate of productivity of the hatchery. The fishery takes wild and hatchery fish at the same catch rate. "Similarly the establishment of a new and highly productive hatchery stock might be responsible for an increase in the rate of exploitation that resulted in elimination of less productive existing stocks harvested in a common fishery" (Paulik et al. 1967). In terms of the portion of the stock taken, small wild stocks often have a greater portion removed than would ever be contemplated if they were the only stocks being managed.

It is the simultaneous fishing of stocks with differing productivities at a high rate suitable for the most productive stock that causes the problem. The result can only be that either the catch rate is diminished so that all stocks survive, or that the diversity of stocks decreases. The principle of the mixed stock management problem applies to other mixed stock situations as well as to salmonids; for example, to mixed species of ocean flatfish.

Multispecies Fishery Models

The Paulik model is a multistock model that is usable as a general multispecies model. The heart of the model is a series of stock-recruitment functions. Either stock or species somatic growth rates can be added and suitable changes can be made so that the stock being modeled is more like a cod as in the model POPSIM (Walters 1969) than a salmon. Several stocks can then be modeled with the Baranov function using catch rates from competing fleets (Ricker 1975); e.g., a longline fleet and a trawler fleet, both fishing the same stock of cod. These multispecies models can be put into a trophic dynamic setting where the target stocks compete with one another for a food resource base of varying complexity (Daan 1987). Statistically determined abiotic relationships can be added so that the rates of natural mortality, reproduction, and somatic growth respond to the biotic and abiotic environment (Pope and Knight 1982, Hilborn and Walters 1992).

I have heard some say that they dislike the single species orientation of the current stock assessment models and that analysts should develop more realistic multispecies models. On the other hand, analysts often respond by saying that they do have multispecies models, and they are developing better ones—that data are not available to either parameterize or drive the models. The limitation is that the statistical nature of species interaction data is uncertain. Often the common wisdom is that "predation has got to be important" or "food supply must be limiting." However, it is usually not possible to show the statistically significant relationships that are the best evidence for these relationships. As a result, multispecies models are generally used for summarizing an understanding of a complex dynamic situation and in gaining insight rather than for making quota estimates.

In our reductionist way, we hope that we will be able to develop natural production system models in computers that will account for species interactions so that we can extend the yield maximization paradigm to the context of a total production system. Though this thesis is attractive, I believe it will be an impossible objective to achieve for many years. The reason that elaborate computer models are not used to calculate potential yields of a species, for example, walleye pollock, is that the information needed is grossly incomplete. Many biologists are working on making this more complete. As scientists gain information and insight on how to apply the information to fishery problems, production system models will be more valuable.

Our partial understanding of variable species interactions and the influence of the physical-chemical world still prevents us from developing predictive models of multispecies yields in a multifactor setting that are predictive, ecosystem-dynamics models. Consequently, there is wisdom in not using those models to make yield estimates, but instead refining the single species models as analysts have done, changing the parameters and driving variables of the models as new information comes on-line.

Large-Scale Changes in Assemblages of Groundfish

It is important to look at the evidence for multispecies system change in case histories. There is evidence of change that goes beyond the reductions in the biomass of the target species that have been overfished. I will review assemblage findings from the Grand Banks of Newfoundland, Canada (western North Atlantic), and for United States fisheries: Georges Bank off New England (western Atlantic), the continental shelf off Washington and Oregon (eastern North Pacific), and the eastern Bering Sea off Alaska.

In terms of species richness, the Grand Banks is the simplest of the four areas. Thirty-five fish species occurred consistently enough in the regional data set to be included in a statistical analysis (Gomes et al. 1995). The dominant species were Atlantic cod (*Gadus morhua*), American plaice (*Hippoglossoides platessoides*), thorny skate (*Raja radiata*), redfish (*Sebastes marinus* and *S. mentella*), Greenland turbot (*Reinhardtius hippoglossoides*), and witch flounder (*Glyptocephalus cynoglossus*). Eleven assemblage groupings of species were identified that had affinities to particular areas (Fig. 1). Species were usually members of more than one group, and the groups were characterized by particular compositions of species. It was particularly interesting that the geographical boundaries of the assemblage groups were consistent over a 15-year period. As Newfoundland's northern cod fishery crisis emerged during the 1980s, it was not only the biomass of cod that changed, the abundances of nontarget species changed as well (Fig. 2). American plaice and broadhead wolffish (*Anarhichus denticulatus*) diminished, while witch flounder became scarce in the more shallow regions but more abundant in deeper assemblages. The multispecies biomass became dominated by thorny skate, apparently resisting the high intensity of trawling.

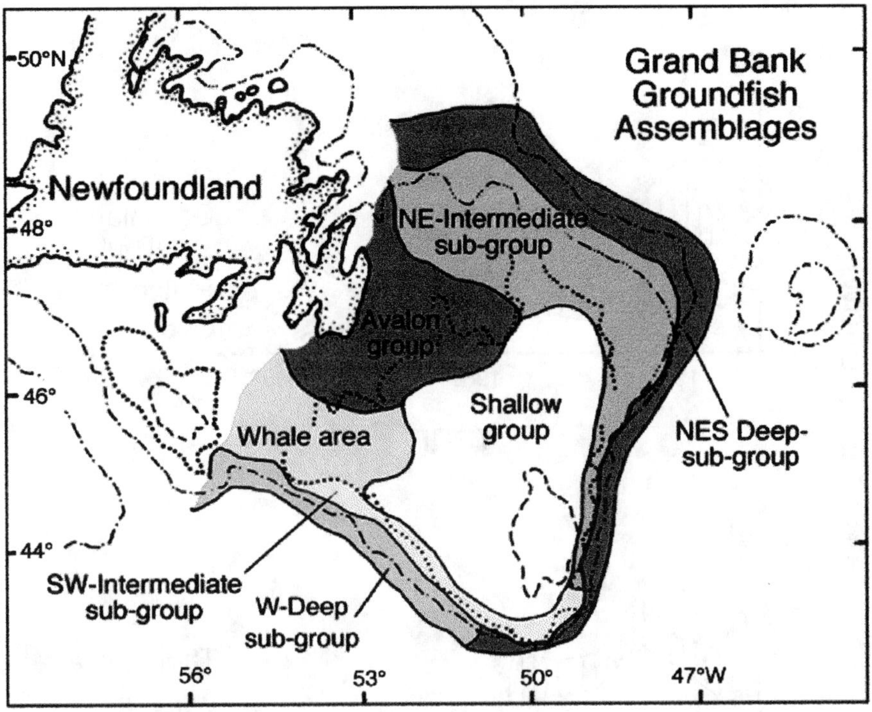

Figure 1. Assemblage map of the Grand Banks, Newfoundland (Gomes et al. 1995).

Five assemblages were identified on Georges Bank (Fig. 3) by using statistical methods similar to the Grand Banks analysis (Overholtz and Tyler 1985, Gabriel 1992). Thirty-six species were included in the assemblage analysis. Dominant species originally included Atlantic cod, haddock (*Melanogrammus aeglefinus*), yellowtail flounder (*Limanda ferruginia*), winter flounder (*Pseudopleuronectes americanus*), and winter skate (*Raja ocellata*). As was the case off Newfoundland, the assemblage boundaries remained fairly constant geographically, but strong distortions occurred in the relative abundance of the component species' abundances concurrent with high fishing rates (Fig. 4). The multispecies biomass became dominated by skates.

On the continental shelf off Washington and Oregon, 53 species were included in the first analysis of assemblages (Gabriel and Tyler 1980), while 33 species were included in a more recent analysis (Jay 1996). The recent assemblage analysis indicates that many of the assemblages have retained their identity to date since the first research survey in 1977, while varying in their geographic distributions. The most abundant species,

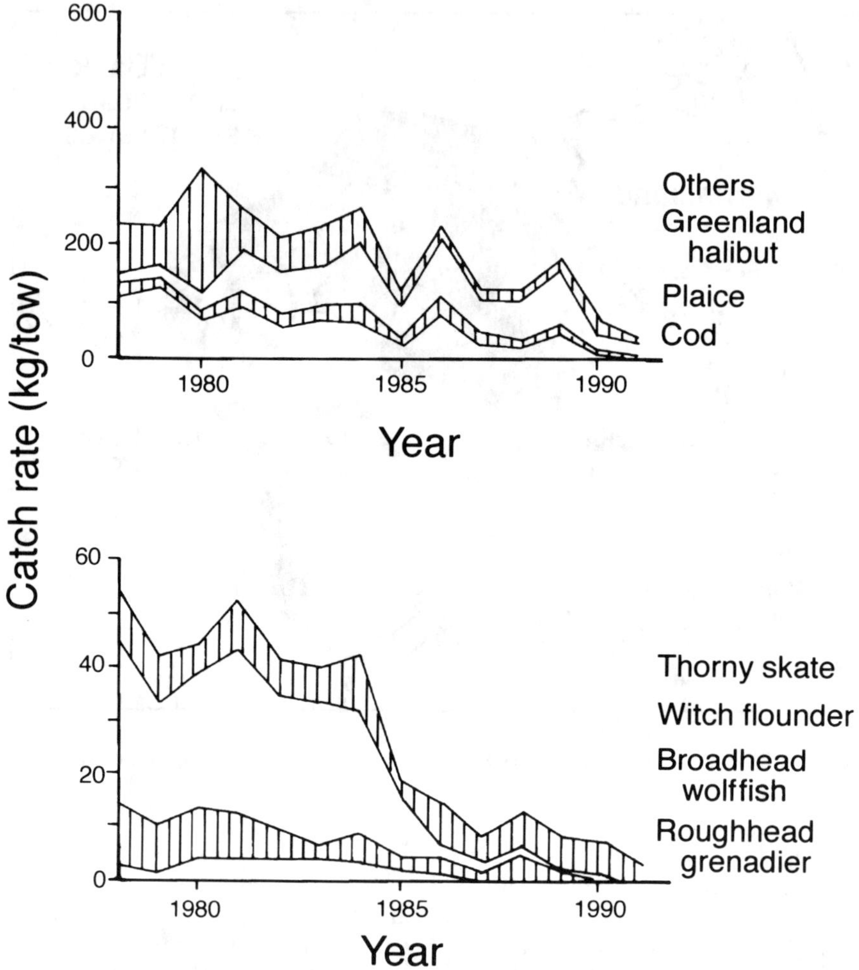

Figure 2. Abundance changes for Grand Banks assemblages (Gomes et al. 1995). The width of the species band, or the species group band, shows the relative abundance. The bands are stacked. The order of the names at the right matches the order of the bands. The first band, marked "Others," is a miscellaneous species category.

Pacific hake (whiting) (*Merluccius productus*), has remained at high levels through the entire period (Pacific Fishery Management Council 1998). English sole (*Pleuronectes vetulus*) has shown a strong increase in its abundance, while petrale sole (*Eopsetta jordani*) has remained fairly constant. The stock assessments for the region indicate, however, that many of the commercially targeted stocks have declined during the past 15 years (Fig. 5), including lingcod (*Ophiodon elongatus*), widow rockfish (*Sebastes entomelus*), yellowtail rockfish (*S. flavidus*), sablefish (*Anoplopoma fimbria*), shortspine thornyhead (*Sebastolobus alascanus*), and longspine thornyhead (*S. altivelis*). Dover sole (*Microstomus pacificus*) showed a decline ending in 1994, and has increased since then.

Five fish assemblages (Fig. 6) have been identified for the continental shelf of the eastern Bering Sea (Walters and McPhail 1982). Regrettably, assemblage analysis has not been carried out on an annual basis, so it is not possible to look at species composition changes by assemblage. The region is by far the most species-rich of the four areas in this comparison. Seventy-eight species of fish were used for the assemblage analysis. In a later publication, Bakkala (1993) reported that there were over 150 species of fish recorded in National Marine Fisheries Service (NMFS) surveys. The number is an estimate because some of the species were identified only to genus. Species biomass levels for the entire area of the annual survey conducted since 1979 by the Alaska Fishery Science Center, NMFS, were available from the computer database (Pers. comm., G. Walters). Striking changes have occurred for all of the dominant species since the first survey (Fig. 7). Many of the dominants have increased over the past decade. Pacific cod (*Gadus macrocephalus*), rock sole (*Pleuronectes bilineata*), and Greenland turbot showed a long-term decline followed by a recent increase in biomass. Note that this last species is also abundant in the Grand Banks fauna. Yellowfin sole (*Pleuronectes asper*) has shown a fairly steady biomass level. The dominant walleye pollock (*Theragra chalcogramma*) has shown increases and decreases, and is now near historical average levels. One high-biomass species that has shown a long-term decline and is still declining is sablefish. The assessments of the 12 main economically important species show that the stocks are in good shape and above the biomasses needed for maximum sustainable yield (MSY), except for sablefish (North Pacific Fisheries Management Council 1997). The stock trends evidenced in the NMFS survey data indicate that the mix of biomasses of less abundant species is also at historical levels with no evidence of deterioration due to fishing.

Adaptive Assemblage Maintenance Programs

What is the alternative to ecosystems dynamics models? I propose an empirically based assemblage maintenance program with extensive, standardized research surveys and fishery observer programs with regulation by area, with two or more levels of fishing effort, one in each of several

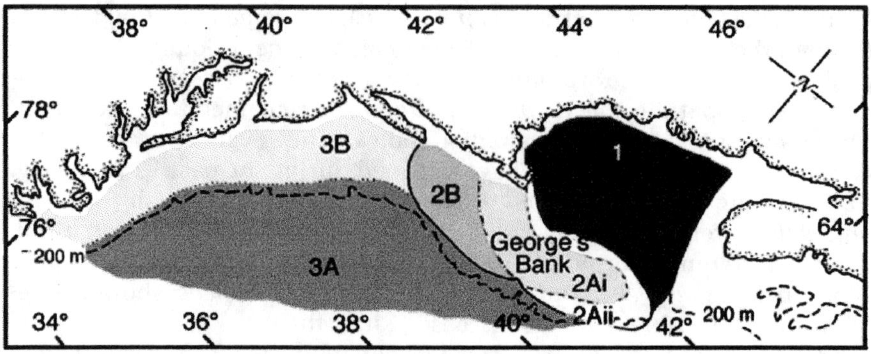

Figure 3. Assemblage map of Georges Bank, New England, U.S. (Gabriel 1992).

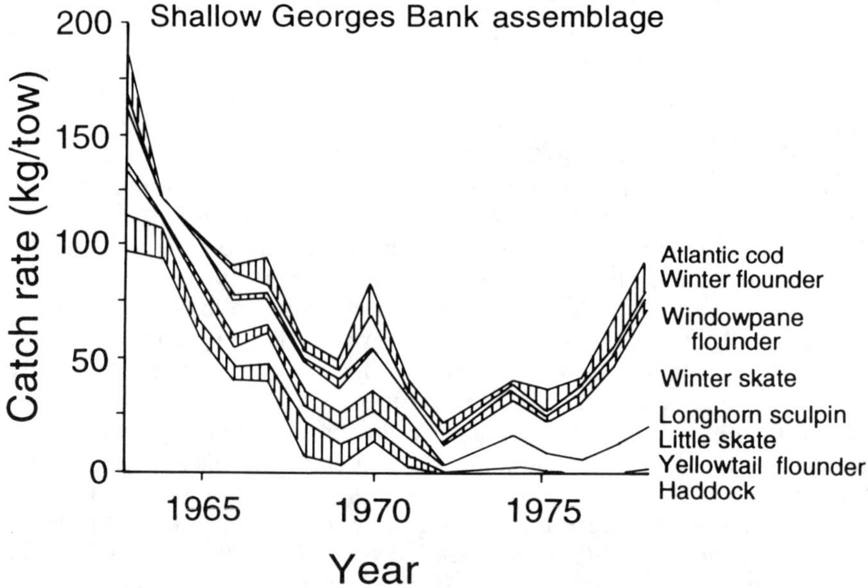

Figure 4. Abundance changes for Georges Bank assemblages (Overholtz and Tyler 1985). The width of the species band, or the species group band, shows the relative abundance. The bands are stacked. The order of the names at the right matches the order of the bands.

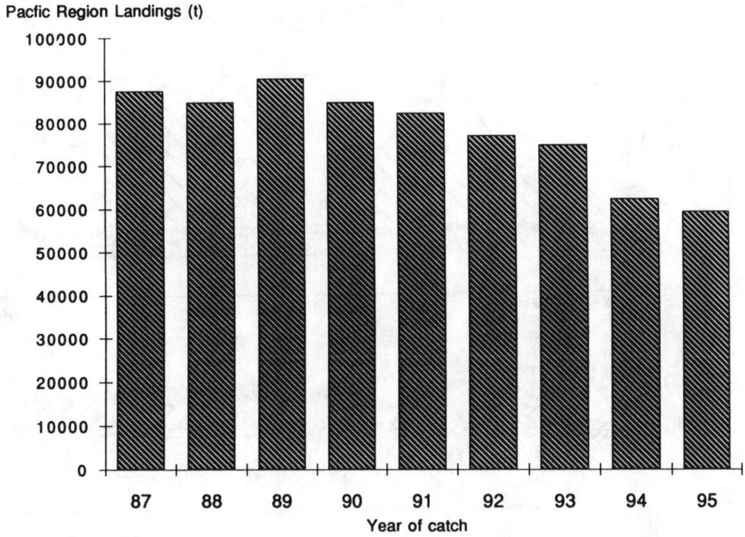

Figure 5. Decline of multispecies landings except for Pacific hake of the Pacific region of the United States (Pacific Fishery Management Council 1998).

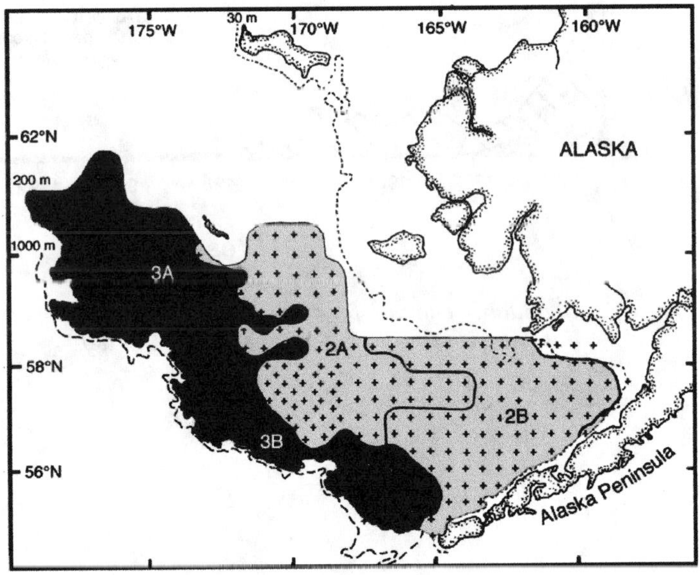

Figure 6. Map of eastern Bering Sea assemblages from Walters and McPhail (1982).

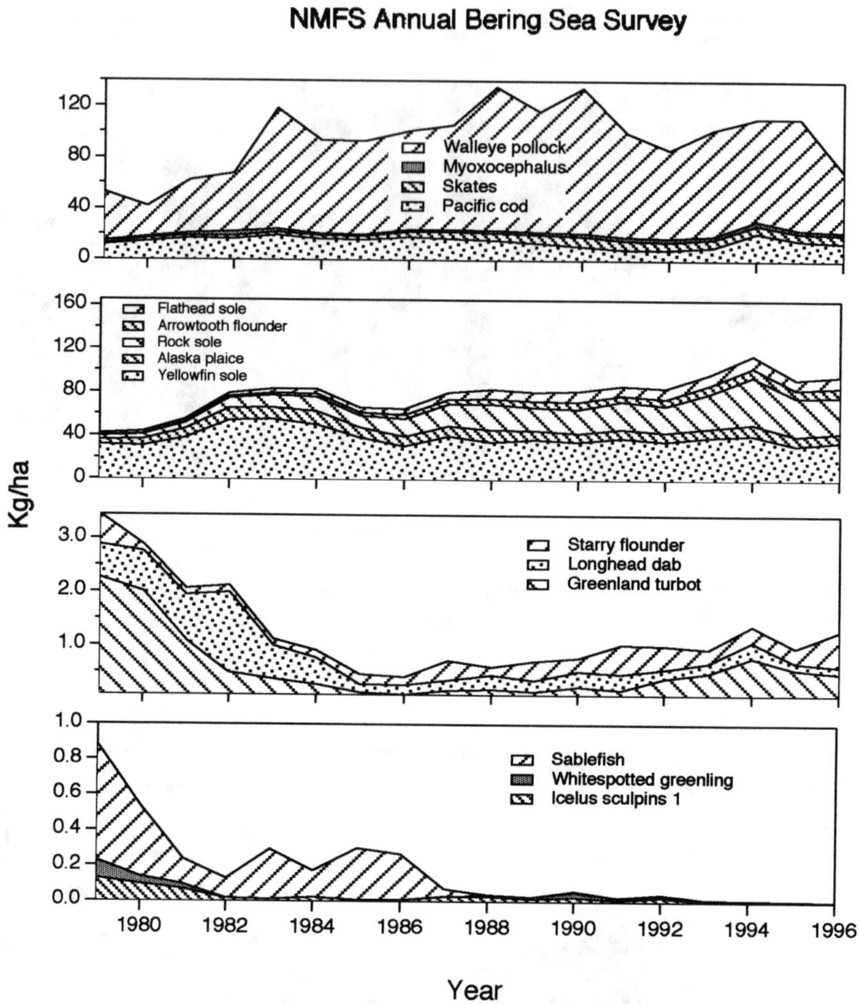

Figure 7. Abundance changes for eastern Bering Sea assemblages from the NMFS database (courtesy of G. Walters, NMFS, Seattle).

specific areas in an experimental or test fishery manner. This is an adaptive approach (Walters and Hilborn 1976) in that it depends on the response of fishery management to trials among test areas. The data would consist of multispecies biomass trends resulting from the contrasting levels fishing effort spaced over large areas of sea bottom. I believe that such a method should be carefully considered by assessment analysts and that it could be applied to many fishery regions.

My example, given below, is of a possible deployment of an assemblage maintenance program to Hecate Strait, the large shelf area between the Pacific mainland of Canada and the Queen Charlotte Islands (Fig. 8). A series of research cruises has been conducted in this area since 1984. These were begun with the Hecate Strait Project of the Pacific Biological Station, Research Branch, Canada Department of Fisheries and Oceans (Westrheim et al. 1984, Tyler 1989). Hecate Strait is bounded to the north by the deep (>200 m) Dixon Entrance and the Alaska Archipelago, and to the south by deep gullies of Queen Charlotte Sound. Consequently the continental shelf fauna within Hecate Strait tends to be separated ecologically from its counterparts elsewhere. Analysis of the research cruises of the 1980s showed that there were four assemblages of fishes in the strait (Fig. 8). Fifty species of fish were used in the statistical analysis of assemblage structure (Fargo and Tyler 1991). The deepwater assemblage in the gullies to the south is quite different from the other Hecate Strait assemblages and will not be discussed in this paper. The three assemblages occupying the 20-100 fathom range (40-200 m) are the subject of this discussion. They are dominated by rock sole, halibut (*Hippoglossus stenolepus*), rex sole (*Glyptocephalus zachirus*), Pacific sanddab (*Citharichthys sordidus*), English sole, Pacific cod, spiny dogfish (*Squalus acathias*), and big skate (*Raja binoculata*). Note that some of the dominant species of the eastern Bering Sea are also members of this fauna. There is evidence that some of the species, for example rock sole, form substocks within Hecate Strait that do not mix between the north and south areas of the strait (Fargo 1990). The differences between the north and south areas have been recognized for years by the traditional fishery there of small trawlers (65-95 ft; 22-32 m). As a result, biologists at the Pacific Biological Station divided the north and south into two fishery statistical areas: 5D (north) and 5C (south) (Fargo and Tyler 1991). This fleet remained the chief fishing effort on groundfish from the 1950s through the 1980s.

A natural fishery experiment, quite apart from fishery management, occurred in Hecate Strait during the 1980s, with the trawler fleet putting two to three time as much fishing effort in the north compared to the south (Fig. 9). The reason for this difference in effort was that the north was closer to the port of Prince Rupert and its processing plants. This effort differential gave the biologists a chance to determine whether the extra fishing effort was causing differences in relative abundance of species between north and south in the assemblages detected by the surveys. They found that catch rates from the research surveys for the groups of

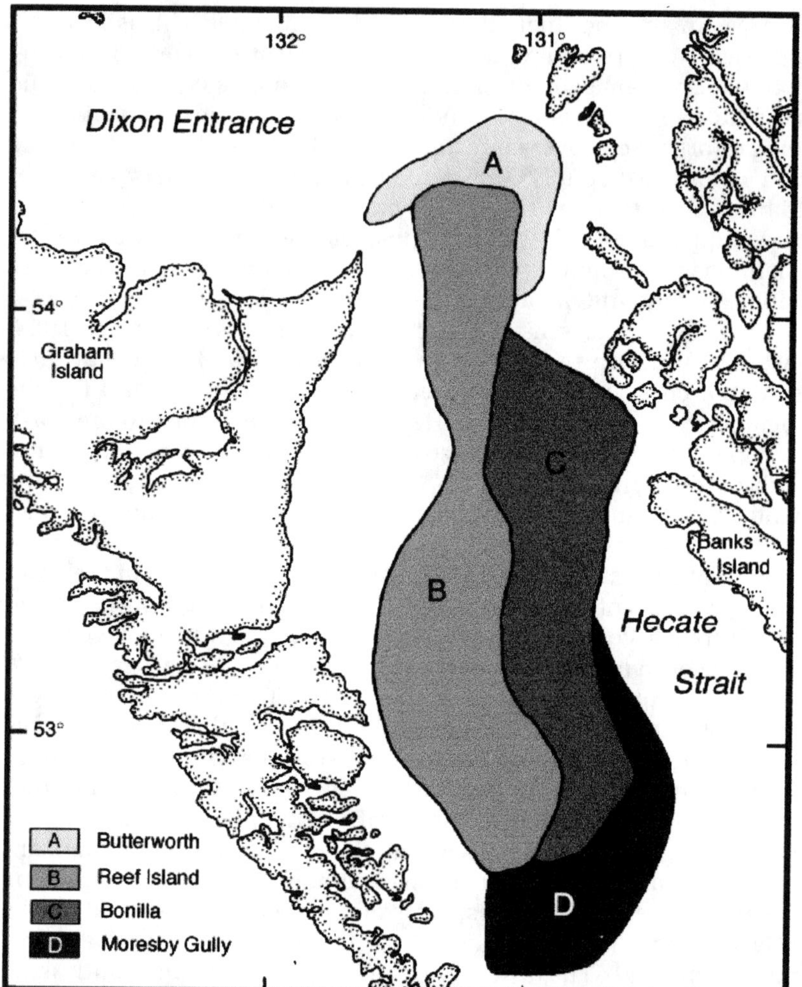

Figure 8. Assemblage map of Hecate Strait, British Columbia (Fargo and Tyler 1991).

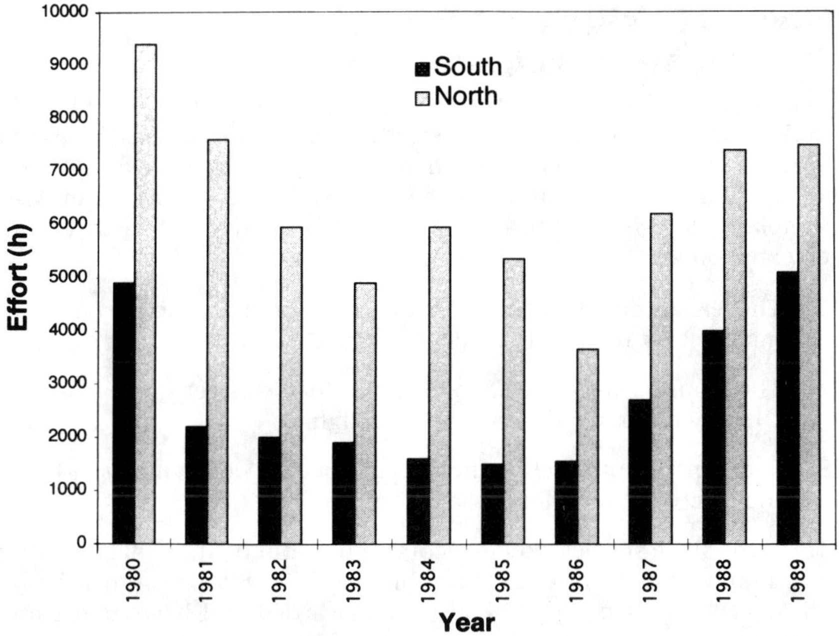

Figure 9. Fishing effort in the northern versus southern fishery statistical areas of Hecate Strait (Fargo and Tyler 1991).

species were unrelated to commercial effort (Fargo and Tyler 1991), and that differences in species diversity, species richness, and evenness were also unrelated. It was clear that fishing practices in Hecate Strait did not bring about the kind of species abundance distortions that were witnessed in the Grand Banks or Georges Bank groundfish assemblages. Skates did not become dominant as they did on Georges Bank. An experiment in fishery management was proposed (Fargo and Tyler 1991), in which the effort level of the south area (5C) would be held at 1988 levels, while levels in the north would be encouraged to double again by way of testing in an adaptive sense (Walters and Hilborn 1976, Sainsbury 1988, Hilborn and Walters 1992) for the level of effort that would just bring about undesirable change. Though this management measure has not been carried out, the analysis can serve as a model for future adaptive management programs and could yet be implemented. The increases could have been implemented through stepwise quota increases every 5 years on key target species such as Pacific cod and English sole. The process of stepwise quota increase could then have been managed over a period of many years.

How to Design an Assemblage Maintenance Program

In looking over the various assemblage responses to heavy fishing effort in the ocean around North America, it is possible to say that there are the specific kinds of multispecies changes that one would expect from an adaptive assemblage maintenance program. I suggest a series of five assemblage-change hypotheses that are testable and based partly on Atlantic experience.

1. The least productive species would become scarce on the basis of the mixed stock model of Paulik et al. (1967).
2. The biomasses of the economically valuable target species would decline by too much if effort was too high.
3. Some of the productive, nontarget species would increase because they were not targeted.
4. Some skate species would become more abundant, perhaps becoming dominant in the mixed-species biomass. The reason for this change is unclear. Perhaps most skates are discarded and have high survival rates.
5. Species diversity indices would be lower because of the many biomass declines without other species replacing them.

If those are the assemblage-level responses one might expect, what would be the management program that would set up a decision-making framework? Change in biomass must be detected that is due to catch rate and distinguished from responses to physical environmental change.

a. For this decision to be made, an adaptive design must include geographic regions with applied fishing rates that are sharply differing—one area having two or three times the level of effort as another area. It would be better if there were more than two areas and levels of effort. The areas would have to be very large, large enough to include subpopulations of several species that do not mix among the areas. Only in this way will differing effects of the effort levels be independently detectable and not smeared through the areas.

b. Fishing effort itself is known to be generally difficult to manage as compared to annual catch quotas. The management of fishing level must be done via annual quotas on key species that would produce approximately a given level of effort.

c. The total allowable catch in any one area would have to be increased in a stepwise manner in 4- or 5-year intervals in order to allow analysis of assemblage change. If no assemblage or individual species response was detected that could be attributed to fishing effort, then either the level would remain in place or an increase could be made.

d. A low catch rate would always be maintained in one area as a control and as a way to sample the assemblage. If similar changes were detected on both the control and the high fishing rate area, it would be assumed that it was not fishing-related but due to an external factor acting on both areas.

e. Specific biological reference points would have to be developed in relation to possible assemblage changes. These reference points would be quantitative thresholds for management action taken from the biological changes described as five hypotheses (above). The reference points would be indicators that changes related to the effects of fishing had progressed too far. When a combination of thresholds was reached, fishing effort would be reduced as a management measure in the high fishing rate area.

f. The program would be carried out as an adjunct to the standard stock assessment programs being conducted on main economic species. If serious changes at the assemblage level were detected, though no overfishing occurred according to individual stock assessments, the level of fishing effort would be cut back to maintain the assemblage. This would be done by reducing the annual quota on one or more of the target species. The assemblage maintenance program would in this way override the individual stock assessments. The operating principle is that assemblages as a whole cannot withstand the high levels of effort appropriate for the most productive species.

g. Assemblage maintenance programs would not apply to very migratory species and to species like walleye pollock, Pacific halibut, and sablefish whose single stocks occupy such vast areas as the continental shelf and slope of the Gulf of Alaska. The program would be applied to groups of stocks that have more restricted distributions and that can be divided into substocks that do not mix rapidly throughout their species distribution. Typically such substocks are not genetically distinct, but have differing somatic growth rates and other vital parameters.

What Next?

Of the cases reviewed here, where could the assemblage maintenance programs be developed in Atlantic or Pacific North America? The strong remedial action that either the Grand Banks or Georges Bank require would mitigate against setting up adaptive programs there. There is a good possibility an area of low effort should be set up with respect to the flatfish fisheries in the Pacific region of either Canada or the United States. A low effort region would be arranged for the flatfish-dominated assemblages described in the published analyses (Fargo and Tyler 1991, Jay 1996). Off Washington and Oregon these lie along the depth contours and tend to be narrow areas running for many miles north and south. The species in these areas would have to have at least two substocks that could be fished with low and high effort levels. These areas would be quasi-replicates in the adaptive management plan. The assemblage areas described by Jay (1996) tend to be very limited, and adjacent assemblages may have to be combined so that the area would be large enough for management. It is well known that it is next to impossible to enforce quota levels on small areas, and so it may be that the low-effort area would have to be closed entirely except for an arranged test fishery that would keep track of abundance changes.

It is likely that an assemblage management program could be developed for the eastern Bering Sea. There are already extensive closures in the southern part put into place to help restore crab fisheries and to prevent trawling in areas frequented by feeding Steller sea lions (Fritz et al. 1995). In addition, some of the on-bottom trawling is limited by the bycatch limitations on catch of red king crabs, Tanner crabs, and Pacific halibut. Because of these restrictions, the yield maximization paradigm has not been fully implemented in the eastern Bering Sea. Perhaps that is why those assemblages seem to be in good condition. A plan might be to set up a specific area for increased fishing rate while simultaneously outlining an area for maintaining the present fishery effort levels consciously into the future. There is evidence from flatfish studies in the continental shelf of the eastern Bering Sea (Wilderbuer et al. 1992) that there are substocks of yellowfin sole that would possibly be a basis for dividing the area into two adaptive management zones (Fig. 10). One might be a high effort zone, the other a zone where present levels would be maintained. The substock structure of other species should be examined to make decisions regarding the boundaries of the areas for assemblages.

The present political climate in fisheries makes it more likely than in previous years that an assemblage maintenance program be established. Never before have so many North American fisheries been closed, or nearly closed, due to overfishing. It is clear to most stakeholders that ocean stocks can be overfished. Fifteen to twenty years ago many were still arguing that overfishing was unlikely. Others will argue that the yield maximization paradigm is sufficient, if only it is applied correctly. I believe the approach is not sustainable and have given arguments here in that direction. In the present way of conducting fisheries, the only proof that it is not working is a failed fishery. The onus of proof must be on the advocates of this intensive approach to resource use, not the other way around (Dayton 1998). If assemblage maintenance programs were instituted, much of the biomass would be fished at levels below those recommended by the yield maximization paradigm. These substocks would be available for

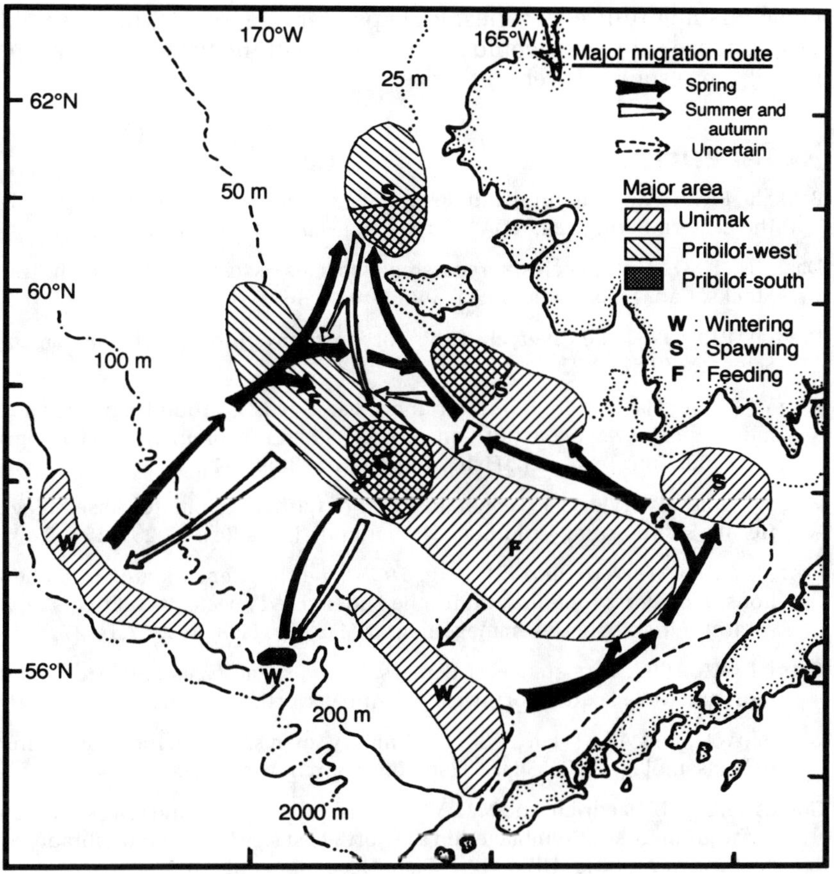

Figure 10. Substock distributions of yellowfin sole in the eastern Bering Sea (Wilderbuer et al. 1992).

reestablishing future fisheries. Perhaps more important, the assemblage maintenance programs would keep a portion of the fishery alive while the rest of the resource is rebuilding.

References

Bakkala, R.G. 1993. Structural and historical changes in the groundfish complex of the eastern Bering Sea. NOAA Tech. Rep. NMFS 114. 92 pp.

Daan, N. 1987. Multispecies versus single-species assessment of North Sea fish Stocks. Can. J. Fish. Aquat. Sci. 44(Suppl. 2):360-370.

Dayton, P.K. 1998. The reversal of the burden of proof in fisheries management. Science 279:821-822.

Fargo, J. 1990. Flatfish. In: A.V. Tyler and J. Fargo (eds.), Groundfish stock assessments for the west coast of Canada in 1989 and recommended yield options for 1990. Can. Tech. Rep. Fish. Aquat. Sci. 1732:95-126.

Fargo, J., and A.V. Tyler. 1991. Sustainability of flatfish-dominated assemblages in Hecate Strait, British Columbia. Netherlands J. Sea Res. 27:237-253.

Fritz, L.W., R.C. Ferrero, and R.J. Berg. 1995. The threatened status of Steller sea lions, *Eumetopias jubatus*, under the Endangered Species Act: Effects on Alaska groundfish fisheries management. NMFS Mar. Fish. Rev. 57:14-27.

Gabriel, W.L. 1992. Persistence of demersal fish assemblages between Cape Hatteras and Nova Scotia, Northwest Atlantic. J. Northwest Atl. Fish. Sci. 14:29-46.

Gabriel, W.L., and A.V. Tyler. 1980. Preliminary analysis of Pacific coast demersal fish assemblages. NMFS Mar. Fish. Rev. (March-April):83-88.

Gomes, M.C., R.L. Haedrich, and M.G. Villagarcia. 1995. Spatial and temporal changes in the groundfish assemblages on the north-east Newfoundland/Labrador Shelf, north-west Atlantic, 1978-1991. Fish. Oceanogr. 4:85-101.

Hilborn, R., and C.J. Walters. 1992. Quantitative fisheries stock assessments: Choice, dynamics and uncertainty. Chapman and Hall, New York. 570 pp.

Jay, C.V. 1996. Distribution of bottom-trawl fish assemblages over the continental shelf and upper slope of the U.S. West Coast, 1977-1992. Can. J. Fish. Aquat. Sci. 53:1203-1225.

North Pacific Fisheries Management Council. 1997. Stock assessment document and evaluation for the Bering Sea and Aleutian Islands. Anchorage, Alaska.

Overholtz, W.J., and A.V. Tyler. 1985. Long-term responses of the demersal fish assemblages of Georges Bank. Fish. Bull., U.S. 83(4):507-520.

Pacific Fishery Management Council. 1998. Status of the Pacific coast groundfish fishery through 1997 and recommended biological catches for 1998: Stock assessment and fishery evaluation-updated. Pacific Fishery Management Council, 2130 SW Fifth Ave., Suite 224, Portland, OR 97201.

Paulik, G.J., A.S. Hourston, and P.A. Larkin. 1967. Exploitation of multiple stocks by a common fishery. J. Fish. Res. Board Can. 24:2527-2537.

Pielou, E.C. 1975. Ecological diversity. John Wiley, New York. 165 pp.

Pope, J.G., and B.J. Knight. 1982. Simple models of predation in multi-age multi-species fisheries for considering the estimation of fishing mortality and its effects. In: M.C. Mercer (ed.), Multispecies approaches to fisheries management advice. Can. Spec. Publ. Fish. Aquat. Sci. 59:64-69.

Ricker, W.E. 1975. Computation and interpretation of biological statistics of fish populations. Canada Fisheries and Marine Service. Bull. 191.

Sainsbury, K. 1988. The ecological basis of multispecies fisheries, and management of a demersal fishery in tropical Australia. In: J. Gulland (ed.), Fish population dynamics. Wiley, Chichester, pp. 349-382.

Tyler, A.V. (ed.). 1989. Hecate Strait Project: Results from four years of multispecies fisheries research. Can. Tech. Rep. Fish. Aquat. Sci. 1675. 60 pp.

Tyler, A.V., W.L. Gabriel, and W.J. Overholtz. 1982. Adaptive management based on structure of fish assemblages of northern continental shelves. In: M.C. Mercer (ed.), Multispecies approaches to fisheries management advice. Can. Spec. Publ. Fish. Aquat. Sci. 59:149-156.

Walters, C.J. 1969. A generalized computer simulation model for fish population studies. Trans. Am. Fish. Soc. 98:505-512.

Walters, C.J., and R. Hilborn. 1976. Adaptive control of fishing systems. J. Fish. Res. Board Can. 33:145-159.

Walters, G., and M.J. McPhail. 1982. An atlas of demersal fish and invertebrate community structure in the eastern Bering Sea. Part I, 1978-81. NOAA Tech. Memo. NMFS F/NWC-35. 122 pp.

Westrheim, S.J., A.V. Tyler, R.P. Foucher, M.W. Saunders, and S.C. Shields. 1984. G B REED groundfish cruise no. 84-3, May 24-June 14, 1984. Can. Data Rep. Fish. Aquat. Sci. 488. 131 pp.

Wilderbuer, T.K., G.E. Walters, and R.G. Bakkala. 1992. Yellowfin sole, *Pleuronectes asper*, of the eastern Bering Sea: Biological characteristics, history of exploitation, and management. NMFS Mar. Fish. Rev. 54:1-18.

Canadian Marine Fisheries in a Changing and Uncertain World

Randall M. Peterman
(Fisheries Panel co-chair) Simon Fraser University, School of Resource and Environmental Management, Burnaby, British Columbia, Canada

Brad deYoung
(Fisheries Panel co-chair) Memorial University, Department of Physics, St. John's, Newfoundland, Canada

A. Rod Dobell
University of Victoria, School of Public Administration, Victoria, British Columbia, Canada

Evelyn Pinkerton
Simon Fraser University, School of Resource and Environmental Management, Burnaby, British Columbia, Canada

Yvan Breton
Université Laval, Department of Anthropology, Laval, Quebec, Canada

Anthony T. Charles
Saint Mary's University, Department of Finance and Management Science, Halifax, Nova Scotia, Canada

Michael J. Fogarty
University of Maryland, Chesapeake Biological Laboratory, Solomons, Maryland

Gordon R. Munro
University of British Columbia, Department of Economics, Vancouver, British Columbia, Canada

Christopher Taggart
Dalhousie University, Department of Oceanography, Halifax, Nova Scotia, Canada

Extended Abstract

In recent years, people have become increasingly aware of the dynamic nature of fishery systems. These systems include not only the fish and their environment, but also people and their associated social and economic institutions and communities. As a result of the challenges created by ever-changing fisheries, the Canadian Global Change Program of the Royal Society of Canada formed a Fisheries Panel of nine people in 1996. The mandate of the panel was to write an authoritative and comprehensive review of the implications of physical, biological, economic, and sociopolitical changes for Canadian marine fisheries and to present options for how to deal with those changes.

The resulting report, "Canadian Marine Fisheries in a Changing and Uncertain World" (deYoung et al. 1999), outlines a vision for future sustainable fisheries for Canada that is applicable to many fisheries elsewhere. This brief extended abstract summarizes the report, which is written for a broad audience, including fisheries scientists and managers, harvesters of fish resources, and residents of fish-dependent communities that rely on renewable aquatic resources. The focus is on capture fisheries, not aquaculture.

The report emphasizes the challenges to biological, economic, and social sustainability of fishery systems: complexity, change, variability, and measurement error. These characteristics lead to uncertainties, which in turn create risks—biological risks for aquatic ecosystems, economic risks for industry, and social risks for coastal communities.

The report develops several guiding principles, which can help attain the overall goals of a biologically productive aquatic system, economically viable fishing industries, and sustainable fishing-dependent communities. These principles, which are intended to be followed by *everyone* involved in fisheries systems, *not* just those in management agencies, evolved from current research in the natural and social sciences and experience not only in Canada but also in fisheries elsewhere in the world. The principles are as follows:

1. Incorporate into decision-making an analysis of structural and dynamic complexities of fisheries systems.

2. Incorporate into decision-making an analysis of change, uncertainty, and risk in all fishery activities.

3. Promote and conserve biological, economic, and social diversity to hedge against variability and uncertainty.

4. Collect, analyze, and openly communicate data and information.

5. Estimate, document, and incorporate into decision-making the social and ecological consequences of decisions and actions.

6. Clearly define the roles, rights, and responsibilities of all fishery participants to align their interests with the overall objectives of sustainability.

Finally, the report identifies strategies for applying these principles:

1. Implement a precautionary approach.
2. Extend risk assessment and risk management procedures.
3. Promote diversity and ecosystem-based management.
4. Ensure adequate collection and communication of information.
5. Inform social decisions through full-cost pricing, accounting, reporting, and charging.
6. Implement institutional reform and incentives to promote compliance.

Within these general strategies, the report makes specific recommendations for how management agencies, industry, fishermen, and coastal fishing-dependent communities should plan and act in order to achieve the goals of biological, economic, and social sustainability *in the presence of change and uncertainty*. The report also notes that some of these strategies are already being implemented in Canada but that they are not being widely or uniformly applied.

The report is expected to be available from the National Research Council Press, Ottawa, Ontario, in the summer of 1999.

Reference

deYoung, B., R.M. Peterman, A.R. Dobell, E. Pinkerton, Y. Breton, A.T. Charles, M.J. Fogarty, G.R. Munro, and C. Taggart. 1999. Canadian marine fisheries in a changing and uncertain world. Can. Spec. Publ. Fish. Aquat. Sci. 129. In press.

A Report on Historical, Human-Induced Changes in Newfoundland's Fisheries Ecosystem

Melanie D. Power and Nathaniel Newlands
University of British Columbia, Fisheries Centre, Vancouver, British Columbia, Canada

Abstract

The groundfish collapse off Canada's East Coast earlier this decade led to a change in the species that are commercially targeted. This recent change is just the latest in a series of shifts within the Atlantic fisheries, which have resulted in a decrease in the mean trophic level at which the fisheries operate. Our study investigates subsequent changes in the mean trophic level, calculated from historical catch statistics, and the changing management priorities of the Newfoundland fishery in the twentieth century. This report reviews the fishery from 1900 through 1995, with an emphasis on the period since Newfoundland joined the Canadian Confederation in 1949. We discuss the changing mean trophic level of fisheries catches in Newfoundland. Annual shifts in mean trophic level are reviewed in the context of social, economic, and political spheres in an effort to determine the management and regulatory decisions that may have contributed to the decline of trophic levels. This analysis aims to provide a direction for fisheries ecosystem management, and considers nonlinear, multispecies interactions encompassing the indirect effects on marine life other than identifiable targeted species in the ecosystem. An awareness of the trophic level impact of management decisions, based on historical trends, can indicate possible future implications of current policy decisions. This work constitutes a component of a comparative examination of Canadian management trends in Atlantic and Pacific fisheries.

Introduction

The Newfoundland fishery has been prosecuted for nearly five centuries, and the cod fishery was a primary reason for the settlement of the island

(Felt and Locke 1995). The cod fishery, which created and sustained Newfoundland society, was the main focus of the fishery until the early 1990s, at which point the well-known and well-documented collapse of the fishery occurred, resulting in a moratorium on fishing. Despite the centrality of the cod to the Newfoundland fishery, other species have attracted interest in the commercial fishery, particularly in the last few decades.

In this paper, we examine how changes in the fishery contributed to the decline of mean trophic level of the Newfoundland fishery catch. Following Pauly et al. (1998), we chart the trend in mean trophic level from 1900 through 1995. We then describe and discuss various factors—social, political, economic—which are likely to have contributed to these trends. In doing so, we aim to highlight important implications of this trend and how ecosystem considerations may prevent such future declines.

The Newfoundland fishery has been selected because of its local and global historical significance. The northern cod fishery, a resource that was long considered impossible to deplete, continues to be closed under a fishing moratorium following its collapse in the early 1990s. In the time since the moratorium was first imposed, the trends of targeting alternative "underutilized" species have become more apparent and pronounced.

This study encompasses catch data from 1900 through 1995, with an emphasis on the period from 1949 onward. It was in 1949 that Newfoundland became the tenth province of Canada and, as will be seen below, the fishery in the second half of the twentieth century was particularly complex.

The last century has witnessed countless changes to the fishery: changes in technology and methods, and changes regarding which species are fished. Furthermore, it was during this period that Newfoundland's political reality was most dramatically altered. A study of this period, from the early 1900s through to the collapse of the cod fishery and to the present, helps to explain the current devastation the fishery now exhibits.

Results

Catch data for Newfoundland for 1900 through 1995 was compiled from catch time series from the following three sources: Statistics Canada (Leacy et al. 1983) for years 1900-1959, Northwest Atlantic Fisheries Organization Statistical Bulletins for 1960 through 1971 (NAFO 1960 through 1971), and the Department of Fisheries and Oceans for 1972 through 1995 (Canada 1998). Catch statistics for years earlier than 1900 have been reported for cod and haddock but were not representative across all major species, and therefore were not included in the data analysis.

The catch data from the above three sources were obtained for years in which the yearly intervals indicated overlap. The catch series from each data source was then scaled proportionally, or "tuned," to remove inconsistencies in catch estimates between each catch series. This produced a full catch time series which was continuous across each of the yearly linked catch segments. The catch statistics were not corrected for discards, bycatch, unreported catch, or wasting of fish. Included in this analysis were groundfish, pelagics and crustaceans; marine mammals were not included as a result of inconsistencies between catch sources in generating a time series across the years 1900-1995.

Following the collapse of many species within the northwest Atlantic and a subsequent closure of the associated fisheries, estimates of catch data later than 1994 are preliminary for selected species. Recent years, including 1996, are also associated with significant changes in atmospheric circulation which resulted in above normal air temperatures over the Labrador Sea region for the first time in a decade. The effect of such large-scale environmental changes on the marine ecosystem remains an important question for consideration in influencing recent yearly catch data. Catch data for 1900 through 1995 are reflected in Fig. 1.

Mean trophic levels were calculated for each of the species (i) reported under Canadian catch by a weighted ratio of yearly catch (C_{ij}) for each species group by total catch (H_j). The weighted average was calculated using the following formula:

$$\overline{TL}_j = \frac{1}{H_j}\sum_i C_{ij} TL_i, \quad H_j = \sum_i C_{ij}$$

The species-specific weights for this ratio comprised estimates for the trophic level (TL_i) of each species obtained from FishBase 97 (FishBase 1997). Where TL information was not available in FishBase, estimates were associated with ancillary statistics for species-groups (Pauly and Christensen 1995).

The calculations of mean trophic level (TL_i) are shown in Fig. 2. While the mean trophic level was remarkably constant through the first six decades of this century, the 1960s marked the beginning of the declining trend. A major decline, unprecedented in this century, is apparent during the 1990s. The mean trophic level ranges from a maximum mean trophic level of 3.79 in 1900 and minimum of 2.78 in 1995. The declining trend is evident.

A plot of mean trophic level versus yearly catch for 1900 through 1995 is provided in Fig. 1. This figure demonstrates that, although the fishery is now being prosecuted at a lower trophic level, catches have declined to roughly the same quantity as that taken before the industrial expansion of the 1950s, indeed to the same volume as at the beginning of the century.

Historical Overview: Pre-Confederation

Newfoundland achieved Responsible Government in 1855 (Neary 1988, Felt and Locke 1995), and as such the British colony held a degree of sovereignty in her affairs, including responsibility for fisheries.

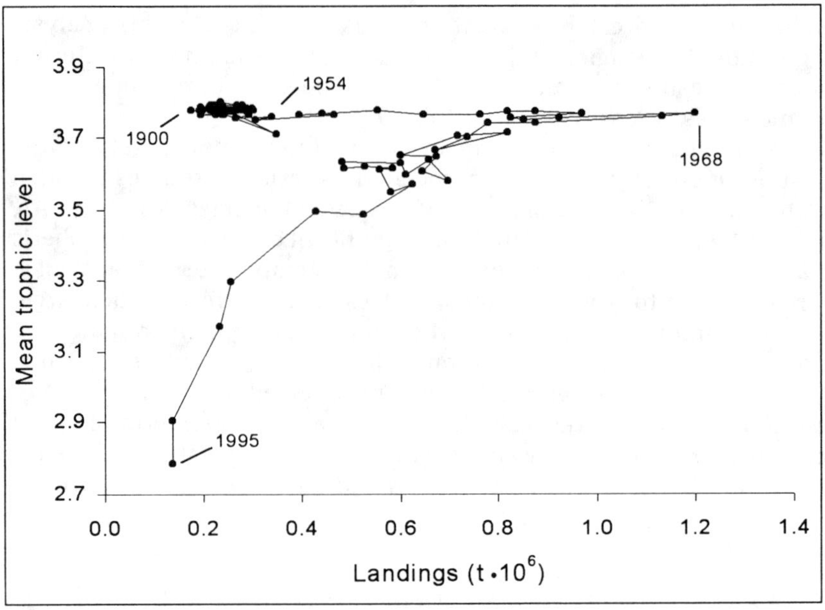

Figure 1. Mean trophic levels in fisheries landings versus landings, Newfoundland, 1900-1995.

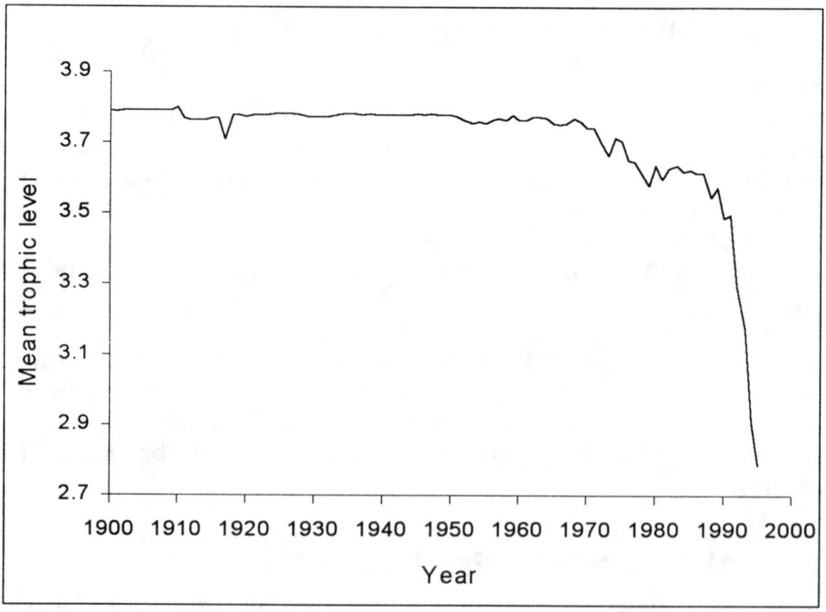

Figure 2. Trend in mean trophic level of fisheries landings, Newfoundland, 1900-1995.

The early decades of the twentieth century saw the colony facing ever-deepening financial difficulties, and during both world wars men and fishing vessels joined the effort, thereby impacting on the fishery. The seal hunt, for example, suffered during the wars as vessels were sent into service (Candow 1989). A more disturbing problem for the fishery was the postwar collapse of fish markets (Rowe 1980). As the value of codfish exports fell dramatically from more than $23 million in 1919 and 1920 to $13.5 million in 1921, the government ran an average annual deficit of $4 million. By 1930, the Great Depression was under way and fish prices plummeted (Rowe 1980).

The economic difficulties prompted Newfoundland to seek a "caretaker receivership" through a Commission of Government (Felt and Locke 1995). Consisting of six members, three of whom were British, the Commission was established in 1934, and Britain assumed responsibility for Newfoundland's finances as part of the arrangement (Rowe 1980).

When World War II broke out, Newfoundland again joined the war effort by sending men and ships to serve. This war proved to be profitable, and by 1940 Newfoundland enjoyed a surplus of revenue which continued through the war years (Rowe 1980). The construction and operation of Canadian and American bases on the strategically located island can be thanked for providing the increased government revenues which brought Newfoundland back to economic respectability.

As the war ended and Newfoundland's much-improved fiscal position became evident, the Commission of Government became somewhat redundant. The Commission was disbanded when, through two referenda, the people of Newfoundland narrowly voted to join with Canada. Confederation occurred on March 31, 1949, ending Newfoundland's stature as a separate country and bringing under Canadian federal jurisdiction the rich fishing banks of Newfoundland (Mackenzie 1986).

During this time of turbulence, the fishery was left remarkably unchanged. The early 1900s saw the continuation of the slow process of industrialization begun in the 1800s. Significantly, fishing gear saw many improvements in the late 1800s, with the development of cod traps in 1866 (Ryan 1986). The new technology was not without controversy over the efficiency of the gear, and faced a ban between 1888 and 1890 (Hutchings and Myers 1995). Motor engines were common in inshore fishing vessels by 1910 (Hutchings and Myers 1995).

The most important technological advances were those that developed through increasing mechanization. For instance, by the 1920s, an aerial spotting service had begun to assist the seal hunters in finding the seal stocks on the ice (Candow 1989). In the 1930s, diesel-powered trawlers were first used in a bank fishery and in 1948 pair trawlers were first used in bank fisheries (Hutchings and Myers 1995).

As noted, there was a period when the cod trap was banned in the late 1800s, following which time there were restrictions on fishing seasons and mesh sizes (Hutchings and Myers 1995). Otherwise, there appear to have been few governmental restrictions on the various fisheries during that time.

Throughout the history of the fishery, catch failures have periodically occurred (Cashin 1993). Furthermore, traditional life in Newfoundland was seasonal, and included fishing, hunting, and other activities. Accordingly, a fisher would target a number of fish species (Felt and Locke 1995). Newfoundland's cod fishery was therefore supplemented by other species, including Atlantic salmon, lobster, mackerel, herring, and seals. Lobster had in fact been a commercial fishery since the late 1800s. On the west coast of the Great Northern Peninsula, lobster fishing began in 1873 with one cannery and expanded quickly to 25 canneries by 1888 (Palmer and Sinclair 1997). During the first half of the twentieth century, fisheries nonetheless remained fairly narrowly focused on groundfish, seals, salmon, pelagics and lobster, with cod still the primary species. The prevalence of cod is evident in Fig. 2.

Historical Overview: Post-Confederation

It was in the second half of the twentieth century, from the time Newfoundland entered into Confederation with Canada through to the present, that the fishery experienced massive changes. With Confederation, Canada gained control of Newfoundland's fisheries, and concurrently, Newfoundlanders received the benefits of a social welfare state. Indeed, "Newfoundland's confederation [sic] with Canada in 1949 brought with it a host of social benefits that undermined the traditional work pattern in the outports" (Candow 1989). The reliance on the fishery was mitigated by the availability of state support. In 1957, fishermen became eligible for unemployment insurance after December 15, which effectively made that date the unofficial close of the fishing season (Kurlansky 1997).

Attempts at diversification of the economy followed Confederation, since the federal government in Ottawa viewed the cod fishery as an economic failure to be rectified (Kurlansky 1997). As part of the effort to modernize, the government began a program of resettlement, causing roughly 200 rural communities to be abandoned (Felt and Locke 1995). Furthermore, the military bases built in Newfoundland during the war accelerated industrialization. "Ironically, the first consequence [of industrialization] was a de-emphasis of the fisheries which had sustained the now-province for several hundred years" (Felt and Locke 1995). As well:

> The combination of the accelerated growth of government and indifferent public support for the fisheries led to a decline in both fishermen and value of fish products during the 1950s. . . . While better technology and expanded effort by the remaining fishermen mitigated reductions in harvest, landings of principal groundfish species (grenadiers, redfish, flatfish, and cod) declined as well, to approximately 200,000 metric tonnes from 246,000 tonnes

in 1946. As in earlier periods, Atlantic cod comprised 70 to 80 per cent of landings (Felt and Locke 1995).

The fact that the number of fishermen declined during the 1950s but the landings increased (as shown in Fig. 1) is indicative of the increased effort during this period of industrialization.

During the 1950s, the foreign fleet fishing off Canada's coast expanded and its catches increased accordingly (Felt and Locke 1995). This increasing foreign presence provoked a degree of concern within Newfoundland by the late 1960s and early 1970s. The federal government was pressured to protect Newfoundland fish (Felt and Locke 1995). Reacting to this pressure, as well as a changing international climate evident during the negotiations for the Third Convention on the Law of the Sea, Ottawa claimed a 12 nautical mile territorial limit in 1970 (Candow 1989). In 1977, Canada claimed a 200 nautical mile exclusive fisheries jurisdiction (Rowe 1980, Blades 1995, Hutchings and Myers 1995).

This second and greater extension was ostensibly aimed to protect the fish. Instead, due to the sense of optimism pervading the fisheries sector, the extension to 200 nautical miles encouraged a build-up of Canadian harvesting and processing capacity (Felt and Locke 1995). This was further encouraged by government, as subsidies and loans were made available to fishers. To policy-makers, "It was apparent that new technologies and increased fishing capacity would put Canada on a firmer competitive footing and boost the lagging economy of Atlantic Canada, so government pulled out all the stops, offering loans and subsidies to competing segments of the industry" (Blades 1995).

The 1980s were a time of boom and expansion, with fishers taking advantage of government loan programs to upgrade their vessels and to make large capital investments in their fishing enterprises. With the development of navigation and fish-finding gear, such as Loran C in the 1980s (Hutchings and Myers 1995), there were plenty of opportunities to spend money on new technologies. Market forces also contributed to the expansion, through "the growing strength of fish prices as health conscious consumers discovered the value of fish as a low-fat staple" (Blades 1995).

Other technological developments had shaped the fishery during this period. For instance, the use of nylon multifilament and later monofilament in nets, rather than cotton, and further developments in mechanization which aided in hauling nets (Hutchings and Myers 1995). Regulations were considerably more numerous than in the first half of the century, such as those regarding net technology: in the 1970s, the gillnet mesh size was 7 inches which was reduced to 5.5 inches during the 1980s. Cod trap mesh size was reduced from 4 inches to 3⅝ inches over this period (Hutchings and Myers 1995).

In this climate of changing regulations and ebullient optimism, the fishery itself began to change, slowly at first and then with surprising rapidity. The cod fishery underwent significant fundamental changes during these few decades, initiated by the expanding offshore foreign fleet. By 1959, and "For probably the first time in the history of the fishery, offshore catches exceeded inshore catches" (Hutchings and Myers 1995). This shift from inshore to offshore continued through the 1960s (Hutchings and Myers 1995).

The increased effort and the new technologies brought constant increases in catches through 1968, whereupon a decline began. Following the extension to 200 miles, catches for all groundfish including cod increased into the 1980s, declined again in the early 1990s, and finally collapsed in 1992. It was in 1992 that the moratorium on Northern cod was first announced (Newfoundland and Labrador 1997). It was subsequently extended and continues yet today. This trend of declining landings is evident in Fig. 1.

What is most interesting about this time period is the change in harvested species. Although cod remained the main fishery of concern, new species were targeted, particularly after the 1992 closure of groundfisheries, as will be discussed below.

Discussion

The changes which occurred in the fishery regarding targeted species are particularly interesting and are presented in Figs. 1 and 2. What were previously considered undesirable species have become foci of the fishing effort. Crab, for instance, was first prosecuted in Trinity Bay in 1968 and has since become one of the most valuable fisheries (Canada 1996a). Yet:

> Traditionally in Newfoundland, crabbing has been of a lower social order and fishermen have resisted it, but the Asian market for snow crab is extremely lucrative and several Newfoundlanders became wealthy in the mid-1990s from it. The landed value of snow crab in 1995 was the highest in dollars of any catch in the history of Newfoundland fisheries (Kurlansky 1997).

Other fisheries were expanded to become commercially oriented, such as capelin. Domestically, capelin were traditionally fished for food, bait, and fertilizer, although by the early 1970s a directed foreign fishery for capelin was established. The fishery was closed in NAFO zone 3L by 1979, and in zone 2J3K in 1993 (Canada 1996c). The capelin fishery boomed in the mid-1980s due to demands for roe in the Japanese market, a lucrative but highly wasteful production (Palmer and Sinclair 1997). This fishery was closed in 1995, but has subsequently resumed (Canada 1996c). Similarly, since the early 1970s lumpfish has been targeted for roe (Felt and Locke 1995).

Northern shrimp were fished for the first time in the mid-1970s. "Annual catches increased to about 9,000 tons [sic] in 1981, but then declined to only 3,000 tons [sic] in 1984. Since then, however, catches have steadily increased, the 1995 catch of approximately 30,000 tons [sic] being the highest recorded" (Canada 1996b). Squid were also a newly targeted species in the late 1960s and early 1970s. Catches climbed and remained high until 1981, when the stock crashed (Blades 1995). Finally, scallops have also been fished in recent decades. However, in the mid- to late 1980s, a price drop depressed the market (and therefore the fishery) for scallops (Palmer and Sinclair 1997).

These shifts in targeted species, or "fishery diversification," are especially obvious in the 1990s. In the report of the Task Force on Incomes and Adjustment in the Atlantic Fishery, chair Richard Cashin explained that:

> Before the collapse, cod made up about two-thirds of the groundfish catch. Shellfish—principally lobster, scallops, shrimp, crab and clams—were 13.5 per cent of the landed volume (in tonnes), but nearly 50 per cent of the landed value. Herring and other pelagic fish were nearly one-quarter of the landed volume, but 8.4 per cent of landed value, as these are relatively low-value species (Cashin 1993).

He further commented that:

> The volume of shellfish landings has increased during the last decade, partly in response to shifts in consumer demand. By landed value, the most important shellfish species to the fishery are lobster, scallop, shrimp and snow crab. The total Atlantic catch of shellfish has grown from 170,000 tonnes in 1982 to 230,000 tonnes in 1992, with catches of lobster almost doubling in that time (Cashin 1993).

His was the only warning that, although catches may seem good and stocks healthy, "we must note that our capacity to predict is always low, given the uncertainties of what happens beneath the ocean surface" (Cashin 1993).

Indeed, the Government of Newfoundland continues to place its hopes on finding new species to harvest. As other groundfish species (specifically redfish, American plaice, flounder, and turbot) have come under moratoria (Newfoundland and Labrador 1997), other species are being sought and harvested. Furthermore:

> Other *positive* developments have also taken place in in [sic] the fishing industry in recent years. Fish harvesters and processors are becoming increasingly innovative in their pursuit of commercial fisheries opportunities. Also, new species are being targeted and a greater emphasis is being placed on the full utilization of existing resources (Newfoundland and Labrador 1997) [emphasis added].

The provincial government is not alone in its eagerness to pursue previously unfished species. Federally, "The Government of Canada has implemented a directed fishery of such underutilized species as capelin, redfish, and turbot species, believing them to be surplus to the country's needs" (Blades 1995). Here history repeats itself. In 1976, a federal Department of Fisheries and Oceans (DFO) paper entitled "Policy for Canada's Commercial Fisheries" devoted a section to the issue of "forage fisheries," in recognition of the complexity of multispecies predator-prey interactions. The section concluded with the promise that "These species interactions and their role in fisheries management will be the object of intensive research in the near future, research which has just commenced during the past couple of years" (Blades 1995). To this end, this paper emphasizes this priority to both government and fishery scientists.

The increased interest in lobster, as well as the relatively new fisheries for crab, lumpfish and capelin roe, and other so-called "underutilized" species stems from a handful of specific causes. First, the increasing industrialization coinciding with Newfoundland's entry into Confederation led to significantly increased catches. As Canada extended its jurisdiction to 200 nautical miles, optimism that those marvelous catches could be taken by Canadians (especially Newfoundlanders) prompted many individuals to make large capital investments into the fishery, often under government-funded loan schemes. While this did not cause undue hardship during the boom times immediately following the extension to 200 miles, once the groundfishery began to decline payments became more difficult. In turn, fishers began to take other nontraditional species in an effort to earn a living and avoid defaulting on loans. Once the groundfishery collapsed in the early 1990s, the situation became desperate. Catch data indicate a refocusing of effort at this time on the other species, especially shellfish.

Similar changes were exhibited by the processing sector. Following the 1977 extension of jurisdiction, processing capability increased rapidly. In fact, during the later 1980s, processing capacity was rarely fully utilized, averaging only 25.11 percent capacity in Newfoundland. "While there is substantial variation, no inshore plant has operated at more than 40 per cent in recent years and no offshore plant at more than 70 per cent. . . . However, it is not unusual for additional capacity to be needed in good years" (Felt and Locke 1995).

With the individual fisher trying to support his capital investment in the fishery by targeting other species, the processing plants also turned to new species in an attempt to keep the plants open and to maintain profits and employment. Even the Task Force on Incomes and Adjustments in the Atlantic Fishery called for regionally centered multispecies processing plants to help reduce the persistent overcapitalization of processing capacity in Atlantic Canada (Cashin 1993).

The other major trend which spurred fishers to target new species is market demand. For instance, the relatively new fisheries for lumpfish and capelin roe developed in response to Japanese market demand. Capelin, for example, boomed during the 1980s due to such demand (Palmer and Sinclair 1997). Snow crab has also increased in value because of the lucrative Asian market (Kurlansky 1997). As prices increase, fishers are naturally (and not unexpectedly) drawn towards the more valuable species.

The two most significant factors which brought about further changes in the fishery were modernization and market pressures, both effectively being economic pressures. Economic pressures (and simple human greed) are seen to have provoked a previously unseen shift in a fishery which had been relatively stable for nearly five centuries. The consequences of this shift are alarming.

Fishery Impacts

From Fig. 2, it is apparent that over the years, the mean trophic level of the fishery has declined. While there was minimal variation in mean trophic level in the first half of the century, in the second half, and particularly from the 1970s through the 1990s, that the decline became most obvious. The trends evident in Fig. 2 are discussed here in the context of the political, socioeconomic, and fisheries developments.

From the establishment of the Commission of Government through Newfoundland's entry into Confederation with Canada and the 1950s, the mean trophic level of the fisheries was remarkably constant. The Commission of Government demonstrated a degree of concern for the viability of the fishery, beginning with the Commission of Inquiry into Fisheries in 1935 (Rowe 1980). One year later, the Newfoundland Fisheries Board was established by the Commission of Government (Rowe 1980, Mackenzie 1986).

To reiterate, by mid-century, new technologies had been and were still being introduced into fisheries. For instance, in 1948, pair trawls were first used in the bank fishery. The development and introduction of such new technologies around the midpoint of the century may have contributed to the slight variations of mean trophic level apparent on Fig. 2.

In 1949, Newfoundland relinquished her sovereignty and became Canada's tenth province. With Confederation, Canada took control of marine fisheries. Interestingly, in 1950 the mean trophic level of the harvest began to demonstrate greater volatility.

Once Newfoundland joined Canada, a greater effort was made to industrialize the province generally. Offshore, foreign fishing fleets increased in size and number and greatly expanded both their capacity and catches. Technologies continued to develop and be introduced to the fishery, especially from the 1960s onward. Radar, echosounders, and nylon multifilament and monofilament were developed and put into use, in a time when it was thought technology and modernization could develop without limits. Foreign fishing off Canada's coast continued, through which the ensuing expansion of fishing jurisdiction was taken as an opportunity to further exploit the existing fisheries and to increase catches.

In addition to increasing effort toward existing fisheries, new species were targeted, especially beginning in the 1970s. This greatly affected the mean trophic level, particularly as many of the species harvested were invertebrates, including snow crab, squid, and shrimp, as well as capelin and lumpfish for roe.

The new lower-order fisheries, combined with a sharply declining seal hunt in the 1980s, led to an increased variability and overall decline in mean trophic level throughout the 1980s. Cod catches were initially high in both the inshore and offshore sectors, but declined mainly in the inshore sector. When the groundfish fishery collapsed in the early 1990s, and the commercial Atlantic salmon fishery was closed in 1995, only the lower-order fisheries remained primarily active, resulting in a particularly acute decline between 1992 and 1993. The 1993 beginning of a Grand Banks fishery for Iceland scallop, with a trophic level of just 2.0, further exacerbated the decline in mean trophic level (Canada 1006d).

The main decline, then, began after 1950 when modernization and development brought new technologies to the fishery. Fear of foreign overfishing of "Canadian" stocks prompted Canada to extend jurisdiction over fisheries to 200 nautical miles and led to the expansion of Canadian fishing effort. As fishers increased their debt to finance new technologies, they needed not only to maintain their incomes but also to increase them in order to meet payments. When fishing for groundfish was no longer adequate, fishers began to search for and exploit other fisheries, especially squid, shrimp, scallop, and other species which are at very low trophic levels. Governments, too, encouraged the development of "underutilized" species, seeing them as a panacea—for low incomes in the fishery, for excess numbers of fishers chasing inadequate numbers of fish, for processing plants operating below capacity. And processing plants, too, sought harvests of other species to supplement the groundfish catch, thereby creating more jobs and for longer periods.

What, then, is the implication of this trend? Simply put, higher trophic level fish species have largely been devastated in their numbers within the area and have, under various moratoria, been left with the chance to rebuild. However, fishers are eager to be at work in their chosen employment, and hence other species have been opened for harvest. What is often overlooked in such a policy decision is that the new species are of a lower trophic level—implying that, either directly or indirectly, they may comprise some part of the diet of the cod and other traditionally preferred species (which are at higher trophic levels). Thus, in addition to having decimated the cod stocks, commercial fishing effort is now making the mistake of catching the cods' prey. In effect, humans have begun to compete with cod and other higher-order fish for their food. If this trend does not end, then there is little hope for the populations of traditionally targeted species to rebuild. Moreover, there will be little food for fish to consume and certainly not enough to support a stock large enough to even be commercially viable.

Future Directions: Fisheries Science and Management

The eventual impact of overexploitive fishing effort globally continues to be overlooked, often despite scientific indications. The inherent uncertainty in spatial stock assessment and predator-prey relationships, and the lack of scientific consensus may have also provided a certain degree of uncertainty in the management and operational advice for fishing within certain regions. The extent to which indirect, nonlinear impacts on marine life have contributed to the rapid decline of identifiable commercially targeted fish species in this region is uncertain from this preliminary ecosystem analysis. However, whether historical fishing levels were justified under the consideration of the nonlinear impact of fishing effort is questionable. This is shown in the yearly mean trophic level changes within this region. Many of the main fisheries in the East Coast region remain either severely depleted and closed or open with small TACs. If a recovery in the size of the fish populations does occur, the awareness of the trophic level impact of management decisions based on historical trends will provide an indication of the possible future implications of policy decisions.

The interactions between species are indeed complex and at present largely beyond the scope of our human understanding. As fisheries engulf more species, what impact will this cause on marine ecosystems? Indeed, "The problems experienced by both the fixed-gear and mobile-gear fishers during the early 1990s also produced the search for "underutilized" species. . . there [is] already talk about today's underutilized species becoming the endangered species of tomorrow" (Palmer and Sinclair 1997).

References

Blades, K. 1995. Net destruction: The death of Atlantic Canada's fishery. Nimbis, Halifax. 184 pp.

Canada DFO. 1996a. Newfoundland and Labrador snow crab. Canada Department of Fisheries and Oceans (DFO) Atlantic Fish. Stock Status Rep. 96/15.

Canada DFO. 1996b. Assessment of northern shrimp off Newfoundland and Labrador. DFO Atlantic Fish. Stock Status Rep. 96/17.

Canada DFO. 1996c. Capelin in Subarea 2 + Div. 3KL. DFO Atlantic Fish. Stock Status Rep. 96/23.

Canada DFO. 1996d. Overview of Canadian marine fisheries resources. Ottawa, Fisheries Research Board.

Canada DFO. 1998. Atlantic commercial landings, by province, 1972-1995. DFO Statistics (http://www.ncr.dfo.ca).

Candow, J. E. 1989. Of men and seals. A history of the Newfoundland seal hunt. National Historic Parks and Sites, Environment Canada, Ottawa. 239 pp.

Cashin, R. (Chairman). 1993. Charting a new course: Towards the fishery of the future. Report of the Task Force on Incomes and Adjustment in the Atlantic Fishery. Communications Directorate, Fisheries and Oceans Canada, Ottawa. 199 pp.

Felt, L.F., and L.W. Locke. 1995. It were well to live mainly off fish: The collapse of Newfoundland's fishery and beyond. In: R. Arnason and A. Felt (eds.), The North Atlantic fisheries: Success, failures and challenges. Island Living Series Vol. 3. Institute of Island Studies, Charlottetown, pp. 197-236.

FishBase. 1997. FishBase 97 CD-ROM. Manila, ICLARM.

Hutchings, J.A., and R. Myers. 1995. The biological collapse of Atlantic Cod off Newfoundland and Labrador: An exploration of historical changes in exploitation, harvesting technology, and management. In: R. Arnason and A. Felt (eds.), The North Atlantic fisheries: Success, failures and challenges. Island Living Series Vol. 3. Institute of Island Studies, Charlottetown, pp. 37-94.

Kurlansky, M. 1997. Cod: A biography of the fish that changed the world. Alfred A. Knopf, Toronto. 294 pp.

Leacy, F.H., M.C. Urquhart, and K.A.H. Buckley (eds.). 1983. Historical statistics of Canada. 2nd edn. Statistics Canada.

MacKenzie, D. 1986. Inside the Atlantic Triangle: Canada and the entrance of Newfoundland into confederation 1939-49. University of Toronto Press, Toronto. 285 pp.

Myers, R.A., J. Bridson, and N.J. Barrowman. 1995. Summary of worldwide spawner and recruitment data. Can. Tech. Rep. Fish. Aquat. Sci. 2020 (http://www.mscs.dal.ca/~myers/).

Neary, P. 1988. Newfoundland in the North Atlantic World, 1929-1949. McGill-Queen's University Press, Kingston, 459 pp.

Newfoundland and Labrador. 1997. The Atlantic groundfish strategy: An analysis of the program on a regional basis. Newfoundland and Labrador Department of Finance, St. John's. 57 pp.

Northwest Atlantic Fisheries Organization (NAFO). 1960 through 1995. NAFO Statistical Bulletin. Vols. 10 through 42. Dartmouth, Canada.

Palmer, C., and P. Sinclair. 1997. When the fish are gone: Ecological disasters and fishers in northwest Newfoundland. Fernwood Publishing, Halifax.

Pauly, D., and V. Christensen. 1995. Primary production required to sustain global fisheries. Nature 74:255-257.

Pauly, D., V. Christensen, J. Dalsgaard, R. Froese, and F. Torres Jr. 1998. Fishing down marine food webs. Science 279:860-863.

Rowe, F.W. 1980. A history of Newfoundland and Labrador. McGraw-Hill Ryerson, Toronto. 563 pp.

Ryan, S. 1986. Fish out of water: The Newfoundland saltfish trade 1814-1914. Newfoundland History Series 2. Breakwater Books, St. John's. 320 pp.

Ecosystem Approaches for Fisheries Management
Alaska Sea Grant College Program • AK-SG-99-01, 1999

Human Effects on the Baltic Sea Ecosystem: Fishing and Eutrophication

Sture Hansson
Stockholm University, Department of Systems Ecology, Stockholm, Sweden

Extended Abstract

The Baltic Sea food web is comparatively very simple, with three species (herring, sprat, and cod) making up about 80% of the total fish biomass (Thurow 1993). Cod (*Gadus morhua*) is the dominant piscivorous species, while herring (*Clupea harengus*) and sprat (*Sprattus sprattus*) dominate among zooplanktivorous species. In some areas, mysids and jellyfish can also be important zooplankton predators. All three fish species, particularly cod, are intensively fished.

Cod is also strongly influenced by humans in another way: eutrophication. Since the turn of the century, phosphorous and nitrogen load to the Baltic has increased 4 and 8 times, respectively (Larsson et al. 1985). The resulting increase in primary production has resulted in an increased sedimentation of organic matter to the deep water, where oxygen conditions have deteriorated and reduced cod reproductive success (Hansson and Rudstam 1990, Sparholt 1996).

As a result of the intensive fishery in combination with the reduced recruitment, the cod population decreased by almost an order of magnitude within 10 years. This resulted in decreased predation pressure on herring and sprat. This effect was particularly evident on sprat, for which predation pressure (consumption divided by production) dropped >90%. The sprat population increased to a level that probably was the "all-time high." Recalling that cod is the dominant piscivorous species in the Baltic, and that the clupeids are the most common zooplanktivors, the changes in the fish community between 1980 and 1990 represent one of the most drastic ecosystem changes in the Baltic in this century. This change was very much caused by a synergy between eutrophication and the fishery, and sustainable management of the Baltic Sea thus requires a new approach, with cooperation between ecologists and fisheries managers.

Acknowledgment

This study was supported by grants from the Nordic Council of Ministers and the European Union (MAST/BASYS).

References

Hansson, S., and L. Rudstam. 1990. Eutrophication and Baltic fish communities. Ambio 19:123-125.

Larsson, U., R. Elmgren, and F. Wulff. 1985. Eutrophication and the Baltic Sea: Causes and consequences. Ambio 14:9-14.

Sparholt, H. 1996. Causal correlation between recruitment and spawning stock size of central Baltic cod? ICES J. Mar. Sci. 53:771-779.

Thurow, F. 1993. Fish and fisheries in the Baltic Sea. ICES Coop. Res. Rep. 186:20-37.

Ecosystem Approaches for Fisheries Management
Alaska Sea Grant College Program • AK-SG-99-01, 1999

Does Harvesting a Million Metric Tons of Sand Lance Per Year from the North Sea Threaten Seabird Populations?

Robert W. Furness
University of Glasgow, Institute of Biomedical and Life Science, Glasgow, United Kingdom

Abstract

The North Sea has for many decades been one of the most heavily fished seas in the world. Most stocks of large predatory fishes are depleted. Stocks of herring and mackerel were greatly reduced during the 1970s. Current annual landings of sand lance are the largest single species catch from the North Sea at up to 1 million t. About 4 million seabirds breed around the North Sea. Most depend largely on sand lance as food during breeding. Over recent decades seabird numbers in the North Sea have increased considerably. Removal of sand lance by the fishery might be expected to have negative effects on seabirds. Seabird populations and breeding productivity in the North Sea are intensively monitored, yet there is no convincing evidence that the sand lance fishery has yet affected seabird numbers or breeding success.

Three factors may explain this. First, seabirds are generally long-lived and have behavioral mechanisms to buffer themselves against effects of short-term fluctuations in fish abundance. Second, seabird consumption of sand lance is partly spatially separated from sand lance fishing effort. Third, the industrial fishery may represent a substitute for removal of sand lance by adult mackerel and some gadoids that occurred previously, such that combined predation imposed by those fishes plus industrial fishing appears to have remained approximately constant. All these possible explanations present important implications for future management of fishing effort in the North Sea if the fisheries and seabirds are to continue to coexist.

Introduction

Although many oceanic or pelagic seabirds feed extensively on cephalopods or crustaceans, most continental shelf and shallow seas seabirds feed predominantly on abundant, small, shoaling pelagic fish, at least during the breeding season (Furness and Monaghan 1987, Montevecchi 1993, Springer and Speckman 1997). Small shoaling fish species are often targets of industrial fisheries for production of fishmeal and oils. The removal of large quantities of these fishes by industrial fisheries might reduce food supply to seabirds. One frequently quoted example of this has been in Peru, where the collapse of the heavily fished Peruvian anchovy (*Engraulis ringens*) stock resulted in a dramatic decrease in numbers of guano seabirds (Duffy 1983). Another is the fishery for capelin (*Mallotus villosus*) in the Barents Sea, which annually harvested 1-3 million metric tons of capelin between 1973 and 1984 until the stock collapsed to 20,000 t in 1987 (Gjøsæter 1997). The cause of the collapse was primarily high predation levels from increased stocks of cod *Gadus morhua* (Bogstad and Mehl 1997) but the low levels of capelin resulted in an 80% decrease in numbers of common murres *Uria aalge* apparently as a result of winter starvation (Vader et al. 1990, Krasnov and Barrett 1995, Anker-Nilssen et al. 1997).

Industrial fishing could hypothetically affect seabird populations by a number of distinct processes. Fishing might reduce stock biomass, so that the prey density available to foraging seabirds, during the seasonal period of the fishery or subsequently, might fall below levels that would support high breeding success or survival. Such an effect would depend on whether seabirds require a certain minimum prey density to be available before foraging would be economic. Over a longer term, fishing might reduce the mean level, or increase variability, of recruitment into the fished stock. Such an effect would come about if fishing reduced spawning stock biomass and this reduction affected the level of recruitment of young fish. Thus the form of any relationship between spawning stock biomass and recruitment is of fundamental interest with regard to possible effects of such fisheries on seabird food supply. Finally, industrial fishing might alter food-web structure by affecting the competitive balance between fished and unfished stocks, or between heavily fished and lightly fished stocks. For example, the relative abundance of two ecologically similar small planktivorous fish species might change such that species A decreased and species B increased over a period of heavy fishing on species A. If species A was the staple prey of seabirds while species B was unavailable to them or uneconomic, this change in community composition could have an adverse effect on seabirds, whereas if species B was the staple prey of seabirds, then a high fishing effort on species A might increase food availability to the seabirds and could result in an increase of seabird numbers as a result.

Seabirds are generally long-lived, mostly producing few fledglings that will only recruit if they survive to several years old (Furness and Monaghan 1987). Thus their populations can only increase slowly, and except in rare instances where mass mortality of adults occurs, they also usually decrease in numbers slowly. In stark contrast, small pelagic fishes exhibit short life-spans, with early and highly variable recruitment. Their populations therefore tend to fluctuate rapidly, and rather unpredictably, in abundance. Seabird population sizes cannot track short-term changes in prey population abundance. Thus, seabirds have a variety of buffering mechanisms to cope with such natural variations in food supply. These vary among species in strength and form (Furness 1996, Phillips et al. 1996b). Theory predicts that the seabird species most vulnerable to reductions in pelagic food supplies would be small surface-feeding seabird species with specialized and energetically expensive foraging methods and little spare time to allow any increase in foraging effort (Furness and Ainley 1984). Obvious examples of such seabirds are the Arctic tern (*Sterna paradisaea*) and the black-legged kittiwake (*Rissa tridactyla*). In Shetland, these species were particularly severely affected by the reduction in sand lance (*Ammodytes marinus*) stocks there in the late 1980s, whereas some of the larger seabirds continued to breed successfully despite the decline in sand lance abundance (Heubeck 1989, Furness and Barrett 1991). Theory also predicts that seabird species would maintain high adult survival rates even in the face of moderately reduced fish abundances, but that breeding success and time budgets would respond to food supply (Cairns 1987). Some data support this (Hamer et al. 1991, 1993; Phillips et al. 1996a; Furness and Camphuysen 1997; Harris and Wanless 1997) but in several cases even adult survival may be affected by reductions in pelagic fish stocks (Vader et al. 1990, Hamer et al. 1991, Heubeck et al. 1991, Harris and Bailey 1992, Krasnov and Barrett 1995, Barrett and Krasnov 1996).

The industrial fishery for sand lance in the North Sea is the largest fishery in the area (Gislason and Kirkegaard 1996). Sand lance is an extremely abundant lipid-rich fish that is a major part of the diet of many seabirds, so seabirds might be in direct competition with the industrial fishery for the resource (Furness and Barrett 1991, Monaghan 1992). The fact that the fishery is well documented and that there are very detailed data on the numbers and breeding success of seabirds in the region, means that this case provides an outstanding opportunity to examine effects of industrial fishing on seabird populations. In the North Sea, many fish stocks are very heavily exploited. An industrial fishery developed during the 1950s, first harvesting young herring (*Clupea harengus*), then taking mackerel (*Scomber scombrus*), and after these stocks had collapsed, fishing mainly for Norway pout (*Trisopterus esmarkii*) and sand lance. The sand lance has become the main target of industrial fishing in the North Sea. The black-legged kittiwake is one of the most abundant and widespread breeding seabirds in Europe, and it feeds very extensively on sand lance during the breeding season (Furness 1990, Harris and Wanless 1997). Many features of its biology suggest that it should be particularly vulnerable to reductions in food supply, such as its surface feeding habits, the fact that one adult is always present to protect the nest site, its relatively small size, high foraging costs (prolonged flapping flight), and specialized diet (Furness and Ainley 1984), and empirical data confirm that black-legged kittiwake breeding success is strongly affected by food abundance (Harris and Wanless 1990, 1997; Hamer et al. 1993). Thus the black-legged kittiwake is the most obvious seabird to study in relation to effects of industrial fishing for sand lance on seabirds in the North Sea. Relationships between black-legged kittiwakes, sand lance stocks, and the sand lance fishery were reviewed by Furness (1998) and details of relationships between kittiwake breeding productivity and measures of sand lance abundance are examined further in this paper.

Methods

Since the sand lance stock of Shetland has been considered to be distinct from the sand lance stock in the rest of the North Sea (Bailey et al. 1991), the data for Shetland have been examined separately from those for the rest of the North Sea. By tradition, and lacking data to allow a disaggregated analysis by separate regional stocks, the sand lance in the rest of the North Sea has been treated as a single stock. This procedure is followed here, although it is recognised that this is likely to be an oversimplification, and that the spatial scale over which breeding seabirds respond to variations in sand lance abundance is much smaller than that of the whole North Sea area. Thus any correlations between seabird breeding success and North Sea areawide sand lance abundance are likely to be weaker than might be the case if spatially resolved sand lance data were available.

Data on the stock biomass and abundances of age classes of sand lance from virtual population analysis (VPA) for the whole North Sea excluding Shetland, on the abundance of each age class of sand lance in Shetland, on landings of sand lance from the North Sea, and on catch per unit of effort (CPUE) by the Danish and Norwegian fleets (responsible for almost the whole catch) were obtained from the literature, especially from the International Council for the Exploration of the Sea (ICES 1997a). VPA and CPUE data are available for years up to and including 1995, but are not yet available for 1997 and are incomplete for 1996. The industrial fishery for sand lance at Shetland was closed for several years in the late 1980s and 1990, so estimation of stock by VPA is impractical. Instead, sand lance abundance and age structure at Shetland have been assessed by counting numbers of fish caught per 30-minute tow during research trawling each summer. In two years this survey work was not carried out, so for those seasons the abundance of each age class was interpolated from the life table data collected in the previous and subsequent seasons, assuming survival rates of each age class to be the mean of those across all years of the surveys.

Data on black-legged kittiwake breeding success at sample colonies on the U.K. coast of the North Sea were obtained from the annual reports of the Joint Nature Conservation Committee (JNCC) Seabird Monitoring Programme. These data are available for all years from 1986 to 1997. Thus correlations between kittiwake performance and sand lance stock estimates can be computed over the period 1986 to 1995 or 1996, giving samples of 10 or 11 data points. Estimates of annual consumption of sand lance by predators in the North Sea were obtained from Tasker and Furness (1996) and Furness and Tasker (1997) for seabirds and from the ICES Multispecies Assessment Working Group (ICES 1997b) for predatory fishes. Relationships between different measures of sand lance stock abundance and between black-legged kittiwake breeding success and sand lance stock sizes were examined by correlation analysis. One-tailed tests of significance were used since the directions of correlations between breeding success and food abundance and between different measures of sand lance stock abundance were anticipated to be positive. Two-tailed tests were used when correlating abundances of 0-group with older sand lance since there was no anticipated direction of correlation in that case.

Results

Different Measures of Sand Lance Abundance in the North Sea

CPUE data were positively correlated with VPA estimates of the abundance of 1 year old and older sand lance or of total stock biomass in all 16 of the possible comparisons across the years 1986-1995 (Table 1). The correlations were mostly between 0.5 and 0.7. Since one would expect CPUE to reflect sand lance abundance, the relatively low correlation coefficients presumably indicate a combination of inaccuracies in the data and spatial variability over the North Sea, among other factors. CPUE during the second half of the year was positively correlated with CPUE in the first half of the same year (northern North Sea $r = 0.74$, $p < 0.05$; southern North Sea $r = 0.51$, $p = 0.05$) despite the fact that catches later in the year include 0-group sand lance whereas this is not the case in January-June. It is noteworthy that the CPUE during the first half of each year was weakly negatively correlated both in the northern and in the southern North Sea with the log of the estimated number of 0-group sand lance on 1 July in that year (Table 1). This suggests that recruitment of 0-group sand lance tends to be lower in years of high sand lance abundance during January-June. To explore this possibility further, using data from ICES (1997a) for the years 1983-1995, the estimated number of 0-group sand lance on 1 July each year was correlated with the number of 1+ sand lance on 1 July and with the total stock by numbers on 1 January of the same year. In both cases

Table 1. Correlations (Pearson product-moment) between different measures of sand lance abundance in the North Sea over the years 1986-1995/1996.

Measure of sand lance abundance	CPUE in northern North Sea Jan-Jun	CPUE in northern North Sea Jul-Dec	CPUE in southern North Sea Jan-Jun	CPUE in southern North Sea Jul-Dec
Log of VPA estimated number of 0-group sand lance on 1 July	−0.34	0.10	−0.45	0.02
Log of VPA estimated number of 1+ group sand lance on 1 July	0.70 $p < 0.05$	0.67 $p < 0.05$	0.68 $p < 0.05$	0.46
Log of VPA estimated sand lance total stock biomass on 1 July	0.71 $p < 0.05$	0.78 $p < 0.05$	0.44	0.52
Log of VPA estimated number of 1+ group sand lance on 1 January	0.61 $p < 0.05$	0.60 $p < 0.05$	0.59 $p < 0.05$	0.44
Log of VPA estimated sand lance total stock biomass on 1 January	0.91 $p < 0.05$	0.65 $p < 0.05$	0.70 $p < 0.05$	0.50

CPUE data are from Danish and Norwegian fisheries data (ICES 1997a: Tables 13.1.3.4-13.1.3.6). VPA data are from the Report of the Working Group on the Assessment of Demersal Stocks in the North Sea and Skagerrak (ICES 1997a: Table 13.1.4.1).

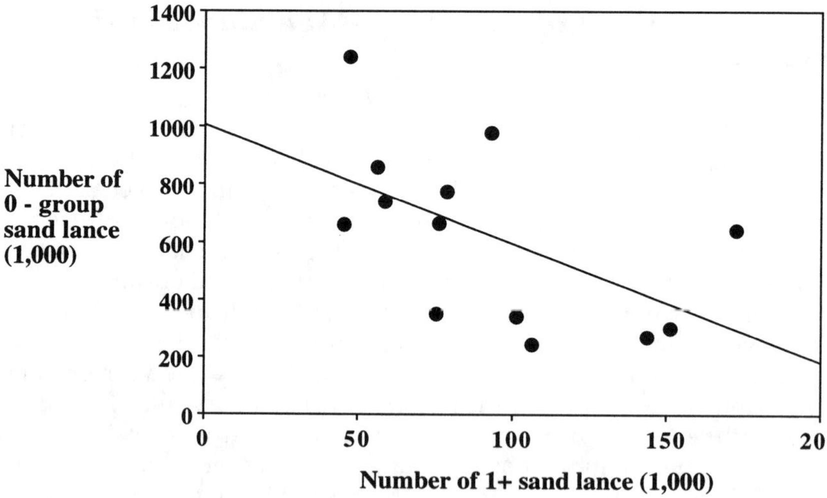

Figure 1. Estimated numbers of 0-group sand lance in the North Sea on 1 July each year 1986-1995 in relation to the estimated numbers of older (1+) sand lance (from VPA). Data from ICES (1997a).

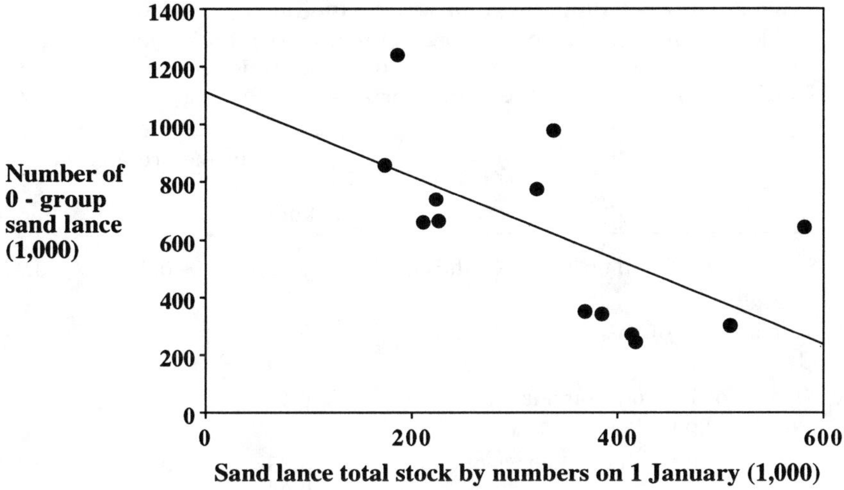

Figure 2. Estimated numbers of 0-group sand lance in the North Sea on 1 July each year 1986-1995 in relation to the estimated sand lance total stock by numbers on 1 January (from VPA). Data from ICES (1997a).

the production of 0-group sand lance was negatively correlated with sand lance abundance (Figs. 1 and 2; correlation with 1+ sand lance on 1 July $r_{11} = -0.55$, $p = 0.05$; correlation with total stock by numbers on 1 January $r_{11} = -0.61$, $p < 0.05$).

Kittiwake Productivity in Relation to Sand Lance Abundance

Breeding productivity of black-legged kittiwakes in three areas of the North Sea (Orkney, east Scotland, and east England) generally showed moderate positive correlations with six different measures of sand lance abundance that excluded 0-group fish (Table 2). Out of 18 correlation coefficients, all 18 were positive (sign test $p < 0.01$), and half were greater than 0.4. The most consistent positive correlations with breeding productivity were correlations with the number of 1+ sand lance on 1 July (Fig. 3). It is particularly worth noting here that over the study period 1986-1995, mean breeding productivity of black-legged kittiwakes never fell below 0.8 chicks per nest in Orkney or below 0.9 chicks per nest in east England and yet in both these areas with predominantly high breeding success, the year-to-year variation in success correlated with sand lance abundance. However, in all three areas, kittiwake breeding productivity was weakly negatively correlated with the number of 0-group sand lance on 1 July. In Shetland, black-legged kittiwake breeding productivity also correlated with sand

Table 2. Correlations (Pearson product-moment) between breeding performance of black-legged kittiwakes (chicks per nest) in monitoring plots at sample North Sea colonies and measures of sand lance abundance in the North Sea for the years 1986-1995/1996.

	Kittiwake productivity		
Parameter	Orkney	East Scotland	East England
Sand lance CPUE in northern North Sea Jan-Jun	0.62 $p < 0.05$	0.01	0.04
Sand lance CPUE in southern North Sea Jan-Jun	0.66 $p < 0.05$	0.16	0.01
Number of 1+ group sand lance in North Sea on 1 Jan	0.41	0.53	0.36
Sand lance total stock biomass on 1 Jan	0.73 $p < 0.05$	0.27	0.31
Number of 1+ group sand lance in North Sea on 1 Jul	0.61 $p < 0.05$	0.41	0.58 $p < 0.05$
Sand lance total stock biomass on 1 Jul	0.62 $p < 0.05$	0.08	0.38
Number of 0-group sand lance on 1 Jul	−0.26	−0.54	−0.41

Data on kittiwake breeding from Thompson et al. (1997). Data on first four measures of sand lance abundance from CPUE data and remaining estimates from VPA data (ICES 1997a), as indicated for Table 1.

Table 3. Correlations (Pearson product-moment) between breeding performance of black-legged kittiwakes (chicks per nest) in monitoring plots at Shetland colonies and measures of sand lance abundance for the years 1986-1995/1996.

Sand lance abundance estimate	Correlation with kittiwake breeding productivity
Log number of 1 group sand lance per 30 minute tow at Shetland	0.72, $p < 0.05$
Log number of 1+ sand lance per 30 minute tow at Shetland	0.70, $p < 0.05$
Log number of 0 group sand lance per 30 minute tow at Shetland	0.20

Data on kittiwake breeding from Thompson et al. (1997). Measures of sand lance abundance from ICES (1997a).

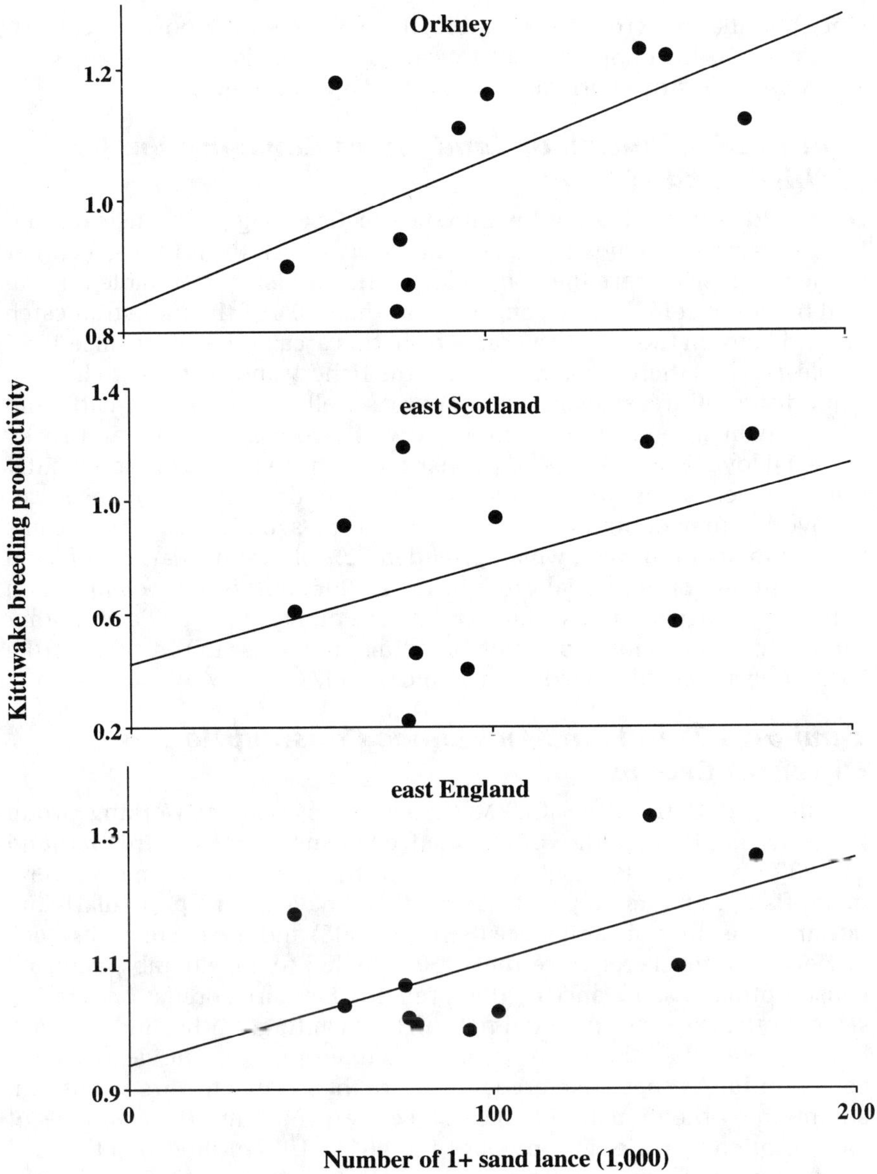

Figure 3. Breeding productivity of black-legged kittiwakes in (a) Orkney, (b) east Scotland, and (c) east England in relation to the numbers of 1+ sand lance in the North Sea stock on 1 July each year 1986-1995 as estimated by VPA.

lance abundance across years (Table 3), showing strong positive correlations with the log of numbers of 1-group or 1+ sand lance, but suggesting a very weak positive correlation with 0-group abundance too.

Spatial Distribution of Sand Lance Consumption by Seabirds and Fishery

Over a broad scale of regions within the North Sea (Fig. 4), the take of sand lance by seabirds is mostly (53%) from ICES IVa west, the northwest corner of the North Sea where there are many cliffs and islands suitable for seabird breeding colonies. By contrast, less than 10% of the industrial catch of sand lance in the North Sea came from this area in the years since 1985 (Table 4). Thus there is a tendency for the fishery and seabirds to harvest from different areas: seabirds from areas well endowed with cliffs and islands (which tend to occur in areas with stronger currents and rocky seabed [Lloyd et al. 1991]), and the fishery from areas with predominantly sandy seabed (areas where cliffs and islands tend to be scarce). The greatest overlap in exploitation by seabirds and the sand lance fishery seems to arise in area IVb west, which provided 22% of seabird harvest of sand lance and 34% of industrial catch in 1991-1996. In this area, competition between the fishery and seabirds may arise on the Wee Bankie and other sandy substrates close to major breeding sites of seabirds such as the Firth of Forth islands (Harris and Wanless 1997).

Temporal Trends in Sand Lance Consumption by Different Groups

Examining data from the ICES Multispecies Assessment Working Group suggests that the quantity of sand lance consumed by seabirds (around 210,000 t per year) is small compared to the quantities taken by several major fish predators. In particular, mackerel have been a particularly important consumer of sand lance (Furness 1998) and the decrease in stock of North Sea mackerel since the 1960s has led to considerably reduced consumption of sand lance by that predator. Similarly, predation rates by several other heavily exploited fish such as whiting, saithe, and haddock have decreased as these stocks have declined in size. The decreases in consumption by these fish predators more than match the growth of consumption by the industrial fishery since the 1960s, and the estimates of consumption by major groups since 1974 (ICES 1997b) show that the total consumption of sand lance has probably decreased slightly over the last 25 years, while the percentage of the consumption due to North Sea mackerel has fallen drastically (Figs. 5 and 6).

Discussion

Black-legged kittiwake breeding numbers have increased on North Sea coasts alongside the large North Sea industrial fishery for sand lance between

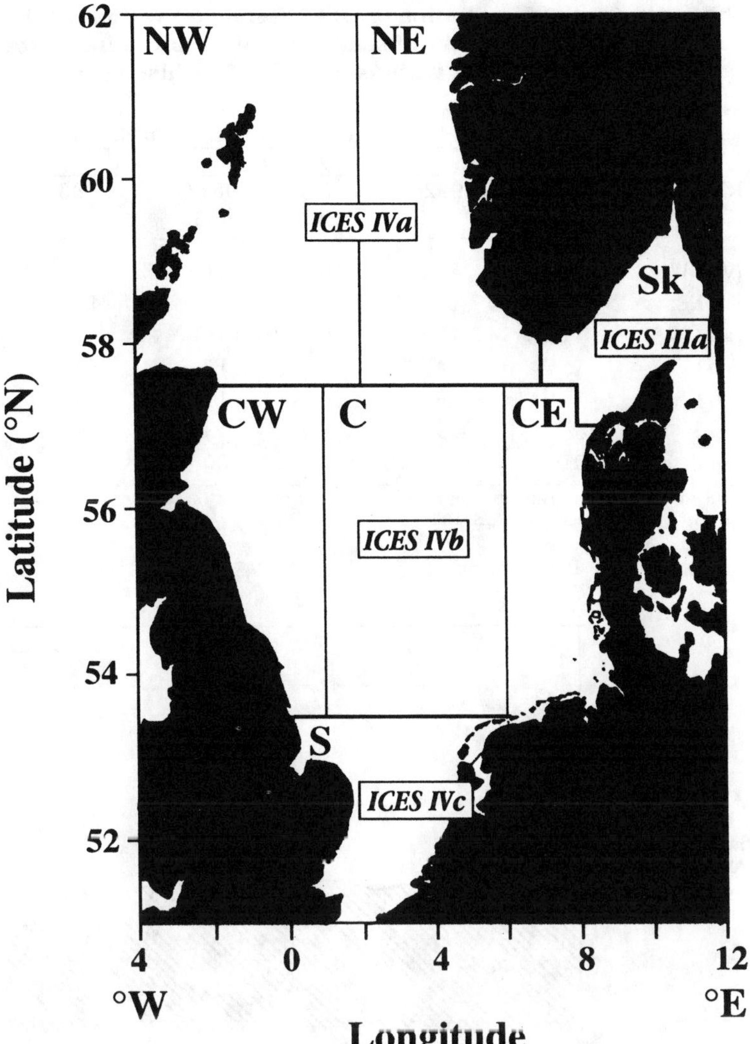

Figure 4. ICES areas of the North Sea as used in this study to compare consumption of sand lance by seabirds and by the industrial fishery.

Table 4. Spatial distribution (by ICES rectangles within the North Sea; see Fig. 1) of the take of sand lance by seabirds and by the industrial fishery during various periods of its history.

	% of total catch of sand lance in North Sea					
		Industrial catch				
ICES area North Sea	Take by seabirds	1972-1975	1976-1980	1981-1985	1986-1990	1991-1996
IVa NW	53	21	16	11	5	9
IVa NE	8	36	23	12	40	35
IVb CW	22	18	26	36	30	34
IVb C	7	4	21	17	10	11
IVb CE	9	8	6	11	10	8
IVc S	1	13	8	13	5	3
Total	100	100	100	100	100	100

Data on sand lance consumption by seabirds from Furness and Tasker (1997). Data on sand lance catches by the fishery from ICES (1997a).

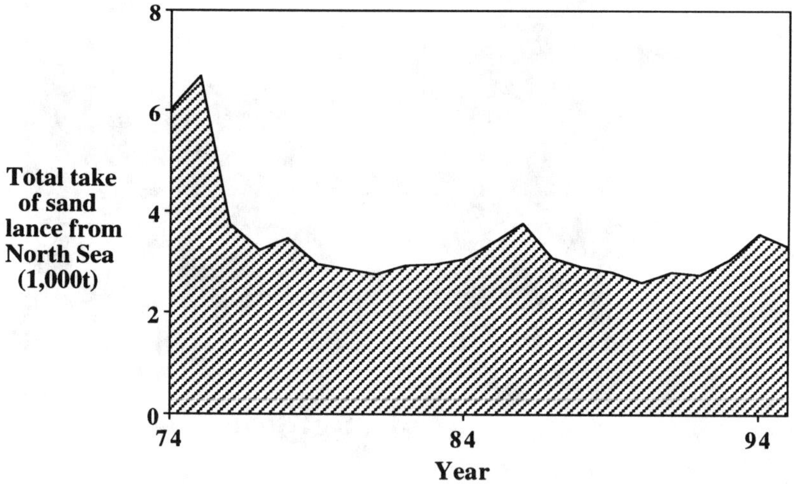

Figure 5. Estimated total consumption of sand lance in the North Sea by major consumer groups showing trends over the study period from 1974 to 1995.

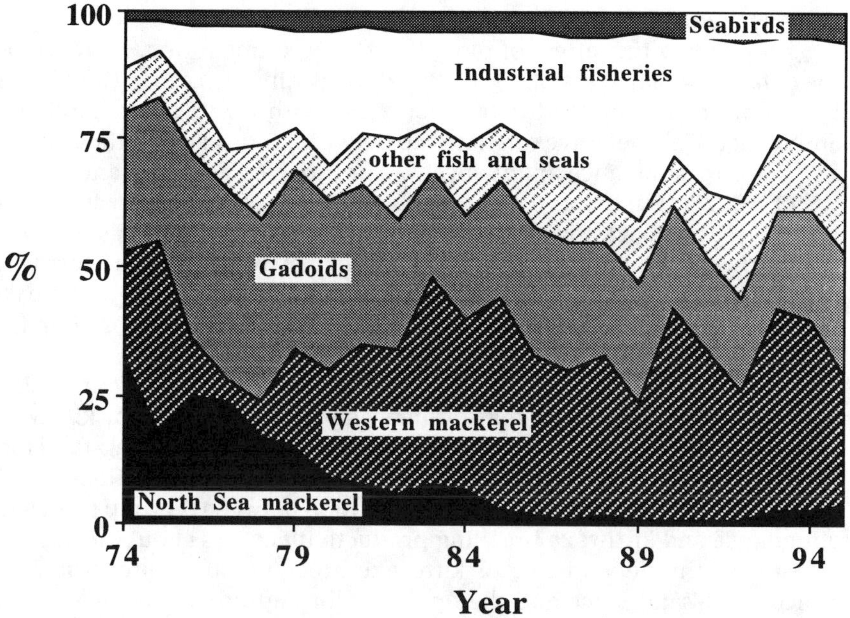

Figure 6. Percentages of annual take of sand lance attributable to particular consumer groups; North Sea mackerel stock, western (Atlantic) mackerel stock, gadoids, other fish and seals, industrial fishery, seabirds. Data were derived from the report of the Multispecies Assessment Working Group of ICES (ICES 1997b).

national censuses in 1969 and 1987 (Lloyd et al. 1991). No national census has been carried out since 1987, but between 1986 and 1995 rates of change in breeding kittiwake numbers at monitoring sites showed no significant difference between sites in Orkney and east England compared to sites in west Britain and Ireland (Thompson et al. 1997). Excepting Shetland, where decreased sand lance abundance in the mid-1980s led to low breeding success (Hamer et al. 1993), mean breeding success of kittiwakes was higher in 1986-1995 at U.K. North Sea colonies from Orkney to east England than at colonies in the Irish Sea or Atlantic coasts of the United Kingdom, and breeding success was higher in years of high sand lance abundance (Furness 1998). This paper expands on that analysis to demonstrate that several measures of sand lance abundance in the North Sea are intercorrelated, and that kittiwake breeding productivity over the years 1986-1995 correlates positively with CPUE and with both winter and summer VPA measures of 1-year-old and older sand lance abundance. However, the analysis suggests that the size of 0-group sand lance production in the North Sea varies as a function of older sand lance abundance. The level of recruitment is highest when preceding abundance of 1+ sand lance was low, suggesting a possible direct effect of older fish on recruitment rate. Furthermore, the data show that years of high kittiwake breeding success in Orkney, east Scotland, or east England tend to be years with low 0-group sand lance abundance. This negative relationship between kittiwake breeding success and 0-group sand lance abundance was unexpected as Harris and Wanless (1990, 1997) argued that 0-group sand lance are "of key importance to the breeding success of kittiwakes" at the Isle of May, east Scotland. They showed that kittiwake breeding success at the Isle of May was higher in years when sand lance formed more of the chick diet, and they concluded that most of the sand lance fed to chicks (all those up to 10 cm long) were 0-group fish. The analysis of the JNCC kittiwake productivity data and ICES North Sea sand lance data over the period 1986-1995 suggest that it is the abundance of older sand lance that influences breeding success of kittiwakes and that the level of production of 0-group sand lance during the chick-rearing period of the kittiwakes is of less, or negligible, influence. The analysis suggests that the negative correlation between 0-group sand lance abundance and kittiwake breeding productivity comes about indirectly as a result of the dependence of kittiwake breeding success on the abundance of 1+ sand lance and the tendency for 0-group sand lance production to be lower when 1+ sand lance abundance is high.

The correlations between 1+ sand lance abundance and kittiwake breeding productivity for Shetland and for North Sea sand lance stocks supports the view that, in the absence of overwhelming effects of predation on kittiwakes (Heubeck et al. 1997, Regehr and Montevecchi 1997), sand lance abundance is a major determinant of kittiwake breeding success. Furthermore, the linear correlations between kittiwake productivity and sand lance abundance imply that any reduction in sand lance abundance caused by the industrial fishery, or by increased rates of predation by mackerel and gadoids, would reduce mean breeding success of kittiwakes. Whether the industrial fishery reduces sand lance abundance is unclear. The fact that the total mortality of sand lance caused by all predators and the industrial fishery combined has probably decreased as a result of the reduction in North Sea mackerel stock may mean that the sand lance population is unlikely to show any overall impact of the industrial fishery over the whole North Sea, since the fishery less than compensates for the reduction in mortality due to decreased stocks of mackerel and other predatory fishes (Fig. 5). Intense local harvesting as seen on the Wee Bankie in recent years (Harris and Wanless 1997), might cause local effects on sand lance abundance, especially if the North Sea population of sand lance is composed of many local stocks on such sandbanks, but no such effects have yet been demonstrated.

Overall, however, the fact that North Sea kittiwake breeding success has remained high until now, despite the harvest of around 800,000 t of sand lance per year from the North Sea, may be due largely to the overwhelming importance of mackerel as a keystone predator in this food web. The reduction of predation by mackerel may have been fundamental in permitting a flourishing industrial fishery for sand lance to exist alongside healthy populations of kittiwakes and other seabirds in the North Sea over the last three decades. However, our understanding of the nature of local sand lance stocks, of the food web interactions affecting sand lance numbers, and of influences of sand lance stock dynamics on seabirds is still very limited. There is a need for further research, for monitoring, and for precautionary management, in order to ensure the continued sustainability of the fishery and the natural predators dependent upon sand lance for food. In particular, recovery of stocks of mackerel or gadoids may have indirect effects on kittiwake productivity as a result of their impact on sand lance stocks. Given that the take of sand lance by predatory fishes has been much greater than that by the industrial fishery, predator-prey interactions among fish stocks may be the most important influences on sand lance availability to seabirds in the future.

Acknowledgments

I thank the International Fishmeal and Oil Manufacturers' Association for supporting this work. The Secretary General of ICES and Dr. J. Rice (Chair, ICES Multispecies Assessment Working Group) kindly gave permission to use ICES data on sand lance stocks and on multispecies assessment. I am most grateful to Dr. K.R. Thompson for helpful comments on a draft of this paper. The analysis was made possible by the fieldwork efforts of many members of The Seabird Group and the support of the seabird productivity monitoring program by the Joint Nature Conservation Committee, Irish Wildbird Conservancy, and Royal Society for the Protection of Birds.

References

Anker-Nilssen, T., R.T. Barrett, and J.V. Krasnov. 1997. Long- and short-term responses of seabirds in the Norwegian and Barents Seas to changes in stocks of prey fish. In: Forage fishes in marine ecosystems. Proceedings of the International Symposium on the Role of Forage Fishes in Marine Ecosystems. University of Alaska Sea Grant, AK-SG-97-01, Fairbanks, pp. 683-698.

Bailey, R.S., R.W. Furness, J.A. Gauld, and P.A. Kunzlik. 1991. Recent changes in the population of the sandeel (*Ammodytes marinus* Raitt) at Shetland in relation to estimates of seabird predation. ICES Mar. Sci. Symp. 193:209-216.

Barrett, R.T., and J.V. Krasnov. 1996. Recent responses to changes in fish stocks of prey species by seabirds breeding in the southern Barents Sea. ICES J. Mar. Sci. 53:713-722.

Bogstad, B., and S. Mehl. 1997. Interactions between Atlantic cod (*Gadus morhua*) and its prey species in the Barents Sea. In: Forage fishes in marine ecosystems. Proceedings of the International Symposium on the Role of Forage Fishes in Marine Ecosystems. University of Alaska Sea Grant, AK-SG-97-01, Fairbanks, pp. 591-615.

Cairns, D.K. 1987. Seabirds as indicators of marine food supplies. Biol. Oceanogr. 5:261-271.

Duffy, D.C. 1983. Environmental uncertainty and commercial fishing: Effects on Peruvian guano birds. Biol. Conserv. 26:227-238.

Furness, R.W. 1990. A preliminary assessment of the quantities of Shetland sandeels taken by seabirds, seals, predatory fish and the industrial fishery in 1981-1983. Ibis 132:205-217.

Furness, R.W. 1996. A review of seabird responses to natural or fishery-induced changes in food supply. In: S.P.R. Greenstreet and M.L. Tasker (eds.), Aquatic predators and their prey. Fishing News Books, Oxford, pp. 166-173.

Furness, R.W. 1998. Are industrial fisheries a threat to seabird populations? In: N. Adams and R. Slotow (eds.), Proc. 22 Int. Ornithol. Congr., Durban, University of Natal. In press.

Furness, R.W., and D.G. Ainley. 1984. Threats to seabird populations presented by commercial fisheries. ICBP Tech. Rep. 2:701-708.

Furness, R.W., and R.T. Barrett. 1991. Seabirds and fish declines. Natl. Geogr. Res. Explor. 7:82-95.

Furness, R.W., and C.J. Camphuysen. 1997. Seabirds as monitors of the marine environment. ICES J. Mar. Sci. 54:726-737.

Furness, R.W., and P. Monaghan. 1987. Seabird ecology. Blackie, Glasgow. 164 pp.

Furness, R.W., and M.L. Tasker. 1997. Seabird consumption in sand lance MSVPA models for the North Sea, and the impact of industrial fishing on seabird population dynamics. In: Forage fishes in marine ecosystems. Proceedings of the International Symposium on the Role of Forage Fishes in Marine Ecosystems. University of Alaska Sea Grant, AK-SG-97-01, Fairbanks, pp. 147-169.

Gislason, H., and E. Kirkegaard. 1996. The industrial fishery and the North Sea sandeel stock. Seminar on the Precautionary Approach to North Sea Fisheries Management, Oslo, 9-10 September 1996.

Gjøsæter, H. 1997. The Barents Sea capelin stock (*Mallotus villosus*): A brief review. In: Forage fishes in marine ecosystems. Proceedings of the International Symposium on the Role of Forage Fishes in Marine Ecosystems. University of Alaska Sea Grant, AK-SG-97-01, Fairbanks, pp. 469-484.

Hamer, K.C., R.W. Furness, and R.W.G. Caldow. 1991. The effects of changes in food availability on the breeding ecology of great skuas in Shetland. J. Zool. (Lond.) 223:175-188.

Hamer, K.C., P. Monaghan, J.D. Uttley, P. Walton, and M.D. Burns. 1993. The influence of food supply on the breeding ecology of kittiwakes *Rissa tridactyla* in Shetland. Ibis 135:255-263.

Harris, M.P., and R.S. Bailey. 1992. Mortality rates of puffin *Fratercula arctica* and guillemot *Uria aalge* and fish abundance in the North Sea. Biol. Conserv. 60:39-46.

Harris, M.P., and S. Wanless. 1990. Breeding success of kittiwakes *Rissa tridactyla* in 1986-88: Evidence for changing conditions in the northern North Sea. J. Appl. Ecol. 27:172-187.

Harris, M.P., and S. Wanless. 1997. Breeding success, diet, and brood neglect in the kittiwake (*Rissa tridactyla*) over an 11-year period. ICES J. Mar. Sci. 54:615-623.

Heubeck, M. 1989. Seabirds and sandeels: Proceedings of a seminar held in Lerwick, Shetland, 15-16 October 1988. Shetland Bird Club, Lerwick. 35 pp.

Heubeck, M., P.V. Harvey, and J.D. Okill. 1991. Changes in the Shetland guillemot *Uria aalge* population and the pattern of recoveries of ringed birds, 1959-1990. Seabird 13:3-21.

Heubeck, M., R.M. Mellor, and P.V. Harvey. 1997. Changes in the breeding distribution and numbers of kittiwakes *Rissa tridactyla* around Unst, Shetland and the presumed role of predation by great skuas *Stercorarius skua*. Seabird 19:12-21.

ICES. 1997a. Report of the Working Group on the assessment of Demersal Stocks in the North Sea and Skagerrak, ICES headquarters, 7-15 October 1996. ICES CM1997/Assess:6. 633 pp.

ICES. 1997b. Report of the multispecies assessment working group, ICES headquarters, 11-19 August 1997. ICES CM1997/Assess:16. 235 pp.

Krasnov, J.V., and R.T. Barrett. 1995. Large-scale interactions among seabirds, their prey and humans in the southern Barents Sea. In: H.R. Skjoldal, C. Hopkins, K.E. Erikstad, and H.P. Leinaas (eds.), Ecology of fjords and coastal waters. Elsevier Science B.V., pp. 443-456.

Lloyd, C., M.L. Tasker, and K. Partridge. 1991. The status of seabirds in Britain and Ireland. Poyser, London. 355 pp.

Monaghan, P. 1992. Seabirds and sandeels: The conflict between exploitation and conservation in the northern North Sea. Biodiv. Conserv. 1:98-111.

Montevecchi, W.A. 1993. Birds as indicators of change in marine prey stocks. In: R.W. Furness and J.J.D. Greenwood (eds.), Birds as monitors of environmental change. Chapman and Hall, London, pp. 217-265.

Phillips, R.A., R.W.G. Caldow, and R.W. Furness. 1996a. The influence of food availability on the breeding effort and reproductive success of Arctic skuas *Stercorarius parasiticus*. Ibis 138:410-419.

Phillips, R.A., R.W. Furness, and R.W.G. Caldow. 1996b. Behavioural responses of Arctic skuas *Stercorarius parasiticus* to changes in sandeel availability. In: S.P.R. Greenstreet and M.L. Tasker (eds.), Aquatic predators and their prey. Fishing News Books, Oxford, pp. 17-25.

Regehr, H.M., and W.A. Montevecchi. 1997. Interactive effects of food shortage and predation on breeding failure of black-legged kittiwakes: Indirect effects of fisheries activities and implications for indicator species. Mar. Ecol. Prog. Ser. 155:249-260.

Springer, A.M., and S.G. Speckman. 1997. A forage fish is what? Summary of the symposium. In: Forage fishes in marine ecosystems. Proceedings of the International Symposium on the Role of Forage Fishes in Marine Ecosystems. University of Alaska Sea Grant, AK-SG-97-01, Fairbanks, pp. 773-794.

Tasker, M.L., and R.W. Furness. 1996. Estimation of food consumption by seabirds in the North Sea. In: G.L. Hunt Jr. and R.W. Furness (eds.), Seabird/fish interactions, with particular reference to seabirds in the North Sea. Int. Counc. Explor. Sea Coop. Res. Rep. 216. Copenhagen, pp. 6-42.

Thompson, K.R., E. Brindley, and M. Heubeck. 1997. Seabird numbers and breeding success in Britain and Ireland, 1996. Joint Nature Conservation Committee, Peterborough. 64 pp.

Vader, W., R.T. Barrett, K.E. Erikstad, and K.B. Strann. 1990. Differential responses of common and thick-billed murres to a crash in the capelin stock in the southern Barents Sea. Stud. Avian Biol. 14:175-180.

Criteria for Designing Ecosystem-Based, Experimental Management: Bottom Trawling and the Bering Sea Ecosystem

Thomas A. Okey
University of British Columbia, Fisheries Centre, Vancouver, British Columbia, Canada

Gretchen A. Harrington
National Marine Fisheries Service, Sustainable Fisheries Division, Juneau, Alaska

Abstract

Detailed knowledge of the status and productivity of stocks has been the primary focus of fisheries management in order to maximize resource extraction. Ironically, maximum extraction is usually constrained only by the most optimistic stock scenarios from detailed single species models. This has led to degradation of marine ecosystems because natural variability and effects on nontarget species have not been adequately considered. Furthermore, the very act of gathering knowledge about ecosystem degradation has the potential to cause further damage due to the system-wide scale of fishery management experiments. However, a strategic approach for identifying trade-offs and risks of management experiments can simultaneously optimize knowledge-gathering, conservation goals, and economic sustainability.

In this paper we propose decision criteria for determining the most appropriate type of experimental management designs within the context of ecosystem-based management. The two "types" of experimental management designs are "unconstrained experimental management" and "precautionary experimental management," distinguished by the presence or absence of "stress treatments," in which extraction or anthropogenic disturbance is increased.

These decision criteria are designed to optimize both resource protection and knowledge-gathering in any experimental management situation.

Initiatives to reveal bottom trawling effects on the Bering Sea ecosystem are an opportunity to design ecosystem-based, experimental management programs wherein conservation objectives can be achieved through a well-designed experimental program for gathering knowledge. Conservation and knowledge-gathering goals have been viewed as conflicting, or "goals to balance," but we argue that a precautionary experimental management design can be optimal for both purposes, especially when using marine protected areas to increase replication, provide ecological controls, and address statistical detection problems by simplifying multitreatment designs. New approaches to spatially explicit ecosystem modeling can be employed to help design such management experiments and help interpret results within an ecosystem context.

Introduction

Bottom trawling occurs throughout broad areas of the sea floor despite a historically poor understanding of its effects on stocks and ecosystems. It is an anthropogenic disturbance (sensu Pickett and White 1985) in that it removes existing organisms, disrupts ecosystem structure and function, creates new (less occupied) space, and changes the physical and chemical environment. It is exotic in that it is new to the system in character, magnitude, frequency, or scale (sensu Sousa 1984). The ecological effects of bottom trawling were poorly known until a recent surge of interest in the topic (see NRC 1994, Dayton et al. 1995, Auster et al. 1996, NPFMC 1998, Auster 1998, Kaiser 1998, Lenihan and Peterson 1998, NRC 1998, Thrush et al. 1998, Auster and Langton1999). Although certain organisms may become more abundant as the result of bottom trawling in some settings (Engel and Kvitek 1998), the default assumption in fisheries management should be that bottom trawling adversely affects marine ecosystems.

Troublesome population trends in the Bering Sea have raised questions about causative links between fishing practices and ecosystem integrity (Bakkala 1993, Ackley et al. 1994, Apollonio 1994, Loher et al. 1995, NRC 1996, NMFS 1999). Despite documented changes in the Bering Sea species composition, current fishing practices are defended by those citing lack of demonstrated relationships between fishing and ecosystem degradation. The default assumption in fisheries management, even in the purportedly progressive North Pacific region, has been that adverse effects of fishing do not exist without demonstrated causation.

Several authors have pointed out the error of this assumption, and that such thinking goes against logic, science, and the public interest (Underwood 1990, Dayton et al. 1995, Dayton 1998). These authors conclude that the burden should be on the users of a public resource to prove that the effects of human activities are negligible, rather than on the public to prove that these activities cause harm. Furthermore, the current (erroneous) default assumption of "no effects when none are demonstrated" violates the Code of Conduct for Responsible Fisheries (FAO 1995) and the precautionary approach as outlined in the United Nations Conference on Straddling Fish Stocks and Highly Migratory Fish Stocks (United Nations 1996; also see review by Garcia 1996).

Fishery managers are beginning to consider the needs of non-target ecosystem components in the decision-making process (e.g., Fritz et al. 1995, Hollowed et al. 1997), thus challenging the classical notion of "surplus production" and recognizing that any removal is a loss to the system. Consideration of nontarget components of ecosystems, has existed as policy since 1980, when the Convention on the Conservation of Antarctic Marine Living Resources (CCAMLR) called for "ecosystem management" of Antarctic living resources. The principle that species do not exist in isolation is highlighted in Peterson's (1996) outline of the Sustainable Biosphere Initiative of the Global Ocean Ecosystems Dynamics program. In the Bering Sea, authors continue to urge that fishery management be based on whole ecosystem considerations (Smith 1995).

A corollary to the shift from the "surplus production" paradigm to the notion that removal of production decreases population resilience (and ecosystem integrity) along a continuum is that anthropogenic disturbances modify natural disturbance regimes and presumably change the resilience of populations of organisms adapted to them. It follows that superimposing anthropogenic disturbance regimes over natural disturbance regimes (see Norse 1990, Watling and Norse 1998) increases stress and degradation of "natural" biotic assemblages along gradients of change, rather than beyond particular thresholds.

Experimental management programs can be designed for optimum conservation, knowledge generation, and economic and cultural sustainability, even before decision makers adopt appropriate logic and burden-of-proof regarding fishery and marine resources. In their review, McAllister and Peterman (1992) argued for the implementation of formal decision analysis, which we provide herein. Huppert (1996) discussed how risk assessment, formal decision models, and risk communication through collaboration can lead to political acceptance of appropriate precautionary measures.

Prior to these authors, Walters (1986) suggested that the management objectives learning and risk-aversion (conservation) are in conflict and therefore should be balanced, in his discussion of adaptive management and feedback policy design. He argued for probing, active adaptive management (experimentation) by employing ranges of conservative through aggressive exploitation to maximize learning (decrease uncertainty) in a feedback policy design framework (also see Sainsbury 1991). By actively probing, managers accept risks to a resource on behalf of the public in order to maximize learning to better achieve management objectives in the future. Walters again (1997) implied an obligate trade-off between knowledge gathering and conservation, based on the belief that "... manipulative experiments always increase at least some ecological risks...."

We diverge from Walters on this point. Learning is sometimes not in conflict with conservation objectives (Okey, in prep.). Highly informative and powerful experiments can be designed using protective treatments (without using treatments that increase extraction or disturbance). This can lead to effective knowledge gathering as well as minimization, even elimination, of ecological risks. A good example is the use of trawl closures, or spatial reserves, in Australia within an experimental context (Smith 1996). An important role of resource managers is to identify situations wherein probing on the aggressive (stressing) end of the experimental continuum is not necessary or appropriate for achieving learning and conservation goals.

The purposes of this paper are to (1) propose general systematic criteria for choosing an appropriate type of experimental management regime by identifying trade-offs, decision points, and risks of experimental actions, (2) present an example of an experimental management design for Bering Sea bottom trawling, and (3) present an approach for evaluating the effects of trawling within an ecosystem, or "interaction web," context (sensu Wootton 1994, Menge 1995).

Methods

We used a working framework to develop criteria for choosing the appropriate type of experimental management for optimizing conservation, knowledge, and sustainability in a given situation. This framework defines experimental and ecosystem-based management, and distinguishes between precautionary and unconstrained experimental management.

Working Framework for Experimental Management

Experimental Management

"Experimental management" means the experimental study of the effects of human activities on an ecosystem (resources use) and the effects of different management regimes on resource use and the ecosystem. Adaptive management entails adaptive modification of management through feedback from monitoring of environmental trends associated with the management, but it does not necessarily denote experimental manipulation, controls, replication, and a priori hypotheses. (Note: Walters [1986] made this distinction using the terms "actively adaptive" and "passively adaptive," but the more explicit "experimental" is preferred here; Walters and Green (1997) have now focused on the logical merits of "deliberate experimentation.") Experimental management is more likely to elucidate causative interactions than adaptive management, though both have the potential to address the values and goals of society through conservation, knowledge-generation, and economic and cultural sustainability.

Ecosystem-Based Management

Ecosystem-based management is the management of human activities (sensu Lotspeich 1998) based on the default assumption that human actions influence all components of a naturally dynamic ecosystem regardless of whether these influences are intentional or deemed biologically significant. This approach departs from single-species fisheries management where concern is focused on the effects of particular fishing regimes on the status and trends of the targeted population or species complex. The effects of human activities are likely to ripple beyond the targeted species, changing ecosystem services elsewhere in the system. In ecosystem-based management, the indirect effects of human activities are explored and considered when making decisions about human interactions with public resources. The need for ecosystem-based management assumes that societal values and goals can be met only by considering these indirect effects; and that more narrowly focused resource exploitation is likely to degrade ecosystems and their services in unpredictable ways.

Precautionary Experimental Management vs. Unconstrained Experimental Management

Experimental management consists of actions, but since these are often actions on preexisting actions (e.g., taking action to change the character of the act of fishing) they can result in either more or less human interaction with an ecosystem. Experimental treatments themselves can be destructive (De Wreede 1985, Foster and Sousa 1985), underscoring the conflict between knowledge gathering and conservation (as well as goals of economic and cultural sustainability). Destructive treatments can be referred to as "stress treatments" in experimental fisheries management. In "unconstrained experimental management," the ecological strain, degradation, or risks associated with stress treatments are deemed acceptable because the knowledge gained provides an increased potential for management success and stewardship in the long term. In "precautionary experimental management," however, stress treatments are deemed inappropriate because the degradation, strain, or risks, are deemed to outweigh their knowledge gathering benefits. A design without stress treatments might provide less knowledge in the short term but a high likelihood of conservation benefits.

In fisheries, there is sometimes no conflict between information-gathering and conservation goals because experimental treatments that limit or remove fishing activities from plots and leave fishing practices unchanged in other areas are left with at least two treatments (fishing and fishing removed). Such an experiment provides both knowledge and conservation benefits, if they occur on appropriate scales with sufficient replication and measurement. Increased replication is a main reason that the notion of marine protected areas is compelling in an experimental context, in addition to the need for ecological baseline controls. Walters (1997) points out that some sensitive species have become reliant on modified habitats and thus adversely affected by restoration, but such reliance is probably rare in marine settings proper.

The alternative situation is one in which conservation goals are in conflict with knowledge-gathering. These are experiments in which damaging human activities are increased or introduced into new places or times to gain information about ecosystem function. Stress treatments (i.e., destructive treatments) are common in marine ecological studies, but these manipulations are undertaken on scales much smaller than system-wide. In these cases, researchers and permitting agencies deem their adverse effects to be small enough and ephemeral enough that the knowledge-gathering benefits outweigh any degradation caused by the experiment (Pers. comm., Mike Foster, Moss Landing Marine Laboratories, Moss Landing, CA 95039, Nov. 1996).

On a fisheries-wide scale, however, such unconstrained experimental management has the potential to do irreversible damage to a system, as in the case of endangered marine species like Steller sea lions. The decision criteria proposed in these guidelines will help managers design programs for optimal knowledge-gathering while preventing degradation of ecosystem integrity and helping to spare such species from the plight of right whales (remnant populations), white abalone (near extinction), Steller's sea cows (extinct), and great auks (extinct).

Decision Criteria for Experimental Management

The following decision criteria provide a systematic approach for managers and the wider public to implement ecosystem-based, experimental management (precautionary vs. unconstrained). The criteria are intended to guide decisions regarding experimental management actions. These criteria are qualitative general guidelines for broadly different situations. The least arbitrary way to use them is to develop and specify quantitative threshold values for each criterion based on the characteristics of the particular management situation (Shaeffer and Cox 1992). However, the criteria can also be used in their qualitative form during informal evaluation of management situations. Precautionary experimental management designs should be considered when one or more of the following criteria is true (managers may wish to modify this critical number, e.g., two or more of the criteria):

1. Critical (vs. non-critical) situations: If endangered species or remnant habitats could be stressed.
2. Low relative proportion of remaining resource: If the amount of remaining resource is a small proportion to that of estimated pristine levels.
3. Low degree of knowledge about biotic associations: If the knowledge of biotic associations is inadequate to predict indirect effects on interdependent species.

4. Limited reversibility: If there is a low potential for reversibility of ecological changes caused by management experiments.
5. Low resiliency: If the degree of resiliency in the resource is considered low.

Figure 1 shows a flow diagram of the decision framework for determining the appropriate type of experimental management. In addition, the following principles should be used when implementing ecosystem-based, experimental management programs:

1. Experimental management should be designed for simultaneous accomplishment of conservation, management, and scientific goals whenever possible.
2. Experimentation should focus on the effects of human activities since we have management control over human activities only.
3. The most positive conservation benefits of the experimental management should be directed towards the most likely explanation of (hypothesis for) degradation.
4. Manager/designer should assume the existence of causative relationships between ecosystem degradation and all reasonable hypotheses, until the hypotheses are removed from the conceptual model through rigorous testing.

An Example: Effects of Bottom Trawling on the Bering Sea Ecosystem

Setting

The predominant bottom types in the Bering Sea are sand, mud, and cobble, though rocky substrate exists along the Aleutian chain (McMurry et al. 1984). The presence or absence of biogenic substrate, such as filamentous bryozoans, erect colonial ascidians, erect sponges, and tubicolous polychaetes, is also a feature of the Bering Sea floor. These bottom types and oceanographic features create complex and dynamic mosaics of physical habitats. The biotic assemblage in a particular setting is not only influenced, and determined, by physical habitat, but the organisms also influence a habitat, adding an additional layer of complexity to a complex mosaic of marine habitats. "Biogenic structure" is the sum of the physical and chemical habitat structure created by organisms. In this example we mean the structural attributes of sessile, attached, benthic invertebrates.

The habitat structure created by organisms is easiest to picture with sessile benthic organisms like gorgonian corals, sponges, bryozoans, tunicates, or sea pens. Removal of biota changes the habitat for other populations. Many organisms are adapted to the habitats provided and modified by particular organisms, but such biotic associations have been described in only a few cases in the Bering Sea. Habitat is modified by many mobile organisms such as gelatinous plankton, sunken drift vegetation, crabs and sea stars, burrowing clams and rays, and walruses and whales. In turn, these mobile biota are directly or indirectly dependent on the habitat that sessile benthic organisms help create.

Management experiments designed to understand the effects of trawling on the Bering Sea floor must account for natural disturbance regimes. It could be predicted that regular trawling in a naturally undisturbed setting would cause a shift from a species-rich assemblage including long-lived and large-bodied species, sessile species, and complex biogenic structure, to a species-poor assemblage of abundant, small-bodied, mobile opportunists, where less biogenic structure exists. Removal of bottom trawling in such a habitat may enable recovery to the species-rich state, depending on long-term habitat modification, species interactions, recruitment, and neighborhood effects. Nevertheless, opinions about recovery are mere speculation until trawling is experimentally excluded from areas of the Bering Sea.

History of Bering Sea Trawling

Trawling has removed over a million metric tons of biomass, on average, from the eastern Bering Sea system every year since the 1960s. Foreign vessels dominated the fisheries from the 1950s through the 1970s, until the Magnuson Fisheries Conservation and Management Act in 1976 phased them out and "Americanized" the fishery (Bakkala 1993). The extensive distribution of bottom trawling effort can be inferred from catch-per-unit-of-effort (CPUE) geographical information system maps in Fritz et al. (1998). Other efforts are currently under way to reconstruct historical patterns of trawling and to begin unraveling the effects of trawling in the Bering Sea and the Gulf of Alaska (reviewed in NPFMC 1998).

Applying the Criteria

A range of management (conservation) problems exists in the Bering Sea, from declining populations of endangered species, such as Steller sea lions, to concerns about essential fish habitat degradation. Thus, step one of our criteria, recognition of the problem (Fig. 1), is often clear to managers, albeit not to all stakeholders. The Steller sea lion issue has a single species focus, yet is clearly connected to broader-ecosystem dynamics (humans included), and would be an appropriate example to use here. However, a multispecies attribute of the system is even more illustrative of an ecosystem approach. Thus we use "biogenic structure" as the assessment endpoint to which we will apply the five criteria in Fig. 1 to determine the appropriate type of experimental management design.

In knowledge-poor situations, the five criteria should be applied qualitatively by managers using a weight-of-evidence approach, but some situations will allow development of quantitative threshold values for some

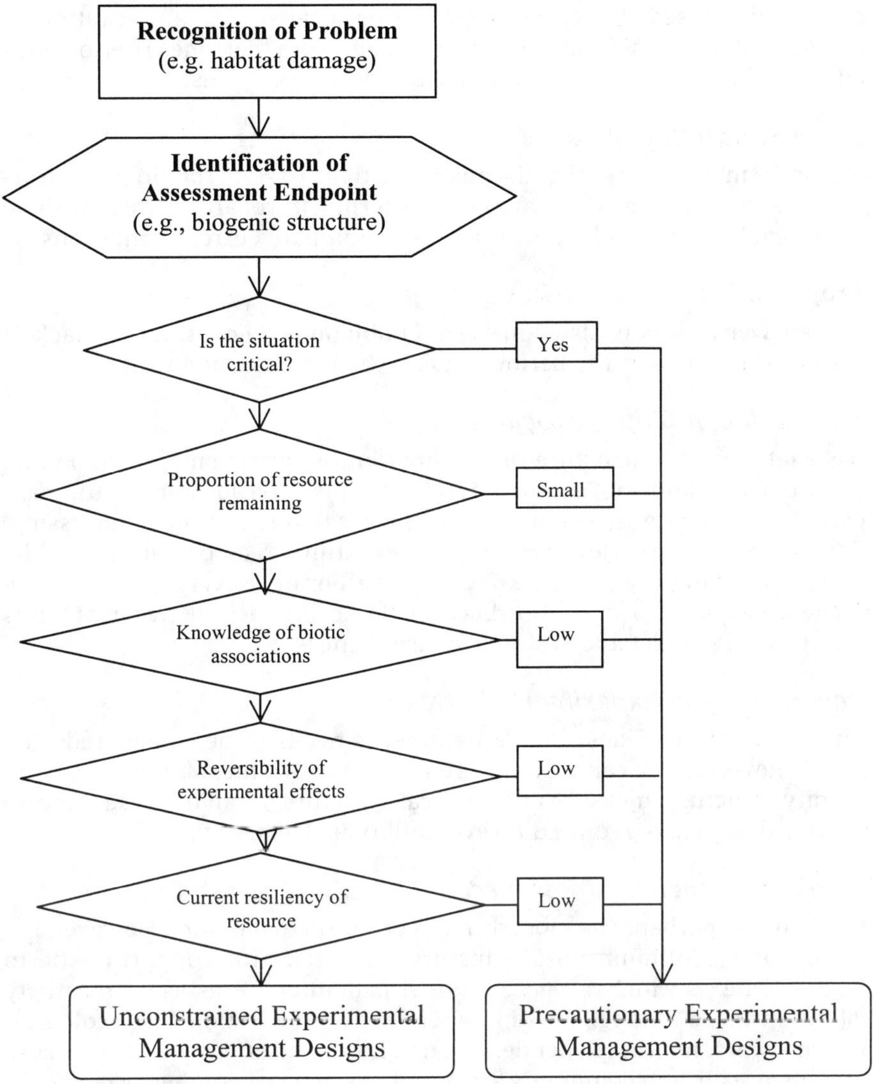

Figure 1. Decision diagram for determining appropriate type of experimental management design—unconstrained experimental management designs in which treatments that increase extraction or anthropogenic disturbance are allowed vs. precautionary experimental management designs in which only treatments with presumed conservation benefits are allowed. The five criteria in the diagram are explained in the text.

criteria (discussed above). Any of the criteria can lead to a precautionary design, but having affirmative answers from more than one criterion adds additional weight to the decision for a precautionary design.

Is the Situation Critical?

Managers may consider that the answer to this question for biogenic structure is ambiguous, as there are no long-term control areas (areas with no historical bottom trawling) with which to compare current conditions.

Proportion of Resource Remaining

The answer to this is also somewhat ambiguous, because of the lack of historical controls in the Bering Sea and the limited studies there.

Knowledge of Biotic Associations

Associations with structure-producing benthic organisms in the Bering Sea are poorly known. The unequivocal low answer leads to precautionary experimental designs, and away from the use of unconstrained (stressing) designs. Some may view this as counter-intuitive, as precautionary designs can hinder the pace of knowledge-gathering. Nevertheless, low answers should lead to precaution because the scale of destructive treatments may have unpredictable systemic consequences.

Reversibility of Experimental Effects

The reversibility of destructive treatments needs to be considered (Roy 1996). Reversibility can be measured in terms of time, abundance, community structure, measures of physical structure, or all of these. Precautionary designs are required if reversibility is deemed low.

Current Resiliency of Resource

Resiliency is perhaps the most difficult criterion to measure. However, the notion of a continuum of life history strategies from opportunistic to "stable-state" is widely accepted. Notwithstanding intraspecies plasticity (shifts in these strategies within species) and the notion that ecological resiliency is somewhat dependent on neighborhood effects (larval sources), organisms can, and commonly are, rated based on where they occur along this "r-K" continuum. We suggest that the resiliency of a species or assemblage can be judged based on these considerations—opportunistic assemblages are generally more resilient to destructive experiments. The most direct way to determine community resiliency is to monitor the trajectory of recovery in replicated, large-scale closed areas.

In cases of ambiguity, as in criteria one and two in this example, proper placement of the burden-of-proof leads to a precautionary experimental design. In addition to leading us to precautionary experimental management designs, application of these criteria to our "biogenic structure" example has repeatedly led us to the need for an experimental design that includes closed areas, i.e., marine protected areas (MPAs). Establishment of MPAs, in an experimental context, is the best way to address four out of five of the questions posed by the criteria. Furthermore, no-take MPAs are necessary to address whole-ecosystem questions in an experimental context. To design an experimental management regime for "biogenic structure" using MPAs in the Bering Sea, it is necessary to identify and describe the existing protected areas (Fig. 2).

Inspection of Fig. 2 reveals four areas in the Bering Sea in which trawling is closed year-round (but the Steller sea lion protection area includes many small trawl exclusion zones around rookeries)(also see map in NPFMC 1996), and there are five additional "areas" that are seasonally closed to bottom trawling. Other types of fishing can occur in trawl exclusion zones. Currently, there are no areas in the Bering Sea designated as year-round "no take" MPAs. In this paper, we suggest that establishment of no take MPSs in an experimental context is needed to help reveal the effects of fishing in general, as well as the status of biogenic structure in the Bering Sea. We also point out that spatially explicit analytical tools are now available to help develop optimal experimental management designs.

Discussion

We agree with Walters (1986) that inclusion of full ranges of treatments in management experiments will result in the maximum knowledge gain. We also agree that such knowledge can help enable future success in fisheries management, including ecosystem protection. However, we explicitly distinguish protective treatments from stress treatments and suggest that it is sometimes inappropriate to include stress treatments in experimental management designs, in fisheries, primarily because experiments are conducted on a large enough scale to do considerable damage to a species, or an ecosystem. Indeed, it is the large scale of management experiments in fisheries that enables a high potential for learning, but the lack of adequate or true replication is a common shortcoming of fisheries management experiments that combines with other constraints to prevent the detection of real trends and processes in marine systems, in which dynamic changes are shaped by multiple factors.

Highly informative management experiments can be designed that have protective conservation effects only, and are truly replicated, through the use of management tools like marine protected areas. Some proponents of marine reserves focus on their "hedging" function as a way to address uncertainty (Fujita et al. 1997). They are careful to make the distinction between minimizing uncertainty through expensive data collection and hedging against uncertainty simply by setting areas aside. The emphasis of this argument is not on knowledge-gathering, but rather, on avoiding the need to understand the details of nature and exhaust the elusive dollars available to fisheries science (also see Ludwig et al. 1993, Lauck 1996, Walters 1998b, Lauck et al. 1998, Sumaila 1998). The underlying assump-

Nearshore Bristol Bay Closure Area: (1996) prohibits all trawling at all times within the area east of 162°W longitude with the exception of an area bounded by 159° to 160°W and 58°43′N that would remain open April 1-June 15. Closure established to protect juvenile red king crab habitat.

Chum Salmon Savings Area: (1994) closed to all trawling August 1-31 with provisional extension to October 5 to reduce bycatch of chum salmon in groundfish trawls.

Bristol Bay Red King Crab Area: (1995) closed seasonally to non-pelagic trawling to protect red king crab population and habitat.

Pribilof Islands Habitat Conservation Area: (1994) closed year-round to all trawling to protect and rebuild blue king crab stocks, marine mammals, seabirds, and prey species.

Crab Protection Zones: (1987) closed to prevent incidental catch of crab and juvenile halibut.

Zone A closed to trawling year-round
Zone B closed to trawling March 15-June 15

Walrus Protection Areas: (1992) closed to all fishing April 1-September 30 to minimize vessel interaction with walrus around haulout sites. (1999) Year-round trawl exclusion 10 nautical miles from haulout sites. Larger year-round pollock trawl exclusion zone.

Steller Sea Lion Protection Areas: (1992 and 1999) combination of 10 nautical mile, year-round all trawl exclustion zones, larger year-round pollock trawl exclusion zones, and other seasonal trawl exclusion zones. Established to reduce disturbance of Steller sea lions (now endangered) around known rookeries.

Herring Savings Areas: (1991) closed to all trawling when trigger reached to control bycatch of herring in groundfish fisheries.

Summer Area closed June 15-July 1
Summer Area 2 closed July 1-August 15
Winter Area closed September 1-March 1

Figure 2. The areas closed in the Bering Sea (NPFMC 1996).

tion of this approach is found in our working definition and guidelines—humans can control human activities (sometimes), but they cannot control nature. This argument also assumes that constraints on our knowledge of ocean processes are real, or expensive, or really expensive.

We agree with this argument for MPAs, but we suggest that scientists and managers will not have purchased conservation at the cost of knowledge. Incorporation of marine reserves in experimental management designs will likely increase knowledge for reasons of replication and, potentially, ecological baseline—a critically important, and often missing, element in marine science. Furthermore, myriad questions about the structure and function of marine ecosystems can be addressed with a combination of precautionary experimental management, using marine reserves, and new whole-ecosystem analytical techniques like Walters' (1998a) Ecospace simulation routine, which utilizes balanced trophic models of whole ecosystems. Such combinations of new analytical techniques and experimental programs can allow optimal learning (CCAMLR 1985, Walters 1986, Walters and Collie 1989, Sainsbury 1991, Wootton 1994), as well as optimal ecosystem protection.

Walters and Green (1997) recently developed a systematic valuation approach to treatment choices in experimental management designs. Their accounting framework is designed to provide a widely useful approach to ". . . improve management performance" by making it easier for stakeholders to reach consensus on experimental management options, but these authors repeatedly point out the glaring problem of uncertainty in long-term valuation trajectories. The approach we outline in the present paper is based on general ecological criteria and the premise that the absolute value of an ecosystem component inherently exceeds its quantifiable aspects (due to the practical impossibility of knowing everything). Ours is a far simpler approach than that of Walters and Green (1997). Our basic criteria approach is useful for choosing one of the two categories of experimental management designs we outline. We suggest the Walters and Green (1997) valuation approach be reviewed once a basic design category has been chosen, but decision makers should be alert that such approaches attempt to encompass only certain values in ecosystems.

Quantitative threshold values, which we advise managers to develop for the criteria we present, should be viewed as decision making conveniences rather than having ecological meaning beyond revelation of detectability thresholds. In reality, the effects of environmental stressors (anthropogenic or natural) are cumulative and, although responses to stressors can be non-linear and interactive (Underwood 1989), they contribute to complex and dynamic gradients of system degradation and integrity rather than manifesting real threshold effects. The notion of cumulative effects gradients has been discussed for decades (Woodwell 1975), but it is only now replacing the often-delusional "threshold effects" paradigm embraced by classical fisheries science and toxicology (see Roberts 1997, Carls et al. 1999, Rice 1999). Therefore, although the goal here is to develop ecologically based criteria for determining experimental management designs, it should be recognized that social and economic factors will contribute to the development of the criteria, and to any "threshold values" designed to operationalize those criteria (Shaeffer and Cox 1992, and others).

Unconstrained experimental management programs have been undertaken on species of known sensitivity to overfishing and known degraded status such as Pacific ocean perch (POP; *Sebastes alutus*). These stocks were intentionally overfished with participation by the commercial fishing industry for a specified period to gain information on stock dynamics and population estimates (Walters and Collie 1989, Leaman and Stanley 1993). Results of this experiment were ambiguous for several reasons that typify some of the constraints of large-scale fishery experiments: (1) the measurement endpoint species (POP) would not be expected to respond quickly relative to the length of the experiment; (2) sampling was somewhat inconsistent, as data collection was partially allocated to commercial fishermen; (3) sampling was not rigorous enough to track changes in POP stocks; (4) replication was low and unequal (one closed area, three quota areas, and three free fishing areas); and (5) treatments were not spatially independent of one another.

This particular experiment would be unlikely to pass the currently proposed criteria for unconstrained experimental management, as it was a depleted and declining stock that was likely below 20% of virgin spawning biomass, and presumably had "unnaturally" low resiliency to additional population stress (e.g., overfishing) (Fujita et al. 1997). Within our framework, experimental programs for this species would be constructed within a precautionary experimental management framework wherein treatments would be limited to those that limit or remove fishing effort rather than adding effort. In this case, this would have meant a simpler, and perhaps more statistically revealing, experimental management design with only two treatments (closed areas, and status quo areas) and higher replication of experimental units. Indeed, when such designs are implemented with marine reserves, they have revealed dramatic impacts of MPAs around the world (e.g., Rowley 1994; Paddack 1996; Russ and Alcala 1996a,b).

Many authors are now calling for establishment of MPAs as a means to achieve ecosystem-based management, usually in a context of rigorous experimentation (Edyvane 1993, Clark 1996, Peterson 1996, Walters et al. 1998). Indeed, tremendous interest in marine protected areas is being generated within the scientific community (Murray et al., In prep.). The establishment of MPAs can serve numerous purposes including experimental treatments and learning. Many other authors have called for fishery managers to move away from a strict fishery production approach towards a habitat-based, ecosystem approach (Wilson et al. 1994, Jones et al. 1996).

The National Marine Fisheries Service and the Federal Fisheries Management Councils in the United States are now required to identify, describe, and protect essential fish habitats in federally managed waters. Many other countries are initiating similar habitat-based approaches to sustaining marine resources. The dilemma of defining and identifying essential fish habitat with extraordinarily limited information about these ecosystems threatened to render the protective goals meaningless, as some resource users advocated policies that incorrectly placed the burden-of-proof for the need for habitat protection on this limited scientific knowledge (see Dayton 1998). One logical irony of such policies is that they decrease opportunities for learning about marine ecosystems, as they continue to be degraded by fishing in ways that are difficult to elucidate without MPAs.

Auster and his colleagues (1997) re-framed the debate with a decision framework in which existing knowledge, however limited, is used with the precautionary approach to designate MPAs, from which new knowledge is generated to modify and refine management actions. These authors correctly stepped back from the discussion about the collection of scientific information for achieving mandated habitat protection, while formulating a management regime that achieves both simultaneously.

The present contribution elaborates on the logical process for determining the appropriate type of experimental management regime for gathering information that is ultimately needed for improving management, similarly pointing out that ecosystem knowledge gathering occurs simultaneously with resource conservation when experimental treatments are protective. The distinction between "unconstrained experimental management" and "precautionary experimental management" will become more important as ecologically sensitive situations become more conspicuous throughout the world's oceans.

Ultimately, the information used for marine resource management must come from sampling and experimentation in the real world. However, ecosystem modeling has reached milestones of usefulness such that considerable learning and some functional predictability is available. We suggest that ecosystem modeling should accompany management experiments throughout their development and implementation to best understand resulting biological changes in an ecosystem context.

Whole-ecosystem, trophic flow models (Polovina 1985, Christensen and Pauly 1992), such as Ecopath, have been constructed all over the world (Christensen and Pauly 1993), including Prince William Sound (Okey and Pauly 1999) and the Bering Sea (Trites et al. 1999). Indeed, several authors have promoted the integration of experimental management with ecosystem modeling to optimally study and manage particular marine settings (CCAMLR 1985, Walters and Collie 1989, Sainsbury 1991), and Walters (1986) presented a framework for such integration. New dynamic temporal and spatial simulation routines—Ecosim and Ecospace—are examples of tools that enable informed policy planning, ecosystem-based analyses of management experiments, and hypothesis formulation (Walters et al. 1997, 1998; Walters 1998a). Ultimately, such empirically based ecosystem models of may help us unravel questions about resilience and stability of ecosystems (Holling 1973, Pimm 1984, 1991, Vasconcellos 1997) as well as helping us develop indices of biotic integrity (Karr 1981) for whole ecosystems (Okey, In prep).

Acknowledgments

We thank L. Thompson, C. Walters, P. Auster, T. Loher, and B. Callahan for helpful comments on the text, and D. Pauly for helpful conversations.

References

Ackley, D., C. Blackburn, K. Brix, D. Clausen, J. DiCosimo, R. Ferrero, L. Fritz, A. Grossman, J. Heifetz, P. Livingston, L.L. Low, R. Merrick, E. Varosi, and D. Witherell. 1994. Ecosystem considerations 1995. Plan teams for the groundfish fisheries of the Bering Sea, Aleutian Islands, and Gulf of Alaska. North Pacific Fishery Management Council, Anchorage, Alaska.

Apollonio, S. 1994. The use of ecosystem characteristics in fisheries management. Rev. Fish. Sci. 2(2):157-180

Auster, P.J. 1998. A conceptual model of the impacts of fishing gear on the integrity of fish habitats. Cons. Biol. 12(6):1198-1203.

Auster, P.J., and R.W. Langton. 1999. The effects of fishing on fish habitat. In: L. Benaka (ed.), Fish habitat: Essential fish habitat (EFH) and rehabilitation. American Fisheries Society, Bethesda, Maryland, pp. 150-187,

Auster, P.J., L. Watling, and A. Reiser. 1997. Comment: the interface between fisheries research and habitat management. N. Am. J. Fish. Manage. 17:591-595.

Auster, P.J., R.J. Malatesta, R.W. Langton, L. Watling, P.C. Valentine, C.L.S. Donaldson, E.W. Langton, A.N. Shepard and I.G. Babb. 1996. The impacts of mobile fishing gear on seafloor habitats in the Gulf of Maine (Northwest Atlantic): Implications for conservation of fish populations. Reviews in Fisheries Science 4(2):185-202.

Bakkala, R.G. 1993. Structure and historic changes in the groundfish complex of the eastern Bering Sea. NOAA Tech. Rep. NMFS 114. Springfield, Virginia.

Carls, M.G., S.D. Rice, and J.E. Hose. 1999. Sensitivity of fish embryos to weathered crude oil: part I. low-level exposure during incubation causes malformations, genetic damage, and mortality in larval Pacific Herring (*Clupea pallasi*). Env. Tox. and Chem. 18(3):481-493.

CCAMLR. 1985. Ecosystem management: Proposal for undertaking a coordinated fishing and research experiment at selected sites around Antarctica. In: Selected papers presented to the Scientific Committee of CCAMLR, 1982-1984. Part 2. Commission Conserv. of Antarctic Marine Living Resources, Hobart, Australia, 1985, pp. 277-284.

Christensen, V., and D. Pauly. 1992. Ecopath II: A system for balancing steady-state ecosystem models and calculating network characteristics. Ecol. Modell. 61:169-185.

Christensen, V., and D. Pauly (eds.). 1993. Trophic models of aquatic ecosystems. International Centre for Living Aquatic Resources Management, ICLARM Conf. Proc. 26. Manila, Philippines. 390 pp.

Clark, C.W. 1996. Marine reserves and the precautionary management of fisheries. Ecol. Appl. 6:369-370.

Dayton, P.K. 1998. Reversal of the burden of proof in fisheries management. Science 279(5352):821-822.

Dayton, P.K., S.F. Thrush, M.T. Agardy, and R.J. Hofman. 1995. Viewpoint: Environmental effects of marine fishing. Aquat. Conserv.: Marine and Freshwater Ecosystems 5:205-232.

De Wreede, R.E. 1985. Destructive (harvest) sampling. In: M.M. Littler and D.S. Littler (eds.). Handbook of phycological methods, ecological field methods: Macroalgae. Cambridge University Press, Cambridge, pp. 147-160.

Edyvane, K.S. 1993. An ecosystem-based approach to marine fisheries management. In: D.A. Hancock (ed.), Sustainable fisheries through sustaining fish habitat. Australian Government Publishing Serv., Canberra, Australia.

Engel, J. and R. Kvitek. 1998. Effects of otter trawling on a benthic community in Monterey Bay National Marine Sanctuary. Cons. Biol. 12(6):1204-1214.

FAO (Food and Agriculture Organization of the United Nations). 1995. Code of conduct for responsible fisheries. FAO, Rome. 41 pp.

Foster, M.S. and W.P. Sousa. 1985. Succession. In: M.M. Littler and D.S. Littler (eds.). Handbook of phycological methods, ecological field methods: Macroalgae. Cambridge University Press, Cambridge, pp. 269-290.

Fritz, L.W., R.C. Ferrero, and R.J. Berg. 1995. The threatened status of Steller sea lions, *Eumetopias jubatus*, under the Endangered Species Act: Effects on Alaska groundfish fisheries management. Mar. Fish. Rev. 57(2):14-27.

Fritz, L.W., A. Greig, and R.F. Reuter. 1998. Catch-per-unit-effort, length, and depth distributions of major groundfish and bycatch species in the Bering Sea, Aleutian Islands, and Gulf of Alaska regions based on groundfish fishery observer data. NOAA Tech Memo. NMFS-AFSC-88. 179 pp.

Fujita, R.M., T. Foran, and K. Honey. 1997. Can no-take marine reserves help rebuild the Pacific coast groundfish fishery? Environmental Defense Fund, 5655 College Ave., Suite 304, Oakland, California.

Garcia, S.M. 1996. The precautionary approach to fisheries and its implications for fishery research, technology and management: An updated review. Precautionary approach to fisheries. Part 2: Scientific Papers. FAO Fish. Tech. Pap. 350(2):1-75.

Holling, C.S. 1973. Resilience and stability of ecological systems. Ann. Rev. Ecol. Syst. 4:1-23.

Hollowed, A.B., B. Brown, J. Ianelli, P. Livingston, B. Megrey, and C. Wilson. 1997. Walleye pollock: Appendix 1. Stock synthesis model incorporating predation mortality. In: Stock assessment and fishery evaluation report for the groundfish resources of the Gulf of Alaska, as projected for 1998. North Pacific Fishery Management Council, P.O. Box 103136, Anchorage, Alaska 99510.

Huppert, D.D. 1996. Risk assessment, economics, and precautionary fishery management. Precautionary approach to fisheries. Part 2: Scientific Papers. FAO Fish. Tech. Pap. 350(2):103-127.

Jones, M.L., R.G. Randall, D. Hayes, W. Dunlop, J. Imhof, G. Lacroix, and N.J.R. Ward. 1996. Assessing the ecological effects of habitat change: Moving beyond productive capacity. In: J.R.M. Kelso (ed.), Workshop on the Science and Management for Habitat Conservation and Restoration Strategies in the Great Lakes. National Research Council of Canada 53(Suppl. 53):446-457.

Kaiser, M.J. 1998. Significance of bottom-fishing disturbance. Cons. Biol. 12(6):1230-1235.

Karr J.R. 1981. Assessment of biotic integrity using fish communities. Fisheries 6(6):21-27.

Lauck, T. 1996. Uncertainty in fisheries management. In: D.V. Gordon and G.R. Munro (eds.), Fisheries and uncertainty: A precautionary approach to resource management, University of Calgary Press, Calgary, Canada, pp. 91-106.

Lauck, T., C.W. Clark, M. Mangel, and G.R. Munro. 1998. Implementing the precautionary principle in fisheries management through marine reserves. Ecol. Appl. Suppl. 8(1):S72-S78.

Leaman, B.M., and R.D. Stanley. 1993. Experimental management programs for two rockfish stocks off British Columbia, Canada. In: J. Smith, J.J. Hunt, and D. Rivard (eds.), Workshop on Risk Evaluation and Biological Reference Points for Fisheries Management, Halifax, Nova Scotia, 19 Nov. 1991. National Research Council of Canada 120:403-418.

Lenihan, H.S. and C.H. Peterson. 1998. How habitat degradation through fishery disturbance enhances impacts of hypoxia on oyster reefs. Ecol. Appl. 8(1):128-140.

Loher, T., P.S. Hill, G. Harrington, and E. Cassano. 1995. Management of Bristol Bay red king crab: A critical intersections approach to fisheries management. University of Washington, School of Marine Affairs/Fisheries Research Institute, SMA/FRI-UW-9510. Seattle.

Lotspeich, F.B. 1998. The blue planet: Mankind's habitat. Bull. Ecol. Soc. Am. 79(3):222-225.

Ludwig, D., R. Hilborn, and C.J. Walters. 1993. Uncertainty, resource exploitation, and conservation: Lessons from history. Science 260:17, 36.

McAllister, M.K., and R.M. Peterman. 1992. Experimental design in the management of fisheries: A review. N. Am. J. Fish. Manage. 12(1):1-18.

McMurray, G., A.H. Vogel, P.A. Fishman, D.A. Armstrong, and S.C. Jewett. 1984. Distribution of larval and juvenile red king crab (*Paralithodes camtschatica*) in Bristol Bay. NOAA, OCSEAP Final Rep. 53(1986):267-477.

Menge, B.A. 1995. Indirect effects in marine rocky intertidal interaction webs: Patterns and importance. Ecol. Monogr. 65(1):21-74.

Murray, S.N., R.F. Ambrose, J.A. Bohnsack, L.W. Botsford, M.H. Carr, G.E. Davis, P.K. Dayton, D. Gotshall, D.R. Funderson, M.A. Hixon, J. Lubchenco, M. Mangel, A. MacCall, D.A. McArdle, J.C. Ogden, J. Roughgarden, R.M. Starr, M.J. Tegner, and M.M. Yoklavich. In preparation. No-take reserve networks: Protection for fishery populations and marine ecosystems. Submitted to Fisheries.

NMFS (National Marine Fisheries Service). 1999. Ecosystem-based fishery management: A report to Congress by the National Marine Fisheries Service Ecosystem Principles Advisory Panel. NMFS, Silver Spring, Maryland, 164 pp. Available at: www.nmfs.gov/sfa/EPArpt.pdf

NMFS. 1999. Fisheries of the exclusive economic zone off Alaska: Steller sea lion protection measures for the pollock fisheries off Alaska. United States Federal Register 64(14):3437-3446, 22 January, 1999.

Norse, E. 1990. Ancient forests of the Pacific Northwest. Island Press, Washington, D.C. 327 pp.

NPFMC (North Pacific Fishery Management Council). 1996. Regulatory and closure areas for the groundfish fisheries in the Bering Sea and Aleutian Islands. North Pacific Fishery Management Council, Anchorage, Alaska.

NPFMC. 1998. Draft Environmental Assessment/Regulatory Impact Review for Amendment 55 to the FMP for BSAI Groundfish, Amendment 55 to the FMP for GOA Groundfish, Amendment 8 to the FMP for BSAI Commercial King and Tanner Crab, Amendment 5 to the FMP for Scallops, and Amendment 5 to the FMP for Salmon Fisheries in the EEZ off the Coast of Alaska: Essential Fish Habitat. North Pacific Fishery Management Council, March 24, 1998, Anchorage, Alaska.

NRC (National Research Council). 1994. Improving the management of U.S. marine fisheries. National Academy Press, Washington, D.C.

NRC. 1996. The Bering Sea ecosystem. National Academy Press, Washington, D.C. 307 pp.

NRC. 1998. Sustaining marine fisheries. National Academy Press, Washington, D.C. 164 pp.

Okey, T.A. In prep. A formulation of biotic integrity for whole ecosystems with a fisheries management example. Will be submitted to Ecol. Appl.

Okey, T.A., and D. Pauly. 1999. A mass-balanced model of trophic flows in Prince William Sound: Decompartmentalizing ecosystem knowledge. In: Ecosystem approaches for fisheries management. University of Alaska Sea Grant, AK-SG-99-01, Fairbanks. (This volume.)

Paddack, M.J. 1996. The influence of marine reserves upon rockfish populations in central California kelp forests. M.S. thesis, University of California, Santa Cruz.

Peterson, C.H. 1996. Ecological research for a sustainable Northeast Shelf ecosystem and biosphere. In: K. Sherman, N.A. Jaworski, and T.J. Smayda (eds.), The Northeast Shelf ecosystem: Assessment, sustainability, and management. Blackwell-Science, Cambridge, Massachusetts, pp. 31-46.

Pickett, S.T.A., and P.S. White. 1985. Patch dynamics: A synthesis. In: S.T.A. Pickett and P.S. White (eds.), The ecology of natural disturbance and patch dynamics. Academic Press, New York, pp. 371-382.

Pimm, S.L. 1981. The balance of nature?: Ecological issues in the conservation of species and communities. University of Chicago Press, Chicago. 434 pp.

Pimm, S.L. 1984. The complexity and stability of ecosystems. Nature 307:321-326.

Polovina, J.J. 1985. An approach to estimating an ecosystem box model. Fish. Bull., U.S. 83(3):457-560.

Rice, S.D. 1999. Lessons learned on the long-term toxicity of oil to fish: Intersection of chance, oil, biology, toxicology, and science. In: Legacy of an oil spill: Ten years after Exxon Valdez, abstracts. March 23-26, 1999. Exxon Valdez Oil Spill Trustee Council, Alaska Department of Fish and Game, Anchorage, Alaska, p. 93.

Roberts, C.M. 1997. Ecological advice for the global fisheries crisis. Trends Ecol. Evol. 12(1):35-38.

Rowley, R.J. 1994. Marine reserves in fisheries management. Aquat. Conserv.: Marine and Freshwater Ecosystems 4(3):233-254.

Roy, N. 1996. What went wrong and what can we learn from it? In: D.V. Gordon and G.R. Munro (eds.), Fisheries and uncertainty: A precautionary approach to resource management. University of Calgary Press, pp. 15-25.

Russ, G.R., and A.C. Alcala. 1996a. Marine reserves: Rates and patterns of recovery and decline of large predatory fish. Ecol. Appl. 6(3):947-961.

Russ, G.R., and A.C. Alcala. 1996b. Do marine reserves export adult fish biomass? Evidence from Apo Island, central Philippines. Mar. Ecol. Prog. Ser. 132(1-3):1-9.

Sainsbury, K.J. 1991. Application of an experimental approach to management of a tropical multispecies fishery with highly uncertain dynamics. In: N. Daan and M.P. Sissenwine (eds.), Multispecies models relevant to management of living resources. ICES Mar. Sci. Symp. 193:301-320.

Schaeffer, D.J. and D.K. Cox. 1992. Establishing ecosystem threshold criteria. In: R. Costanza, B.G. Norton, and B.D. Haskell (eds.), Ecosystem health: New goals for environmental management. Island Press, Washington, D.C., pp.157-169.

Smith, A. 1996. Adaptive fisheries management strategies. In: R. Thackway (ed.), Developing Australia's representative system of marine protected areas. Proceedings of a technical meeting held at the South Australian Aquatic Sciences Centre, West Beach, Adelaide, 22-23 April 1996. Australian Nature Conservation Agency, Canberra. Available on the worldwide web at http://www.erin.gov.au/portfolio/esd/coast_marine/or2000/mpa/Httoc.htm

Smith, T.D. 1995. United States practice and the Bering Sea: Is it consistent with a norm of ecosystem management? Ocean Coast. Law J. 1(2):141-186.

Sousa, W.P. 1984. The role of disturbance in natural communities. Ann. Rev. Ecol. Syst. 15:353-391.

Sumaila, U.R. 1998. In: T. J. Pitcher, D. Pauly and P. Hart (eds.), Reinventing fisheries management, Chapman and Hall Fish and Fisheries Series, Kluwer Academic Publishers, Boston, pp. 303-309.

Thrush, S.F., J.E. Hewitt, V.J. Cummings, P.K. Dayton, M. Cryer, S.J. Turner, G.A. Funnell, R.G. Budd, C.J. Milburn, and M.R. Wilkinson. 1998. Disturbance of the marine benthic habitat by commercial fishing: Impacts at the scale of the fishery. Ecol. Appl. 8(3):866-879.

Trites, A.W., P. Livingston, M.C. Vasconcellos, S. Mackinson, A.M. Springer, and D. Pauly. 1999. Ecosystem considerations and the limitations of ecosystem models in fisheries management: Insights from the Bering sea. In: Ecosystem approaches for fisheries management. University of Alaska Sea Grant, AK-SG-99-01, Fairbanks. (This volume.)

Underwood, A.J. 1989. The analysis of stress in natural populations. Biol. J. Linn. Soc. 37:51-78.

Underwood, A.J. 1990. Experiments in ecology and management: Their logics, functions and interpretations. Aust. J. Ecol. 15:365-389.

United Nations. 1996. Agreement for the Implementation on the Provisions of the United Nations Convention on the Law of the Sea of 10 December 1982 Relating to the Conservation and Management of Straddling Fish Stocks and Highly Migratory Fish Stocks. United Nations General Assembly, New York.

Vasconcellos, M., S. Matkinson, K. Sloman and D. Pauly. 1997. The stability of trophic mass balance models of marine ecosystems: A comparative analysis. Ecol. Modell. 100:125-134.

Walters, C.J. 1986. Adaptive management of marine resources. Macmillan Publishing Company, New York. 374 pp.

Walters, C.J. 1997. Challenges in adaptive management of riparian and coastal ecosystems. Conserv. Ecol. [online]1(2):1. Available from the Internet: http://www.consecol.org/vol1/iss2/art1

Walters, C.J. 1998a. Ecosim and Ecospace: Basic considerations. In: D. Pauly (ed.), Use of Ecopath with Ecosim to evaluate strategies for sustainable exploitation of multi-species resources. University of British Columbia, Fisheries Centre Research Report 6(2):11-14.

Walters, C.J. 1998b. Designing fisheries management systems that do not depend on accurate stock assessments. In: T.J. Pitcher, D. Pauly, and P. Hart (eds.), Reinventing fisheries management. Chapman and Hall Fish and Fisheries Series, Kluwer Academid Publishers, Boston, pp. 280-287.

Walters, C.J., and J.S. Collie. 1989. An experimental strategy for groundfish management in the face of large uncertainty about stock size and production. Can. Spec. Publ. Fish. Aquat. Sci. 108:13-25.

Walters, C.J., and R. Green. 1997. Valuation of experimental management options for ecological systems. J. Wildl. Manage. 61(4):987-1006.

Walters, C.J., V. Christensen, and D. Pauly. 1997. Structuring dynamic models of exploited ecosystems from trophic mass-balance assessments. Rev. Fish Biol. Fish. 7(2):139-172.

Walters, C.J., D. Pauly, and V. Christensen. 1998. Ecospace: Prediction of mesoscale spatial patterns in trophic relationships of exploited ecosystems, with emphasis on the impacts of marine protected areas. Presented at the Annual Science Conference, International Council for the Exploration of the Sea. Session S: Visualization of Spatial (including Survey) Data. Cascais, Portugal. CM 1998/S:4. 20 pp.

Watling, L. and E.A. Norse. 1998. Disturbance of the seabed by mobile fishing gear: A comparison to forest clearcutting. Cons. Biol. 12(6):1180-1197.

Wilson, J.A., J.M. Acheson, M. Metcalfe, and P. Kleban. 1994. Chaos, complexity and community management of fisheries. Marine Policy 18(4):291-305.

Woodwell, G.M. 1975. The threshold problem in ecosystems. In: S.A. Levin (ed.), Ecosystem analysis and prediction. Society for Industrial and Applied Mathematics, Philadelphia, pp. 9-23.

Wootton, J. T. 1994. Predicting direct and indirect effects: An integrated approach using experiments and path analysis. Ecology 75(1):151-165.

Ecosystem Approaches for Fisheries Management
Alaska Sea Grant College Program • AK-SG-99-01, 1999

"Back to the Future": A Method Employing Ecosystem Modeling to Maximize the Sustainable Benefits from Fisheries

Tony J. Pitcher, Nigel Haggan, David Preikshot, and Daniel Pauly
University of British Columbia, Fisheries Centre, Vancouver, British Columbia, Canada

Abstract

Single-species approaches to fishery assessment are not sufficient to avoid the changes to ecosystems caused by fishing. These changes include the replacement of high-trophic-level, high-value species, by low-trophic-level, low-value species. Commercial fisheries practices and ecosystem effects interact, producing serious depletion that is usually detected too late, resulting in fishery collapses and the failure of stocks to rebuild. Multispecies management is required. Whole ecosystem modeling can evaluate the impacts of fishing on nonharvested groups. Ecosystems and their embedded fisheries can now be effectively modeled using the mass balance system of Ecopath, and new routines, Ecosim and Ecospace, which allow simulation of "what if" questions related to changes in time and space by a fishery sector. Beyond the limits of numerical simulation, a new method termed "Back to the Future" draws up ecosystem models of current, past, unexploited, and alternative ecosystems. Archival and archaeological combined with traditional environmental knowledge may be used in the construction of models of past ecosystems. Ecoval comprises a new set of techniques for the valuation of alternative ecosystems and their benefits to society. In natural aquatic ecosystems, past states provide the alternatives: in human-made systems, a wide range of possible ecosystems, including present and unfished systems, may be compared. Benefits evaluated can include total catch, economic value, diversity of fishery products, employment, biodiversity, and intersectoral conflict. Using Ecoval, the ecosystem and associated fisheries that maximize total benefit to society may be adopted as a policy goal, taking into account the costs of restoration, monitoring, and enforcement when shifting from the present system.

Introduction

Fisheries management has a history of unpleasant surprises that have dismayed its founders (e.g., Larkin 1996, Beverton 1998, Holt 1998) and has lead some fishery scientists to consider no fishery as having ever been managed in a sustainable fashion (Ludwig et al. 1993). At one time vaunted by the Canadian government as the "best managed fishery in the world" (Finlayson 1994), cod were historically one of the most prolific fish stocks in the world (Kurlanski 1997). Serious mistakes have been identified in the assessment of the cod stock (e.g., errors in biomass estimation: Walters and Maguire [1996]; failure to use spatial data: Hutchings [1996]; inexplicit and unachievable policy goals and instruments: Matthews [1995]). There is even suspicion that political goals compromised scientific management of the fishery (Hutchings et al. 1997). Even today, the Newfoundland cod resource has not begun to recover despite 5 years of closure, suggesting that management has missed the vital factor of ecosystem change. Moreover, the lesson has not have been fully learned by other fisheries managers as North Sea cod stocks are now in danger of collapse (Cook et al. 1997). Three explanations lay behind collapses such as that in Newfoundland.

Fisheries are embedded in aquatic ecosystems that are very imperfectly understood. One contributory reason underlying fishery disasters is that management has generally not been designed to adapt and learn in the face of errors in data, uncertain assessment, and ineffective control instruments despite the long availability of sophisticated quantitative methods for adaptive management (Walters 1986, Punt and Hilborn 1997, Bundy 1998, Hilborn and Liermann 1998).

Moreover, overcapacity has been identified as a major worldwide bioeconomic problem that is out of control, for which no one can propose a method to arrest (Mace 1997). Generating overcapacity (Ludwig et al. 1993) has been termed "Ludwig's Ratchet" (Pitcher, In press). Unfortunately, human responses to these difficulties in terms of management actions and commercial fishing decisions tend to be maladaptive (Haggan 1998; Hart and Pitcher 1998), despite advances in understanding how co-management may alleviate some of these problems (Pinkerton 1989, Pinkerton and Weinstein 1995).

There is a third, and more fundamental, reason for fishery collapses, that suggests future disasters will likely occur at an increasing rate. This is the long-term impact of fishing on the species composition of aquatic ecosystems. Through several direct and indirect effects, fishing alters niches towards generalist, k-selected species. The result is "simpler" ecosystems and higher volatility and lower trophic levels. The ecological processes leading to these changes, termed "Odum's Ratchet" (Pitcher, In press), are difficult to reverse and are, as yet, imperfectly understood. However, recent evidence supports this conjecture. There has been a progressive decline in the trophic level of fish caught in all areas of the world, including freshwaters (Pauly et al. 1998b), as fishers maintain their income by switching to species lower down the food web when higher-trophic-level fish become depleted. Fisheries thus sequester increasing proportions of primary production (Christensen and Pauly 1995) and large high-value species with specialized niches are rapidly lost (Christensen and Pauly 1997). A consequence is that "trash" fish come to replace high-value table fish, a process which has reached disaster levels in many areas of the world, e.g., the Gulf of Thailand (Christensen 1998), and is proceeding unchecked almost everywhere else (Pauly et al. 1998b). Moreover, the emergence of new cephalopod fisheries independently supports the notion of such a worldwide shift in the nature of exploited marine ecosystems (Caddy and Rodhouse 1998).

Avoiding the profound changes in aquatic ecosystems that are wrought by fisheries requires a major change in the philosophy underlying fisheries management (Pitcher and Pauly 1998). Traditional single-species fish stock assessment, although necessary for computing the details of age structure and population biomass, is simply incapable of providing the information to remedy or reverse this process (Pitcher, In press). What is needed is an evaluation of the impacts of fishing on aquatic ecosystems, and the adoption of policy goals that aim to maximize profits, or total benefits to society, by comparing the fisheries in alternative exploited ecosystems (Pitcher 1998a, In press, Pitcher and Pauly 1998). This agenda requires multispecies, ecosystem-based stock assessment models.

The essential features of these techniques are, first, to model the reconstruction of past and alternative ecosystems, and second, to evaluate their present economic value if they were to be restored, including the costs and uncertainties of restoration. The policy goal for management becomes the restoration of the ecosystem that maximizes net benefits to society. We term this the "Back to the Future" (BTF) policy process. This is a fundamentally different process from the conventional use of sustainability as a policy goal, which, at worst, may serve only to sustain the present misery (Pitcher and Pauly 1998). Adopting the BTF method counters the tendency to use as a baseline the state of things as they were at the start of our careers: a cognitive impediment to comprehending the full effects of fishing on aquatic abundance and biodiversity (Pauly 1995) that has been termed "Pauly's ratchet" (Pitcher, In press). In this paper we outline the methodological basis for the BTF process, along with two examples of pilot work.

Ecosystem Modeling

Although multispecies methods for assessing the exploitation status of fisheries have long been recognized as being critically needed (e.g., Larkin et al. 1984), there are as yet no methods that are generally accepted by the fisheries research community. The contributions in Sissenwine and Daan (1991) comprehensively review multispecies methods that showed promise at that time. In the past, most forms of ecosystem modeling have generally been of theoretical interest rather than practical relevance. Most are complex, involve many hundreds of parameters, do not encompass the whole ecosystem, and do not usually describe the fishery in sufficient detail, for example, by gear sector, to be used directly in management. Some examples are: Ulltang (1995) on the Barents Sea; Andersen and Ursin (1977) and Ursin (1979) on the North Sea; and Laevastu and Larkins (1981) and Low (1983) on the North Pacific. These models usually only include part of the ecosystem. They also tend to confound physical oceanographic (or limnological) processes not fundamental to understanding the fishery with essential biological features. In general, there seems to have been little advance on Larkin and Gazey's (1982) pioneering attempt to model fish communities by multiple predator/prey models. Published models tend to be highly site specific and so are difficult to generalize to new sites, and, moreover, are not robust enough to make predictions or understand ecosystem processes.

In recent years, a series of developments in ecosystem models based on the simplifying assumption of mass balance have improved the situation. Ecopath is a practical trophic mass-balance model developed at ICLARM (Christensen and Pauly 1992, 1993), based on original work by J.J. Polovina at the National Marine Fisheries Service in Hawaii (Polovina 1984, 1985). Recent extensions to this comprise Ecosim (Walters et al. 1997), a dynamic simulation technique, and Ecospace (Walters et al. 1998), a spatial ecosystem modeling tool. These developments are described below.

Ecopath tallies the flows of matter within the major components of a system, defines the trophic level of each component, and can be used to estimate biomass (*B*) given information on diet, mortality (Production, *P*, to biomass, *B* ratio, which equals the total mortality rate, *Z*), and consumption rates (consumption, *Q*, to *B* ratio). It is not the intention here to provide full details of the method, but the basic Ecopath approach is to solve a set of simultaneous linear equations, one for each group, *i*, in the system:

$$\text{[Production of } i] - [\text{all predation on } i] - \\ [\text{non predation losses of } i] - [\text{export of } i] = 0$$

This may be expressed more explicitly as:

$$[\{B_i \times (P/B)_i \times EEi\} - \Sigma\{(B_j \times (Q/B)_j + DC_{ji}\} - EX_i - Bacc(i)] = 0$$

where,

B_i	=	mean biomass of group i;
$(P/B)_i$	=	production/biomass, or (at equilibrium) total mortality rate of group i;
EE_i	=	ecotrophic efficiency, the fraction of the production of i ($=P_i(P/B)_i$) that is either consumed within the system or exported out of the system;
Σ	=	summation for $j = 1$ to k, predator trophic groups;
$(Q/B)_j$	=	food consumption/biomass of the predator j;
B_j	=	mean biomass of predator j;
DC_{ji}	=	fraction of i in the diet of predator j;
EX_i	=	exports (catches + emigration) of group i; and
$Bacc(i)$	=	accumulated biomass of i.

An equation like this is drawn up for each functional group in the system (up to 50 groups may be considered, including different types of detritus, and discarded bycatch), and the set of simultaneous equations solved using a robust algorithm (Mackay 1981). The model uses four parameters for each group (derivation of the model equations from the general principle set out above may be found in Christensen and Pauly [1992] and in the "help" files associated with the Ecopath software package): the production to biomass ratio, P/B; the consumption to biomass ratio, Q/B; the biomass, B, usually expressed in tonnes per square kilometer; and the ecotrophic efficiency, EE, expressing the proportion of a group's production that is utilized within the system. In addition, exports from each group, and the proportion of each group's diet represented by each other group in the ecosystem, are estimated. Three of these four parameters are required as input and the remaining parameter is estimated for each group. As methods exist for the field estimation of biomasses and P/B, and Q/B can be estimated from laboratory experiments, or comparative studies, it is usually EE which is left unknown, to be estimated when the above system of equations is solved by Ecopath. These estimated values of EE, which must range between 0 and 1, can then be used as diagnostic for groups that are not balanced within the system, leading to adjustments of the input data until the system is balanced (i.e., all the EE values range between 0 and 1). Overall, drawing up a preliminary ecosystem model is not so hard as might be imagined, given that data required as input exist in unpublished form in most research laboratories, or have been published.

Many ecologists worry about the validity of the mass-balance assumption, yet it is consistent with the work of most aquatic biologists, whose state and rate estimates represent averages applied to a certain period.

Here state and rate estimates are applied during an arbitrary period. In many cases, the period considered will be a typical season, or a typical year, but the state and rate estimates used for model construction may refer to different years. For example, models may represent a decade or more, during which little change has occurred. When ecosystems have undergone large changes, two or more models may be needed, representing the ecosystem before, during, and after the changes, such as before and after the advent of large-scale industrial fishing.

Ecopath mass balance models can be built without requiring vast amounts of survey data: input variables can be estimated by back calibration, i.e., by using the match of independently observed quantities, such as trophic levels, with Ecopath estimates of the same quantities (Kline and Pauly 1998), to adjust input values. The resulting models are robust in that even preliminary, incomplete models are still helpful and useable and may be improved when more accurate and extensive data become available. They are quite easy to explain to nonscientists, and relatively easy to build and validate by junior and student scientists through the software.

Ecologists have also worried that uncertainty was not taken account of in earlier versions of Ecopath. The Ecoranger module in the most recent releases of the Ecopath software now remedies this by allowing input of error distributions (uniform, triangular, normal, or log-normal) to be attached to all the model inputs. Random values are drawn from these distributions in a Monte Carlo fashion for one model run, and the process repeated thousands of times. Of the models that pass a selection criteria (such as mass balance, and match between observed and predicted trophic levels), the best-fitting one is chosen using a least square criterion, and its error bounds evaluated from the runs completed, a process involving prior and posterior distributions, analogous to Bayesian estimation. Note that among the selection criteria, Ecopath can also include observed levels of persistent pollutants in the tissues of the organisms in different functional groups, in which case Ecopath functions as a "fate model" (Dalsgaard et al. 1998a).

In addition to its use in evaluating fisheries and ecosystems, Ecopath has a number of heuristic advantages:

1. Requiring the modeler to review and standardize all available data on a given ecosystem, and identify information gaps.

2. Requiring the modeler to identify estimates (of states, or rates) that are mutually incompatible.

3. Requiring the modelers to interact with disciplines other than their own.

In addition, the Ecopath files for a model act as a consolidated database, accessible to all, that stores all references and notes about the sources and methods of estimating values (see www.ecopath.org).

Simulation of Changes in Ecosystems—Ecosim

The need for a dynamic version of the static Ecopath mass-balance model of an ecosystem has long been recognized. Ecosim is based upon a simple idea, the substitution of rate of change of biomass for the mass-balance equation of Ecopath, but the resulting simultaneous differential equations are both conceptually and computationally challenging. Walters et al. (1997) have provided a practical solution and their paper lists a range of examples of the successful validation of the technique. Ecosim, which is based on Ecopath files, has proven capable of reproducing changes that have been observed in a number of ecosystems, such as Lake Victoria (Walters et al. 1997) and the Gulf of Thailand (Christensen 1998). The first published examples of the use of Ecosim are an investigation of the influence of small pelagic fish and their fisheries on ecosystems by Mackinson et al. (1997), and an exploration of the stability of exploited ecosystems by Vasconcellos et al. (1997).

Ecosim is capable of evaluating the impact of changes in fishing rates on the ecosystem for each of the modeled fisheries. Clearly, Ecosim has great potential to be used in fisheries management in this way, although it is premature and risky to place great reliance on exact results (see contributions in Pauly 1998). Management plans have to be devised in such as way as to learn from the monitoring of changes that result from new policies, and to be flexible enough to adapt to unexpected outcomes.

Simulation of Spatial Dynamics of Ecosystems—Ecospace

Another recent development, Ecospace, introduces the facility to spatially model ecosystems that have been summarized by Ecopath (Walters et al. 1998). This will be especially useful in the evaluation of the benefits of closed areas.

The model enables the layout of a gridded map including coastlines, areas of higher productivity, habitat preferences of ecopath groups, and shapes and sizes of areas closed to fishing. It includes the movement characteristics of each Ecopath group and the bioeconomic spatial behavior of the fishers.

Reconstructing Past Ecosystems

In the Back to the Future (BTF) approach, these tools are used to construct and evaluate present and alternative ecosystems. Ecosystems that existed in the past are reconstructed using historical documents, archaeology and traditional or local environmental knowledge (TEK and LEK) (Pitcher 1998a, 1998b, In press). For some human-made ecosystems like freshwater or brackish impoundments (e.g. Lake Nasser, Pitcher 1999, 1997), BTF is clearly not appropriate, but in these cases a range of alternative ecosystems can be constructed.

Although direct simulation modeling of major ecological shifts is likely to remain infeasible for the foreseeable future, the Ecopath approach has made it possible to bypass this problem. The method makes it credible to use other information to construct models of past and alternative ecosystems. The first steps are to draw up likely species compositions and the resulting predation matrix. For example, data can be used from similar systems that may exist elsewhere, or may have existed in the recent past. Alternatively, information on likely diet and metabolic parameters may be borrowed from models of similar systems, or from FishBase (see Froese and Pauly [1998] and www.fishbase.com). By 1998 nearly 100 Ecopath models have been, or will soon be published (see, for example, contributions in Christensen and Pauly [1993]; see also Ecopath files posted in www.ecopath.org). The ability to run checks on the internal consistency of the model, by comparing respiration rates, is valuable here. Where species are globally extinct, modeling may be more approximate, but still feasible (e.g., Steller's sea cow; Pitcher 1998b).

Reconstructing past systems adds to the heuristic advantages of Ecopath by requiring the modeler to interact with the holders of TEK, archaeologists, and historians. Evaluation of a series of reconstructed ecosystems since ancient times can illustrate how past marine catches have foreclosed our future economic options. The BTF methodology supplies a practical direct use for the knowledge of maritime historians, archaeologists, ecological economists, fisheries ecologists, and the TEK of indigenous peoples. Interestingly, TEK, often denied a voice or trivialized in conventional fisheries science, is strengthened in the BTF process by a cross-validation with ecological science (Johannes 1981, 1987), and may thus be endowed with a real and valuable role in shaping future fisheries policy.

Through use of carefully designed questionnaires and interviews, it is possible to entrain TEK into ecological science. For example, presence and absence, place and time, and abundance can be scored relative to other times or organisms. This partnership between TEK and Ecopath can produce a powerful description of past ecosystems.

Figure 1 is a diagrammatic representation of the BTF method. The triangles represent Ecopath models for the present day and for specified past times. The Ecopath model is conventionally represented by triangles, the vertex angle and height of the triangle reflecting overall system transfer efficiency and size of biomass flux, respectively (see contributions in Christensen and Pauly [1993]). The dotted lines represent limits to Ecosim "what if" simulations employing species present at each stage only. The boxes represent species, and the size of boxes, the relative abundance. Note that boxes could equally well represent genetically distinct lineages within a species. Arrows between boxes show the time line of species, ending when local extinction occurs. Information about what species were present comes from archaeology, traditional knowledge, or archival documents.

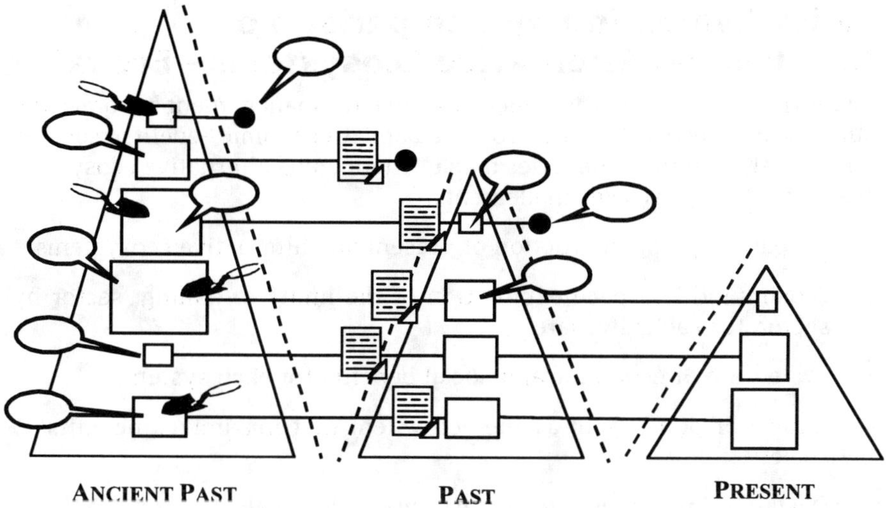

Figure 1. Diagram illustrating the Back to the Future methodology for the evaluation of past ecosystems. Triangles represent three Ecopath models, constructed at appropriate past times, where vertex angle and height is inversely related to biodiversity and internal connectance. Timing of models depends on the locality, the dawn of quantitative documentary evidence, and major shifts in resource history; a fourth model might be drawn up for a pre-modern human, late Pleistocene era. Broken lines next to triangles represent limits to Ecosim simulation modeling of "what if?" scenarios based on the Ecopath models. Time lines of some representative species in the models are indicated, where the sizes of boxes indicate relative abundance. Solid circles indicate local extinction. Sources of information for constructing and tuning the Ecopath models are illustrated by the symbols for historical documents (document), archaeological data (trowel) and the traditional environmental knowledge of indigenous Peoples (voice balloon). For further details, see text.

Policy Evaluation by Comparison of Constructed Alternative Ecosystems—Ecoval

Under the BTF method, the policy objective for management becomes the building of the ecosystem that would maximize economic benefit to society. The Ecoval method for fisheries management using alternative ecosystem evaluation comprises seven elements:

1. Ecopath model construction of present and alternative ecosystems.
2. Ecosim and Ecospace exploration of the limits to fishing, sector by sector, for each alternate.
3. Evaluation of economic and social benefits for each system.
4. Choice of policy goal as the ecosystem that maximizes benefits to society.
5. Design of instruments to achieve this policy goal.
6. Evaluation of costs of these management measures.
7. Adaptive implementation and monitoring of management measures.

Figure 2 illustrates the comparative policy evaluation process. The triangles represent Ecopath models, which are drawn up for the current ecosystem and its fishery alongside several alternative ecosystems with different fishing regimes. One model, for example, might represent an unfished ecosystem.

The vertical axis of the pyramid represents the trophic level of resources that might be exploited. Schematic resources are illustrated at three trophic levels. Vertical arrows leading to ovals at the top represent fishery catches. In the ecosystem at the left, mainly top carnivores are exploited. In the center, there are diverse fishery sectors with most of the catch coming from middle trophic levels. In the ecosystem at the right, catches are mainly from lower trophic levels. In addition, moving from left to right represents a gradient of ecosystem depletion so that organisms available for exploitation are at lower trophic levels. (This need not necessarily be the case, but represents what we know has happened to aquatic resources throughout the world [Pauly et al. 1998b].)

The table below the diagram summarizes suggested evaluation criteria. Evaluation of the relative benefits to be gained from alternative ecosystems is rather different from conventional stock assessment. First, the potential yield estimation procedure using an ecosystem model automatically provides sustainable yields, since anything greater will alter the nature of the ecosystem as detected by the Ecosim modeling. Second, in addition to estimating the total catch and catch value, values may also be put upon the diversity of the fishery products, the amount of employment, and secondary social and economic benefits of the fishery to local

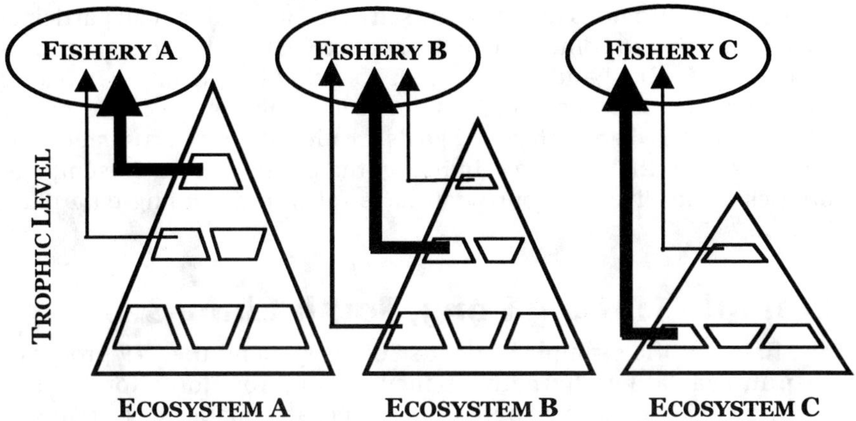

EVALUATION CRITERIA			
Total Catch	A1	B1	C1
Catch Value	A2	B2	C2
Product	A3	B3	C3
Employment	A4	B4	C4
Sectoral Conflict	A5	B5	C5
Social Benefits	A6	B6	C6
Ecosystem	A7	B7	C7

Figure 2. A schematic illustrating the Ecoval comparative policy evaluation process. The triangles represent Ecopath models, which are drawn up for the current ecosystem and its fishery alongside several alternative ecosystems with different fishing regimes. The vertical axis of the pyramid represents the trophic level of resources that might be exploited. Schematic resources are illustrated at three trophic levels. Vertical arrows leading to ovals at the top represent fishery catches. The table below the diagram summarizes suggested evaluation criteria, including costs of implementation. For further details, see text.

communities. Evaluation may further encompass conservation values such as biodiversity, the preservation of endangered species, and benefits and costs for multiple users of the aquatic resource.

Many will likely be surprised by the reconstruction of what the system was like prior to heavy industrial fishing. This state might be described as primal abundance and diversity (PAD), a concept that strikes a chord with aboriginal peoples (Haggan 1998). Ability to determine the PAD provides an important baseline for future simulations.

The BTF/Ecoval procedure has social benefits too. Workshops held to help build models of alternative systems can act as a neutral forum where intersectoral discussions can proceed. Comparing species levels predicted by the model provides talking points. It is recognized that all participants have something to contribute to the knowledge base for conservation and wise use (see Okey and Pauly 1999). Focusing on the alternatives highlights what could be achieved, as opposed to fighting over present scarcity. Moreover, when such policy goals are identified, an ecosystem–based agenda means that, during rebuilding, the public can act as sentinels of progress, and diverse groups can have roles in providing data (Pitcher 1998a).

Example 1: Hong Kong, South China Sea

The first specific example of the use of Ecosim and the BTF process has been in evaluating alternative policy options for Hong Kong fisheries (Pitcher et al. 1998). The South China Sea is an aquatic ecosystem that is among the most devastated by uncontrolled fishing (see Christensen [1998], Saeger [1993], and contributions in Silvestre and Pauly [1998]). Ecosim has been used to predict the impact of different scenarios representing changes in the management of fishing gear sectors on the relative abundance of species in Hong Kong (Pitcher et al. 1998).

For example, halving the current fishing mortality from trawlers provides considerable benefits for all fishery sectors, and those elements of the ecosystem with a conservation aspect such as marine mammals. The model suggests that the full benefits of such a policy may, however, take a decade to be realized. Associated modeling work in Hong Kong is evaluating the benefits to the resource of an artificial reef/closed areas system that is being established (Watson and Walters 1998, Pitcher et al. 1999).

Data from the 1950s suggest that large, high-value demersal croakers, groupers, and snappers were common in the fishery. Comparison of ecosystem models of the past state with that of the present day provides a clear policy choice: restoration of just some elements of the past system, even allowing for the establishment and monitoring costs, would increase the existing resource value. Restoration that included the reestablishment of heavily depleted species no longer present in the present-day catch, and the reintroduction of high-value species that have become extinct, would more than double this again. Moreover, social and governance advantages will likely accrue from the successful co-management of the reserve/artificial reef system (Haggan et al. 1998b).

Example 2: Strait of Georgia, British Columbia

The historical times at which it may be appropriate to draw up models will vary among sites. For example, in the Strait of Georgia, British Columbia, Ecopath models have been constructed for the 1990s; for the late 1900s, i.e., before the huge modern expansion of fisheries; and for a period 500 years before the present, i.e., before contact of native peoples with Europeans and the expansion of the fur trade, in the context of a project devoted to cross-validate TEK and ecosystem modeling (see contributions in Pauly et al. 1998a). Interviews were conducted with groups of natives by the ecosystem modelers to help them establish baselines for the historic ecosystem model (Haggan et al. 1998a, Salas et al. 1998). By such investigations it became possible to synthesize information from archival and archaeological sources pertaining to the abundance and ranges of many species historically. This inclusive process was extended to a workshop attended by members of the native community, commercial fishers, government scientists, academics, and historians. The team of modelers was thus able to create a framework of past and present abundance and distributions by synthesizing the information that came from these different sources (Preikshot et al. 1998).

Ecopath models of the Strait of Georgia 100 years ago suggest that for many species of economic importance, the last century has been one of significant declines. Figure 3 indicates the magnitude of biomass change over the past 100 years for the groups modeled in the Strait of Georgia (Dalsgaard et al. 1998b). Two results of this exercise catch the eye. The first is that more groups have lost biomass than gained. Secondly, the biomass decreases were experienced mainly by species of high economic value. Conversely, hake, the only group whose biomass strongly increased, is a very low-value fish, now dominating catches in British Columbia.

The exercise in the Strait of Georgia also provided a local confirmation of a phenomenon already mentioned, that of fishing down marine food webs (Wallace 1998). One result that might be expected from fishing down food webs would be larger catches, since more biomass appears to be available at lower trophic levels. However, this has not been the experience in the Strait of Georgia. Using data from British Columbia fisheries in total, the trend was increasing catches up to the early 1960s. After 1963 catches declined sharply to almost half their peak level. Although there have been some periods with minor increases the general trend has been stagnation (Wallace 1998). This supports the case of Pauly et al. (1998a) that long-term trends of fishing down food webs lead to reduced catches.

Discussion: Advantages, Problems, and Pitfalls in the BTF Process

Ecosystem modeling has to be an integral part of the methodology required for this new BTF and comparative ecosystem evaluation approach. Ecopath, Ecosim, and Ecospace modeling have advantages in making clear the impacts of catches, comparing the effects of alternative gear types, and being able to provide estimates of unknown biomasses. A disadvantage is that in its present state of development, ecosystem modeling is not itself able to provide single-species quotas because, among other things, several species are usually combined into one "box" in the model. Conventional stock assessment methods will continue to be needed, but biomass values for single species will have to be constrained by the results of the ecosystem model.

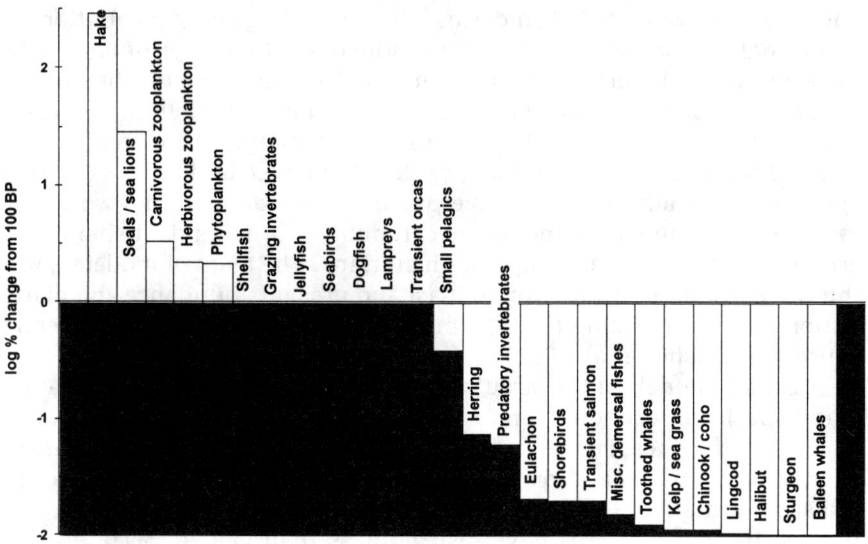

Figure 3. Changes in biomass of modeled functional groups of organisms in the Straight of Georgia, British Columbia, from 100 BP to the present day, estimated using the Back to the Future ecosystem simulation approach (see text). Bars indicate percentage change of present day biomass compared to that in 1900 (log scale calculated by the transform {sign(p) × log[abs(p+1)]} where p is the percentage change from 1900 biomass values in per square km. Shaded area represents biomass loss area down to a minimum of 100%.

Ways of merging current sophisticated single-species stock assessment methods with ecosystem modeling of the impacts of harvest need to become an active research area. The BTF approach proposes rebuilding aquatic ecosystems as the proper goal of fisheries management, and it is to be expected that the fishing industry and management agencies will resist this idea at first until its advantages become transparent.

Dayton (1998) calls for a "reversal of the burden of proof" to oblige fisheries to demonstrate that their activities cause no ecologically significant long-term changes, but too late to do this for past depletion of the world's oceans. Many fisheries ecologists call for ecosystem management (see, e.g., Larkin 1996). The BTF and Ecoval methods provide an opportunity to avoid foreclosing future options to profit from the seas, and to recover the lost opportunity costs of collapsed resources (Pitcher 1998c). The same goal is now well established in terrestrial ecology (Sinclair et al. 1995) and a similar approach is advocated to solve the problems of forestry (Marchak 1995), which in many respects parallel those of fisheries. In conclusion, a rebuilding and restoration objective for aquatic resources is neither fanciful, preservationist, nor emotive, because the policy produces tangible economic gains in tomorrow's markets. Restored systems have higher biodiversity, allowing more market niches for seafood products. Moreover, restored marine ecosystems maximize benefit to society because they have higher value from multiple users while satisfying conservation objectives.

References

Andersen, K.P., and E. Ursin. 1977. A multispecies extension to the Beverton and Holt theory of fishing, with accounts of phosphorus circulation and primary production. Medd. Dan. Fisk. Havunders. N.S. 7:319-435.

Beverton, R.J.H. 1998. Fish, fact and fantasy: A long view. Rev. Fish Biol. Fish. 8(3).

Bundy, A. 1998. The red light and adaptive management. In: T.J. Pitcher, P.J.B. Hart, and D. Pauly (eds.), Reinventing fisheries management. Chapman & Hall, London, pp. 361-368.

Caddy, J., and P. Rodhouse. 1998. Cephalopod and groundfish landings: Evidence for ecological change in global fisheries? Rev. Fish Biol. Fish. 8:1-14.

Christensen, V. 1998. Fisheries-induced changes in a marine ecosystem: Insights for models of the Gulf of Thailand. J. Fish Biol. 53(Suppl. A).

Christensen, V., and D. Pauly. 1992. ECOPATH II: A system for balancing steady-state ecosystem models and calculating network characteristics. Ecol. Modell. 61:169-185.

Christensen, V., and D. Pauly (eds.). 1993. Trophic models of aquatic ecosystems. ICLARM Conf. Proc. 26. 390 pp.

Christensen, V., and D. Pauly. 1995. Primary production required to sustain global fisheries. Nature (374):255-257.

Christensen, V., and D. Pauly. 1997. Changes in models of aquatic ecosystems approaching carrying capacity. Ecol. Appl. 8(1):104-109.

Cook, R.M., A. Sinclair, and G. Stefansson. 1997. Potential collapse of North Sea cod stocks. Nature 385:521-522.

Dayton, P. 1998. Reversal of the burden of proof in fisheries management. Science 279:821-822.

Dalsgaard, J., A. Jarre-Teichmann, C. Walters, and D. Pauly. 1998a. An approach to the modelling of persistent pollutants in marine ecosystems. ICES CM 1998/V:10. 16 pp.

Dalsgaard, J., S.S. Wallace, S. Salas, and D. Preikshot. 1998b. Mass balance models of the Strait of Georgia: The present, one hundred and five hundred years ago. In: D. Pauly, T. Pitcher, and D. Preikshot (eds.), Back to the future: Reconstructing the Strait of Georgia ecosystem. University of British Columbia, Fisheries Centre Res. Rep. 6(4):72-89.

Finlayson, A.C. 1994. Fishing for truth. Memorial University, Institute of Social and Economic Research, Newfoundland. 176 pp.

Froese, R., and D. Pauly (eds.). 1998. FishBase 98: Concepts, design and data sources. ICLARM, Manila. 293 pp. (Distributed with 2 CD-ROMs.)

Haggan, N. 1998. Re-inventing the tree: Reflections on the organic growth and creative pruning of fisheries management structures. In: T.J. Pitcher, P.J.B. Hart, and D. Pauly (eds.), Reinventing fisheries management. Chapman & Hall, London, pp. 64-90.

Haggan, N., J-A. Archibald, and S. Salas. 1998a. Knowledge gains power when shared. In: D. Pauly, T.J. Pitcher, and D. Preikshot (eds.), Back to the Future: Reconstructing the Strait of Georgia Ecosystem. University of British Columbia, Fisheries Centre Res. Rep. 6(5):8-13.

Haggan, N., R. Brown, and K. Vodden. 1998b. Public consultation on the establishment and management of artificial reefs and protected areas in Hong Kong. Report to Environmental Resource Management Ltd., Hong Kong. 27 pp.

Hart, P.J.B., and T.J. Pitcher. 1998. Conflict, cooperation and consent: The utility of an evolutionary perspective on individual human behaviour in fisheries management. In: T.J. Pitcher, P.J.B. Hart, and D. Pauly (eds.), Reinventing fisheries management. Chapman & Hall, London, pp. 215-225.

Hilborn, R., and M. Liermann. 1998. Standing on the shoulders of giants: Learning from experience in fisheries. Rev Fish Biol. Fish. 8(3):273-283.

Holt, S.J. 1998. Fifty years on. Rev. Fish Biol. Fish. 8(3):357-366.

Hutchings, J.A. 1996. Spatial and temporal variation in the density of northern cod and a review of hypotheses for the stock's collapse. Can. J. Fish. Aquat. Sci. 53:943-962.

Hutchings, J.A., C.J. Walters, and R.L. Haedrich. 1997. Is scientific inquiry incompatible with government information control? Can. J. Fish. Aquat. Sci. 54:1198-1210.

Johannes, R.E. 1981. Words of the lagoon: Fishing and marine lore in the Palau district of Micronesia. University of California Press, Berkeley.

Johannes, R.E. 1987. The value today of traditional management and knowledge of coastal marine resources in Oceania. NAGA (ICLARM Quarterly) 10(4):5.

Kline Jr., T.C. and D. Pauly. 1998. Cross-validation of trophic level estimates from a mass-balance model of Prince William Sound using $^{15}N/^{14}N$ data. In: F. Funk, T.J. Quinn II, J. Heifetz, J.N. Ianelli, J.E. Powers, J.F. Schweigert, P.J. Sullivan, and C.-I. Zhang (eds.), Fishery stock assessment models. University of Alaska Sea Grant, AK-SG-98-01, Fairbanks, pp. 693-702.

Kurlanski, M. 1997. Cod: A biography of the fish that changed the world. Knopf, Toronto. 294 pp.

Laevastu, T., and H.A. Larkins. 1981. Marine fisheries ecosystems: Quantitative evaluation and management. Fishing News Books, U.K. 176 pp.

Larkin, P.A. 1996. Concepts and issues in marine ecosystem management. Rev. Fish Biol. Fish. 6:139-164.

Larkin, P.A., and W. Gazey. 1982. Applications of ecological simulation models to management of tropical multispecies fisheries. In: D. Pauly and G.I. Murphy (eds.), Theory and management of tropical fisheries. ICLARM Conf. Proc. 9:123-140.

Larkin, P.A., C.W. Clark, N. Daan, S. Dutt, V. Hongskul, S.A. Levin, G.G.Newman, and D. Pauly. 1984. Strategies for multispecies management. In: R.M. May (ed.), Exploitation of marine communities. Dahlem Workshop Reports 32. Springer-Verlag, Berlin, pp. 287-301.

Low, L.L. 1983. Application of a Laevastu-Larkins ecosystem model for Bering Sea groundfish management. In: G.D. Sharp and J. Csirke (eds.), Proceedings of the Expert Consultation to Examine Changes in Abundance and Species Composition of Neritic Fish Resources. FAO Fish. Rep. 291(3):1161-1175.

Ludwig, D., R. Hilborn, and C.J. Walters. 1993. Uncertainty, resource exploitation, and conservation: Lessons from history. Science 260:17-18.

Mace, P.M. 1997. Developing and sustaining world fisheries resources: The state of science and management. In: D.A. Hancock, D.C. Smith, A. Grant, and J.P. Beumer (eds.), Developing and sustaining world fisheries resources: The state of science and management. CSIRO, Collingwood, Australia, pp. 1-20.

Mackay, A. 1981. The generalized inverse. Practical Computing 1981 (September):108-110.

Mackinson, S., M. Vasconcellos, T.J. Pitcher, C.J. Walters, and K. Sloman. 1997. Ecosystem impacts of harvesting small pelagic fish using a dynamic mass-balance model. In: Forage fishes in marine ecosystems. University of Alaska Sea Grant, AK-SG-97-01, Fairbanks, pp. 731-750.

Marchak, P. 1995. Logging the globe. McGill-Queen's University Press, Montreal. 440 pp.

Matthews, D.R. 1995. Commons versus open access: The collapse of Canada's east coast fishery. The Ecologist 25(2/3):86-104.

Okey, T.A., and D. Pauly. 1999. A mass-balanced model of trophic flows in Prince William Sound: Decompartmentalizing ecosystem knowledge. In: Ecosystem approaches for fisheries management. University of Alaska Sea Grant, AK-SG-99-01, Fairbanks. (This volume.)

Pauly, D. 1995. Anecdotes and the shifting baseline syndrome of fisheries. Trends Ecol. Evol. 10(10):430.

Pauly, D. (ed.). 1998. Use of Ecopath with Ecosim to evaluate strategies for sustainable exploitation of multi-species resources. University of British Columbia, Fisheries Centre Res. Rep. 6(2). 49 pp.

Pauly, D., T.J. Pitcher, and D. Preikshot (eds.). 1998a. Back to the future: Reconstructing the Strait of Georgia ecosystem. University of British Columbia, Fisheries Centre Res. Rep. 6(5). 99 pp.

Pauly, D., V. Christensen, J. Dalsgaard, R. Froese, and F. Torres Jr. 1998b. Fishing down marine food webs. Science 279:860-863.

Pinkerton, E. (ed.). 1989. Co-operative management of local fisheries. University of British Columbia Press, Vancouver. 299 pp.

Pinkerton, E., and M. Weinstein. 1995. Fisheries that work. Sustainability through community-based management. The David Suzuki Foundation, Vancouver, B.C. 199 pp.

Pitcher, T.J. 1997. The fisheries resources of African lakes. In: J.H. Annala (ed.), Fisheries and aquaculture research planning needs for Africa and West Asia. ICLARM Conf. Proc. 50, pp. 49-51.

Pitcher, T.J. 1999. The use of ecosystem modelling in comparative policy analysis: Maximising sustainable benefits from Lake Nasser's aquatic resources. In: J. Craig (ed.), ICLARM Workshop on Lake Nasser's Fisheries. In press.

Pitcher, T.J. 1998a. "Back to the Future": A novel methodology and policy goals in fisheries. In: D. Pauly, T.J. Pitcher, and D. Preikshot (eds.), Back to the future: Reconstructing the Strait of Georgia ecosystem. University of British Columbia, Fisheries Centre Res. Rep. 6(5):4-7.

Pitcher, T.J. 1998b. Pleistocene pastures: Steller's sea cow and sea otters in the Strait of Georgia. In: D. Pauly, T.J. Pitcher, and D. Preikshot (eds.), Back to the future: Reconstructing the Strait of Georgia ecosystem. University of British Columbia, Fisheries Centre Res. Rep. 6(5):49-52.

Pitcher, T.J. 1998c. Ecosystem simulation models and the new "Back to the Future" approach to fisheries management. In: D. Pauly (ed.), the use of Ecopath with Ecosim to evaluate strategies for sustainable exploitation of multi-species resources. University of British Columbia, Fisheries Centre Res. Rep. 6(2):21-23.

Pitcher, T.J. In press. Rebuilding as a new goal for fisheries management: Reconstructing the past to salvage the future. Environ. Conserv.

Pitcher, T.J., and D. Pauly. 1998. Rebuilding ecosystems, not sustainability, as the proper goal of fishery management. In: T.J. Pitcher, P.J.B. Hart, and D. Pauly (eds.), Reinventing fisheries management. Chapman & Hall, London, pp. 311-329.

Pitcher, T.J., R. Watson, A.M. Courtney, and D. Pauly. 1998. Assessment of Hong Kong's inshore fishery resources. University of British Columbia, Fisheries Centre Res. Rep. 6(3). 168 pp.

Pitcher, T.J., R. Watson, N. Haggan, S. Guénette, R. Kennish, and R. Sumaila. 1999. Marine reserves and the restoration of fisheries and marine ecosystems in the South China Sea. Bulletin of Marine Science. In press.

Polovina, J.J. 1984. Model of a coral reef ecosystem. Part I: The ECOPATH model and its application to French Frigate Shoals. Coral Reefs 3:1-11.

Polovina, J.J. 1985. An approach to estimating an ecosystem box model. Fish. Bull. U.S. 83(3):457-460.

Preikshot, D., S.S. Wallace, S. Salas, and J. Dalsgaard. 1998. Combining knowledge of the Strait of Georgia: Report from workshop discussions. In: D. Pauly, T.J. Pitcher, and D. Preikshot. (eds.), Back to the future: Reconstructing the Strait of Georgia ecosystem. University of British Columbia, Fisheries Centre Res. Rep. 6(4):65-71.

Punt, A.E., and R. Hilborn. 1997. Fisheries stock assessment and decision analysis: The Bayesian approach. Rev. Fish Biol. Fish. 7:35-63.

Saeger, J. 1993. The Samar Sea, Philippines: A decade of devastation. NAGA (ICLARM Quarterly) 16(4):4-6.

Salas, S., J-A. Archibald, and N. Haggan. 1998. Aboriginal knowledge and ecosystem reconstruction. In: D. Pauly, T.J. Pitcher, and D. Preikshot (eds.), Back to the Future: Reconstructing the Strait of Georgia Ecosystem. University of Biritish Columbia, Fisheries Centre Res. Rep. 6(5):22-27.

Silvestre, G., and D. Pauly (eds.). 1998. Status and management of tropical coastal fisheries in Asia. ICLARM Conf. Proc. 53. 208 pp.

Sinclair, A.R., S. Hik, O.J. Schmitz, G.G.E. Scudder, D.H. Turpin, and N.C. Larter. 1995. Biodiversity and the need for habitat renewal. Ecol. Appl. 5:579-587.

Sissenwine, M.P., and N. Daan. 1991. An overview of multispecies models relevant to management of living resources. Multispecies models relevant to management of living resources. In: N. Daan, and M.P. Sissenwine, M.P. (eds.), ICES Mar. Sci. Symp. 193:611.

Ulltang, O. 1995. Multispecies modelling and management with reference to the Institute of Marine Research's multispecies model for the Barents Sea. In: A.S. Blix, L. Walloe, and O. Ulltang (eds). Whales, seals, fish and man. Elsevier, Holland.

Ursin, E. 1979. Multispecies fish stock and yield assessment in ICES. In: M.C. Mercer (ed.), Multispecies approaches to fisheries management advice. ICES Mar. Sci. Symp. 59:39-47.

Vasconcellos, M., S. Mackinson, K. Sloman, and D. Pauly. 1997. The stability of trophic mass-balance models of marine ecosystems: A comparative analysis. Ecol. Modell. 100:125-134.

Wallace, S. 1998. Changes in human exploitation of marine resources in British Columbia pre-contact to present day. In: D. Pauly, T.J. Pitcher, and D. Preikshot (eds.), Back to the future: Reconstructing the Strait of Georgia ecosystem. University of British Columbia, Fisheries Centre Res. Rep. 6(4):58-64.

Walters, C.J. 1986. Adaptive management of renewable resources. Reprinted facsimile edition, 1997. University of British Columbia, Fisheries Centre, Vancouver.

Walters, C.J., and J-.J. Maguire. 1996. Lessons for stock assessment from the northern cod collapse. Rev. Fish Biol. Fish. 6:125-137.

Walters, C.J., V. Christensen, and D. Pauly. 1997. Structuring dynamic models of exploited ecosystems from trophic mass balance assessments. Rev. Fish Biol. Fish. 7:139-172.

Walters, C., D. Pauly, and V. Christensen. 1998. Ecospace: A software tool for predicting mesoscale spatial patterns in trophic relationships of exploited ecosystems, with special reference to impacts of marine protected areas. ICES C.M. 1998/S:4. 20 pp.

Watson, R., and C.J. Walters. 1998. Ecosim and MPAs: A quasi-spatial use of Ecosim. In: D. Pauly (ed.), Use of Ecopath with Ecosim to evaluate strategies for sustainable exploitation of multi-species resources. University of British Columbia, Fisheries Centre Res. Rep. 6(2):15-20.

The Effect of Urbanization on Fish Community Structure in a Large Michigan Watershed

Kurt R. Newman, Daniel B. Hayes, and William W. Taylor
Michigan State University, Department of Fisheries and Wildlife, East Lansing, Michigan

Abstract

Management of stream fish populations depends on management of the stream's watershed. This is because land use modifications occurring within a watershed ultimately influence the quality of the stream environment and fish habitat. Documenting changes in land cover over time and comparing them to quantitative changes in fish community composition is important to allow us to assess and predict the consequences of proposed or observed changes in land use for fish communities. This study evaluates changes in the fish community between 1938 and 1996 within a large Michigan watershed undergoing extensive urbanization. Of 65 fish species observed in 1938, 24 species have disappeared, the distribution of 35 species has been reduced, and the distribution of only 6 species showed no change or increased by 1996. Mean fish species richness declined significantly ($P < 0.0001$) from 13.7 species per site in 1938 to 3.7 species per site in 1996. Continued urbanization of the watershed is expected to adversely affect the ability of the fish community to persist in its present form if left unchecked.

Introduction

Changes in land use are thought to affect aquatic ecosystem health, species richness, diversity, and productivity (Larimore and Bayley 1996, Williams et al. 1997). Changes in human land use generally alter the composition of aquatic communities by modifying physical processes that affect stream habitat and water quality. Impacts on stream habitat and water quality are particularly apparent in watersheds undergoing extensive urbanization.

A problem that commonly occurs in watersheds experiencing shifts in human land use from agricultural or more "natural" covers to an urbanized landscape is accelerated runoff caused by an increase in the surface area of impermeable structures (Wood et al. 1997). In urbanized landscapes, wetlands are often filled and small tributaries are often channelized or converted to storm drains. Other components of the urban environment such as roads, parking lots, rooftops and gutters also contribute to accelerated runoff. These developments, combined with the loss of vegetative cover that allows precipitation to slowly percolate through the soils and recharge groundwater systems, result in increased frequency and magnitude of flood events, and reduced summertime base flows. Such changes in flow regime can have direct impacts on fish survival. For example, reduced base flows can have a seasonal effect on survival by decreasing the amount of stream habitat available through changes in water temperature, velocity, depth, and wetted stream width (Williams et al. 1997). Changes in flow regime also have longer-term effects on stream habitats by altering the geomorphology of the stream itself (Williams et al. 1997).

In addition to changes in flow regime, rivers and streams draining urbanized landscapes often suffer from poor water quality. Frequently, these systems receive substantial annual loads of urban pollutants (e.g., PCBs, trace metals, pesticides), sediment, and debris (Shepp and Cummins 1997). Combined sewer overflow events can also lead to reductions in dissolved oxygen levels below that required by many fish species. Such degradation of water quality can have serious ramifications for the structure of aquatic communities.

Although there is good evidence demonstrating changes in stream ecosystem structure and function in urbanized watersheds, there is little empirical evidence quantifying changes in the fish community due to urbanization. The goal of this study is to determine the quantitative changes in the fish community of a watershed undergoing extensive urbanization. To accomplish this goal we used historical data on fish community composition and land cover from the Huron River watershed.

The Huron River watershed, located in southeastern Michigan, is an example of a drainage that has undergone extensive urbanization. The watershed is situated on the edge of the Detroit metropolitan area (Hay-Chmielewski et al. 1995); it has been projected that between 1990 and 2010 the human population of southeastern Michigan will increase by 6% and that urban land use will expand by 40% (Southeast Michigan Council of Governments 1991). Much of this expansion is expected to take place within the Huron River watershed.

Our a priori hypotheses were: (1) fish species richness at individual sites will decrease as urbanization in the watershed increases; and (2) the ubiquity (defined as the number of sites occupied by a species throughout the watershed) of species intolerant to human development (Whittier and Hughes 1998) will decrease as urban land cover increases. To evaluate these hypotheses, we had the following main objectives: (1) to quantify

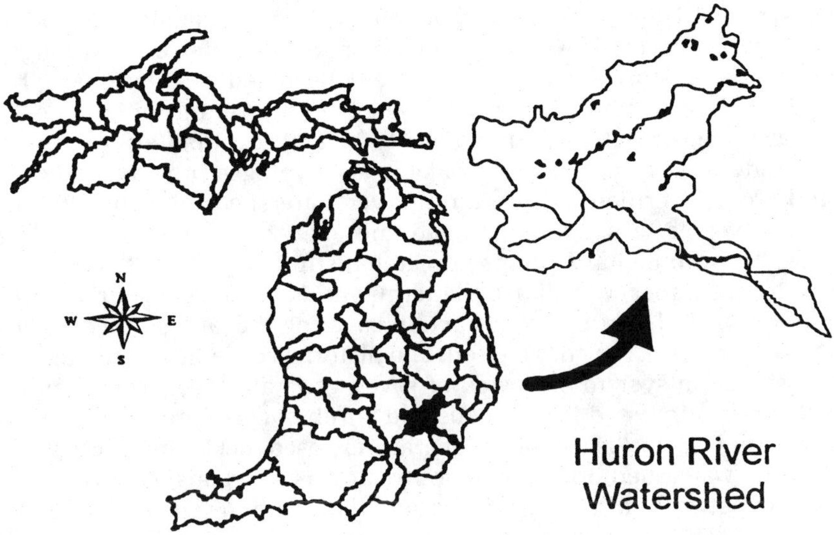

Figure 1. Location of the Huron River watershed study area in southeastern Michigan.

the land cover in the Huron River watershed in 1938 and 1996, (2) to determine the extent of changes in cover that took place over this time period, (3) to quantify fish species richness and ubiquity for 1938 and 1996 at 90 sites in the Huron River watershed, and (4) to evaluate the extent of changes in fish species richness and ubiquity over this same time period.

Study Area

The Huron River watershed is located in southeastern Michigan (Fig. 1), and includes portions of seven counties within the state. The watershed drains approximately 2,300 km^2 and eventually empties into the northwest corner of Lake Erie. The headwaters of the main stem originate in the northeast lobe of the watershed, and continue 218 km to the outlet into Lake Erie. The main stem has 24 tributaries that contribute an additional 587 km of streams within the basin. The elevation of the Huron River ranges from 310 m at the headwaters to 174 m above sea level at Lake Erie. Although there are some areas of relatively high gradient, the overall gradient of the Huron River is quite low, at 0.62 m/km.

Methods

The first objective of this study was to quantify the land cover in the Huron River watershed in 1938 and 1996. Watershed boundaries were determined from 7.5-minute topographic maps available from the U.S. Geological Survey. The 1938 land cover patterns (referred to throughout the text as historical land cover) were determined by interpreting black and white aerial photographs obtained from the Center for Remote Sensing at Michigan State University. The 1996 land cover patterns (referred to as modern land cover) were available as coverages previously digitized and ground verified by the Huron River Watershed Council (HRWC). All land cover patterns were digitized and analyzed using ARC/INFO software (Environmental Systems Research Institute).

In this paper we will use the term land cover as a description of the vegetation and artificial construction covering the land surface (Osborne and Wiley 1988). For both modern and historical coverages, our land cover classification system consisted of seven general categories: urban, agriculture, nonforested (herbaceous and shrub cover), forested, water, wetlands, and barren (beach, sand dune, and exposed rock). Land cover patterns and the watershed boundary were treated as polygons (enclosed areas), stream networks as line coverages, and fish sampling sites as point coverages. All digitized coverages were converted to a standardized geographic reference scale (Michigan State Plane 1927) so that spatial overlays of any coverage type (i.e., polygon, line, or point) could be accomplished during analyses.

The total area of each land cover category for each era (historical and modern) was determined using ARC/INFO software. The data from the two time periods were compared to assess the net change for each land cover category in the Huron River watershed between 1938 and 1996. Maps depicting the land cover categories experiencing the largest absolute gains and losses were developed for visual comparisons between the two time periods.

The next objective of this study was to analyze changes in fish species richness and ubiquity. To accomplish this, we based our sampling on a previous survey that characterized the fish species assemblage in the watershed in 1938 (Brown and Funk 1945). In the 1938 survey, samples were collected using a 3 m by 1.2 m seine net with a 0.42 cm^2 mesh. Although Brown and Funk (1945) identified 121 sites sampled in the previous survey, we found clear fish species and location information for only 90 of those sites on the original data sheets. In their survey, Brown and Funk recorded information on fish species caught and the size range of individuals caught. As such, we were limited to analyses on species richness and ubiquity. In the 1996 survey, we resampled the 90 sites identified above, employing the same sampling methods and sampling the same stream area as reported on the original data sheets. This allowed us to directly compare results from the two surveys under the assumption that each survey was subject to the same limitations imposed by the gear used. A paired *t*-test was performed to determine the significance of differences in species richness between the two surveys.

Maps of the 90 sites sampled in the Huron River watershed were generated to depict the species richness observed at each site in each survey. In order to evaluate the spatial arrangement of the severest losses in relation to those areas in which we observed increased urbanization, we developed maps showing the percent change in species richness at each site.

Results and Discussion

Land Cover

In 1938, agriculture was the predominant land cover, encompassing 131,469 hectares (ha) or 55.8% of the Huron River watershed. Forested lands comprised the next largest percentage of the watershed, covering 38,617 ha or 16.4%. Forests were followed by nonforested cover (27,047 ha, 11.5%), and wetlands (17,102 ha, 7.3%). Urban cover comprised only 12,955 ha of the Huron River drainage in 1938 or 5.5% of the watershed. Approximately 8,576 ha or 3.6% of the watershed were covered with water. Barren lands comprised only 1 ha or less than 1% of the watershed. From 1938 to 1996, the acreage in five of our land cover categories increased while two decreased (Table 1). The greatest absolute change between 1938 and 1996 was in the agricultural land cover category, which decreased by more than 69,000 ha or by about 53% of the 1938 value (Fig. 2). As agricultural lands decreased within the watershed, urban land cover increased by 51,159 ha. That increase represents nearly a 4-fold increase in urban land cover between 1938 and 1996 (Table 1). We observed the greatest concentration of increased urbanization in the upper or northern portion of the watershed that sits nearest the expanding Detroit metropolitan area, and in the extreme southern arm of the watershed that encompasses the growing Ann Arbor and Ypsilanti urban areas (Fig. 3). Nonforested cover gained more than 15,000 ha, while about 12,000 ha of forests were lost during the period. A notable increase was that of the combined area of water and wetland coverages observed in 1996. These two coverages increased a combined total of 14,421 ha since 1938. One reason for this increase is due to increased reservoir surface area and the associated wetlands that resulted from increased dam construction throughout the drainage since 1938. The observed change in wetland area may also be due in part to the difficulty in identifying wetlands from the black and white aerial photographs available for 1938. Changes in barren land covers were small relative to the changes described above.

Fish Community

We expected the increased urbanization of the Huron River watershed to have significant influences on the fish community. For the watershed as a whole, Brown and Funk (1945) observed 65 fish species. In 1996 we observed 47 fish species, a net loss of 18 species (Table 2).

Table 1. Total area (hectares) encompassed by each of the major land cover categories in the Huron River watershed during the historical and modern era, along with the net absolute change observed for each category.

Land cover	1938 (ha)	1995 (ha)	Net absolute change (ha)
Urban	12,955	64,113	51,159
Agriculture	131,469	62,094	-69,375
Nonforested	27,047	42,642	15,595
Forested	38,617	26,814	-11,803
Water	8,576	10,784	2,208
Wetland	17,102	29,315	12,213
Barren	1	3	2
Total	235,766	235,766	0

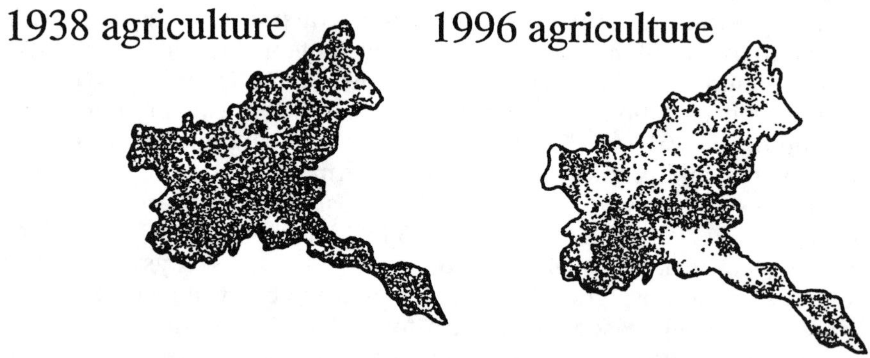

Figure 2. Change in agricultural land use within the Huron River watershed between 1938 and 1996. Areas shaded in black represent the acreage encompassed by agriculture land uses; all other land use and land cover categories are in white.

Table 2. Changes in the ubiquity of fish species observed in the Huron River between the 1938 and 1996 faunal surveys.

Fish species newly observed since 1938 survey	6
Fish species with expanded ubiquity since 1938 survey	3
Fish species showing no change in ubiquity since 1938 survey	3
Fish species with reduced ubiquity since 1938 survey	35
Fish species lost since 1938 survey	24
Total fish species richness 1938	65
Total fish species richness 1996	47

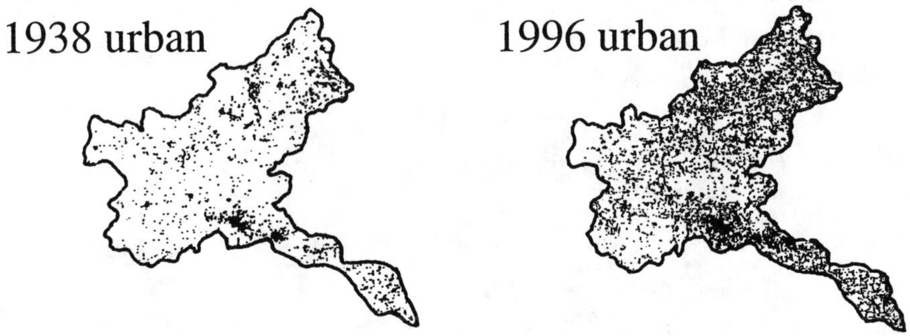

Figure 3. Change in urban land use within the Huron River watershed between 1938 and 1996. Areas shaded in black represent the acreage encompassed by urban land uses; all other land use and land cover categories are in white.

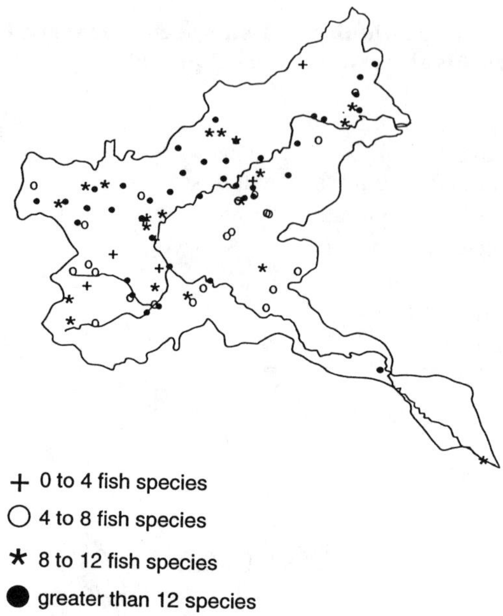

Figure 4. *Fish species richness among sites sampled in 1938.*

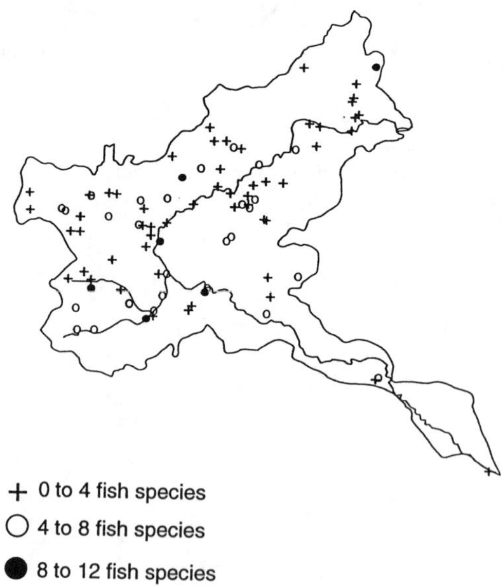

Figure 5. *Fish species richness among sites sampled in 1996.*

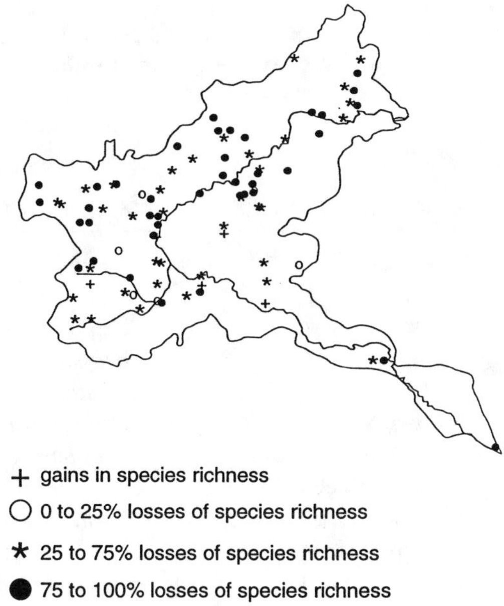

Figure 6. Percent change in fish species richness among sites sampled between 1938 and 1996.

Species richness among the sites sampled in each survey was highly variable, but changes from 1938 to 1996 show an alarming trend. In 1938, fish species richness at individual sites ranged from 3 to 45 species, with a mean species richness of 13.7 species ±0.80 (1 standard deviation) across all sites. In 1996, fish species richness ranged from 0 to 12 species, with a mean of 3.7 species ±0.32 across all sites. Results of a paired t-test indicate that the decline in species richness that occurred with the changing land use practices from 1938 to 1996 was highly significant ($t = 10.6$, df = 89, $P < 0.0001$).

Maps of fish species richness at individual sites show that although the loss in species richness was spread across the watershed, some areas experienced greater losses than others. In 1938, the highest species richness values were observed in the northeastern lobe of the watershed, and in the extreme southern arm of the drainage (Fig. 4). In contrast, relatively moderate or low species richness values were observed in the central and southwestern lobe of the watershed. Just the opposite pattern was observed in 1996; relatively low species richness values were recorded in the northeastern lobe and extreme southern arm of the watershed, while higher species richness values were recorded in the central and south-

Table 3. Fishes recorded from the Huron River in 1938 and 1996 faunal surveys, and the number of sites at which individual species were observed.

Common name	Scientific name	Number of sites sampled	
		1938	1996
Rock bass	*Ambloplites rupestris*	62	6
Bowfin	*Amia calva*	6	1
Central stoneroller	*Campostoma anomalum*	21	8
Goldfish	*Carassius auratus*	2	0
White sucker	*Catostomus commersoni*	25	15
Mottled sculpin	*Cottus bairdi*	44	19
Lake chub	*Couesius plumbeus*	9	0
Brook stickleback	*Culaea inconstans*	3	11
Common carp	*Cyprinus carpio*	3	2
Gizzard shad	*Dorosoma cepedianum*	1	0
Creek chubsucker	*Erimyzon oblongus*	17	0
Lake chubsucker	*Erimyzon sucetta*	22	0
Grass pickerel	*Esox americanus*	74	4
Northern pike	*Esox lucius*	11	2
Greenside darter	*Etheostoma blennioides*	12	6
Rainbow darter	*Etheostoma caeruleum*	54	9
Iowa darter	*Etheostoma exile*	16	3
Barred fantail darter	*Etheostoma flabellare*	26	0
Least darter	*Etheostoma microperca*	21	0
Johnny darter	*Etheostoma nigrum*	29	19
Banded killifish	*Fundulus diaphanus*	2	0
Blackstripe topminnow	*Fundulus notatus*	24	3
Northern hogsucker	*Hypentelium nigricans*	18	12
Northern brook lamprey	*Ichthyomyzon fossor*	1	0
Black bullhead	*Ictalurus melas*	2	1
Yellow bullhead	*Ictalurus natalis*	59	3
Brown bullhead	*Ictalurus nebulosus*	3	4
Brook silverside	*Labidesthes sicculus*	9	4
Longnose gar	*Lepisosteus osseus*	2	0
Green sunfish	*Lepomis cyanellus*	41	13
Pumpkinseed	*Lepomis gibbosus*	63	20
Warmouth	*Lepomis gulosus*	4	2
Bluegill	*Lepomis macrochirus*	63	30
Longear sunfish	*Lepomis megalotis*	53	0
Smallmouth bass	*Micropterus dolomieu*	15	5
Largemouth bass	*Micropterus salmoides*	63	25
Black redhorse	*Moxostoma duquesnei*	1	0

Table 3. Continued.

Common name	Scientific name	Number of sites sampled 1938	1996
Golden redhorse	*Moxostoma erythrurum*	5	0
Greater redhorse	*Moxostoma valenciennesi*	1	0
Horneyhead chub	*Nocomis biguttatus*	36	5
River chub	*Nocomis micropogon*	0	1
Golden shiner	*Notemigonus crysoleucas*	7	8
Emerald shiner	*Notropis atherinoides*	3	0
Striped shiner	*Notropis chrysochephalus*	0	18
Common shiner	*Notropis cornutus*	66	9
Pugnose minnow	*Notropis emillae*	1	0
Blackchin shiner	*Notropis heterodon*	10	0
Blacknose shiner	*Notropis heterolepis*	9	0
Spottail shiner	*Notropis hudsonius*	3	2
Silver shiner	*Notropis photogenis*	1	0
Rosyface shiner	*Notropis rubellus*	5	3
Spotfin shiner	*Notropis spilopterus*	0	13
Sand shiner	*Notropis stramineus*	1	1
Mimic shiner	*Notropis volucellus*	2	0
Stonecat	*Noturus flavus*	1	0
Tadpole madtom	*Noturus gyrinus*	15	0
Brindled madtom	*Noturus miurus*	13	0
Rainbow trout	*Oncorhynchus mykiss*	0	1
Yellow perch	*Perca flavescens*	30	8
Logperch	*Percina caprodes*	20	11
Blackside darter	*Percina maculata*	33	1
Northern redbelly dace	*Phoxinus eos*	2	2
Bluntnose minnow	*Pimephales notatus*	32	21
Fathead minnow	*Pimephales promelas*	0	6
White crappie	*Pomoxis annularis*	3	1
Black crappie	*Pomoxis nigromaculatus*	6	6
Blacknose dace	*Rhinichthys atratulus*	19	6
Brook trout	*Salvelinus fontinalis*	1	0
Creek chub	*Semotilus atromaculatus*	71	23
Pearl dace	*Semotilus margarita*	0	1
Central mudminnow	*Umbra limi*	76	11

western lobe of the drainage (Fig. 5). Overwhelmingly, we observed high losses in species richness (75-100%) in those areas where urban cover increased the most (Fig. 6). Conversely, small gains or relatively little changes in species richness were observed near the center and in the southwestern lobe of the watershed where land cover changed little from the agricultural landscape of 1938.

Several changes in the ubiquity of individual fish species were observed across the Huron River watershed between the two surveys (Table 2). While six new fish species were observed in the 1996 survey and six other fish species either expanded their range or remained unchanged since 1938, the majority of the changes observed were either complete losses or reductions in the ubiquity of fish species since 1938 (Table 2).

The six new species found in the 1996 survey include river chub, striped shiner, spotfin shiner, rainbow trout, fathead minnow, and pearl dace (Table 3). Of these, river chub, rainbow trout, and pearl dace were found only at a single site each during the 1996 survey. Three fish species expanded their ubiquity within the watershed in 1996: brown bullhead, brook stickleback, and golden shiner (Table 3). Each of these species has been categorized as being tolerant to human activity (Scott and Crossman 1973, Whittier and Hughes 1998). Three other fish species showed no change in ubiquity: sand shiner, black crappie, and northern redbelly dace (Table 3). Of these, sand shiner and black crappie are considered tolerant to the observed changes in human land use, but the northern redbelly dace is considered intolerant (Whittier and Hughes 1998). However, neither the sand shiner nor the northern redbelly dace were widespread within the watershed, being found in only one and two sites, respectively.

Of the 35 fish species showing reductions in ubiquity, rock bass, grass pickerel, rainbow darter, blackside darter, and common shiner are examples of species that experienced dramatic reductions in the number of sites where they were detected (Table 3). Each of these species are moderately affected by human disturbance in the watershed (Scott and Crossman 1973, Whittier and Hughes 1998). If the trends in urbanization continue unchecked, these fish species may be subject to complete elimination from the watershed. Notable fish species that disappeared since the 1938 survey include barred fantail and least darters, blackchin and blacknose shiners, tadpole and brindled madtoms, and longear sunfish (Table 3). Of this group, darters, shiners, and madtoms are generally considered to be sensitive to human disturbance within the watershed.

Our results suggest that urbanization in the Huron River watershed has had a significant influence on the composition and distribution of the fish community. If we assume that fish are the ultimate integrators of human activities within the watershed (Taylor et al. 1998), our results suggest that the watershed is not ecologically healthy as defined by Williams et al. (1997). Clearly, the watershed has lost ecological structure and function as human demands on the landscape have changed, and the prognosis is not good in the face of the projected increase in demands over the next several decades. We need to consider the long-term consequences of societal decisions on the health, diversity, and productivity of the land to ensure the ability of human society to prosper into the future (Wood et al. 1997). Fortunately, local watershed councils are working with state and federal agencies to protect and restore the Huron River watershed. It is hoped that restoration efforts can reverse the trends observed in this study. Ultimately, the effectiveness of any restoration effort will depend upon our ability to integrate an ecosystem perspective with a better understanding of the needs of society as they relate to land use.

References

Brown, C.J.D., and J.L. Funk. 1945. A fisheries survey of the Huron River, its tributaries and impounded waters. Mich. Dep. Nat. Resour. Fish. Res. Rep. 1003.

Hay-Chmielewski, E.M., P.W. Seelbach, G.E. Whelen, and D.B. Jester Jr. 1995. Huron River assessment. Mich. Dep. Nat. Resour. Fish. Div. Spec. Rep. 16.

Larimore, R.W., and P.B. Bayley. 1996. The fishes of Champaign County, Illinois, during a century of alterations of a prairie ecosystem. Ill. Nat. Hist. Surv. Bull. 35(2):53-183.

Osborne, L.L., and M.J. Wiley. 1988. Empirical relationships between land use/cover and stream water quality in an agricultural watershed. J. Environ. Manage. 26:9-27.

Scott, W.B., and E.J. Crossman. 1973. Freshwater fishes of Canada. Fish. Res. Board Can. Bull. 184.

Shepp, D.L., and J.D. Cummins. 1997. Restoration in an urban watershed: Anacostia River of Maryland and the District of Columbia. In: J.E. Williams, C.A. Wood, and M.P. Dombeck (eds.), Watershed restoration: Principles and practices. American Fisheries Society, Bethesda, Maryland, pp. 297-317.

Southeast Michigan Council of Governments. 1991. The environment, 1990/2010: Regional development initiative, briefing paper #1. Detroit.

Taylor, W.W, K.D. Lynch, and K.W. Taylor. 1998. Integrating fisheries management with watershed processes. Verh. Internat. Verein. Limnol. In press.

Whittier, T.R., and R.M. Hughes. 1998. Evaluation of fish species tolerances to environmental stressors in lakes in the northeastern United States. N. Am. J. Fish. Manage. 18:236-252.

Williams, J.E., C.A. Wood, and M.P. Dombeck (eds.). 1997. Watershed restoration: Principles and practices. American Fisheries Society, Bethesda, Maryland. 561 pp.

Wood, C.A., J.E. Williams, and M.P. Dombeck. 1997. Learning to live within the limits of the land: Lessons from the watershed restoration case studies. In: J.E. Williams, C.A. Wood, and M.P. Dombeck (eds.), Watershed restoration: Principles and practices. American Fisheries Society, Bethesda, Maryland, pp. 445-458.

Ecosystem Approaches for Fisheries Management
Alaska Sea Grant College Program • AK-SG-99-01, 1999

Will Reduced Discarding Help or Harm Seabird Populations?

Robert W. Furness
University of Glasgow, Institute of Biomedical and Life Science, Glasgow, United Kingdom

Abstract

About 27 million t of fish was discarded worldwide in 1993. Particularly high discarding levels occur in heavily exploited mixed fisheries where possibility for gear selectivity is constrained. Clearly reduction of discarding is desirable. The FAO (Food and Agriculture Organization of the United Nations) estimates that a reduction of at least 60% in bycatch is possible by the year 2000. However, discards now form a major food supply for scavenging seabirds. In the North Sea, discard consumption by seabirds varies, being highest in winter, highest in the northwest North Sea, and high for offal (94-100%) and roundfish (70-92%) but low for flatfish (10-35%), elasmobranchs (12%), and benthic invertebrates (3-17%). Larger scavenging seabirds steal fish from smaller seabirds and obtain higher intake rates. Technical measures to reduce discarding are likely to increase competition, reducing scavenging success for smaller seabirds. Technical measures that increase the mean size of discards (such as increased net mesh size) will particularly reduce food supply to the smaller scavenging seabirds. Severe effects on breeding success or population size of scavenging seabirds following reductions in discarding have already been documented.

Discards form more than half the breeding season diet of skuas. Skua colony sizes in the North Sea are many times larger than those in the sub-Antarctic where skuas feed predominantly by predation on seabirds. The increase in skua numbers in the North Sea seems to have been made possible by the provision of discards. Reducing discards available to seabirds, leading to switching diet, may have a severe impact on other seabird populations, since skuas are likely to switch to predation on other seabirds. The large populations of skuas in the North Sea might extirpate kittiwakes and other seabirds before their own populations fall to sustainable levels.

Introduction

The term "discards" describes the animal waste, generated by fishing operations, which is jettisoned at sea. This therefore includes undersized fish and shellfish, fish which cannot be taken to market because quotas are exceeded or the catch is of little or no commercial value, and offal and waste from cleaning fish at sea. A recent report to the FAO estimated that 27 million t of fish was discarded worldwide in 1993 (Alverson et al. 1994). Particularly high discarding levels occur in heavily exploited mixed fisheries where possibility for gear selectivity is constrained (Furness 1992, Reeves et al. 1992). Clearly reduction of discarding is desirable (FAO 1995). The FAO estimates that a reduction of at least 60% in bycatch is possible by the year 2000 (Alverson et al. 1994). However, discards now form a major food supply for very large numbers of scavenging seabirds. This has been documented for many different fisheries: in Australia (Blaber and Wassenberg 1989, Blaber et al. 1995), in the western Mediterranean trawl fishery (Gonzalez-Solis et al. 1997, Oro and Ruiz 1997), in the shrimp fishery of the Wadden Sea (Berghahn and Rosner 1992, Walter and Becker 1994, 1997), in the finfish fishery at the Falkland Islands (Thompson and Riddy 1995), in the cod (*Gadus morhua*) fishery of the Gulf of St. Lawrence (Chapdelaine and Rail 1997), in the hoki (*Macruronus novaezelandiae*) fishery off New Zealand (Freeman 1997, Freeman and Smith 1998), in the trawl fisheries off southern Africa (Abrams 1983, Ryan and Moloney 1988) and in greatest detail in the trawl fisheries for fish and for crustacea in the North Sea (Furness and Hislop 1981; Furness et al. 1988, 1992; Hudson and Furness 1988, 1989; Hudson 1989; Camphuysen et al. 1993, 1995; Garthe and Huppop 1994; Tasker and Furness 1996). A sudden end to discarding would deprive these birds of a major source of food. Not only would this affect their breeding success and population size (Oro 1995, 1996), but these birds would be forced to switch, where possible, to alternative food sources (Castilla and Perez 1995, Arcos and Oro 1996). Such switching could have major impacts on other wildlife populations, since many of the seabirds that feed on discards are large, aggressive birds that could switch to kleptoparasitism (Oro and Martinez-Vilalta 1994) or predation (Furness 1997) on smaller birds. Sudden reductions in discarding could seriously affect the ecological balance in wildlife communities. Here I report on one well-documented example of this problem, but point out that similar examples are likely to arise in many parts of the world.

Use of Discards by Seabirds in the North Sea

In the North Sea, discard consumption by seabirds varies, being highest in the northwest North Sea and high for offal (94-100%) and roundfish (70-92%) but low for flatfish (10-35%), elasmobranchs (12%), and benthic invertebrates (3-17%) (Garthe et al. 1996). Discards not consumed by seabirds eventually sink, and may be eaten by large fish or by benthic invertebrates (Ramsay et al. 1997). Larger scavenging seabirds also steal fish from smaller seabirds and obtain higher intake rates (Hudson 1989). The high level of competition among scavenging seabirds for discarded offal and roundfish (Hudson 1989, Camphuysen et al. 1995) and the high proportion of the diet that these foods can represent (Furness and Hislop 1981, Oro and Ruiz 1997), suggests that the scavenging seabirds are dependent on these resources to maintain their breeding success and population sizes.

Evidence in support of this inference comes from studies of scavenging seabirds in areas where discarding has suddenly stopped. For example, in the western Mediterranean Sea a trawling moratorium had profound effects on the breeding success and interspecific interactions of scavenging seabirds there, including the endangered species Audouin's gull (Oro et al. 1995, 1996; Oro and Ruix 1997). Current and impending measures to reduce discarding (CEC 1993, 1995) are likely to increase competition, particularly reducing scavenging success for smaller seabirds (Furness 1992). Technical measures that increase the mean size of discards (such as increased net mesh size) will also particularly reduce food supply to the smaller scavenging seabirds. One species that is highly dependent on discards is the great skua *Catharacta skua*, a rare seabird found breeding only in the Northeast Atlantic, and especially at northern North Sea islands (Furness 1987, Hamer et al. 1991).

Potential Effects of Reduced Discarding on Great Skuas

Discards currently form more than half the breeding season diet of great skuas (Furness 1997). Great skua breeding numbers at North Sea colonies have increased remarkably, at a sustained rate of about 7% per year from 1900 to 1990 (Furness 1987, 1997), and it seems probable that the provision of discards has had an important influence on this increase. Great skua colony sizes in the North Sea are many times larger than those of the closely related brown skua (*Catharacta antarctica*) which breeds on sub-Antarctic islands. Brown skuas feed predominantly by predation on burrow-nesting seabirds. Brown skua breeding numbers are closely related to the numbers of their bird prey. Considering the 19 sub-Antarctic islands or archipelagoes where there are census data for brown skuas and for their burrow-nesting petrel prey (prions, storm petrels, diving petrels, and gadfly petrels), there is a strong linear relationship (on a log-log scale) between the number of breeding pairs of brown skuas and the number of breeding pairs of burrow-nesting petrels (log pairs of skuas = −1.15 + 0.608 log pairs of burrow-nesting petrels, $r^2 - 73\%$, $p < 0.001$). It seems that the numbers of breeding brown skuas are determined by their food supply.

Table 1. Numbers of pairs of prey seabirds of great skuas (kittiwakes, auks, and storm petrels) at colonies or groups of colonies, in relation to the predicted number of pairs that could be sustained by predation on seabirds as the sole feeding method according to the brown skua subantarctic model (see text) and in relation to the predominant diet of these birds in recent years.

Region	Pairs of prey seabirds available (kittiwakes, auks, and storm petrels)	Pairs of great skuas present (most recent census)	Pairs of skuas sustainable according to subantarctic brown skua model	Main diet of great skuas in this region during breeding season
Shetland	300,000	5,500	151	Discards and sand lance
Orkney	280,000	2,500	145	Discards and sand lance
St. Kilda	400,000	250	180	Discards and seabirds
Handa	120,000	100	87	Seabirds
Faeroe Islands	1,500,000	250	403	Seabirds

Census data are from Lloyd et al. (1991) and the Joint Nature Conservation Committee Seabird Colony Register.

We can guess that great skua numbers would be similarly related to the numbers of seabird prey available to them if they were not able to feed on discards and sand lance. If we compare numbers of great skuas at colonies in the North Atlantic with the numbers of potential seabird prey at these sites and look at these data in relation to the regression of brown skua numbers against their seabird prey population sizes, we can see (Table 1) that there appear to be far too many great skuas to be sustained only by predation on seabirds, at North Sea colonies (Shetland and Orkney) where great skuas currently feed predominantly on sand lance and discards, but that numbers at colonies in St. Kilda, and Handa (both northwest Scotland) and the Faeroe Islands where seabirds are important prey for great skuas, are more closely in agreement with the predictions from the brown skua sub-Antarctic regression. Reduction of discarding in the northwestern North Sea would almost certainly lead to large decreases in great skua numbers at the large North Sea colonies of Shetland and Orkney. It would also in the short term increase the level of predation by skuas, which can affect sheep farming (Furness 1987) as well as other seabird populations (Furness 1997).

Potential Effects of Reduced Discards for Skuas on the Seabird Community

Reducing discards available to seabirds, leading to large scavenging seabirds switching diet, may have a severe impact on other seabird populations, since great skuas (and great black-backed gulls, *Larus marinus*) are likely to switch to predation on other seabirds. The large populations of skuas in the North Sea might extirpate black-legged kittiwakes (*Rissa tridactyla*), Atlantic puffins (*Fratercula arctica*), black guillemots (*Cepphus grylle*), and other seabirds before their own populations fall to sustainable levels. There is already evidence of this starting to occur. Great skuas are estimated to have eaten about 200,000 seabirds in the northwestern North Sea in the summer of 1996 (Furness 1997). Kittiwake numbers in Shetland have fallen by 60% over the last 15 years as a result of predation impact of great skuas (Heubeck et al. 1997, Thompson et al. 1997). Further reductions in discarding will probably increase this impact. There may be a need to manage scavenging seabird numbers to protect other seabirds from such impacts.

Conclusion

There is a clear need to reduce discarding in many fisheries. However, this may have severe impacts in the short term on populations of scavenging seabirds that have increased during decades of widespread discarding and depend on discards for food. Many of these large scavengers are likely to switch to predation as an alternative feeding method, and so could cause severe disruption to seabird communities and other wildlife. There is a conservation problem here that needs to be addressed. In this paper I have emphasised the problem of skuas at North Sea colonies, but a similar situation may apply in many other parts of the world with artificially enlarged populations of large scavenging seabirds such as large gulls and giant petrels.

Acknowledgments

I thank the International Fishmeal and Oil Manufacturers' Association for supporting the preparation of this paper.

References

Abrams, R.W. 1983. Pelagic seabirds and trawl-fisheries in the southern Benguela Current region. Mar. Ecol. Prog. Ser. 11:151-156.

Alverson, D.L., M.H. Freeberg, S.A. Murawski, and J.G. Pope. 1994. FAO Fish. Tech. Pap. 339.

Arcos, J.M., and D. Oro. 1996. Changes in foraging range of Audouin's Gulls *Larus audouinii* in relation to a trawler moratorium in the western Mediterranean. Colon. Waterbirds 19:128-131.

Berghahn, R., and H.-U. Rosner. 1992. A method to quantify feeding of seabirds on discard from the shrimp fishery in the North Sea. Neth. J. Sea Res. 28:347-350.

Blaber, S.J.M., and T.J. Wassenberg. 1989. Feeding ecology of the piscivorous birds *Phalacrocorax varius, P. melanoleucos* and *Sterna bergii* in Moreton Bay, Australia: Diets and dependence on trawler discards. Mar. Biol. 101:1-10.

Blaber, S.J.M., D.A. Milton, G.C. Smith, and M.J. Farmer. 1995. Trawl discards in the diets of tropical seabirds of the northern Great Barrier Reef, Australia. Mar. Ecol. Prog. Ser. 127:1-13.

Camphuysen, C.J., K. Ensor, A. Follestad, R.W. Furness, S. Garthe, O. Huppop, G. Leaper, H. Offringa, and M.L. Tasker. 1993. Seabirds feeding on discards in winter in the North Sea. NIOZ Rapport 1993-8, Netherlands Institute for Sea Research, Texel.

Camphuysen, C.J., B. Calvo, J. Durinck, K. Ensor, A. Follestad, R.W. Furness, S. Garthe, G. Leaper, H. Skov, M.L. Tasker, and C.J.N. Winter. 1995. Consumption of discards by seabirds in the North Sea. NIOZ Rapport 1995-5, Netherlands Institute for Sea Research, Texel.

Castilla, A.M., and J.J. Perez. 1995. Relationships between fishery activities and presence of the Audouin's gull *Larus audouinii* in the Columbretes Islands. Colon. Waterbirds 18:108-112.

Chapdelaine, G., and J.-F. Rail. 1997. Relationship between cod fishery activities and the population of herring gulls on the North Shore of the Gulf of St. Lawrence, Quebec, Canada. ICES J. Mar. Sci. 54:708-713.

CEC (Commission of the European Communities). 1993. The new components of the Common Fisheries Policy and their practical implementation. COM(93)664 final. Brussels.

CEC (Commission of the European Communities). 1995. Annual report to the Council and the European Parliament on the progress of the multi-annual guidance programmes for the fishing fleets at the end of 1994. COM(95)463 final. Brussels.

FAO. 1995. Code of conduct for responsible fisheries. FAO, Rome.

Freeman, A.N.D. 1997. The influence of Hoki fishing vessels on Westland petrel (*Procellaria westlandica*) distribution at sea. Notornis 44:159-164.

Freeman, A.N.D., and P.J. Smith. 1998. Iso-electric focusing and the identification of fisheries' waste in the diet of Westland petrels (*Procellaria westlandica*). N. Z. J. Mar. Freshwater Res. 32:177-180.

Furness, R.W. 1987. The skuas. T. and A.D. Poyser, Calton. 363 pp.

Furness, R.W. 1992. Implications of changes in net mesh size, fishing effort and minimum landing size regulations in the North Sea for seabird populations. JNCC Report No. 133. Joint Nature Conservation Committee, Peterborough.

Furness, R.W. 1997. The impact of predation by great skuas on other seabird species, with particular reference to special protection areas in Shetland. Report to Scottish Natural Heritage. Applied Ornithology Unit, Glasgow. 99 pp.

Furness, R.W., and J.R.G. Hislop. 1981. Diets and feeding ecology of great skuas *Catharacta skua* during the breeding season in Shetland. J. Zool. (Lond.) 195:1-23.

Furness, R.W., K. Ensor, and A.V. Hudson. 1992. The use of fishery waste by gull populations around the British Isles. Ardea 80:105-113.

Furness, R.W., A.V. Hudson, and K. Ensor. 1988. Interactions between scavenging seabirds and commercial fisheries around the British Isles. In: J. Burger (ed.), Seabirds and other marine vertebrates: Competition, predation and other interactions. Columbia University Press, New York, pp. 240-268.

Garthe, S., and O. Huppop. 1994. Distribution of ship-following seabirds and their utilization of discards in the North Sea in summer. Mar. Ecol. Prog. Ser. 106:1-9.

Garthe, S., C.J. Camphuysen, and R.W. Furness. 1996. Amounts of discards by commercial fisheries and their significance as food for seabirds in the North Sea. Mar. Ecol. Prog. Ser. 136:1-11.

Gonzalez-Solis, J., D. Oro, L. Jover, X. Ruiz, and V. Pedrocchi. 1997. Trophic niche width and overlap of two sympatric gulls in the southwestern mediterranean. Oecologia 112:75-80.

Hamer, K.C., R.W. Furness, and R.W.G. Caldow. 1991. The effects of changes in food availability on the breeding ecology of great skuas *Catharacta skua* in Shetland. J. Zool. (Lond.) 223:175-188.

Heubeck, M., R.M. Mellor, and P.V. Harvey. 1997. Changes in the breeding distribution and numbers of kittiwakes *Rissa tridactyla* around Unst, Shetland and the presumed role of predation by great skuas *Stercorarius skua*. Seabird 19:12-21.

Hudson, A.V. 1989. Interspecific and age-related differences in the handling time of discarded fish by scavenging seabirds. Seabird 12:40-44.

Hudson, A.V., and R.W. Furness. 1988. Utilization of discarded fish by scavenging seabirds behind whitefish trawlers in Shetland. J. Zool. (Lond.) 215:151-166.

Hudson, A.V., and R.W. Furness. 1989. The behaviour of seabirds foraging at fishing boats around Shetland. Ibis 131:225-237.

Lloyd, C.S., M.L. Tasker, and K. Partridge. 1991. The status of seabirds in Britain and Ireland. Poyser, London. 355 pp.

Oro, D. 1995. The influence of commercial fisheries in daily activity of Audouin's Gull *Larus audouinii* in the Ebro Delta, NE Spain. Ornis Fennica 72:154-158.

Oro, D. 1996. The effects of trawler discard availability on the egg-laying and the breeding success of the lesser black-backed gull *Larus fuscus* in the western Mediterranean. Mar. Ecol. Prog. Ser. 132:43-46.

Oro, D., and A. Martinez-Vilalta. 1994. Factors affecting kleptoparasitism and predation rates upon a colony of Audouin's Gull (*Larus audouinii*) by Yellow-legged Gulls (*Larus cachinnans*) in Spain. Colon. Waterbirds 17:35-41.

Oro, D., and X. Ruiz. 1997. Exploitation of trawler discards by breeding seabirds in the north-western Mediterranean: Differences between the Ebro Delta and the Balearic Islands areas. ICES J. Mar. Sci. 54:695-707.

Oro, D., M. Bosch, and X. Ruiz. 1995. Effects of a trawler moratorium on the breeding success of the yellow-legged gull *Larus cachinnans*. Ibis 137:347-349.

Oro, D., L. Jover, and X. Ruiz. 1996. Influence of trawling activity on the breeding ecology of a threatened seabird, Audouin's gull *Larus audouinii*. Mar. Ecol. Prog. Ser. 139:19-29.

Ramsay, K., M.J. Kaiser, P.G. Moore, and R.N. Hughes. 1997. Consumption of fisheries discards by benthic scavengers: Utilization of energy subsidies in different marine habitats. J. Anim. Ecol. 66:884-896.

Reeves, S.A., D.W. Armstrong, R.J. Fryer, and K.A. Coull. 1992. The effect of mesh size, cod-end extension length and cod-end diameter on the selectivity of Scottish trawls and seines. ICES J. Mar. Sci. 49:279-288.

Ryan, P.G., and C.L. Moloney. 1988. Effect of trawling on bird and seal distributions in the southern Benguela region. Mar. Ecol. Prog. Ser. 45:1-11.

Tasker, M.L., and R.W. Furness. 1996. Estimation of food consumption by seabirds in the North Sea. In: G.L. Hunt Jr. and R.W. Furness (eds.), Seabird/fish interactions, with particular reference to seabirds in the North Sea. ICES Coop. Res. Rep. 216:6-42.

Thompson, K.R., and M.D. Riddy. 1995. Utilization of offal and discards from "finfish" trawlers around the Falkland Islands by the black-browed albatross *Diomedea melanophris*. Ibis 137:198-206.

Thompson, K.R., E. Brindley, and M. Heubeck. 1997. Seabird numbers and breeding success in Britain and Ireland, 1996. Joint Nature Conservation Committee, Peterborough. 64 pp.

Walter, U., and P.H. Becker. 1994. The significance of discards from the brown shrimp fisheries for seabirds in the Wadden Sea: Preliminary results. Ophelia, Suppl. 6:253-262.

Walter, U., and P.H. Becker. 1997. Occurrence and consumption of seabirds scavenging on shrimp trawler discards in the Wadden Sea. ICES J. Mar. Sci. 54:684-694.

Mixed Stock Pollock and Herring Fishery in the Northern Sea of Okhotsk During 1997-1998

Vladimir Radchenko
Pacific Scientific Research Fisheries Centre (TINRO-Centre), Vladivostok, Russia

Introduction

The results of the first experiment on a large-scale mixed stock fishery in the Russian far-eastern seas are given in this report. Until recently, all large-scale fisheries traditionally remained single-species fisheries with regard to target species (pollock, herring, salmon, saury, etc.). The mixed stock fishery approach has been applied in the near coastal groundfish fishery for food production from small vessels. The total harvest of this groundfish fishery did not exceed 50,000 tons for each fisheries region in the Okhotsk and Bering seas. Experiments on the application of the mixed stock fishery approaches to a large-scale pollock fishery in the Okhotsk Sea was planned after the 1996-1997 fishing season. We from TINRO reported on the results of this experiment to interested specialists during the two-day Alaska Seas Marine Conservation Biology Workshop in 1997 in Anchorage, Alaska (Center for Marine Conservation 1998).

Premises

During the 1990s, significant changes appeared in pollock and herring stock abundance in the northern Sea of Okhotsk. The herring stock increased 2-2.5 times (from a biomass level of about 1 million tons) and the pollock stock has declined twice (Shuntov et al. 1997). In this regard, a problem appeared in the traditional single-species pollock fishery during the fishing season of 1996-1997. In March and early April of 1997, the pollock fleet moved from western Kamchatka due to significant herring bycatch which exceeded the allowable 8% of the total catch almost everywhere.

The situation became complicated by light ice cover conditions during that winter. Vessels fishing for pollock were obliged to undertake a search for pollock aggregations in the vast ice-free waters. As areas of pollock prespawning migrations and herring wintering widely overlap on the upper slope and outer shelf zones of the northern Okhotsk Sea, herring aggregations often have been caught instead of pollock. Despite the "pollock" target of this fishery, many vessels got pure herring catches. Such unallowable catches cannot be sent to processors. Areas with the highest rate of herring bycatch were temporarily closed to fishing. This led to additional costs for fishing vessels; to pollock quota underutilization of 103,400 tons in this region; and to losses of herring, because unsanctioned herring bycatch was released due to fear of penalties. The fish conservation authorities gradually issued permission for herring bycatch in amounts of 8,000 and then 10,000 and 15,000 tons, which was quickly caught by the fleet. According to official fisheries statistics, the Okhotsk herring bycatch was at 15,200 tons from December 1996 to April 1997.

The problem of precise estimation of the Okhotsk herring stock abundance has been discussed since 1994 (Radchenko and Glebov 1996). According to one viewpoint, in 1994-1997 the spawning stock of the Okhotsk herring was between 398,000 and 608,000 tons. These estimates were based on data from air and diving observations of the herring spawning grounds. Large-scale trawl surveys covering the shelf and offshore zones of the northern Okhotsk Sea during herring feeding migrations showed that the stock was significantly underestimated. Herring was widely distributed through the northern sea and adjacent areas, up to 48°N in the summer of 1995. The spawning stock of Okhotsk herring was assessed as 2.45 ± 0.25 million tons in 1997. Herring MSY was increased to 400,000 tons. Harvest reached 220,700 tons during the traditional fall season "fat" herring fishery. The annual Okhotsk herring harvest totaled 312,800 tons in 1997.

Before the 1997-1998 fishing season, it became clear that the pollock fishery lost single-species consideration in the northern Okhotsk Sea. It must be considered as a bi-species (or mixed stock) pollock and herring fishery and cannot be interrupted due to bycatch of herring or pollock of commercial body size before the herring or pollock quotas are caught. The quota for the pollock fishery was corrected to 460,000 tons (previously 360,000 tons). The allowable Okhotsk herring quota was 50,000 tons for the pollock fishery in 1998 (from January to April). In the northeastern Okhotsk Sea (also called the western Kamchatka fisheries region) the quota for the pollock fishery was established at 850,000 tons. The remaining 1997 quota of 89,000 tons were used in December 1997.

Results

The winter ice conditions in the northern Okhotsk Sea in 1997-1998 differed sharply from the two previous winters. Due to heavy icing of the region, the fishing fleet had to concentrate on the narrow outer shelf area which was free from ice. At of late February, fishing efforts were mainly distributed along the western Kamchatka shelf edge from 52°30′ to 55°30′N (Figure 1). Some herring bycatch occurred there as in previous years (Figures 2 and 3). Local herring aggregations were formed by the Gizhyginsky herring population with their spawning area centered in Shelikhov Bay. This herring stock was not included in consideration of the mixed stock fishery experiment because herring bycatch had not previously been a problem there. However, it became clear in 1998 that the same principles must also be applied in the northeastern Okhotsk Sea now that the Gizhyginsky herring stock abundance has been increasing.

In the northern Okhotsk Sea region (Figures 2 and 3) the pollock fishery mainly began in late March 1998. Fishery conditions were the most stable on the western and northwestern slopes of the TINRO Basin (between 55° and 57°N). Herring appeared there almost in all trawl catches (91.8% from catches of the R/V *Borodino*). It would be impossible to fish pollock there under the previous fishery regulation. Nevertheless, in early April the northern area was closed to the fishery due to a large bycatch of juvenile pollock under commercial body size (32 cm). The pollock harvest totaled 474,600 tons in the northern Okhotsk Sea fisheries region, or 36.5% of the total catch of the western Kamchatka region—1,299,100 tons during the 1997-1998 fishing season.

The total harvest of Okhotsk herring was unexpectedly low during the 1997-1998 fishery season (2,800 tons) and did not reach the established quota. The main reasons were the early closing of the fishing areas due to the large juvenile pollock bycatch and heavy winter ice conditions in the Okhotsk Sea. Vast ice covering partly protected the wintering Okhotsk herring aggregations from fishing. The fishing fleet did not spend so much time searching for fish aggregations due to the narrow ice-free area (i.e., suitable for trawl operation and potential for pollock aggregation formation). The heavy ice condition also prevented herring schools from intermingling with pollock. As a result, pure herring catches rarely occurred despite herring often showing up in pollock catches.

It must be noted that only pure herring catches were usually sent for processing and were included in fishery statistics. It is a traditional procedure to account for fish catches according to the total volume. In this case, mixed pollock and herring catches being difficult to separate are usually destined for fish flour production. Accounting for fish flour value is made in accordance with norms for pollock, the herring catch is included in total harvest statistics as pollock-only catch. All this leads to underestimating the herring harvest and its portion in the total catch of the mixed stock fishery.

During the trawl survey conducted by R/V *Borodino*, average herring catch was 4,394 fish per hour of trawl haul. The fishing fleet spent 5,368 vessel-days in the northern Okhotsk Sea fisheries region during the 1997-1998 season. If the average herring catch was the same for commercial vessels, and they conducted trawl operations 18 hours per day, the underestimated Okhotsk herring harvest would amount to about 5,300 tons

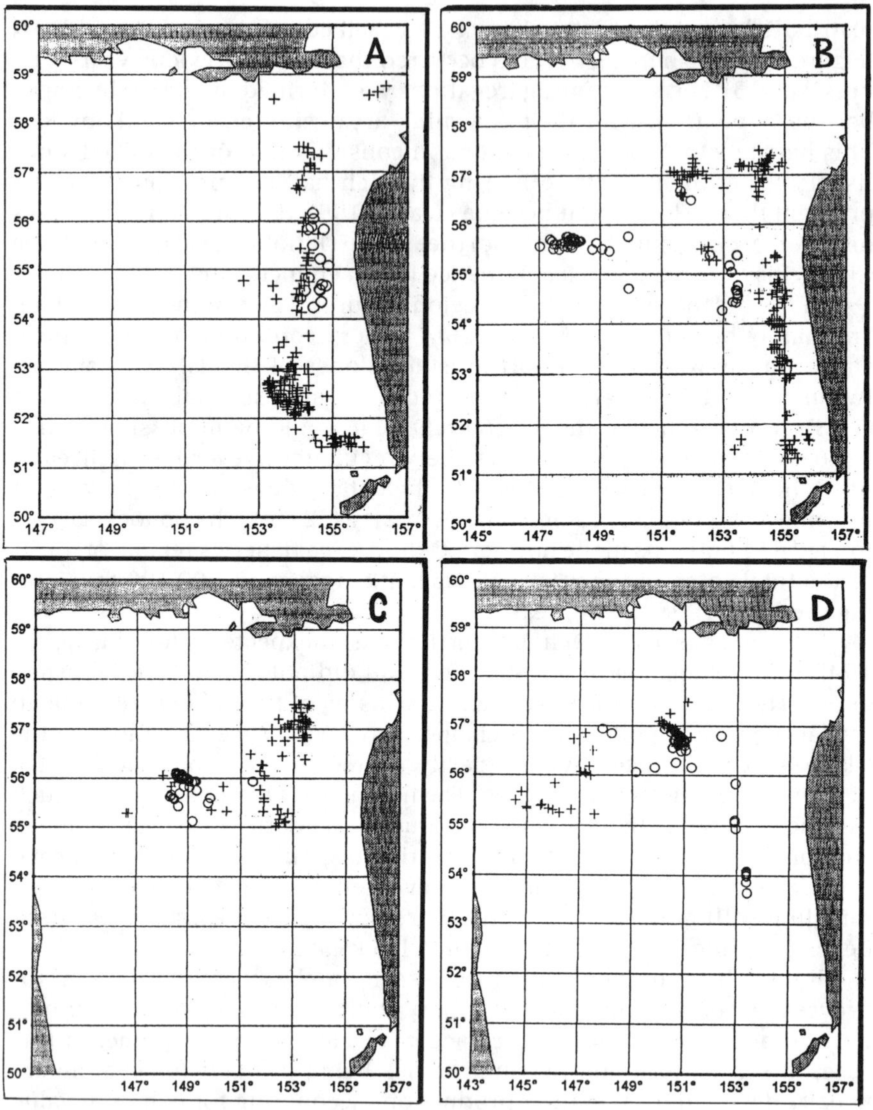

Figure 1. Fishing fleet distribution during the 1997-1998 fishery season in the Sea of Okhotsk: A, 11-19 January; B, 25-31 March; C, 4-5 April; D, 19-30 April 1998.

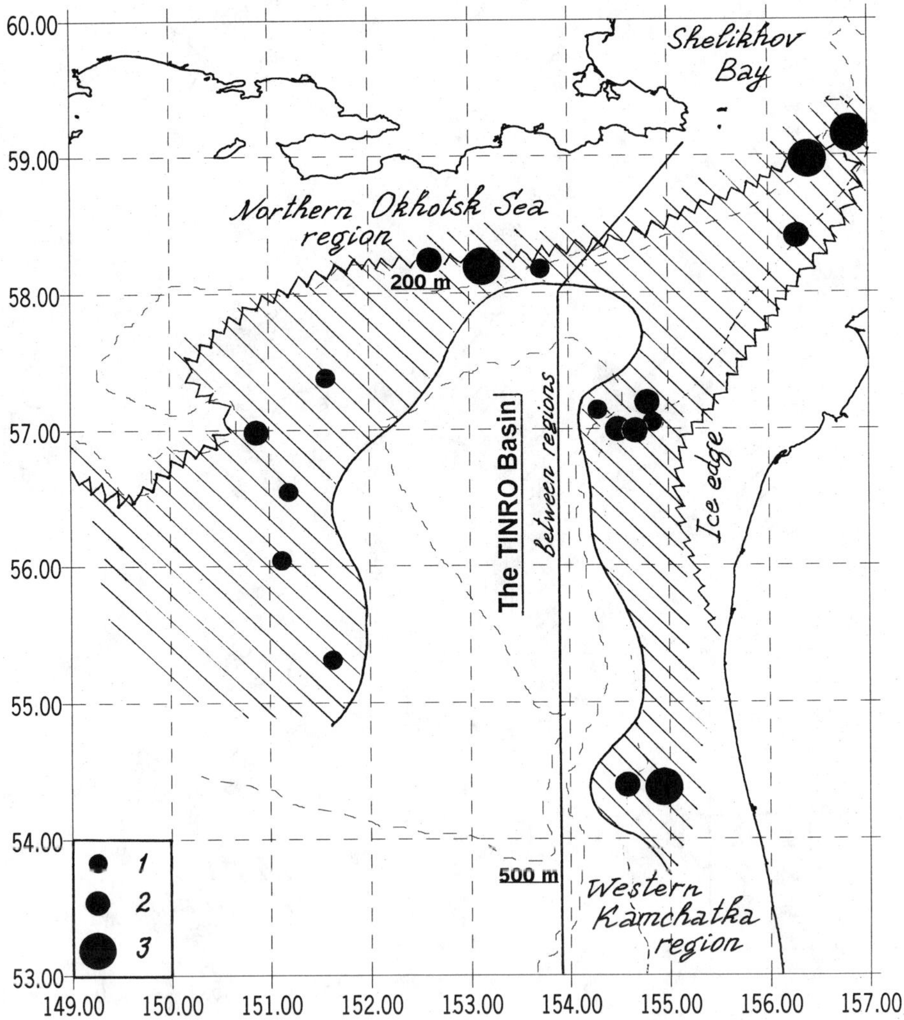

Figure 2. Areas of potential herring bycatch (shaded) during the prespawning pollock fishery in the Okhotsk Sea (combined data of 1995-1996 and 1996-1997 fishery seasons). Points of noticeable herring bycatch by research vessels are given by dots: 1, from 0.1 to 1 tons; 2, from 1 to 5 tons; 3, more than 5 tons per hour of trawl haul. The 500 and 200 m contours, and ice edge for late March are given.

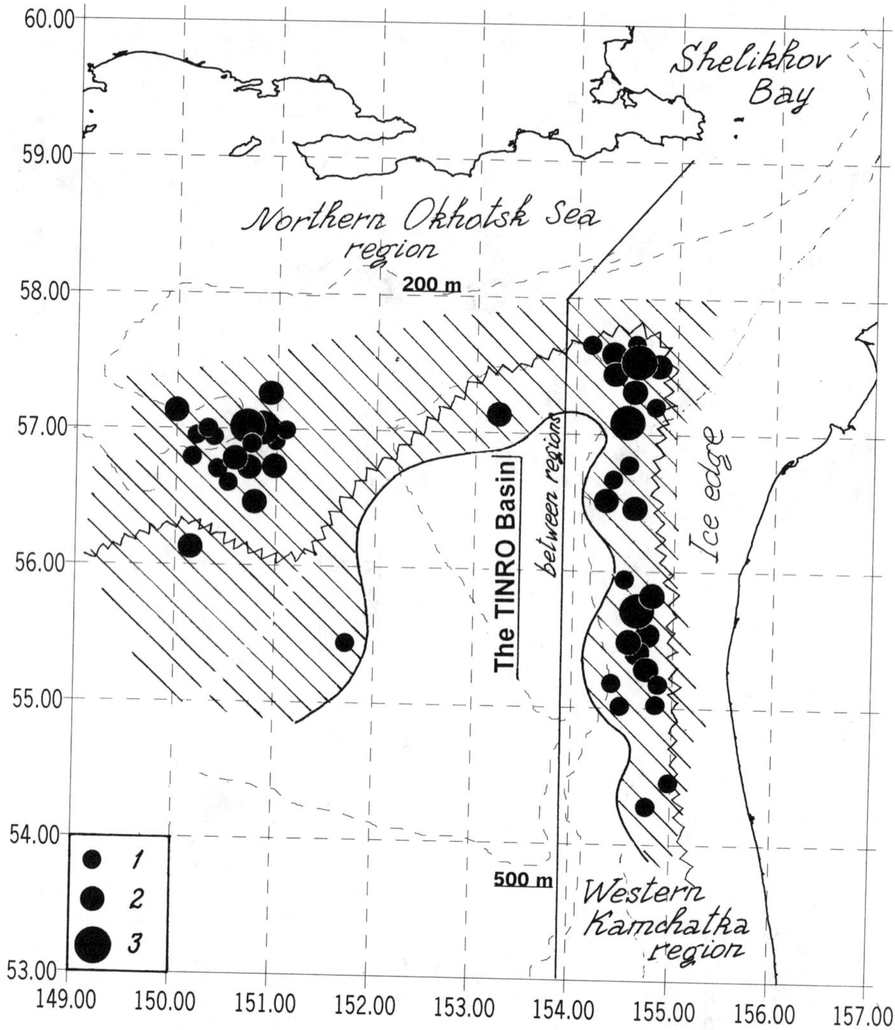

Figure 3. Areas of potential herring bycatch (shaded) during the mixed stock (in the northern Okhotsk Sea) and prespawning pollock (in the western Kamchatka) fisheries during the 1997-1998 fishery season. Points of noticeable herring bycatch by research vessels are given by dots: 1, from 0.1 to 1 tons; 2, from 1 to 5 tons; 3, more than 5 tons per hour of trawl haul. The 500 and 200 m contours, and ice edge for late March are given.

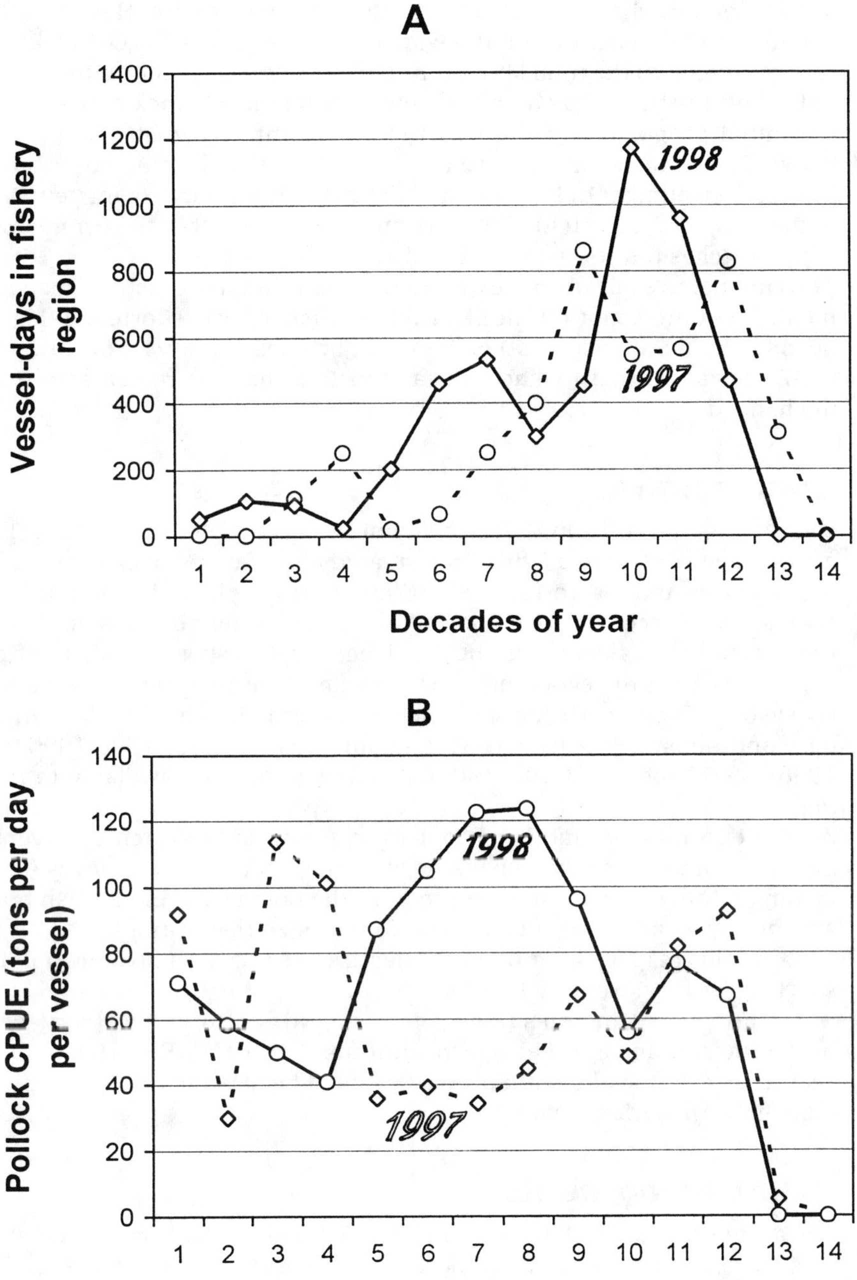

Figure 4. Number of fishery efforts (vessel-days in fishery region) (A) and pollock CPUE (B) in the northern Okhotsk Sea in 1997 and 1998.

in addition to 2,800 tons reflected in the fishery statistics. However, this amounts to an insignificant stock-specific harvest rate for Okhotsk herring compared to the total MSY value. Okhotsk herring stock abundance is only slightly affected by a pollock and herring mixed stock fishery since the initial proportion of the fishery catch comprised of herring is low (Lloyd 1996). Based on data from the R/V *Borodino*, herring contributed only 8.1% in number of fish caught during the survey (approximately 2.5% of the weight of the catch). This portion could be higher in the commercial vessel catches but it scarcely exceeded 10%.

The main result of this experiment was to improve fishery management. The time spent by the fleet for the fishery was shortened (Fig. 4) despite the number of vessels being slightly less (255 vs. 290 in 1996-1997) in the northern Okhotsk Sea, and total harvest remained almost unchanged.

Conclusions

Measures undertaken in the experiment on large-scale mixed stock fishery management were useful. They created more favorable conditions for the fisheries and led to full MSY utilization for pollock in the northern Okhotsk Sea. Accordingly, it allowed a decrease in number of vessels and time spent by the fleet to pursue the fishery. Decreasing the fishery effort is good for both the economic efficiency and conservation of the marine ecosystem. The effectiveness of such an experiment could be higher under conditions of low ice covering (as in the winters of 1995-1996 and 1996-1997), when wintering Okhotsk herring are more available to fishing.

We encountered another problem: a significant bycatch of juvenile pollock prohibited to the fishery. As was suggested by N.S. Fadeyev (Pers. comm., TINRO), it is possible not to limit the pollock fishery by fish body size but by establishing a larger trawl mesh size than was allowed previously. In this respect, continued studies are necessary on juvenile pollock survival for fish swept by trawls.

Improvement in collecting fishery statistics also is necessary, with new principles and approaches. In addition, during the 1998-1999 fishing season mixed stock fishery approaches must be expanded to the western Kamchatka fisheries region.

Acknowledgments

The author thanks Dr. E.P. Karedin, who instilled basic ideas to the mixed stock fishery organization in the northern Okhotsk Sea, and Dr. N.S. Fadeyev, who collected data on fishing fleet and distribution of effort during the fishery season (both from TINRO, Vladivostok, Russia).

References

Center for Marine Conservation. 1998. Ecosystem-Based Management in the Bering Sea. Proceedings of Alaska Seas Marine Conservation Biology Workshop. October 6 and 7, 1997. Anchorage, Alaska. 102 pp.

Lloyd, D.S. 1996. Relative effect of mixed stock fisheries on specific stocks of concern: A simplified model and brief case study. Alaska Fish. Res. Bull. 3(1):21-31.

Radchenko, V.I., and I.I. Glebov. 1996. Present state of the Okhotsk herring stock and fisheries outlook. In: Proceedings of the Workshop on the Okhotsk Sea and Adjacent Areas. PICES Sci. Rep. 6:384-390.

Shuntov, V.P., V.I. Radchenko, E.P. Dulepova, and O.S. Temnykh. 1997. Biological resources of the Russian far-eastern economic zone: Structure of pelagic and bottom communities, present states, trends of long-term dynamics. Izvestiya TINRO 122:3-15. (In Russian.)

Ecosystem Approaches for Fisheries Management
Alaska Sea Grant College Program • AK-SG-99-01, 1999

Identifying Seasonal Spatial Scale for the Ecological Analysis of Herring and Other Forage Fish in Prince William Sound, Alaska

Evelyn D. Brown
University of Alaska Fairbanks, Institute of Marine Science, Fairbanks, Alaska

Jia Wang
University of Alaska Fairbanks, International Arctic Research Center, Fairbanks, Alaska

Shari L. Vaughan
Prince William Sound Science Center, Cordova, Alaska

Brenda L. Norcross
University of Alaska Fairbanks, Institute of Marine Science, Fairbanks, Alaska

Extended Abstract

Recently there has been increasing interest in the distribution, abundance, and ecology of forage fish populations because of the crucial role they play in nearshore and pelagic ecosystems. However, relatively little is known about the ecology of forage fishes in Alaska. From 1995 to 1997 extensive ecosystem studies were conducted in response to the *Exxon Valdez* oil spill in 1989. Monthly broadscale aerial surveys were included in those studies for the purpose of determining distribution and abundance of juvenile Pacific herring (*Clupea pallasi*) and other surface-schooling forage fishes (Brown and Norcross 1997). Many other types of physical and biological data were also available for the same area and dates that could be used for ecological analyses of forage fishes. In order to proceed with hypothesis-driven science, we recognized that basic descriptive life history parameters and the spatial overlap with other ecological parameters needed to be documented first. Therefore, a general research goal was established to perform an ecological analysis of forage fish distribution and abundance using the appropriate spatial scale. The research objective for this effort was to identify and define the appropriate seasonal spatial scale. In order to address the objective we used geographic information system (GIS) methodology since it is a powerful tool for examining spatial processes. Because of the profusion of data available we focused on Prince William Sound (PWS), situated at the northern boundary of the Gulf of Alaska in southcentral Alaska.

A critical starting place for ecological analyses is to define the spatial scale at which processes affect the species of interest. Fish affected mainly by ocean conditions and zooplankton (bottom forced), such as herring, may be affected by one scale during spawning, yet another during the larval stage, and possibly a third during the juvenile stage. Therefore, seasonal effects on spatial scale should be defined first. In addition, traditional nested quadrat analyses may not truly reflect the scale of bottom forcing events. We therefore chose to start by looking at the spatial distribution of forcing events. For this examination of scale, we evaluated some of the available spatial data by season and removed interannual variability by pooling all years for which data were available. In order to gain a better understanding of the timing of physical and biological processes affecting forage fish, we plotted events along a linear time scale pooled over the years of data availability (Fig. 1).

Timing of Events

Most of the ecological activity associated with the timing of forage fish activities in the surface waters (upper 100 m) occurred during the spring and summer months from April to August. In addition, we found that biological processes affecting forage fish during this period were concentrated in the upper 20 m of the water column. The activity was initiated by the formation and strengthening of the stratified layer (data source CLAB [Continuous-Linked Automated Buoy] Buoy 1991-1997; Eslinger 1998; central PWS); this is accompanied by a steady increase of temperature in the upper 20 m of the water column (Fig. 1A).

The phytoplankton bloom peaked in April during the formation of the stratified layer at approximately 20 m at a temperature of approximately 5°C (chlorophyll *a*, CLAB Buoy and southwestern PWS, 1994-1997; Unpublished data, D.L. Eslinger, University of Alaska Fairbanks; McRoy et al. 1997; Eslinger 1998; Fig. 1B). The offshore zooplankton bloom began in April and peaked in June (mid-sound and southwestern PWS, 1981-1997; Cooney 1997, 1998; Unpublished data, R.T. Cooney, University of Alaska Fairbanks). The inshore zooplankton bloom (near the juvenile herring rearing sites; Stokesbury et al. 1997) followed similar timing to the offshore bloom but was apparently more highly concentrated (1996 data; Unpublished data, R.J. Foy, University of Alaska Fairbanks; Fig. 1B).

Pacific herring spawning commenced after temperatures started rising (above 4°C) and peaked when the stratified layer first formed (15 d

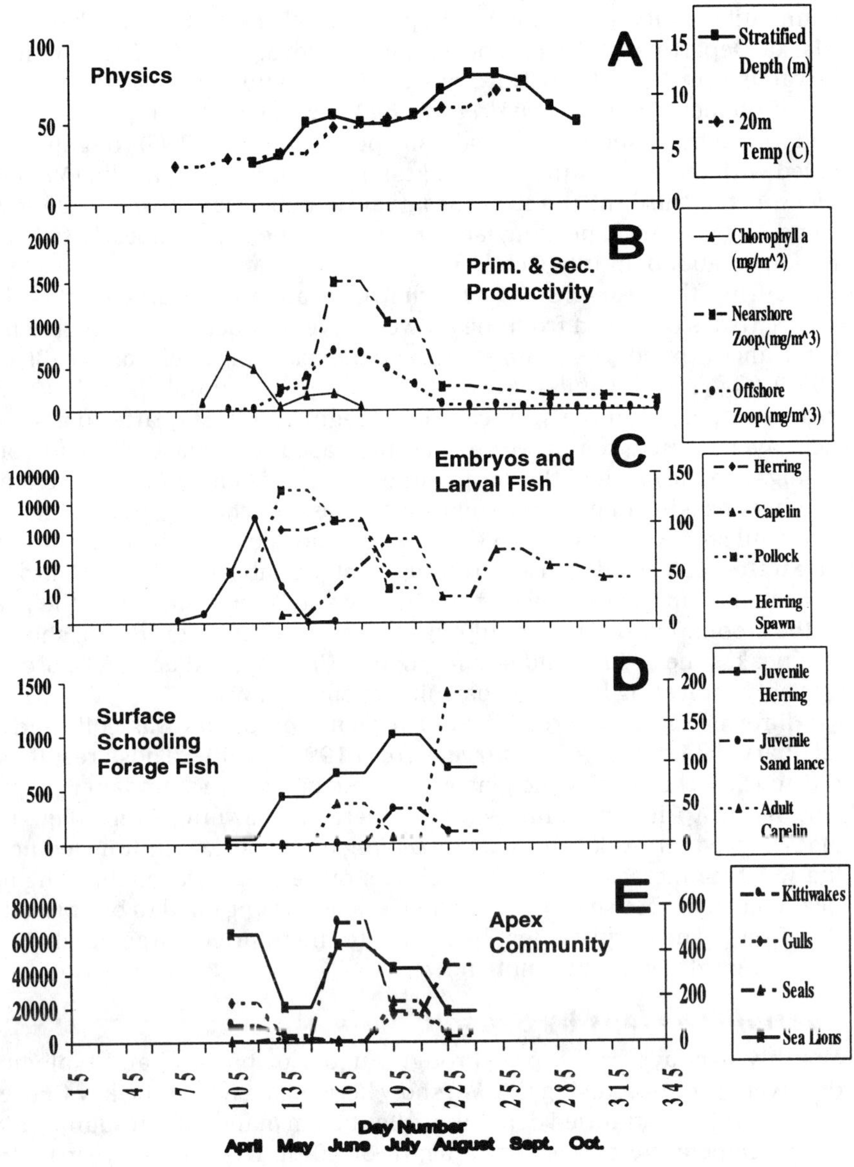

Figure 1. The timing of key ecological events in Prince William Sound, Alaska, including: (A) the formation of the stratified layer (depth in meters, left axis) and ocean temperatures at 20 m of depth (°C, right axis); (B) primary and secondary production (mg/m² or mg/m³); (C) herring spawning (cumulative miles of spawn, right axis) and larval fish (number of larvae/m³, left axis); (D) surface-schooling forage fish (total m² school surface areas; capelin on right axis, sand lance and herring on the left); and (E) apex predators (total number of individuals; kittiwakes and gulls on left axis, sea lions and seals on right).

cumulative estimates of miles of spawn, 1985-1997; Unpublished data, Alaska Department of Fish and Game, Cordova; Fig. 1C). From a single larval fish survey in 1989 (Norcross and Frandsen 1996), peak densities of larval pollock (*Theragra chalcogramma*) and Pacific herring (the second and fourth most abundant larval fish species found in PWS) roughly coincided with the peak of the zooplankton bloom (Fig. 1C). Capelin (*Mallotus villosus*; the third most abundant larval fish species) showed a bimodal trend in abundance, peaking later in July and then again in early fall, and remained abundant into October.

During the years 1995-1997, the most abundant surface-schooling forage fishes observed from the air were juvenile Pacific herring, juvenile sand lance (*Ammodytes hexapterus*), and prespawning adult capelin (Brown 1997) (Fig. 1D). Juvenile herring (mainly age 1; Unpublished data, E.D. Brown) appeared in the surface waters beginning in May after the stratified layer formed, temperatures had increased to at least 5°C, and coinciding with the onset of the zooplankton bloom. Their abundance steadily climbed, peaking in July and August after the age-0 herring joined the age-1 juveniles in the nursery bays. The appearance of sand lance in the surface waters coincided with a recruiting event (Unpublished data, E.D. Brown) of age-0 fish in the nearshore zone in July and August well after the peak of the bloom. The occurrence of large prespawn schools of capelin showed two peaks, one in June and one in August (Unpublished data, E.D. Brown) possibly indicating bimodal spawning events in PWS.

Information on aggregations of foraging predators was collected as auxiliary data during aerial surveys from 1995 to 1997. Those results indicated Steller sea lions and glacous-winged gulls were abundant in aggregations in April, coinciding with the herring-spawning event (Fig. 1E). Black-legged kittiwake and sea lion aggregations peaked in June, coinciding with the increase in surface schools of herring and capelin. The appearance of harbor seals peaked while sea lions continued to be abundant during July and August, most likely due to the return of large numbers of spawning salmon in PWS (not shown).

Spatial Overlays by Season

Monthly overlays from April through August of physical and biological data were produced using GIS. We show here only a few of the key figures.

In April, the stratified depth is at about 25 m in the central sound with a mean temperature of 4.2°C at a depth of 20 m. Over the 20-year hydrographic data set available (but not shown), the warmest regions occurred in northeast, north, and northwest PWS (up to 5.7°C); the coolest (2.9°C) occurred in the central sound associated with anti-cyclonic gyre and at the entrance and exit points to PWS from the Gulf of Alaska (GOA) (Fig. 2). The densest water was associated with the central gyre and the least dense water was associated with warmer regions and downstream from western glacial-filled fjords. The residual currents were generally low, ranging from 0.001 to 0.210 m/s with a mean flow of 0.04 m/s (Fig. 2). The flow into and out of

PWS during this month was minimal. Herring spawning regions occurring during the 1980s were generalized. Spawning regions occurred adjacent to, south, north, and east of the central gyre, but never west; all spawning occurred well inside the "oceanographic boundaries" of PWS. During this month all the surface-schooling fish (all herring), gulls, and aggregations of sea lions were associated with the herring spawn. Larval fish abundance was generally low and only pollock larvae were observed (Fig. 2).

In May, the stratified depth had increased to 30 m with an average temperature at 20 m of 5.5°C. The temperatures all over PWS showed a large increase from April (not shown), but the northeastern region of PWS remained the warmest (up to 6.9°C) and the regions associated with high flow and gyres the coolest (down to 4.2°C). Some of the "hot spots" in PWS were now associated with exchange of GOA water at the entrance and exits. The overall density of the water was lower during May with the least dense water still associated with higher temperatures. Ocean flow increased from April to a range of 0.004-0.284 m/s with a mean flow of 0.042 m/s (Fig. 3). The central gyre intensified, reversed direction (becoming anticyclonic), and extended farther north. In addition, the net flow into and out of PWS greatly increased during this month. Topography steered cross sound flow was also observed in the north region of PWS. Larval fish peaked in abundance during May and were broadly distributed (except capelin that was only observed in the northeast region) with and outside of the gyre-related currents. Note that herring larvae were as abundant west of the spawning regions as well as adjacent to them.

By June, temperatures at 20 m had risen significantly (mean at 7.3°C) with the stratified layer depth approaching 50 m. The temperatures during June were more uniform over the sound, with a range of 5.4-9.8°C; however, "hot spots" were still observed in the eastern sound and adjacent to the GOA. Density declined slightly from May, but apparently in response mainly to changes in salinity rather than temperature and salinity. Although the range of ocean flow lessened slightly from May (to 0.001-0.204 m/s), the mean flow was higher at 0.054 m/s. The flow into and out of the sound intensified from May with the formation of smaller topography-steered cyclonic gyres east and west of the inflow; overall, the exchange of surface water appeared to have increased over May. Larval fish were still relatively abundant and widespread (not shown), with an increase in the relative abundance of herring over pollock. No capelin larvae were observed. The most notable feature in June was the increase in appearance of surface-schooling forage fishes (Fig. 4). The most abundant species inside the sound was juvenile herring and outside the sound was prespawn adult capelin. A few sand lance schools were also observed inside the sound. Black-legged kittiwakes were highly associated with surface schooling forage fishes, with notably large flocks associated with aggregations of spawning capelin. Gulls and sea lions (not shown) did not appear to be significantly associated with the juvenile herring; however, aggregations of both were observed near spawning capelin.

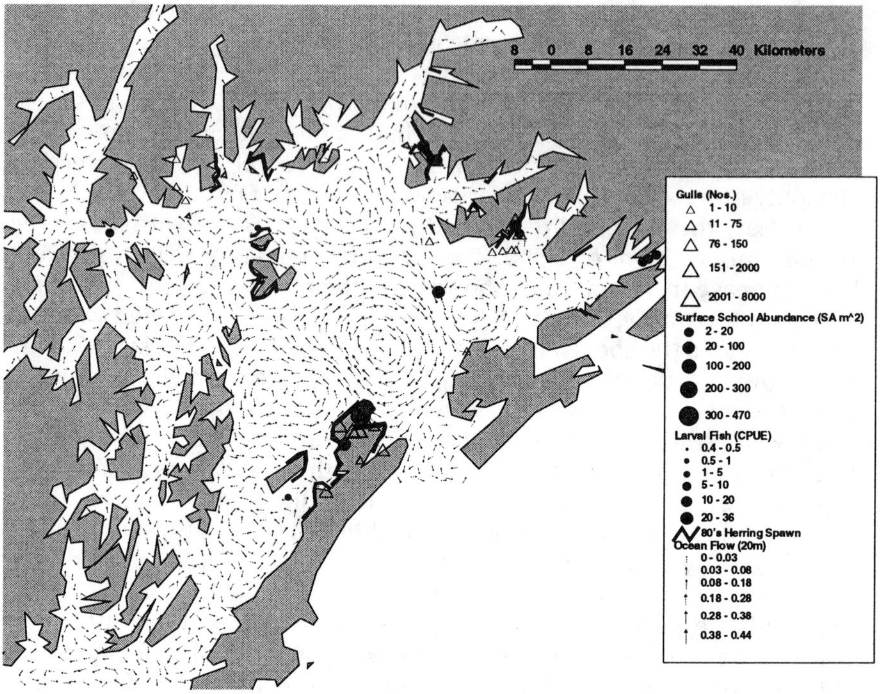

Figure 2. Ecological features within Prince William Sound, Alaska, during the month of April.

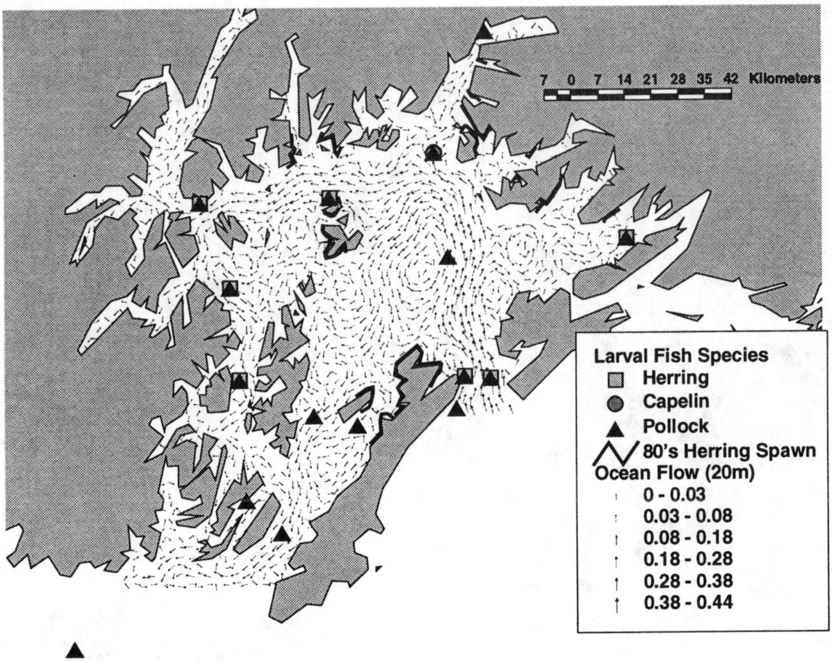

Figure 3. Ecological features within Prince William Sound, Alaska during the month of May.

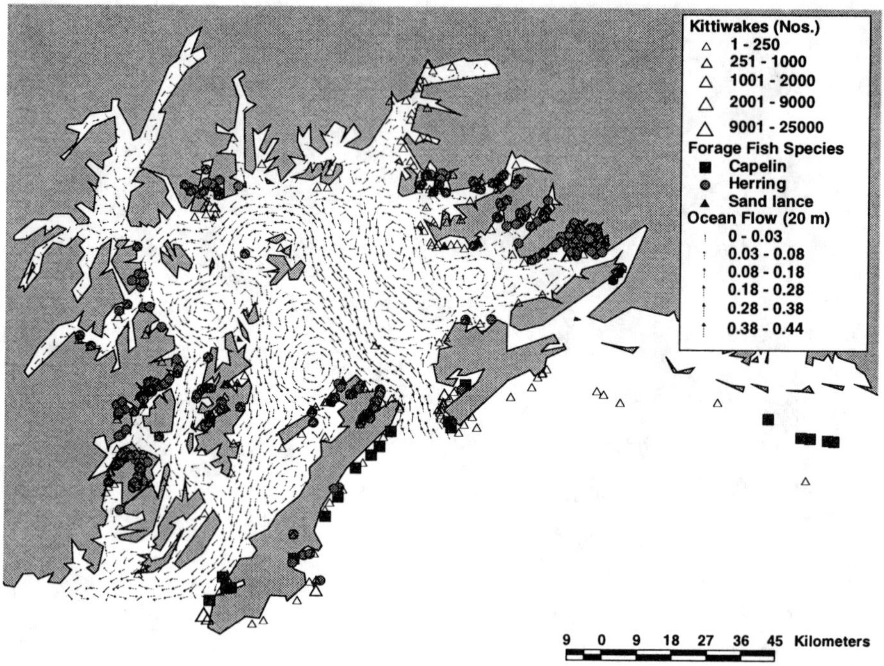

Figure 4. Ecological features within Prince William Sound, Alaska during the month of June.

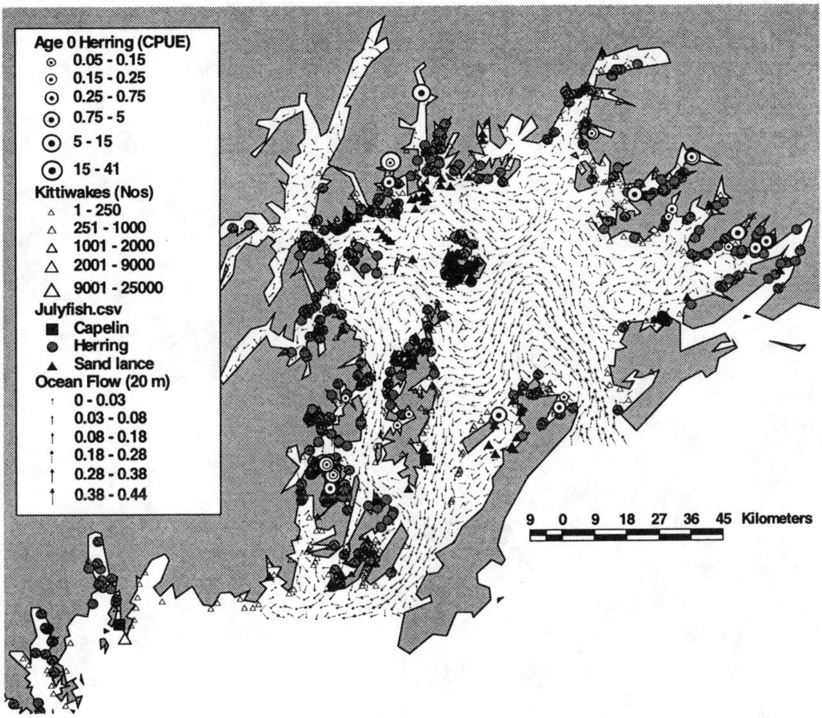

Figure 5. Ecological features within Prince William Sound, Alaska during the month of July.

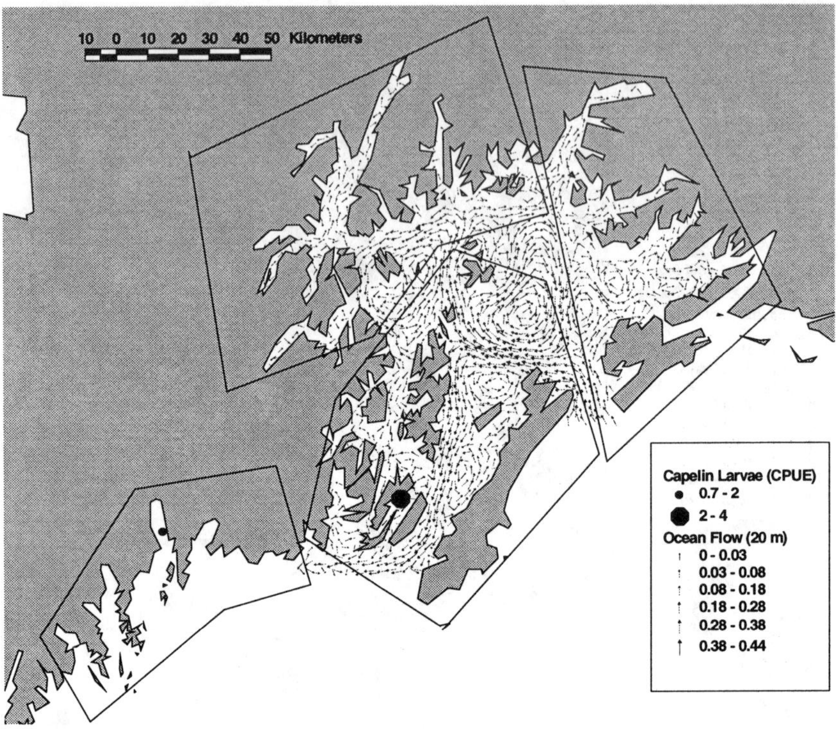

Figure 6. Ecological features within Prince William Sound, Alaska during the month of August and delineation of proposed ecological "regions."

In July, the temperatures at 20 m climbed to 10.1°C but the depth of the stratified layer was similar to its depth in June. The overall temperature range was 6.0-13.6°C, with the warmest regions occurring again in the eastern and, additionally, the northern sound. The coolest region remained in the central sound associated with strong currents. Densities were similar to those in June. Ocean flows were also similar to those in June, ranging from 0.0002 to 0.315 m/s with a mean flow of 0.054 m/s (Fig. 5). Larval fish were overall less abundant than in June but still widespread, with a decrease in the occurrence of both pollock and herring; the notable feature was the occurrence of large numbers of capelin broadly distributed within the sound (probably resulting from the spawning events observed in June; not shown). The numbers of surface-schooling fish increased significantly over June largely due to an increase in both herring and sand lance (Fig. 5). This distribution of herring and sand lance extended well west of PWS along the outer Kenai Peninsula. We believe this increase is due to recruiting events of age-0 fish for both species (based on unpublished data of E.D. Brown). As in June, kittiwakes were largely associated with the surface-schooling fishes while aggregations of associated gulls and sea lions were not observed.

In August, both the stratified depth (over 50 m) and temperatures at 20 m (mean 10.4°C) peaked over this 5-month period. Temperatures ranged from 4.3 to 12.7°C with the same pattern of higher temperatures around the "rim" of PWS but cooler temperatures associated with, presumably, ice-melt in the northern fjords (not shown). Ocean density and salinity remained very similar to those in July, although ocean flow peaked during this month (over the 5-month series) with a range of 0.0002-0.435 m/s and a mean of 0.082 m/s (Fig. 6). The flow into and out of the sound remained strong but the central gyre reversed direction to generally anticyclonic flow. Winds within PWS determined to a great degree the strength and directions of the gyres and surface currents. All the biological activity observed in the surface waters declined greatly during this month (not shown). Although capelin larvae were still present, they appeared to have migrated or advected out of PWS (Fig. 6).

The conclusion of this exercise is that there is variability in the seasonal spatial scale of bottom-driven forcing events in PWS. In an analysis of spatially explicit ecological processes in PWS, there appear to be three major regions to be considered (Fig. 6). The eastern portion of the sound is generally warmer, east of but influenced biologically by the central gyre and inflow, and seasonally highly variable. The northern and northwestern part of the sound is generally quieter, less variable, and less influenced by the inflow. The central region adjacent to the GOA is very dynamic, affected by the most intense flows, and is very likely responding to events outside of PWS to a much greater degree than the other two regions. This last region is also probably associated with greater losses of larval fish (advection of larvae out of PWS). The area off the outer Kenai Peninsula may be an "overflow" area for biological processes occurring within the sound.

References

Brown, E.D., and B.L. Norcross. 1997. Assessment of forage fish distribution and abundance using aerial surveys: Survey design and methodology. In: R.T. Cooney, Sound Ecosystem Assessment (SEA): An integrated science plan for the restoration of injured species in Prince William Sound, Appendix I, Chapter 11. FY96 Annual Report for the Exxon Valdez Trustee Council, Anchorage, Alaska, pp. 25-53.

Brown, E.D. 1997. Preliminary documentation of temporal and spatial variability of Pacific herring, other forage fish, and seabirds in Prince William Sound, Alaska. In: R.T. Cooney, Sound Ecosystem Assessment (SEA): An integrated science plan for the restoration of injured species in Prince William Sound, Appendix II, Chapter 10. FY96 Annual Report for the Exxon Valdez Trustee Council, Anchorage, Alaska, pp. 44-79.

Cooney, R.T. 1997. The role of zooplankton. Restoration Project 97320-H. In: R.T. Cooney, Sound Ecosystem Assessment (SEA): An integrated science plan for the restoration of injured species in Prince William Sound, Chapter 4. FY96 Annual Report for the Exxon Valdez Trustee Council, Anchorage, Alaska. 35 pp.

Cooney, R.T. 1998. The role of zooplankton. Restoration Project 97320-H. In: R.T. Cooney, Sound Ecosystem Assessment (SEA): An integrated science plan for the restoration of injured species in Prince William Sound, Chapter 4. FY97 Annual Report for the Exxon Valdez Trustee Council, Anchorage, Alaska. 22 pp.

Eslinger, D.L. 1998. Biophysical modeling and validation through remote sensing. In: R.T. Cooney, Sound Ecosystem Assessment (SEA): An integrated science plan for the restoration of injured species in Prince William Sound, Chapter 5. FY97 Annual Report for the Exxon Valdez Trustee Council, Anchorage, Alaska. 37 pp.

McRoy, C.P., A. Ward, E.P. Simpson, K. Tamburello, J. Cameron, S. McCullough, and P. Cassidy. 1998. Phytoplankton and nutrients. Restoration Project 97320G. In: R.T. Cooney, Sound Ecosystem Assessment (SEA): An integrated science plan for the restoration of injured species in Prince William Sound, Chapter 3. FY97 Annual Report for the Exxon Valdez Trustee Council, Anchorage, Alaska. 85 pp.

Norcross, B.L., and M. Frandsen. 1996. Distribution and abundance of larval fisheries in Prince William Sound, Alaska, during 1989 after the *Exxon Valdez* Oil Spill. American Fisheries Society Symposium 18:463-486.

Stokesbury, K.D.E., E.D. Brown, R.J. Foy, and B.L. Norcross. 1997. Juvenile herring growth and habitats. Restoration Project 95320T Annual Report. In: R.T. Cooney, Sound Ecosystem Assessment (SEA): An integrated science plan for the restoration of injured species in Prince William Sound, Chapter 11. FY96 Annual Report for the Exxon Valdez Trustee Council, Anchorage, Alaska. 75 pp.

Development of a Marine Habitat Protection Area in Bristol Bay, Alaska

David Ackley
Alaska Department of Fish and Game, Juneau, Alaska

David Witherell
North Pacific Fishery Management Council, Anchorage, Alaska

Abstract

An extensive area in Bristol Bay, Alaska, was proposed as a no-trawling zone to protect red king crab habitat from potential impacts. The Bristol Bay red king crab stock has remained at low abundance levels since a stock collapse in 1980. A literature review indicated that the proposed area encompassed all available habitat essential for red king crab juveniles of this stock. Published studies have shown that trawling could potentially degrade high-relief living habitat used extensively by red king crab juveniles, yet portions of the proposed area were also important and lucrative areas for trawl fisheries targeting yellowfin sole. Analysis indicated that a sub-area within the proposed closure area could be open to trawling on a seasonal basis and not severely impact crab habitat or increase crab bycatch. Closure of the remaining larger area to trawling could result in improved crab recruitment and subsequent benefits to crab fisheries.

Introduction

In January 1995, the North Pacific Fishery Management Council (NPFMC) reviewed a proposal to institute a trawl closure in the northeast and eastern sections of Bristol Bay. In combination with existing closures, all of Bristol Bay was proposed to be closed to trawling. It was felt that such a closure could reduce crab bycatch and protect red king crab (*Paralithodes camtschaticus*) habitat from potential degradation due to trawling. The directed red king crab pot fishery was closed in 1994 due to low abundance, and crab fishing representatives had requested that the NPFMC do what they could to rebuild this red king crab stock. Proponents of the closure believed that trawl closure areas would protect crabs from unobserved mortality by contact with passing gear and prevent habitat losses they believed were associated with trawl gear. Trawl fishing representatives agreed that actions should be taken to rebuild crab populations, but were concerned that a large-scale area closure could force them out of local areas in northern Bristol Bay where flatfish catch rates were reputed to be high and crab bycatch rates were low.

Several different fisheries occur in various areas of Bristol Bay. Pacific herring (*Clupea pallasi*) and Pacific salmon (*Oncorhynchus* sp.) gillnet fisheries take place in the spring and summer months within state waters (within 3 nautical miles from the coast). Commercial fisheries for red king and Tanner crabs (*Chionoecetes bairdi*) are prosecuted with pot gear largely during November in recent years. Trawl fisheries generally target yellowfin sole (*Pleuronectes asper*) in an area within the northern part of the bay after the ice cover melts (April-June). Large-scale trawl, longline, and pot fisheries directed at groundfish occur just outside Bristol Bay (west of 162°W). Prior to implementation of earlier area closures, some trawling was directed at Pacific cod (*Gadus macrocephalus*) and rock sole (*Pleuronectes bilineatus*) within the western part of Bristol Bay.

Area closures have been established in Bristol Bay for a number of reasons. In 1959, the Japanese prohibited their trawl vessels from fishing in a king crab pot sanctuary comprising the southern portion of Bristol Bay to minimize interaction with its crab pot and tanglenet fisheries. This area remained off-limits to trawling until 1983, when the NPFMC Bering Sea Groundfish Fishery Management Plan was amended to enhance a developing domestic trawl fishery by allowing fishing in this productive area. A crab protection zone was implemented by the NPFMC in 1987 to prevent the incidental catch of adult male and female red king crabs in the domestic trawl fisheries. This area, coincidental with the current National Marine Fisheries Service (NMFS) statistical area 512, was closed to trawling year-round and covered a substantial portion of the red king crab mating area (Fig. 1). In 1989, the current NMFS statistical area 516 was selected for closure during the period April 15-June 15 annually specifically to help protect mating and molting red king crabs. State waters bordering most of Bristol Bay were closed to trawling to protect mating and molting crabs. In 1967, much of Bristol Bay was declared a halibut nursery by the International Halibut Commission, and all commercial halibut longlining was prohibited from this area. In 1990, 12-mile buffer zones were established in NMFS regulations around three walrus haulouts in northern Bristol Bay. It was felt that prohibiting all vessels from entering these areas would minimize human interactions with these marine mammals.

Landings of red king crabs peaked at 60 million kg in 1980, followed by a stock collapse to very low abundance by 1983. No fishery occurred in 1983, but signs of improved abundance allowed fisheries to be prosecuted at low levels (roughly 2-7 million kg per year) through 1993 (Otto et al.

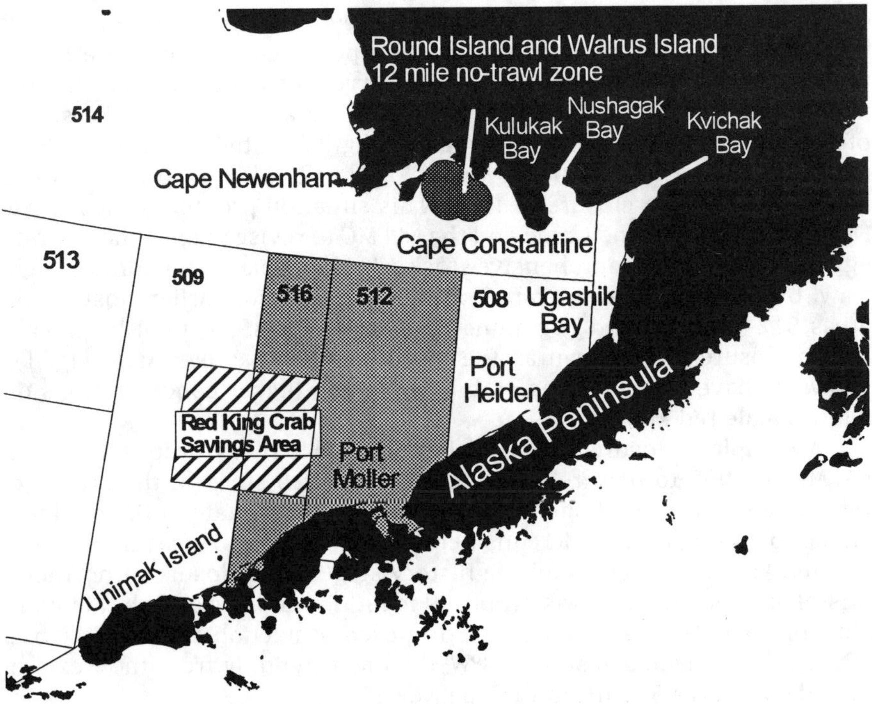

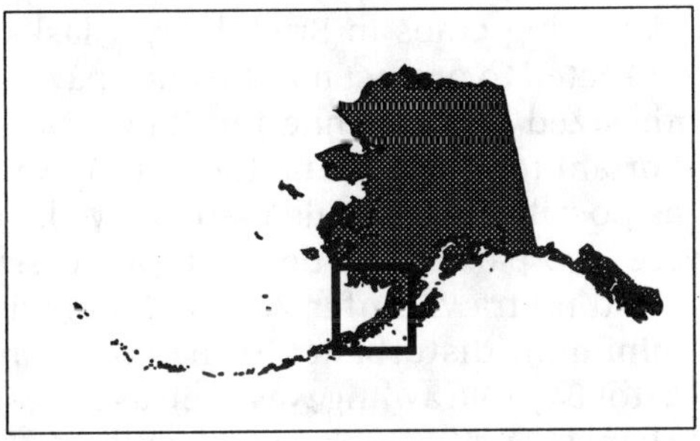

Figure 1. *Map of Bristol Bay, Alaska, with NMFS three-digit statistical areas and place names. Trawl closure areas are variously shaded and include the Round Island no-trawl zone, areas 512 and 516, and the Red King Crab Savings Area. For reference the western edge of area 512 is 163°W.*

1997). In 1994, the NMFS bottom trawl survey revealed low abundance of prerecruit red king crabs, indicating that poor year classes were being produced compared to the 1970s. The survey also detected that the female red king crab stock in Bristol Bay was below an established threshold of 8.4 million females >90 mm carapace length (CL), the mean size at 50% maturity. As a result, the fishery was closed in 1994. Similar survey estimates resulted in a closure in 1995. This situation prompted the Alaska Department of Fish and Game and the NPFMC to revise management strategies and develop comprehensive stock rebuilding plans. In addition, high trawl bycatch of red king crabs continued in spite of earlier closures of areas 512 and 516, and beginning with actions in 1994, the NPFMC created a closure area known as the Red King Crab Savings Area (Fig. 1), shown to have high red king crab bycatch and higher concentrations of adult female red king crabs.

A complete closure of Bristol Bay to trawling was proposed to the NPFMC in 1995 to help rebuild the red king crab stock in the area and revitalize the directed fishery. Due to the depressed state of the red king crab population, the NPFMC considered these additional measures to protect red king crabs across all life history stages. The closure of northern Bristol Bay and area 508 was proposed to reduce impacts on habitat favorable for juvenile crab survival. The proposed protection area would close all waters to trawling east of 163°W, the line extending from the western boundary of area 512 north to Cape Newenham.

This paper is a retrospective look at the information the NPFMC considered in 1995 and 1996, and the actions taken in response to continued low abundance of red king crabs in Bristol Bay, Alaska. The NPFMC had recently (1994-1996) acted to protect mature and spawning red king crabs. It had been hypothesized that juvenile red king crab settlement stages were the most important to survival, and the NPFMC wanted to provide as much protection as possible to juvenile crabs as well. Further ecosystem considerations were also part of the decision process. Bristol Bay is important to walrus, and no-trawl buffer zones had previously been established to afford minimum disturbance to haulouts and rookeries. The closure of all of Bristol Bay to trawling was seen as a means to broaden the areas of nondisturbance. Further, annual spawning of herring is concentrated in portions of the northern shore of Bristol Bay, and the perceived impacts that trawling could have on the migration paths and schools of herring were seen to be diminished by a closure. Bristol Bay is also important to migrating and nesting seabirds, and although the impacts of trawling on seabirds has been little studied, reducing the co-occurrence of seabirds and fishing activity was postulated as being a positive side-benefit.

The information presented to the NPFMC included a literature search on juvenile crab habitat requirements and effects of trawling on bottom habitat. An expanded literature search is provided in this paper. Where possible, data were analyzed that could verify or augment information from the literature and from public testimony. A Windows NT-based PC running ARC/INFO and ArcView geographic information system (GIS) programs was used to create the maps in this document and perform the spatial analyses for the NPFMC.

Area Description

Bristol Bay comprises the southeast corner of the Bering Sea shelf, including the waters north and east of Port Moller and south and east of Cape Newenham (Fig. 1). The shelf in this area is characterized by relatively smooth topology in waters <100 m. The shallow nearshore shelf deposits consist of gravel and coarse sand along the open shore, with clay and silt also found in some bays or coastal indentations including the shallow flat expanses near the mouths of the Kvichak and Nushagak rivers. The size of sediment decreases with distance from shore, moving from coarse sand through fine sand to mud as depth increases (Sharma 1979). Figure 2 provides a composite of the various bottom sediments graded on a sand-silt continuum as presented in Sharma (1979), and the general distribution of gravel coverage as presented in McDonald et al. (1981). The more coarse sediments are located to the north of the Alaska Peninsula and in the inner portion of Bristol Bay. Northern Bristol Bay, west and north of Cape Constantine, generally has finer sand and silt nearer the shoreline. Trawl locations in the commercial fishery for yellowfin sole overlap the areas of finer sand coverage (indicated in Fig. 2 and discussed below).

Crab Biology, Distribution, and Habitat Requirements

Bering Sea red king crab females bear an egg clutch for 11 months, and the eggs hatch prior to the female molting and extruding a new batch of eggs. Hatching generally occurs in April and May (Pereyra et al. 1976, Armstrong et al. 1981, McMurray et al. 1984). Larvae are subsequently in the water column for approximately 10 weeks (Pereyra et al. 1976, Wainwright et al. 1992) before settling to the bottom in July or August (Armstrong et al. 1981) and metamorphizing to a megalops stage and a subsequent benthic existence. The timing, locations, and movement of larval crabs have been studied in the Outer Continental Shelf Environmental Assessment Program (OCSEAP) which was undertaken to investigate possible impacts of oil exploration and drilling in the eastern Bering Sea and Bristol Bay. Armstrong et al. (1981) and McMurray et al. (1984) investigated larval and juvenile red king crab distributions and hypothesized on larval drift as part of this series of studies, and Hebard (1959) and Takeuchi (1962) reported on initial investigations into current and larval drift. In general, larval drift is with the current into Bristol Bay.

Habitat preferences for juvenile red king crabs were investigated by McMurray et al. (1984), Rice et al. (1988), and Rounds et al. (1990). In summary, juveniles prefer a solitary existence in high-relief habitat, pref-

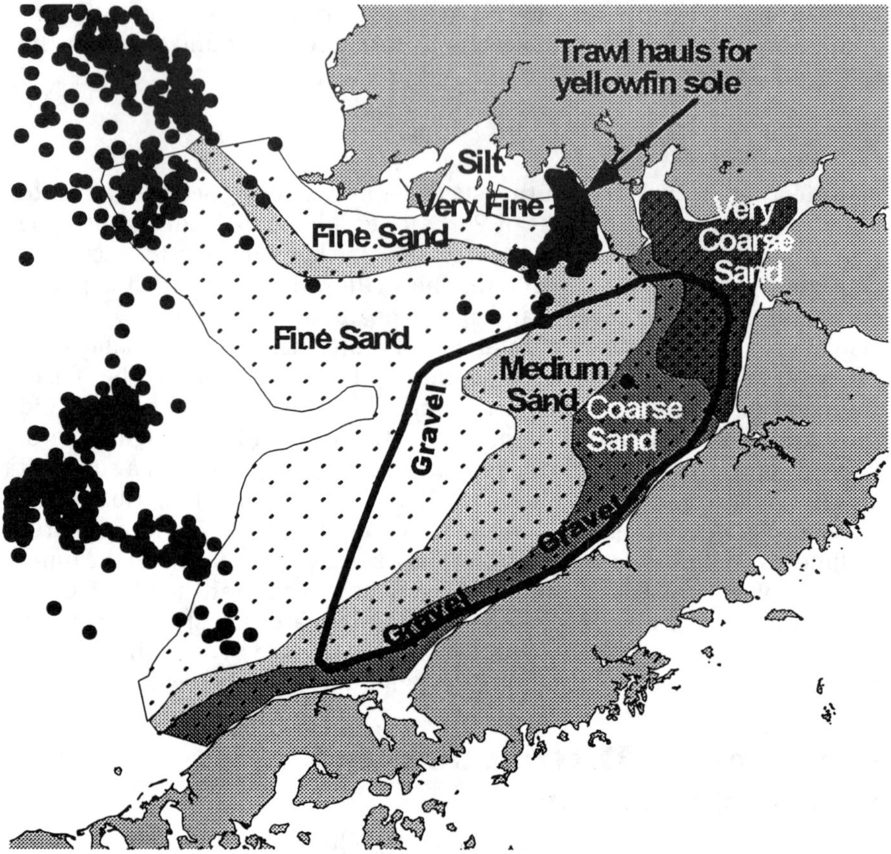

Figure 2. Bottom sediment gradations in Bristol Bay (shaded areas) as provided by Sharma (1979). The general distribution of gravel is as indicated by McDonald et al. (1981) and portrayed by a heavy dark line. The locations of trawl hauls targeting yellowfin sole in 1994 are indicated by large dots.

erably with abundant biota for protection and food. Specifically, juveniles are mainly found among biogenic assemblages, such as tube-building polychaete worms, sea onion (*Boltenia ovifera*), bryozoans, mussels, kelp, and ascidians (McMurray et al. 1984, NMFS 1991, Armstrong et al. 1993). If no epifaunal community exists, juveniles can be found on rocky or gravel substrate, but it is considered to be inferior habitat (McMurray et al. 1984). A shallow-water survey for juvenile red king crabs in Bristol Bay found most small crabs (4-11 mm CL) in stands of mussels, attached to bryozoans or hydroids at the base of sea onion stalks, among clumps of polychaete tubes, or in tows with abundant *Asterias* or *Evasterias* (NMFS 1991).

Juvenile red king crabs are found in shallow waters <30 m in depth and into the littoral zone (Pereyra et al. 1976), or in the eulittoral zone between 20 m and 60 m (Armstrong et al. 1993). In a review of NMFS annual survey data, Stevens (1990) found that most small crabs (<50 mm CL) were captured in depths <45 m and the major stations encountering crabs of this size were within 30 nautical miles of estuaries along the Alaska Peninsula. A survey of Port Moller specifically for red king crab juveniles (Wainwright et al. 1992) found few juveniles in either soft bottoms or in rock surveyed with a small rock dredge. Nearly all of the juveniles encountered were found in a foot survey of the intertidal area. There is a tendency of juveniles between 20 and 60 mm to form pods (Armstrong et al. 1993), and as red king crabs near adulthood, they join the adults in deeper waters. Adult crabs annually migrate to shallower nearshore waters to spawn in the winter and spring and back to deeper waters in summer and autumn (Takeshita et al. 1990).

Survey Information on Crab Abundance and Distribution

Data from the annual summer NMFS Bering Sea trawl survey (in this case, 1975-1997) was used to examine locations of crab abundance. The annual trawl survey covers the intersection points of a 20 × 20 mile grid throughout lower Bristol Bay and along the Bering Sea shelf. Crabs are identified by sex, measured for carapace length and other biological data are collected. Surveys to assess king crab stocks in the Bering Sea began in 1955 (Hayes 1983) and annual assessments of the red king crab stock in Bristol Bay have been conducted since 1968 (e.g., Otto et al. 1997). In addition to overall abundance estimates, the NMFS survey reports estimates for three size groups of male crabs (juveniles <110 mm CL, prerecruits ≥110 mm CL and <35 mm CL, and legals ≥135 mm CL) and two size groups of female crabs (small females <90 mm CL and large females ≥90 mm CL). While the annual trawl survey is capable of assessing larger crabs, the trawls are generally unable to retain small (<30 mm CL) crabs (Reeves and Marasco 1980, Armstrong et al. 1981), and the survey does not occur in the shallow waters believed to be the principal habitat for young crabs (Armstrong et al. 1981, Armstrong et al. 1993).

A supplementary survey cruise was conducted prior to the annual survey in 1991 to specifically sample for juvenile red king crabs and juvenile flatfish (Stevens and MacIntosh 1991). Seven transects were conducted in waters shallower or with rougher bottom type than is normally included in the annual trawl survey. Haul locations, numbers, and sizes of juvenile crabs encountered in the survey were contained in the cruise report. Similarly, in 1995 a supplementary cruise was made of shallow-water locations in Togiak and Kulukak bays in northern Bristol Bay to assess yellowfin sole and other flatfish in the area (NMFS 1995).

Although representing only seven transects along the northern Alaska Peninsula and into Kvichak Bay, the survey for juvenile red king crabs in 1991 (Stevens and MacIntosh 1991) yielded some interesting information which helps to confirm the location and habitat preferences of juvenile crabs. The station data provided in the cruise report are reproduced in Fig. 3. The station furthest into Kvichak Bay was located over very rough substrate, and even a modified scallop dredge met with limited success. The results of the survey do confirm, however, that the smallest crabs (4-11 mm CL in this case) are located in the shallower stations, and in the vicinity of estuaries as had been reported in Stevens (1990). A large portion of these small crabs were found to be in close association with the biota in the area as discussed above. Crabs in the middle size group (15-49 mm CL) appeared to be distributed similarly to the small size class. However, a haul nearest the shore in the vicinity of Port Moller at a depth of 20 m had a high catch of small crabs with no crabs from larger size categories (Fig. 3). The larger crabs (>49 mm CL) were strongly segregated from the smaller size groups and found in deeper water (Stevens and MacIntosh 1991).

McMurray et al. (1984) reported juvenile red king crabs (age 0+) within Togiak Bay. Additional NMFS survey hauls were made into these shallow waters in 1989 and 1991 because of the trawl fishing activity, and no red king crabs were found in the hauls within Togiak Bay (Stevens and MacIntosh 1989, Stevens et al. 1991). A cruise to sample the shallow waters of Togiak and Kulukak bays specifically to assess abundant flatfish such as yellowfin sole was conducted in May of 1995 (NMFS 1995). This survey found

Figure 3. (Facing page.) Catch of crab in a supplemental NMFS trawl survey (NMFS 1991) specifically to locate juvenile crab near Port Moller, Port Heiden, and Kvichak Bay in Bristol Bay. Size of symbols is proportional to the catch in the tow. Three size groups are presented as follows: small juvenile red king crab (4-14 mm) as an asterisk; medium juvenile crab (15-49 mm) as a box; and large juvenile crab (>49 mm) as a circle.

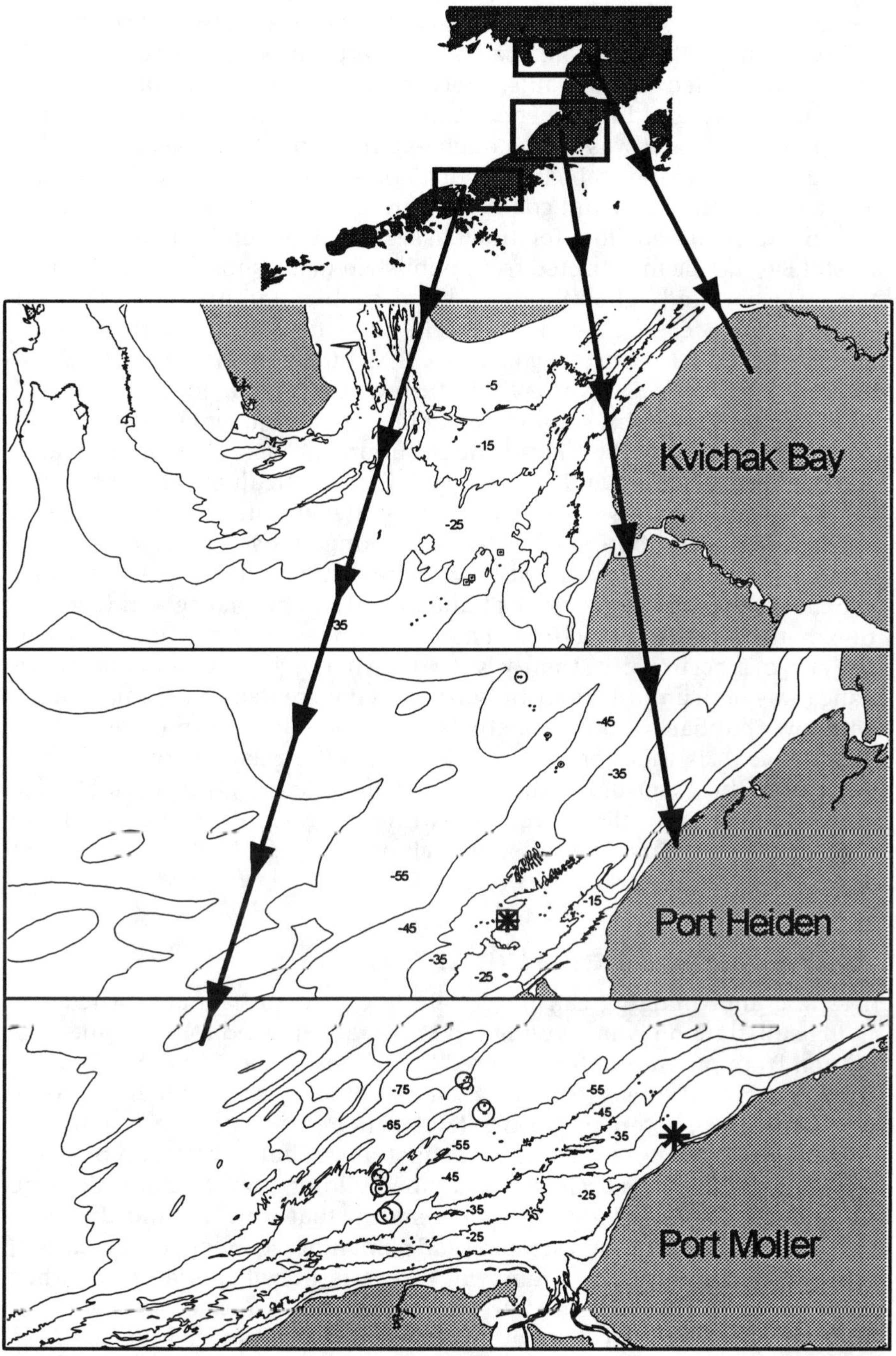

Kvichak Bay
Port Heiden
Port Moller
-5
-15
-25
-35
-45
-35
-55
-45
-15
-35
-25
-75
-55
-45
-65
-35
-55
-45
-25
-35
-25

resident juvenile yellowfin sole and mature adults in spawning condition, verifying that yellowfin sole spawn in the area. Among the dominant 22 species identified in the hauls, there were no red king crabs reported, although these could have been recorded in the miscellaneous category. The mesh of the net was fine enough (4.0 mm) to retain juvenile crabs.

Juvenile red king crabs live among epifaunal communities, which are associated with gravel and cobble substrate. Much of this habitat type is too rough or too shallow for trawl assessment. Juvenile distribution in Bristol Bay can be interpreted from published maps showing the distribution of associated substrate (gravel and cobble) and areas sampled for young crabs (McMurray et al. 1984, Stevens and MacIntosh 1991, Wainwright et al. 1992, Armstrong et al. 1993). Suitable juvenile habitat is "extremely patchy" in Bristol Bay (McMurray et al. 1984, Jewett and Onuf 1988). Areas shown by surveys to contain age 0-2 juvenile crabs do not adequately describe their actual distribution because: (1) the entire area has not been sampled; and (2) young crabs are difficult to catch with sampling gear, particularly in cobble habitats (Stevens and MacIntosh 1991). Furthermore, distribution of juvenile red king crabs may be affected by year-class strength. Juvenile surveys were conducted in 1984 and 1991, which were years of low stock abundance. The abundance and distribution of juveniles might have been more encompassing and conclusive had the surveys occurred in the early 1970s when overall population abundance was high and the distribution covered a greater area. Sample areas with low abundance could nonetheless be viable habitat in times of high abundance. It is therefore not possible from the existing survey data to determine the exact distribution of juveniles in any given year. Studies have combined juvenile survey data with substrate information to create a general map of juvenile red king crab habitat (Hsu 1986, Armstrong et al. 1993).

Trawl Impacts on Habitat

Trawling and dredging can negatively impact crab habitat, particularly living substrate on which young red king crabs depend for food and protection from predators. Juvenile red king crabs in the Bering Sea depend on both physical substrate and biogenic assemblages for settlement, food, and protection from predators (McMurray et al. 1984, Stevens and MacIntosh 1991). Both the physical substrate (cobble, shell) and biogenic assemblages (such as ascidians and tube-building polychaete worms) are vulnerable to trawling. Studies have shown that trawling and dredging impact the seabed through scraping and ploughing, sediment resuspension, and physical destruction, removal, or scattering of nontarget benthos (Messieh et al. 1991, Jones 1992). In the Wadden Sea, scientists have observed destruction and elimination of erect epifaunal species (Reise 1982). If habitat is impacted by trawling and dredging, crab settlement and survival could be reduced, thereby lowering recruitment.

Armstrong et al. (1993) and Cassano et al. (1995) proposed more extensive trawl closures to protect both spawning females and juvenile habitat from impacts of trawling. They conjectured that the lack of suitable habitat could be a population constraint, and habitat protection should be considered as a means to increase red king crab populations.

Groundfish Fisheries

Trawling was fairly limited in the proposed area with the exception of a yellowfin sole trawl fishery in April and May which targeted the flatfish when they migrated to the shallower waters of northern Bristol Bay to spawn. This fishery is prosecuted in Kulukak and Togiak bays (the vicinity of Walrus Island, Fig. 1) on a fine sand substrate in shallow waters (Fig. 2). The fishery is typically located in a confined area as represented by observed hauls in 1994 (Fig. 2), and occurs in mid-April to mid-June. Trawlers argued, and an analysis of the observer data confirmed, that few crabs were taken as bycatch in the fairly confined area this fishery utilized. An area proposed to remain open for this fishery was between 159°W and 160°W and between 58°N and 58°43′N (Fig. 4).

Data from individual fishing vessel trawl hauls were obtained from the NMFS observer program. Observers collect specific catch information from randomly selected hauls on vessels requiring an observer. A comparison of the directed yellowfin sole catch and red king crab bycatch in this area with the catch and bycatch in the rest of the Bering Sea for the yellowfin sole fleet indicated that a high of 13% of the red king crab bycatch from the yellowfin sole fishery was taken in this area in 1991. Similarly, 12% was taken in 1993. However, in most years, this percentage ranged between 1% and 5%. An analysis of the observer data indicated that most of the crab bycatch is from tows along the southern boundary of the area to remain open, or along the northern edge of area 512. In contrast, between 10% and 40% of the directed yellowfin sole catch came from this area in any given year.

Discussion and Conclusion

Fishery managers usually must focus upon measures that will monitor and/or control the portions of the exploited population affected by fishing gear. Setting seasons, catch limits, or the size or sex to be taken are examples of management options which can control total fishing mortality. Also, fishery managers must assure that critical nursery habitat is protected and that all sources of mortality (including bycatch) are taken into account. Specifically, red king crabs in the Bering Sea have been at low population levels since the early 1980s, and at the time of this action mature female crabs had declined in numbers to a level below an established threshold believed to be necessary to the continuing health of the population. In addition, recruitment to the population is variable and has

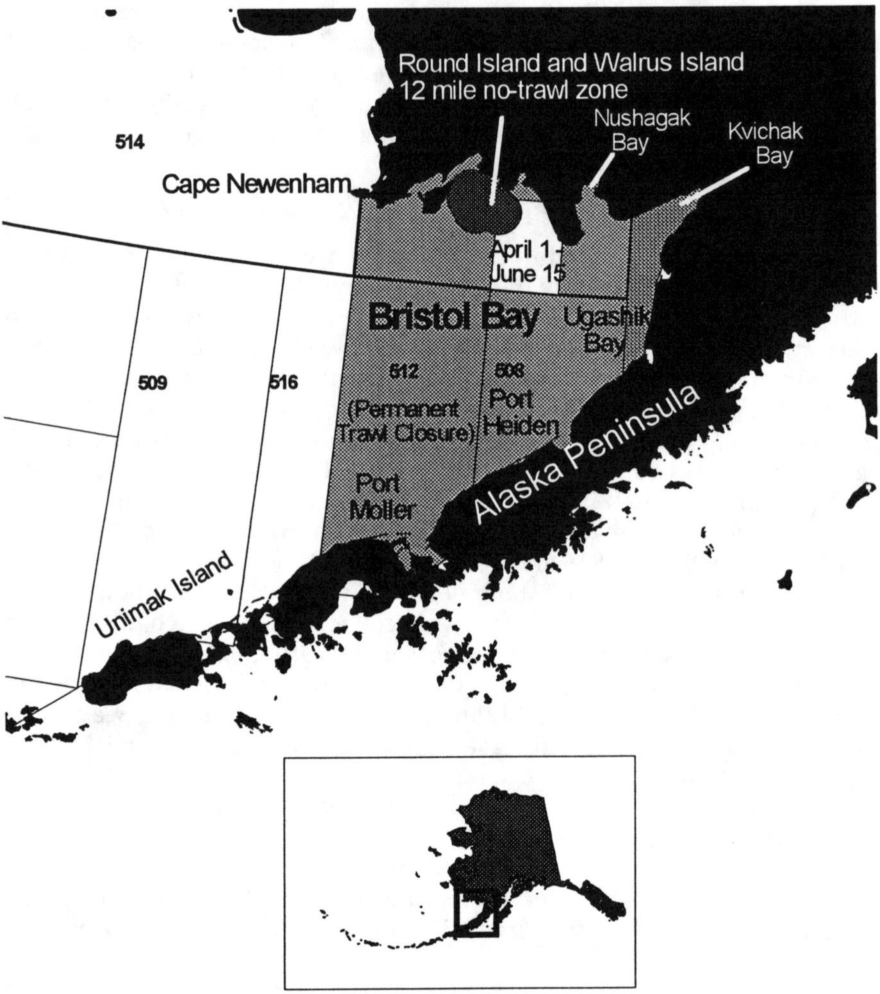

Figure 4. Map of the Bristol Bay trawl closure area. The unshaded area within Bristol Bay is an area open to trawling for yellowfin sole from April 1 to June 15 annually.

mostly been low in recent years. With the closure of Bristol Bay and the Red King Crab Savings Area to trawling, managers and policy makers have moved to increase protection to include all possible life-history stages.

The effectiveness of any trawl/dredge closure designed to protect juvenile red king crabs hinges on our understanding of recruitment dynamics and the distribution of juveniles and their habitat. Red king crab year-class strength depends both on the number of spawners and on environmental conditions such as temperature and currents (Tyler and Kruse 1995). It has been hypothesized that the availability of appropriate red king crab habitat at the settlement and juvenile stages constrains juvenile abundance, in turn affecting year-class strength and recruitment (McMurray et al. 1984, Armstrong et al. 1993, Cassano et al. 1995). Larvae drift with the current before settling to begin life on the benthos. Larval settlement and survival patterns vary according to ocean currents and availability of appropriate substrate. If the current transports the larvae to an area without suitable habitat, the chances of survival are slim (McMurray et al. 1984, Jewett and Onuf 1988).

In his analysis of the life history of major red king crab stocks, Rodin (1990) established the following four basic habitat components for a viable population: (1) conditions experienced by hatching larvae in the coastal zone must be combined with transport of larvae to favorable habitats for survival; (2) a well-developed sessile community (dense concentration and large areas of hydroids, bryozoans, and sponges) is needed for refuge from predators, and a food base is necessary where the massive settlements of larvae occur; (3) a broad continental shelf is necessary with a rich food base for adults; and (4) no physical barriers to the movement of juvenile crabs into the adult spawning population and no barriers to the migration of adults into or out of spawning areas should exist. Rodin (1990) concluded the list with the following observation: "The most important, judging from everything, is the presence of favorable habitat conditions for the post larval stages."

Crab stocks will rebuild only when recruitment increases. Managers can affect recruitment by addressing the following three components: (1) ensure there are adequate numbers of spawners, (2) provide adequate habitat available for settlement, and (3) reduce fishing-related impacts on juvenile crabs. The closure of Bristol Bay to trawling should provide measures to reduce impacts on all three components necessary to recruitment. Mature adults, although protected by previous closure measures such as areas 512 and 516, and the Red King Crab Savings Area, should enjoy further protection, especially along the northern section of Bristol Bay. Cessation of trawling in the nearshore areas of Bristol Bay is expected to allow sessile benthic fauna to thrive. And, although not readily retained by trawl gear, juveniles in the closure area will not experience injuries or mortality from the passing trawl gear.

An area within northern Bristol Bay to seasonally permit trawling for yellowfin sole was established at the same time the larger Bristol Bay closure was implemented. An analysis of NMFS observer data indicated this was a highly productive area for yellowfin sole, and relatively low bycatch of red king crabs. The substrate within this subarea is relatively fine sand and silt, and does not appear to have the high relief habitat and biota necessary to juvenile king crabs. Surveys in the area were not able to locate red king crabs. The NPFMC determined that the area could be opened during the spawning season of yellowfin sole without appreciable effects on king crab juveniles or their habitat. In summary, the closure of Bristol Bay, with a sub-area available for yellowfin sole trawling, ensures that the possible impacts by trawl gear on the benthos are minimized.

Further benefits can be expected for marine mammals that inhabit Bristol Bay and already have had no-trawl zones established around rookeries and haulouts to minimize trawl and marine mammal interactions. Bristol Bay is also an important area for migrating and nesting seabirds to which disturbances would be reduced. Finally, the northern border of the area open to yellowfin sole trawling was purposefully established south of the inner reaches of Kulukak Bay to minimize trawl interactions with spawning herring which utilize the first few hundred meters from the shore for spawning.

Acknowledgments

The authors thank Gordon Kruse, Fritz Funk, and an anonymous reviewer for providing review and useful comments in the preparation of this manuscript. They would also like to thank the National Marine Fisheries Service Observer Program for the use of observer data, and would like to extend thanks to the National Marine Fisheries Service Alaska Fisheries Science Center–Kodiak for information from the annual trawl surveys. This paper was funded in part by the Pacific States Marine Fisheries Commission through the Alaska Fisheries Information Network (AKFIN).

This document is contribution PP-177 of the Alaska Department of Fish and Game, Division of Commercial Fisheries, Juneau.

References

Armstrong, D.A., L.S. Incze, D.L. Wencker, and J.L. Armstrong. 1981. Distribution and abundance of decapod crustacean larvae in the southeastern Bering Sea with emphasis on commercial species. NOAA, Office of Marine Pollution Assessment. OCSEAP Final Report, Research Unit 609.

Armstrong, D.A., T.C. Wainwright, G.C. Jensen, P.A. Dinnel, and H.B. Andersen. 1993. Taking refuge from bycatch issues: Red king crab (*Paralithodes camtschaticus*) and trawl fisheries in the eastern Bering Sea. Can. J. Fish. Aquat. Sci. 50:1993-1999.

Cassano, E., G. Harrington, P.S. Hill, and T. Loher. 1995. Management of Bristol Bay red king crab: A critical intersections approach to fisheries management. SMA/FRI-UW-9510. University of Washington, Seattle.

Hayes, M.L. 1983. Variation in the abundance of crab and shrimp with some hypotheses on its relationship to environmental causes. In: W.S. Wooster (ed.), From year to year: Interannual variability of the environment and fisheries of the Gulf of Alaska and the eastern Bering Sea. Washington Sea Grant Program, University of Washington, Seattle, pp. 86-101.

Hebard, J.F. 1959. Currents in southeastern Bering Sea and possible effects on king crab larvae. NOAA Tech. Rep. NMFS SSRF 293.

Hsu, C.C. 1986. Spatial and temporal distribution patterns of female red king crabs in the southeastern Bering Sea. Ph.D. thesis, University of Washington, Seattle. 298 pp.

Jewett, S.C., and C.P. Onuf. 1988. Habitat suitability index models: Red king crab. U.S. Fish Wildl. Serv. Biol. Rep. 82(10.153).

Jones, J.B. 1992. Environmental impact of trawling on the seabed: A review. N. Z. J. Mar. Freshwater Res. 26:59-67.

McDonald, J., H.M. Feder, and M. Hoberg. 1981. Bivalve mollusks of the southeast Bering Sea. In: D. W. Hood and J.A. Calder (eds.), The eastern Bering Sea shelf: Oceanography and resources. Vol. 2. National Oceanic and Atmospheric Administration, Office of Marine Pollution Assessment. University of Washington Press, Seattle.

McMurray, G., A.H. Vogel, P.A. Fishman, D.A. Armstrong, and S.C. Jewett. 1984. Distribution of larval and juvenile red king crabs (*Paralithodes camtschatica*) in Bristol Bay. NOAA OCSEAP Final Rep. 53(1986):267-477.

Messieh, S.N., T.W. Rowell, D.L. Peer, and P.J. Cranford. 1991. The effects of trawling, dredging and ocean dumping on the eastern Canadian continental shelf seabed. Continental Shelf Res. 2:1237-1263.

NMFS (National Marine Fisheries Service). 1995. Cruise results. Cruise 95-1 Crawdad, 1995 Togiak nearshore beam-trawl survey, May 18-May 29, 1995. NMFS, Alaska Fisheries Science Center, Resource Assessment and Conservation Engineering Division, Seattle, Washington.

Otto, R.A., J.A. Haaga, and R.A.MacIntosh. 1997. Report to industry on the 1996 eastern Bering Sea survey. NMFS, Alaska Fisheries Science Center, Processed Rep. 97-02. Seattle, Washington.

Pereyra, W.T., J.E. Reeves, and R.G. Bakkala. 1976. Demersal fish and shellfish resources of the eastern Bering Sea in the baseline year 1975. NMFS, Northwest Fisheries Center, Processed Rep. Seattle, Washington.

Reeves, J.E., and R. Marasco. 1980. An evaluation of alternative management options for the southeastern Bering Sea king crab fishery. NMFS, Northwest and Alaska Fisheries Center, Processed Rep. 80-6. Seattle, Washington.

Reise, K. 1982. Long-term changes in the macrobenthos invertebrate fauna of the Wadden Sea: Are polychaetes about to take over? Netherlands J. Sea Res. 16:29-63.

Rice, S.D., T.C. Shirley, and S.M. Shirley. 1988. Coastal fisheries oceanography of the southern Bering Sea and north Aleutian Basin: Habitat requirements of red king crab larvae and juveniles. NOAA, National Ocean Service, Ocean Assessments Division, Alaska Office. Interim Final Report, OCSEAP Research Unit 693.

Rodin, V.E. 1990. Population biology of the king crab *Paralithodes camtschatica* Tilesius in the north Pacific Ocean. In: Proceedings of the International Symposium on King and Tanner Crabs. University of Alaska Sea Grant, AK-SG-90-04, Fairbanks, pp. 133-144.

Rounds, P.M., C. Brodersen, and M.M. Babcock. 1989. Effects of cohort density and habitat on survival and growth of juvenile red king crab, *Paralithodes camtschatica.* In: Proceedings of the International Symposium on King and Tanner Crabs. University of Alaska Sea Grant, AK-SG-90-04, Fairbanks, pp. 209-218.

Sharma, D.S. 1979. The Alaska shelf: Hydrographic, sedimentary, and geochemical environment. Springer-Verlag, New York.

Stevens, B.G. 1990. Temperature-dependent growth of juvenile red king crab (*Paralithodes camtschatica*), and its effects on the size-at-age and subsequent recruitment in the eastern Bering Sea. Can. J. Fish. Aquat. Sci. 47:1307-1317.

Stevens, B.G., and R.A. MacIntosh. 1989. Report to the industry on the 1989 Eastern Bering Sea crab survey. NMFS, Alaska Fisheries Science Center, Processed Rep. 89-18, Seattle, Washington.

Stevens, B.G., and R.A. MacIntosh. 1991. A survey of juvenile red king crabs with dredge and beam trawl in Bristol Bay, Alaska. Cruise results supplement OH-91-1 (Ocean Hope 3). Available from National Marine Fisheries Service, P.O. Box 1638, Kodiak, AK 99615.

Stevens, B.G., R.A. MacIntosh, and J.A. Haaga. 1991. Report to the industry on the 1991 Eastern Bering Sea crab survey. NMFS, Alaska Fisheries Science Center, Processed Rep. 91-17, Seattle, Washington.

Takeshita, J., H. Fugita, and S. Matsuura. 1990. A note on population structure in the eastern Bering Sea adult red king crab, *Paralithodes camtschaticus.*In: Proceedings of the International Symposium on King and Tanner Crabs. University of Alaska Sea Grant, AK-SG-90-04, Fairbanks, pp. 427-434.

Takeuchi, I. 1962. On the distribution of zoea larvae of king crab, *Paralithodes camtschatica,* in the southeastern Bering Sea. Bull. Hokkaido Reg. Fish. Res. Lab. 24:163–170.

Tyler, A.L., and G.H. Kruse. 1995. Report of the modeling workshop on year-class strength formation of red king crab. Regional Information Rep. 5J95-11. Alaska Department of Fish and Game, Commercial Fisheries Division, Juneau, Alaska.

Wainwright, T.C., D.A. Armstrong, H.B. Andersen, P.A. Dinnel, D.W. Herren, G.C. Jensen, J.M. Orensanz, and J.A. Shaffer. 1992. Coastal fisheries oceanography of the southern Bering Sea and northern Aleutian Basin: Port Moller king crab studies. Minerals Management Service, Alaska OCS Region, OCS Study MMS 92-0040.

Comparing Models of Trophic Flows in the Northern and Southern Benguela Upwelling Systems During the 1980s

Lynne Jane Shannon
Marine and Coastal Management, Cape Town, South Africa

Astrid Jarre-Teichmann
Danish Institute for Fisheries Research, North Sea Centre, Hirtshals, Denmark

Abstract

Since 1981, the Benguela Ecology Programme has aimed at increasing ecological understanding of a large marine upwelling region of importance to fisheries. In an attempt to summarize results of this program, experts brought together both published and unpublished data at a workshop held in Cape Town in 1989. This was in an attempt to construct simple input-output carbon budgets of the Northern and Southern Benguela ecosystems. Based on these results, and together with more recently published data, ECOPATH models of the two systems during the 1980s have been prepared. In the Northern Benguela system, the dominant pelagic fish species during the 1980s was horse mackerel, whereas in the Southern Benguela, anchovy were dominant. In both systems, sardine were at low levels and hake were commercially important. These, together with different structures in the zooplankton communities and the different fishing levels in the respective systems, are used as a basis upon which to compare the trophic functioning of the two parts of the Benguela ecosystem during the period 1980-1989. Preliminary EcoSim analysis highlights differences in the impact of fisheries on selected groups in the two systems.

Introduction

There are four major marine upwelling systems in the world (Fig. 1). The Benguela upwelling system is subdivided into relatively distinct upwelling systems: the Northern Benguela off Namibia, and the Southern Benguela off South Africa. The trophic functioning of the Northern Benguela system has been examined in two periods, namely 1971-1977 and 1978-1983 (Jarre-Teichmann et al. 1998; Jarre-Teichmann and Christensen 1998a, 1998b). In addition, an ECOPATH model of the Northern Benguela has also been developed for the 1990s (Heymans 1996). Preliminary models of trophic flows in both the Northern (Shannon and Jarre-Teichmann 1999) and Southern (Jarre-Teichmann et al. 1998) Benguela systems during the 1980s have been constructed. The present paper serves as a preliminary comparison of the trophic functioning of the upwelling ecosystems of the Northern and Southern Benguela during this decade. The paper aims to stimulate and encourage further revision of the data and models, so that more detailed interdecadal and intersystem comparisons can be made. It is planned that this work will form part of a fuller comparison of the trophic functioning of both the Northern and Southern Benguela ecosystems during periods dominated by different species. Using an approach such as this, it would be possible to compare the way in which harvest and conservation strategies impact the northern and southern systems under different species dominance regimes. This would be of benefit to both South Africa and Namibia, particularly as the fisheries have been managed in different ways in the two systems over the last few decades. Researchers and managers in each system could learn from the other, and both could work together towards improving fisheries management and conservation in the future.

Methods

Trophic flow budgets for the Northern and Southern Benguela ecosystems during the 1980s were constructed using ECOPATH. The latter is a widely used ecosystem modeling tool that has been under development since 1987 by the International Centre for Living Aquatic Resources Management (ICLARM). Christensen and Pauly (1992, 1995) describe the ECOPATH model in detail. ECOPATH is based on a model by J.J. Polovina of the U.S. National Marine Fisheries Service, Honolulu, Hawaii (Polovina 1984). The steady state approach was recently extended to a dynamic simulation tool called ECOSIM by scientists of the Fisheries Centre, University of British Columbia, Canada (Walters et al. 1997).

At a workshop held in 1989, experts on the different species groups in the Benguela upwelling region brought together both published and unpublished data to construct a simple input-output carbon budget. These data, together with those which have since become available, have been used to construct preliminary Ecopath models of both the Southern (Jarre-

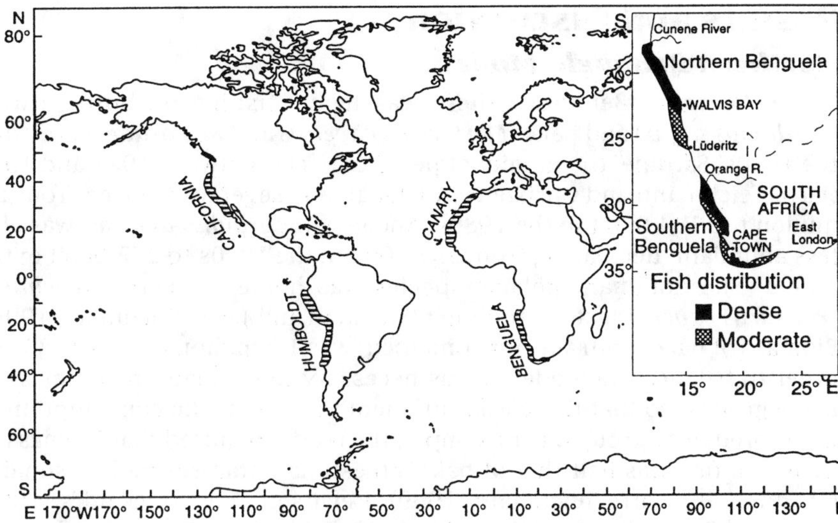

Figure 1. Map showing the four major upwelling regions of the world, with the inset showing the Northern and Southern Benguela systems (after Jarre-Teichman et al. 1998).

Teichmann et al. 1998) and the Northern (Shannon and Jarre-Teichmann 1999) Benguela upwelling systems. The model of the Northern Benguela comprises the shelf region to about 500 m depth, extends from 15°S to 29°S and covers an area of 179,000 km^2. The model of the Southern Benguela comprises the shelf region to about 500 m depth, extends from 29°S, around the southern tip of Africa and eastwards to 28°E, covering an area of 220,000 km^2. The Southern Benguela model, presented by Jarre-Teichmann et al. (1998), has been slightly modified to separate some components for clarity, and to incorporate some new hake data, which only recently became available. Parameters used are tabulated in Shannon et al. (1999). The two systems are compared in terms of flow diagrams, primary system statistics, relative consumption by components, primary production equivalents, and so on. In addition, analyses such as mixed trophic impact assessment (Ulanowicz and Puccia 1990) are used to highlight overall competition between species and the effects of species on others within an ecosystem.

ECOSIM has only been released in an alpha form at this stage. We show here how it can be used to perform preliminary analyses of the impacts of changing fishing mortality in the two systems. All settings were default, with interactions assumed to be of the mixed control type (0.5). Once thorough testing and validating of ECOSIM have been done, this work will be taken further.

Results and Discussion

Northern Benguela Model

In the Northern Benguela, there was heavy fishing on both sardine (*Sardinops sagax*) and hake (*Merluccius capensis* and *M. paradoxus*) during the 1970s. Sardine, the dominant pelagic species between 1971 and 1977 (Jarre-Teichmann and Christensen 1998a), was severely reduced from 1.5 million t to 134,000 t in the 1980s. Anchovy (*Engraulis capensis*) was also less abundant, decreasing from 593,000 t in the 1970s to 252,000 t in the 1980s. The dominant pelagic species was horse mackerel (*Trachurus trachurus capensis*), at 2.5 million t (Shannon and Jarre-Teichmann 1999). Biomass of hake (both species combined) was 1.3 million t in the 1980s. In order to balance the model, it was necessary to estimate the biomass of some groups, so that there was sufficient to support the consumption of other predatory groups. For example, the model required that biomass of demersal fish was half that of hake, an estimate that seemed reasonable (Shannon and Jarre-Teichmann 1999). Hewitson and Cruickank's (1993) estimate of biomass of pelagic goby (*Sufflogobius bibarbatus*) was doubled in the model. The proportion of cephalopods in the diets of many species was reduced to avoid consumption of cephalopods exceeding production. Estimates of benthic fauna and production are lacking. Therefore the model was used to estimate required benthic biomass. That the benthic biota in the northern ecosystem are less abundant (at ony 25% of that in the southern Benguela) is to be expected (Shannon and Jarre-Teichmann 1999); the northern model is of an upwelling area whereas the southern model extends over shelf (Agulhas Bank) and upwelling areas. Further, low oxygen levels are frequently found in shelf waters off Namibia (Bailey 1991).

Southern Benguela Model

During the 1980s, anchovy was the dominant pelagic fish, at a biomass of 1.1 million t off South Africa. Sardine was at a low level of only 129,000 t, mesopelagic fish (lanternfish, dominated by *Lampanyctodes hectoris* and lightfish, *Maurolicus muelleri*) at 1.7 million t, and hake at about 624,000 t (both species combined). In order that the model balance, some of the parameters of a few groups had to be reevaluated. Hake and other demersal fish caused major problems in the model, particularly related to diet composition of the various size groups. The proportion of these groups in the diet of others was reduced in many cases and it was necessary to allow the model to estimate the predatory biomass of these groups, based on the amount required as prey and catches. Hake biomass was required to be about 64% higher than estimated, and that of other demersal fish, 7.5 times higher. These adjustments can send ripple effects through the system as a result of prey species being consumed in larger amounts. As in the Northern Benguela, cephalopods were not abundant enough to support demands by other components, and the proportion in the diets of some groups had to be reduced dramatically.

Table 1. Comparison of summary statistics in the Northern and Southern Benguela upwelling systems in the 1980s. Flows are in tons per km² per year.

	Northern Benguela	Southern Benguela
Sum of all consumption	11,743	14,196
Sum of all exports	1,254	4,300
Sum of all flows to detritus	4,265	10,170
Total system throughput	23,327	36,435
Sum of all production	10,034	15,427
Net system production	1,254	4,300
Sum of all respiratory flows	6,065	7,769
Total net primary production	7,319	12,068
Total catches	7	3

Comparison of Northern and Southern Benguela Upwelling Systems

Summary Statistics

Total biomass (excluding detritus) was higher in the Northern Benguela (360 t per km^2) than in the Southern Benguela (297 t per km^2). With the exception of total catches, all major flows were higher in the Southern Benguela (Table 1), indicating that the Southern Benguela is a more productive system. Mean path lengths (sensu Finn 1976) in the two systems were similarly short, at 3.02 in the Southern and 3.19 in the Northern Benguela.

Flow Diagrams

Trophic levels of the components of the systems, biomasses, and flows through the two systems are compared (Fig. 2). Many components in the Southern Benguela occupy higher trophic levels than the same groups in the Northern Benguela.

Consumption

Consumption by anchovy, redeye, hake, and other demersal fish and mesopelagic fish are most important in the Southern Benguela (Fig. 3b). In the Northern Benguela (Fig. 3a), horse mackerel and goby also consume large portions of production.

Primary Production Required

Use of primary production equivalents enables the effects of fishing at different trophic levels to be compared (Pauly and Christensen 1995). The

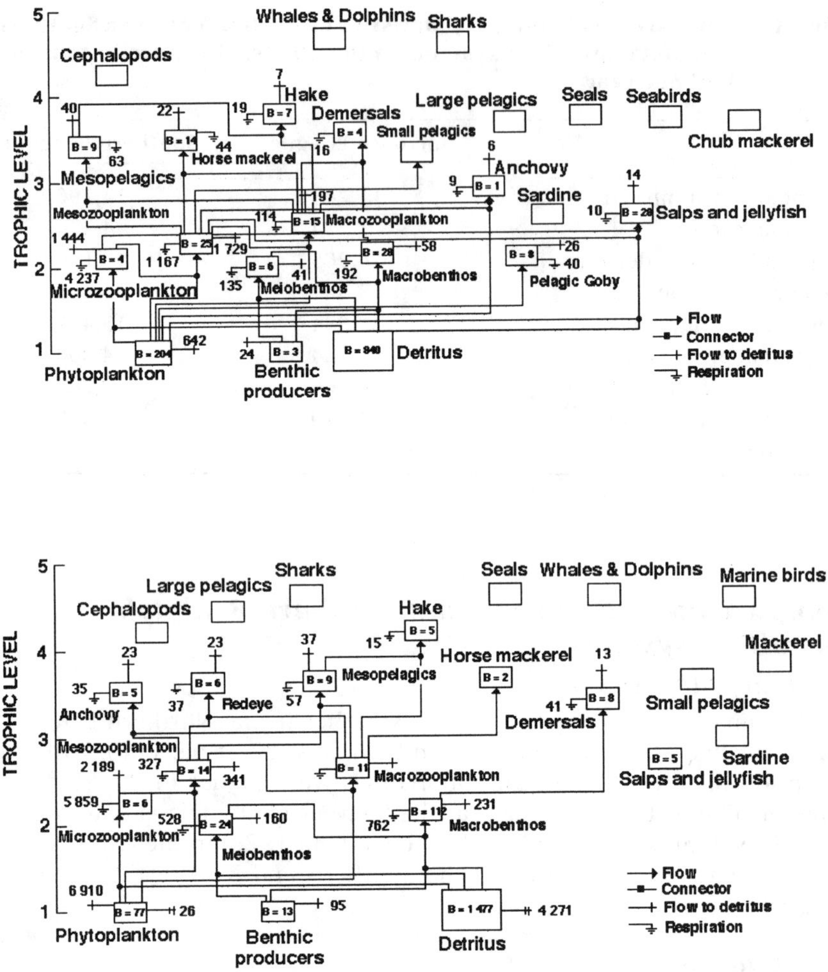

Figure 2. *Diagrams of the trophic flows in the Northern (top) and Southern (bottom) Benguela during the 1980s. Boxes arranged along vertical axis by trophic level. Flows are tons wet mass per km² per year. Biomass of components denoted by B where B >1 t per km². Flows leave boxes in upper half, enter in lower half. Values are indicated for respiration and flows to detritus, with those less than 5 t per km² per year (i.e., around 0.3-0.4 ppt of total consumption) omitted for clarity.*

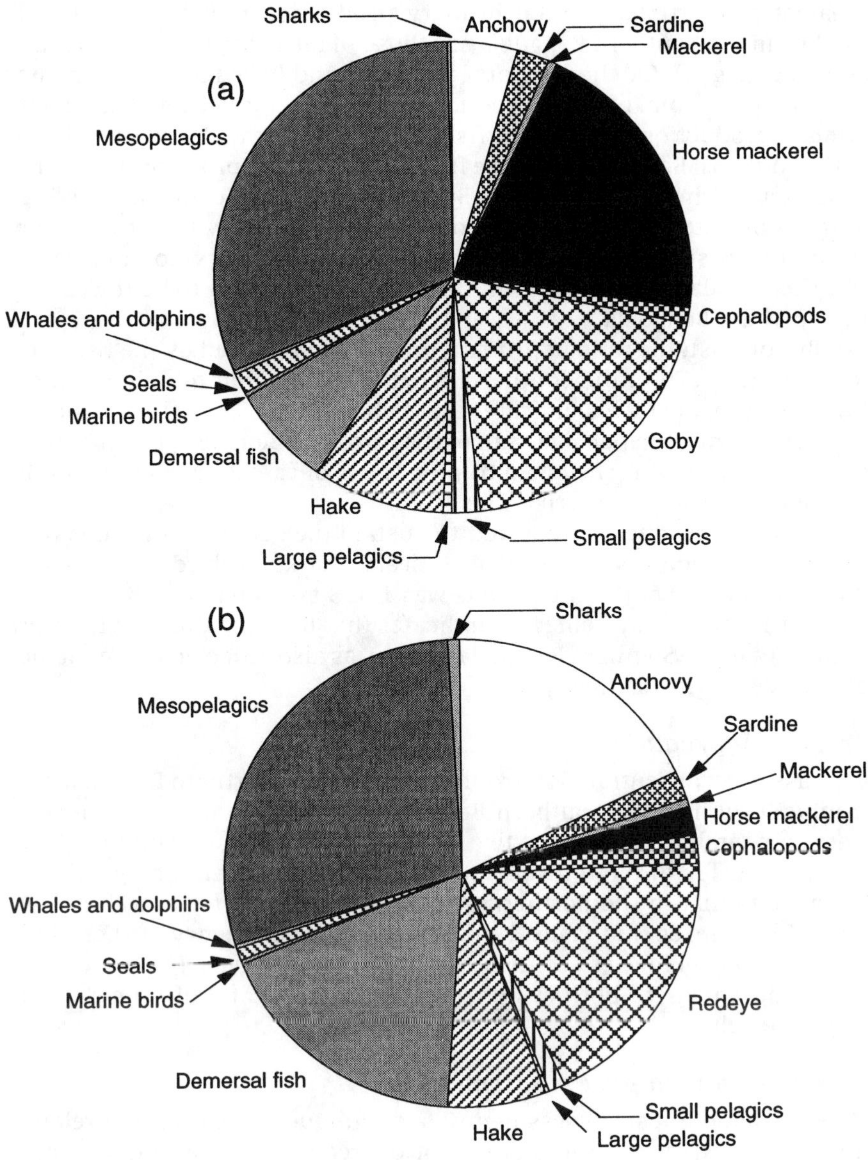

Figure 3. Production consumed by predators in the Northern (a) and Southern (b) Benguela during the 1980s.

greatest proportions of total primary production required to support catches in the Southern Benguela are those of anchovy, large pelagic fish, and hake (Fig. 4). Catches of horse mackerel and hake require far greater proportions of total primary production in the Northern Benguela system. Primary production required for fisheries is 5.8% in the Northern and 3.6% in the Southern Bengeula. That the fraction of primary production required to sustain catches in the Southern Benguela is so small, could be explained by the mismatch of phyto- and zooplankton productivity and fish consumption, as suggested by Hutchings (L. Hutchings, pers. comm., Marine and Coastal Management, South Africa). This seems not to be the case in the Northern Benguela.

By contrast, the ratio of primary production required by the fisheries, to harvest, is 179 in the Southern Benguela, compared to only 96 in the Northern Benguela. This indicates that fishing in the Southern Benguela is ecologically more expensive than in the Northern Benguela, despite lower catches in the south. The mean trophic levels of catches are similar in the southern (4.7) and northern (4.6) systems.

Primary production required to sustain (i.e., support the consumption) of top predators (large pelagics, seals, sharks, seabirds, whales, and dolphins) in the Southern Benguela was 1,553 t, compared to 657 t in the Northern Benguela system. This indicates that top predators are far more important in the Southern Benguela system, as also reflected in the catches of species serving as their prey.

Trophic Aggregation

Biomass was concentrated in trophic level I in the Northern Benguela and trophic level II in the Southern Benguela (Table 2). Transfer efficiencies were comparable between the two systems at most levels, with the exception of transfer through level IV in the southern system being twice as efficient as that in the northern system. As expected for upwelling systems (Christensen and Pauly 1993), the transfer efficiency at these high trophic levels is low, as are those at the herbivore level. There are some long food chains in both systems, resulting in a total of nine discrete trophic levels.

Mixed Trophic Impact Assessment

Mixed trophic impact assessment is a technique measuring the relative impact of a change in the biomass of one component on other components of the ecosystem (Ulanowicz and Puccia 1990). It assumes that trophic structure is constant; i.e., the technique cannot be used for predictive purposes, but should rather be used as sensitivity analysis. In the Southern Benguela, microzooplankton benefited mesozooplankton, anchovy, sardine, redeye (*Etrumeus whiteheadi*), and other small pelagic fish, such as saury (Scomberesocidae) and flying fish (Exocoetidae), by providing a food source for these groups and by serving as food for others upon which

Table 2. Trophic aggregation in the Benguela upwelling systems during the 1980s.

Trophic level	Biomass (t per km^2)		Transfer efficiency, % (all flows)		Total throughput (all flows)	
	Northern	Southern	Northern	Southern	Northern	Southern
I	206.87	89.60			11,584.02	22,238.61
II	86.91	147.78			9,670.81	10,936.25
III	48.83	40.77	10.6	7.5	1,028.39	823.36
IV	14.55	14.58	12.3	22.1	123.15	180.62
V	3.02	4.27	17.5	20.5	19.63	35.99
VI	0.22	0.46	9.2	8.8	1.27	2.80
VII	0.01	0.02	7.3	6.3	0.06	0.13
VIII	0.00	0.00	3.0	2.9	0.00	0.00
IX	0.00	0.00	2.9	1.5	0.00	0.00
X	0.00	0.00	0.2	0.4	0.00	0.00

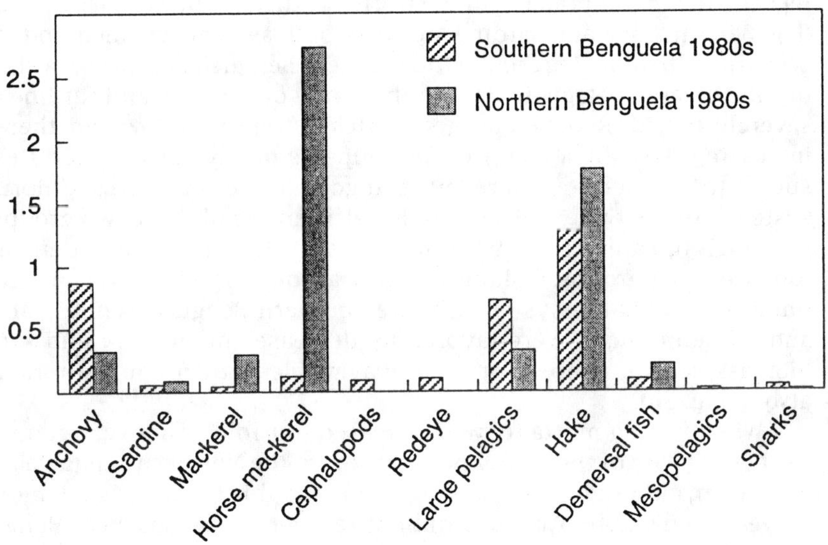

Figure 4. Percent of total primary production required to sustain catches in the Northern and Southern Benguela ecosystem during the 1980s.

these groups prey (Fig. 5). By contrast, in the northern system, microzooplankton competed with these groups (goby replaces redeye in the northern system) for phytoplankton food, thereby having a net negative impact on them. Mesozooplankton, through consuming phytoplankton, negatively affected pelagic goby in the Northern Benguela. However, mesozooplankton was an important prey item for its counterpart in the Southern Benguela, the redeye. Owing to its greater biomass in the south, anchovy had more pronounced impacts on other groups in the Southern Benguela. Similarly, horse mackerel, the dominant pelagic species in the Northern Benguela, had greater effects on its competitors in the north. Sardine showed small net effects in both systems. As expected from higher catches in the Northern Benguela, groups in this system were more severely affected by fishing.

ECOSIM Analysis of the Effects of Altered Levels of Fishing Mortality

Altered fishing mortality has different effects in the two systems (Fig. 6a-c). Increasing fishing mortality (F) on the three most abundant small pelagic fish in each system (namely anchovy and sardine in both systems, and redeye in the south and goby in the north) had less pronounced effects on other components in the Northern than in the Southern Benguela (Fig. 6a). Anchovy and sardine stocks in both systems crashed under this scenario of fishing. Further, in the Southern Benguela, there was a sharper decline in large pelagic fish when their prey of anchovy and sardine were severely fished. Redeye and goby catches were very low and therefore increasing these by a factor of four appears to have little effect. Despite such low fishing effort on redeye and goby in the southern and northern systems, respectively, the ecotrophic efficiencies of these two groups are very high (greater than 0.9), indicating that they are in high demand as food for other groups. Fishing at higher absolute levels would have major implications for the systems. In the Southern Benguela, chub mackerel and horse mackerel were favored by decreases in anchovy and sardine biomass, leaving more zooplankton available to their competitors (refer also to Fig. 5b).

When fishing mortality was increased fourfold as above, but only for the first 4 years of the simulation, sardine in the Northern Benguela began to recover, although only reaching about one-third of its original biomass by year 10 (Fig. 6b). Anchovy did not recover. In the Southern Benguela, both species recovered by year 10.

Increasing fishing mortality of hake fourfold in the first four years of simulation reduced hake biomass, thereby favoring some of the prey species of hake, such as mesopelagic fish and squid (Fig. 6c). However, in the Northern Benguela, other small pelagic fish (e.g., saury) were reduced and did not recover once fishing on hake was restored to its original level. In both systems, horse mackerel were negatively affected by increased fishing on hake, stabilizing at a lower level of abundance once fishing on hake

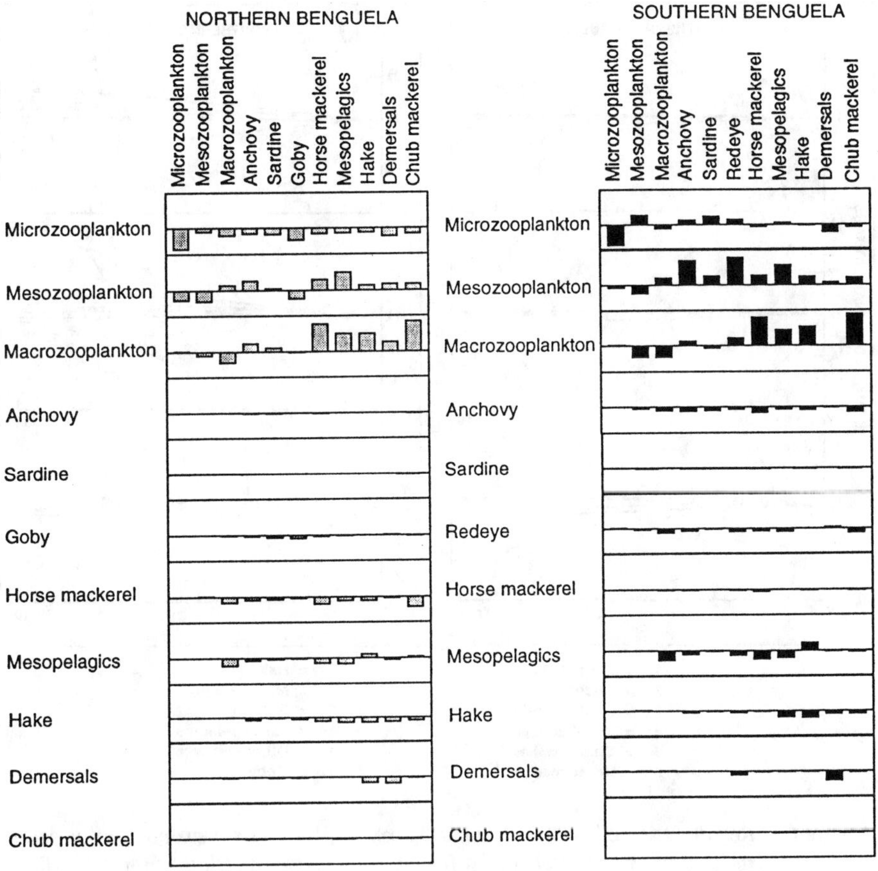

Figure 5. *Mixed trophic impacts of selected groups in the Northern and Southern Benguela during the 1980s. Bars extending above the line of zero impact for each species represent net positive impacts. Conversely, bars extending below the zero line represent net negative impacts.*

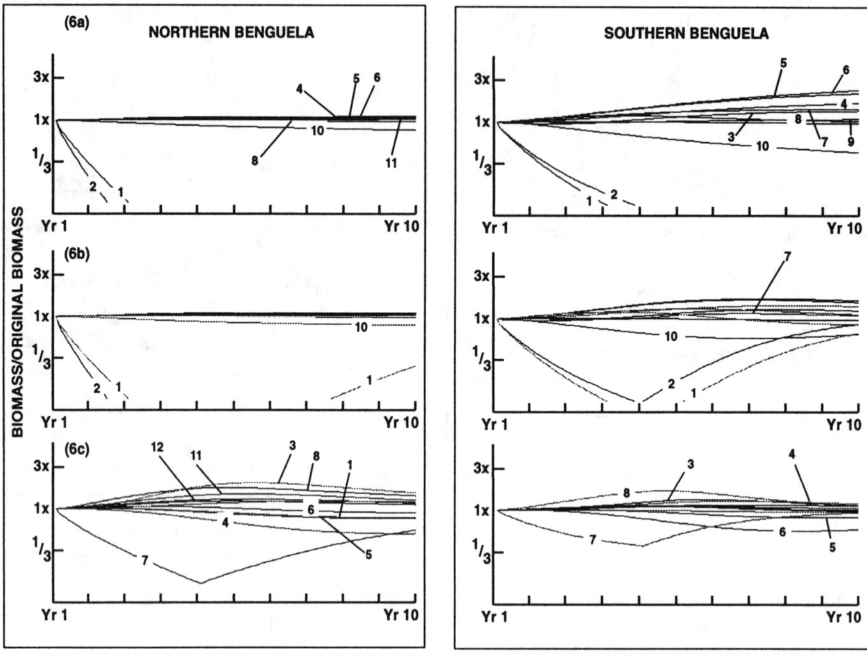

1 = Sardine
2 = Anchovy
3 = Cephalopods
4 = Small pelagics
5 = Chub mackerel
6 = Horse mackerel
7 = Hake
8 = Mesopelagics
9 = Redeye
10 = Large pelagics
11 = Other demersals
12 = Goby

Figure 6. Results of ECOSIM simulations. Biomass plotted as a proportion of original biomass. (a) Effects of a fourfold increase in fishing mortality (F) of anchovy, sardine, and redeye (Southern Benguela) or goby (Northern Benguela) over 10 years. (b) Effects of a fourfold increase in fishing mortality (F) of anchovy, sardine, and redeye (Southern Benguela) or goby (Northern Benguela) for the first 4 years, after which Fs are set back to original levels. (c) Effects of a fourfold increase in fishing mortality (F) of hake for the first 4 years, after which F is set back to original level.

was restored to its original level. The effect was more severe in the Southern Benguela. This is explained by the greater net negative impact which mesopelagic fish had on horse mackerel, by competing with them for zooplankton food (Fig. 5b).

As horse mackerel were lightly fished in the Southern Benguela, increasing F by a factor of 4, for example, had little effect. In the Northern Benguela, increasing F on horse mackerel fourfold for 4 years reduced horse mackerel biomass dramatically, so that by year 10, biomass was only at half its original level. Horse mackerel had a negative effect on chub mackerel by competing with them for zooplankton prey (Fig. 5a). Therefore chub mackerel were favored by this scenario, leveling off at a biomass factor of 1.7 that at the start.

Conclusions

There are clearly differences in the functioning of the Northern and Southern Benguela systems, related to abundances of the various components, as well as to the level of harvesting on these. Catches were higher in the Northern Benguela, despite the higher overall production in the Southern Benguela. However, fishing was more ecologically expensive in the southern system. Top predators were more important components in the Southern Benguela, as indicated by twice the primary production required to support consumption by top predators in the south. Demersal fish biomass required to support consumption in the Southern Benguela was double that in the Northern Benguela. However, when hake biomass is taken into account, the overall demersal component was only slightly higher in the Southern Benguela, and is explained by the fact that the Southern Benguela system is a continuum from upwelling to shallow bank. Although only lightly fished in both systems, mesopelagic fish, redeye, and pelagic goby were required to sustain the production of other components within the systems, and their heavier exploitation should be considered with caution. Zooplankton was more abundant in the Northern Benguela, but not as efficiently transferred to higher trophic levels.

Preliminary ECOSIM analyses show that increasing fishing mortality (in proportion to its original value) has different effects in the two systems, and that these are complex responses to indirect and direct competition. In the Southern Benguela, large pelagic fish were more severely affected by fishing strategies that reduce their small pelagic prey. Also, horse mackerel and hake were more closely linked, so that reducing hake biomass had a more pronounced effect on horse mackerel in the Southern Benguela.

Acknowledgments

We would like to thank Coleen Moloney, Patti Wickens and John Field for their foresight in organizing the workshop held in 1989, for which much of the basic data used in this model were prepared and presented by local experts. We thank Coleen Moloney for her encouragement and support in this first comparative attempt. We are grateful to the European Union Framework Programme IV, Cooperation with Third Countries and International Organisations, Part C, Scientific and Technological Cooperation with the Developing Countries "Putting fishing effects in the ecosystem context" for funding which enabled us to work together on this project. We thank Villy Christensen for his efforts to make available to us the alpha version of EcoSim used in this paper. Intentions to model this system for the period 1980-1989 were presented at the International Symposium on Environmental Variability in the South-east Atlantic, held 30 March-1 April 1998 in Swakopmund, Namibia, and parts of the manuscript presented were incorporated into this paper.

References

Bailey, G.W. 1991. Organic carbon flux and development of oxygen deficiency on the modern Benguela continental shelf south of 22°S: Spatial and temporal variability. In: R.V. Tyson. and T.H. Pearson (eds.), Modern and ancient continental shelf anoxia. Geological Society Special Publication 58:171-183.

Christensen, V., and D. Pauly. 1992. ECOPATH II: A software for balancing steady-state ecosystem models and calculating network characteristics. Ecol. Modell. 61:169-185.

Christensen, V., and D. Pauly (eds.). 1993. Trophic models of aquatic ecosystems. ICLARM Conference Proceedings 26. 390 pp.

Christensen, V., and D. Pauly. 1995. Fish production, catches and the carrying capacity of the world oceans. NAGA (ICLARM Quarterly) 15(4):26-30.

Finn, J.T. 1976. Measures of ecosystem structure and function derived from analysis of flows. J. Theor. Biol. 56:363-380.

Hewitson, J.D., and R.A. Cruickshank. 1993. Production and consumption by planktivorous fish in the northern Benguela ecosystem in the 1980s. S. Afr. J. Mar. Sci. 13:15-24.

Heymans, J.J. 1996. Network analysis of the carbon flow model of the northern Benguela Ecosystem, Namibia. Ph.D. thesis, University of Port Elizabeth, South Africa. 213 pp.

Jarre-Teichmann, A., and V. Christensen. 1998a. Comparative modelling of trophic flows in four large upwelling ecosystems: Global vs. local effects. In: M.-H. Durand, P. Cury, R. Medelssohn, C. Roy, A. Bakun, and D. Pauly (eds.), Global vs. local changes in upwelling ecosystems. Proceedings of the First International CEOS Meeting, ORSTOM, Paris, pp. 423-443.

Jarre-Teichmann, A., and V. Christensen. 1998b. Modelling trophic flows in large eastern ocean ecoystems: Temporal and spatial comparisons. ICLARM Studies and Reviews 24. In press.

Jarre-Teichmann, A., L.J. Shannon, C. Moloney, and P. Wickens. 1998. Comparing trophic flows in the Southern Benguela to those of other upwelling ecosystems. In: S.C. Pillar, C. Moloney, A.I.L. Payne, and F.A. Shillington (eds.), Benguela dynamics: Impacts of variability on shelf-sea environments and their living resources. S. Afr. J. Mar. Sci. 19:391-414.

Pauly, D., and V. Christensen. 1995. Primary production required to sustain global fisheries. Nature 374:255-257.

Polovina, J.J. 1984. Model of a coral reef ecosystem. I. The ECOPATH model and its application to French Frigate Shoals. Coral Reefs 3:1-11.

Shannon, L.J., and A. Jarre-Teichmann. 1999. A model of the trophic flows in the Northern Benguela upwelling system during the 1980s. S. Afr. J. Mar. Sci. 21. In press.

Shannon, L.J., A. Jarre-Teichmann, and P. Cury. 1999. Comparing trohpic flows and effects of fishing on the Southern Benguela ecosystem during two decades. Submitted to ICES Symposia (ICES/SCOR Symposium "Ecosystem effects of fishing," March 1999, Montepelier, France).

Ulanowicz, R.E., and C.J. Puccia. 1990. Mixed trophic impacts in ecosystems. Coenoses 5:7-16.

Walters, C., V. Christensen, and D. Pauly. 1997. Structuring dynamic models of exploited ecosystems from trophic mass-balance assessments. Rev. Fish Biol. Fish. 7:139-172.

Modeling of Chub Mackerel *(Scomber japonicus)* and Sardine *(Sardinops sagax melanosticta)* During the Annual Migration Cycle

Mikhail V. Pavlychev
Far East State Technical University, Natural Science Institute, Vladivostok, Russia

V.A. Belayev
Pacific Scientific Research Fisheries Centre (TINRO-Centre), Vladivostok, Russia

E.Y. Frisman
Institute for Automation and Control Processes, Vladivostok, Russia

Vadim P. Pavlychev
Pacific Scientific Research Fisheries Centre (TINRO-Centre), Vladivostok, Russia

Abstract

In this work we have attempted to mathematically describe chub mackerel *(Scomber japonicus)* and sardine *(Sardinops sagax melanostricta)* migrations, and construct a model and verify it by retrospective forecast. Three models are described. The first is constructed based on a linear migration model excluding any environmental factors. The second is constructed by taking into account the thermocline depth. The third model adds the additional characteristic of a temperature gradient in the thermocline. Positive results permit predicting migrations one month in advance for April-June and September-October for chub mackerel and for January-April and September-October for sardine.

Introduction

Fish of the genus *Scomber* are one of the most numerous fishes found in the pelagic waters of temperate and subtropical zones of the Pacific Ocean. Chub mackerel is widely distributed and occurs in the temperate waters around Japan, where it has periods of high abundance. Large aggregations of Pacific sardine are also observed in this area. The least understood aspects of the biology of these commercially targeted fish is their foraging and spawning migrations (Figs. 1 and 2). Therefore, forecasting these occurrences is very important for the fishing fleet.

The mathematical modeling of the spatial-temporal distribution of biological populations and communities is not well developed, as is the modeling of dynamics of their biomass (Bocharov 1990). Therefore the problem of modeling the migrations of pelagic fishes is quite practical.

Based on the scientific research of the distributions and migrations of chub mackerel and Pacific sardine, a problem arose in describing the seasonal migrations of these fishes by mathematical methods; i.e., constructing a model to obtain model coefficients and verify them with actual data.

We constructed three models based on a chamber model, where the considered biological population is divided into subpopulations (chambers), which are connected only by migrations between chambers. The first model is constructed with no environmental factors being considered. The second model is constructed on the basis of the scientific assumption that chub mackerel and sardine, as well as many other pelagic fishes, inhabit the upper layer of water, i.e., in a layer above the thermocline. The third model is constructed by including in the second model the additional characteristic of a vertical gradient in the thermocline.

Chamber Model

In construction of a chamber model (Tuzinckevich 1989) the area of habitation of the biological object is divided into p fragments (chambers), which are independent and connected to each other only by migration flows:

$$\frac{du_{ik}}{dt} = f_{ik}(u_{1k},...,u_{nk}) + M_{ik}(u_{i1},...,u_{ip})$$

where

u_{ik} is biomass (abundance) of species i in fragment k;

f_{ik} is intrapopulation interaction of species i in fragment k;

M_{ik} is inflow of migrants of species i in fragment k; and

n is number of species.

The main advantage of the chamber approach is the simplicity of the model. Considering the use of ordinary differential equations, chamber

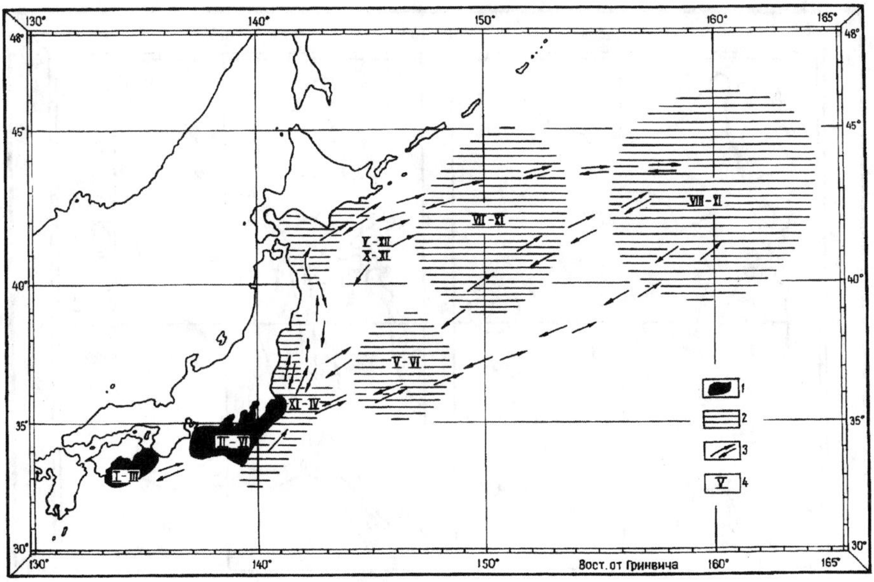

Figure 1. *Distribution of chub mackerel in the northwest Pacific (Belayev 1984). 1 = Spawning region, 2 = Foraging region, 3 = Migrations, 4 = Month.*

models carry out the analysis rather easily, which was one of the factors for selecting them for the mathematical description of migrations of the studied biological objects.

In considering the construction of the migration models, we divide the ocean surface into equal squares and attach each square to a coordinate grid $[i, j]$. Use data on catch per effort (in kilograms) as density of fish concentration in each square in the given month and designate it as $X_{i,j}$. By considering one of the squares in the given month (n) we take four adjacent squares on the horizontal and vertical, and assume at the current time the fish migrate from square to square; i.e., movement between squares (subpopulations) occurs.

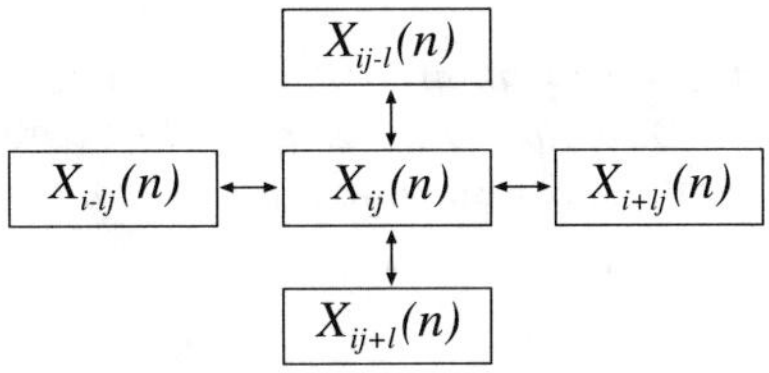

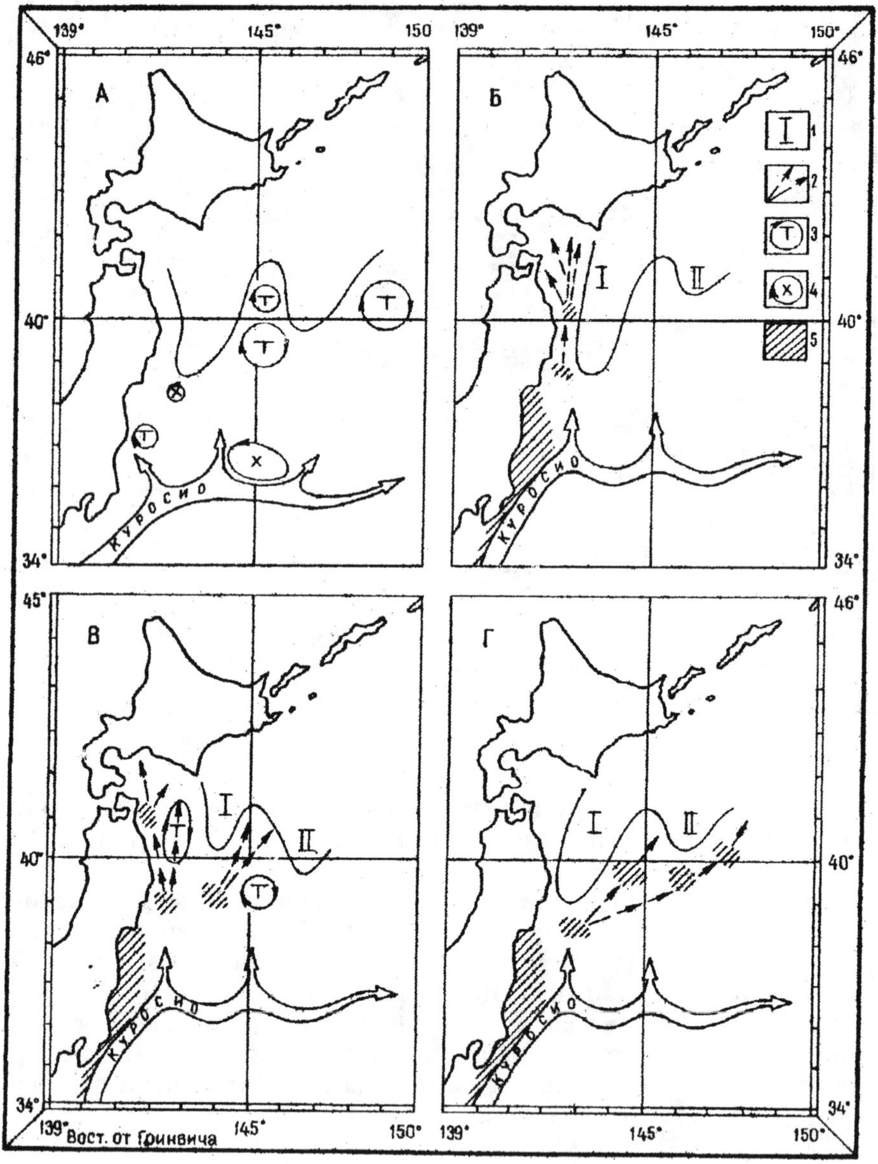

Figure 2. Scheme of northern migrations of Pacific sardine (Kenya 1982). 1 = Oyashio branches, 2 = Direction of moving sardine assemblage, 3 = Warm waters, 4 = Cold waters, 5 = Sardine aggregations.

For the simplicity in model appearance, we designate fish concentrations in the adjacent squares as X_1, X_2, X_3, and X_4.

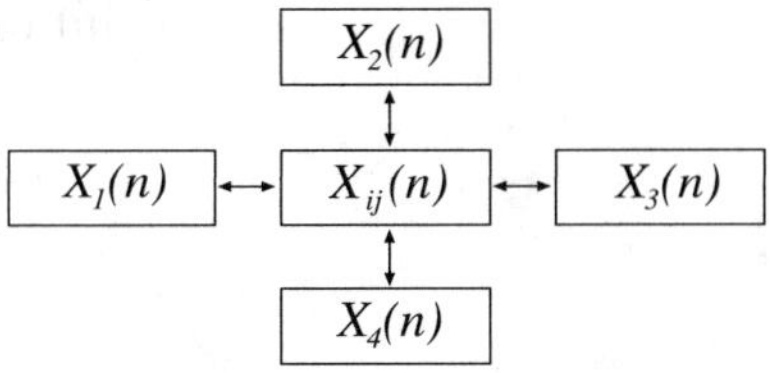

Model 1

We designate a coefficient a in our model, and obtain a linear forecast equation for the following month (n+1) for the given square X_{ij}^*:

$$X_{ij}^*(n+1) = X_{ij}(n)(1-4a) + X_1(n)a + X_2(n)a + X_3(n)a + X_4(n)a.$$

Group items with respect to the coefficient a:

$$X_{ij}^*(n+1) = X_{ij}(n) + aX_1(n) + X_2(n) + X_3(n) + X_4(n) - 4X_{ij}(n).$$

For simplicity, let:

$$c = X_1(n) + X_2(n) + X_3(n) + X_4(n) - 4X_{ij}(n),$$

then the predicted fish concentration for month n+1 for the given square is:

$$X_{ij}^*(n+1) = X_{ij}(n) + ac.$$

Write the residual sums of squares:

$$U = \sum_{n}\left[X_{ij}^*(n+1) - \overline{X}_{ij}\right]^2, \text{ where } \overline{X}_{ij} \text{ are observed data,}$$

then minimize the obtained function:

$$U \to min.$$

Equating the partial derivative with respect to parameter a to zero, we obtain equation:

$$\frac{\partial U}{\partial a} = 2\sum_{n} (X_{ij}^{*}(n+1) - \overline{X}_{ij})c = 0 \cdot$$

After some transformations we obtain the solution, i.e., coefficient of the model:

$$a = \frac{\sum_{n} c[\overline{X}_{ij} - X_{ij}(n)]}{\sum_{n} c^2}.$$

Model II

Through extensive statistical analyses, scientists have revealed that oceanographic factors, such as the depth of a thermocline and its gradients of temperatures (Sokolovskiy 1971), greatly influenced the distribution of chub mackerel in waters of 11-16°C.

Therefore in our model we limit fish migrations to depth of the thermocline; i.e., assume that fish will migrate in the direction where the thermocline is deeper. Given the differences of thermocline depth between one square and its adjacent squares r_1, r_2, r_3, and r_4, and designating coefficients a and b, we obtain a linear forecast equation for the following month (n+1) for the given square X_{ij}^{*}:

$$\begin{aligned} X_{ij}^{*}(n+1) = X_{ij}(n)[1 - 4a - 4b(\rho_1 + \rho_2 + \rho_3 + \rho_4)] + X_1(n)(a + b\rho_1) \\ + X_2(n)(a + b\rho_2) + X_3(n)(a + b\rho_3) + X_4(n)(a + b\rho_4). \end{aligned}$$

Group items with respect to coefficients a and b:

$$\begin{aligned} X_{ij}^{*}(n+1) = X_{ij}(n) + a[X_1(n) + X_2(n) + X_3(n) + X_4(n) - 4X_{ij}(n)] \\ + b[X_1(n)\rho_1 + X_2(n)\rho_2 + X_3(n)\rho_3 + X_4(n)\rho_4 \\ - 4X_{ij}(n)(\rho_1 + \rho_2 + \rho_3 + \rho_4)].. \end{aligned}$$

Let:

$$\begin{aligned} c &= X_1(n) + X_2(n) + X_3(n) + X_4(n) - 4X_{ij}(n), \\ d &= X_1(n)\rho_1 + X_2(n)\rho_2 + X_3(n)\rho_3 + X_4(n)\rho_4 - 4X_{ij}(n)(\rho_1 + \rho_2 + \rho_3 + \rho_4), \end{aligned}$$

the predicted fish concentration for month n

$$X_{ij}^{*}(n+1) = X_{ij}(n) + ac + bd.$$

Write the residual sums of squares:

$$U = \sum_n [X_{ij}^*(n+1) - \overline{X}_{ij}]^2 .$$

Then minimize the obtained function:

$$U \to min \cdot$$

Equating the partial derivative with respect to parameters *a* and *b* to zero, we get a closed system of two equations with two unknown parameters:

$$\begin{cases} \dfrac{\partial U}{\partial a} = 2\sum_n (X_{ij}^*(n+1) - \overline{X})c = 0, \\ \dfrac{\partial U}{\partial b} = 2\sum_n (X_{ij}^*(n+1) - \overline{X})d = 0 \end{cases}$$

Having made some transformations, we obtain the solutions to our system, i.e., coefficients of the model, where *a* is expressed as an implicit form:

$$\begin{cases} b = \dfrac{\sum_n c^2 \sum_n d[\overline{X}_{ij} - X_{ij}(n)] - \sum_n cd \sum_n c[\overline{X}_{ij} - X_{ij}(n)]}{\sum_n c^2 \sum_n d^2 - (\sum_n cd)^2}, \\ a = \dfrac{\sum_n c[\overline{X}_{ij} - X_{ij}(n)] - b\sum_n cd}{\sum_n c^2} \end{cases}$$

Model III

To have a model more flexible in response to hydrological conditions, we add one more coefficient *c* with regard to gradient in the thermocline. Assigning g_1, g_2, g_3, and g_4 as the differences of gradients in the thermocline between one square and its adjacent squares, we obtain a linear forecast equation for the following month (n+1) for the given square X_{ij}^*:

$$\begin{aligned} X_{ij}^*(n+1) = X_{ij}(n)[1 - 4a - 4b(\rho_1 + \rho_2 + \rho_3 + \rho_4) \\ - 4c(g_1 + g_2 + g_3 + g_4)] + X_1(n)(a + b\rho_1 + cg_1) \\ + X_2(n)(a + b\rho_2 + cg_2) + X_3(n)(a + b\rho_3 + cg_3) \\ + X_4(n)(a + b\rho_4 + cg_4) \end{aligned}$$

Group items with respect to coefficients *a*, *b* and *c*:

$$
\begin{aligned}
X^*_{ij}(n+1) = X_{ij}(n) &+ a[X_1(n) + X_2(n) + X_3(n) + X_4(n) \\
&- 4X_{ij}(n)] + b[X_1(n)\rho_1 + X_2(n)\rho_2 + X_3(n)\rho_3 + X_4(n)\rho_4 \\
&- 4X_{ij}(n)(\rho_1 + \rho_2 + \rho_3 + \rho_4)] + c[X_1(n)g_1 + X_2(n)g_2 + X_3(n)g_3 + X_4(n)g_4 \\
&- 4X_{ij}(n)(g_1 + g_2 + g_3 + g_4)]
\end{aligned}
$$

Again let:

$$
\begin{aligned}
l &= X_1(n) + X_2(n) + X_3(n) + X_4(n) - 4X_{ij}(n), \\
m &= X_1(n)\rho_1 + X_2(n)\rho_2 + X_3(n)\rho_3 + X_4(n)\rho_4 - 4X_{ij}(n)(\rho_1 + \rho_2 + \rho_3 + \rho_4), \\
n &= X_1(n)g_1 + X_2(n)g_2 + X_3(n)g_3 + X_4(n)g_4 - 4X_{ij}(n)(g_1 + g_2 + g_3 + g_4).
\end{aligned}
$$

The predicted fish concentration for month n+1 for the given square is:

$$X^*_{ij}(n+1) = X_{ij}(n) + al + bm + cn.$$

Minimize the residual sums of squares:

$$U = \sum_n [X^*_{ij}(n+1) - \overline{X}_{ij}]^2 .$$

Equating the partial derivative with respect to parameters a, b, and c to zero, we obtain a closed system of three equations with three unknowns:

$$
\begin{cases}
\dfrac{\partial U}{\partial a} = 2\sum_n \left[X^*_{ij}(n+1) - \overline{X}\right] l = 0, \\
\dfrac{\partial U}{\partial b} = 2\sum_n \left[X^*_{ij}(n+1) - \overline{X}\right] m = 0, \\
\dfrac{\partial U}{\partial c} = 2\sum_n \left[X^*_{ij}(n+1) - \overline{X}\right] n = 0
\end{cases}
$$

Make replacement:

$$X = \overline{X}_{ij} - X_{ij}(n)\cdot$$

Put the system into matrix:

$$\begin{cases} a\sum_n l^2 + b\sum_n lm + c\sum_n ln = \sum_n lX, \\ a\sum_n lm + b\sum_n m^2 + c\sum_n mn = \sum_n mX, \\ a\sum_n ln + b\sum_n mn + c\sum_n n^2 = \sum_n nX \end{cases}$$

Solve the system using the Gauss method. For this purpose let:

$$R_1 = \left(-\frac{\sum_n l^2}{\sum_n lm} \right), \qquad R_2 = \left(-\frac{\sum_n l^2}{\sum_n ln} \right), \text{ then}$$

$$\begin{cases} a\sum_n l^2 + b\sum_n lm + c\sum_n ln = \sum_n lX, \\ b\left(\sum_n lm + R_1\sum_n m^2 \right) + c\left(\sum_n ln + R_1\sum_n mn \right) = \sum_n lX + R_1\sum_n mX, \\ b\left(\sum_n lm + R_2\sum_n mn \right) + c\left(\sum_n ln + R_2\sum_n n^2 \right) = \sum_n lX + R_2\sum_n nX \end{cases}$$

Again make replacement:

$$R_3 = \left(-\frac{\sum_n lm + R_1\sum_n m^2}{\sum_n lm + R_2\sum_n mn} \right),$$

$$R_4 = \sum_n ln + R_1\sum_n mn + R_3\left(\sum_n ln + R_2\sum_n n^2 \right),$$

$$R_5 = \sum_n lX + R_1\sum_n mX + R_3\left(\sum_n lX + R_2\sum_n nX \right)$$

then:

$$\begin{cases} a\sum_n l^2 + b\sum_n lm + c\sum_n ln = \sum_n lX, \\ b\left(\sum_n lm + R_1\sum_n m^2 \right) + c\left(\sum_n ln + R_1\sum_n mn \right) = \sum_n lX + R_1\sum_n mX, \\ c(R_4) = R_5 \end{cases}$$

After some transformations, we obtain the implicit solutions to our system, i.e., coefficients of the model:

$$\begin{cases} c = \dfrac{R_5}{R_4}, \\ b = \dfrac{\sum_n lX + R_1 \sum_n mX - c\left(\sum_n ln + R_1 \sum_n mn\right)}{\sum_n lm + R_1 \sum_n m^2}, \\ a = \dfrac{\sum_n lX - b\sum_n lm - c\sum_n ln}{\sum_n l^2} \end{cases}$$

Data

As the basis of our work we used materials collected in TINRO surveys in 1984-1986, including chub mackerel catch per unit of effort data (kg per hour) and water temperature in a 0-600 m layer at 34°15′-44°30′N and 141°-160°E. Data on chub mackerel and sardine catches (SDD [ship daily data] from work of the Soviet-Russian fishing fleet) from 1980 to 1992 were used. A file of long-term, deepwater data, created by the employees of DVNIGMI was also used, where the temperature data were arranged on 1 degree squares in each month of a year for a layer of 0-500 m. Data processing for the above three sources is as follows.

Survey Data

Survey data for 1984-86 were interpolated in units of a regular square grid with steps of 15 minutes (0.25°) with the help of the graphic package Surfer (Win32) (Version 6.01, Surface Mapping System, Copyright 1993-95, Golden Software Inc.). (Survey data were not averaged.) Various methods of interpolation of data in this software are widely available. For interpolation of our data we used the method of "Inverse Distance to a Power." This gridding method is a weighted average interpolator, and can be either an exact or a smoothing interpolator. The power parameter controls how the weighting coefficients drop off as distance from a grid node increases. For a larger power, closer data points are given a higher fraction of the overall weight; for a smaller power, the weights are more evenly distributed among the data points.

The weight given to a particular data point when calculating a grid node is proportional to the inverse of the distance to the specified power of the observation from the grid node. When calculating a grid node, the assigned weights are fractions, and the sum of all the weights is equal to 1.0.

When an observation is coincident with a grid node, the observation is given a weight of essentially 1.0, and all other observations are given a weight of almost 0.0. In other words, the grid node is assigned the value of the coincident observation. This is an exact interpolator. Graphic representation of data on chub mackerel is shown in Fig. 3a and 3b. For an example of interpolated data for chub mackerel in September see Table 1a-c.

As the duration of the survey was lengthy, coefficients of migrations for some months were not obtained for the models.

Commercial Data

Commercial data for 1980-1992 were averaged for the fishing period by months. For this purpose the data array was divided by months, and the region of fishing for each month was broken into a regular square grid node in 3-minute segments (0.05°). Data in each square were averaged for the center of the square. An example of a long-term data array on sardine fishing is shown in Fig. 4.

These data were then also interpolated in nodes of a regular grid by steps of 6 minutes (0.1°) by the same method of interpolation in the Surfer32 software. An example of interpolated data for sardine in October is in Table 2a-c.

Hydrological Data

Hydrological data provided from the files of DVNIGMI were averaged and interpolated in nodes of a regular grid with steps of 6 minutes (0.1°) by the same method of interpolation in the Surfer32 software.

A program was written in the language of a high level Turbo Pascal (Version 7.0, Copyright 1983-92, Borland International, Inc.) to compare data from two nearest months, calculate coefficients of the models, and make prediction using the obtained coefficients.

Conclusions

For chub mackerel on scientific surveys we obtained coefficients for models II and III, some of which are indicated below.

September-October:

$a = 0.3987151$, $b = 0.0235035$;

$a = 0.3851383$, $b = 0.0334592$, $c = 0.00280627$;

April-May:

$a = 0.4358854$, $b = 0.0051921$;

$a = -0.8897466$, $b = 0.0028711$, $c = 0.08901928$;

May-June:

$a = 0.0579618$, $b = 0.0027537$;

$a = 0.0888050$, $b = 0.0010291$, $c = 0.03524888$.

Coefficients were also obtained using three models on monthly distribution data averaged over many years for sardine fishing, some of which are indicated below:

January-February:

a = 0.3220480;

a = 0.2684861, b = 0.0017835;

a = 0.2743469, b = 0.0033472, c = 0.0740538;

February-March:

a = 0.4763302;

a = 0.3867460, b = 0.0020080;

a = 0.3579247, b = 0.0001139, c = 0.0670430;

March-April:

a = 0.3945235;

a = 0.3208751, b = 0.0007745;

a = 0.3131046, b = 0.0001545, c = 0.0727093;

September-October:

a = 0.2780193;

a = 0.2337730, b = 0.0479562;

a = 0.2171924, b = 0.0264501, c = 0.0014328.

Based on the work we have done, we conclude:

1. Models of chamber type give rather reliable results pertaining to the actual data; i.e., the retrospective forecast errors are insignificant, taking into account the quantity of information used.

2. Data arrays on chub mackerel and sardine fishing, obtained from scientific surveys and during fishing, as well as hydrological data are well interpolated using the Surfer32 graphic package.

3. Figures 5-9 drawn with the help of the Surfer32 graphic package show that predicted fish concentrations are smoother than the actual data, but the general trend in the distribution of biological objects is maintained.

4. Based on the coefficients of migrations obtained from the survey data on chub mackerel, it is seen that the beginning of spawning (southern) migrations is rather heavily influenced by the change of thermocline depth, and the beginning of foraging (northern) migrations by vertical temperature gradients.

5. The same conclusions can be made on coefficients of migrations, obtained through the application of long-term fishing data on sardine.

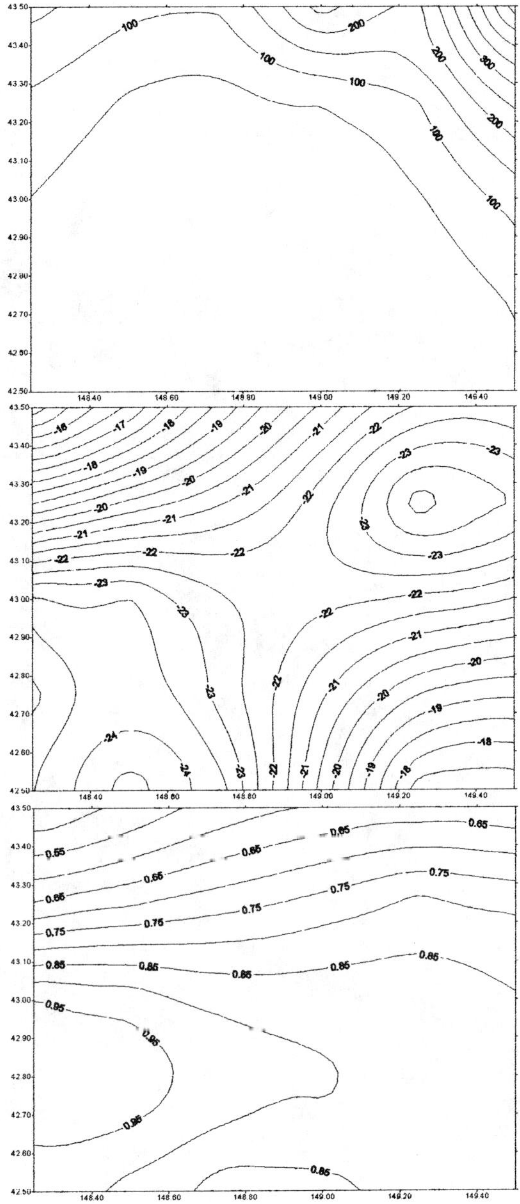

Figure 3a. *Graphic representation of data on chub mackerel for September. Top = catch per effort (kg per hour), Middle = the depth of thermocline (m), and Bottom = temperature gradient (°C per m) in the thermocline for September.*

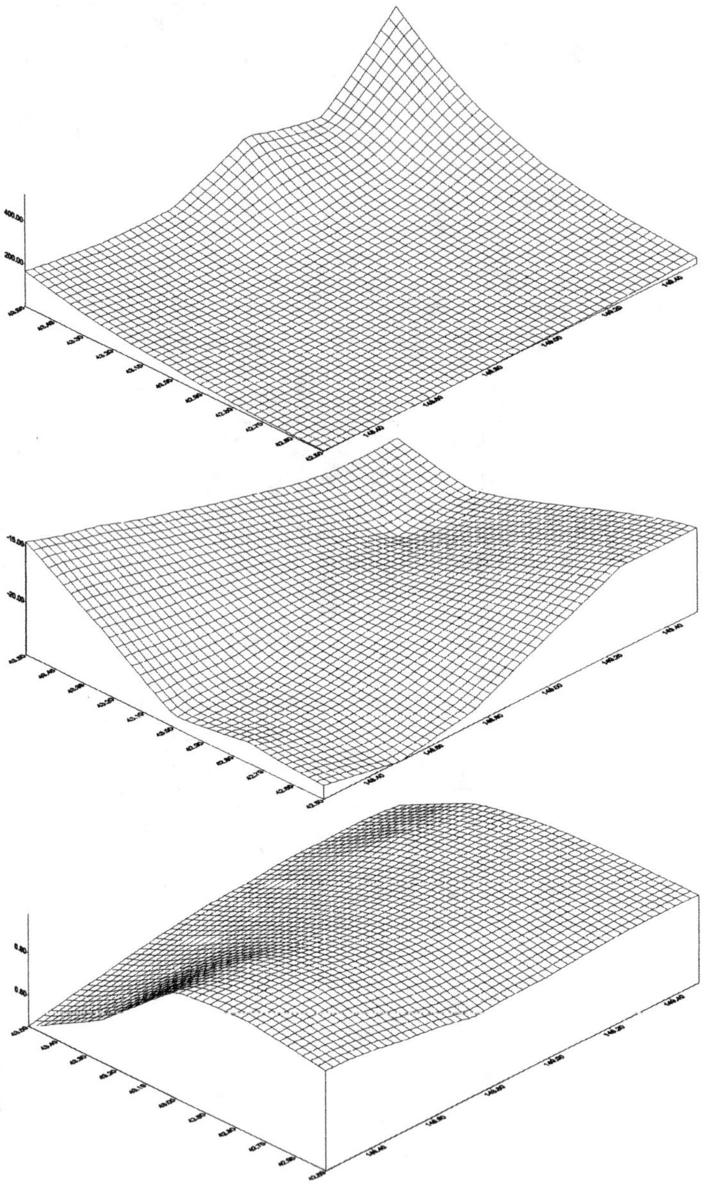

Figure 3b. Graphic representation of data on chub mackerel for September. Top = catch per effort (kg per hour), Middle = the depth of thermocline (m), and Bottom = temperature gradient (°C per m) in the thermocline for September.

Table 1a. Data on catch per effort (kg per hour).

42.50°N	4.49	0.82	2.00	4.91	12.35	30.61
42.75°N	16.07	4.86	4.77	9.17	22.15	57.34
43.00°N	49.24	17.66	13.40	20.73	45.75	121.33
43.25°N	90.50	42.74	29.87	52.40	97.34	262.54
43.50°N	164.40	114.46	107.08	272.87	186.68	528.74
	148.25°E	**148.50°E**	**148.75°E**	**149.00°E**	**149.25°E**	**149.50°E**

Table 1b. Data on depth of thermocline (m).

42.50°N	-23.47	-24.73	-23.60	-20.33	-17.24	-17.20
42.75°N	-22.8283	-23.6674	-22.7371	-21.0879	-19.5075	-19.0869
43.00°N	-23.6603	-23.5470	-22.5199	-22.1602	-21.8740	-21.4924
43.25°N	-18.9556	-20.1278	-20.9953	-22.2936	-24.2375	-23.4953
43.50°N	-14.6758	-16.5767	-18.7258	-20.5734	-22.0299	-22.0828
	148.25°E	**148.50°E**	**148.75°E**	**149.00°E**	**149.25°E**	**149.50°E**

Table 1c. Data on a gradient in thermocline (°C/m).

42.50°N	0.923147	0.882312	0.822160	0.834828	0.864296	0.867404
42.75°N	0.976460	0.976767	0.923929	0.902199	0.887763	0.869506
43.00°N	0.954640	0.927858	0.882868	0.877402	0.870742	0.853594
43.25°N	0.653358	0.693884	0.731696	0.768550	0.815047	0.787706
43.50°N	0.454835	0.513467	0.571861	0.603753	0.621295	0.609358
	148.25°E	**148.50°E**	**148.75°E**	**149.00°E**	**149.25°E**	**149.50°E**

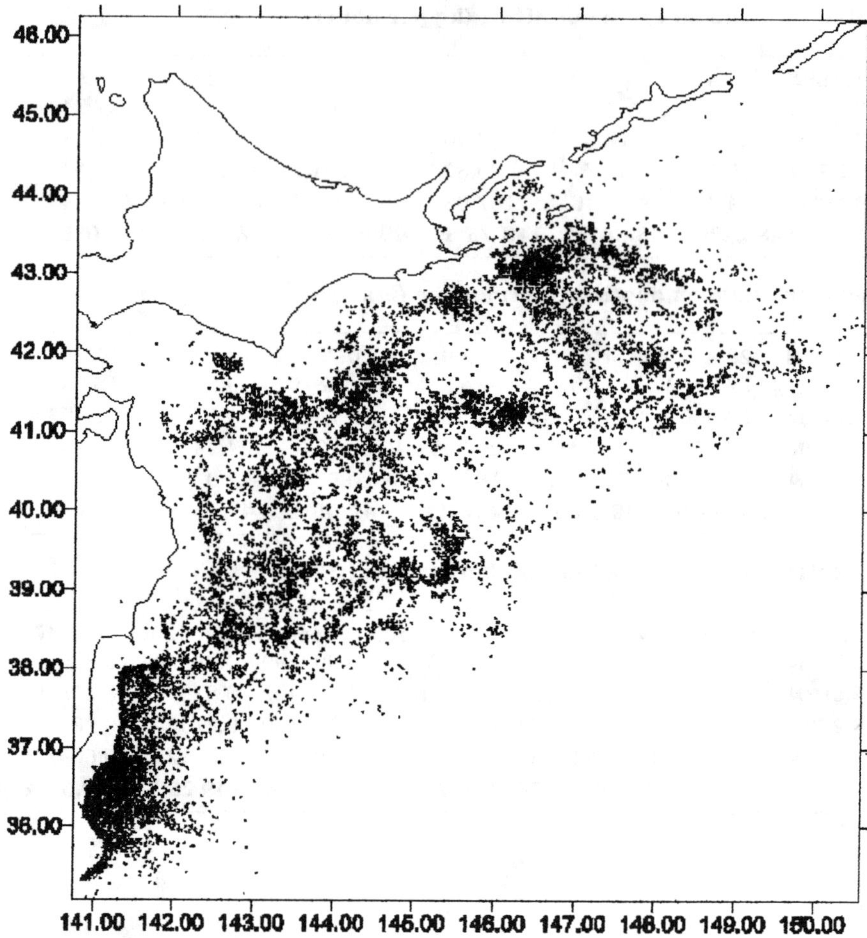

Figure 4. Data array from daily ship data from the Soviet-Russian fleet fishing for sardine 1980-1992.

Table 2a. Data on catch per effort (kg per hour).

37.00°N	9153.38	9253.43	9248.12	8685.04	14792.3	14761.5
37.10°N	10660.3	9962.00	10011.6	9389.48	17330.0	16617.0
37.20°N	9069.67	8575.42	8573.56	8089.74	9254.16	10638.2
37.30°N	9200.62	9094.62	8915.80	8343.98	6971.42	7912.60
37.40°N	9620.74	9453.39	9307.78	8979.86	8534.27	9765.96
37.50°N	10764.0	10399.3	10308.8	10241.4	11017.5	11727.7
	141.00°E	**141.10°E**	**141.20°E**	**141.30°E**	**141.40°E**	**141.50°E**

Table 2b. Data on depth of thermocline (m).

37.00°N	−77.325	−77.340	−76.936	−76.327	−75.408	−74.053
37.10°N	−75.875	−76.253	−75.900	−74.364	−73.614	−72.514
37.20°N	−76.531	−76.907	−75.575	−75.171	−74.545	−73.613
37.30°N	−77.270	−76.587	−76.480	−76.265	−75.875	−75.256
7.40°N	−78.091	−77.483	−77.570	−77.613	−77.575	−77.400
37.50°N	−78.974	−78.479	−78.805	−79.168	−79.559	−79.960
	141.00°E	**141.10°E**	**141.20°E**	**141.30°E**	**141.40°E**	**141.50°E**

Table 2c. Data on a gradient in thermocline (°C per m).

37.00°N	0.090768	0.090016	0.090233	0.090452	0.090668	0.090454
37.10°N	0.093703	0.092767	0.092902	0.093424	0.093442	0.093368
37.20°N	0.094328	0.093122	0.093991	0.094061	0.094069	0.093989
37.30°N	0.094960	0.094231	0.094352	0.094434	0.094747	0.094694
37.40°N	0.095600	0.094893	0.095021	0.095115	0.095166	0.095451
37.50°N	0.096245	0.095563	0.095701	0.095807	0.095879	0.095917
	141.00°E	**141.10°E**	**141.20°E**	**141.30°E**	**141.40°E**	**141.50°E**

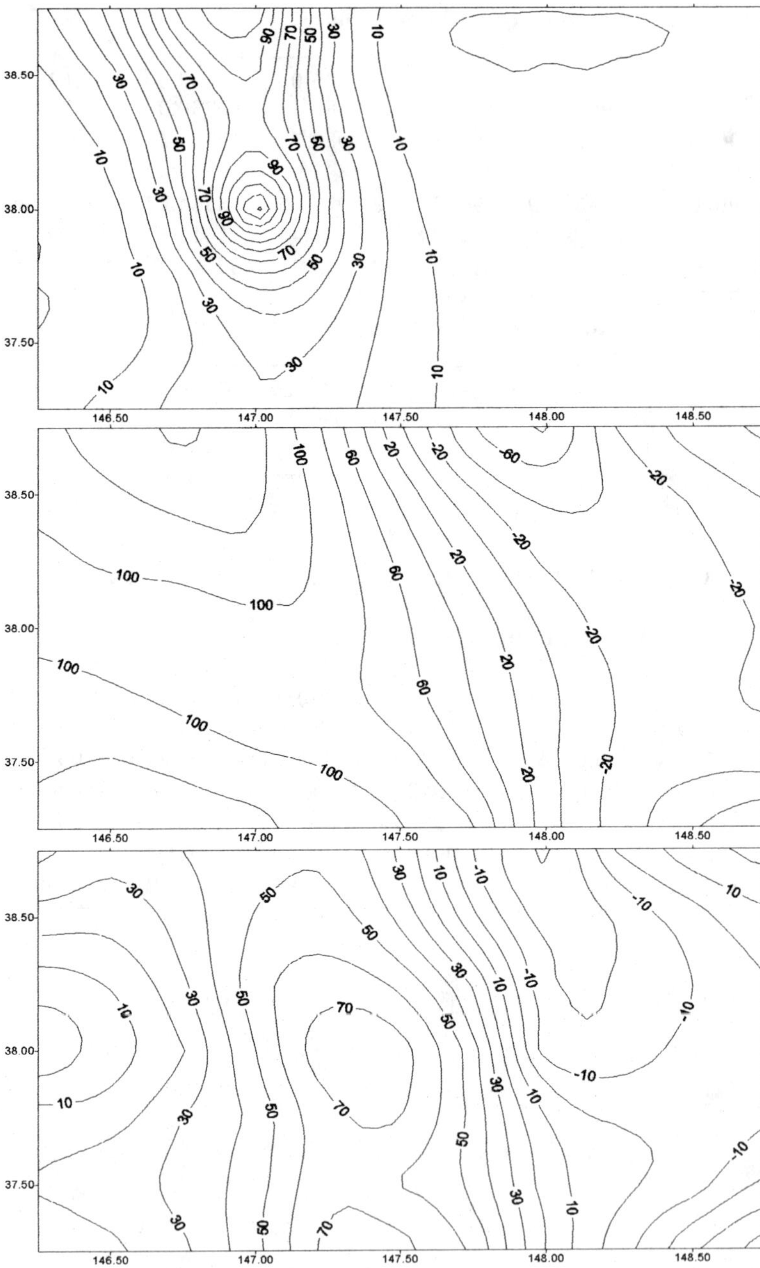

Figure 5a. Distribution of chub mackerel in May. Top = actual data forecasted from April, Middle = Model II, Bottom = Model III.

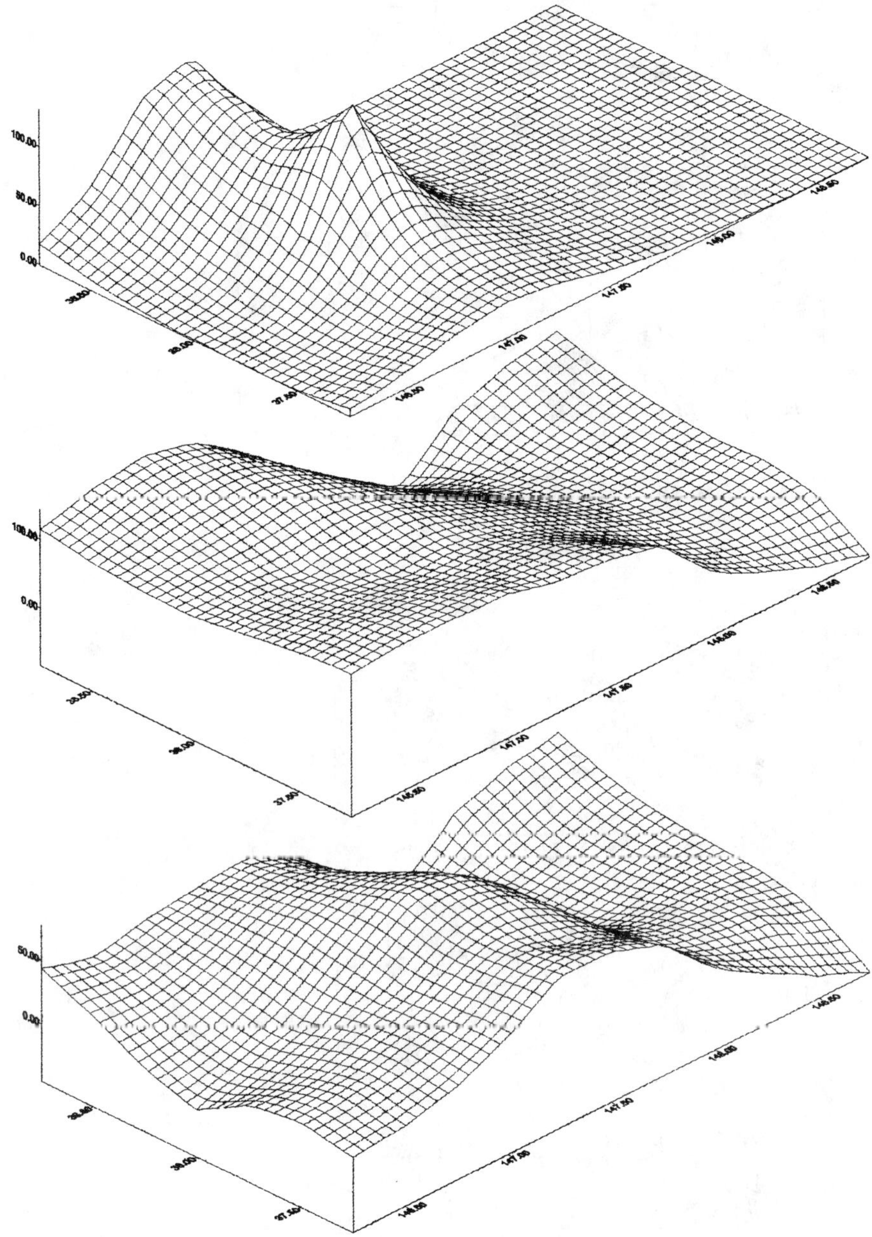

Figure 5b. Distribution of chub mackerel in May. Top = actual data forecasted from April, Middle = Model II, Bottom = Model III.

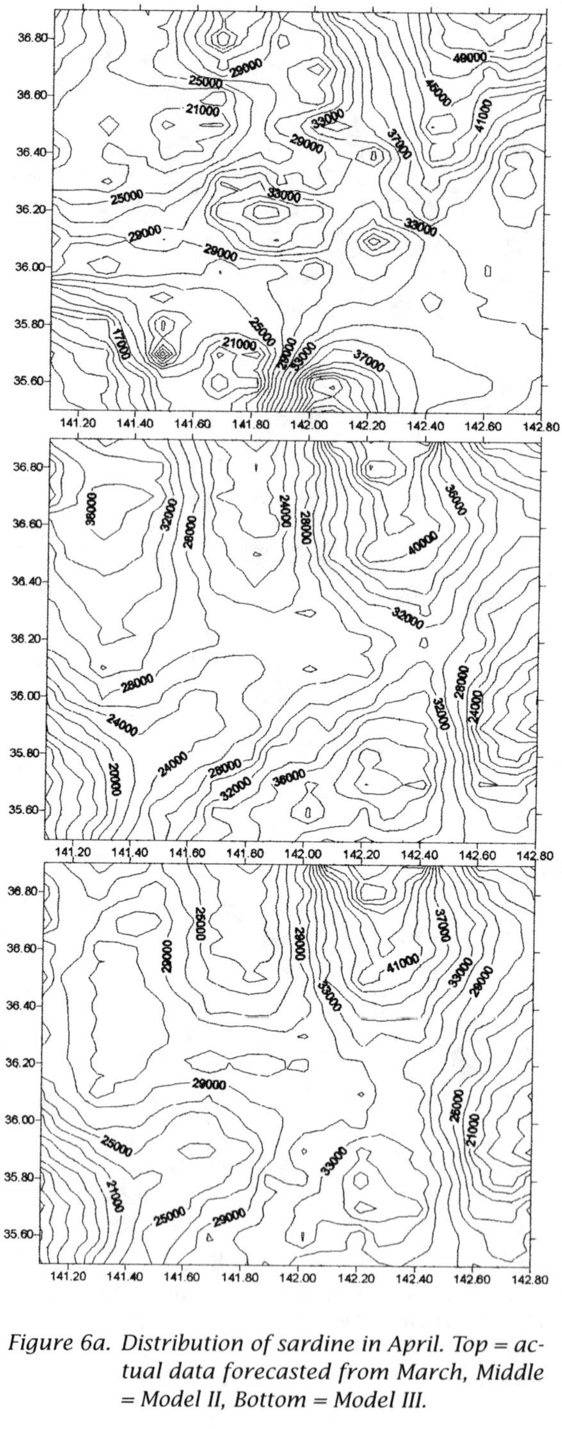

Figure 6a. Distribution of sardine in April. Top = actual data forecasted from March, Middle = Model II, Bottom = Model III.

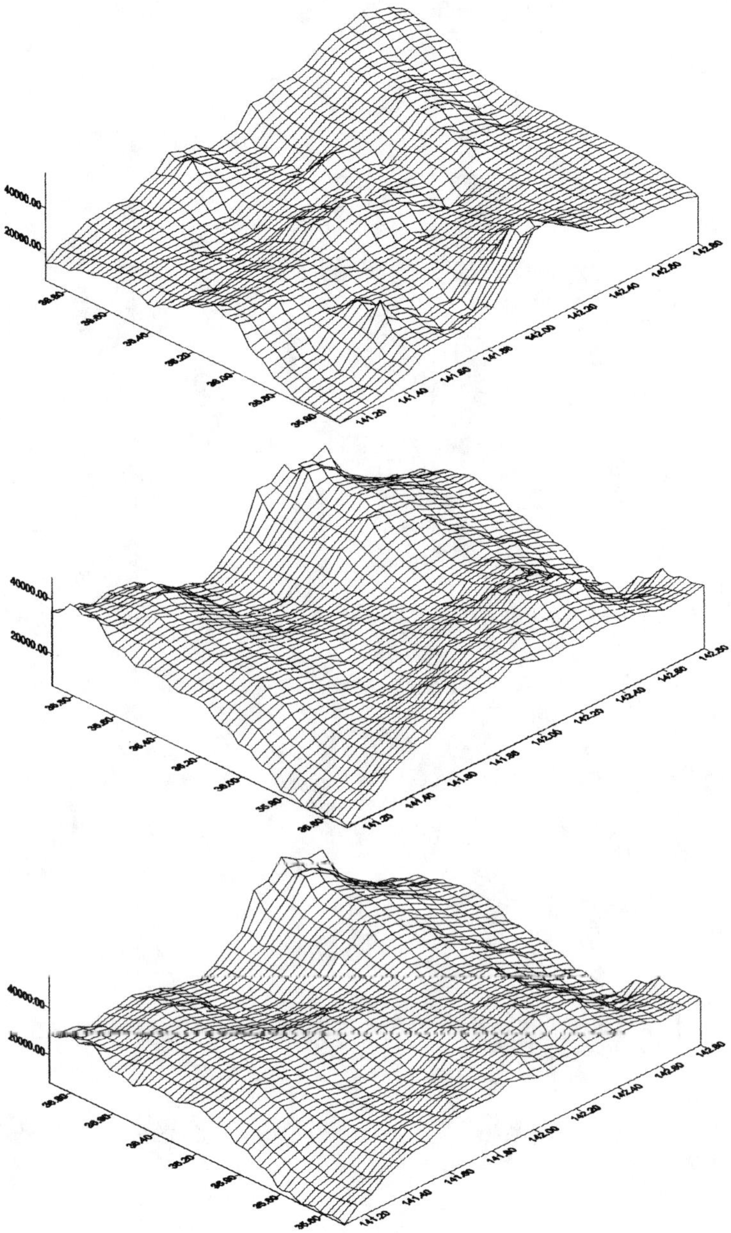

Figure 6b. Distribution of sardine in April. Top = actual data forecasted from March, Middle – Model II, Bottom = Model III.

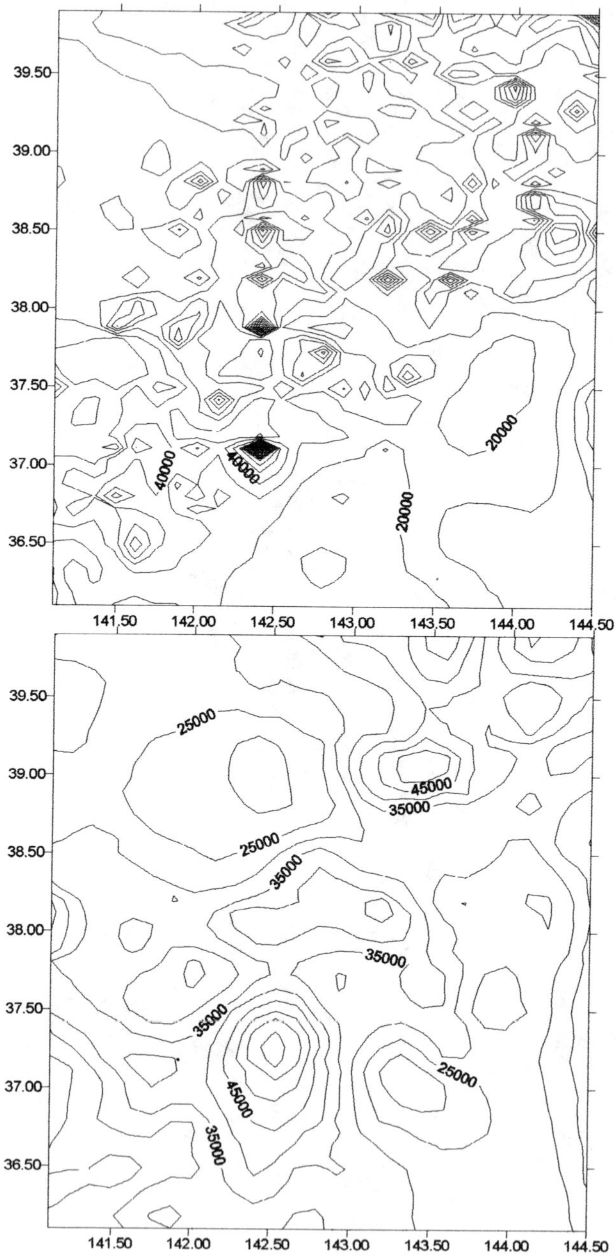

Figure 7a. Distribution of sardine in May. Top = actual data forecasted from April, Bottom = Model III.

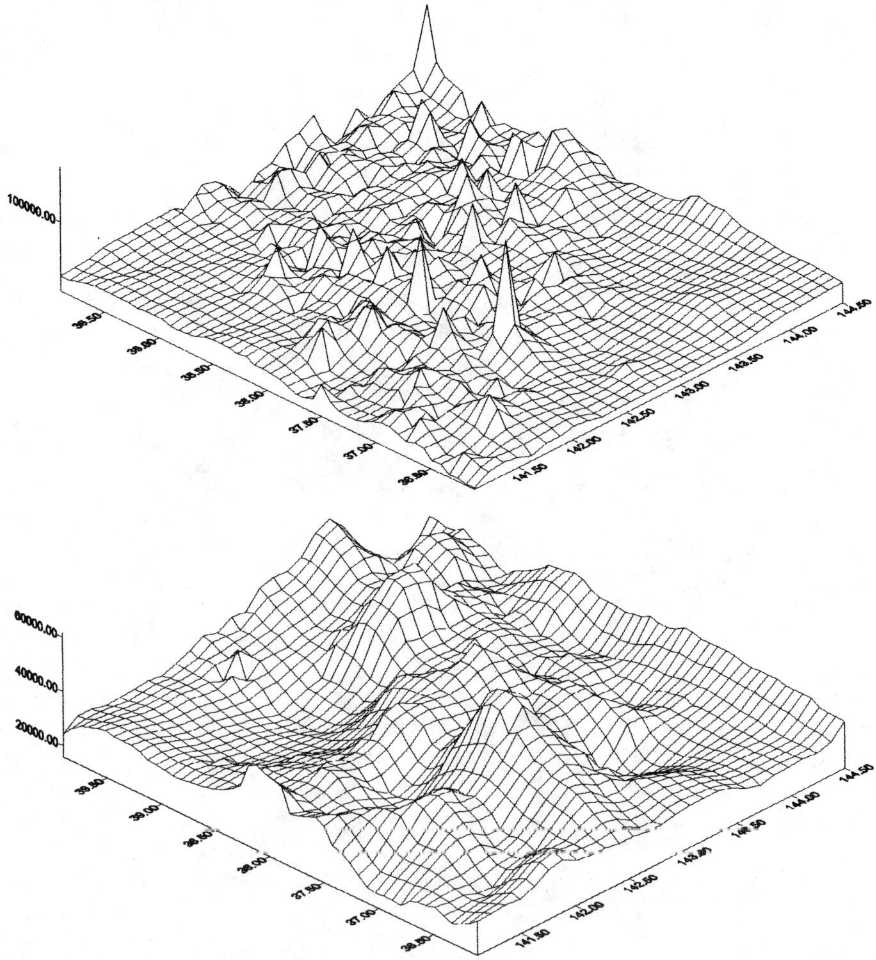

Figure 7b. Distribution of sardine in May. Top = actual data forecasted from April. Bottom = Model III.

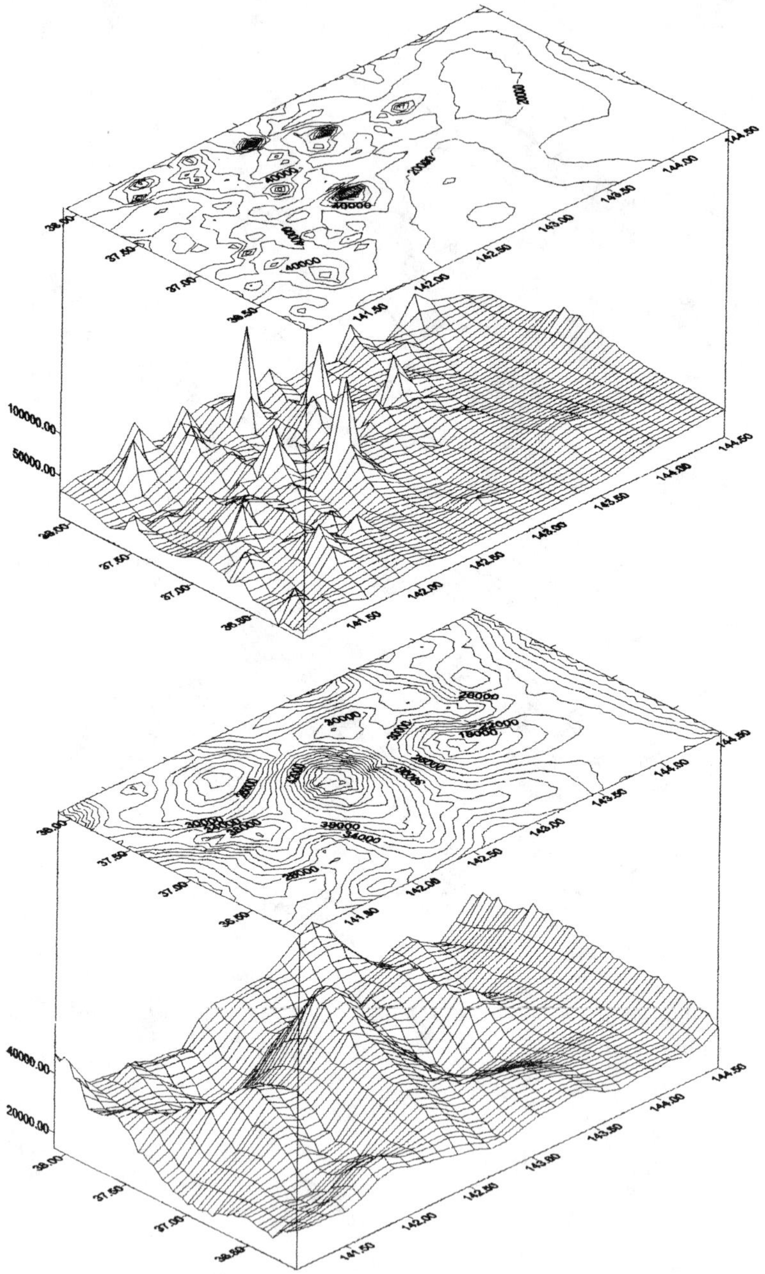

Figure 8. *Distribution of sardine in May. Top = actual data forecasted from April, Bottom = Model III.*

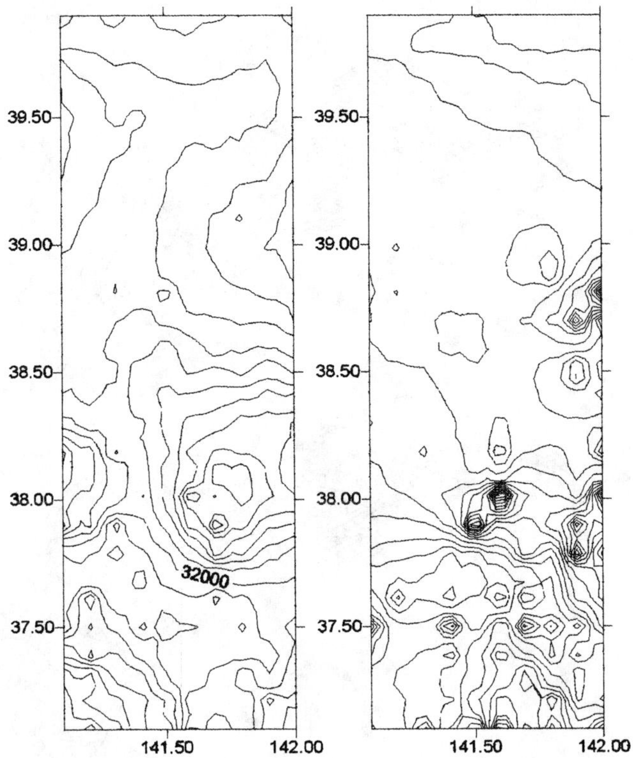

Figure 9a. Distribution of sardine in May. Right = actual data forecasted from April, Left = Model III.

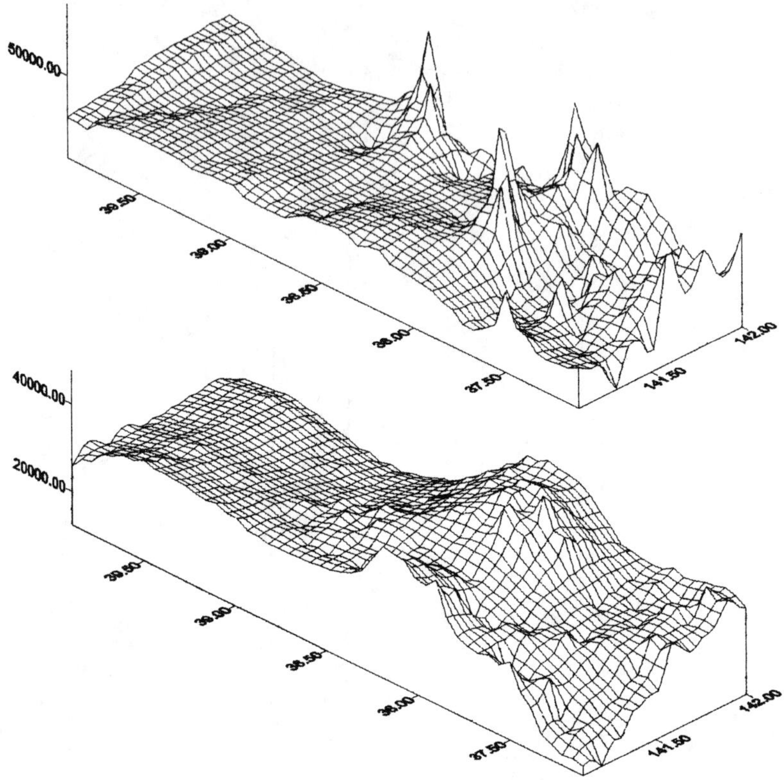

Figure 9b. Distribution of sardine in May. Right = actual data forecasted from April, Left = Model III.

6. In view of the above it is possible to conclude that the assumptions stated by the Sokolovskiy (1971) and Belayev (1984) that chub mackerel and sardine distribution is dependent on the depth of the thermocline, and the temperature gradient in the thermocline are confirmed.

References

Belayev, V.A. 1984. Structure of chub mackerel arial in the north-west Pacific. Okeanologiya 14(4):689-695. (In Russian.)

Bocharov, L.N. 1990. System analysis in short-term fishery forecasts. Nauka, Leningrad. 208 pp. (In Russian.)

Kenya, V.S. 1982. New data on migrations and distribution of far eastern sardine in the north-west Pacific. Biol. Morya (Vladivost.) 1:44-51. (In Russian.)

Sokolovskiy, A.S. 1971. To question on influence of thermocline on distribution of chub mackerel accumulations. Rybn. Khoz. 10:5-6. (In Russian.)

Tuzinckevich, A.V. 1989. Integral models of spatial-temporal dynamics of ecosystem. IAPU Akad. Nauk SSSR. 184 pp. (In Russian.)

(Re)Constructing Food Webs and Managing Fisheries

Jason Link
National Marine Fisheries Service, Northeast Fisheries Science Center, Woods Hole, Massachusetts

Abstract

The construction and analysis of a food web is an ecosystem approach that augments classical fisheries management. Food web analysis delineates important ecosystem linkages representing species interactions such as predator-prey relationships; from these linkages emerges a better understanding of ecosystem resiliency, resistance, connectivity, energy transfer efficiency, mass flux, and energy partitioning. In turn, with an understanding of food web dynamics, critical fisheries issues, including the relative importance of fishery and natural mortality, identification of critical life stages, production surplus and partitioning, multispecies yield dynamics, and forecasting, the impact of fishery management scenarios can be more effectively examined. Unfortunately, elucidating a food web is not a trivial task. One of the more parsimonious, cost-effective, and fisheries-amenable methods of reconstructing (at least portions of) a food web is diet analysis. Incorporated as part of standard resource surveys, analysis of stomach composition can not only qualify the linkages of a food web, but can also quantify the magnitude and rate of energy and mass exchange. Examples of statistical analyses from northwest Atlantic diet data demonstrate the utility of this approach in constructing a food web that produces information useful in addressing key fisheries issues.

Introduction

Food webs are the skeletal and circulatory analogues of ecosystem "anatomy and physiology." That is, the predator-prey interactions that comprise food webs can ultimately determine the fate and flux of every population in an ecosystem, particularly upper level consumers of fiscal importance (Hairston et al. 1960, Oksanen et al. 1981, Carpenter et al. 1985, Carpenter and Kitchell 1993, Christensen 1996). Numerous examples demonstrate the utility and insight that examining multispecies interactions have had on fisheries management around the world (e.g., Andersen and Ursin 1977, Mercer 1982, Daan and Sissenwine 1991, Payne et al. 1992, Christensen and Pauly 1993, Walters et al. 1997, Pauly et al. 1998). While assessing fish populations with disregard to other species is known to be unrealistic, intellectually unsatisfying, and potentially misleading (sensu May et al. 1979, Murawski 1996), there persist challenges, be they scientific, institutional, philosophical or otherwise, that impede the implementation of multispecies and ecosystem approaches to fisheries management. Perhaps this is best expressed in a quote from Hilborn and Walters (1992: p. 448), which captures a pervasive attitude among fisheries scientists regarding food webs:

> We believe the food web modelling approach is hopeless as an aid to formulating management advice; the number of parameters and assumptions required are enormous. Such large model-building exercises are perhaps of some utility in designing research programs, but will not help manage fisheries. The use of trophic relationships to estimate general yield potentials also seems unlikely to produce anything that fishery managers can use in the near future, although such relationships may become clearer in the long term as data are accumulated from a wide variety of fisheries.

I hope to demonstrate the contrary in this work, and will attempt to elucidate the utility that a classical food web approach can provide for fisheries management.

What Is a Food Web?

Food webs have been defined literally as road maps of species interactions, or as the relationship or connection between species, particularly trophic interactions (Pimm 1982, Cohen et al. 1993, Paine 1988). Implied are diagrams that graphically depict which species in a community interact, or serve as a schematic of energy flow in an ecosystem. DeAngelis (1992: p. 2) correctly separates the two major aspects of ecological systems into strict food web dynamics (the interactions between populations), in an ecological perspective, and energy and material flux (rates or magnitude of these interactions, expressed as units of energy or mass), in a thermodynamically balanced perspective. In this work I will primarily focus on the strict food web aspect of ecosystem dynamics, although this is highly interrelated with the flow of mass and energy. Although similar to the ECOPATH and ECOSIM (reviewed in Walters et al. 1997) approaches familiar to many fishery scientists, this approach, championed by May (1973), Cohen (1978), and Pimm (1982, 1991), focuses on summary statistics from a formal, ecologically oriented analysis that elucidates the structure of a food web and highlights major properties peculiar to a particular ecosystem. My goal is to relate these key ecological parameters to metrics and concepts familiar to fishery managers.

Materials and Methods

There are several ways to reconstruct and assess food webs, described in Table 1. Many of these methodologies are understandably cost- and time-prohibitive for most fishery agencies. However, one cost-effective and assessment-amenable method of reconstructing a food web is diet analysis. Augmented by other directed studies, this approach can provide a robust data set with which to analyze food webs. When incorporated as part of standard resource surveys, stomach analysis can qualify linkages and quantify the magnitude of these linkages across broad taxonomic, spatial, and temporal scales. If an agency is examining selected individual fish lengths, weights, ages, and maturities, then examining stomach contents is a natural extension of the existing sampling protocol.

The approach I used is based primarily upon stomach content evaluation, although I also relied on literature reports and body size and general biological inferences, particularly for prey of most invertebrates. I used the database on selected fish stomachs collected by the Northeast Fisheries Science Center (NEFSC) from 1973 to 1998. These data were collected in the spring and fall NEFSC surveys, covering the entire northeast U.S. continental shelf (ranging from Cape Hatteras to the Gulf of Maine). Stomachs were analyzed for total weight, prey volume (during 1981-present, weights prior to 1981), percent prey composition, and prey lengths if feasible to do so. These data are the primary source for the following analyses, and most species had more than 250 stomachs sampled.

I selected 75 "species" (i.e., groups of organisms at taxonomically feasible and functionally related levels) common in the northwest Atlantic. I included humans in this web. The lowest trophic level is highly aggregated (i.e., phytoplankton, detritus), yet there are 33 groups per species of invertebrates at the next trophic levels, as well as 35 species per groups of fish. Admittedly, this food web is vertebrate centric, and is, like all food webs, by default incomplete (DeAngelis 1992).

I first constructed an interaction matrix of the 75 pair-wise interactions, and assigned a –1, 0 or +1 if species A preyed on, did not interact with, or was eaten by species B, respectively. The presence or absence of a species in the diet of another species was the primary determinant of these values. A food web diagram was also constructed from these trophic linkages.

I then constructed an index of interaction strength matrix. This was based upon the assumption that percent frequency of occurrence of a prey item integrated predator spatial overlap, selectivity, and feeding rates for these prey items at varying abundances over the 25-year period across this broad geographic scale. This parameter was used to index the strength of trophic links between each of the 75 pairs. I calculated the mean interaction strength to assess how strongly connected the species in this community were.

I calculated standard food web metrics from these interaction matrices.

Table 1. Common methods for reconstructing food webs, particularly species interactions and interaction strengths.

Approach	Comments
In situ observations	*Observing or inferring a predatory event.* Video/photographic Spatial overlap-acoustics, trawls, etc. Anecdotal evidence Directed observations
Experimental studies	*Identifying and quantifying predation impacts.* Laboratory rates-e.g. selection, consumption, etc. Unplanned field manipulations Enclosures, translocations, removals, etc.
Diet analysis	*Examining what has been eaten by a predator.* Stomach content evaluation-mass, volume, prey number Composition (i.e. proportion) per prey type Selectivity indices, overlap indices Mean stomach content, related statistics
Body size inferences	*Relating morphometry with known feeding patterns.* e.g. Gill rakers Mouth gape Stomach length Aspect ratio
Microchemistry	*Stoichiometry to infer diet.* e.g. Fatty acids Bomb calorimetry Microconstituent analysis DNA, RNA probes Immuno-assays
Stable isotopes	*Reconstructing trophic level position from predator tissue.* e.g. N^{15} C^{13}
Ad hoc	*Educated guessing in a situation with a lack of information.* Related literature values Evidence from similar species, systems Try to "think like a fish"

Food Web Concepts

The series of "standard" food web metrics merit revisiting to, at the least, clarify terminology. The first metric noted was simply the number of species (S). The number of interactions (or trophic links; L) was also counted. This number was then divided by the total number of possible interactions to determine the connectivity (C) of a system:

$$C = \frac{L}{S(S-1)/2} \tag{1}$$

The number of interactions per species (L/S) was also calculated. Both C and L/S index the level of interactions, potential competition, trophic specialization, and overall inter-connectedness of a food web (May 1973, Pimm 1982, 1991; Cohen et al. 1993). The product $S \times C$, when coupled with an assessment of overall web interaction strength (either mean interaction strength [May 1973] or eigenvalues of the interaction matrices [Pimm 1982]) gives an assessment of mathematical stability for a system. Here stability is defined as whether a population will return to equilibrium after a perturbation, realizing that there are both global and local stabilities (DeAngelis 1992, Pimm 1991). Associated with the idea of stability are concepts of: (1) equilibrium, defined as when time approaches infinity, a system becomes relatively constant (given local variability) near a "carrying capacity"; even though it is contentious as to whether equilibrium truly exists in biological systems, yet is often assumed for distinct time scales; (2) resilience, defined as how fast a system returns to equilibrium after a perturbation; (3) persistence, which is a measure of the strength of a system against departure from equilibrium, or how long component populations persist at certain levels before being changed to new equilibria; and (4) resistance, defined as the relative magnitude that a change in one population has on other populations in a system, measured as the ratio or percent change before and after a perturbation for all component populations (Pimm 1991, DeAngelis 1992). Recognizing the contentious nature of these concepts (e.g., Hastings 1988, Paine 1988, Pimm and Kitching 1988, Warren 1990), I will only address these items qualitatively.

The number of basal species (no trophic levels supporting these organisms), the number of top predators (defined here as a species having less than three predators), the number of species at trophic levels 2 and 3, and the total number of intermediate species (non-basal and non-top-predator species) were enumerated, as were the number of cannibals (species known to prey upon themselves), the number of cycles (when species A preys upon species B and vice versa), and the number of omnivores (defined as the number of species eating prey from more than one trophic level; Pimm and Lawton 1978). The modal food chain length was calculated to determine the most common number of trophic levels in this food web. Finally, the number of predators for a prey item, the number of prey items for a predator, and the predator to prey ratio were calculated. This set of metrics details the specific properties of a particular food web and gives insight into the rate and flow of material and energy distribution in an ecosystem. Identifying the links between species is required to assess energy/mass partitioning, energy transfer efficiency, and surplus production of an ecosystem.

Results and Discussion

What Does This Food Web Information Tell Us, and How Does It Relate to Fisheries?

The first thing to note is that this is a relatively large food web, as most food webs average <20 species, and 95% of webs have <40 species (Table 2; Cohen 1989). I recognize the high degree of aggregation at lower trophic levels in this web, but 33 invertebrate groups is larger than most whole food webs (discussed in Goldwasser and Roughgarden 1993). The level of observation on the feeding habits for this number of species is unusually high, yet by no means exhaustive. This is a complicated system that has been observed rather frequently, yet needs continued attention, including focus on the multispecies system attributes and better quantification of species linkages.

The values for the connectivity and number of interactions per species (*C* and *L/S*, Table 2) demonstrate that this is a highly connected food web, as depicted in Fig. 1. If one examines the level of connectivity in this system in relation to the number of species, this food web is approaching the theoretical maximal connectivity (ca. 50%) and is an extremely distinct outlier in the decreasing hyperbolic curve of *C* vs. *S* (Pimm 1982, Warren 1990). Most food webs that have >50 species have a connectance of approximately 10%. I recognize that I was conservative in my assessment of interactions, and the high connectivity I report for this system may be an artifact of the grouping for invertebrates, the focus on fish, construction by using all known interactions (disregarding interaction strength when ascertaining linkages), and the inclusion of humans in the web. However, this invertebrate grouping is no worse than lower trophic levels in other webs, and again the number of groups at lower trophic levels is relatively more exhaustive than most other whole webs. The ratio of predators to prey is lower than the ECOWEB mean (Table 2), also suggesting that this web is not as vertebrate (or top predator) centric as others. The results from this food web may be an artifact of our sampling protocol, but more likely reflect true features of this system. Certainly, a higher threshold of species interaction other than presence in the diet of another species will lower the overall connectivity of this web, but unless this threshold is relatively high the connectivity will still be distinct from other food webs examined. The scale of the data, both temporally and spatially, is larger and longer than most food web analyses. The orders of magnitude in size

Table 2. Summary of food web statistics from the northwest Atlantic continental shelf community.

Parameter	NW Atlantic	ECOWEB mean
# Species (*S*)[a]	75	19, 95% of webs <40
# Interactions (*L*)	2,557	–
# Possible interactions	5,550	–
Connectivity (*C*)	46.1%	31.2%
$S \times C$	34.6	3.744
# Interactions/Spp. (*L*/*S*)	34.1	1.72
# Cannibals	20	–
% Cannibals	27%	
# Omnivores	48	
% Omnivores	64%	27%
# Top predators	8	
% Top predators	11%	30.1%
# Basal species[b]	2	
% Basal species	3%	20.9%
# Species at trophic level 2	33	–
% Species at trophic level 2	44%	
# Species at trophic level 3	31	
% Species at trophic level 3	41%	
# Intermediate species	66	
% Intermediate species	88%	49%
# Cycles	70	–
% Cycles	3%	
Modal chain length	4	2-3 (2.59)
Maximum	9	3.74
Minimum	3	–
Mean # Prey/predator	17.5	2.21
Maximum	61	
Minimum	1	
Mean # Predator/prey	15.7	2.46
Maximum	38	
Minimum	0	
Predator/prey ratio	0.894	1.18
Mean interaction strength	0.0697	–

The ECOWEB mean values are from 213 food webs analyzed in a similar fashion; – indicates not examined (Cohen 1989, cited in Goldwasser and Roughgarden 1993). See text for description of interaction strength.

[a]Some grouping, particularly in lower trophic levels.

[b]Taxonomic resolution very low.

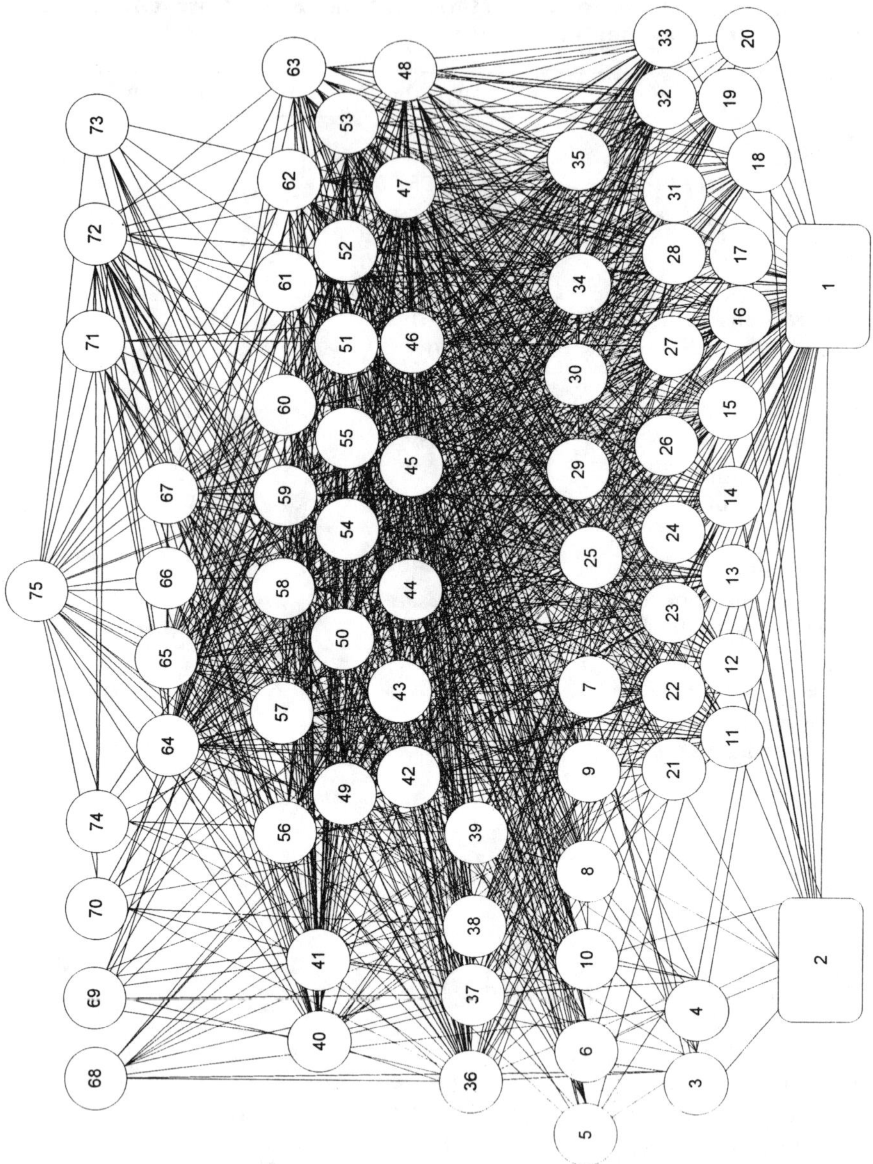
1
2
3
4
5
6
7
8
9
10
11
12
13
14
15
16
17
18
19
20
21
22
23
24
25
26
27
28
29
30
31
32
33
34
35
36
37
38
39
40
41
42
43
44
45
46
47
48
49
50
51
52
53
54
55
56
57
58
59
60
61
62
63
64
65
66
67
68
69
70
71
72
73
74
75

Figure 1 *Species and links of the northwest Atlantic food web. This tangled "bird's nest" represents interactions at the approximate trophic level of each species, with increasing trophic level towards the top of the web. The left side of the web generally typifies pelagic organisms, and the right to middle represents more benthic/demersally oriented organisms.*

1. Detritus.
2. Phytoplankton.
3. Calanus sp.
4. other copepods.
5. Ctenophores.
6. Chaetognatha (i.e. arrow worms).
7. Jellyfish.
8. Euphasiids.
9. Crangon sp.
10. Mysids.
11. Pandalids.
12. other decapods.
13. Gammarids.
14. Hyperiids.
15. Caprellids.
16. Isopods.
17. Pteropods.
18. Cumaceans.
19. Mantis Shrimps.
20. Tunicates.
21. Porifera.
22. Cancer crabs.
23. other crabs.
24. Lobster.
25. Hydroids.
26. corals, anenomes.
27. Polychaetes.
28. other worms.
29. Starfish.
30. Brittle stars.
31. Sea cucumbers.
32. Scallops.
33. clams, mussels.
34. Snails.
35. Urchins.
36. Sand lance.
37. Atlantic Herring.
38. Atlantic Mackerel.
39. Butterfish.
40. Loligo.
41. Illex.
42. Pollock.
43. Silver Hake.
44. Spotted Hake.
45. White Hake.
46. Red Hake.
47. Atlantic Cod.
48. Haddock.
49. Sea Raven.
50. Longhorn Sculpin.
51. Little Skate.
52. Winter Skate.
53. Thorny Skate.
54. Ocean Pout.
55. Redfish.
56. Yellowtail Flounder.
57. Windowpane Flounder.
58. Summer Flounder.
59. Witch Flounder.
60. Four-spot Flounder.
61. Winter Flounder.
62. American Plaice.
63. American Halibut.
64. Spiny Dogfish.
65. Goosefish.
66. Weakfish.
67. Bluefish.
68. Baleen Whales.
69. Toothed Whales, Porpoises.
70. Seals.
71. Migratory Scombrids.
72. Migratory Sharks.
73. Migratory Billfish.
74. Birds.
75. Humans.

across the ontogeny of these organisms will increase the number of connections relative to other systems. Finally, a limited number of marine food webs have been analyzed, thus the high connectivity of this system relative to theoretical expectations remains an intriguing aspect for further research.

Regardless of whether this food web revises connectance theory, two major points arise from the high connectivity observed for this food web: (1) there are a lot of generalists in this ecosystem, and (2) there is a high degree of diet overlap for these component organisms. A cursory glance at Fig. 1 shows the "bird's nest" of interactions in this system, and the almost opaque quality of this diagram enforces the idea of ubiquitous overlap and connectedness in this food web. For simplicity, Fig. 2 depicts the food web for the major groups from a fishery perspective, yet even this simplification exemplifies the complexity of these interactions.

The mean number of predators per prey and mean number of prey per predator are also higher than the ECOWEB means (Table 2). In addition, 64% of the species in this food web feed omnivorously. Even splitting out the lower trophic levels would not ameliorate this high level of overlap and omnivory. Omnivory indicates opportunistic feeding. Thus, there is high potential for competition in this food web, particularly if fish feed opportunistically and switch diet preferences based upon relative abundance of prey. For example, there are 38 predators of sand lance (species 36 in Fig. 1) in this system. Therefore, many of the fish in this ecosystem are ecologically interchangeable (at least from the perspective of prey populations).

There also exists a high potential for indirect effects on other species in this system, given an impact on a particular species. When a system is as connected as this northwest Atlantic system (Fig. 1, Table 2), changes in predator or competitor populations may release a particular species or group of species from interaction pressure (i.e., the interaction strengths decline), allowing that species to drastically increase. For example, Fogarty and Murawski (1998) hypothesize that the notable increase in elasmobranchs on Georges Bank is due in large part to a decrease in competitive pressure from the gadoids that have been overfished. The potential for Markovian spirals increases with this level of connectance. As in the previous example, impacts on species *A* influence species *B* which directly impacts species *C* and also indirectly releases species *K*, a competitor of both species *C* and *D*, ad infinitum. That is, from a stability perspective, resistance may be low for this web. The challenge is then to predict what the future portions of biomass will be in this system, at least in qualitative trends, for various management scenarios. The large number of interactions contributes to the stochastic nature of this ecosystem, to the consequent uncertainty in the predictions, and hence to the magnitude of this challenge. The northwest Atlantic food web demonstrates the need to consider these species as a whole and not as individual populations, particularly noting the complexity and ubiquity of these interspecific interactions.

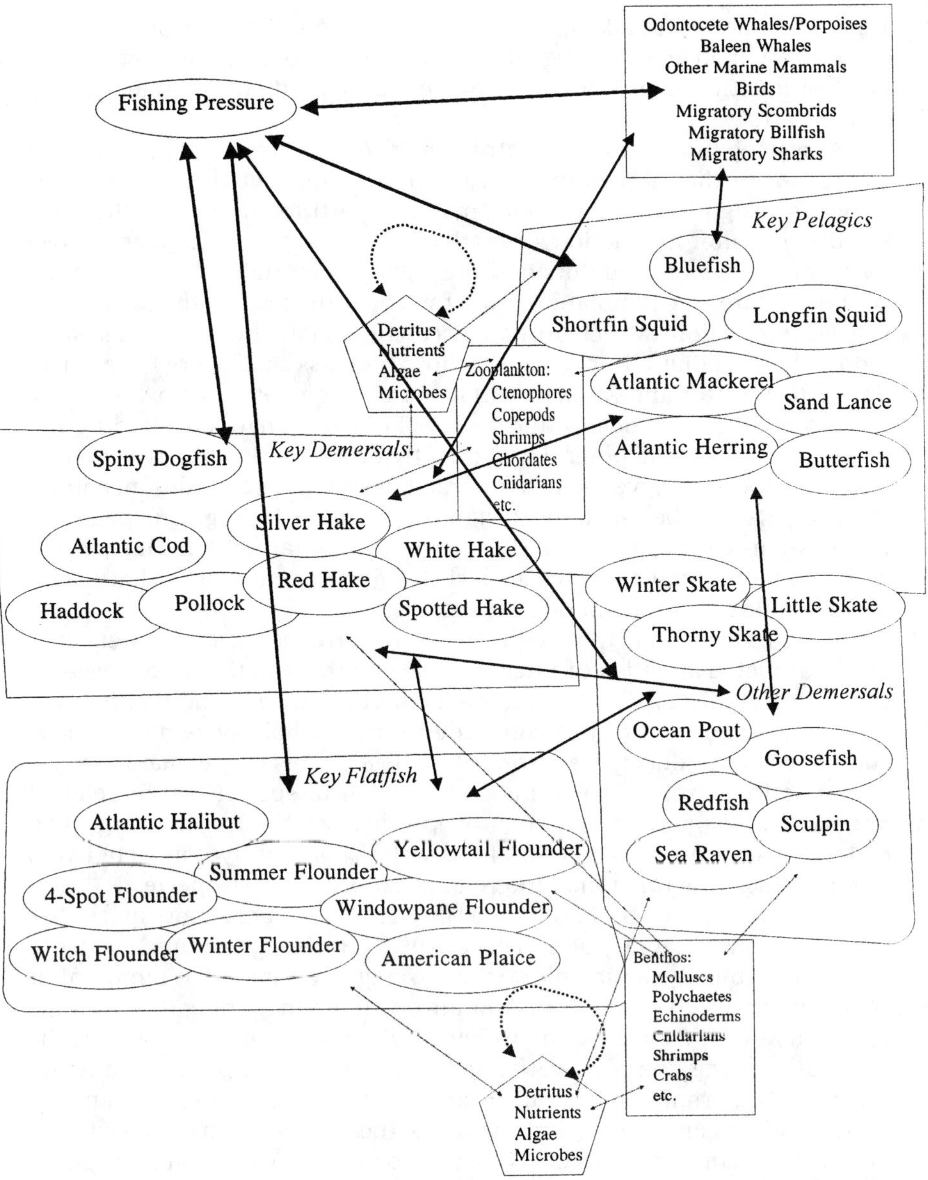

Figure 2. Simplified schematic of the northwest Atlantic food web depicting the major fish groups and their interactions with other major prey and predator groups.

Yield dynamics must take into consideration a multispecies perspective, else fishery managers will be surprised at the amount of predicted "surplus" production that is actually consumed by predators or inhibited by competitors.

Assessing interaction strengths for a food web is difficult, definitely an area for further improvement (Paine 1988, Cohen et al. 1993), and yet provides insight into energy and material partitioning within the web. Factors that alter interaction strength are abundance, overlap, and selectivity/avoidance of predators and prey. Fishing mortality (F) impacts the abundance of these populations, and thus directly impacts the strength of species interactions, which ultimately can propagate throughout a system indirectly affecting all species and how biomass and energy are partitioned (Carpenter and Kitchell 1993, Christensen 1996, Murawski 1996, Walters et al. 1997, Fogarty and Murawski 1998). That is, cascading impacts from F have the potential to directly or indirectly alter every population in a food web given a set of interaction strengths. In addition, natural mortality (M) can be much more significant than F for a given species in this system (Overholtz et al. 1991; Overholtz et al. 1999). Thus, if the major portion of mortality for a species is due to predation, fishery managers should be cognizant of this fact.

If we assume that the frequency of occurrence of a prey item in a predator diet is an index of interaction strength, then this web is weakly connected (Table 2). That is, although there may be numerous interactions, few are of sufficient magnitude to drive whole-system dynamics, similar to the results of Sissenwine et al. (1982). This may explain the lack of competitive exclusion for this high number of species, as the relative importance of many interactions drop out from the major energy and material pathways through this food web. This loosely connected web also reiterates the functional interchangeability of these species.

The structure of this food web, in terms of co-occurring species, is static (Sissenwine et al. 1984, Gabriel 1992). However, the functioning of this food web is dynamic. Interaction strengths examined at temporal or spatial intervals give greater insight into system functioning than a static, cumulative view (Schoenly and Cohen 1991). Species interactions are the pathways for the flux of material and energy from one population to another, and determine how biomass and energy are partitioned in an ecosystem. Addressing the dynamic nature of these interactions is not a trivial task, yet explains the "physiology" of an ecosystem. To illustrate this point, in the 1970s the major portion of this food web was comprised of gadoids and flounders (reviewed in Backus 1987, NEFSC 1995, and Fogarty and Murawski 1998). If one blacks out species 36-41, 49-55, and 64 in Fig. 1, the result is a historical representation of this ecosystem as depicted in Fig. 3a. That is, this was a vertically oriented system, with the preponderance of biomass in the gadoid and flounder species. Shift forward in time to the removal of foreign trawlers, the increase in the domestic groundfish fleet, and the subsequent decline of historically abundant gadoids

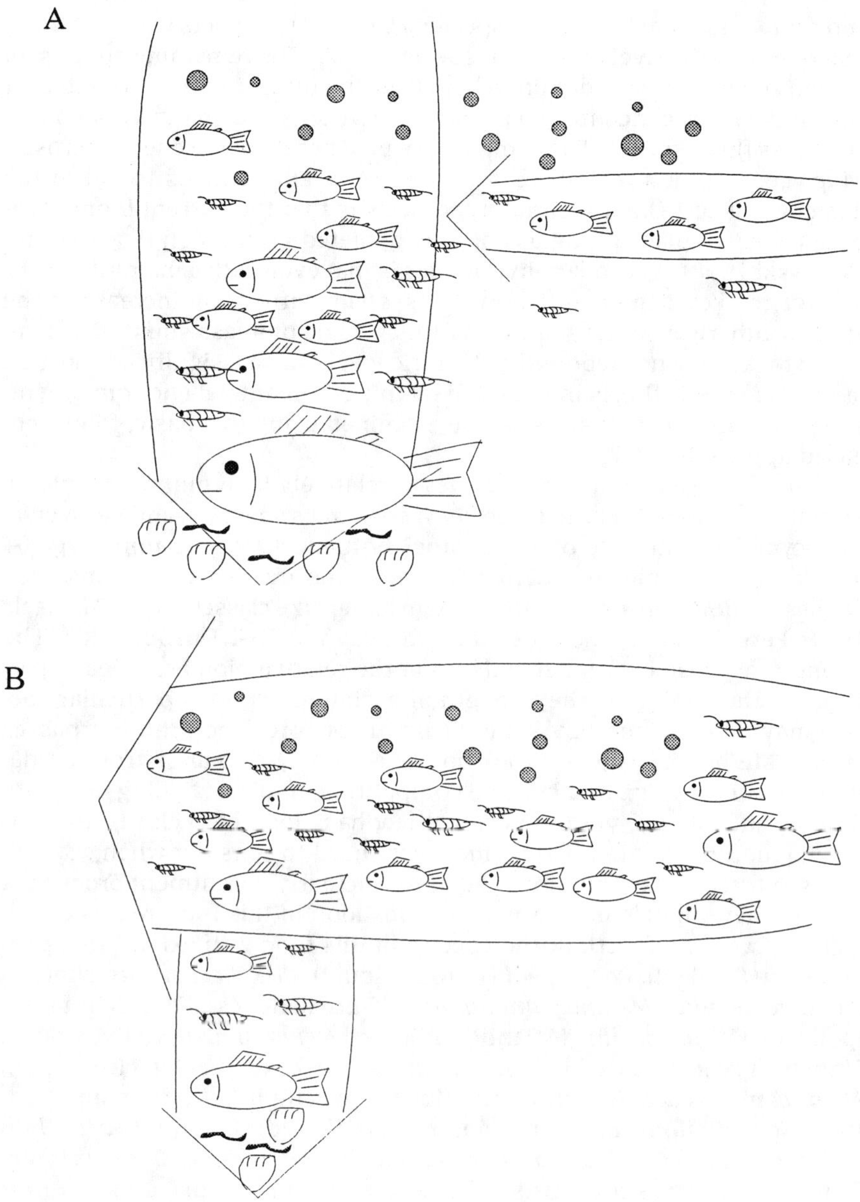

Figure 3. *(A) Depiction of a "vertical" food web, where the majority of the energy and mass flux is to demersal and benthic organisms. Representative of the northwest Atlantic ecosystem in the 1970s. (B) Depiction of a horizontal food web, where the major pathway of energy and mass flux is captured in migratory pelagics. Representative of the current northwest Atlantic ecosystem.*

and flounders. That is, human (species 75 in Fig. 1) impacts were so strong that one can effectively black out species 42-63. The remaining species on the left of Fig. 1 are predominately pelagic, resulting in the food web as it currently exists, concentrated primarily in a horizontal pathway of energy and mass flux (Fig. 3b). These migratory pelagics disperse energy across a wider geographic range instead of a more localized focus found in the demersal fishes. One potential hypothesis is that the current biomass of pelagics has replaced biomass of now-depleted groundfish (Fogarty and Murawski 1998). A corollary hypothesis is that even with a major decrease in F for the key demersal fish in this system, without an increase in the yield or otherwise overt removal of this pelagic biomass shunt, the demersal stocks may not rebound to historic levels since a significant portion of their energetic flux is intercepted by this dominant pelagic component. Similar qualitative scenarios can be examined using this basic, food web-based approach.

This food web contains 70 cycles, a relatively high number (Table 2). Recalling that a cycle is when species A consumes species B and vice versa, each cycle may indicate potential stock bottlenecks. For example, age 0+ gadoids are consumed by Atlantic mackerel, who are themselves consumed by these gadoids once the gadoids attain larger size classes (Fig. 1, Michaels 1991; Pers. comm., W.M. Michaels, NEFSC, and E.G. Durbin, URI). The strength of these cycles depends upon the relative biomass of each participant. Depending on the strength of each interaction, a particular species may be consumed by some of its major "prey" before it surpasses critical life history stages. Stock bottlenecks may also arise from predation, lack of prey due to increased competition, or direct fishing mortality on these life stages, yet the negative feedback loop of cycles is an overlooked mechanism that fishery managers need to consider when examining age structured population models, particularly recruitment processes.

Another important negative feedback loop in this food web is cannibalism. Over one-fourth of the species in this food web exhibit this phenomenon (Table 2). *Loligo pealii* (longfin squid), *Illex illecebrosus* (shortfin squid), copepods, *Melanogrammus aeglefinus* (haddock), *Pollachius virens* (pollock), *Gadus morhua* (Atlantic cod), *Merluccius bilinearis* (silver hake), *Urophycis regia* (spotted hake), *U. tenuis* (white hake), *U. chuss* (red hake), *Myoxocephalus octodecemspinosus* (longhorn sculpin), *Sebastes* spp. (redfish), *Scophthalmus aquosus* (windowpane flounder), *Squalus acanthias* (spiny dogfish), *Cynoscion regalis* (weakfish), *Pomatomus saltatrix* (bluefish), migratory scombrids, migratory sharks, birds, and *Lophius americanus* (goosefish) all exhibit cannibalism. Of these, silver hake (11%), pollock (9%), longfin squid (2.6%; 16% all cephalopods), shortfin squid (2.7%; 53% all cephalopods), goosefish (3.7%), and bluefish (7%) have a significant portion of their diet composed of individuals of their own species. Cannibalism is generally not a positive phenomenon for a stock other than as a density-regulating mechanism, and is particularly noteworthy when >10% of a fish's diet consists of conspecifics. Aside from the voracious and opportunistic feeding nature of these fish, this implies that different life stages of a fish are functionally different species. Most fish in this community exhibit ontogenetic shifts towards increasing piscivory, particularly after 40-50 cm, and the cannibalism demonstrated by these fish is by larger size classes on 10-25 cm individuals. Fishing mortality is known to target larger size classes (Hilborn and Walters 1992). Thus, with both cannibalism and cycles impacting smaller size classes, I hypothesize that when both cannibalism and cycles are operating on small size classes, the negative effects on already depleted stocks are intensified. A few individuals may reach sizes large enough to be piscivorous, in large enough numbers to reduce incoming cohorts through cannibalism but in too few numbers to limit their other prey populations. If those prey populations consume the larval phases of the depleted stock, the negative feedback becomes more intense. A cohort must then incur predation by its eventual prey plus predation by its parents before it reaches maturity. Once it reaches maturity, the cohort is smaller than its predecessor, and is even less likely to reduce the prey population. The number of ensuing larvae is then consumed by a larger number of prey, but is still subject to cannibalism. Again, the interaction strength determines the importance of cannibalism and the negative feedback loop described here. This secondary mechanism of overfishing could have occurred in the gadoids on Georges Bank after the early 1980s, and may still impacting part of this species complex.

The modal chain length for this web is reasonable compared to the ECOWEB average, considering that I included humans. This is slightly larger than the estimate Pauly et al. (1998) present for this region, yet is still lower than the five or six levels that were historically observed. When fishing pressure removes larger size classes and alters the dynamics of species interactions, forcing fish to feed upon lower trophic levels, yet another negative feedback loop is initiated. Economic value of fish is typically inversely proportional to its trophic level, thus this food web impact has far-reaching implications. That humans are also consuming more forage fish of lower trophic levels (Pauly et al. 1998) is alarming.

Quantitatively addressing the stability of the northwest Atlantic food web is a moot point given the continual perturbations it has experienced over the past several decades. Assuming this system is at least locally stable (i.e., composition of community biomass is relatively static on a short time scale; e.g., 2-3 years), then two points stand out. One is that with the high degree of interactions in this web, the resilience is going to be very high. That is, to return to historical equilibrium, presuming that (1) it existed, (2) it resembled a state where cod and other gadoids were predominant in this community, and (3) we can manipulate it to do so, will take a long time. Second, even if we reduce the human-induced perturbations of the system in the future, we may find that the system has already been perturbed beyond its historical limits and may have shifted to a new local equilibrium. Measuring the persistence of this system is difficult, but we have clearly exceeded earlier persistence limits as indicated by changes in component populations ranging across several orders of magnitude (NEFSC 1995). How the populations in this, and similar, ecosystems will fluctuate stochastically from one local equilibrium to another remains a major, if not the key management challenge for our nation's resource managers. Enhanced understanding of complex systems via a lucid examination of food webs is one useful way to meet this challenge.

Acknowledgments

I thank all members of the Food Web Dynamics Program of the NEFSC, past and present, for their dedicated effort at collecting, auditing, and maintaining this food habits database of unprecedented scale and scope. I thank F. Almeida, W. Gabriel, L. Garrison, M. Fogarty, M. Terceiro, and other reviewers for constructive comments on earlier versions of this manuscript. I also thank W. Overholtz, F. Almeida, M. Fogarty, R. Brown, G. Begg, P. Rago, M. Sissenwine, S. Murawski, J. Collie, L. Maden, R. Rountree, and L. Garrison for stimulating conversations about this food web that sparked many of the ideas presented in this paper. This work was supported by NOAA, NMFS, NEFSC, and NOAA's Coastal Ocean Program, Georges Bank Study. Finally, I assert *Solo Deo Gloria*.

References

Andersen, K.P., and E. Ursin. 1977. A multispecies extension of the Beverton and Holt theory of fishing with accounts of phosphorus circulation and primary production. Medd. Dan. Fisk. Havunders. N.S. 7:319-435.

Backus, R.H. 1987. Georges Bank. MIT Press, Cambridge, Massachusetts.

Carpenter, S.R., and J.F. Kitchell. 1993. The trophic cascade in lakes. Cambridge University Press, New York. 385 pp.

Carpenter, S.R., J.F. Kitchell, and J.R. Hodgson. 1985. Cascading trophic interactions and lake productivity. Bioscience 35:634-639.

Christensen, V. 1996. Managing fisheries involving predator and prey species. Rev. Fish Biol. Fish. 6:417-442.

Christensen, V., and D. Pauly (eds.). 1993. Trophic models of aquatic ecosystems. ICLARM Conf. Proc. 26. 390 pp.

Cohen, J.E. 1978. Food webs and niche space. Princeton University Press, Princeton.

Cohen, J.E. 1989. Ecologist' co-operative web bank. Version 1.0. Machine-readable data base of food webs. Rockefeller University, New York.

Cohen, J.E., R.A. Beaver, S.H. Cousins, D.L. DeAngelis, L. Goldwasser, K.L. Heong, R.D. Hold, A.J. Kohn, H.H. Lawton, N. Martinez, R. O'Malley, L.M. Page, B.C. Patten, S.L. Pimm, G.A. Polis, M. Rejmanek, T.W. Schoener, K. Schoenly, W.G. Sprules, J.M. Teal, R.E. Ulanowicz, P.H. Warren, H.M. Wilbur, and P. Yodzis. 1993. Improving food webs. Ecology 74:252-258.

Daan, N., and M.P. Sissenwine (eds.). 1991. Multispecies models relevant to management of living resources. ICES Mar. Sci. Symp. 193. 358 pp.

DeAngelis, D.L. 1992. Dynamics of nutrient cycling and food webs. Chapman and Hall, New York. 270 pp.

Fogarty, M.J., and S.A. Murawski. 1998. Large-scale disturbance and the structure of marine systems: Fishery impacts on Georges Bank. Ecol. Appl. 8(Suppl. 1):S6-S22.

Gabriel, W. 1992. Persistence of demersal fish assemblages between Cape Hatteras and Nova Scotia, northwest Atlantic. J. Northwest Atl. Fish. Sci. 14:29-46.

Goldwasser, L., and J. Roughgarden. 1993. Construction and analysis of a large Caribbean food web. Ecology 74:1216-1233.

Hairston, N.G., F.E. Smith, and L.B. Slobodkin. 1960. Community structure, population control and competition. Am. Nat. 94:421-425.

Hastings, A. 1988. Food web theory and stability. Ecology 69:1665-1668.

Hilborn, R., and C.J. Walters. 1992. Quantitative fisheries stock assessment. Choice, dynamics and uncertainty. Chapman and Hall, New York. 570 pp.

May, R.M. 1973. Stability and complexity in model ecosystems. Princeton University Press, Princeton. 235 pp.

May, R.M., J.R. Beddington, C.W. Clark, S.J. Holt, and R.M Laws. 1979. Management of multispecies fisheries. Science 205:267-277.

Mercer, M.C. (ed.). 1982. Multispecies approaches to fisheries management advice. Can. Spec. Publ. Fish. Aquat. Sci. 59.

Michaels, W.L. 1991. The impact of mackerel predation on the survival of pelagic age-0 sand lance, cod, and haddock on Georges Bank during spring of 1986. M.S. thesis, Southern Massachusetts University. 154 pp.

Murawski, S.A. 1996. Can we manage our multispecies fisheries? In: K. Sherman, N.A. Jaworski, and T.J. Smayda (eds.), The Northeast shelf ecosystem: Assessment, sustainability and management. Blackwell Science, Cambridge, Massachusetts, pp. 491-510.

NEFSC. 1995. Status of the fishery resources off the northeastern United States for 1994. NOAA Tech. Memo. NMFS-NE-108. 140 pp.

Oksanen, L.S., D. Fretwell, J.A. Arruda, and P. Niemela. 1981. Exploitation ecosystems in gradients of primary productivity. Am. Nat. 118:240-262.

Overholtz, W.J., J.S. Link, and L.E. Suslowicz. 1999. Consumption and harvest of pelagic fishes and squids in the Gulf of Maine–Georges Bank ecosystem. In: Ecosystem Approaches for Fisheries Management, University of Alaska Sea Grant, AK-SG-99-01, Fairbanks. (This volume.)

Overholtz, W.J., S.A. Murawski, and K.L. Foster. 1991. Impact of predatory fish, marine mammals, and seabirds on the pelagic fish ecosystem of the northeastern USA. In: N. Daan and M.P. Sissenwine (eds.), Multispecies models relevant to management of living resources. ICES Mar. Sci. Symp. 193:198-208.

Paine, R.T. 1988. Food webs: Road maps of interactions or grist for theoretical development? Ecology 69:1648-1654.

Pauly, D., V. Christensen, J. Dalsgaard, R. Froese, and F. Torres. 1998. Fishing down marine food webs. Science 279:860-863.

Payne, A.I.L., K.H. Brink, K.H. Mann, and R. Hilborn (eds.). 1992. Benguela trophic functioning. S. Afr. J. Mar. Sci. 12. 1108 pp.

Pimm, S.L. 1982. Food webs. Chapman and Hall, London. 219 pp.

Pimm, S.L. 1991. The balance of nature? University of Chicago Press, Chicago. 434 pp.

Pimm, S.L., and R.L. Kitching. 1988. Food web patterns: Trivial flaws or the basis of an active research program? Ecology 69:1669-1672.

Pimm, S.L., and J.H. Lawton. 1978. On feeding on more than one trophic level. Nature 275:542-544.

Schoenly, K., and J.E. Cohen. 1991. Temporal variation in food web structure:16 empirical cases. Ecol. Monogr. 61:267-298.

Sissenwine, M.P., E.B. Cohen, and M.D. Grosslein. 1984. Structure of the Georges Bank ecosystem. Rapp. P.-V. Reun. Cons. Int. Explor. Mer 183:243-254.

Sissenwine, M.P., B.E. Brown, J.E. Palmer, R.J. Essig, and W. Smith. 1982. Empirical examination of population interactions for the fishery resources off the northeastern USA. In: M.C. Mercer (ed.), Multispecies approaches to fisheries management advice. Can. Spec. Publ. Fish. Aquat. Sci. 59:82-94.

Walters, C., V. Christensen, and D. Pauly. 1997. Structuring dynamic models of exploited ecosystems from trophic mass-balance assessments. Rev. Fish Biol. Fish. 7:139-172.

Warren, P.H. 1990. Variation in food-web structure: The determinants of connectance. Am. Nat. 135:689-700.

A Multivariate Approach to Monitoring Changes in Species Composition of a Demersal Fish Community

Franz J. Mueter and Brenda L. Norcross
University of Alaska Fairbanks, Institute of Marine Science, Fairbanks, Alaska

Extended Abstract

While extensive biomass surveys routinely gather information on the abundance of commercial and noncommercial fishes, information from these surveys is often analyzed on a species-by-species basis only. A more "holistic" analysis of community data is usually the domain of community ecology. Various approaches have been used to describe communities at different sites, to relate community structure to underlying explanatory factors, and to assess changes in community structure over time.

We describe a multivariate approach to derive indices of species com position based on trawl catch data from any number of stations. Our primary goal is to describe if and how the species composition of a community changes across some environmental gradient, across space, or over time, based on samples from the community at different points in space and time. Specific objectives are (1) to develop robust indices which summarize the major patterns of variation in species composition, (2) to model the indices as a function of environmental variables, (3) to identify the major trends over time in these indices, and (4) to interpret the indices in terms of species composition by relating them back to underlying species abundances.

The suggested approach follows a sequence of steps which keeps species abundance data and environmental data separate until appropriate indices of species composition are obtained. The indices are subsequently related to environmental gradients:

1. Sampling sites are selected from a clearly defined area and time frame in accordance with the goals of the study.

2. Species to be included in the analysis are selected and possibly combined into species groups at an appropriate level of aggregation. Rare species may be eliminated or, alternatively, can be aggregated at higher taxonomic levels.

3. After constructing a site-by-species matrix of abundances the data are transformed in order to reduce the influence of a few very abundant species. Field et al. (1982) recommend a fourth-root transformation which in its effect is similar to a log-transformation, but has the advantage that it is invariant to scale changes when used with the Bray-Curtis index of dissimilarity.

4. From the matrix of transformed abundances, Bray-Curtis dissimilarities among each pair of sites are computed.

5. Based on the Bray-Curtis dissimilarities an ordination using non-metric multidimensional scaling (NMDS) is performed. We recommend choosing a criterion for goodness of fit in advance and do the ordination using as many axes as is necessary to achieve the criterion. Goodness of fit is assessed using Kruskal's stress criterion, a measure of discrepancy between the distances in ordination space and the underlying Bray-Curtis dissimilarities.

6. Because the orientation of axes is arbitrary, the solution is rotated using principal components such that the first axis corresponds to the axis of maximum variation. We suggest using the scores for each site along each rotated ordination axis as indices of species composition. Each index can be interpreted as representing a different, independent aspect of species composition.

7. The indices are subsequently related to appropriate explanatory variables, either in an exploratory sense to identify variables that are most strongly related to species composition, or to test hypotheses regarding species composition. The principal advantage of using indices of this kind is that they are statistically well behaved and allow us to conduct hypothesis tests regarding species composition. Furthermore, multivariate indices provide more sensitive indicators of change in community structure and can identify general trends that may not be apparent when using univariate measures (Austen and Warwick 1989).

8. Apart from relating the indices to environmental variables, we suggest a straightforward approach to interpret indices in terms of the underlying species compositions using scatterplots and rank correlations. Plotting the abundance of a species against each index is used to identify individual species that are strongly related to an index, and to assess the nature of the relationship. We found that species were almost always monotonically related to indices. In such cases we

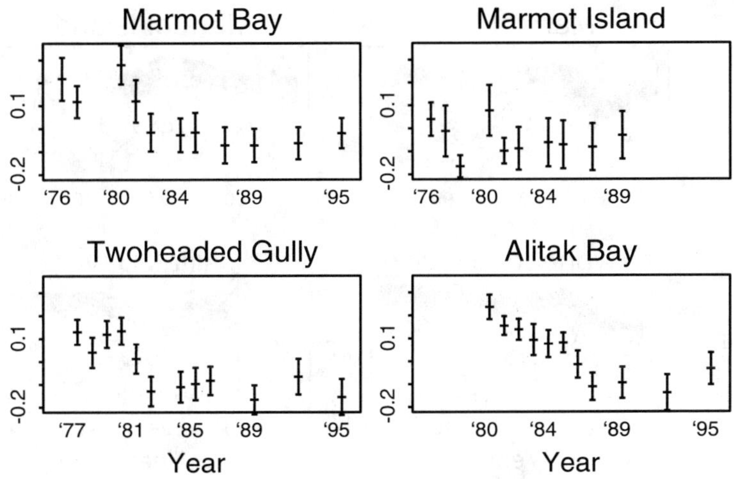

Figure 1. Estimated mean values for the first index of species composition against year by area, adjusted for depth effects, with 95% confidence intervals. The y-axis is dimensionless. For interpretation of the index, see text.

recommend that important associations be defined simply in terms of the magnitude of the observed rank correlations.

The approach is illustrated using data from shrimp trawl surveys conducted by the Alaska Department of Fish and Game and the National Marine Fisheries Service in nearshore areas of the central Gulf of Alaska. For this example we used data collected between 1976 and 1995 in four nearshore areas around Kodiak Island. Species abundance data were transformed as described above and the NMDS ordination was conducted in 5 dimensions to achieve a satisfactory fit (Kruskal's stress = 12%). Only results for the first axis, the major mode of variation in species composition, are described here.

The first axis primarily reflected interannual variation; i.e., most of the variation in species composition in this data set occurred over time. There was a much stronger difference over time than among areas or along other gradients. The index, adjusted for depth effects, showed a strong decreasing trend with a sharp drop after 1980, but there were significant differences among the four areas (Fig. 1).

A group of five taxa (shrimp, Pacific herring, capelin, Pacific sand lance, and Pacific tomcod) showed a strong positive association with the first index, as defined by rank-correlations greater than 0.35 (plus group). In contrast, a minus group consisting of arrowtooth flounder, flathead sole,

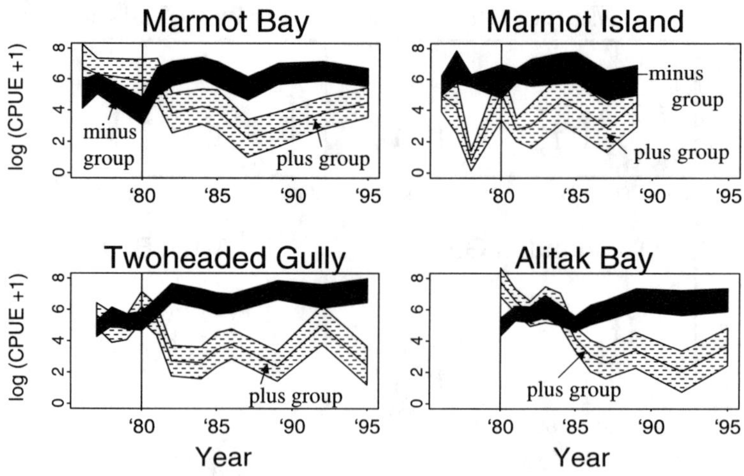

Figure 2. Estimated trends over time by area of the combined CPUE of shrimp, Pacific herring, capelin, Pacific sand lance, and Pacific tomcod (plus group), and the combined CPUE of arrowtooth flounder, flathead sole, Pacific cod, and walleye pollock (minus group). Estimated means and 95% confidence intervals, adjusted for depth effects.

Pacific cod, and walleye pollock was negatively correlated with the first index. Thus, the sharp decrease in the first index was shown to be related to a strong decrease in catch per unit of effort (CPUE) of the plus group, and a simultaneous increase of the flatfish and gadid species in the minus group (Fig. 2).

Using the outlined approach we were able to clearly demonstrate a pronounced shift in species composition that occurred in the nearshore areas of Kodiak Island in the early 1980s as well as significant differences in timing among different areas. The proposed index of species composition based on Bray-Curtis dissimilarities and NMDS ordination provides a sensitive indicator of variation in community composition and can provide a powerful link for statistically relating species composition to environmental variation and for testing hypotheses regarding species composition.

References

Austen, M.C., and R.M. Warwick. 1989. Comparison of univariate and multivariate aspects of estuarine meiobenthic community structure. Estuarine Coastal Shelf Sci. 29:23-32.

Field, J.G., K.R. Clarke, and R.M. Warwick. 1982. A practical strategy for analyzing multispecies distribution patterns. Mar. Ecol. Prog. Ser. 8:37-52.

OSMOSE: A Multispecies Individual-Based Model to Explore the Functional Role of Biodiversity in Marine Ecosystems

Yunne-Jai Shin
ORSTOM-LIA, Bondy cedex, France

Philippe Cury
University of Cape Town, Oceanography Department, Rondebosch, South Africa

Abstract

Considering the management of marine resources as the management of an ecosystem integrating different interactive components has enriched our understanding of fisheries dynamics. Indeed, the dynamics of marine exploited communities do not only reflect the effects of fishing mortality and of hydroclimatic conditions but is also the result of multiple interspecific interactions. An individual-based model OSMOSE (Object-oriented Simulator of Marine biOdiverSity Exploitation) is developed in Java programming language to explore the functional role of biodiversity in the exploitation of multispecies systems. By its flexibility, individual-based modeling permits implementation of different levels of biodiversity. In OSMOSE, species are age- and size-structured and modeled as interacting in the bosom of a spatial trophic web. Each species can be the predator or the prey of another species depending on its stage of its life cycle. Two simple rules form the basis of the trophic model. There is first a criterion of body length for the selection of prey and secondly a law of spatio-temporal co-occurrence. Thus, the fishes would prey regardless of the taxonomic identity of their prey and, moreover, predation and competition can vary with time, according to species abundances and changes

Introduction

With the increase of the number of both species and trophic levels being exploited, multispecies approaches have gone through a large expansion in fisheries science (e.g., Mercer 1982, May 1984, Daan and Sissenwine 1989, Pauly et al. 1998). Two main types of approaches have so far been developed.

The first multispecies investigation consisted of extending to several species the monospecies models that were classically used in fisheries science. This approach encompasses the multispecies Schaefer model (May et al. 1979, Kirkwood 1982, Ströbele and Wacker 1991) and the Multispecies Virtual Population Analysis (MSVPA; Andersen and Ursin 1977, Helgason and Gislason 1979, Pope 1979). The multispecies Schaefer model is a system of several differential equations of Schaefer including a few linear terms of interspecies interactions. Some variants exist in the expressions of species dynamics and interactions, in particular for the case of a two species predator-prey system, the pioneer study of which is the one of Lotka-Volterra. Indeed, three functions are investigated: the growth rate of a predation-free prey population, the functional and the numerical responses of the predator (Holling 1959, Hassel and Varley 1969, Arditi and Ginzburg 1989, Yodzis 1994). The MSVPA lies on the estimation of the predation rate mortality of exploited species by partitioning the food of each stock into the different potential preys in relation to their biomass and a suitability coefficient. This latter has to take into account the size suitability of the prey, the degree of species habitats overlap, and the chance a predator has to encounter a prey.

The second type of multispecies approach is more concerned with the emergent properties of multispecies assemblages or ecosystems without explicitly detailing the species processes that are at stake. Three reasons which could motivate this approach are: (1) Practical reasons since investigating aggregated levels requires a less exhaustive information. It is illustrated by the use of the MSY concept for the total exploited fish biomass in a given ecosystem (Brown et al. 1976, FAO 1978); i.e., the equilibrium production of a multispecies assemblage is a parabolic function of fishing effort and the MSY corresponds to the exploitation of this assemblage up to half its virgin biomass. (2) There is a need to search for other ways than classical fisheries approaches to estimate the marine biomass production and to understand the trophic structure of marine ecosystems. To this end, Christensen and Pauly (1992) and Pauly et al. (1998) developed the trophic model ECOPATH. In this compartment approach, the species are segregated into a certain number of groups that are linked together with matter flows. Two mass conservative equations describing the equilibrium production and consumption of each species group form the basis of ECOPATH. (3) The analysis of both catch and scientific trawl survey data has allowed to bring to the fore the existence of noteworthy conservative patterns at the scale of the ecosystem. A widespread observation in fisheries science, and in other ecological fields as well, is the relative stability of ecosystems total production compared to individual species production (e.g., Sutcliffe et al. 1977, Beddington 1984, Murawski et al. 1991). Another observed property is the relative uniformity in the shape of marine ecosystem size spectra; i.e., the distribution of biomass concentration is a slightly decreasing function of organism size (Sheldon et al. 1972, Pope and Knights 1982, Murawski and Idoine 1992, Rice and Gislason 1996). This conservative property, despite the fluctuations observed in species abundance, suggests that the interactions within a species assemblage tend to spread the biomass over the whole range of organism sizes, that is to fully occupy the production potential of a given system (Kerr and Ryder 1989).

Two main multispecies approaches can then be distinguished, the first lying on the coupling of species dynamics with interaction modules, the second paying particular attention to the macroscopic properties of the ecosystem. Two main ideas can be drawn from these different approaches: First, they almost all refer to trophic interactions, and second, the different studies agree to conclude that interspecific interactions help the key to understanding the emergent features at the ecosystem level. Nonetheless, the processes that link the dynamics at the levels of the species and the ecosystem are not clearly defined. On the one hand, the coupled species dynamics provide a partial vision of the ecosystem (few species are considered) and on the other hand, the aggregated approaches do not make clear the interspecies interactions that give rise to the investigated macroscopic behavior. The complexity of fisheries systems can be at the origin of both the modeling difficulty and the difficulty in linking both approaches. This complexity has been defined by Pimm (1979) and Kikkawa (1986) as the number of pairs of interspecific interactions in the system. In addition, for Goodman (1975), the complexity of ecosystems is synonymous with trophic diversity which can be measured by species richness. Accordingly, it appears that the notion of biodiversity and its derivatives can link the different scales of multispecies studies and could be the key to understanding ecosystem dynamics. For these reasons, an alternative multispecies approach has been developed that enables one to take into account ecosystem biodiversity and to explore its influence on the macroscopic attributes which are of interest for the management and the assessment of marine systems, namely their stability, resistance, and resilience.

Functional Role of Biodiversity

Relationship Between Ecosystem Biodiversity and Stability

In terrestrial ecology, the nature of the relationships between the diversity of the organisms living in an ecosystem and the properties of this latter, in particular the relationships between diversity and stability, are among the most investigated, experimentally and theoretically as well. MacArthur (1955), Odum (1969) or more recently Hammer et al. (1993) and Tilman (1996) draw a parallel between a better stability of the ecosystem with a greater species richness, thereby assuming that the energy flow in a complex trophic web, i.e., which presents a great number of interspecies links, would be less sensitive to disruptions since it can use more different pathways through the system. In contrast, at the local level, a higher species richness would decrease the stability of species abundances by these phenomena of abundance compensations following, for example, climate perturbations (Tilman 1996). Fisheries ecologists have also been aware of the close links that could exist between the biodiversity of the resource and its stability, in the sense defined by MacGillivray and Grime (1995); i.e., the capacity of an ecosystem (1) to resist to a change while maintaining a function to a certain level (resistance), and (2) to recover normal levels of a function after perturbation (resilience). In their simulations of multispecies systems, Wilson et al. (1991) have observed that in the case where the biological and demographical parameters of the populations are similar, the species dynamics are chaotic but such that the production peaks are in phase: thereby, the system biomass varies greatly. In contrast, when the population parameters are such that there is a functional diversity within the species, the total biomass of the system becomes stable. This "diversity-stability" hypothesis is the most common in ecology although some experimental studies have led to its rejection. With the idiosyncratic hypothesis, Lawton (1994) discusses the possibility that no particular relationship exists between the species diversity and the properties of the system. The resulting controversies raise some questions about the relevance of species richness as the study variable and consequently about the functional equivalence of the species in an ecosystem. Other hypotheses have then been proposed as alternatives to the "diversity-stability" hypothesis. The "rivets" hypothesis suggests that the resistance of an ecosystem can decrease with the number of species even though the system performance (biomass production, for example) does not seem to be affected. But the consequences of a perturbation can be sudden and drastic. This hypothesis has been introduced by Ehrlich and Ehrlich (1981) who compare the species of an ecosystem to the rivets of a plane. The progressive removal of the rivets up to a certain threshold would cause the ecosystem to collapse. The authors suppose that some species extinctions can go unnoticed in terms of system performance because redundancy may exist within the system, thus generating a nonlinear relationship between diversity and stability.

Redundancy Hypothesis

The redundancy hypothesis proposed by Walker (1992) is in a way the logical development of the rivets hypothesis and considers that some species are capable of extending their functions in the ecosystem in order to compensate certain decreasing abundance of similar species. In the core of this concept is the idea that the ecosystem is segregated into a certain number of functional groups, within which species would have functional analogs. Some species can disappear with little effects on ecosystem processes as long as each functional group is represented. An important role of redundancy is to provide an insurance to the system. During an environmental change, a species can disappear since the biotic and climatic conditions do not correspond any more to its tolerance range. But other species belonging to the same functional group may be resistant to the perturbation and even increase their production as a consequence of competition. By extending this redundancy hypothesis, a keystone species would be a functional group without redundancy. Its loss would cause massive changes in ecosystem structure and may give rise to many other extinctions (Fig. 1). It is worth noticing that the keystone status can be attributed to or removed from a species following the evolution of the system. A redundant species can indeed become a keystone species if it becomes the last representative of its functional group and conversely.

Ecosystems containing the same number of species can have different functional structurations, thereby resulting in difficulties of analyzing, comparing ecosystems and in establishing clear relationships between species richness and the stability properties of ecosystems. Indices used for characterizing biodiversity also have to be nonambiguous. The number of species of a community, for example, is an unsufficient measure to characterize the composition of a community. At equal density and species richness, two communities can have very different structures. The concept of "species diversity" thus substitutes for the simple one of species richness to allow for the relative species abundances. It corresponds to the cardinal measures of biodiversity (Cousins 1991), both Shannon and Simpson indices of which are the most used. The interest of these indices is to permit global comparison of either different communities or the successive states of the same community. But they do not convey the functional structures, namely the different modes of community organization. New measures are proposed, called ordinal measures (Cousins 1991), which do not consider the species in an equivalent way. The "taxonomic distance" effect can, for instance, be taken into account (Vane-Wright et al. 1991) with the hypothesis that the phylogenetic classification of the species reflects anatomical and morphological differences between species. Previously evoked is also a measure based on the size of individuals, namely the size spectra of the ecosystem. The underlying hypothesis is that the trophic organization of a system is linked to morphometric differences between species. The observed gap between the prey species ranges of different predators would most often stem from similar gaps in the dimensions of the organs involved in food intake (e.g., body size, jaws size). The existence of such relationships would thus allow viewing the size spectrum of a community as a global indirect expression of its functional structure.

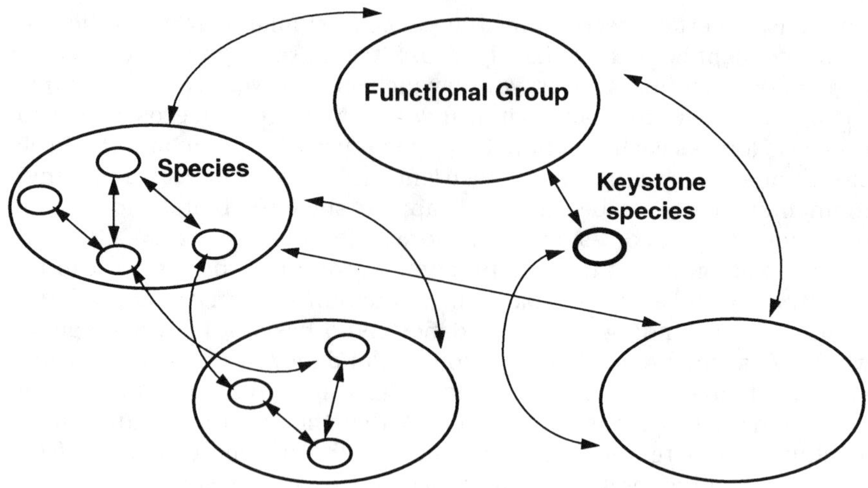

Figure 1. Theoretical structuration of ecological communities.

To conclude, the notion of biodiversity can have major consequences in fisheries management. It seems indeed indissociable from ecosystems properties of stability since it can allow this latter to dispose of a certain margin of adaptability while facing some perturbations, especially those due to fishing mortality. An important investigation area would be to define some criteria enabling determination of the functional role of each taxonomic entity in the ecosystem (redundant or keystone species).

OSMOSE: An Alternative Model for Diversified and Dynamic Trophic Web

Underlying Hypotheses

In this context, OSMOSE (Object-oriented Simulator of Marine biOdiverSity Exploitation) has been developed to possibly account for two aspects which are supposed to be important in the functioning of marine multispecies communities: (1) the diversity of the components (e.g., species richness, biological parameter diversity, size diversity). Indeed, OSMOSE is devoted to investigating the functional role of biodiversity in ecosystem stability and subsequently to exploring the consequences of fishing activity on this biodiversity. (2) the possibility for the ecosystem trophic structure to vary. In the multispecies models previously mentioned, there is little place for adaptation and change, as the elements of the system can hardly learn or modify their behavior. As an example, the classical models represented by a more or less complex system of differential equations and describing various dynamics (stable, cyclic, or chaotic) are capable of functioning but not of evolving (Allen 1990). Indeed, the species dynamics are modeled from preestablished relationships of predation or competition. In this context, an attempt is made to model a dynamic trophic web by implementing the possibility for the predators to shift species in relation to species abundance conditions. As stated by Rice (1995), life-history omnivory has profound effects on system dynamics. For Cousins (1991), any species can be the prey of a predator if the predator encounters some individuals of this species, if the prey is not too large (size inadequacy for possible predation, escapement) and finally if the prey is not too small (null energy gain). In OSMOSE, the predation behavior is continuously conditioned by the first two conditions. The third condition will be respected only in the case where prey of adequate size are in sufficient abundance in the predator environment. Thus fishes would prey regardless of the taxonomic identity of their prey (see for example Bond 1979, Laevastu and Larkins 1981, Sissenwine 1984). This opportunism seems realistic in the case of animal preys for which the catches are mainly due to physical and mechanical processes. More specificity exists, due to chemical processes, for parasitism relationships or for example the search of food of terrestrial herbivores (Cousins 1991).

An Individual-Based Simulation Tool

The individual-based approach we have adopted enables, by its flexibility, taking into account a great number of species and, by its principle, managing predator opportunism by a simple law of spatio-temporal co-occurrence. Two biological principles form, indeed, the basis of this approach (DeAngelis and Gross 1992), namely: (1) the physiological and behavioral unicity of each individual which results from the combination of both its genetic inheritance and its environment, and (2) the localization principle which states that an individual is mainly affected by the only organisms that are located in its spatio-temporal vicinity. The biological entity which should correspond to a computer individual had first to be chosen. We had the choice between the species, the cohort, the shoal or the fish level. As individual size is potentially an important variable to set the predation process and to address the functional role of ecosystem biodiversity, the species is too aggregated a level to represent the unit of action and interaction of the model. Besides, for the individual entity chosen, the variables of interest for our study (biological parameters, fish size, predation behavior) have to be relatively homogeneous. As shoals are generally composed of fishes having the same size, the same food, the same spatial coordinates, and belonging to the same species it was not relevant to consider the fish level. In OSMOSE, the unit of interest is then the fish school. In object-oriented programs, as many objects as desired can be created from some molds called classes and can interact by means of messages. Written in Java, OSMOSE is an artificial ecosystem containing a certain number of object species (up to a hundred species) which are a set of age classes which are in turn a collection of fish schools (Fig. 2). This hierarchical structure permits following some key features at different levels of aggregation. The object species are characterized by some input parameters as survival, growth, and reproduction ones (Table 1). These are parameters that can be easily collected from the literature or from the FishBase database (Froese and Pauly 1997).

Description of the Modeled Processes

The model represents annual dynamics of interactive species which are heterogeneously distributed in a rectangular lattice. To initiate the system, an abundance value is given to the age class 0 of each species. Then, a pseudo-cohort is formed from the following classical survival equation and is randomly distributed in the lattice:

$$N_{a+1} = N_a e^{-M} \quad with \quad \begin{cases} N_a = \text{abundance of age class } a \\ M = 1 \qquad \text{for } a = 0 \\ M = 0,2 \quad \text{for } 1 \leq a < \text{longevity} \end{cases}$$

The mean lengths and weights at age for each species are calculated from the input von Bertalanffy parameters. Thereby, at time 0 and within an age class, all the shoals have the same values for the length and weight attributes. Once the system is initiated, the whole life cycle of fishes is modeled following the processes that occur in the course of one simulated year (Table 2). First the fishes are submitted to the so-called diverse mortality. For fishes older than age 1, this mortality is independent of feeding activity (predation and starvation). For age class 0, on the contrary, the starvation mortality is implicitly taken into account in this diverse mortality rate, the interannual variability of which, follows the one supposed for the ecosystem primary production (stable, cyclic, or chaotic). The subsequent stages are only applied to age-1 and older fishes. Following the foraging activity, the predation process occurs such that a predator can potentially feed on any species, provided that there is spatio-temporal co-occurrence and that the prey size does not exceed a certain threshold value. Laevastu and Larkins (1981) and Gislason and Helgason (1985) estimate the upper predator/prey size ratio at about 10%. Thus, as all fish go through the larval-juvenile stage, they can be at a given moment at the basis of the carnivore trophic levels. Thereby, two species can be simultaneously the predator and the prey of each other, according to their stage in the life cycle. On the other hand, fishes should consume about 3.5 grams of food per gram of fish and per year. This value represents an average calculated from different species values provided by Longhurst and Pauly (1987). From the ratio between the quantity of food the shoals achieve to consume and the one the shoals should consume, a predation success rate is calculated. One can notice that during this stage,

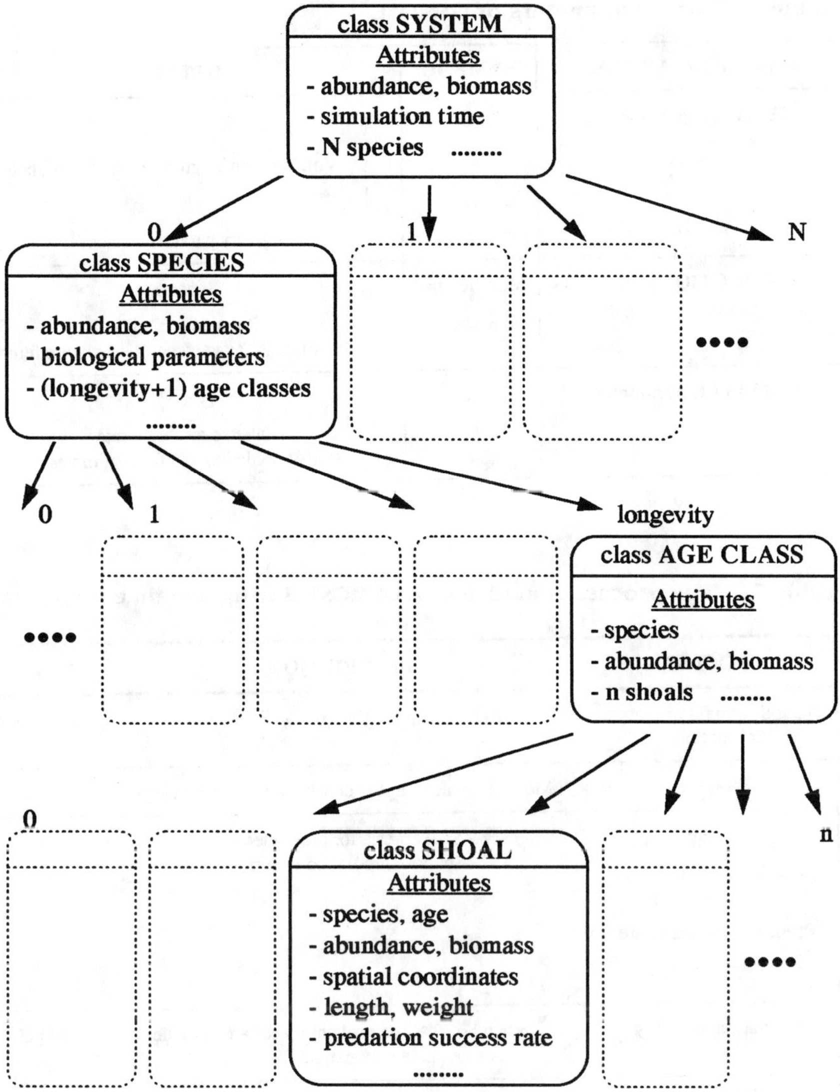

Figure 2. Hierarchical organization of OSMOSE.

Table 1. Input parameters of OSMOSE.

INPUT PARAMETERS	SYMBOL	DETAIL
GROWTH parameters	L_∞ a_0 K	Von Bertalanffy growth model parameters
	c	Allometry parameter (g.cm^{-3})
REPRODUCTION parameters	Maturity age	(years)
	α	Relative fecundity (number of eggs spawned per g. of mature female)
SURVIVAL parameters	Longevity	(years)
	D	Diverse mortality rate (free of predation and fishing, in year^{-1})

Table 2. The processes modeled in OSMOSE during one time step (year).

PROCESS	DESCRIPTION	LEVEL
Natural mortality without predation	$N_{a+1,t+1} = N_{a,t}\, e^{-D}$	Age class
Foraging	Moving towards higher densities of potential preys	Shoal
Predation ↓ Predation success rate S	- Size criteria for prey selection - Spatio-temporal co-occurrence	Shoal
Starvation mortality	When S<30%, the probability for a fish to die is a negative exponential function of S	Shoal
Growth	Von Bertalanffy model: $\Delta L_a = L_\infty (1 - e^{-K})\, e^{-K(a-a_0)}$ Depending on S, $\Delta L_a \pm 10\%$.	Shoal
Reproduction	$N_0 = \alpha\, SB$	Species

(t=simulation time in years, a=age in years, N=fish number, D=diverse mortality rate in year^{-1}, L=fish length in cm, α=relative fecundity in number of eggs per g. of mature female, SB=spawning biomass in g).

the predation mortality of fish is taken into account in a completely explicit manner. Depending on the predation success rate, a phenotypic plasticity is then implemented each year for fish mortality (starvation), growth and indirectly for reproduction (Table 2). In other words, the more the fish feeds, the more it grows, the less it dies and the more it reproduces. It is worth noticing that in OSMOSE, the key stages of the fish life cycle are explicitly modeled. For example, the stock-recruitment relationship observed in the simulator stems from explicit density-dependent processes such as predation or starvation mortalities. Also, the three functions investigated in Lotka-Volterra type predator-prey systems (namely the growth rate of a predation-free prey population, the functional and the numerical responses of the predator) emerge in OSMOSE from simple microscopic predation behaviors.

Conclusion

OSMOSE provides a possible alternative to other multispecies models by enabling investigation of the spatial dynamics of a great number of interactive species without making any a priori hypothesis on the food web architecture. OSMOSE is still in progress but different simulations can from now on be envisaged. The model will be generically implemented with the study of different types of exploitation inspired from the main world fisheries. The fisheries can be classified on the basis of the type of exploited resource as it was proposed by Larkin et al. (1984): the number of species caught (in the order of magnitude of about ten or of a hundred), habitat (e.g., pelagic, demersal), the number of exploited trophic levels. Besides, the existence of different strategies of exploitation allows establishment of a transverse typology. For example, target fisheries can be distinguished from nondiscriminating fisheries. The first exploitation type is illustrated by the North Sea fishery which can be represented by a set of diversified fleets targeting one or a small set of species of commercial interest (Gulland and Garcia 1984). For the second fishing strategy, a great number of species is exploited by a uniform fishing fleet. These fisheries would possess a more important operational flexibility (Garrod 1973). They can seasonally or annually shift species assemblages in order to stabilize their income despite the resource variability. By means of a comparative approach of these different typologies, OSMOSE will permit testing of different hypotheses concerning the functional role of marine ecosystem biodiversity. For this purpose, the main interest of this model is to provide as output the ecosystem size spectra, which is suspected to be an important indicator of ecosystem structuration, accordingly to the various observations and theoretical developments that have been mentioned in this article. With OSMOSE, an attempt will be made to link the notion of species redundancy with the distribution of fish sizes in an ecosystem. Moreover, the spatialized size spectrum is provided in OSMOSE and can be used to study the effect of marine refuges on fish population dynamics. Consisting of closing some fishing areas in order to preserve ecosystem (functional?) biodiversity, this management strategy is increasingly advocated but we still lack, on the one hand, criteria to determine the spatio-temporal limits of the refuges, and on the other hand, experience to judge the consequences of this type of measure. The virtual laboratory OSMOSE may permit exploration of the conditions within which these spatio-temporal restrictions can be useful.

References

Allen, P.M. 1990. Why the future is not what it was. New models of evolution. July/August:555-570.

Andersen K.P., and E. Ursin. 1977. A multispecies extension to the Beverton and Holt theory of fishing, with accounts of phosphorus circulation and primary production. Medd. Dan. Fisk. Havunders. 7:319-435.

Arditi, R., and L. Ginzburg. 1989. Coupling in predator-prey dynamics: Ratio-dependence. J. Theor. Biol. 139:311-326.

Beddington, J.R. 1984. The response of multispecies systems to perturbations. In: R.M. May (ed.), Exploitation of marine communities. Springer Verlag, Berlin, pp. 209-225.

Bohnsack, J.A., and J.S. Ault. 1996. Management strategies to conserve marine biodiversity. Oceanography 9(1):73-82.

Bond, C.E. 1979. Biology of fishes. Saunders College Publishing, Philadelphia. 514 pp.

Brown, B., J. Brennan, M. Grosslein, E. Heyerdhal, and R. Hennemuth. 1976. The effect of fishing on the marine fish biomass in the Northwest Atlantic from The Gulf of Maine to Cape Hatteras. ICNAF Res. Bull. 12:49-68.

Christensen, V., and D. Pauly. 1992. ECOPATH II: A software for balancing steady-state ecosystem models and calculating network characteristics. Ecol. Modell. 61:169-185.

Cousins, S.H. 1991. Species diversity measurement: Choosing the right index. TREE 6(6):190-192.

Daan, N., and M.P. Sissenwine (eds.). 1989. Multispecies models relevant to management of living resources. ICES Mar. Sci. Symp. 193:1-385.

DeAngelis, D.L., and M.J. Gross (eds.). 1992. Individual-based models and approaches in ecology: Populations, communities, and ecosystems. Chapman & Hall, New York. 525 pp.

Ehrlich, P.R., and A.H. Ehrlich. 1981. Extinction. The causes and consequences of the disappearance of species. Random House, New York.

FAO. 1978. Some scientific problems of multi-species fisheries. Report of the Expert Consultation on Management of Multi-species Fisheries. FAO Fish. Tech. Pap. 181. 42 pp.

Froese, R., and D. Pauly (eds.). 1997. FishBase 97: Concepts, design and data sources. ICLARM, Manila, Philippines. 256 pp.

Garrod, D.J. 1973. Management of multiple resources. Optimal exploitation of multiple stocks by a common fishery: A new methodology. J. Fish. Res. Board Can. 30:1977-1985.

Gislason, H., and Th. Helgason. 1985. Species interaction in assessment of fish stocks with special application to the North Sea. Dana 5:1-44.

Goodman, D. 1975. The theory of diversity-stability relationships in ecology. Quart. Rev. Biol. 50:237-266.

Gulland, J.A., and S. Garcia. 1984. Observed patterns in multispecies fisheries. In: R.M. May (ed.), Exploitation of marine communities. Springer-Verlag, Berlin, pp. 155-190.

Hammer, M., A. Jannson, and B.O. Jansson. 1993. Diversity change and sustainability: Implications for fisheries. Ambio 22(2-3):97-105.

Hassel, M.P., and G.C. Varley. 1969. New inductive population model for insect parasites and its bearing on biological control. Nature 223:1133-1136.

Helgason, T., and H. Gislason. 1979. VPA analysis with species interaction due to predation. CIEM C.M. 1979/G. 52 pp.

Holling, C.S. 1959. The components of predation as revealed by a study of small mammal predation of the European pine sawfly. Can. Entomol. 91:293-320.

Kerr, S.R., and R.A. Ryder. 1989. Current approaches to multispecies analyses of marine fisheries. Can. J. Fish. Aquat. Sci. 46:528-534.

Kikkawa, J. 1986. Complexity, diversity and stability. In: J. Kikkawa and D.J. Anderson (eds.), Community ecology. Pattern and process. Blackwell, Oxford, pp. 41-62.

Kirkwood, G.P. 1982. Simple models for multispecies fisheries. In: D. Pauly and G.I. Murphy (eds.), Theory and management of tropical fisheries. ICLARM Conference Proceedings, Manila, Philippines, 9:83-98.

Laevastu, T., and H. Larkins. 1981. Marine fisheries ecosystem. Its quantitative evaluation and management. Fishing News Books Ltd., Farnham, Surrey, England. 162 pp.

Larkin, P.A., C.W. Clark, N. Daan, S. Dutt, V. Hongskull, S.A. Levin, G.G. Newman, D.M. Pauly, G. Radach, and H.K. Rosenthal. 1984. Strategies for multispecies management. In: R.M. May (ed.), Exploitation of marine communities. Springer Verlag, Berlin, pp. 287-301.

Lawton, J. 1994. What do species do in ecosystems. Oikos 71:367-374.

Longhurst, A.R., and D. Pauly 1987. Ecology of tropical oceans. Academic Press, London. 406 pp.

MacArthur, R. 1955. Fluctuations of animal populations and a measure of community stability. Ecology 36:533-536.

MacGillivray, C.W., and J.P. Grime. 1995. Testing predictions of the resistance and resilience of vegetation subjected to extreme events. Funct. Ecol. 9:640-649.

May, R.M. (ed.). 1984. Exploitation of marine communities. Springer-Verlag, Berlin. 370 pp.

May, R.M., J.R. Beddington, C.W. Clark, S.J. Holt, and R.M. Laws. 1979. Management of multispecies fisheries. Science 205:267-277.

Mercer, M.C. (ed.). 1982. Multispecies approaches to fisheries management advice. Can. Spec. Publ. Fish. Aquat. Sci 59:1-169.

Murawski, S.A., and J.S. Idoine. 1992. Multispecies size composition: A conservative property of exploited fishery systems? J. Northwest Atl. Fish. Sci. 14:79-85.

Murawski, S.A., A.M. Lange, and J.S. Idoine. 1991. An analysis of technological interactions among Gulf of Maine mixed-species fisheries. ICES Mar. Sci. Symp. 193:237-252.

Odum, E.P. 1969. The strategy of ecosystem development. Science 164:262-270.

Pauly, D., V. Christensen, J. Dalsgaard, R. Froese, and F. Torres Jr. 1998. Fishing down marine food webs. Science 279:860-863.

Pimm, S.L. 1979. Complexity and stability: Another look at MacArthur's original hypothesis. Oikos 33:351-357.

Pope, J.G. 1979. A modified cohort analysis in which constant natural mortality is replaced by estimates of predation levels. ICES C.M. 1979/H. 16 pp.

Pope, J.G., and J. Knights. 1982. Simple models of predation in multi-age multispecies fisheries for considering the estimation of fishing mortality and its effects. In: M.C. Mercer (ed.), Multispecies approaches to fisheries management advice. Can. Spec. Publ. Fish. Aquat. Sci. 59:64-69.

Rice, J. 1995. Food web theory, marine food webs, and what climate change may do to northern marine fish populations. In: R.J. Beamish (ed.), Climate change and northern fish populations. Can. Spec. Publ. Fish. Aquat. Sci. 121:561-568.

Rice, J., and H. Gislason. 1996. Patterns of change in the size spectra of numbers and diversity of the North Sea fish assemblage, as reflected in surveys and models. ICES J. Mar. Sci. 53:1214-1225.

Sheldon, R.W., A. Prakash, and W.H. Sutcliffe. 1972. The size distribution of particles in the ocean. Limnol. Oceanogr. 17:327-340.

Sissenwine, M.P. 1984. Why do fish populations vary? In: R.M. May (ed.), Exploitation of marine communities. Springer-Verlag, Berlin, pp. 59-94.

Ströbele, W.J., and H. Wacker. 1991. The concept of sustainable yield in multispecies fisheries. Ecol. Modell. 53:61-74.

Sutcliffe, W.H., K. Drinkwater, and B.S. Muir. 1977. Correlations of fish catch and environmental factors in the Gulf of Maine. J. Fish. Res. Board Can. 34:19-30.

Tilman, D. 1996. Biodiversity: Population versus ecosystem stability. Ecology 77(2):350-363.

Vane-Wright, R.I., C.J. Humphries, and P.H. Williams. 1991. Biol. Conserv. 55:235-254.

Walker, B. 1992. Biodiversity and ecological redundancy. Conserv. Biol. 6:18-23.

Wilson, J.A., J. French, P. Kleban, S.R. McKay, and R. Townsend. 1991. Chaotic dynamics in a multiple species fishery: A model of community predation. Ecol. Modell. 58: 303-322.

Yodzis, P. 1994. Predator-prey theory and management of multispecies fisheries. Ecol. Appl. 4(1):51-58.

Ecosystem Considerations and the Limitations of Ecosystem Models in Fisheries Management: Insights from the Bering Sea

Andrew W. Trites
University of British Columbia, Fisheries Centre, Vancouver, British Columbia, Canada

Patricia A. Livingston
National Marine Fisheries Service, Alaska Fisheries Science Center, Seattle, Washington

Marcello C. Vasconcellos and Steven Mackinson
University of British Columbia, Fisheries Centre, Vancouver, British Columbia, Canada

Alan M. Springer
University of Alaska Fairbanks, Institute of Marine Science, Fairbanks, Alaska

Daniel Pauly
University of British Columbia, Fisheries Centre, Vancouver, British Columbia, Canada

Abstract

Over the past 10 years there has been increasing criticism of management decisions that are based on single-species approaches and a call for the implementation of ecosystem approaches. The major criticism of single-species models is that they cannot predict changes in community structure. Unfortunately, our experience in modeling the Bering Sea shows that these same criticisms can also be leveled against ecosystem models.

We employed trophic mass-balance models (Ecopath and Ecosim) to examine some possible explanations for the changes that occurred in the Bering Sea between the 1950s and 1980s. We removed fish and mammals from the modeled system and tracked how other components of the ecosystem responded. Our mass-balance models indicate that neither whaling nor commercial fisheries were sufficient to explain the 400% increase in pollock biomass and other changes that may have occurred between the two time periods. The simulations further suggest that environmental factors, affecting recruitment or primary production, may be more important in determining the dynamics of the Bering Sea ecosystem than predator-prey interactions alone. These findings illustrate that mass balance models that do not account for the impact of climate variability on year-class strength cannot provide reliable estimates of trends in marine fish production. However, our models can show how predation and fishing can affect trophic interactions among species. As such, ecosystem models are a useful scientific tool to identify gaps in understanding and data needs, but are unlikely to ever replace single-species models. They may instead complement and provide parameters to single-species models. Ecosystem models such as ours are still in the early stages of development and will become increasingly more important as a management tool as they begin to incorporate spatial and oceanographic/climatic information.

Introduction

The Magnuson-Stevens Fishery Conservation and Management Act (1996) is the basis of fisheries management in U.S. jurisdictional waters. It maintains that fisheries are to be conserved and maintained to protect the marine ecosystem and to ensure the greatest overall benefit. It further encourages the application of ecosystem principles in fishery conservation and management.

To date, fisheries have not had good tools for understanding how fishing for one species will affect other components of the ecosystem. Fisheries scientists and managers are not yet able to replace single-species approaches with multispecies ecosystem approaches that can reliably predict shifts in species composition, abundance, and productivity. This is changing, however, as people begin to pool their collective knowledge into multispecies and ecosystem models. A number of models have recently been constructed for marine ecosystems around the world (e.g., see contributions in Christensen and Pauly 1993, Pauly and Christensen 1996), and their numbers keep increasing (see http://www.ecopath.org). However, ecosystem modeling is still in its infancy, leaving it unclear to what extent such models can guide ecosystem-based management.

A fundamental question for fisheries concerns how ecosystem models can be used to guide management decisions. Ecosystem models provide information on trophic linkages in a system and have the potential to show how fishing on one species might have unintended effects on other species in a system. Such models were applied to various regions of the North Pacific more than 20 years ago (Laevastu and Larkins 1981, Laevastu et al. 1982). However, in order to make progress in marine ecosystem management, scientists need to consider more fully how to sufficiently validate these models and communicate the results to managers in a meaningful way that can guide management decisions. This is the challenge we face in our attempt to model the large-scale changes that occurred in the eastern Bering Sea between the 1950s and 1980s.

An Ecosystem Model of the Eastern Bering Sea

We constructed an ecosystem model of the eastern Bering Sea using the Ecopath and Ecosim approach. Ecopath is a software package that describes an ecosystem at steady state using a mass-balance approach (Christensen and Pauly 1992, 1995). It is a relatively simple, but powerful method to model the biomass and fluxes of an ecosystem, and gain some insights into the relationships of the various species in the ecosystem. Ecosim can be applied to the Ecopath files, turning them into fully parameterized simulation models (Walters et al. 1997).

The area we delineated in the eastern Bering Sea encompasses the region covered by the Alaska Fisheries Science Center's bottom trawl surveys of the shelf and slope down to 500 m (Fig. 1). It included a wide range of marine habitats but was treated as a single homogenous area. (A spatially explicit model is being developed using Ecospace software.) Nearshore fauna and the northern portion of the Bering Sea were not considered. We constructed two models: one for the 1950s before large-scale commercial fisheries were under way, and another for the 1980s after many marine mammal populations had declined. Both were annual average models, which means that the biomass and species composition of summer and winter were averaged to provide a year-round annual average.

The main motivation for developing these models was to test the commercial whaling cascade hypothesis (i.e., that pelagic whaling in the eastern Bering Sea in the 1960s set up a chain of events that released prey, reduced predation, and increased the Bering Sea's walleye pollock carrying capacity; Merrick 1995, NRC 1996). How the ecosystem differed between the pre-whaling and post-whaling periods, and how the effects of fishing and whaling altered the dynamics of the Bering Sea were among the questions we posed. We also used the model to consider how changing current fishing quotas for pollock might alter the dynamics of other species in the ecosystem.

A complete description of the model and parameters used is contained in Trites et al. (In press). We split the hundreds of species that make up the Bering Sea ecosystem into 25 groups (Fig. 2). Information gathered from published sources included the numbers or biomass of each group of species living in the Bering Sea; their diets, rates of consumption, and production. The two primary data sources for groundfish biomass estimates were surveys or stock assessments conducted by the Soviet Union during the 1950s and the United States National Marine Fisheries Service

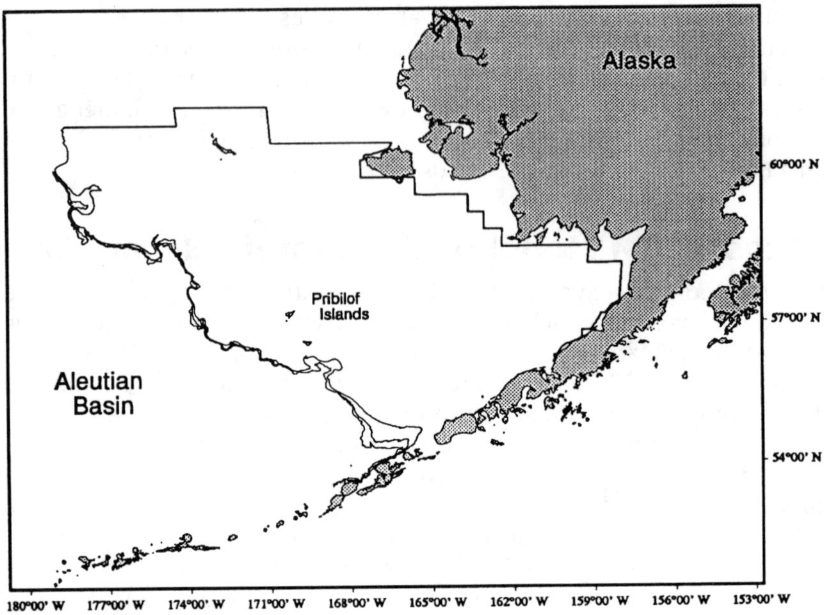

Figure 1. The eastern Bering Sea as defined in the ecosystem model. Total area is approximately 500,000 km².

(NMFS) during the 1980s. Population estimates for marine mammals came from Perez (1990), the NRC (1996) report, and the 1987-1988 Marine Mammal Protection Act annual report (NMFS 1988). Estimates for other species were drawn largely from stock assessments performed by NMFS scientists. Diet data for groundfish come primarily from the groundfish food habits database of NMFS while diet data for marine mammals come primarily from the 1996 NRC report and references therein. Several assumptions were made with respect to the 1950s model. For groups that had no abundance information from that time period, we assumed they had the same biomass as the 1980s period. We also made assumptions about the diet compositions of animals during that time period, particularly assuming that pelagic fish (herring, capelin, and others) were a more important prey than pollock in the 1950s.

The biomass estimates that we input to our mass-balance ecosystem models reflected our assumptions that most of the top predators (trophic level IV) declined from the 1950s to the 1980s (Fig. 3). They included Steller sea lions, seals, sperm whales, deepwater fishes, and other demersal fishes. The only top predators that were assumed to increase were large flatfish such as arrowtooth flounders. At the mid-trophic level (III), baleen whales and pelagic fishes were assumed to decline while small

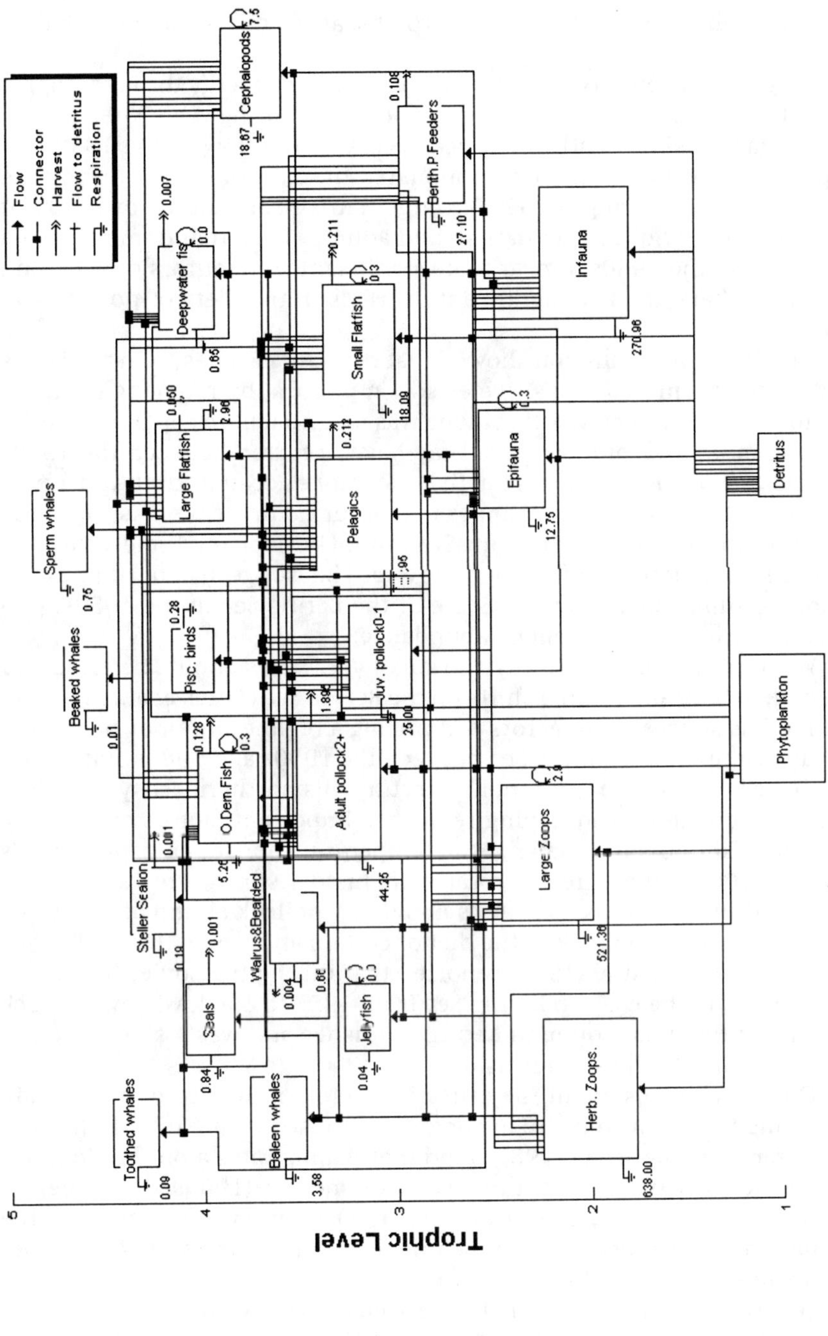

Figure 2. *Flow chart of trophic interactions in the eastern Bering Sea during the 1980s. All flows are in t per km² per year. Minor flows are omitted as are all backflows to the detritus.*

flatfish, pollock, and walrus and bearded seals were assumed to have increased.

Pollock contributed over 50% of the total flow of energy through trophic level III during the 1980s compared to only 10% during the 1950s. In contrast, pelagic fishes contributed nearly 50% of the flow in the 1950s. At trophic level IV, no one species dominated the flow of energy during the 1950s. However, during the 1980s, large flatfish contributed over 60% of the total energy flow. Large flatfish and adult pollock that dominated the Bering Sea in the 1980s appear to be significant competitors of seals and sea lions. There are also significant overlaps in the diets of pollock and baleen whales.

Our dynamic simulation showed that removing whales from the 1950s model ecosystem had a positive effect on pollock, by reducing competition for food. However, whaling alone was insufficient to explain the 400% increase in pollock biomass that may have occurred between the 1950s and 1980s. Nor could commercial fisheries alone account for the observed changes. The magnitude of changes that occurred in the biomass of all the major groups in the eastern Bering Sea cannot be explained solely through trophic interactions and fishing removals. This suggests a need to examine our assumptions about the state of the Bering Sea in the 1950s, particularly with regard to pollock abundance.

We used our models to test various hypotheses about the effects of whaling, fishing and regime shifts on the Bering Sea (Trites et al., In press). Our models suggest that factors comprising a regime shift may have been at play in altering the Bering Sea between the 1950s and 1980s states, and may have been more important in affecting this northern ecosystem than trophic interactions and fishing removals. Important factors comprising the regime shift could be changes in temperature and/or current patterns that have differential effects on early life history survival of various species. In addition, our assumptions about low pollock abundance and the low importance of pollock in the diet of certain animals in the 1950s may be inaccurate. Shuntov (1972) reported that during the Soviet fishery investigations in the eastern Bering Sea from 1957 to 1964, walleye pollock was one of the most common Bering Sea fishes and was a staple food of large flatfish as well as other fishes.

Our conclusions about the causative factors influencing the shifts in the Bering Sea ecosystem are in contrast to those for the Gulf of Thailand ecosystem. Christensen (1998) found that fishing rates alone could move this tropical ecosystem from one state to another (1960s-1980s: before and after the development of trawl fisheries). Future work to improve the Bering Sea 1950s model should include testing the possibility of a higher dominance of pollock than we assumed.

We examined three pollock fishing scenarios to explore how the Bering Sea ecosystem might be changed from its 1980s state. We considered the effects of reducing the biomass of adult pollock by 50%. We also considered what might happen if pollock were overfished to the point that all

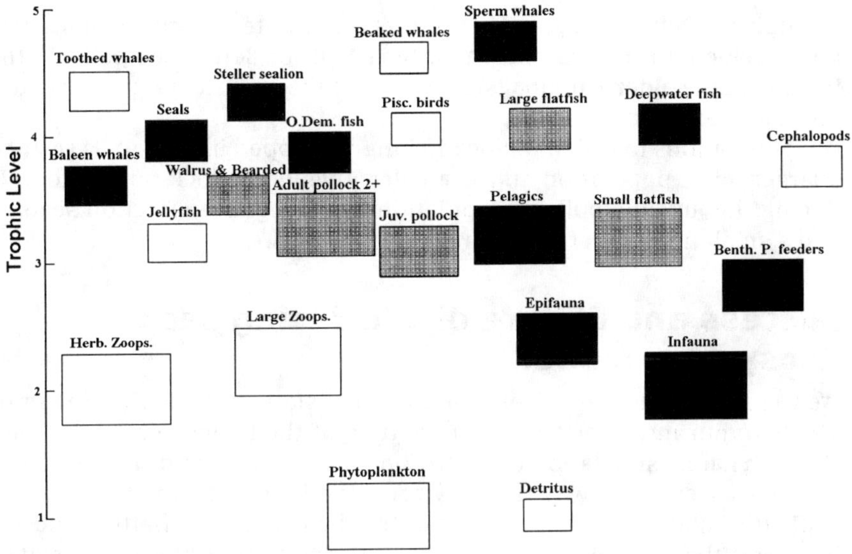

Figure 3. Trophic levels and relative abundance of species in the eastern Bering Sea during the 1980s. Black boxes indicate groups that had lower estimated abundance in the 1980s than in the 1950s, and shaded boxes show species with higher estimated biomass in the 1980s than in the 1950s. Major flows of energy between the boxes are shown in Fig. 2.

the juvenile and adult pollock were removed. Obviously, these two scenarios are hypothetical and do not reflect fishing policies that would be considered under the present fishery management regime. Finally, we considered how the Bering Sea ecosystem might look if pollock were not caught at all. All simulations were run over 30 years.

Our model suggests that increased fishing pressure on pollock has only a small effect on the equilibrium estimates of adult biomass due to a continuous replenishment from the juvenile stock (Fig. 4). At certain fishing levels, juvenile pollock may benefit from reduced cannibalism. System-wide effects are minimal, with the biomass changes of individual groups changing from less than 1% to about 30%, because the adult pollock stock does not change appreciably. However, the model predicts that equilibrium biomass levels of seals, sea lions, and piscivorous birds would increase due to an increase in the abundance of juvenile pollock.

Dynamic simulations of overfishing pollock to the point of extinction in the 1980s, predict the decline of seabirds that consume juvenile pollock. However, reducing the adult biomass of adult pollock by 50% would have a positive effect on seals, sea lions, and piscivorous birds because the abundance of juvenile pollock, which they consume, increases as cannibalism by adult pollock is reduced. This result corroborates the conclusion previously made by Laevastu and Favorite (1988) for Bering Sea pollock: that fishing on the older, cannibalistic portion of the stock might increase juvenile survival and abundance.

Simulations in which pollock fishing is stopped in the 1980s result in a larger adult population and a smaller juvenile pollock population. Reducing the juvenile pollock population has a negative effect on seabirds and a small negative effect on marine mammals.

Success and Failure of the Bering Sea Ecosystem Model

We were unable to move from the assumed state of the Bering Sea in the 1950s to our more certain understanding of the Bering Sea in the 1980s using dynamic simulation of the trophic interactions and fishing removals. Our assessment was that environmental change (which we did not explicitly model) is one explanation for the changes in flatfish and pollock, and the decline of pelagic fishes. Uncertainty in the 1950s state is the other explanation.

We do not know at this point how well our models represent the Bering Sea ecosystem, nor do we have a straightforward means of quantifying the uncertainty of our results. However, we can do further testing to examine our assumptions of the 1950s model. We can also try to project the 1980s model forward in a way that matches present observations. Our models are based on the best available data and have been used in a simple and rigorous modeling framework (Ecopath) that has a number of checks and balances to ensure consistency. Thus, we feel that we can make inferences about some of the general and fundamental properties of the Bering Sea, despite our incomplete knowledge of the past and present. For example, we can identify some of the major flows of energy through the Bering Sea and better understand the trophic relationships of the different species living in the Bering Sea. Our model suggests that the Bering Sea is a system where cascading effects of changing one component tend to be sluggish and may be small compared to the magnitude of change that the environment can cause via its impact on recruitment or primary production. Whether fishing for pollock or other species results in the unexpected changes predicted by our model remains to be verified. It nevertheless highlights the strength of ecosystem approaches to understanding the whole system.

Our conclusions are broad and general and may not be particularly useful to people charged with setting fishing quotas or making other management decisions. However, our models are the first step in a series of models that will come later. Splitting more groups of species into immature and mature stages, and adding habitat and spatial/migratory relationships (Walters et al. 1998) will all improve this model. Ecosystem

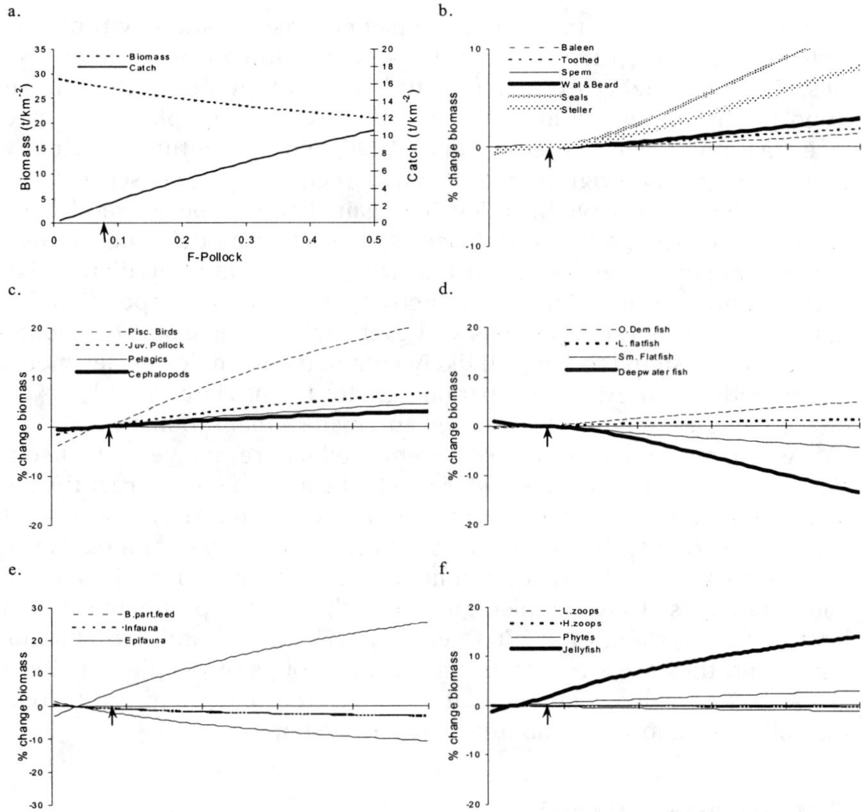

Figure 4. Equilibrium biomass for Bering Sea species following changes to the fishing mortality of adult pollock in the 1980s model. Arrows mark the instantaneous rate of fishing (F per year) during the 1980s. The top left panel shows changes in the biomass and catch of baleen whales under different levels of F. The other five panels show the relative change (%) that would occur to other species in the ecosystem model to compensate for changes in the abundance of adult pollock.

models will also need to explicitly consider seasonal and environmental factors influencing carrying capacity and recruitment of dominant species. These are being extensively studied by the National Oceanic and Atmospheric Administration's (NOAA) Fisheries Oceanography Coordinated Investigations program (Kendall et al. 1996) and the Southeast Bering Sea Carrying Capacity program (http://www.pmel.noaa.gov/sebscc/).

Criticisms that have been leveled against single species models can be equally raised against multispecies ecosystem models such as ours. Neither type of model does a particularly good job yet of predicting large decadal shifts in abundance, productivity, or species composition. It is unlikely that ecosystem models will ever replace single-species models. Instead, ecosystem models will likely complement single-species models and provide a context and/or parameters for them. They should prove to be another addition to the scientific and management toolbox.

As a management tool, ecosystem models are not yet very useful. They still need to prove themselves by being able to reconstruct the past or by making realistic predictions about the future. They also need to convey uncertainty. However, as a scientific tool, ecosystem models such as ours are very useful. They can help to identify gaps in understanding and data needs. They can also guide the choice of experiments to highlight our understanding (Walters et al. 1997). Finally, and perhaps most important, they are a tool to bring diverse groups of people together to share their knowledge about small pieces of the ecosystem and increase the collective knowledge about the whole system.

Acknowledgments

This study was supported in part by a grant from the David and Lucile Packard Foundation, and from the North Pacific Marine Science Foundation through the North Pacific Universities Marine Mammal Research Consortium. We gratefully acknowledge the scientific reviews of this paper provided by Anne Hollowed and Beth Sinclair, and the technical review by Gary Duker.

References

Christensen, V. 1998. Fishery-induced changes in a marine ecosystem: Insight from models of the Gulf of Thailand. J. Fish Biol. 53:128-142.

Christensen, V., and D. Pauly. 1992. The Ecopath II: A software for balancing steady-state ecosystem models and calculating network characteristics. Ecol. Modell. 61:169-185.

Christensen, V., and D. Pauly (eds.). 1993. Trophic models of aquatic ecosystems. ICLARM Conference Proceedings 26. 390 pp.

Christensen, V., and D. Pauly. 1995. Fish production, catches and the carrying capacity of the world oceans. Naga (ICLARM Quarterly) 18(3):34-40.

Kendall Jr., A.W., J.D. Schumacher, and S. Kim. 1996. Walleye pollock recruitment in Shelikof Strait: Applied fisheries oceanography. Fish. Oceanogr. 5(Suppl. 1):4-18.

Laevastu, T., and F. Favorite. 1988. Fishing and stock fluctuations. Fishing News Books, Ltd., Surrey, England. 239 pp.

Laevastu, T., and H.A. Larkins. 1981. Marine fisheries ecosystem: Its quantitative evaluation and management. Fishing News Books, Ltd., Surrey, England. 161 pp.

Laevastu, T., F. Favorite, and H.A. Larkins. 1982. Resource assessment and evaluation of the dynamics of the fisheries resources in the Northeastern Pacific with numerical ecosystem models. In: M.C. Mercer (ed.), Multispecies approaches to fisheries management advice. Can. Spec. Publ. Fish. Aquat. Sci. 59:70-81.

Merrick, R. 1995. The relationship of the foraging ecology of Steller sea lions (*Eumetopias jubatus*) to their population decline in Alaska. Ph.D. thesis, University of Washington, Seattle. 171 pp.

NMFS (National Marine Fisheries Service. 1988. Marine Mammal Protection Act of 1972 Annual Report 1987-88. NOAA, NMFS, Office of Protected Resources, Silver Spring, MD 20910.

NRC (National Research Council). 1996. The Bering Sea ecosystem. National Academy Press, Washington, D.C. 207 pp.

Pauly, D., and V. Christensen (eds.). 1996. Mass-balance models of north-eastern Pacific ecosystems. University of British Columbia, Fisheries Centre Research Rep. 4(1). 131 pp.

Perez, M.A. 1990. Review of marine mammal and prey population information for Bering Sea ecosystem studies. NOAA Tech. Memo. NMFS F/NWC-186.

Shuntov, V.P. 1972. Seasonal distribution of black and arrow-toothed halibuts in the Bering Sea. In: P.A. Moiseev (ed.), Soviet fisheries investigations in the north-eastern Pacific, Part V. (Translated from Russian by Israel Program for Scientific Translations, Jerusalem, NTIS document TT71-50127, pp. 397-408.)

Trites, A.W., P. Livingston, M.C. Vasconcellos, S. Mackinson, A.M. Springer, and D. Pauly. In press. Ecosystem change and the decline of marine mammals in the Eastern Bering Sea: Testing the ecosystem shift and commercial whaling hypotheses. University of British Columbia, Fisheries Centre Rep. 7.

Walters, C., V. Christensen, and D. Pauly. 1997. Structuring dynamic models of exploited ecosystems from trophic mass-balance assessments. Rev. Fish Biol. Fish. 7(2):139-172.

Walters, C., D. Pauly, and V. Christensen. 1998. Ecospace: Prediction of mesoscale spatial patterns in trophic relationships of exploited ecosystems, with emphasis on the impacts of marine protected areas. ICES C.M. 1998/S:4. 20 pp.

Ecosystem Approaches for Fisheries Management
Alaska Sea Grant College Program • AK-SG-99-01, 1999

A Mass-Balanced Model of Trophic Flows in Prince William Sound: Decompartmentalizing Ecosystem Knowledge

Thomas A. Okey and Daniel Pauly
University of British Columbia, Fisheries Centre, Vancouver, British Columbia, Canada

Abstract

Just as real-world food webs contain complex interactions among species, so too must scientists and others interact to describe food webs in realistic ways. The most useful ecosystem models are constructed through collaboration among a wide range of experts. Collaboration among Prince William Sound (PWS), Alaska, researchers resulted in a mass-balanced Ecopath model of trophic flows including all ecosystem components (explicitly or implicitly). This study was conducted to describe functional interactions among components, and to reveal thermodynamic constraints of these interactions, thus enabling further refinement of contributed estimates as well as dynamic simulations of ecosystem perturbations.

Since the 1989 *Exxon Valdez* oil spill (EVOS) in Prince William Sound, and adjacent areas, research teams focused on the status of particular biological components of the PWS ecosystem, changes to those components, and the effects of EVOS. Some research groups investigated the effects of EVOS on larger segments of the PWS ecosystem, but a system-wide analysis was not undertaken until the current model was constructed. Estimates of basic population parameters were lacking for several biological components of the ecosystem, but the Ecopath approach enables refinement of knowledge of uncertain groups based on the constraints of interrelationships among groups. The whole-system model described herein can be used by managers, schools, or local communities for learning, knowledge refinement, or simulation of environmental disturbances such as oil spills or increased fishing. Ecopath modeling through multiway collaboration is presented as a broadly accessible tool for restoration and resource planning with the potential to be highly community-based.

Only fluency across the boundaries will provide a clear view of the world as it really is.

E.O. Wilson, Consilience: The Unity of Knowledge, 1998

Introduction

It is increasingly apparent that single-species approaches to fisheries management fail in all but the most fortuitous circumstances (Wilson et al. 1994, Roberts 1997; also see Laevastu et al. 1996). This realization provides increased impetus for resource managers to adopt a more ecologically sophisticated logic. For example, components of ecosystems interact with each other, the actions of resource users affect nontarget components of ecosystems, resource users can influence the "biotic integrity" of an ecosystem (NMFS 1998), and natural changes in certain ecosystem components can cause changes in other components, which are often unpredictable. Incorporation of such whole-ecosystem considerations into resource decision-making is called "ecosystem-based management," in which human activities are managed within the context of a naturally dynamic and integral ecosystem (Langton and Haedrich 1997, NMFS 1998; also see Juda 1996, Okey and Harrington 1999).

Such nice ideas are not challenging to discuss; the real key to achieving ecosystem-based management in decision-making forums is the application of integrative analytical tools. The first tool needed is one that describes interactions of the components of a defined ecosystem, and the constraints that might exist among those components. For example, thermodynamic (energy flow) constraints can limit the sizes of populations, either absolutely or in conjunction with other limiting factors like predation or recruitment. The second analytical tool needed for the transition to ecosystem-based management is one that can be used to predict the effects of changes in one ecosystem component on other components. These tools should be comprehensive enough to provide a cohesive picture of the defined ecosystem, in as accurate a manner as possible, while retaining adequate simplicity to enable a wide range of interested parties to comprehend and use the model. Most importantly, these tools can function optimally when parallel collaborative structures enable maximum flow of ecosystem knowledge.

A mass-balanced trophic model of the Prince William Sound ecosystem was constructed using the user-friendly Ecopath software, with the collaborative contributions of a working group of experts from the region. This is a static model that includes all biotic components of the PWS ecosystem, either implicitly or explicitly, in a possible scenario of interrelationships (trophic energy flows among components). The data in the model were then analyzed in dynamic simulation routines called Ecosim and Ecospace to predict indirect effects of simulated perturbations on the biotic system. This modeling approach is discussed in the methods section below.

The purpose of this paper is to point out the natural necessity for broad collaboration to achieve (1) realistic descriptions of whole ecosystems, and (2) functional policies and "ecosystem-based management" of human activities (defined in Okey and Harrington 1999). We illustrate this necessity for collaboration by describing the methods used to construct the PWS model. This paper is about collaboration, not Ecopath modeling per se.

Our premise is that human knowledge of the ecosystem is, to a large degree, compartmentalized among individuals, research teams, institutions, and other groups. This is especially true within the modern milieu of western culture and science. We suggest that the most realistic description of an ecosystem can be constructed by de-compartmentalizing knowledge through collaborative efforts such as the one described herein, and through everyday communications and working relationships. Furthermore, traditional ecological knowledge of native communities may prove invaluable for achieving a fuller understanding of ecosystems, as this knowledge may be less compartmentalized, albeit generally less quantifiable.

The science conducted subsequent to the EVOS has been criticized for failing to maximize opportunities for knowledge gathering as the result of political and legal constraints (Keeble 1991, Wheelwright 1994, Paine et al. 1996). Our efforts to construct a balanced trophic model of PWS revealed that reliable estimates of basic information such as biomass, production and consumption rates, and diet composition are lacking for many groups. One explanation for this paucity of information is that EVOS research was not guided by a system-wide analytical framework. Another is that some components of the ecosystem have simply not been studied in depth because the ecosystem is complex. Notwithstanding these alternative explanations, a considerable amount of information has been collected about the biota within Prince William Sound during the years since the spill (Spies et al. 1996). Moreover, significant efforts have been made to describe, in detail, larger functioning segments of the PWS ecosystem to reveal system-level effects of EVOS (Cooney 1997, Duffy 1997, Holland-Bartels et al. 1997). These programs are the source of much of the available knowledge of the PWS ecosystem, but a comprehensive, system wide synthesis had not been undertaken until the current model was constructed (see contributions in Okey and Pauly [1998] for a more detailed description of the model).

The mass-balanced trophic model of Prince William Sound was constructed to integrate and synthesize what is being learned from the various research and monitoring projects within the *Exxon Valdez* oil spill (EVOS) restoration program, and to enable insight into "the effects of the oil spill and the long-term restoration and management of injured resources and services from an ecosystem-level perspective," as desired by the EVOS Trustee Council (1996:53). A collaborative synthesis of compartmentalized information can optimize ecosystem-level insights into the impacts of EVOS and other anthropogenic stressors. Moreover, the EVOS Trustee

Council stated that existing PWS data sets "need to be integrated in a simple [cost-effective] model to benefit long-term resource management," and "the restoration program will increasingly focus on an integrated, ecological approach." The goals of our study are to achieve these stated objectives.

Methods

The Ecopath Model of Prince William Sound

The balanced trophic model of PWS was constructed to describe the most likely flow scenario during the period from 1994 to 1996. A quantitative description of the whole trophic structure of PWS and adjacent waters and the relationships among the different species and groups inhabiting the area will place the results of individual EVOS projects into a realistic context and enable marine resource policy planning on an ecosystem level (multispecies as opposed to single species). The PWS model has unique potential as its 50 defined ecosystem components makes it, by far, the most explicit Ecopath model to date. There are many possible examples of its use; a PWS Ecopath model can be used to reveal of shifts in trophic structure in the wake of the oil spill that might be hindering the recovery of seabirds and marine mammals. Likewise, a quantitative analysis of the relationships between seabird foraging and hatchery-released fish will help to identify the ecological role of the hatchery program. Also, it may help track pollutants as they move through the food web (Dalsgaard et al. 1998). The versatility of the Ecopath system allows it to produce a fast and cost-effective overview of any part of the system. The basic idea of this project is that the use of a mass balance model such as Ecopath will allow easy identification of areas of trophic flux that will be of interest to those involved in policy making and restoration.

Constructing the PWS Model

The collaborative process of constructing a balanced trophic model of the PWS ecosystem consisted of four components: (1) a scoping period to identify ecosystem components and experts, (2) workshops, (3) coordinated e-mail and telephone communications, and (4) an edited volume that provided a venue for authored contributions. Initial identification of components and contributors was accomplished through inputs from EVOS program scientists, conversations with other experts, and our knowledge of the PWS marine ecosystem and the scientific literature.

Three meetings occurred over a 9-month period: a preparatory working lunch held in conjunction with the 1998 EVOS restoration workshop in January 1998; a model specification workshop during March 2-4, 1998, at which invited experts provided initial estimates of biomass, production, consumption, diet composition, migration, and spatial distributions; and an evaluation workshop on October 5, 1998, at which participants evaluated the balanced model and initial analyses to refine strategies of future model iterations and analyses.

At the model specification workshop, parameters were contributed within the context of modifying a pre-existing, preliminary model of PWS, constructed from existing literature sources (Dalsgaard and Pauly 1997, Pauly et al. 1998b). This format of building upon an existing, simpler model served three purposes: (1) anticipating skepticism regarding construction of a realistic whole-ecosystem model, (2) avoiding pressure for commitment by contributors early in the process, and (3) ameliorating a seemingly daunting challenge by refining an existing model rather than starting from scratch.

Facilitated communication among working group participants was crucial to keep participants coordinated within the context of the whole interactive ecosystem, and the edited volume of authored sections enabled many experts to contribute data and invest time without sacrificing recognition. Table 1 shows that contributors from a broad range of affiliations contributed to the integration of knowledge about a broad range of ecosystem components.

Contributors provided estimates for each of the following parameters:

1. Biomass in wet weight units and expressed as density (t per km^2) for PWS as a whole (9,059 km^2).

2. The P/B ratio (production/biomass). In Ecopath-type models, this is equivalent to an instantaneous rate of total mortality (i.e., Z; per year).

3. The Q/B ratio (consumption/biomass). This is a population-weighted estimate of food consumption per unit biomass (per year), or *ration* for an average-sized individual.

4. Exports from the system consist of catches (here in t wet weight per year) and animals leaving the system.

5. Information for a group in which diet fractions add up to 1.

When available, contributors provided seasonal means allowing for consideration of seasonal oscillations. They were also asked to quantitatively indicate increasing or decreasing trends as well as uncertainty by providing confidence intervals or likely minimum and maximum values.

Temporal Simulations with Ecosim

Beyond the uses of static representations of the PWS ecosystem, the data in Ecopath files were used in dynamic simulations using the Ecosim approach of Walters et al. (1997). Ecosim models allow rapid exploration of the predicted consequences of natural or anthropogenic disturbances on all components of an ecosystem simultaneously over a specified time period (typically 10 years). These could include changes in fishing, anthropogenic disturbances like another oil spill, or natural changes in agents of physical forcing. Other changes in resource use or potential management actions can likewise be simulated.

Hypothetical Scenarios for Simulating Perturbations

After the balanced trophic model of PWS was constructed, hypothetical "what if" scenarios were simulated using Ecosim. These scenarios were contributed by B. Spies, B. Wright, and A. Gunther at the model specification workshop:

1. What if fishing pressure on herring increases or decreases; what if there is one stock of herring? two? three?
2. What if somebody decides to fish sandlance or capelin? this is probably far-fetched, but model simulations would likely show important trophic impacts of removing important forage fishes.
3. What if an earthquake raises the upper 10 m of intertidal above sea level?
4. What if PWSAC goes broke and the hatcheries close?
5. What if there is another oil spill?
6. What if human impacts from the road to Whittier result in damage to intertidal habitats in the western part of PWS?
7. What if recreational fishing pressure removes 90% of the rockfish from PWS?
8. What if there is a major warm-water episode for 2 years with the upper 200 m of water over the shelf in the GOA is elevated by 2°C ?
9. What if the bloom and sustained productivity lasts only for 3 weeks instead of the usual 12 weeks in PWS ?
10. What if the harbor seals continue to decline at 8% per year ?
11. What if Dungeness crab return to PWS?
12. What if salmon prices drop or increase?
13. What if pollock disappear from PWS?
14. What if salmon farming were allowed in PWS?
15. What if a road were established to Cordova?
16. What if cruise ship traffic increases into Cordova?

Spatial Simulations with Ecospace

The recently developed Ecospace routine (Walters 1998, Walters et al. 1998) was used to simulate changes in spatial distributions of Prince William Sound groups starting with information on habitat preferences and spatial distributions of habitats and organisms provided by contributors (also see Okey 1998). Ecospace simulates dynamic, two-dimensional redistribution of ecosystem components based on trophic interactions (flow) among organisms, their relative preferences for spatially specified habitats,

Table 1. Some Prince William Sound ecosystem component groupings and associated contributors.

Ecosystem component	Contributor	Affiliation
Cetaceans	Craig Matkin	North Gulf Oceanic Soc., Homer
Sharks	Lee Hulbert	NMFS Auke Bay Laboratory
Pinnipeds	Kathy Frost	ADF&G, Fairbanks
Cetaceans and pinnipeds	Rod Hobbs	NMML, NMFS, Seattle
Pacific halibut	Bob Trumble	IPHC, Seattle
Adult arrowtooth flounder	Mark Willette	ADF&G, Cordova
Shallow large epibenthos	Tom Dean	Coastal Resources Associates, Vista, CA
Walleye pollock	Mark Willette	ADF&G, Cordova
Miscellaneous fish groups	Tom Okey	UBC Fisheries Centre, Vancouver
Seabirds and raptors	Bill Ostrand, David Irons	USFWS, Anchorage
Adult salmon	Slim Morestead (consulted)	ADF&G, Cordova
Juvenile salmon fry	Tom Kline	PWS Science Center, Cordova
Nearshore demersal fishes	Tom Dean	Coastal Resources Associates, Vista, CA
Sea otter	J. Bodkin, D. Monson, G. Esslinger	USGS-BRD, Anchorage
Squid	Jay Kirsh	PWS Science Center, Cordova
Forage fishes	Evelyn Brown	UAF Institute of Marine Science
Deep epibenthos	Tom Okey	UBC Fisheries Centre, Vancouver
Adult Pacific herring	John Wilcock (consulted)	ADF&G, Cordova
Sea ducks	Dan Esler	USGS-BRD, Anchorage
Juvenile Pacific herring	Robert Foy	UAF Institute of Marine Science
Jellyfish	Jennifer Purcell	Horn Point Lab, Cambridge, MD
Small benthic infauna	Stephen Jewett	UAF Institute of Marine Science
Nearshore zooplankton	Robert Foy	UAF Institute of Marine Science
Offshore zooplankton	Ted Cooney	UAF Institute of Marine Science
Deep large infauna	Tom Okey	UBC Fisheries Centre, Vancouver
Shallow small epibenthos	Tom Dean	Coastal Resources Associates, Vista, CA
Shallow large infauna	Tom Dean	Coastal Resources Associates, Vista, CA
Phytoplankton	Peter McRoy (consulted)	UAF Institute of Marine Science
Macroalgae and eelgrass	Tom Dean	Coastal Resources Associates, Vista, CA
Forage fish diets	Molly Sturdevant	NMFS Auke Bay Laboratory
Birds eating herring eggs	Mary Anne Bishop	Pac. NW research station, USFS, Cordova
Recreational catches	Scott Meyer	ADF&G, Homer
Commercial catches	Bill Bechtol	ADF&G, Homer
Preliminary model of PWS	J. Dalsgaard, D. Pauly	UBC Fisheries Centre, Vancouver
"What if" scenarios	B. Spies, B. Wright, A. Gunther	AMS; NMFS Juneau; AMS

and their movement rates and vulnerability to predators in the various specified habitats.

Results

Workshop participants helped define the ecosystem and its components, they contributed estimates for the input parameters listed above, and they provided information on spatial and temporal changes. Figure 1 shows the trophic levels and the relative biomasses of biotic components of PWS based on the contributed estimates to the trophic flow model (trophic flows are left out of Fig. 1, as they are too numerous to display in this format). This model and its graphical representation represents a likely scenario of energy flow, based on the assumption of equilibrium in the system.

The structure of trophic flows provides an indicator of knowledge flow and a guide for researcher interactions when refining the parameter estimates for a group. Figure 2 identifies all of the direct flows connected to a single group in the model (in this case, adult Pacific herring). Ideally, the parameter estimates for a given group are derived empirically and have a high degree of accuracy and precision, but in reality, there are varying degrees of uncertainty among groups. Inconsistencies in energy flow between connected groups are indicated when the Ecopath model is unbalanced. These highlighted inconsistencies enable researchers to revisit the data and refine the model in a systematic way. This refines the realism of the overall model as well as estimates for individual groups. If a particular group is "unbalanced" within the model (i.e., when the ecotrophic efficiency is greater than 1; "ecotrophic efficiency" is the proportion of production by a group that is either consumed or exported), this may indicate that biomass or production/biomass values for the group are underestimates, or that consumption by other groups has been overestimated. Thus, the researcher for a group can use trophic flow connections as a guide to collaborative interactions in order to refine the information for a group.

Ecosystem components extracted from other trophic positions reveal unique trophic flow patterns that indicate different knowledge flow structures for refinement and learning. The relative magnitudes of trophic flows are not shown in these figures, but optimal knowledge flow structures would also account for magnitude of flows.

One Ecosim simulation is presented as an example of the usefulness of the collaborative approach; Fig. 3 shows functional responses of some groups to an aggressive fishery on sandlance in PWS, as trajectories of biomass changes. The model predicts that both seabirds and avian predators (birds that eat seabirds) would decline in response to such a fishery, while other groups would increase.

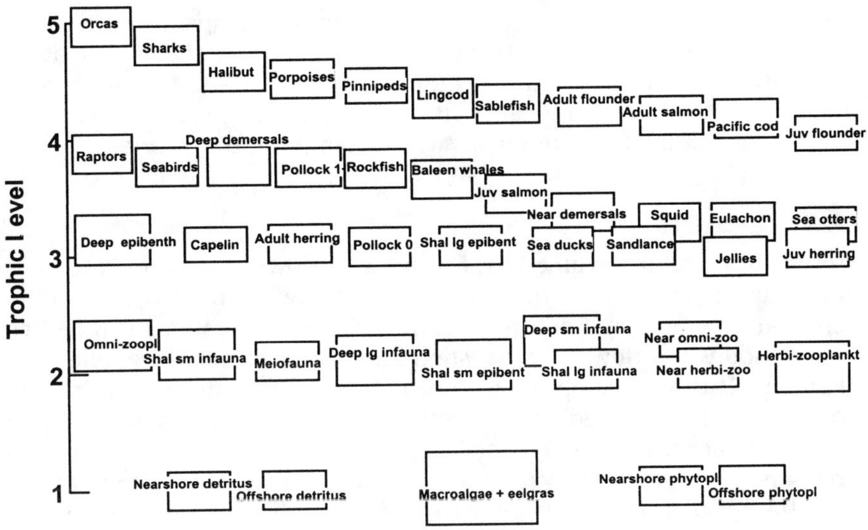

Figure 1. Components of the balanced trophic model of Prince William Sound, Alaska, displayed on a trophic level scale. Box size represents the log relative standing biomass of each component. Trophic flows are not displayed here, as there are too many connections for this format.

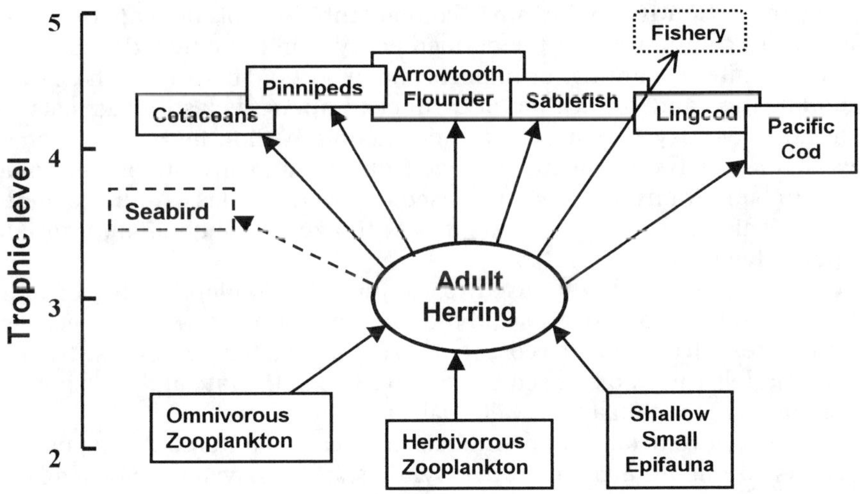

Figure 2. A mid-trophic-level component (adult Pacific herring) extracted from the overall food web along with directly connected components. Known trophic flows among components serve as indicators of knowledge flow among experts to optimize refinement of information about the group and its interrelationships (e.g., refine model input parameters).

When the same Ecosim scenario is used in the Ecospace routine, components are redistributed in two-dimensional, habitat-specified space at the relative biomasses revealed by the Ecosim run. Thus, a comparison of the "equilibrium" and "sand lance fishery" scenarios reveals different predictions of spatial distributions as well as temporal trajectories.

Discussion

Trophic interactions in an ecosystem are also flows of energy, which vary in rate and magnitude among connections (between the various components) and in time and space. We contend that ecosystems can be best understood when flows of ecosystem knowledge closely resemble flows of energy. Pathways to optimize knowledge flow for one ecosystem component should be based on the trophic flows immediately surrounding that component. Ecosystem-based trophic models, like Ecopath, integrate previously disparate ecosystem information through a system of knowledge flow that resembles trophic flow. This method was used to synthesize a cohesive picture of an apparently compartmentalized ecosystem.

Several EVOS-funded projects, notably the Alaska Predator Ecosystem Experiment (APEX), the Nearshore Vertebrate Predators (NVP) project, and the Sound Ecosystem Assessment (SEA) project, are devoted to the biology and ecology of distinct groups of organisms, sometimes including their prey, their predators, or both. As a result of these programs, the resolution of information is high for some ecosystem components. However, resolution is low for other components. Use of an Ecopath model allows all components of the defined ecosystem to be included and balanced (while accounting for imports and exports). This approach enables modification and verification of distinct components, as well as insights into whole ecosystem structure and function. Within the Ecopath framework, the precise information gained from the large investments in some research programs results in increased knowledge of less-studied components. This knowledge refinement is optimized through an appropriate knowledge flow structure.

This approach can be used to gain a better understanding of individual resource components and their potential trajectories in an ecosystem context, in addition to ecosystem structure and function. These trophic flow models can also be used to accurately map the fate and transport of contaminants within a food web (Dalsgaard et al. 1998).

A mix of reluctance and enthusiasm was encountered during our attempts to initiate a collaborative synthesis of ecosystem knowledge in Prince William Sound using the Ecopath approach. Some of the initial reluctance was linked with skepticism about the Ecopath approach. However, feedback from participants indicates a higher degree of enthusiasm and acceptance of the approach, as well as increased interaction and discussion among research groups, now that a face-to-face workshop was conducted.

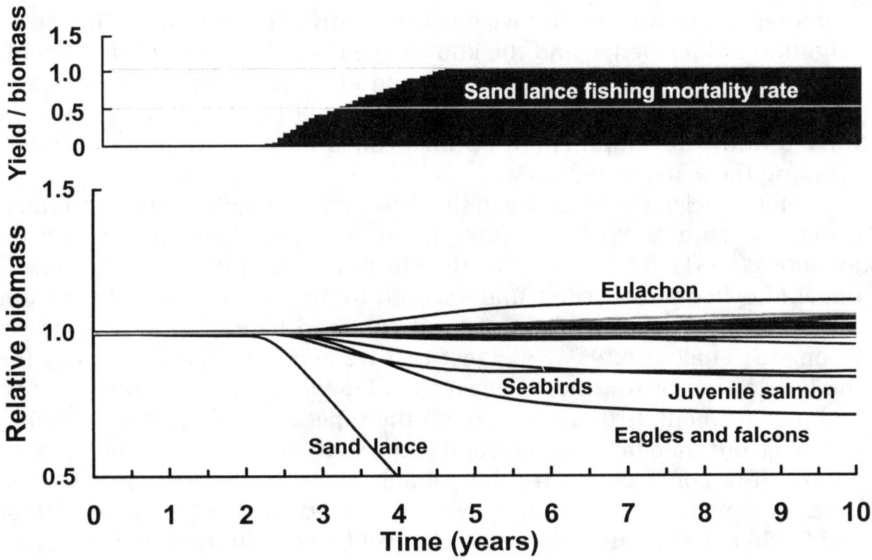

Figure 3. Ecosim simulation of the response to an aggressive sand lance fishery showing biomass declines in seabirds (which prey on sand lance), juvenile salmon, and avian raptors (which depend on seabirds and salmon). Eulachon, a competitor of juvenile salmon and a prey of seabirds, increases as seabirds decline. Aggregation of seabirds into one group masks the magnitude of declines in seabird species that are particularly dependent on sand lance. The functional responses predicted by Ecosim integrate the relative trophic forces in an ecosystem. These functional responses, rather than the absolute magnitude of responses, are useful for resource decision making and research planning. This sand lance example is a relatively simple one; some simulations predict more broad and dramatic responses.

Aside from technical skepticism about a particular analytical approach, reluctance to collaborate in an ecosystem context, in general, is linked to aspects of human behavior, such as social group dynamics and perceptions of territory. Behaviors reinforcing reluctance to collaborate are natural and adaptive, but offset by other motivations that reinforce collaboration. Based on our experience with this synthesis, we suggest that reluctance to collaborate can be overcome by redefining working communities and territories to resemble ecosystem structure. We also suggest that this can be done through workshops, collaborative reports, and other venues of knowledge flow patterned after energy flows in ecosystems.

The initial PWS model was constructed through the collaboration of members of the scientific community in the region, and it is planned to also integrate traditional knowledge from native communities. The representation of knowledge and the knowledge flow structures of these communities, however, differ in fundamental ways from the knowledge structure of the scientific community. Examination of these differences will aid in the development of optimal collaborative structure for understanding the ecosystem.

Some overlap exists between these two knowledge systems, and thus, some opportunity exists for integration, as elegantly demonstrated by Johannes (1981) in the South Pacific. The potential of incorporating traditional knowledge is further underscored by the richness of information attainable by examining lists of local common names, as discussed by Palomares et al. (1999). We suggest that tremendous potential exists for the integration of traditional ecosystem knowledge into current science and management structures through the type of collaborative modeling approach outlined in this paper, and using analytical tools like the Ecopath approach to constructing trophic models of historical ecosystems to be used as benchmarks. This has been discussed by Haggan (1996, 1998) and Pitcher (1998), and has been achieved by contributors to Pauly et al. (1998a). Failure to include traditional knowledge in such a process certainly limits the realism of the model, as well as its ultimate effectiveness as a management tool.

One potential outcome of Ecopath modeling activities in PWS is community-based resource management and policy development. By this we mean resource management that functions well because it is appropriately, and constructively, influenced by stakeholder communities and other trustees. The future success of oil spill restoration and resource management planning will be optimized through the participation of stakeholders, especially when venues for participation are functional, educational, and collaborative. We suggest that community-based management naturally occurs in parallel to ecosystem-based management that is achieved through knowledge flow structures that resemble ecosystem structures, as explored during this collaborative approach.

A great deal can be learned from Ecosim simulation models to aid resource managers in making decisions that affect the development of these communities. Perhaps even more importantly, the outputs and implications of Ecosim model runs are easily grasped by anyone because of the friendly user interface and graphics that are clear and intuitive. To encourage the process of ecosystem-based management, the model of PWS will be widely disseminated among the public as well as among managers. A CD-ROM version of the PWS model is available for distribution to interested organizations and institutions, including schools through the *Exxon Valdez* Oil Spill Trustee Council and appropriate trustee agencies. This CD-ROM also includes a local/traditional language database of the marine organisms of PWS and beyond. In addition, a locally enriched, customized version of "FishBase," the global, computerized encyclopedia of fishes, will also be made available on the worldwide web (see MacCall and May 1995 and www.fishbase.com).

In this paper we have outlined an iterative process. Knowledge about trophic interactions must exist before optimal knowledge flow structures can be indicated by trophic structure. Knowledge flow structures suggested by this process can then be used to refine the input parameter estimates, and thus increase the realism of the model. As the realism of the model increases, the tool becomes more useful for resource assessment and management, as well as research scoping and planning. At the same time, interest in the approach increases within the communities that are aware of the model. This in turn continues to increase model realism and applications. In this sense, the Ecopath approach has the potential of becoming a focal point for a living synthesis of ecosystem information.

Acknowledgments

We thank the *Exxon Valdez* Oil Spill Trustee Council for support of project BAA 98330. We are grateful to L.C. Thompson for comments that helped clarify the manuscript, and C.J. Jevons for inspiration.

References

Cooney, R.T. 1997. Sound Ecosystem Assessment (SEA): A science plan for the restoration of injured species in Prince William Sound, Alaska. *Exxon Valdez* Oil Spill Restoration Project Annual Report (Restoration Project 96320). University of Alaska Fairbanks, Institute of Marine Science.

Dalsgaard, J., and D. Pauly. 1997. Preliminary mass-balance model of Prince William Sound, Alaska, for the pre-spill period, 1980-1989. Fisheries Centre Research Report 5(2). University of British Columbia, Vancouver.

Dalsgaard, J., A. Jarre-Teichmann, C. Walters, and D. Pauly. 1998. An approach to modeling persistent pollutants in marine ecosystems. Presented at the Annual Science Conference, International Council for the Exploration of the Sea. Session V: Recovery and Protection of Marine Habitats and Ecosystems from Natural and Anthropogenic Impacts. Cascais, Portugal. CM 1998/S:4. 20 pp.

Duffy, D.C. 1997. APEX project: Alaska Predator Ecosystem Experiment in Prince William Sound and the Gulf of Alaska. *Exxon Valdez* Oil Spill Restoration Project Annual Report (Restoration Project 96163 A-Q). University of Alaska Anchorage, Alaska Natural Heritage Program and Department of Biology.

EVOS Trustee Council. 1996. Invitation to submit restoration proposals for federal fiscal year 1997. *Exxon Valdez* Oil Spill Trustee Council, Anchorage.

Haggan, N. 1996. Integration of local environmental knowledge. In: D. Pauly and V. Christensen (eds.), Mass balance models of north-eastern Pacific ecosystems. Fisheries Centre Research Report 4(1). University of British Columbia, Vancouver, pp. 88-89.

Haggan, N. 1998. Back to the future with Ecopath and Ecosim. In: D. Pauly (ed.), Use of Ecopath with Ecosim to evaluate strategies for sustainable exploitation of multi-species resources. University of British Columbia, Vancouver, Fisheries Centre Research Report 6(2):26.

Holland-Bartels, L. 1997. Mechanisms of impact and potential recovery of near-shore vertebrate predators. *Exxon Valdez* Oil Spill Restoration Project Annual Report (Restoration Project 96025). Alaska Science Center, Biological Resources Division, 1011 East Tudor Road, Anchorage, Alaska 99503.

Johannes, R.E. 1981. Words of the lagoon: Fishing and marine lore in the Palau district of Micronesia. University of California Press, Berkeley. 245 pp.

Juda, L. 1996. Developing international law and ecosystem-based fisheries management. In: K. Sherman, N.A. Jaworski, and T.J. Smayda (eds.), The Northeast shelf ecosystem: Assessment, sustainability, and management. Blackwell-Science, Cambridge, pp. 31-46.

Keeble, J. 1991. Out of the channel: The *Exxon Valdez* oil spill in Prince William Sound. Harper Collins, New York. 290 pp.

Laevastu, T., D.L. Alverson, and R.J. Marasco. 1996. Exploitable marine ecosystems: Their behaviour and management. Fishing News Books, Oxford. 321 pp.

Langton, R.W., and R.L. Haedrich. 1997. Ecosystem-based management. In: J. Boreman, B.S. Nakashima, J.A. Wilson, and R.L. Kendall (eds.), Northwest Atlantic groundfish: Perspectives on a fishery collapse. American Fisheries Society.

MacCall, R.A., and R.M. May. 1995. More than a seafood platter. Nature 376:735.

NMFS (National Marine Fisheries Service). 1999. Ecosystem-based fishery management: A report to Congress by the National Marine Fisheries Service Ecosystem Principles Advisory Panel. NMFS, Silver Spring, Maryland. 164 pp.

Okey, T.A. 1998. The Prince William Sound model. In: D. Pauly (ed.), Use of Ecopath with Ecosim to evaluate strategies for sustainable exploitation of multi-species resources. University of British Columbia, Vancouver, Fisheries Centre Research Report 6(2):32-33.

Okey, T.A., and D. Pauly (eds.). 1998. A trophic mass-balance model of Alaska's Prince William Sound ecosystem, for the post-spill period 1994-1996. University of British Columbia, Vancouver, Fisheries Centre Research Report 6(4). 144 pp.

Okey, T.A., and G.A. Harrington. 1999. Criteria for designing ecosystem-based, experimental management programs: Effects of bottom trawling on the Bering Sea ecosystem. In: Ecosystem considerations for fisheries management, University of Alaska Sea Grant, AK-SG-99-01, Fairbanks. (This volume.)

Paine, R.T., J.L. Ruesink, A. Sun, E.L. Soulanille, M.J. Wonham, C.D.H. Harley, D.R. Brumbaugh, and D.L. Secord. 1996. Trouble on oiled waters: Lessons from the *Exxon Valdez* oil spill. Ann. Rev. Ecol. Syst. 27:197-235.

Palomares, M.L.D., C.V. Garilao, and D. Pauly. 1999. On the biological information content of common names: A quantitative case study of Philippine fishes. Cybium. In press.

Pauly, D., T.J. Pitcher, and D. Preikshot (eds.). 1998a. Back to the future: Reconstructing the Strait of Georgia ecosystem. University of British Columbia, Vancouver Fisheries Centre Research Report 6(5). 100 pp.

Pauly, D., J. Dalsgaard, T.A. Okey, R. Powell, and S. Pimm. 1998b. Mass-balance food web ecosystem models as an alternative approach for combining multiple information sources in fisheries. In: F. Funk, T.J. Quinn II, J. Heifetz, J.N. Ianelli, J.E. Powers, J.F. Schweigert, P.J. Sullivan, and C.-I. Zhang (eds.), Fishery stock assessment models. University of Alaska Sea Grant, AK-SG-98-01, Fairbanks, pp. 815-816.

Pitcher, T.J. 1998. Rebuilding as a new goal for fisheries management: Reconstructing the past to salvage the future. In: K. Ekschmidt et al., Proceedings of INTECOL 1998. Florence, Italy, July 1998.

Roberts, C.M. 1997. Ecological advice for the global fisheries crisis. Trends Ecol. Evol. 12(1):35-38.

Spies, R.B., S.D. Rice, D.A. Wolfe, and B.A. Wright. 1996. The effects of the *Exxon Valdez* oil spill on the Alaskan coastal environment. In: Proceedings of the *Exxon Valdez* Oil Spill Symposium. American Fisheries Society Symposium 18:1-16.

Walters, C. 1998. Ecosim and Ecospace: Basic considerations. In: D. Pauly (ed.), Use of Ecopath with Ecosim to evaluate strategies for sustainable exploitation of multi-species resources. University of British Columbia, Vancouver, Fisheries Centre Research Report 6(2):11-14.

Walters, C., D. Pauly, and V. Christensen. 1998. Ecospace: Prediction of mesoscale spatial patterns in trophic relationships of exploited ecosystems, with emphasis on the impacts of marine protected areas. Presented at the Annual Science Conference, International Council for the Exploration of the Sea. Session S: Visualization of Spatial (including Survey) Data. Cascais, Portugal. CM 1998/S:4. 20 pp.

Walters, C., V. Christensen, and D. Pauly. 1997. Structuring dynamic models of exploited ecosystems from trophic mass-balance assessments. Rev. Fish Biol. Fish. 7(2):139-172.

Wheelwright, J. 1994. Degrees of disaster. Prince William Sound: How nature reels and rebounds. Simon and Schuster, New York. 348 pp.

Wilson, J.A., J.M. Acheson, M. Metcalfe, and P. Kleban. 1994. Chaos, complexity and community management of fisheries. Marine Policy 18(4):291-305.

Applying Ecosystem Management to Fisheries in the Strait of Georgia

Richard J. Beamish and Gordon A. McFarlane
Department of Fisheries and Oceans, Pacific Biological Station, Nanaimo, British Columbia, Canada

Abstract

The Strait of Georgia is located between Vancouver Island and mainland British Columbia. It is one of the most important marine ecosystems in the Pacific, as most British Columbians live within 10 km of its shores and it is a major rearing and nursery area for a number of important species of fishes. Various components of the dynamics of this ecosystem have been identified over the past 50 years. However, the connection between fish dynamics and ecosystem dynamics remains poorly understood. Recent studies have shown that major changes in the composition of fish species can occur naturally in response to shifts in the climate and ocean environment. In the Strait of Georgia, these regime shifts affect the matching of copepod movement into the surface waters with the first feeding stage of larval hake and herring. As these two species are the dominant fishes in the Strait of Georgia, changes in their biomass will affect the dynamics of the populations of other species. We propose that the key process that regulates fish abundance is related to the factors that regulate the timing of the vertical migrations and abundance of copepods. According to our hypothesis, copepods are the component of the zooplankton community that link the fundamental processes of microscopic plant production through the association of light and nutrients with the survival of larval fish. We suggest that effective management of the fish populations in the Strait of Georgia requires an understanding of how climate and ocean changes affect ecosystem dynamics. A first step toward ecosystem management could be the production of species ecosystem management charts. We provide examples of these charts for Pacific herring, Pacific hake, and coho salmon. The charts included in this paper may ultimately be shown to require major changes. It is the process of identifying the key factors affecting species abundance that is important in the initial stages of ecosystem management.

It may be as challenging to find methods to communicate the information among experts as it is to acquire the relevant information. There is considerable speculation in our approach, but we propose that it provides a framework to begin to manage single species in the context of their ecosystems.

Introduction

The Strait of Georgia is the center of recreational and commercial activity in British Columbia. The Strait, which is located between Vancouver Island and the British Columbia mainland (Fig. 1), is one of the most important juvenile salmon rearing areas in the Pacific Ocean. The Canadian portion of the strait is 220 km long, 33 km wide, and has a surface area of about 6,900 km^2 (Thomson 1981). The maximum depth is 421 m and the average depth is 155 m. The area south of the Strait that is in United States water is known as the San Juan Islands and Puget Sound. This area is shallower and influenced oceanographically by winds and tides. The Fraser River contributes about 80% of the freshwater runoff into the strait (Thompson 1981). The surface salinity averages about 31‰ which is 2.8‰ less than the average salinity (33.8‰) found off the west coast. The Strait of Georgia is a semi-enclosed sea that is connected to the Pacific Ocean in the north by Johnstone Strait and in the south by Juan de Fuca Strait. These two narrow passages affect the movement of water in and out of the Strait of Georgia, resulting in reduced salinity and a temperature profile that is influenced on the surface by the Fraser River and on the bottom by inflowing deep water from offshore. The total volume of freshwater discharge reduces the surface salinity particularly in the summer. Water circulation, mixing, and stratification in the strait are largely influenced by its freshwater discharge, together with tides and wind. The more general estuarine circulation (Thomson 1981) was described about 35 years ago. The fresh water that is added to the Strait must eventually leave on the surface. The replacement by bottom water from offshore is about 1,000 times richer in nitrates (NO_3) than surface water (Harrison et al. 1994). This nutrient-rich bottom water is eventually mixed into the surface, providing nutrients for phytoplankton that become food for other organisms. The Fraser River discharge starts to increase in March; reaches a maximum in June; gradually decreases in July, August, and September; and remains near minimum levels throughout the rest of the year. From March to September, the discharged fresh water is mixed with seawater in the Strait and most of the mixture forms as a large plume extending over as much as one-third of the southern area of the Strait. This plume, and specifically the edge of the plume, is an area of high productivity that attracts large numbers of young salmon and herring as well as their predators.

Wind mixing is the dominant physical mechanism that entrains nitrates from the nitrate-rich deep water into the surface layer in the Strait of Georgia. Modeled impact of wind events on nitrate fluxes (St. John et al.

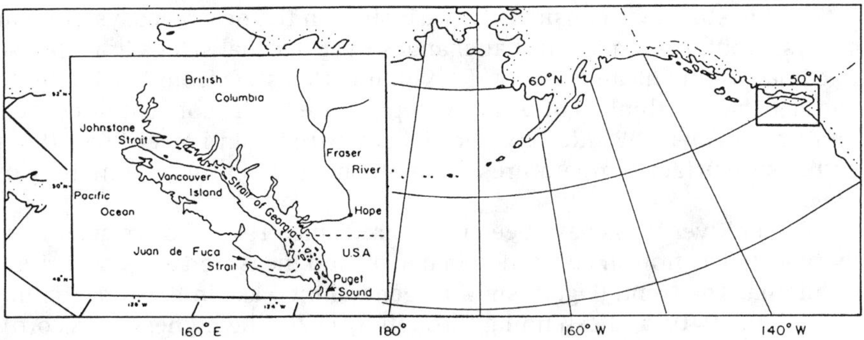

Figure 1. The Strait of Georgia, located off the west coast of Canada between Vancouver Island and mainland British Columbia.

1993) in the southern Strait, showed that nitrate entrainment in the surface layer was reduced when runoff from the Fraser River was high as greater energy was required to break down the more buoyant surface layer. Increases in primary production lagged behind the wind mixing events and were lower and occurred earlier when runoff was high. An earlier spring bloom occurred because the increased stability of the more buoyant surface layer resulted in a shallower depth of mixing for plankton and effectively increased the light available for photosynthesis at a time light limits plankton growth (St. John et al. 1993). Variations in river runoff have also been shown to affect the nutrient supply to the surface waters (Harrison et al. 1994). When Fraser River flows were reduced, wind mixing was more effective and productivity increased. It is this period from April to September when production is periodically nutrient-limited (Parsons et al. 1981) that we believe is important for salmon survival (Beamish and Mahnken 1998).

Li et al. (1999) modeled the salinity and volume fluxes in the Strait of Georgia as a function of Fraser River runoff. The model calculations identified a rapid response in the circulation trends to changing trends in Fraser River flows. The influence of sea level wind patterns was not incorporated into the model but the authors noted that short-term persistent wind events would result in interannual variability in the modeled response to Fraser River runoff. A dramatic decrease in river flow would, according to the model, result in an increase in surface salinity within about 1 year. Other studies of the factors affecting the daily and seasonal oceanographic conditions within the Strait of Georgia have been described by a number of investigators (Hutchinson and Lucas 1931; Waldichuck 1957; Thomson 1981, 1994; LeBlond 1983; Crean et al. 1988; Harrison et al. 1994), and all identify the Fraser River as having a major influence on the oceanography of the Strait of Georgia.

In this study, we consider the variability in the ocean ecosystem that occurs about a mean trend that changes over decadal scales in response to climate shifts (Beamish et al. 1999b). It is this shift from one "state" to another that we think is of primary importance in the concept of ecosystem management. We identify decadal-scale trends in Fraser River flows and in sea surface temperatures, both of which affect the productivity of fishes.

Fraser River flows have been monitored since 1912 and are customarily reported as flow in cubic meters per second measured at Hope, British Columbia. The trend (Fig. 2) shows a general increase in flows beginning in the late 1940s and extending through to 1976 (the highest on record). Since 1976 the flows have generally remained low. In contrast, 1997 was the second highest annual discharge this century but 1998 will be one of the historic low values. Major shifts in trends occurred in the mid-1940s and in 1977. If the annual Fraser River discharge, expressed as m^3/s, is shown as ± 1 standard deviation from the time series mean (Fig. 2), the period beginning in the late 1940s through to 1976 is a period of extreme high flows. Periods before and after were characterized by average to below average flows (Fig. 2). The years of extreme discharge tend to occur the year following a La Niña (Table 1) but there were years in the 1960s of higher discharges that did not follow a prominent La Niña. There also is a tendency for extreme low discharge years to follow El Niño years, but again there were extreme low discharge years that did not follow El Niño years.

There was not a major change in the trend of annual flows in the 1990s, but there was a significant increase in the volume of discharge in April (Beamish et al. 1999a; Fig. 3). Beginning about 1989, flows in April increased which indicated an earlier beginning of the spring freshette (Fig. 3). There also has been an increase in sea surface temperatures (Fig. 4). Our data series indicates a distinct period from 1969 to 1977 with an average of about 10.4°C; from 1978 to 1991 with an average of about 11.3°C; and a recent warming beginning about 1992 with an average of 11.6°C. Thus the overall surface temperature has increased about 1.2°C in the last 2 decades. In general the pattern of flows of the Fraser River and the patterns of temperatures (Beamish et al. 1999a) reflect the trends seen in the Aleutian Low Pressure Index (Beamish et al. 1999b) which indicates that global climate processes affect the ocean environment in the Strait of Georgia.

Moore (1991), Moore and McKendry (1996), and Mantua et al. (1997) showed that there were trends in the amount of runoff from large rivers on the west coast of North America that were related to trends in climate. Moore (1991) showed that the depth of the snowpack in the Fraser River basin declined by 22% over the same period that Fraser River flows declined by 28%. Beamish et al. (1999a, 1998c) showed that decadal scale changes in climate were related to shifts in the migratory behavior of coho salmon and synchronous decreases in marine survival. These previous

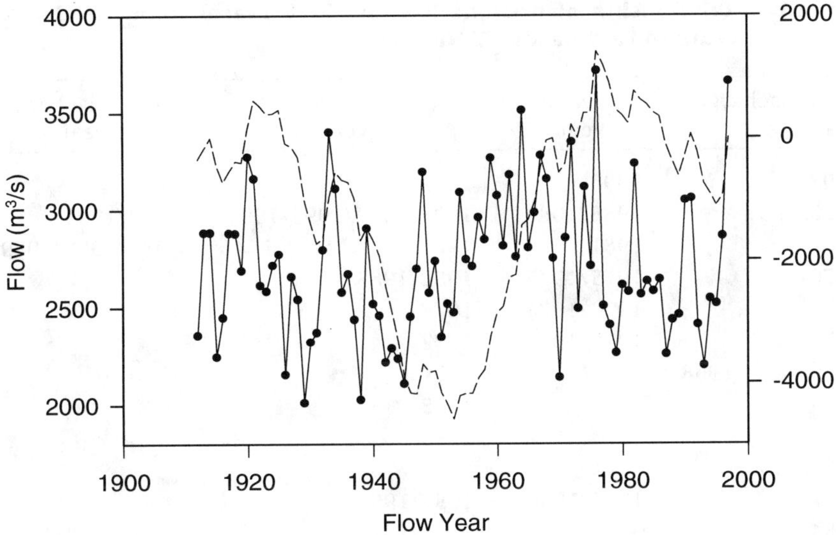

Figure 2. Average annual discharge from the Fraser river (m³/s). Flow year is April 1 to March 31 of the following year. The dashed line is the cumulative sum trend (Beamish et al. 1998d). The cumulative sum of the anomalies provides a simple visual picture of the trends in the index. The method of accumulating sums or CuSum (Murdock 1979) is a way of studying trends by including information from past data points. The calculation is a simple addition of a data point to the sum of all previous data points. Each data point may have a constant subtracted from it, to enhance visualization of trends in the series. The constant is usually the mean of the entire series. If there is a trend in the data there will be a trend in the CuSum graph. In recent years there was an increasing trend in discharge that changed in 1977 to a declining trend.

Table 1. Relationship of extreme low and high Fraser River discharge years to La Niña and El Niño events.

High discharge year	La Niña year	El Niño year	Low discharge year
1997	1996	1997	1998
1990/1991	1988/1989	1991/1992/1993/1994	1992/1993
1982	1981	1987	1987/1988/1989
1976	1975/1976	1982/1983	
1974	1973/1974	1977	1978/1979
1972	1970/1971	1972	
1966/1967/1968		1969	1970
1964		1965	
1962		1963	
1959/1960			
1957	1955/1956	1957/1958	
1954		1951/1953	1951/1952/1953

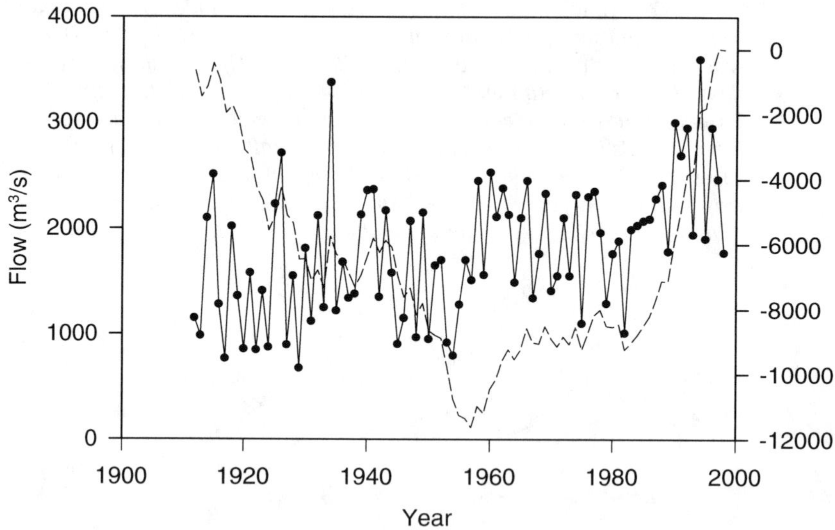

Figure 3. Average April discharges from the Fraser River (m^3/s). The CuSum trend (dashed line) shows an abrupt change in the late 1980s.

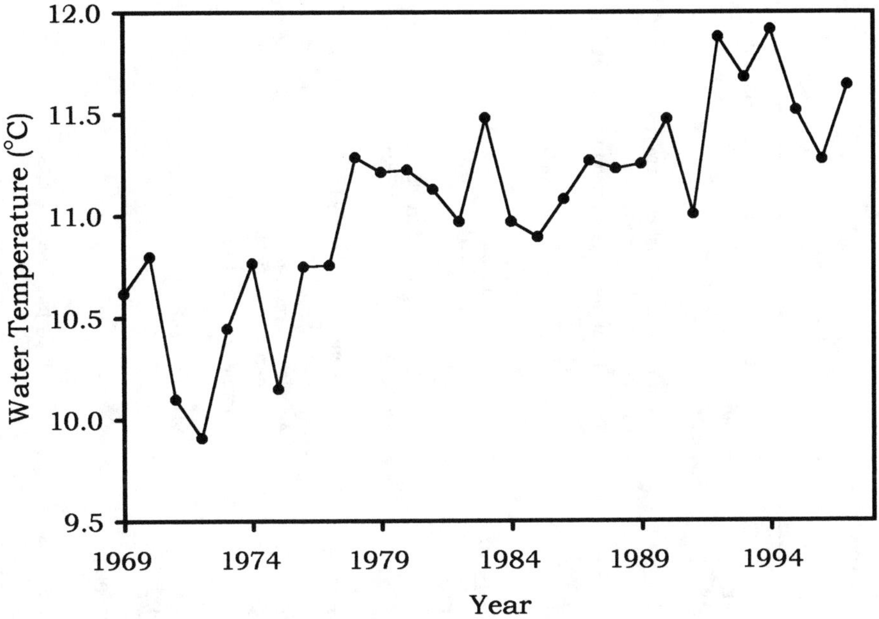

Figure 4. Average annual sea surface temperatures in the central Strait of Georgia, showing a distinct warming trend since the early 1970s.

studies identify a relationship between the dynamics of the ocean ecosystem in the Strait of Georgia and climate. The long-term or regime changes that may alter the survival pattern of a particular year class are important, as these impacts need to be distinguished from fishing effects. In this report, we cannot examine the relationships of the population dynamics of all major species, but we briefly review the state of the key species and relate the dynamics of three species to the dynamics of the ecosystem.

Major Fisheries

The current state of the major fisheries in the Strait of Georgia is summarized in Table 2 and Fig. 6. Other fisheries and other important biota have not been included in this report but additional descriptions are available in Ketchen et al. (1983).

The most abundant resident fish species is Pacific hake (Table 2). The abundance has increased slightly in recent years (Saunders and McFarlane 1997) and may now exceed 65,000 t or more than 300 million fish. This estimate is considerably lower than the earlier estimates because revisions in target strength have reduced all previous estimates (Saunders and McFarlane 1997). Most Pacific hake spawn in the Strait of Georgia

Table 2. Key fish species and condition of stocks in the Strait of Georgia.

Species	Condition	Comments
Coho salmon *Oncorhynchus kisutch*	Wild stocks are in very low abundance. In recent years about 3 million juveniles rear in the strait until fall, then leave. Ocean age 1 juveniles have not returned in recent years.	Major recreational and commercial fisheries now closed.
Chinook salmon *O. tshawytscha*	Wild stocks are in low abundances but stable.	Major recreational fisheries are restricted and the commercial fishery is closed.
Sockeye salmon *O. nerka*	Declining from historical high levels to below long-term average.	Transitional-juveniles remain in the strait less than 2 months.
Pink salmon *O. gorbuscha*	Declining from historical high levels to below long-term average.	Transitional-juveniles remain in the strait less than 3 months.
Pacific herring *Clupea pallasi*	Historical high levels of abundance but individual size declining. Recent catches in the last 5 years averaged 16,000t with a stock size of 80,000 t.	Largest commercial fishery for roe of all British Columbia herring stocks.
Pacific hake *Merluccius productus*	Historical high levels of abundance and largest resident biomass in the Strait of Georgia. May be over 1 billion fish or 75,000 t.	Small commercial fishery during 1980s; averaging 8,000 t during 1990s.
Walleye pollock *Theragra chalcogramma*	Probably at high levels of abundance, possibly 15,000 t or 65 million fish.	Small commercial fishery (less than 2,000 t).
Lingcod *Ophiodon elongatus*	Very low abundance.	A restricted recreational fishery and the commercial fishery closed.
Pacific cod *Gadus macrocephalus*	Very low abundance.	Commercial fishery closed.
Inshore rockfishes *Sebastes* sp.	This group of rockfish consists of about 6 species of importance for the fisheries. The combined abundance is low and declining.	Valuable commercial and recreational species.
Spiny dogfish *Squalus acanthias*	Probably at average levels of abundance of 60,000 t or 35 million fish.	Small commercial fishery.
English sole *Pleuronectes vetulus*	Very low abundance relative to pre-1980s abundances.	Not actively managed.

from late February to late March. Most spawn for the first time at an age of 3 years and live for up to approximately 15-20 years. Hake in the Strait of Georgia are resident and feed mainly on euphausiids. In April, they also can be a major predator of herring (McFarlane and Beamish 1985); however, there is no evidence that they are an important predator of salmon. The increase in abundance in the 1990s has been associated with a large decline in individual length and a continuous series of strong year classes.

Associated with Pacific hake populations are walleye pollock (Table 2). They are considerably less abundant than Pacific hake but feed on similar organisms. The Strait of Georgia stock is at the southern limit of pollock distribution in the subarctic Pacific. Abundance was relatively stable during the early 1980s and declined during the late 1980s, but has increased since 1994 (Saunders and Andrews 1997). There currently is a small fishery at approximately 1,000 t.

Pacific herring stocks in the Strait of Georgia are currently at or near historical high levels of abundance (Schweigert et al. 1998). Most Pacific herring spawn in the strait in late winter. The juveniles remain for about 1 year and migrate out of the strait to the west coast of Vancouver Island. Thus, relatively large numbers of juveniles feed in the Strait of Georgia in their first year of life. The number that leaves the Strait has been speculated to be about one-half of the year class. These fish return to spawn in the Strait at age 3, but there is considerable doubt about the location of the migrants when they are not in the Strait. The current estimated abundance of mature fish of approximately 75,000 t (at 100 g/fish) is equivalent to about 750 million adult fish. However, only approximately half of the fish remain and feed over the summer.

Lingcod and Pacific cod abundances are extremely low relative to past abundances. Both species feed primarily on other fishes, although the juveniles feed on plankton in the first few months after hatching. Rockfish species are aggregated into groupings labeled "inshore rockfish" (about six species). These fish tend to be slow-growing and long-lived. In general, they are in low abundance, relative to previous abundances. They are important for the recreational and commercial fisheries, but in recent years the total catch of "inshore rockfish" species was less than 500 t. Spiny dogfish are in large abundance and are predators of a number of other fishes (Beamish et al. 1992). Despite their long life and predatory habit, they are extremely slow-growing and do not consume large numbers of other fish. They also tend to feed mainly on plankton for the first 10-20 years of life. They are an important component of the ecosystem, but their impact on the dynamics of other species remains uncertain.

Historically, the flatfish fishery was a major groundfish fishery in the Strait of Georgia (Ketchen et al. 1983). Although several species were harvested, English sole was the main species in the catch. Catches declined abruptly in 1981 and have remained at low, but stable levels ever since (Fargo 1994). There is no regulation of catch levels, but there are restrictions on fishing locations and fishing effort.

Pacific salmon are the group of fishes of most interest to managers and biologists. Six species of salmon, including steelhead (*Oncorhynchus mykiss*), enter the Strait of Georgia. Historically, only coho and chinook salmon remained in the Strait over the winter and only a small percentage of these survived the first winter (Healey 1980, Sandercock 1991). The movement into salt water and movement out of the Strait varies among species, with sockeye salmon spending the least amount of time (Groot and Cooke 1987). Chum salmon now remain in the Strait longer than in the past (Beamish and Folkes 1998), and probably in greater numbers. The greater abundance and longer resident time may be a result of releases from hatcheries in Canada and the United States. Despite the low abundance of coho and chinook salmon older than ocean age 0, there may be more juveniles entering the Strait now than in the past as a result of releases from hatcheries (Beamish et al. 1995a,b).

It is important to have some approximation of the abundance of juvenile salmon (ocean age 0) of all species that feed in the Strait of Georgia to compare to abundances of other species. A difficulty with producing this estimate is the very high early marine mortality and the tendency for juveniles to migrate quickly out the Strait (Healey 1980). We used the procedures described in Beamish et al. (1998b) to produce abundance estimates of the various species. The swept volume estimates are believed to be minimal, but they provide a general indication of abundance that can be used to compare to the estimated abundances of other species during the summer feeding period. We estimated the abundance of coho, chinook, and chum salmon in September of 1997 and pink and sockeye salmon in June and July of 1998. Recognizing that most sockeye salmon and possibly a large proportion of pink salmon may have left the strait by June and July (Groot and Cooke 1987) these estimates will be low. In general, the total abundances of all five species of Pacific salmon range around 10 million individuals (Table 3) and possibly as high as 20 million. Abundances decline substantially by late fall (Beamish et al. 1999a), but the abundances in Table 3, probably approximate juvenile salmon abundance during the summer. The important observation is that the abundances of juvenile salmon are substantially smaller than the abundance of hake and herring.

Plankton and Larval Feeding

We collected fish larvae in 50 m or 200 m oblique plankton tows in the central portion of the Strait of Georgia, using 56 cm diameter bongo nets, with a 250 μ nitex mesh towed at 1 m/s while the vessel speed was 2 knots. Each sample was preserved in 5% formaldehyde solution and sorted in the laboratory at a later date. Volume of water strained was determined from a flow meter within the net. The complete sample was examined for fish larvae. In this paper, we report the gut contents of the first 30 larvae that had stomach contents. It was necessary to open the stomachs of hake

Table 3. Approximate abundance of juvenile Pacific salmon in the Strait of Georgia in the summer.

Species	Sampling period	Habitat depth (m)	Abundance	Error limits
Coho	Sep 1997	0-45	2,762,000	1,451,000-4,073,000
Chinook	Sep 1997	0-45	3,097,000	2,379,000-3,816,000
Chum	Sep 1997	0-15	3,635,000	1,134,000-6,137,000
Pink	Jun-Jul 1998	0-15	3,434,000	1,836,000-5,032,000
Sockeye	Jun-Jul 1998	0-15	684,000	173,000-1,120,000

to determine if there were contents even though the body wall appeared to be transparent. Calanoid copepods found in the gut were generally transparent and fragmented, making the identification difficult. Often, prosome length and width were the only parts measured. Whole copepods were counted and identified when possible, but identifications were difficult because some parts were missing. We confirmed our identifications by exchanging samples with another laboratory.

In 1997 and in 1998, virtually all larval hake had absorbed their yolk sacs by April 1. Over 90% of the gut contents of hake and herring in both years were copepods including copepod eggs (Tables 4 and 5). In 1997, the dominant food items were juvenile and adult *Pseudocalanus minutus* (Table 4). In 1998, there were more *Neocalanus plumchrus* (about 70%) than *P. minutus*, and all *N. plumchrus* were juveniles (Table 5). Many of the *P. minutus* had well-developed genital segments averaging approximately 1.3 mm in total length. Juvenile stages (C3, C4, and C5) were also present in the guts. It is possible that in 1997, our ability to detect larvae with stomach contents was not as reliable as in 1998. The results, therefore, show the contents and are not used to estimate the percentage that were feeding. We recorded the mean size of the larval fish to confirm that these specimens were at the first feeding stage.

Plankton samples were collected either separately or in association with the larval fish diet study using the nets and procedures described earlier. Plankton analysis was carried out by subsampling catches and estimating the numbers of organisms per 100 m^3 filtered, using the flow meter in the plankton net to estimate the total volume filtered. In 1996 and 1997, the abundance of *P. minutus* in the surface waters increased about 10 days before the maximum abundance of *N. plumchrus* (Fig. 5). For both these species, the increase in abundance was greatest at the end of March and the first week of April. Adult *P. minutus* were present in the first samples collected in February with maximum abundance from mid-March to early April. *N. plumchrus* abundance maximums followed *P. minutus* but *N. plumchrus* reached much higher densities. After June, adult

Table 4. Contents of stomachs of larval Pacific hake and Pacific herring at first feeding, April 1997.

Date	Species	Total	Number examined	Number of stomachs	Average length (mm)	*Neocalanus plumchrus*	*Pseudo-calanus minutus*	Copepod eggs	Copepod parts or other species	Other items (non-copepods)
Apr 8	Hake	13	13	3	4.2	0	0	1	2	0
	Herring	74	74	37	8.5	0	0	5	≤50	0
Apr 26	Hake	164	30	5	4.8	0	0	5	0	0
	Herring	8	4	2	17.9	0	1	8	2	2
Apr 26	Hake	264	40	14	5.1	0	7	50	10	0
Apr 26[a]	Hake	253	48	26	4.7	0	5	58	15	1

[a] 200 m tow.

Table 5. Contents of stomachs of larval Pacific hake and herring at first feeding, April 1998.

Date	Species	Total	Number examined	Number of stomachs	Average length (mm)	*Neocalanus plumchrus*	*Pseudo-calanus minutus*	Copepod eggs	Copepod parts or other species	Other items (non-copepods)
Apr 6	Hake	154	37	30	4.4	9	1	0	91	1
	Herring	NC	79	30	11.3	0	0	36	33	8
Apr 6	Hake	566	27	27	4.4	4	1	0	37	0
	Herring	NC	489	30	10.8	2	0	28	8	2
Apr 23	Hake	329	123	30	4.1	10	7	2	19	0
	Herring	NC	169	30	12.4	0	1	87	2	0
Apr 23	Hake	271	89	30	4.3	0	2	0	30	0
	Herring	NC	146	16	12.8	1	1	21	13	0

NC = Not counted.

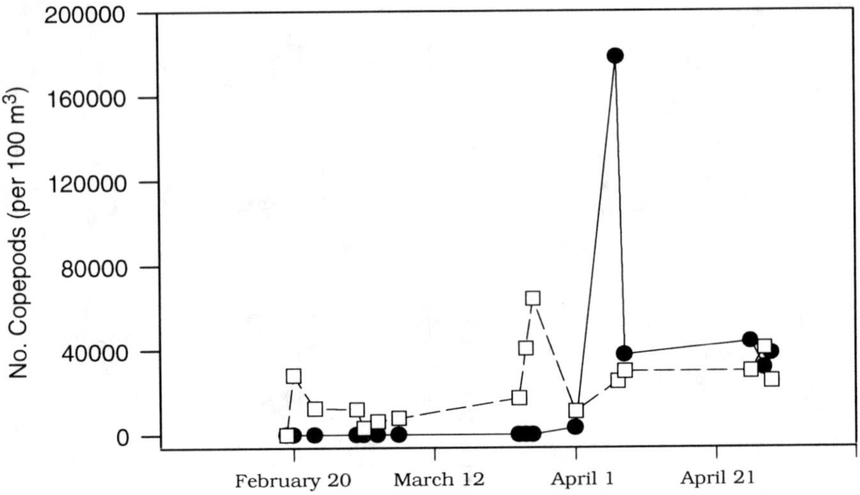

Figure 5. The average number of N. plumchrus (solid circles) and P. minutus (open squares) in the top 50 m of the Strait of Georgia, from eight sampling sites throughout the strait. The increases in abundance occur in late March and early April, with P. minutus increases preceding N. plumchrus.

N. plumchrus were not found in the surface waters, but there was a second peak in the abundance of *P. minutus* in the southern Strait in June and in the north in August. We show average densities for the Strait in Fig. 5. In 1997 *P. minutus* density was higher than *N. plumchrus* in the north and the reverse was true in the south. The copepods *Calanus pacificus* and *Metridia pacifica* were also present and in some sampling locations their abundance in 1997 was similar to *P. minutus* and *N. plumchrus.* In 1996, small numbers of *M. pacifica* were found in the samples.

Bornhold et al. (1998) studied the juvenile development of *N. plumchrus* in the Strait of Georgia. As the naupliar stages migrate to the surface they develop through to the first copepodite stage about the time they reach the surface. The copepodite stage C5 is the stage of maximum biomass and the stage that signals a return to deeper waters. In addition to sampling the surface 50 m for plankton composition in the mid-1990s, Bornhold et al. (1998) reexamined data from three other plankton studies, dating from 1967. They observed that the date that 50% of the *N. plumchrus* were at C5 stage in the surface 50 m changed from about May 17 in 1967 to April 21 in 1997. From 1981 to 1997, the change was from about April 26 to April 21, or less than 1 week. Overall, they identified a change in timing of 25-30 days in 30 years, with the greatest change occurring in the first 15 years between 1967 and 1981.

We noted that in our study, *P. minutus* was an important food item for larval hake and herring in both 1997 and 1998. *N. plumchrus* was an important item in 1998, but less common in our identifications in 1997. However, as mentioned, the representativeness and accuracy of the 1997 analyses may not be equivalent to the 1998 analyses. It is clear that *P. minutus* is an important component of the diet, and its earlier appearance than *N. plumchrus* in the plankton samples may indicate that it also is more abundant in the surface waters earlier in the year.

Spawning and Abundance Changes for Hake and Herring

Hake biomass remained stable at approximately 60-70 kt during the 1980s and the 1990s. From the late 1970s through the 1980s the average size was 42 cm for males and 44 cm for females age 4 and older. Since 1990, the numbers of hake increased as a consequence of five strong year classes. Associated with the increase in numbers has been a decrease in length to 36 and 37 cm (age 4 and older) for males and females, respectively. Therefore, although biomass has remained relatively stable, numbers have increased threefold. Pacific hake spawn late February to early April with peak spawning occurring in the midwater depths exceeding 200 m in mid- to late March. Large midwater spawning concentrations are found in the south-central Strait of Georgia. Eggs hatch in shallower water, often in about 5-6 days. Larvae begin feeding 4-6 days after hatching. Extensive studies of the timing of hake spawning have not been conducted, thus we do not know if the mean spawning date has changed.

Herring spawn in tidal and subtidal waters around the Strait of Georgia. Eggs hatch within 8 days and larvae start to feed within 4 days of hatching. Herring spawn throughout the Strait of Georgia from mid-February to late March (Hay and McCarter 1998). The duration of spawning has decreased since the mid-1980s, but the date of maximum spawning has not changed from mid-March (Hay and McCarter 1998). Herring abundance has approached historical high levels since 1990 and is estimated to be 75 kt. This high abundance has been maintained throughout the 1990s by above-average year classes in 1990, 1991, 1992, 1995, and 1996 (Schweigert et al. 1998).

Natural Regulation of the Strait of Georgia Ecosystem: A Hypothesis for Ecosystem Management

The details of the mechanisms that affect the dynamics of the Strait of Georgia ecosystem are obviously complex. However, we propose that the key process that regulates fish abundance is related to the factors that regulate the timing of the vertical migrations and abundance of copepods.

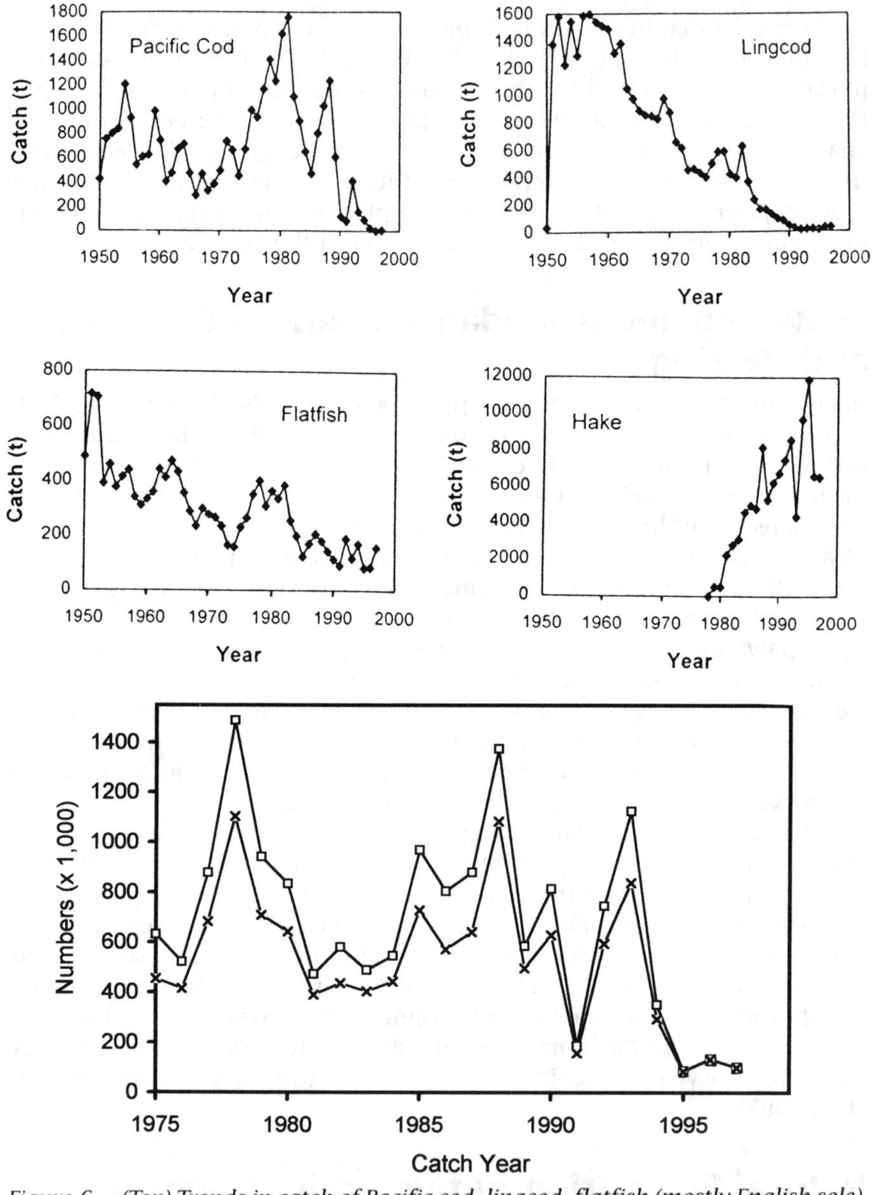

Figure 6. *(Top) Trends in catch of Pacific cod, lingcod, flatfish (mostly English sole), and Pacific hake in the Strait of Georgia. (Bottom) Catches of coho salmon in the Strait of Georgia from the recreational fishery (Xs) and the commercial and recreational fishery combined (open squares). Actual abundances were not determined; however, fishing effort was virtually unrestricted over this period.*

According to our hypothesis, copepods are the component of the zooplankton community that link the fundamental processes of microscopic plant production through the association of light and nutrients with the survival of larval fish. It is no coincidence that most fishes in the North Pacific reproduce in the winter immediately before copepods move from the diapause state in deeper water (Conover 1988) into the surface waters, which ensures that eggs hatch about the time copepods are available for food.

The availability of copepods immediately after yolk-sac resorption by larval hake and herring is the fundamental mechanism that we propose is the lever that shifts the dynamics of this ecosystem. The dynamics of the fish community in the Strait of Georgia respond to the timing shifts in copepod vertical migration through the abundance shifts of Pacific hake and herring, as these are the dominant fish species in the Strait.

Throughout the North Pacific, including the Strait of Georgia, the calanoid copepods are dominant within the zooplankton biomass (Mackas and Tsuda 1999). The life history of copepods places them in the key position in the Strait of Georgia ecosystem. Copepods that dominate the zooplankton biomass have a seasonal migration from a winter residence in deeper water to the surface water in the spring. The juvenile copepods complete their growth in the surface waters before returning to deep water in the late spring and summer (Miller et al. 1984). In the Strait of Georgia, most *N. plumchrus* reproduce in these deepwater areas in January and February (Fulton 1973). As the fecundity appears to be around 1,000 eggs (Mackas and Tsuda 1999), it can be argued that their abundance is regulated more by the habitat conditions than number of offspring as is common with plants and animals that produce large numbers of seeds and babies (Colinveaux 1978).

We propose that the connection to the decadal shifts in climate and the changes in the abundance trends of fishes in the Strait of Georgia is through changes in the vertical migration of copepods into the surface. Changes in the timing have been shown for the North Pacific (Mackas et al. 1998, Mackas and Tsuda 1999) and as previously mentioned for the Strait of Georgia (Bornhold et al. 1998). Off the west coast of British Columbia, the movement into and out of the surface waters by *N. plumchrus* has advanced by 50-60 days (Mackas et al. 1998) and in the Strait of Georgia by 25-30 days (Bornhold et al. 1998). Thus, the timing of copepod migration becomes an important reference point for the management of fishes as it provides information about the relative success of hake and herring spawning.

Ecosystem Management Charts

If the dynamics of the Strait of Georgia ecosystem is regulated as we propose, it is possible to identify the key processes that regulate abundance in a simplified chart. We call these ecosystem management charts. These charts are developmental, as the relationships are both poorly studied and poorly understood. However, the charts identify the associations of the dynamics of fish populations with environmental factors and other fishes. The charts are intended to provide a broader scope for management by specifying the possible causes of abundance changes and helping to separate fishing impacts from natural changes. The charts are also useful for research planning as they highlight important relationships that need clarification. We suggest that these charts facilitate the development of ecosystem management. Initially, the chart may greatly oversimplify relationships, but the long-term objective should be to maintain simplicity. The difference between the initial and more advanced charts may not be the number of associations, rather a better appreciation of the importance of the key relationships. In this first attempt we identify the key linkages for three species in the Strait of Georgia.

Coho Salmon Ecosystem Management Chart

Coho salmon abundance is regulated naturally by the carrying capacity in the ocean and by the number of juveniles (smolts) produced in fresh water. The number of smolts produced is a function of freshwater habitat conditions and the number of spawning adults. The number of spawning adults is related to the amount of fishing mortality and to the marine survival. Currently, in the Strait of Georgia, about 77% of the population is from hatchery-reared smolts and fry (Beamish et al. 1998a). Thus, the total abundance is only partly affected by fishing effects on wild stocks. Marine survival is a function of early predation-based mortality and growth-related mortality that is proposed to occur in the fall and winter (Beamish and Mahnken 1998). According to our proposed relationship, stock and subsequent recruitment of wild stocks is linear for small stock sizes but recruitment is not a function of stock size at larger abundances. At these larger abundances stock size is regulated by the ocean environment.

The management chart for coho salmon (Fig. 7) identifies only the principal mechanisms currently regulating coho salmon abundance. According to the chart, coho salmon abundance is regulated primarily through the timing of copepod production, fishing effects, and the availability of suitable freshwater habitat for wild coho salmon. Because the wild spawning biomass or escapement is supplemented by large additions of hatchery fish, the interannual changes in total abundance are currently regulated by the ocean environment and not by fishing. Favorability of the ecosystem for coho salmon will be reflected in the "condition" of coho salmon in the fall which is a reflection of availability of food in summer according to the critical size and critical period hypothesis of Beamish and Mahnken (1998). Persistent poor condition would indicate poor overwinter survival. The regime response which determines the mean carrying capacity of the system for coho salmon is reflected in the timing of copepod production

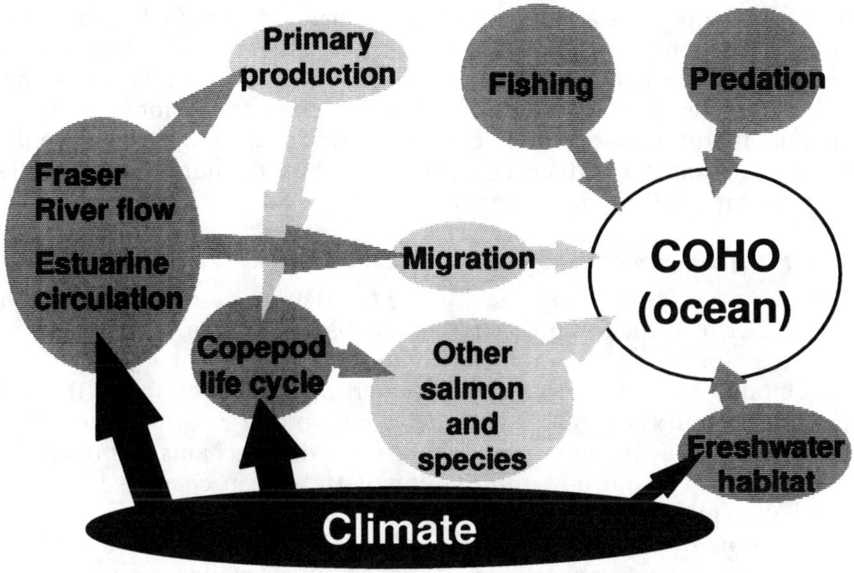

Figure 7. Coho salmon ecosystem management chart. The chart identifies the relationships that may be most influential in the regulation of abundance. The three key areas are fishing, climate-ocean environment, and associated species. Climate, as a regime, affects Fraser River flows, which may affect primary productivity in the Strait of Georgia. Climate also affects the freshwater habitat, but the principal impact of climate is on the timing of copepod migration into the surface waters. The timing of this migration affects the survival of key competitors for food of coho salmon durinq their summer feedinq period. Growth during the summer affects survival in the winter and fall and eventually determines brood-year strength. Predation is important but adult abundance is determined through growth. Fishing is important if removals restrict the ability of a stock to replenish itself. A manager needs to know if the regime is productive or less productive for coho salmon and this would be determined using climate indices and timing of copepod migration. In favorable regimes, more smolts would be required to achieve the maximum productivity. In less favorable regimes, the manager needs to be sensitive to the impacts of associated species and other coho salmon. In less favorable regimes, adding more coho salmon smolts or other potential competitors may also increase competition for food resources, which are less available.

as this will affect the abundance of Pacific hake and Pacific herring which compete for food with juvenile coho salmon. A suggested ecosystem management approach by managers would be to determine the relative strength of hake and herring year classes which affect competition for food of coho salmon, monitor the condition of coho salmon in the fall, and ensure that fishing and hatchery effects combine to produce the number of juveniles appropriate for the current regime.

Herring Ecosystem Management Chart

Herring in the Strait of Georgia migrate to offshore waters after spending about 1 year in the Strait (Hourston and Haegele 1980). There are some resident stocks and the timing and amount of movement apparently varies among years (Hourston 1982). As herring mature offshore, they migrate back into the Strait of Georgia to spawn and return offshore after spawning. It may be important to determine the mechanisms involved in order to understand how future climate and ocean changes will affect stock dynamics. It is clear that this migration complicates any interpretation of how the dynamics of the Strait of Georgia affect the abundance trends of herring. Our ecosystem management chart incorporates the offshore residence as a general predation component.

We propose that the year-class strength is determined initially by the availability of copepods for the larval herring (Fig. 8). It is also possible that salinity and temperature may affect survival as spawning and hatching occur in the shallow, intertidal areas. However, at this time we are not considering egg mortality to be a major source of interannual larval mortality. There is no question that larval mortality is a function of predation, but again, we propose that it is copepod availability that is the key factor regulating the initial year-class strength. We consider that predation in the first marine year becomes important during the summer and fall and during the offshore residence. It is well known that herring are a principal food item of a number of species (Hourston and Haegele 1980). It is less well known that it has not been possible to relate surveys of age-0 herring with subsequent recruitment, possibly indicating that predation mortality is also an important factor regulating the abundance of adults. Fishing impacts directly affect the abundance of spawning adults, but we do not consider that fishing currently affects recruitment as long as a minimal or "cut-off" biomass is maintained.

Ecosystem management of herring would require knowledge of the vertical migration timing of two copepods: *N. plumchrus* and *P. minutus*. A general classification such as early, average, or late may be appropriate in the initial stages of ecosystem management. Late copepod migration would indicate the possibility of poor larval herring survival and an early migration to the surface would indicate that larval survival will be high.

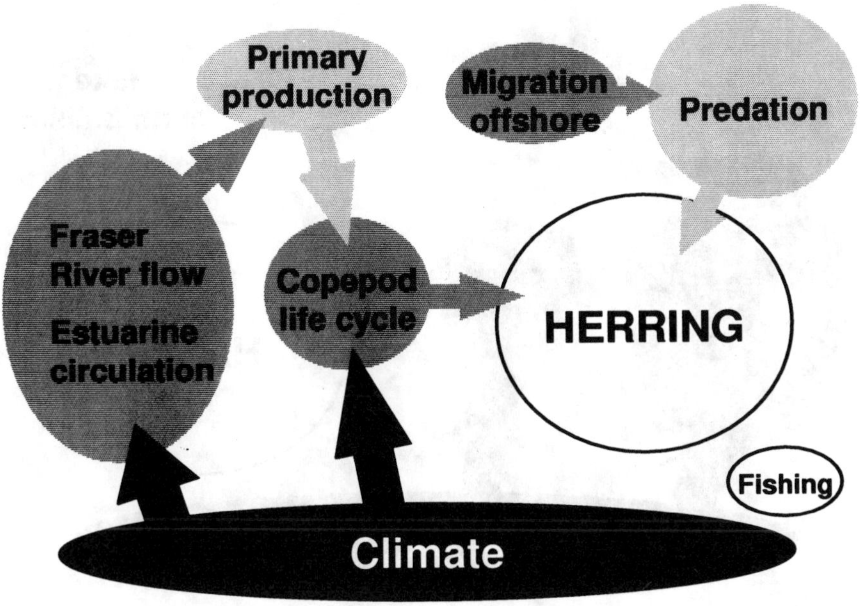

Figure 8. *Herring ecosystem management chart. The chart identifies the key relationships that currently affect the abundance of herring in the Strait of Georgia. Climate will affect Fraser River flows and possibly the level of primary production in the strait, but the main impact of climate is the timing of copepod migration into the surface waters. Earlier movement into the surface waters makes food more available for larval herring at the time of first feeding. Herring abundance is regulated naturally through predation on all age classes, but predation does not restrict the ability of the population to replenish itself. The migration of herring out of the Strait of Georgia increases the predation impact. The current level of fishing would not affect abundance levels as indicated by the increasing abundance trends. Managers would need to know the favorability of the regime as indicated by the timing of copepod migration and any changes to a regime as indicated by the climate indices. Some form of abundance survey would identify trends, but declines could be for natural reasons if there was a change from a favorable to a less favorable regime.*

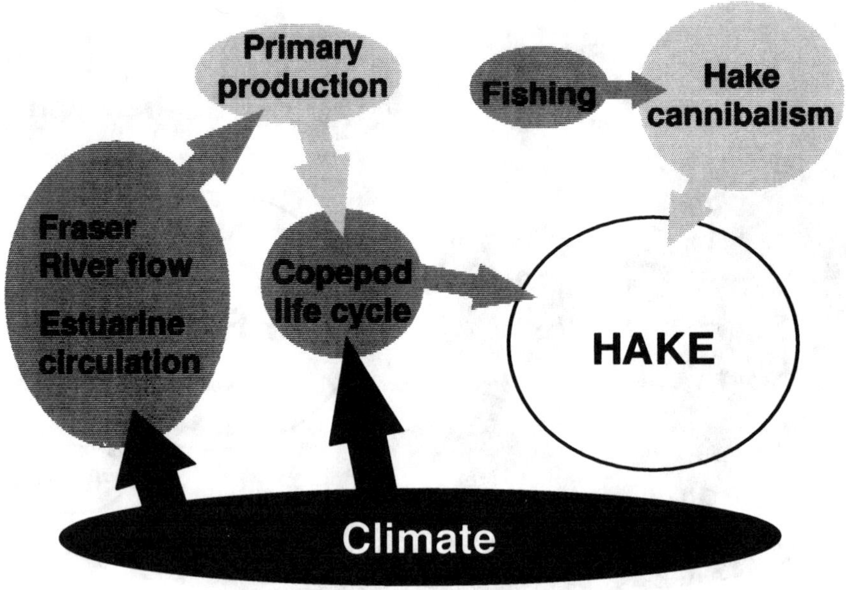

Figure 9. Hake ecosystem management chart. This is the simplest of the three charts. Hake abundance is regulated by timing of copepod migration, similar to herring. As earlier migration is more favorable, more food is available at the time of first feeding of larval hake. The large abundance and high fecundity of hake ensures that a very large number of eggs are produced. Thus abundance is closely related to larval survival. Hake abundance is regulated naturally through cannibalism. Currently, fishing levels have an impact on the number of adults that consume juvenile hake. Increasing fishing effort would probably result in a higher survival of juveniles if the current regime persists.

Hake Ecosystem Management Chart

Hake, as the dominant species in the Strait of Georgia, has the simplest ecosystem management chart. Hake abundance is entirely dependent on the timing of the copepod reproduction cycle and cannibalism (Fig. 9). Early migration of copepods to the surface favors improved hake larval survival resulting in a strong year class. The natural regulation of hake abundance is proposed to be through cannibalism. Fishing at the current levels, in the current regime, would have little impact on the abundance of hake. Consequently, increasing fishing in an attempt to reduce competition for a species such as coho salmon would have no effect unless high amounts of hake were removed. However, because the natural regulation of hake abundance is through cannibalism the removal of large numbers of adults may reduce cannibalism resulting in even larger adult biomass a few years later and even greater competition.

Conclusion

This study combines a basic understanding of the oceanography, productivity process, and life history strategies of key fishes in the Strait of Georgia to develop an approach to ecosystem management. The oceanography of the Strait of Georgia is affected by Fraser River flows which are strongly affected by decadal-scale trends that we call regimes. Within these regimes there is considerable interannual variability (Moore and McKendry 1996, Mantua et al. 1997). The Strait of Georgia is also affected directly by wind patterns and intensities. All of these influences alter the ecosystem dynamics but according to our proposal, the principal factor regulating the dynamics of fishes is the timing of the copepod life cycle. Copepods migrating to the surface in the spring provide food for first-feeding larval fish and have immediate impacts on their marine survival. The early, average, or late migration of copepods to the surface alters the relative abundances of hake and herring and their abundances affect the survival of the resident fish species through the changes in availability of food. Changes in the abundance of the fishes reported in the paper have in the past been thought to be the direct result of fishing effects. Although excessive fishing pressures would be expected to reduce abundances, the regime concept implies that trends in abundances will change for natural reasons. Thus, the recorded changes in abundance of all species in the Strait of Georgia may not simply be associated with fishing. The fish abundances at any particular time may be adjusting to new states, as there is inertia within the population dynamics that affects the rate of change. Thus, changes in fish abundance may appear to be gradual. Fishing will have direct impacts on the ability of a stock to replenish itself if the abundance is lower than a minimal level. Abundances higher than this level are mostly regulated by environmental factors. This minimum level needs to be determined for each regime and for each species. The concept of regimes means that the dynamics of the ecosystem can change. By identifying when large-scale climate shifts occur (Beamish et al. 1998d), it is possible to warn managers to reassess the dynamics of particular species. Regime shifts are not cycles or oscillations, but changes in state. The response to changes may have to be measured, rather than forecasted. Modeling attempts are useful as long as it is recognized that key components may be missing.

In the Strait of Georgia there were changes in the pattern of Fraser River flow that would affect the dynamics of the ecosystem in 1977, 1989, and possibly in 1996. It is difficult to study the changes earlier than the mid-1970s, but the trend of Fraser River flow also changed in the late 1940s, at the same time that indices of global climate changed (Mantua et al. 1997, Minobe 1997, Beamish et al. 1999b). The importance of the regime concept is the requirement to check ecosystems after a regime change to ensure that past relationships remain unchanged. For example, since 1989, there has been a trend to earlier Fraser River flows and higher sea levels (Beamish et al. 1999a) that have had profound effects on the behavior of coho salmon in the Strait of Georgia.

We propose that the key to starting ecosystem management is the ecosystem management chart. These charts identify the major mechanisms that affect the abundance of a particular species. The charts are clearly developmental, but they do show that changes in abundance can be a function of several variables, all of which must be considered when management is proposed. The charts in association with models such as Ecopath 3.0 (Christensen and Pauly 1996) also identify the key research issues required to improve management. Having discussions about the relative importance of the relationships in the charts is an important part of the process. The charts included in this paper may ultimately be shown to require major changes. However, it is the process of identifying the key factors affecting species abundance that is important in the initial stages of ecosystem management. It may be as challenging to find methods to communicate the information among experts as it is to acquire the relevant information. The charts are intended to provide a focus for the process of ecosystem management. They retain the identity of single species, but they attempt to incorporate essential impacts of the environment and associated species.

Acknowledgments

We very much appreciate the assistance of a number of colleagues over the years who have studied the Strait of Georgia ecosystem. The list is too extensive to identify everyone, but we want to mention Ray Scarsbrook, Bill Andrews, Mike Smith, and Chrys-Ellen Neville. Carol Cooper identified plankton and larval stomach contents. The captain, officers, and crew of the *W.E. Ricker* provided excellent support. The project was funded from the high priority project fund of DFO management, Ottawa.

References

Beamish, R.J., and M. Folkes. 1998. Recent changes in the marine distribution of juvenile chum salmon off Canada. In: D.W. Welch, D.M. Eggers, K. Wakabayashi, and V.I. Karpenko (eds.), Assessment and status of Pacific Rim salmonid stocks. Bull. No. 1. North Pacific Anadromous Fish Commission. Vancouver, B.C., pp. 443-453.

Beamish, R.J., and C. Mahnken. 1998. Natural regulation of the abundance of coho and other species of Pacific salmon according to a critical size and critical period hypothesis. North Pacific Anadromous Fish Commission Doc. 319.

Beamish, R.J., G.A. McFarlane, and R.E. Thomson. 1999a. Recent declines in the recreational catch of coho salmon (*Oncorhynchus kisutch*) in the Strait of Georgia are related to climate. Can. J. Fish. Aquat. Sci. 56. In press.

Beamish, R.J., R. Sweeting, and Z. Zhang. 1998a. Estimating the percentage of hatchery-reared juvenile coho salmon in the Strait of Georgia in 1997. PSARC Working Paper S98-8. Cited with permission.

Beamish, R.J., B.L. Thomson, and G.A. McFarlane. 1992. Spiny dogfish predation on chinook and coho salmon and the potential effects on hatchery-produced salmon. Trans. Am. Fish. Soc. 121(4).

Beamish, R.J., C.M. Neville, J. Rice, and Z. Zhang. 1995b. Factors affecting the marine survival of coho salmon in the Strait of Georgia. Pacific Stock Assessment Review Committee (PSARC) Working Paper S95-4. Pacific Biological Station.

Beamish, R.J., J. King, D. Noakes, G.A. McFarlane, and R. Sweeting. 1998b. Evidence of a new starting in 1995 and the relation to Pacific salmon catches. North Pacific Anadromous Fish Commission Doc. 321. 21 p.

Beamish, R.J., B.E. Riddell, C.M. Neville, B.L. Thomson, and Z. Zhang. 1995a. Declines in chinook salmon catches in the Strait of Georgia in relation to shifts in the marine environment. Fish. Oceanogr. 4:243-256.

Beamish, R.J., D. Noakes, G.A. McFarlane, L. Klyashtorin, V.V. Ivanov, and V. Kurashov. 1999b. The regime concept and natural trends in population of Pacific salmon. Can. J. Fish. Aquat. Sci. 56. In press.

Beamish, R.J., D.J. Noakes, G.A. McFarlane, W. Pinnix, R. Sweeting, and J. King. 1998c. Trends in coho marine survival in relation to the regime concept. Fish. Oceanogr. (In review).

Bornhold, E., D. Mackas, and P. Harrison. 1998. Interdecadal variations in developmental timing of the copepod *Neocalanus plumchrus* (Marukawa) in the Strait of Georgia. Eos 79(1). (Abstract.)

Burgner, R.L. 1991. Life history of sockeye salmon (*Oncorhynchus nerka*). In: C. Groot and L. Margolis (eds), Pacific salmon life histories. UBC Press, Vancouver, B.C., pp. 1-117.

Christensen, V., and D. Pauly. 1996. Ecological modelling for all. Naga (ICLARM Quarterly) 19(2).

Colinveaux, P. 1978. Why big fierce animals are rare. Princeton University Press. 256 pp.

Conover, R.J. 1988. Comparative life histories in the general *Calanus* and *Neocalanus* in high latitudes of the Northern Hemisphere. Hydrobiologia 167/168:127-142.

Crean, P.B., T.S, Murray, and J.A. Stronach. 1988. Mathematical modelling of the tides and estuarine circulation: The coastal seas of southern British Columbia and Washington State. In: M.J. Bowman et al. (eds.), Lecture Notes on Coastal and Estuarine Studies 30. Springer Verlag. 471 pp.

Fargo, J. 1994. Flatfish stock assessments for 1993 and recommended yield options for 1994. In: M. Stocker (ed.), Groundfish stock assessments for the west coast of Canada in 1993 and recommended yield options for 1994. Can. Tech. Rep. Fish. Aquat. Sci. 1975:93-120.

Fulton, J. 1973. Some aspects of the life history of *Calanus plumchrus* in the Strait of Georgia. J. Fish. Res. Board Can. 30:811-815.

Groot, C., and K. Cooke. 1987. Are the migrations of juvenile and adult Fraser River sockeye salmon (*Oncorhynchus nerka*) in near-shore waters related? In: H.D.Smith, L. Margolis, and C.C. Wood (eds.), Sockeye salmon (*Oncorhynchus nerka*) population biology and future management. Can. Spec. Publ. Fish. Aquat. Sci. 96:53-60.

Harrison, P.J., D.L. Mackas, B.W. Frost, R.W. Macdonald, and E.A. Crecelius. 1994. An assessment of nutrients, plankton and some pollutants in the water column of Juan de Fuca Strait, Strait of Georgia and Puget Sound, and their transboundary transport. In: R.C.H. Wilson, R.J. Beamish, F. Aitkens, and J. Bell (eds.), Review of the marine environment and biota of Strait of Georgia, Puget Sound and Juan de Fuca Strait. Proceedings of the B.C./Washington Symposium on the Marine Environment, January 13-14, 1994. Can. Tech. Rep. Fish. Aquat. Sci. 1948. 398 pp.

Hay, D.E., and P.B. McCarter. 1998. Distribution and timing of herring spawning in British Columbia. Pacific Stock Assessment Review Committee Working Paper H98-5. Cited with permission.

Healy, M.C. 1980. The ecology of juvenile salmon in Georgia Strait, British Columbia, In: W.S. McNeil and D.C. Himsworth (eds.), Salmonid ecosystems of the North Pacific. Oregon State University Press, Corvallis, pp. 203-229.

Hourston, A.S. 1982. Homing by Canada's west coast herring to management units and divisions as indicated by tag recoveries. Can. J. Fish. Aquat. Sci. 39:1411-1422.

Hourston, A.S., and C.W. Haegele. 1980. Herring on Canada's Pacific coast. Can. Spec. Publ. Fish. Aquat. Sci. 48. 23 pp.

Hutchinson, A.H., and C.C. Lucas. 1931. Epithalassa of the Strait of Georgia. Can. J. Res. 5:231-284.

Ketchen, K.S., N. Bourne, and T.H. Butler. 1983. History and present status of fisheries for marine fishes and invertebrates in the Strait of Georgia, British Columbia. Can. J. Fish. Aquat. Sci. 40:1095-1119.

LeBlond, P.H. 1983. The Strait of Georgia: Functional anatomy of a coastal sea. Can. J. Fish. Aquat. Sci. 40(7):1033-1063.

Li, M., A. Gargett, and K. Denman. 1999. Seasonal and interannual variablilty of estuarine circulation in a box model of the Strait of Georgia and Juan de Fuca Strait. Atmosphere-Ocean. In press.

McFarlane, G.A., and R.J. Beamish. 1985. Biology and fishery of Pacific whiting, *Merluccius productus*, in the Strait of Georgia. Mar. Fish. Rev. 47(2):23-34.

Mackas, D.L., and A. Tsuda. 1999. Mesozooplankton in the eastern and western subarctic Pacific: community structure, seasonal life histories, and interannual variability. Progress in Oceanog. In press.

Mackas, D.L., R. Goldblatt, and A.G. Lewis. 1998. Interdecadal variation in developmental timing of *Neocalanus plumchrus* populations at Ocean Station P in the subarctic North Pacific. Can. J. Fish. Aquat. Sci. 55:1-16.

Mantua, N.J., S.R. Hare, Y. Zhang, J.M. Wallace, and R.C. Francis. 1997. A Pacific interdecadal climate oscillation with impacts on salmon production. Bull. Am. Meteor. Soc. 78:1069-1079.

Miller, C.B., B.W. Frost, H.P. Batchelder, M.J. Clemons, and R.E. Conway. 1984. Life histories of large, grazing copepods in a subarctic ocean gyre: *Neocalanus plumchrus, Neocalanus cristatus*, and *Eucalanus bungii* in the Northeast Pacific. Prog. Oceanogr. 13:201-243.

Minobe, S. 1997. A 50-70 year climatic oscillation over the North Pacific and North America. Geophys. Res. Lett. 24:683-686.

Moore, D.R. 1991. Hydrology and water supply in the Fraser River basin. In: H.J. Dorsey (ed.), Water in sustainable development: Exploring our common future in the Fraser River basin. Westwater Research Centre, University of British Columbia, Vancouver, B.C., pp. 21-40.

Moore, R.D., and I.G. McKendry. 1996. Spring snowpack anomaly patterns and winter climatic variability, British Columbia, Canada. Water Resour. Res. 32:623-632.

Murdock, J. 1979. Control charts. MacMillan Press Ltd., London. 150 pp.

Parsons, T.R., J. Stronach, G.A. Borstad, G. Louttit, and R.I. Perry. 1981. Biological fronts in the Strait of Georgia, British Columbia, and their relation to recent measurements of primary production. Mar. Ecol. Prog. Ser. 6:237-242.

Sandercock, F.K. 1991. Life history of coho salmon (*Oncorhynchus kisutch*). In: C. Groot and L. Margolis (eds.), Pacific salmon life histories. UBC Press, Vancouver, B.C., pp. 394-445.

Saunders, M.W., and W. Andrews. 1997. Walleye pollock stock assessment for 1997 and recommended yield options for 1998. Pacific Stock Assessment Review Committee Working Paper G97-7. Cited with permission.

Saunders, M.W., and G.A. McFarlane. 1997. Pacific hake stock assessment for 1997 and recommended yield options for 1998. Pacific Stock Assessment Review Committee Working Paper G97-6. Cited with permission.

Schweigert, J., C. Fort, and R. Tanasichuk. 1998. Stock assessment for British Columbia herring in 1998 and forecasts of the potential catch in 1999. Pacific Stock Assessment Review Committee Working Paper H98-1. Cited with permission.

St. John, M.A., S.G. Marinone, J. Stronach, P.J. Harrison, J. Fyfe, and R.J. Beamish. 1993. A horizontally resolving physical-biological model of nitrate concentration and primary productivity in the Strait of Georgia. Can. J. Fish. Aquat. Sci. 50:1456-1466.

Thomson, R.E. 1981. Oceanography of the British Columbia coast. Can. Spec. Publ. Fish. Aquat. Sci. 56. 291 pp.

Thomson, R.E. 1994. Physical oceanography of the Strait of Georgia-Puget Sound-Juan de Fuca system. In: R.C.H. Wilson, R.J. Beamish, F. Aitkens, and J. Bell (eds.), Review of the marine environment and biota of Strait of Georgia, Puget Sound, and Juan de Fuca Strait. Proceedings of the B.C./Washington Symposium on the Marine Environment, January 13-14, 1994. Can. Tech. Rep. Aquat. Sci. 1948. 398 pp.

Waldichuk, M. 1957. Physical oceanography of the Strait of Georgia, British Columbia. J. Fish. Res. Board Can. 14:321-486.

Keystone Predators in the Central Pacific

James F. Kitchell
University of Wisconsin, Center for Limnology, Madison, Wisconsin

Christofer H. Boggs
National Marine Fisheries Service, Southwest Fisheries Center, Honolulu, Hawaii

Xi He
CSIRO (Commonwealth Scientific and Industrial Research Organisation) Marine Research, Hobart, Tasmania, Australia

Carl J. Walters
University of British Columbia, Fisheries Centre, Vancouver, British Columbia, Canada

Abstract

We evaluated the potential of keystone predator effects among the guild of sharks, tunas, and billfishes at the apex of pelagic food webs in the central North Pacific ecosystem. The Ecopath approach was used as the basis for a dynamic model (Ecosim) that simulates the ecological effects of environmental change and/or fishery exploitation patterns. No single fish species of the highest trophic levels appears to have a profound and uniquely important role in the organization and structure of the central Pacific ecosystem. Instead: (1) The most important components among the guild of apex predators appear to be those such as yellowfin and skipjack tunas which have greater biomass than other apex predators, highly diverse diets, and rapid turnover rates. (2) Intraguild cannibalism and substantial diet overlap are common in these systems. The apparent complexity and importance of cannibalism invites more extensive data collection. (3) Simulated depletions of shark or tuna populations exhibit ecological changes similar to those observed when management or fishery exploitation effects are intense. (4) The ecological effects of longline fisheries are diffuse and diverse. If we are to affix a keystone predator label in this ecosystem, it is that due to fisheries exploitation.

Introduction

> By removing so many of the sea's keystone predators, we are weakening an entire tier at the top of the food chain. This may have dire biological consequences throughout the ecosystem, far beyond the social, economic and moral costs of depleted fisheries. —*Hinman 1998*

The quote above summarizes a question and a concern expressed in a variety of reports (Apollonio 1994, Botsford et al. 1997). The problem was addressed by Pauly et al. (1998), whose analysis of fishery effects demonstrates a general reduction in the average trophic level for both marine and freshwater fishes harvested over the past several decades. The conceptual framework for this issue derives from keystone predator effects in a rocky intertidal community (Paine 1966), followed by a wide application in community ecology (Power et al. 1996) and extended to the ecosystem level in development of the trophic cascade hypothesis (Carpenter and Kitchell 1993). In other words, there is a growing body of evidence that changes in the tops of food webs are expressed at all trophic levels in a wide variety of aquatic ecosystems.

A recent review (Power et al. 1996) of the keystone predator and keystone species literature sought precise and predictable descriptions of keystone effects but concluded that a unifying ecological principle was not apparent and more research was necessary. However, that review did produce the general, operational definiton of a keystone predator as: "one whose impact on its community or ecosystem is large, and disproportionately large relative to its abundance." That concept and definition will suffice for the present.

Our study focused on the ecological role of apex predators in food webs of the Central North Pacific. As suggested by Hinman (1998), we wondered if one or more of those fishes served as a keystone predator. Unlike Atlantic stocks, central Pacific tuna and billfish stocks do not yet appear to be overfished (Stroud 1989, Shomura et al. 1994). Ecological effects stemming from the development of these fisheries have not been evaluated.

In this paper we develop an Ecopath model for the food web characteristic of the central Pacific pelagic system. Harvests in the model are those of the Asian and U.S. longline, purse seine, pole and line, and drift gill-net fisheries circa 1988 which include an assemblage of sharks, billfishes, tunas, mahimahi, and flying squid (Ueyanagi et al. 1989, Boggs and Ito 1993). The Ecopath model is employed in Ecosim to evaluate the relative importance or keystone predator role for each of a suite of predator species sited at the apex of this food web.

Methods

Model Structure and Inputs

Our analyses are based on the Ecopath model initially developed by Polovina (1984) and subsequently modified (Christensen and Pauly 1992) in applications to scores of ecosystems (Christensen and Pauly 1993). At this writing, Ecopath models have been developed for approximately 90 different systems (Pers. Comm., D. Pauly, Fisheries Centre, University of British Columbia, Vancouver, B.C.). Recent applications include analyses of the linkage between primary production and fish production for the world's oceans (Pauly and Christensen 1995) and estimates of changes in trophic structure due to fishery exploitation in both marine and freshwater systems (Pauly et al. 1998).

Derivation and application of the Ecopath model are detailed in Pauly et al. (1993) and will be only briefly summarized herein. The pertinent software and documentation are available without cost at the University of British Columbia Fisheries Centre website (http://www.fisheries.com). The core component of an Ecopath model is the balance of production and losses for each component of an ecosystem and expressed in the equation:

$$\text{Production} + \text{Immigration} - \text{Predatory losses} - \text{Non-predatory mortality} - \text{Harvest} - \text{Emigration} = 0 \quad (1)$$

In other words, the system is set to steady state based on estimates of rates for each of the terms in Equation 1. This equilibrium condition allows analyses of flows between trophic levels and a means for estimating the relative importance of individual state variables (e.g., species) or processes (e.g., detritivory, selective predation). An important element of this process is that the ratio of production to biomass (P/B) is equal to the sum of all sources of mortality: those due to predation, exploitation, and other causes. Thus, in Ecopath, the system is set to equilibrium based on the levels of mortality, biomass, and production at one point in time.

Walters et al. (1997) developed the Ecosim model as a way to simulate dynamic effects of changes away from the initial steady-state condition defined by an Ecopath model. In Ecosim, any change in exploitation rates is reflected in development of a new steady-state condition. User inputs can alter the initial conditions for exploitation rates, predator-prey interactions, and/or external changes in the environment as applied to any of the state variables in the Ecopath model. Ecosim then calculates the time course for consequent change in all other state variables and the new equilibrium condition. Derivation of the Ecosim model is detailed in Walters et al. (1997). The software is also available at http://www.fisheries.com, including a version with population size structure and density-dependent or compensatory responses in growth rates and a new model (Ecospace) that sets the Ecosim dynamics in a spatial context.

In its simplest configuration, Ecosim represents dynamic change for each biomass state variable (B) with a differential equation of the form:

$$dB/dt = \text{Food Consumption} - \text{Predatory losses} - \text{Non-predatory losses} - \text{Harvest} \qquad (2)$$

The consumption and loss rates are predicted over time as functions of prey and predator biomasses, using functions that explicitly account for "bottom-up" versus "top-down" control effects by representing prey as exchanging between behaviorally unavailable versus available states. Low exchange rates into predator-vulnerable states cause the model to exhibit bottom-up trophic control structure, while high exchange rates result in classic predator-prey control patterns (e.g., trophic cascades).

Because all Ecopath variables are embedded in a food web, effects of changes in one may be expressed in any and all others. Their dynamics depend on the linkages stated in the trophic flow matrix and in the parameters used to define each. For example, organisms with large P/B values (e.g., phytoplankton or zooplankton) respond rapidly to perturbation and achieve new steady-state values quickly. Those with low P/B values (e.g., sharks), which are indicative of K-strategy life histories or low turnover rates, respond slowly and with substantial lags. Both Ecopath and Ecosim include a diversity of options and alternative functions available to the user. In both cases, we employed the initial or default conditions provided with the software.

The trophic structure developed for the central Pacific model is presented in Fig. 1. Resolution is high at the top of the food web but aggregation is required for lower trophic levels where hundreds of species may be represented by one state variable (e.g., phytoplankton). We distinguish between adult and juvenile stages for all apex predators because intraguild cannibalism is a common feature of diets for these fishes. Separation of adult and juvenile stages was based on sizes at maturity or sizes at which each of the adult stages becomes vulnerable to longline gear. We scaled all state variables in biomass units per area and bounded the system as that represented for epipelagic waters from latitudes of 0°N to 40°N and longitudes of 140°E to 140°W.

The basic inputs for an Ecopath include estimates of biomass, P/B, total annual consumption/biomass (Q/B), fishery harvests and other exports from the system, and ecotrophic efficiency (EE) which is defined as the proportion of the total annual production that is harvested, exported, or preyed upon. If one of those is unknown, the Ecopath software balances the ecosystem budget and provides an estimate for that value. For this analysis, we used biomass as the unknown and estimated values for

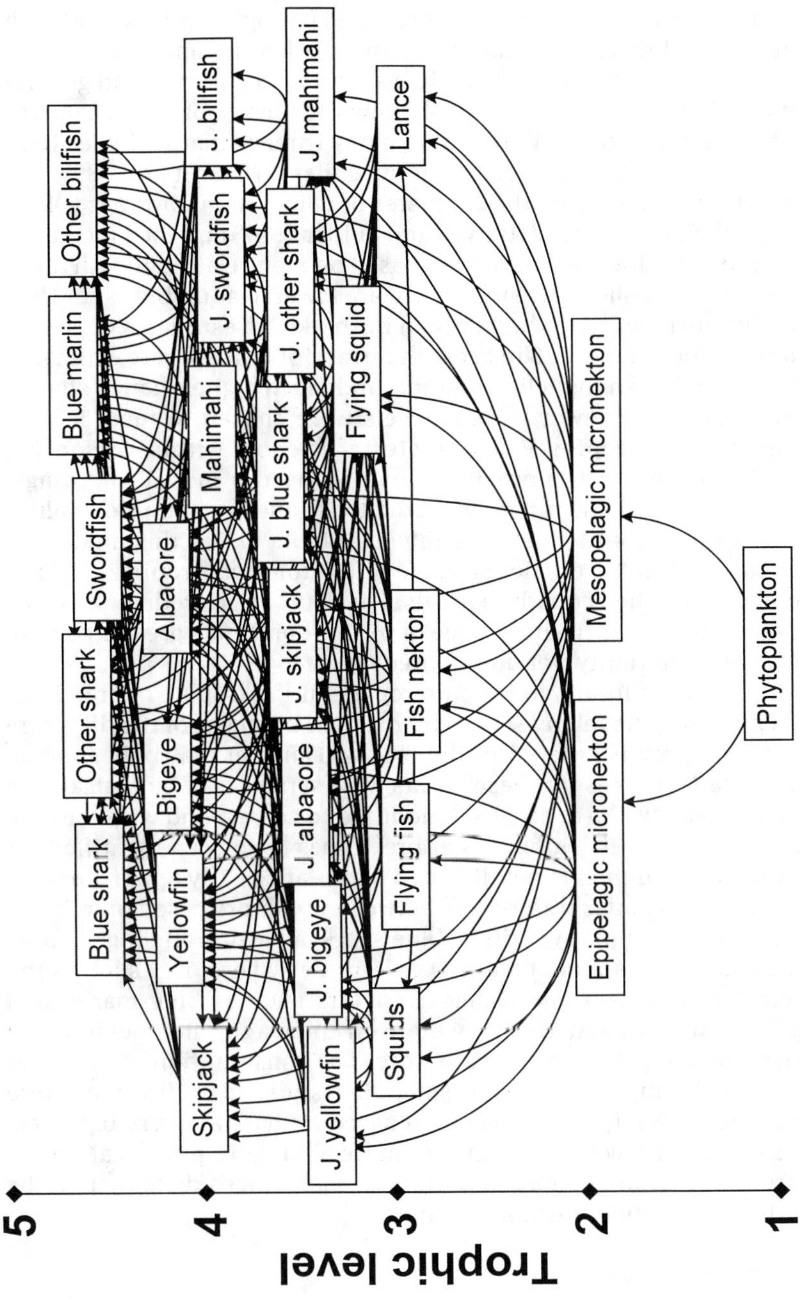

Figure 1. A food web diagram of the Ecopath model for the central Pacific.

each of the other Ecopath inputs. Values for P/B roughly correspond with estimates of total mortality as derived from stock assessments (Shomura et al. 1994) or were estimated from life history characteristics and generation times which are reasonably well known for many of the organisms represented in this model. We note that many tropical marine fishes have very rapid growth rates, particularly in juveniles (Prince et al. 1989).

For Q/B values, we used the estimates of daily ration for tunas (Olson and Boggs 1986) and sharks (Stillwell and Kohler 1992) and then increased or decreased those for other predators based on their relative growth rates. For example, metabolic costs and growth are lower in adult billfishes than in yellowfin tuna and much higher than in sharks, so estimates of annual prey consumption rates (Q/B in Ecopath terms) for billfish were estimated to be much lower than for yellowfin tuna, and much higher than in sharks. Parameter values for lower trophic levels were chosen to meet P/B requirements while keeping P/Q (transfer efficiency) within reasonable bounds. These informal approximations produced a model for pelagic predators with a realistic range of productivity and consumption values related to a variety of life history strategies (Table 1). For this model, we assumed an ecotrophic efficiency value of 0.9 for most trophic levels as recommended by the Ecopath manuals. For adult apex predators, we assumed an ecotrophic efficiency value of 0.5 to represent a greater component of natural mortality not due to predation.

A second set of inputs to Ecopath represent the trophic interactions. For each predator, the relative components (% of biomass) of the diet composed of each prey is entered in the matrix. Our diet matrix is summarized in Table 2. Sources for these inputs were synopses of biological data (e.g., Bayliff 1980, Palko et al. 1982) modified to correspond with the simplified model food web. Hundreds of prey types have been identified for these predators and their generalist food habits are clearly opportunisitic. The diet matrix includes differences between predators based on relative habitat use. For example swordfish, blue shark, albacore, and bigeye tuna are seasonally more abundant at higher latitudes where prey such as flying squid are dominant. Epipelagic predators such as blue marlin and skipjack tuna were assumed to prey more on epipelagic micronekton than deep-swimming bigeye tuna and albacore. The consumption of juveniles by adults (including cannibalism, since diet study identifications were often limited to family levels) was also based on habitat overlap. For example, skipjack and yellowfin tuna are more abundant in tropical waters and were assumed to be more important in each other's diets than in the diet of the more temperate albacore tuna.

Modeling Protocol

Based on diet information, Ecopath constructs a food web and assigns each component to an average trophic level (Fig. 1). For this analysis, each

Table 1. Paramenter estimates for the central Pacific Ecopath model.

State variable	Biomass	P/B	Q/B	*F*
Blue marlin	3.3	0.3	4.0	0.15
Other billfishes	1.9	0.5	5.0	0.16
Other sharks	1.2	0.1	1.5	0.05
Swordfish	3.6	0.4	5.0	0.15
Blue shark	16.0	0.15	2.1	0.08
Albacore tuna	7.9	0.5	12.0	0.18
Bigeye tuna	7.7	0.8	16.0	0.32
Yellowfin tuna	14.4	1.2	18.0	0.48
Skipjack tuna	12.2	2.0	20.0	0.74
Juvenile billfishes	3.7	7.0	40.0	0.02
Mahimahi	1.1	5.0	20.0	0.90
Juvenile swordfish	6.9	6.0	35.0	0.01
Juvenile blue shark	5.8	0.5	4.0	–
Juvenile other sharks	6.0	0.5	4.0	–
Juvenile skipjack	79	5.0	25.0	0.04
Juvenile albacore	34	2.0	25.0	0.01
Juvenile bigeye	23	3.0	30.0	0.01
Juvenile mahimahi	46	10.0	50.0	0.02
Juvenile yellowfin	89	3.5	35.0	0.01
Flying squids	67	10.0	50.0	0.15
Lance	879	0.5	2.9	–
Squids	294	5.0	25.0	–
Flying fishes	399	2.0	10.0	–
Fish nekton	2,090	1.5	6.0	–
Epipelagic micronekton	2,190	10.0	50.0	–
Mesopelagic micronekton	2,742	7.0	35.0	–
Phytoplankton	4,108	100.0	0.0	–

Units for biomass estimates are kg per km^2. Ratios of production/biomass (P/B) and consumption/biomass (Q/B) are estimated per year. Instantaneous annual fishery mortality is represented by F.

Table 2. Summary of the trophic interactions matrix used in the central Pacific Ecopath model. For each predator, diet composition (as percentage by mass) is listed in descending order of importance for the top four prey types.

Predator	Dominant prey types
Blue marlin	Squids, juv. and adult skipjack, juv. yellowfin
Other billfishes	Squids, juv. and adult skipjack, juv. yellowfin
Other sharks	Squids, fish nekton, juv. skipjack and juv. yellowfin
Swordfish	Flying squid, squids, fish nekton, juv. apex predators
Blue shark	Flying squid, squids, fish nekton, juv. apex predators
Albacore tuna	Flying squid, squids, micronekton, fish nekton
Bigeye tuna	Flying squid, squids, fish nekton, mesopel. micronekton
Yellowfin tuna	Fish nekton, flying fishes, micronekton, juv. skipjack
Skipjack tuna	Micronekton, fish nekton, juv. apex predators, squids
Juvenile billfishes	Juvenile skipjack, micronekton, fish nekton, juv. mahimahi
Mahimahi	Flying fishes, fish nekton, micronekton, juv. tunas
Juvenile swordfish	Micronekton, fish nekton, squids, juv. apex predators
Juvenile blue shark	Micronekton, fish nekton, squids, flying squids
Juvenile other sharks	Micronekton, fish nekton, squids, juv. sharks
Juvenile skipjack	Micronekton, fish nekton, squids, juv. tunas
Juvenile albacore	Micronekton, fish nekton, squids, flying squid
Juvenile bigeye	Micronekton, fish nekton, squids, flying squid
Juvenile mahimahi	Flying fishes, micronekton, fish nekton, squids
Juvenile yellowfin	Micronekton, fish nekton, squids, juv. mahimahi
Flying squids	Micronekton, fish nekton, squids, flying squid
Lance	Micronekton, fish nekton, squids
Squids	Micronekton, fish nekton
Flying fishes	Micronekton
Fish nekton	Micronekton
Epipelagic micronekton	Micronekton, phytoplankton
Mesopelagic micronekton	Micronekton, phytoplankton
Phytoplankton	–

of the fishes represented by trophic level 4 and above was considered as a potential keystone predator.

Paine (1966) derived the keystone predator concept by evaluating responses of a rocky intertidal community to experimental removal of a predaceous starfish species. We conducted an analogous test through Ecosim by simply creating a sudden and sustained tenfold increase in the fishing mortality (*F*) for each individual apex predator compartment at year 2, then allowing the model to run for a total of 30 years. In all cases, this was sufficient to allow establishment of a new steady state or an obvious, new trajectory. In all cases, the manipulated predator biomass dropped to very low levels (ecological extinction). As a final test, we removed fishery exploitation from the entire apex assemblage. This simulation tested the role of the fishery as the keystone predator.

We judged the relative responses to each test by examining the model output for qualitative and quantitative changes based on two general questions: Did one or more other predators increase in proportion to that one decreased by exploitation (competition effects)? Did other species change substantially and in response to the altered apex predator assemblage (predation effects and indirect effects)? In a general sense, we simply asked how much change occurred in the Ecopath food web when one of the apex predators was removed. Because this is a modeling study with all of its attendant assumptions and problematic parameter estimates, we were conservative in the interpretation of "change." We aggregated responses at the most general level: no substantial change (<50%) and major increase or decrease (>50%). To facilitate interpretation, we present the simulation results by normalizing the response of each state variable. We divided the simulated biomass by the initial biomass level specified in the Ecopath model (i.e., 1.0), then graphed responses in time as they depart from the initial condition. This facilitates interpretation of relative responses, but the reader should be cautioned that initial biomass levels can be substantially different for individual state variables. For example, initial biomass of blue marlin is estimated at 3.3 kg per m^2 while that of blue shark is 16 kg per km^2 (Table 2), yet both are at the apex of the food web (Fig. 1).

Results

Food Web Structure

The food web diagram presented in Fig. 1 indicates the extensive connectedness in this system, the compromise of presenting a food web that represents as many as 1,000 species arrayed across many trophic levels, and interactions that summarize cumulative, annual feeding rates over the entirety of a large foraging arena. In keeping with the caution of Walters et al. (1997) we attempted to aggregate prey types with a minimum of vertical trophic overlap. Individual apex predator species and their juveniles were separated based on the practical grounds that understanding of basic biological characteristics (e.g., diets, age at maturity, growth rates) and records of fishery harvest are not all-inclusive. Each of those selected at the species level includes sufficient knowledge that would allow us to recognize its unique characteristics. Other groupings represent separation of the epipelagic from mesopelagic forms and the fact that diel migration crosses those habitat boundaries. Aggregates called epipelagic or mesopelagic micronekton basically include organisms of the size range from small zooplankton to euphausids and includes both herbivores and omnivores. Fish nekton includes all forms small enough to serve as prey and not identified by a more specific recognition as juvenile apex predators. The category labeled "lance" includes mesopelagic fishes of similar, small size. Squids are aggregated as large (flying squids) and small (all other squids).

The highest trophic levels of the food web (Fig. 1) are occupied by adult blue marlin (*Makaira mazara*), blue shark (*Prionace glauca*), swordfish (*Xiphias gladius*), and two aggregated components that include rare species of similar trophic position and life history. Other shark includes forms such as the thresher shark (*Alopias vulpinus*) and mako shark (*Isurus oxyrinchus*), while the category of other billfish includes forms such as the black marlin (*Makaira indica*), striped marlin (*Tetrapturus audax*), and sailfish (*Istiophorus platypterus*). Together, these forms are characterized by trophic positions between level 4 and level 5, relatively low P/B values, and modest biomass, except for blue shark which is very abundant.

The next tier of apex predators, at about trophic level 4, includes adult tunas such as skipjack (*Katsuwonus pelamis*), yellowfin (*Thunnus albacares*), bigeye (*Thunnus obesus*), and albacore (*Thunnus alalunga*), plus adult mahimahi (*Coryphaena hippurus*), although the latter has a trophic position similar to that of juvenile swordfish and juvenile other billfish (Fig. 1). Each of this group has higher P/B ratios and, in the case of skipjack and yellowfin tunas, a substantial biomass.

Juveniles of apex predators occupy the region between trophic levels 3 and 4. As stated above, they are treated as separate entities in this model because of substantial evidence of intraguild predation and diet overlap with adult forms. Their P/B ratios, biomasses, and prey consumption rates are generally high. In estimating biomass for the juvenile tunas, we included the assemblage of smaller tuna species (e.g., *Auxis* sp.) and mackerels. Although the food web does not identify them as individual categories, the life history characteristics of smaller Scombridae are included as their surrogate, juvenile tunas.

Diets of central Pacific apex predators typically contain a huge diversity of prey types and substantial overlap among species. For example, the diet of adult blue shark includes 19 of the forms represented in Fig. 1 and those range in size from small squids to adult billfishes. The two most abundant tunas, skipjack and yellowfin, exhibit almost complete overlap in the diet proportions for each of the 12 prey types in their stomachs. Only adult mahimahi has a substantially different diet from that of other apex predators in that flying fish are its primary prey. While the general similarity of diets is one of the fascinating features of this system, summarizing this diversity is a challenge because we see only modest evidence of the resource partitioning so elegantly elaborated in studies of freshwater systems and terrestrial food webs (Polis and Winemiller 1995). Instead of reporting the complete diet matrix developed in this Ecopath model, we present only the major prey types representative of the diets for each predator and their overlap with others (Table 2).

Modeling Results

Simulated deletion of each, individual apex predator produced a range of results that we aggregated as two general kinds of responses. One group of results evidenced strong effects on only a few other components of the food web. Another group of results indicated a much more extensive and diverse set of effects. For the highest trophic levels (blue marlin, blue shark, other billfish, other shark, and swordfish), removal of the apex predator produced only modest responses among other members of the food web. The response to blue shark removal is representative (Fig. 2). In this case, rapid (within 5 years) and sustained increases occured in the other billfish group, in swordfish, and in mahimahi, while juvenile blue shark gradually declined. The strongest response was a delayed but exponential and continuing increase in juveniles of other sharks. Other components of the food web were little affected.

Removal of blue marlin or other billfishes produced similar effects: skipjack tuna, all juvenile tunas, and mahimahi increased while juvenile billfish and flying fishes decreased. Although these responses are apparent by our criteria (i.e., > 50%), they were modest and established a new equilibrium biomass within less than 10 years. Swordfish removal caused increases in adult albacore and mahimahi, but no substantial declines in other components. Removal of albacore or bigeye tunas produced similar effects: juvenile swordfish and adult mahimahi increased while flying fishes decreased in direct response to the increase in their major predator, mahimahi. On the other hand, removal of adult mahimahi produced no strong responses in other food web components as compensatory consumption by other predators prevented increase in flying fish abundance.

The second group of results are represented by strong responses expressed in a diversity of food web components. The keystone predator test for yellowfin tuna is indicative of that and is presented in Fig. 3. Removal of adult yellowfin caused rapid, strong (>2×), and sustained increases in the biomass of juvenile yellowfin, flying fishes, mahimahi, and juvenile billfishes. Juvenile skipjack and fish nekton declined, but the latter exhibited a slow and continuing response. Other components were little affected. Removal of adult skipjack tuna produced responses very similar to those from adult yellowfin.

Although not pictured in Fig. 1, the apex of this food web is actually that represented by longline fisheries. As a test of that prospect, we used

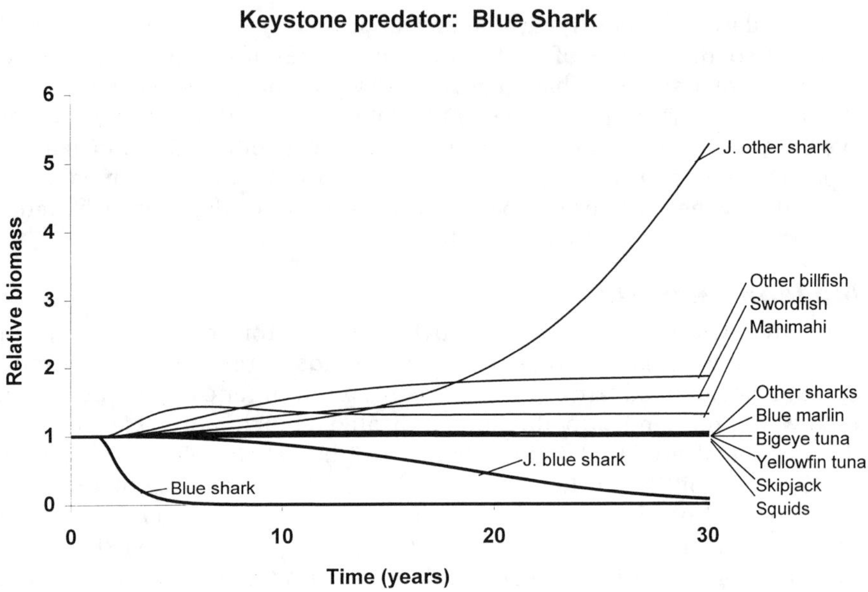

Figure 2. Responses of selected food web components to removal of blue sharks. Biomasses for each are normalized to their initial condition. The Ecosim model was perturbed at 5 years into a 30-year simulation by a tenfold increase in fishery mortality for blue shark.

Ecosim to remove the estimated levels of fishing mortality from each and all of the apex predators. The results of that trial are summarized in Fig. 4 and exhibit more substantial responses that any of those derived from removing any of the individual predators.

Most, but not all, of the adult apex predators increased in response to removal of the fishery. Rates of response corresponded with life history characteristics (rapid maturity, high P/B values) with the tuna species responding most quickly. Over time, blue marlin, swordfish, and blue shark biomasses came to dominate the assemblage with the most substantial change (>8×) exhibited by blue marlin, with blue shark continuing to increase at 30 years. Conversely, adult mahimahi declined quickly and remained at low levels. Juvenile skipjack and juvenile yellowfin demonstrated opposite responses. The latter increased initially, then declined to sustained and low levels. Other food web components were less affected. In the absence of fishing, this food web would have a greater overall abundance of large fishes and would be dominated by those of highest trophic position.

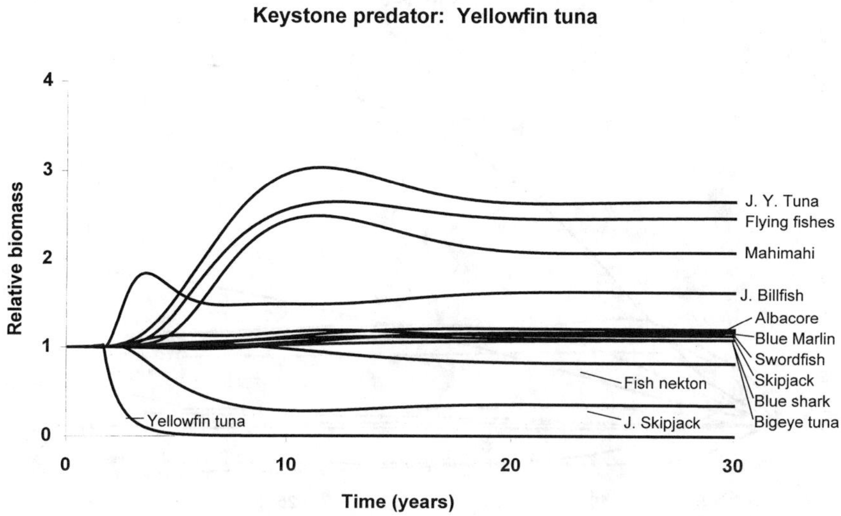

Figure 3. Responses of selected food web components to removal of yellowfin tuna. Biomasses for each are normalized to their initial condition. The Ecosim model was perturbed at 5 years into a 30-year simulation by a tenfold increase in fishery mortality for yellowfin tuna.

Discussion

Contrary to expectation, no one member of this apex predator guild can be considered a clearly demonstrable keystone predator. Simulations of removal demonstrate complex compensatory responses in this ecological context. None of the simulations produced results indicative of simple replacement processes. For example, removal of the abundant blue shark did not cause an increase in the group of other adult sharks which have similar life history characteristics and diets. Instead, the juveniles of other sharks became most abundant. Two explanations may be offered. One is to simply note that the complexity of this food web context makes simple expectations subject to question. That may be true, but has no explanatory power and does not advance our understanding. The second is that we have provided inadequate or insufficient inputs to the Ecopath model that underlies these simulations. Most likely among those prospects is the insufficient accounting for intraguild cannibalism that would control survival of young sharks. Although our diet matrix does include some cannibalism (juvenile sharks are 0.5% of diets for blue shark and other sharks),

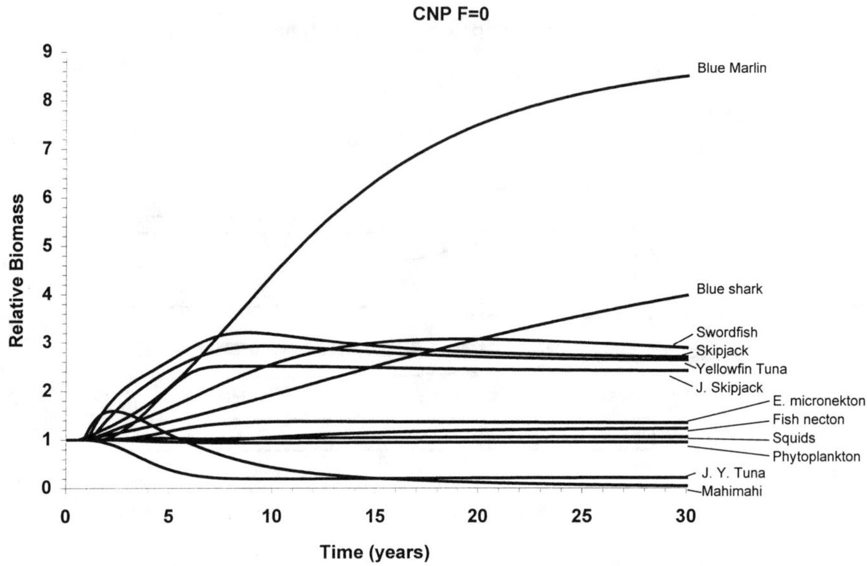

Figure 4. Responses of selected food web components to removal of fishing mortality for all apex predators.

this may underrepresent the importance of cannibalism because shark diets are not well studied. Given the low fecundity and slow maturation of these fishes, a modest increase in cannibalism rates could have profound ecological effects.

An alternative approach is to seek independent evidence of ecological changes that may have transpired in places where shark removal programs or intensive fisheries have selectively decreased the populations of adult sharks (Bonfil 1994). Shark removals or shark exclusions (large-mesh nets) have been conducted in a number of places with a universal, general result: increase in abundance of small sharks. As simulated in the keystone predator test using Ecosim, real world removal of large sharks produces increases in small sharks. This comparison provides the encouragement of independent evidence, but does not explain why juvenile blue sharks should decline. This version of the model treats adult and juvenile members of a species as independent entities. Adults may cannibalize their juveniles and they may compete for some prey types, but there is no direct stock-recruit linkage. Therefore, first principles would have it that removal of adult blue sharks should produce more juvenile blue sharks. The opposite occurred and its explanation is not obvious and, again, calls for reevaluation of the adequacy of biological information used to develop Ecopath inputs for these fishes. This conclusion (more study is necessary!) should not overshadow that fact that the model and its embedded assumptions produced a general result remarkably analogous to ecological changes documented in several places where shark removal programs have been conducted.

Two members of the apex predator guild, adult yellowfin and adult skipjack tunas, apparently have critically important roles in this food web. Their removal evoked substantial and sustained changes in the structure of the system. Their importance corresponds with the fact that they are major biomass components of the apex guild and with their relatively high P/B ratios. They are, by definition, important and highly responsive members of the system. Their diets and trophic position also assure an important role as they are among the prey for higher-order predators such as billfishes and sharks, competitors of all members of the guild, and predators on a diversity of lower trophic levels. In this complex system with high degrees of trophic interaction and diet overlap, these tunas are the most effective generalists. They are opportunistic predators and capable of rapid biomass increase. Their abundance confirms the advantages of those traits.

Independent evidence of the keystone role for yellowfin tuna is also available in those systems where exploitation has been and continues to be more intense than in the central Pacific. Longline fishing has been intense in the Gulf of Mexico and Caribbean Sea for many years with the large tunas, especially yellowfin tuna, as major target species. Commensurate with a decline in yellowfin landings, catch reports from those fisheries include a general increase in the abundance of smaller tunas and a recent, rapid increase in the landings of mahimahi (Hinman 1998). The Ecosim simulation (Fig. 3) forecasts similar responses for the central Pacific food web.

As made apparent by Fig. 4, the real keystone component of this system is that owing to fishing. We cannot evaluate the structure of these systems prior to the development of exploitation because of the conundrum that what we know of these fishes and their ecosystem derives from fisheries. Blue marlin demonstrates the most dramatic response to removal of exploitation. We could expect, therefore, that the abundance of large billfishes such as the marlin species and swordfish would be the first to show rapid declines as the fisheries began to develop.

Again, there is some independent evidence in support of that hypothesis. As the Japanese longline fleet developed and expanded into the central Pacific in the 1950s, catch rates for large billfishes were the first to decline (Stroud 1989). As the Hawaii-based swordfish fleet developed, the catch per unit of effort dropped very quickly to less than one-third of the initial rates (Boggs and Ito 1993) while total catches rose to a sustained level as new ships were added to the fleet. In other words, the advent of new fisheries had its first and most profound effect on large billfishes. Removal of fisheries should produce the opposite effect, and that is seen in Fig. 4.

Our analyses indicate that knowing more about the intensity of cannibalism, especially that by sharks, would provide major returns in our understanding of food web dynamics. We also assumed that the juvenile tuna groups would represent the equivalent of life histories for other, smaller members of the tuna group (Scombridae). Organisms such as the bullet mackerel (*Auxis rochei*) are at the nexus of food webs in the Eastern Tropical Pacific (Olson and Boggs 1986). Their analogues play an essential role in our model, but we have no direct measures of their actual abundance and dynamics. That will remain an important unknown in this and similar pelagic systems until an adequate stock assessment technology is developed. As an immediate alternative, our simulations suggest that increases in mahimahi catches may be a signal of changing ecological interactions derived from intensified exploitation of larger and less resilient species such as marlin, other billfishes, and/or yellowfin and bigeye tunas.

Trophic cascade effects are well known from lakes (Carpenter and Kitchell 1993), were recently documented for the North Pacific (Shiomoto et al. 1997), and appear in some applications of the Ecosim model (Walters et al. 1997). We see no evidence of trophic cascade effects in these manipulations of the central Pacific model. Two conclusions derive from that observation. First, the Ecopath model structure includes high resolution at the upper trophic levels but substantial aggregation for the lower trophic levels. Thus a manipulation at the apex predator level may be dispersed by compensatory responses among the many other components at high trophic levels. Second, the previous and current levels of fishery exploitation are such that a manipulation of exploitation levels of apex predators is not sufficiently strong to be transmitted from trophic levels 4 or 5 and through the food web to the phytoplankton. Trophic cascades derive from strong and sustained manipulation of food webs (Carpenter and Kitchell 1993). Although there were strong responses to simulated removal of fisheries from the central Pacific food web, they were obvious only among components at or above trophic level 3.

Our conclusions about keystone predator effects are based on a model that works in the singular currency of energetics. This is a unifying method, but it may be insufficient in that it underrepresents the prospect that alternative currencies (e.g., nutrients, behavioral interactions that apportion space) and alternative mechanisms (e.g., migrations, predator avoidance) may play major roles in the structure of this community. We do not know that and, given the spatial scales of the key interactions, we probably will not see better alternatives or experimental tests of those for some time to come. Better models may be a quicker way to create and test alternatives. Comparative studies (e.g., Atlantic vs. Pacific) offer another opportunity for rapid learning.

Like all models, Ecopath and Ecosim are simplifications that effect a compromise between the merits of simple, general theory and the constraints of overwhelming ecological detail. Ecopath has the merit of being based on the need for a balanced energy budget. That forces an empirical accounting in its parameterization and the experience of Ecopath modelers has accumulated to the point where the software includes challenges and warnings about unrealistic estimates. This creates a stable basis for using Ecosim as a tool to evaluate alternative scenarios.

Obviously, there is substantial opportunity for increasing the extent and accuracy of the basic data requirements of the central Pacific Ecopath. But, we cannot and should not use "waiting for better data" as an excuse. The lessons from the Atlantic Ocean (Hinman 1998) and from global fisheries development (Pauly et al. 1998), teach us that ecological changes have and will continue to occur more rapidly than can be documented by field studies. There is some urgency to these matters because resource management policies for highly migratory stocks and those based on ecosystem principles must be developed in the near future (Botsford et al. 1997). The rate of change in these ecosystems is substantially greater than our rate of learning based solely on accumulation of empirical evidence.

The main lesson for fisheries managers is that the effects of longline gear are ecologically complex. Models such as that described herein afford one tool for evaluating that complexity in an ecosystem context. An important lesson for conservationists is that the status of these apex predator populations and their ecological context can only be known through monitoring of the fisheries. One consequence of the 1992 United Nations ban on drift gillnets in the North Pacific was that the Asian fleets no longer provide ready access to data from their longline fleets. Those fleets account for approximately 95% of fishery exploitation in the Central North Pacific. Conservation concerns for the future of billfishes, sharks, and some tuna stocks are important, but must be expressed with the understanding that restriction of longline fishers registered under the U.S. flag accounts for a very small fraction of the exploitation pressure and, more importantly, constrains or eliminates our most important source of information about the status of these apex predators and the pelagic ecosystem that supports them. The Hawaii-based longline fleet is our essential monitoring tool in the central Pacific. Research must rise to the challenge of understanding how fishery-derived data can be extended to the ecosystem context required for sustainable management of these fisheries.

Acknowledgments

This work was conducted as part of the Apex Predators in Marine Systems Working Group supported by the National Center for Ecological Analysis and Synthesis (NCEAS), a center funded by the National Science Foundation (Grant #DEB-94-21535), the University of California-Santa Barbara, and the State of California. The basic Ecopath and Ecosim models for the central Pacific were developed while J.F.K. was on sabbatical leave and jointly supported by NCEAS and the College of Letters and Science, University of Wisconsin-Madison. More recent support for this project derives from a grant to J.F.K. through the Biological Oceanography Program of the National Science Foundation.

References

Apollonio, S. 1994. The use of ecosystem characteristics in fisheries management. Rev. Fish. Science 2:157-180.

Bayliff, W.H. (ed.). 1980. Synopses of biological data on eight species of scombrids. Inter-American Tropical Tuna Commission Special Report No. 2.

Boggs, C.H., and R.Y. Ito. 1993. Hawaii's pelagic fisheries. Mar. Fish. Rev. 55:69-82.

Bonfil, R. 1994. Overview of world elasmobranch fisheries. FAO Fish. Tech. Paper 341. Rome, Italy. 119 pp.

Botsford, L.W., J.C. Castilla, and C.H. Peterson. 1997. The management of fisheries and marine ecosystems. Science 277:509-515.

Carpenter, S.R., and J.F. Kitchell (eds.). 1993. The trophic cascade in lakes. Cambridge University Press, Cambridge, England.

Christensen, V., and D. Pauly. 1992. ECOPATH II: A software for balancing steady-state models and calculating network characteristics. Ecol. Modell. 61:169-185.

Christensen, V., and D. Pauly (eds.). 1993. Trophic models of aquatic ecosystems. ICLARM Conf. Proc. 26. International Center for Living Aquatic Resources Management, Manila. 390 pp.

Hinman, K. 1998. Ocean roulette: Conserving swordfish, sharks and other threatened pelagic fish in longline-infested waters. National Coalition for Marine Conservation, Leesburg, Virginia. February 1998.

Olson, R.J., and C.H. Boggs. 1986. Apex predation by yellowfin tuna (*Thunnus albacares*): Independent estimates from gastric evacuation and stomach contents, bioenergetics, and cesium concentrations. Can. J. Fish. Aquat. Sci. 43(9):1760-1775.

Paine, R.T. 1966. Food web complexity and species diversity. Am. Nat. 100:65-75.

Palko, B.J., G.L. Beardsley, and W.J. Richards. 1982. Synopsis of the biological data on dolphin-fishes, *Coryphaena hippurus* Linneaus and *Coryphaena equiselis* Linneaus. NOAA Tech. Rep. NMFS Circular 443, FAO Fish. Synopsis 130.

Pauly, D., and V. Christensen. 1995. Primary production required to sustain global fisheries. Nature 374:255-257.

Pauly, D., M.L. Soriano-Bartz, and M.L.D. Palomares. 1993. Improved construction, parameterization and interpretation of steady-state ecosystem models. In: V. Christensen and D. Pauly (eds.), Trophic models of aquatic ecosystems. ICLARM Conf. Proc. 26. International Center for Living Aquatic Resources Management, Manila, pp. 1-13.

Pauly, D., V. Christensen, J. Dalsgaard, R. Froese, and F. Torres Jr. 1998. Fishing down marine food webs. Science 279:860-863.

Polis, G., and K.O. Winemiller (eds.). 1995. Food webs: Integration of patterns and dynamics. Chapman Hall, New York.

Polovina, J.J. 1984. Model of a coral reef ecosystem I. The ECOPATH model and its application to French Frigate Shoals. Coral Reefs 3:1-11.

Power, M.E., D. Tilman, J.A. Estes, B.A. Menge, W.J. Bond, L.S. Mills, G. Daily, J.C. Castilla, J. Lubchenco, and R.T. Paine. 1996. Challenges in the quest for keystones. BioScience 46:609-620.

Prince, E.D., D.W. Lee, J.R. Zweifel, and E.B. Brothers. 1989. Ageing young Atlantic blue marlin (*Makaira nigricans*) from otolith microstructure. In: R.H. Stroud (ed.), Planning the future of billfishes: Research and management in the 90s and beyond. Part 2, Contributed papers. National Coalition for Marine Conservation, Inc., Savannah, Georgia, pp. 295-308.

Shiomoto, A., K. Tadokoro, K. Nagasawa, and U. Ishida. 1997. Trophic relations in the subarctic North Pacific ecosystem: Possible feeding effect from pink salmon. Mar. Ecol. Prog. Ser. 150:75-85.

Shomura, R., J. Majkowsi, and R.F. Harman (eds.). 1995. Status and interactions of Pacific tuna fisheries in 1995. FAO Fish. Tech. Paper 365. Rome, Italy. 612 pp.

Shomura, R., J. Majkowsi, and S. Langi (eds.). 1994. Interactions of Pacific tuna fisheries. Volume 2. Papers on biology and fisheries. FAO Fish. Tech. Paper 336/2. Rome, Italy. 439 pp.

Stillwell, C.E., and N.E. Kohler. 1992. Food habits of the sandbar shark *Carcharhinus plumbeus* off the U.S. northeast coast, with estimates of daily ration. Fish. Bull., U.S. 91:138-150.

Stroud, R.H. (ed.). 1989. Planning the future of billfishes: Research and management in the 90s and beyond. National Coalition for Marine Conservation, Inc., Savannah, Georgia. 321 pp.

Ueyanagi, S., R.S. Shomura, Y. Watanabe, and J.L. Squire. 1989. Trends in fisheries for billfishes in the Pacific. In: R.H. Stroud (ed.), Planning the future of billfishes: Research and management in the 90s and beyond. Part 1. Fishery and stock synopses, data needs, and management. National Coalition for Marine Conservation, Inc., Savannah, Georgia, pp. 31-45.

Walters, C.J., V. Christensen, and D. Pauly. 1997. Structuring dynamic models of exploited ecosystems from trophic mass-balance assessments. Rev. Fish Biol. Fish. 7:1-34.

Ecosystem Modeling of Becharof Lake, a Sockeye Salmon Nursery Lake in Southwestern Alaska

Ole A. Mathisen
University of Alaska Fairbanks, School of Fisheries and Ocean Sciences, Juneau, Alaska

Norma Jean Sands
Alaska Department of Fish and Game, Juneau, Alaska

Abstract

Ecosystem management is dependent on knowledge of species interrelationships within an ecosystem. For sockeye salmon (*Oncorhynchus nerka*) nursery lakes such as Becharof Lake in southwestern Alaska, we are interested in the food competitors, predators, and prey of the juvenile sockeye salmon. Two methods were used to examine the structure of the ecosystem and determine trophic levels of species groups: (1) the use of marine-derived nitrogen levels in organisms determined from stable isotope ratios, and (2) the use of an ecosystem model based on biomass, production, and consumption estimates of each species group. In spite of the many assumptions needed to develop the ecosystem model, the two methods gave similar estimates of trophic level per species group. The development of the ecosystem models helps identify what data are available for the system under study and what data are needed to improve understanding of the system.

Introduction

Management of sockeye salmon (*Oncorhynchus nerka*) runs during the last three decades has been directed mainly toward achieving escapement goals that, on the average, give the highest surplus yield. Predicting returns can be done in two ways: either (1) just by combining the freshwater

Contribution PP-178 of the Alaska Department of Fish and Game, Division of Commercial Fisheries, Juneau.

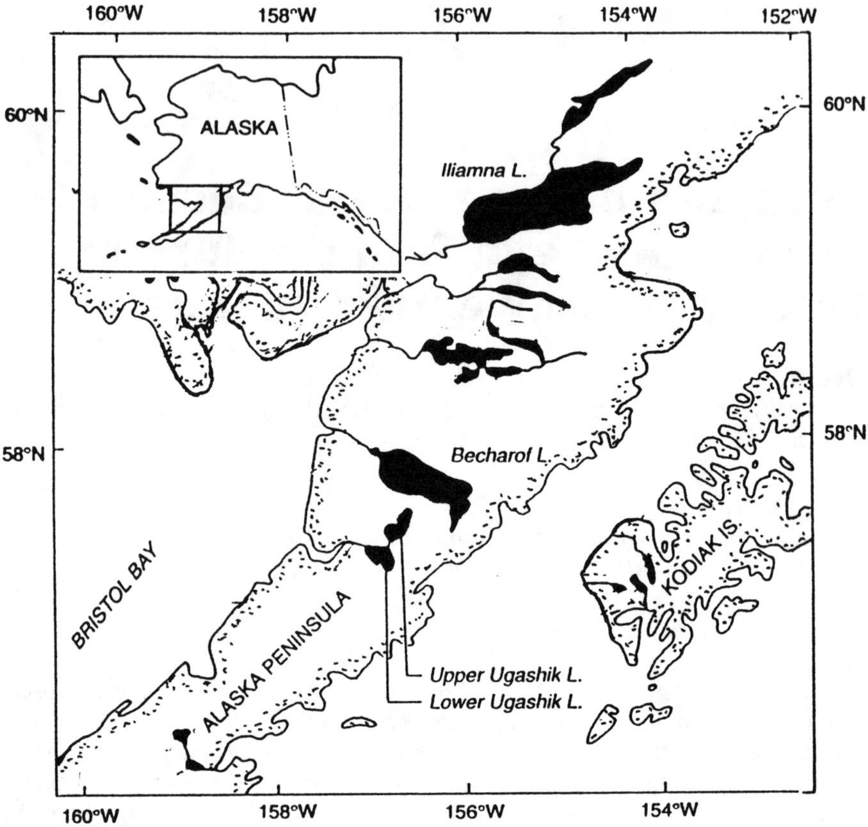

Figure 1. Map of the Becharof Lake area, Alaska.

and marine mortalities to predict adult returns from escapements (standard spawner-recruit analysis); or (2) by additionally considering environmental and biological factors during the freshwater stage to predict the number of smolts produced with expected ages and average lengths. The latter approach is followed in this paper through the use of an ecosystem study. A next step in developing good management is to determine ways to improve survival and growth of juvenile sockeye salmon. This leads to ecosystem management, i.e., managing the system, not just the fishery, and may involve control of predators in the nursery lakes or lake fertilization to improve smolt survival.

In recent years, the emphasis in ecosystem studies has been to look at the correlation between environmental factors such as temperature and air pressure and salmon abundance (Wooster 1983, Beamish and McFarlane 1989). In our study, we examined the trophic structure of Becharof Lake, a

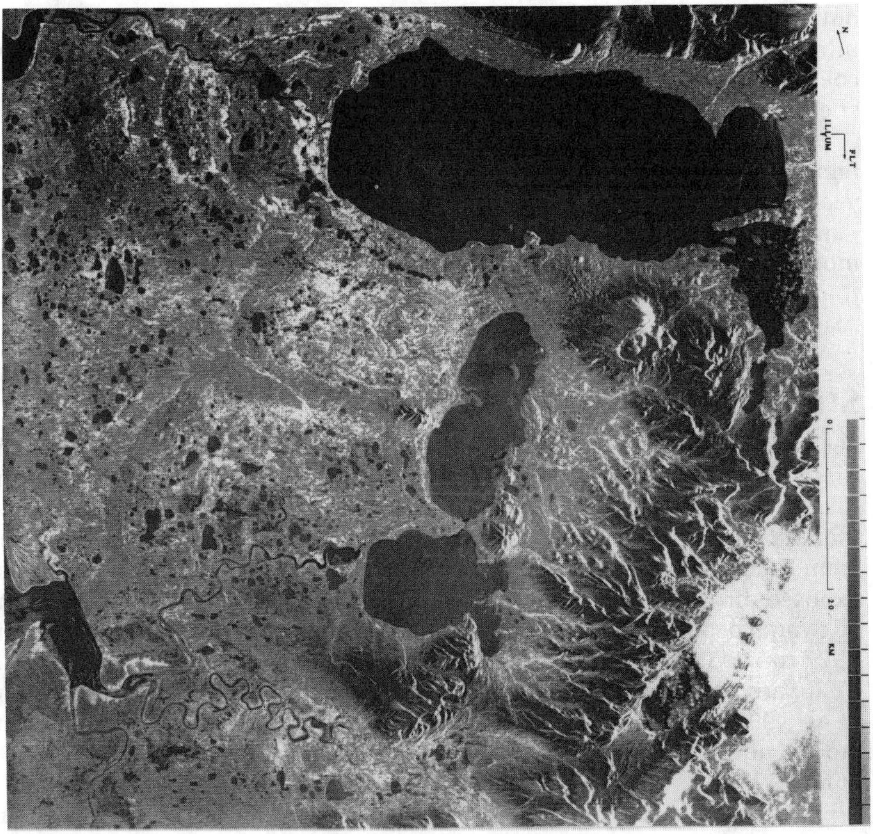

Figure 2. Satellite photograph from an altitude of 785 km showing a swath of 100 km with Becharof Lake on the left and the Ugashik Lakes on the right. Courtesy of Alaska SAR, University of Alaska Fairbanks, received from a satellite operated by the European Space agency (©ESA-1992).

salmon nursery lake in southwestern Alaska, for possible use in management decisions that will improve smolt survival. Lake systems provide relatively simple ecosystems in relation to marine systems. Ecosystem models, such as the ECOPATH II model developed by Christensen and Pauly (1992), help to determine the trophic relationships between species or species groups in the ecosystem using mass balancing under steady-state conditions. As many input parameters for such a model must be assumed or taken from the literature, it is imperative to have a second method to help verify the results. In anadromous lake systems, trophic structure analysis takes advantage of the marine derived nutrients that spawning salmon transport into the lake system and that are then dispersed throughout the system by decomposing carcasses. The ratio of isotope marine nitrogen, ^{15}N, to ^{14}N is specific to trophic level and, using known transfer coefficients from one trophic level to another, allows the trophic level of a species to be estimated (Minagawa and Wada 1984).

Apparently Krogius et al. (1964) were the first to develop an ecosystem model for a salmon lake. Their modeling effort culminated about 30 years of observations of physical and biological parameters in Lake Dalnee, Kamchatka Peninsula. To the best of our knowledge, no similar ecosystem model has been attempted for salmon nursery lakes in North America. In this paper, we provide an ecosystem model developed for Becharof Lake, Alaska.

Study Site—Becharof Lake

Becharof Lake is located in the Egegik River watershed in southwestern Alaska and feeds into Bristol Bay (Fig. 1). It is the second largest lake in Alaska with a surface area of 1,142 km^2 and a mean depth of 56.7 m. Low mountains to the southeast on the Pacific side provide limited shelter from Pacific winds and the main body of the lake is situated in lowlands exposed directly to winds coming from the Bering Sea (Fig. 2). Most salmon spawning occurs in the upper end of the lake (personal observation). From 1977 to 1993, there was a sharp increase in the sockeye salmon runs (Fig. 3), apparently related to an increase in air and water temperatures (Mathisen et al. 1998). The last few years have seen a decline of sockeye salmon stocks, not only in Egegik, but also in all river systems in Bristol Bay (Hart et al. 1998).

The juvenile sockeye salmon migrate to the sea usually either in their second (age-I) or third (age-II) year and, occasionally, in their fourth (age-III) year (Bue 1984). The number of outmigrating smolts is a function, not only of the size of the parent escapements, but also of the amount of predation and food competition the juveniles have been exposed to in the nursery lake. Smolt outmigration from Becharof Lake, as determined by hydroacoustic methods (Crawford and Cross 1998), varies from year to year (Fig. 4) and does not appear to be correlated with returns (Fig. 3). The age ratio (age I to age II) of the outmigrating smolts also various greatly from year to year (Fig. 4). Finally, the average length of the smolts varies (Fig. 5) and is governed by growing conditions in the lake and freshwater age at outmigration. Becharof Lake consistently produces the largest sockeye smolts of any system in Bristol Bay (Bue 1984). In addition to sockeye salmon in Lake Becharof, there is a run of coho salmon (*Oncorhynchus kisutch*) numbering around 25,000 spawners (ADF&G database).

Becharof Lake and other salmon lakes feeding into Bristol Bay have relatively few species of fish. For ecosystem modeling purposes, they can conveniently be grouped on the basis of their feeding behavior. Apex predators include Dolly Varden (*Salvelinus malma*), Arctic char (*Salvelinus alpinus*), and northern pike (*Esox lucius*). Demersal fish include pygmy

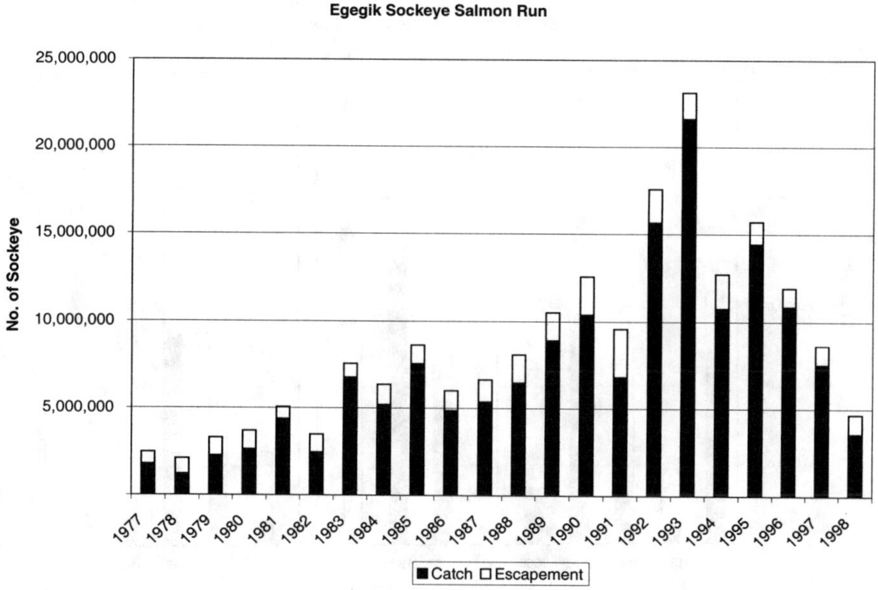

Figure 3. Sockeye salmon returns to Egegik drainage from 1977 to 1998, showing catch and escapement.

whitefish (*Prosopium coulteri*), round whitefish (*Prosopium cylindraceum*), humpback whitefish (*Coregonus pidschian*), and sculpins (genus *Cottus*). Small pelagic fish include the threespine stickleback (*Gasterosteus aculeatus*) in addition to juveniles of sockeye and coho salmon.

Bosmina sp., *Daphnia* sp., *Cyclops* sp., and *Diaptomus* sp. are the common zooplankters in Becharof Lake. The dominant form of the zoobenthos is an isopod that was identified for us by Professor T.C. Shirley as *Saduria entomon*. It is a circumpolar species found in many of the larger lakes in Norway and Sweden. Some marine crustaceans were trapped after the last glacial period when the land started to lift, and these animals became landlocked. The same probably happened in the lakes with outflow into Bristol Bay; these lakes are all less than 50 m above the high tide level in the outlet streams.

Material and Methods

Data Collection

Fish and plankton samples from Becharof Lake were collected and processed by the U.S. Fish and Wildlife Service, King Salmon Fishery Assistance Office. Resident fish samples were provided by Rendan Scalan, a

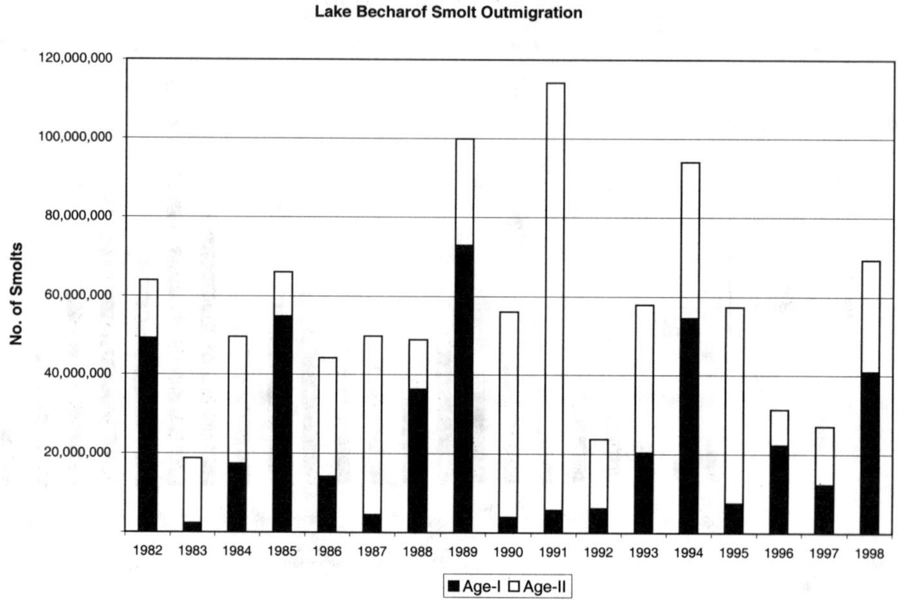

Figure 4. Sockeye smolt outmigration from Becharof Lake from 1982 to 1998, showing the age composition.

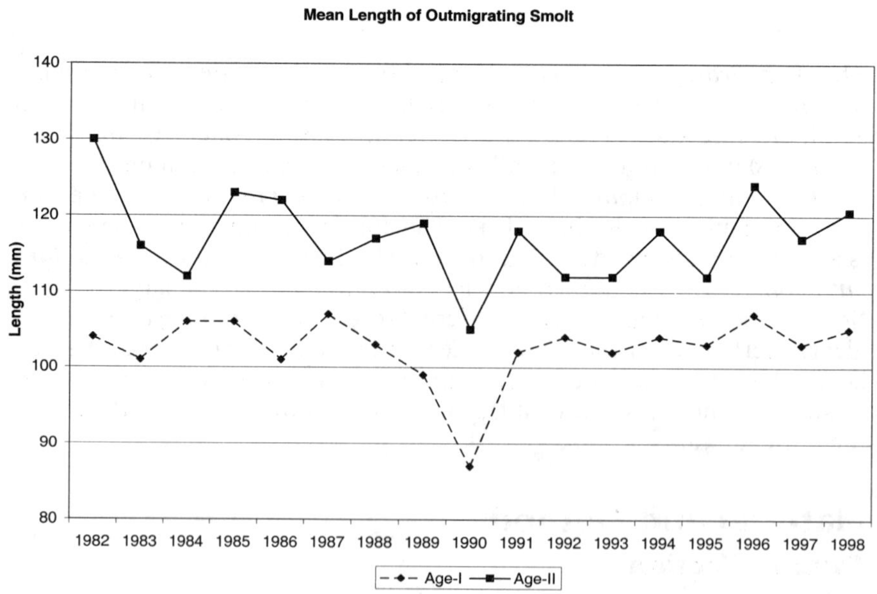

Figure 5. Mean length of age-I and age-II outmigrating sockeye smolts from Becharof Lake from 1982 to 1998.

graduate student at the School of Fisheries and Ocean Sciences, University of Alaska Fairbanks. Salmon escapement and smolt counts were obtained from Alaska Department of Fish and Game (ADF&G) databases. Plankton abundance estimates were obtained from the ADF&G Limnology Lab. Nitrogen isotope analysis was conducted by the Water and Environmental Research Center using mass spectrometry according to the methods of Kline et al. 1993.

Estimating Trophic Levels from Isotope Data

Measurements of stable isotope of nitrogen (^{15}N) are used to trace assimilated material and provide excellent definition of relative trophic level (Wada et al. 1991). The conventional reporting method for isotope measurements is to express them as a ratio of the isotope to the normal form of the element in the sample and then express this ratio relative to the ratio found in a defined international standard. For nitrogen the isotope to normal ratio is ^{15}N to ^{14}N and air is the standard used for comparison. By definition then:

$$\delta^{15}\mathrm{N} = \left(\frac{R_{sample}}{R_{standard}} - 1\right)1{,}000‰ \tag{1}$$

where:

R_{sample} = the ratio $^{15}N/^{14}N$ of the sample, and

$R_{standard}$ = the ratio $^{15}N/^{14}N$ of air.

The heavy isotope ratio of nitrogen, $\delta^{15}N$, is enriched about 3.4‰ with each trophic level (DeNiro and Epstein 1981). For any group above the herbivore level, the trophic level may be found by the following equation (Hobson and Welch 1992, Kline and Pauly 1998):

$$TL_i = \left(\frac{\delta^{15}\mathrm{N}_i - \delta^{15}\mathrm{N}_H}{EF}\right) + TL_H \tag{2}$$

where:

TL_i = trophic level of group *i*,

$\delta^{15}N_i$ = isotopic ratio for group *i*,

$\delta^{15}N_H$ = isotopic ratio for herbivore group in system,

EF = enrichment factor of 3.4‰, and

TL_H = trophic level of herbivores = 2.

Trophic Levels Estimated from Ecosystem Model

The ecosystem model used in this study is based on the ECOPATH model of Christensen and Pauly (1992), but it is constructed in EXCEL so that adult spawners may be treated as a source of nutrients entering the lake annually and are neither a primary producer nor consumer. In addition, in our EXCEL version, salmon juveniles are either consumed or exported from the lake system during smolt outmigration; none are assumed to die and contribute to detritus. The model is based on solving simultaneous equations expressing surplus production of a group as equaling consumption by all predator species plus decomposition to detritus plus any export from the system. In our case, export from the system may be due to predation by bears and birds or, in the case of plankton, lake discharge. We used units of g/m^2 per year, which is equivalent to t/km^2 per year. The production equations are expressed as:

$$B_i(P/B)_i - \sum_j [B_j(Q/B)_j DC_{ij}] - [B_i(P/B)_i(1 - EE_i)] - EX_i = 0$$

or

$$B_i(P/B)_i EE_i - \sum_j [B_j(Q/B)_j DC_{ij}] - EX_i = 0 \quad (3)$$

where:

B	=	biomass for subscripted prey i or predator j;
$(P/B)_i$	=	production per unit biomass for prey i;
$(Q/B)_j$	=	consumption per unit biomass for predator j;
DC_{ij}	=	diet composition of predator group j on prey group i;
EE_i	=	ecotrophic efficiency or proportion of production of group i that is utilized (i.e., consumed) by other groups or removed from the system; thus $(1-EE_i)$ is the proportion of production that goes to detritus; and
EX_i	=	the amount of production of group i that is exported from the system (e.g., by runoff or any form of harvest).

The trophic level for a species group is determined by calculating the weighted mean trophic level of all its prey and adding 1. The model represents a steady state system with values averaged over a year. Input data used are presented in Table 1 with sources of the data. The model was run with a phytoplankton value of 270 and then, for sensitivity study, was run again with a phytoplankton *P/B* value of 175 (used in Alaska Gyre, Pauly and Christensen 1996) and with a value of 365 (used in Lake Ontario model, Halfon and Schito 1993). Sensitivity runs were also made by increasing the biomasses of apex, pelagic, demersal, and zoobenthic groups, groups for which we had no estimates for biomass levels in Becharof Lake.

For both the coho and sockeye juvenile groups, it was assumed that practically all production went to either predation by other groups or was exported from the system as outmigrating smolts; i.e., *EE* was set to 98%

for these two groups. For the apex predator group, the only predation on these fish was assumed to come from bears and birds outside the lake ecosystem and was assumed to be low; therefore, *EE* was set to 1.2% to account only for the export. An initial diet composition matrix was determined based on knowledge of dietary preferences of the species groups. Predation on insects constitutes an import to the system. In addition, since nothing is known about benthic algae, consumption of this group by demersal fish and zoobenthos is classified as an import. Once a full set of initial input parameters was established the model was run; diet composition proportions were adjusted as necessary for all the biomass production equations to balance and efficiencies to fall within reasonable ranges (near 90% for most groups but low for apex predators, phytoplankton, and spawners).

Results

Isotope Analysis

The result of the isotope analysis is summarized in Table 2. Some of the sample sizes are small and should be augmented by future research. A single isopod, *Saduria entomon*, dominates the zoobenthic community. It is an herbivore and its $\delta^{15}N$ value is used as the reference herbivore value in equation 2. The trophic ranking of the examined groups of fish (Table 2) is as one would expect from their food habits, with the three apex predators all having a trophic level equal to 3.0 or greater. The trophic level for each group, corresponding to the ecosystem model groups, is presented as the average, having no relative abundances of species within groups to use as weights.

Ecosystem Model

The ecosystem model was run and the diet composition matrix (D_{ij}) was adjusted until the model balanced. The resulting diet composition matrix is given in Table 3. The solution is not unique, but appears to give a plausible balance in consumption and availability of prey. The resulting balanced model input and output parameters are given in Table 4, with the estimated trophic level for each species group. This information is also presented in a diagram of the trophic flow in Becharof Lake (Fig. 6). The different components of the Becharof Lake ecosystem have been placed on the trophic axis (y axis) according to their trophic levels as calculated by the model and given in Table 4.

The trophic levels estimated per species group are fairly similar between the stable isotope method and the ecosystem model, with a p of 0.056 for a two-tailed paired t-test. The trophic levels estimated from the isotope analysis corroborate those determined from the model and this strengthens the validity of the diet composition matrix for the Becharof Lake ecosystem model.

Table 1. Model components and sources of data for the Becharof Lake ecosystem model.

Species group/Process	Data values	Source
1. Apex predators		
Biomass	Unknown; guess 0.1 g/m^2.	Value used for Schlei Fjord (Christensen & Pauly 1992)
Production/Biomass	0.7	Schlei Fjord
Consumption/Biomass	6.7	Schlei Fjord
Export from system	Eaten by bears and birds, assume 1.2%.	
2. Demersal fish		
Biomass	Unknown; guess 1 g/m^2.	
Production/Biomass	1	Lake Ontario sculpins (Halfon & Schito 1993)
Consumption/Biomass	10.8	Schlei Fjord
3. Pelagic fish		
Biomass	Unknown; guess 1 g/m^2.	
Production/Biomass	1.4	Schlei Fjord
Consumption/Biomass	14.2	Schlei Fjord
4. Coho juveniles		
Biomass	Est. from outmigration, assumed fry survival to smolt 16%, emergent fry 0.3 g; smolts 20 g.	Weights from Marianne McNair, ADF&G, Juneau
Production/Biomass	3.65 calc. from growth curve.	
Consumption/Biomass	10.6; determined during model runs so that gross efficiency same as for sockeye juv.	
Ecological efficiency	Assumed 100%, i.e., all production goes to predation or outmigration.	
Export	Outmigration of 1 million smolts.	ADF&G database
5. Sockeye juveniles		
Biomass	Est. from outmigration, assumed fry survival to smolt 16%; emergent fry 0.2 g, smolts 12.5 g.	Weights from Marianne McNair, ADF&G, Juneau
Production/Biomass	2.49 calculated from growth curve.	
Consumption/Biomass	7.3	Eats 2% of body weight/day, Eric Prestigard, Douglas Island Pink and Chum, Inc.
Export	Outmigration of 56 million smolts.	Crawford & Cross 1998

Table 1. (Continued.)

Species group/Process	Data values	Source
6. Zooplankton		
Biomass	1.7	ADF&G Limnological Lab, Soldotna
Production/Biomass	63	Alaska Gyre (Pauly & Christensen 1996)
Consumption/Biomass	210	
Export	Assumed to be half of that of passive export from lake discharge.	
7. Zoobenthos		
Biomass	Unknown, guess 10.	
Production/Biomass	3.5	Georgia Strait (Pauly & Christensen 1996)
Consumption/Biomass	23	
8. Spawners		
Biomass	Sockeye escapement average 1 M with average weight 2.7 kg; Coho escapement average 15,000 with average weight 3.3 kg.	ADF&G database, Juneau
Production/Biomass	Biomass enters lake in fall, all is available by system; therefore, we used P/B = 1 so available production = annual escapement.	
Consumption/Biomass	Do not feed in lake.	
Ecological efficiency	Low, most goes to detritus.	
9. Phytoplankton		
Biomass	0.9 μg/L Chl *a*	ADF&G Limnological Lab, Soldotna
	600, conversion factor Chl a to wet weight.	Pauly & Christensen 1996
Production/Biomass	270, average of	
	365	Lake Ontario
	175	Alaska Gyre
Export	Assume passive export from lake discharge.	
10. Detritus	No information	
Physical environment	Becharof Lake	
Size	1,142 km^2; 57.6 m mean depth.	
Discharge	Average of 2,760.25 cfs	USFWS, Mitch Linne, Anchorage
	(78 m^3/s) 1997;	
	125.36 m^3/s (average 84-97).	Crawford & Cross 1998

Table 2. Sample size (n), $\delta^{15}N$ mean values, and standard deviation for species from Becharof Lake and the estimated trophic level (TL) based on those values.

Species	n	Mean $\delta^{15}N$	SD	TL	Group TL
Apex predators					3.26
Dolly Varden	49	11.98	1.390	3.5	
Northern pike	6	11.38	1.196	3.3	
Arctic char	13	10.15	1.372	3.0	
Demersal fish					2.65
Pygmy whitefish	6	9.71	0.406	2.8	
Sculpins	14	9.34	0.952	2.7	
Humpback whitefish	9	8.96	0.527	2.6	
Round whitefish	2	8.64	0.530	2.5	
Pelagic fish					2.60
Threespine stickleback	9	9.09	0.261	2.6	
Salmon					
Coho juveniles	10	11.78	1.562	3.4	3.40
Sockeye juveniles	1	7.70	na	2.2	2.20
Zooplankton					2.26
Cyclops	1	8.29	na	2.4	
Cladocera	1	7.87	na	2.3	
Zooplankton	2	7.33	0.113	2.1	
Reference herbivore					2.00
Saduria entomon	10	6.91	0.511	2.0*	
Total	133				

*Used as reference trophic level in calculation of trophic levels of other groups.

na = not applicable

Table 3. Diet composition of each of the predator groups within Becharof Lake ecosystem.

Prey/Predator	1	2	3	4	5	6	7
1. Apex predators	–	–	–	–	–	–	–
2. Demersal fish	–	0.07	–	–	–	–	–
3. Pelagic fish	0.22	–	0.08	0.15	–	–	
4. Coho juveniles	0.02	–	–	–	–	–	
5. Sockeye juveniles	0.62	–	–	0.32		–	
6. Zooplankton	–	0.05	0.60	0.24	0.70	0.20	0.03
7. Zoobenthos	–	0.40	–	–	–	–	0.12
8. Spawners	–	0.01		–	–	–	0.001
9. Phytoplankton	–	–	0.17	0.04	0.05	0.80	
10. Detritus	–	0.30	–	–	–	–	0.30
Import (e.g., insects, benthic plants)	0.15	0.17	0.15	0.25	0.25	–	0.55
Sum	1.00	1.00	1.00	1.00	1.00	1.00	1.00

Table 4. Results for the Becharof Lake ecosystem model giving steady state biomass, surplus production per unit biomass (P/B), consumption per unit biomass (Q/B), ecotrophic efficiency (EE%), export from the lake, and trophic level (TL) for each species group.

Species group	Biomass g/m^2	P/B	Q/B	EE%	Export g/m^2	TL
1. Apex predators	**0.10**	<u>0.7</u>	<u>6.7</u>	1.2%	0.001	**4.0**
2. Demersal fish	**1.00**	<u>1.0</u>	<u>10.8</u>	**77%**	–	**2.7**
3. Pelagic fish	**1.00**	<u>1.4</u>	<u>14.2</u>	**93%**	–	**3.1**
4. Coho juveniles	0.008	3.6	**10.6**	98%	0.018	**3.6**
5. Sockeye juveniles	0.43	2.5	7.3	98%	0.610	**3.1**
6. Zooplankton	1.70	<u>63.0</u>	<u>210.0</u>	**84%**	0.051	**2.3**
7. Zoobenthos	**10.00**	<u>3.5</u>	<u>23.0</u>	**91%**	–	**2.2**
8. Spawners	2.39	1.0	–	**88%**	0.056	1.0
9. Phytoplankton	30.62	<u>270.0</u>	–	**4%**	1.852	1.0
10. Detritus	**7,928**	–	–	**1%**	–	

Notes: Normal font = input data collected from Lake Becharof

Underline = input values taken from literature

Bold font = guesses that work to balance model for biomass except detritus, otherwise model output

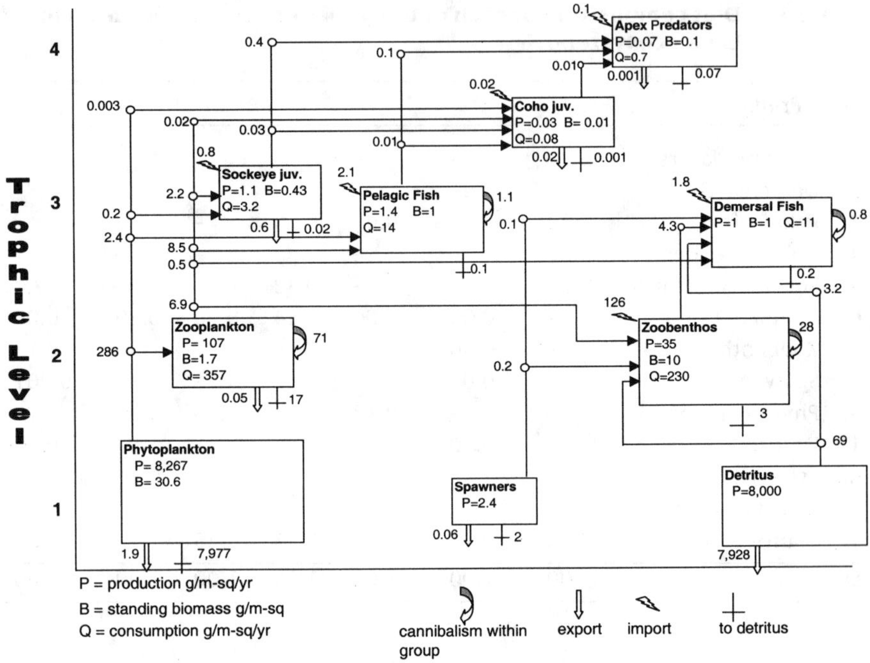

Figure 6. Trophic flow in Becharof Lake as determined by the steady state ecosystem model.

The model does show that neither phytoplankton nor zooplankton are limiting factors to the growth of juvenile salmon, with only 4% of the phytoplankton and 84% of the zooplankton being utilized, the rest decomposing to detritus. Sensitivity runs of the model were made testing the biomass guesses for apex predators, demersal fish, pelagic fish, and zoobenthos. First the pelagic fish biomass was increased from 1 to 2 g/m^2. This increased the consumption on zooplankton and phytoplankton, increasing utilization (EE%) of zooplankton from 84% to 91%, but not noticeably changing that of phytoplankton. Other species groups, except detritus, were not affected. The net detritus accumulation decreased slightly. Next both pelagic and apex fish biomass were doubled. Since we are working with a steady state model and given biomass and production levels for the various species groups, consumption of salmon juveniles remained the same and additional prey for the apex predators had to come from imports to the system, or insects. The trophic level for the apex predator, thus, decreased from 4.0 to 3.6. The trophic level for the other groups did not change. The pelagic fish consumed more zooplankton and phytoplankton, but not more than was available.

Next, demersal fish biomass was increased from 1 to 2 g/m^2 (pelagic and apex fish biomasses were restored to original values). This resulted in an unbalanced model with more zoobenthos being consumed than was available. By doubling the zoobenthos biomass as well, a balanced model was again achieved. There was an increase in consumption of both zooplankton and phytoplankton with utilization of zooplankton being 91% and the utilization of spawners increasing from 17% to 31%.

Another sensitivity test was made changing only the phytoplankton P/B value, trying both 175 and 365 (the two literature values that were averaged to obtain 270, see Table 1). The only change in the model was to change the amount of phytoplankton that accumulated to detritus.

Discussion

The stable isotope analysis for estimating trophic levels relies on having a good estimate of $\delta^{15}N$ for a reference group with a known trophic level, usually a plant group with $TL = 1$ or an herbivore group with $TL = 2$. For this study we assumed that *Saduria entomon* was an herbivore (i.e., $TL = 2.0$). *Saduria entomon* has a worldwide northerly distribution and a depth range from surface down to at least 100 m depth. It displays predatory or cannibalistic behavior depending upon depth and location. In our case, we found it to have the lowest $\delta^{15}N$ and, therefore, classified it provisionally as an herbivore. If future studies in Becharof Lake should show it to forage a wider range of food and its trophic level were slightly greater than 2.0, then this analysis would underestimate the trophic levels of the other groups. However, the similar estimates from the two methods, isotope analysis and ecosystem model, suggest that any error due to this is not large.

The mass-balancing ecosystem model developed here describes a steady state system with parameter values representing annual averages. Thus, diet composition parameter values presented in Table 3 represent yearly averages under conditions of stable salmon production. Clearly, diets change seasonally depending on prey availability; however, annual estimates may be calculated. In this study, our main interest is the production of sockeye salmon in the nursery lake. It is known that there is predation on the developing eggs by species such as sculpins (Foote and Brown 1998); however, in this study we omitted egg mortality and started with consideration at the fry emergence stage. Upon emergence the fry are exposed to various predators as they leave the spawning beds and enter the littoral areas of the nursery lake. When they finally move into the pelagic zone the number of predators is reduced, but at the time of smolt migration the juvenile sockeye salmon again becomes highly vulnerable, both to resident fish species and to external predators such as birds. Northern pike are found only in sheltered inshore bays and, thus, opportunities to prey on sockeye juveniles are limited. The most significant predator on juvenile sockeye salmon in Becharof Lake appears to be from the apex predator group and is probably Dolly Varden. While researchers have found that coho salmon can take up to 7 times as many juvenile sockeye salmon as the Dolly Varden (Roos 1960), the small numbers of coho salmon make it an unimportant predator on sockeye juvenile in this lake. The model shows that apex predators take 94% (0.4 g/m^2) of the consumed sockeye juveniles and coho juveniles take the remaining 4% (0.03 g/m^2) (see Fig. 6).

Ecosystem management implies that some factors of the ecosystem can be manipulated to increase production of the target species group, in this case sockeye salmon. A goal of lake ecosystem management might be to increase the number of smolts with the greatest probability of surviving during their ocean residence prior to return as mature adults. It is commonly assumed that ocean survival is a function of average length of smolts at time of outmigration from the nursery lake. This would favor a greater percentage of age-II than age-I smolts; however, this means an additional year in fresh water with exposure to predators and food competitors reducing the number of outmigrating smolts. While age-II juveniles may be less susceptible to predators than age-I juveniles, this model does not make the distinction.

The principal tool available to the manager is regulation of the size of the escapements, but escapements to Becharof Lake have been relatively stable over the last 20 years (Fig. 3). The decline in total run since 1992 is steeper and more consistent than one would expect from the smolt production (Fig. 4). This implies changing climate or marine conditions as a possible driving force.

Mathisen et al. (1998) demonstrated, using stable nitrogen and carbon isotope ratio measurement from the Egegik smolts, that smolts do not change food habits with increasing age or length but that they compete for the same food base during their entire freshwater residence. If too large an escapement is allowed in a year or a series of years, the juvenile salmon will migrate out with a smaller average length than in years with an average escapement, although their numbers might be greater. However, these changes in numbers of smolts produced each year and the variation in the ratio of age-I to age-II smolts (Fig. 4) are influenced by changes in other elements or species of the ecosystem. Mathisen (1997) suggested that the apex predators control the numbers of smolt produced. In order to understand this relationship better, one would like to be able to vary the population sizes of both predator and prey. The only feasible population control for the apex predators is an indirect one by reducing their principal food, juvenile salmon, through variable escapements. Currently, this is difficult because the idea of a fixed optimum escapement is so ingrained among managers and fishers. The next best approach is constant monitoring of predator populations, or at least the most important, Dolly Varden, to relate its numbers to the variable salmon escapement levels.

Coho salmon escapement can be controlled through changes in the rate of exploitation, usually after the sockeye salmon fishery. Coho salmon is a commercial species with its own monetary value to a more restricted group of fishers; the desired relative abundance of the two salmon species is ultimately economic in nature.

Another form of lake enhancement is through lake fertilization. However, as pointed out by Mathisen et al. (In press), the contribution of marine derived nitrogen in the Egegik smolts amounts to only 13.4%. Hence, lake fertilization is not a viable alternative in this particular ecosystem. Becharof Lake is surrounded by Permian rocks, which contribute sufficient amount of minerals for a high food production such that the system not nutrient limited. Lake fertilization is not viable because the smolts are already large sized and plankton is apparently not limiting the sockeye production. The spawned out salmon carcasses add nutrients to the system, which increase plankton production. Unfortunately, the Ecopath model at this time does not allow production of primary producers to be dependent on another species group in the system. Due to the increased awareness of the importance of carcasses to the nutrient balance of lake systems (Kline et al. 1997), this would be a valuable expansion to the model.

Ecosystem management, that is, management of all components of the system, can be beneficial when it is possible to manipulate numbers of different species in order to obtain age composition and average length of salmon smolts that result in the best subsequent marine survival. While it is easy to speak about ecosystem management, our present knowledge is insufficient for such manipulation of component species in the system. A most important task seems to be monitoring of abundance of major predators, such as Dolly Varden in Becharof Lake. If its numerical abundance is to a large extent dependent upon the supply of juvenile salmon, a strong case can be made for a variable escapement pattern for the salmon. Meanwhile, the vital ecosystem parameters accumulated yearly allow better estimation of returns as functions of smolt numbers, age, and average lengths, which will guide the manager in setting escapement goals.

Acknowledgments

This study of Becharof Lake has only been possible through the assistance of a great number of people and organizations. A special thanks goes to James P. Larson, U.S. Fish and Wildlife Service, King Salmon Fishery Assistance Office, who together with his assistants provided samples over several years and also arranged for monetary support and in kind assistance. Brendan Scalan, a graduate student at the School of Fisheries and Ocean Sciences, University of Alaska Fairbanks, who is studying the char populations in Becharof Lake, was instrumental in securing samples of a number of the resident fishes in Becharof Lake. Norma Haubenstock of the Water and Environmental Research Center did all the isotope analyses in a quick and exact manner for which we are grateful. We wish to thank Professor T.C. Shirley for his identification of the isopod from Becharof Lake. And finally we extend our thanks to the Lake and Peninsula Borough for its continued economic support during these investigations.

References

Beamish, R.J., and G.A. McFarlane (eds.). 1989. Effects of ocean variability on recruitment and an evaluation of parameters used in stock assessment models. Can. Spec. Publ. Fish. Aquat. Sci. 108.

Bue, B.G. 1984. 1982 Egegik river sockeye salmon smolt studies. In: D.M. Eggers and H.J. Yuen (eds.), 1982 Bristol Bay sockeye salmon smolt studies. Alaska Department of Fish and Game Technical Data Report No. 103.

Christensen, V., and D. Pauly. 1992. ECOPATH II: A software for balancing steady-state ecosystem models and calculating network characteristics. Ecol. Modell. 61:169-185.

Crawford, D.L., and B.A. Cross. 1998. Bristol Bay sockeye salmon smolt studies for 1997. Alaska Department of Fish and Game, Division of Commercial Fisheries, Regional Information Report No. 2A98-13.

DeNiro, M.J., and S. Epstein. 1981. Influences of diet on the distribution of nitrogen isotopes in animals. Geochim. Cosmochim. Acta 45:341-353.

Foote, C.J., and G.S. Brown. 1998. Ecological relationship between freshwater sculpins (genus *Cottus*) and beach-spawning sockeye salmon (*Oncorhynchus nerka*) in Iliamna Lake, Alaska. Can. J. Fish. Aquat. Sci. 55:1524-1533.

Halfon, E., and N. Schito. 1993. Lake Ontario food web, an energetic mass balance. In: V. Christensen and D. Pauly (eds.), Trophic models of aquatic ecosystems. ICLARM Conf. Proc. 26:29-39.

Hart, D., D. Petree, and H.J. Geiger. 1998. Run forecasts and harvest projections for 1998 Alaska salmon fisheries and review of the 1997 season. Alaska Department of Fish and Game, Division of Commercial Fisheries, Regional Information Report No. 5J98-04.

Hobson, K.A., and H.E. Welch. 1992. Determination of trophic relationship within a High Arctic marine food web using $\delta^{13}C$ and $\delta^{15}N$ analysis. Mar. Ecol. Prog. Ser. 84:9-18.

Kline Jr., T.C., and D. Pauly. 1998. Mass-balance ecosystem model validation using $^{15}N/^{14}N$ data. In: Fishery stock assessment models. University of Alaska Sea Grant, AK-SG-98-01, Fairbanks, pp. 693-702.

Kline Jr., T.C., J.J. Goering, and R.J. Piorkowski. 1997. The effect of salmon carcasses on Alaskan freshwaters. In: A.M. Milner and M.W. Oswood (eds.), Freshwaters of Alaska: Ecological syntheses. Ecological Studies 119. Springer-Verlag, New York, pp. 179-204.

Kline Jr., T.C., J.J. Goering, A.A. Mathisen, P.H. Poe, P.L. Parker, and R.S. Scalan. 1993. Recycling of elements transported upstream by runs of Pacific salmon: II. δ^{15}N and δ^{13}C evidence in the Kvichak River watershed, southwestern Alaska. Can. J. Fish. Aquat. Sci. 50:2350-2365.

Krogius, F.V., E.M. Krokhin, and V.V. Menshutkin. 1964. Soobshchestvo pelagicheskikh ryb ozera dalnego. Nauka, Leningrad. 85 pp. (In Russian.)

Mathisen, O.A. 1997. Prey-predator relationship in the Ugashik Lakes. University of Alaska Fairbanks Juneau Center, School of Fisheries and Ocean Sciences Report 1997-04(III). 6 pp.

Mathisen, O.A., J.J. Goering, and E.V. Farley. 1999. Nitrogen and carbon isotope ratios in sockeye salmon smolts. Verh. Int. Verein. Limnol. 27. In press.

Mathisen, O.A., J.J. Goering, and R. Russel. 1998. Sockeye production in Egegik and Ugashik watersheds. Verh. Int. Verein. Limnol. 26:2280-2286.

Minagawa, M., and E. Wada. 1984. Stepwise enrichment of ^{15}N along food chains: Further evidence and the relation between δ^{15}N and animal age. Geochim. Cosmochim. Acta 48:1135-1140.

Pauly, D., and V. Christensen (eds.). 1996. Mass-balance models of north-eastern Pacific Ecosystems: Proceedings of a workshop held at the Fisheries Centre, University of British Columbia, Vancouver, B.C., Canada, November 6-10, 1995. Fisheries Centre Research Reports 4(1). 131 pp.

Roos, J.F. 1960. Feeding habits of the Dolly Varden, *Salvelinus malma* (Wahlbaum), at Chignik, Alaska. Trans. Am. Fish. Soc. 88:253-260.

Wada, E., H. Mizutani, and M. Minagawa. 1991. The use of stable isotopes for food web analysis. Crit. Rev. Food Sci. Nutr. 30:361-371.

Wooster, W.S. (ed.). 1983. From year to year: Interannual variability of the environment and fisheries of the Gulf of Alaska and the eastern Bering Sea. Washington Sea Grant Program, WSG-WO 83-3, Seattle. 208 pp.

Bering Sea Ecosystem—A Call to Action: An Interagency Collaborative White Paper

Steven Pennoyer
National Marine Fisheries Service, Juneau, Alaska

Frank Rue
Alaska Department of Fish and Game, Juneau, Alaska

Deborah Williams
U.S. Department of the Interior, Anchorage, Alaska

This paper (written June 1998) represents an unusual gathering of representatives of organizations and agencies, to establish a common vision on the Bering Sea ecosystem and to coherently put together a concept paper. Such inter-agency collaboration is unusual, but on this critical issue a common goal is shared.

Abstract

The Bering Sea is one of the most productive large marine ecosystems in the world. This vast area, almost 2.6 million square kilometers, is located prominently between Alaska and Russia and supports more than 450 species

---

Contributors to this interagency collaborative white paper:
Doug Eggers, Alaska Department of Fish and Game, Juneau, Alaska
Bill Hines, National Marine Fisheries Service, Juneau, Alaska
Gordon Kruse, Alaska Department of Fish and Game, Juneau, Alaska
Pat Livingston, National Marine Fisheries Service, Seattle, Washington
Loh-Lee Low, National Marine Fisheries Service, Seattle, Washington
John Martin, U.S. Fish and Wildlife Service, Homer, Alaska
Ron McCoy, U.S. Department of the Interior, Anchorage, Alaska
Susan McNeil, U.S. Department of the Interior, Anchorage, Alaska
Clarence Pautzke, National Marine Fisheries Service, Anchorage, Alaska
Phil Rigby, Alaska Department of Fish and Game, Juneau, Alaska
William Seitz, U.S. Geological Survey, Anchorage, Alaska
Joe Sullivan, Alaska Department of Fish and Game, Anchorage, Alaska

of fish, crustaceans, and mollusks; 50 species of seabirds; and 25 species of marine mammals. The breeding seabird population of 25 million represents 43% of breeding seabirds in the United States. It provides a world class fishery; the pristine waters contribute 56% of the U.S. fishery production. The total landed value of the catch was one billion dollars in 1997. In recent years the Bering Sea ecosystem has been showing signs of stress. Both marine mammals and seabirds are experiencing population declines. Scientists' knowledge of the various relationships in the ecosystem is limited. Managers need vital and complex information to properly carry out their responsibilities in future years. The ecosystem demands a long-term commitment to assure its protection. It is an extremely valuable treasure of the United States. A showcase ecosystem of the United States and the world, its long-term protection will require comprehensive and coordinated research.

The Bering Sea Ecosystem: A World Class Resource

The Bering Sea is one of the most productive large marine ecosystems in the world. This vast area of almost 2.6 million square kilometers and is located prominently between Alaska and Russia. It is bounded on the north by the Bering Strait, a narrow pass between the Chukotka Peninsula in Russia and the Seward Peninsula in Alaska that once was a land bridge between the two continents thousands of years ago. To the south is the graceful arc of the Aleutian Islands.

This third largest semi-enclosed sea in the world has one of the most extensive continental shelves. The broad shelf, enhanced by nutrient upwelling and intermixing of Pacific Ocean and Bering Sea waters along the Aleutian Chain, provides extremely favorable habitats for a host of marine birds, mammals, and fish that are of international and domestic importance.

The Bering Sea ecosystem supports more than 450 species of fish, crustaceans, and mollusks, more than 50 species of seabirds, and at least 25 species of marine mammals. It provides for a world class fishery; the cold, pristine waters contribute 56% of the U.S. fishery production. A variety of fish species such as salmon, flatfish, cod, walleye pollock *(Theragra chalcogramma)*, Pacific halibut *(Hippoglossus stenolepis)*, herring, sablefish *(Anoplopoma fimbria)* and shellfish has provided the cornerstones of the fishery over the past half century. The total landed value of the catch was approximately one billion dollars in 1997. The North Pacific Fishery Management Council (NPFMC), the National Marine Fisheries Service (NMFS), and the Alaska Department of Fish and Game (ADFG) jointly manage the fisheries.

Fish are just one part of this extraordinary ecosystem. It also has one of the largest internationally shared marine mammal populations, the world's largest eelgrass *(Zostera marina)* beds, and the largest international aggregation of seabirds in the world. The breeding seabird population of

25 million individuals represents 43% of all breeding seabirds in the United States. The Bering Sea is the crossroads for international waterfowl coming from wintering areas in Mexico and Japan, with flights heading to Alaska, Canada, and Russia. Anchoring the American side of the Bering Sea are four national wildlife refuges managed by the Fish and Wildlife Service (FWS): Yukon Delta, Togiak, Izembek, and the Alaska Maritime. A host of other critical federal and state wildlife reserves and wilderness areas border the Bering Sea.

The resources and animals of the Bering Sea ecosystem have been, and continue to be, used extensively by the indigenous peoples as a subsistence resource for thousands of years. The Bering Sea also plays an important role in the cultural and religious traditions of many indigenous peoples in the United States and Russia. In short, the Bering Sea is recognized as one of the most productive and important of the high-latitude seas in the world.

Signs of Stress? A Brief History and Some Concerns

Throughout time, populations of fish, shellfish, birds, and marine mammals that inhabit the Bering Sea have fluctuated in abundance. Fish populations, for example, were lightly exploited and at relatively high levels in the early 1950s, after the lapse in fishing during World War II. Foreign fleets heavily exploited many of the species in the 1960s and early 1970s. It was typical of that era for foreign vessels to find a concentration of fish and fish it down, before moving on to areas of higher abundance. This practice led to cases of stock depletion (e.g., Pacific ocean perch [*Sebastes alutus*]) and provided much of the catalyst for the United States to extend its fisheries jurisdiction out to 322 km (200 miles) in the Bering Sea and elsewhere along the American coast in 1976.

Since then, most, but not all, of the fish populations have been restored to high levels with some species (e.g., many flatfishes) at highest recorded levels of abundance. Pollock remains the dominant fishery, but flatfish and Pacific cod *(Gadus macrocephalus)* also are very abundant. The harvests of all commercial species are closely monitored and controlled by federal and state agencies. Fisheries managers place biologically safe harvest and bycatch limits on each species using the best scientific information available. Fishing seasons are closed or shortened when those catch limits are reached. In general, the abundance of major fish species are based on a few years of very high survival of young fish, that survival being influenced by a host of environmental and biological factors, many of which are outside the control of the managers. The managers can control the catch by how and when it is taken, but often the production of young of the year has a weak to moderate relationship to parental fish abundance.

Shellfishes also are a valuable resource that have fluctuated widely. In the late 1970s, red king crab *(Paralithodes camtschaticus)* made up over 90% of the commercial crab catch, and by 1980, the Bering Sea produced 59 million kg of king crab. After 1980, the population declined precipitously with the commercial catch dropping to 1.4 million kg by 1982. Since then, the fishery has been limited to an annual average of 3.6 million kg (including 3 years of no fishing) as managers allow the stocks to rebuild. Other Bering Sea crab fisheries have experienced cycles with different time trends. In 1997, as the red king crab stock shows promising signs of recovery, no fishing was permitted on the Tanner crab *(Chionoecetes bairdi)* stock which recently fell to record low levels, while at the same time a boom in the snow crab *(C. opilio)* stock led to a harvest of 109 million kg—the largest crustacean fishery in the world. Some scientists believe crab productivity depends on an optimum combination of ice cover, circulation, and seawater temperatures. They think the abundant flatfish stocks in the Bering Sea may be playing a key role through predation and competition. The potentially complex roles of size- and sex-selective fishing on population dynamics continue to be explored as contributing factors to declines.

Some higher trophic level species also are experiencing population declines. Steller sea lion *(Eumetopias jubatus)* abundance has declined by over 80% in the past 30 years in the southeastern Bering Sea. Northern fur seal *(Callorhinus ursinus)* populations have declined by 50% since the 1950s. In the central Aleutians, sea otters *(Enhydra lutris)* are declining rapidly. According to USGS scientists, the Adak sea otter population has experienced an annual decline of 40%. Aleut residents of Atka report similar sea otter declines.

Populations of seabirds such as common murres *(Uria aalge)*, thick-billed murres *(U. lomvia)*, and red-legged *(Rissa brevirostris)* and black-legged kittiwakes *(R. tridactyla)* have declined as much as 50% in some regions of the Bering Sea. Spectacled eiders *(Somateria fisheri)* declined from 50,000 pairs in 1971 to about 1,700 pairs in 1992. In 1993, they were listed as threatened under the Endangered Species Act (ESA). The Steller's eiders *(Polysticta stelleri)* population numbering about 150,000 is also declining dramatically and was listed as threatened in July 1997 under the ESA. Other species such as oldsquaw *(Clangula hyemalis)* and emperor goose *(Chen canagica)* are in decline too.

Many reasons are offered for these declines. The National Research Council has proposed a cascade hypothesis in which ongoing ecosystem changes were triggered by extensive whaling in the late 1800s. Others have speculated that heavy fishing of Bering Sea stocks over the past 40 years has intensified the population declines observed in the Bering Sea ecosystem and that if commercial fishing was reduced, most species with any trophic relationship to the commercially fished species would quickly recover. But most likely, this is an overly simplistic answer. In general, harvest rates applied to Bering Sea fisheries have been very conservative when compared to fisheries virtually everywhere else in the world. However, despite setting overall safe harvest limits, the potential for localized depletion and other unforeseen effects exist. In many cases, managers and scientists' knowledge of the relationship between fishing pressures and a specific marine mammal or seabird population is limited. In other cases scientists know there are many interrelated factors that interact to determine a species' abundance. Additional research is required to understand these relationships. Managers will need to expand their scope of information and decision-making and move toward multispecies, ecosystem-based management.

The broader scope of information needed to characterize the linkages in the ecosystem is for the most part missing. For example, consider the status of the Steller sea lions in the southeastern Bering Sea, listed as endangered in the Federal Register, May 5, 1997. Their population has continued to decline despite limits on pollock harvests and closures to fishing around rookeries and haulouts. Scientists have raised many questions. Is it better to leave more pollock on the fishing grounds? Do pollock have low nutritional value for a sea lion? Are the buffer zones around rookeries and haulouts effective? What are the critical stages in the sea lion life history that affect population changes? These questions must be answered, and that will require comprehensive information on the trophic relationships of the ecosystem.

Contaminants are an emerging issue in the circumpolar region and in the Bering Sea ecosystem. Even though commercially important fish stocks remain safe to consume, some upper trophic level animals that typically bioaccumulate pollutants are now at risk. A study of Aleutian green-winged teal *(Anas crecca)* revealed that 25% of the eggs collected had mercury levels high enough to cause deformities in chicks hatched in laboratory tests. Over 25% of the samples had high enough PCB contamination to cause reduced hatchability in laboratory eggs. All samples of teal eggs had PCBs present as well as DDE, a derivative of DDT.

PCB contamination is not limited to the Aleutian green-winged teal. Bald eagle *(Haliaeetus leucocephalus)* eggs collected in the Aleutian Chain also showed high levels of PCBs. Our national symbol, nesting on one of the remotest national wildlife refuges in the United States, is contaminated with PCBs and also DDT.

Scientists found that sea otters from southeast Alaska had trace levels of DDT while those from around Adak had 36 times greater concentrations. PCB levels in Adak sea otters are even higher than sea otters from elsewhere. The mean liver concentrations for contaminants were the equivalent to the level that causes reproductive failure in mink *(Mustela vison)* and that caused population declines or extinctions of Eurasian otters *(Lutra lutra)*.

Studies of blue mussels *(Mytilus trossulus)* have shown that Adak and Dutch Harbor have high levels of PCB concentrations with the Adak mussels ranking in the top ten of contaminated sites in the United States mussel watch program conducted by the National Oceanic and Atmospheric Administration (NOAA).

The worst examples of contamination are due to local sources, but global atmospheric conditions have transported some pollutants from lower latitudes in to Arctic regions. The degree to which pollution impacts the Bering Sea resources now or will impact it in the future bears investigation. Pollution is the primary reason for establishment of the Arctic Council's Arctic Monitoring and Assessment Programme, and radionuclide contamination from Russia is one of their major concerns.

Finally, and of considerable concern, the environmentally important winter ice cover in the Bering Sea is retreating; it has retreated 5% in 40 years, but principally since the late 1970s. Concurrently, in the late 1970s, a measurable climate change contributed to a major shift in the abundance of many Bering Sea fish species. It is unclear the extent to which these changes are the result of natural processes or anthropogenic effects. Additional global warming could cause extensive ecosystem modifications including flooding, coastal inundation, permafrost loss, and changes in ecosystem productivity. The degree to which human activities have become drivers of ecosystem changes must be evaluated while changes in management may still be effective.

Compared to many regions of the world, the Bering Sea appears to be relatively healthy, a case study of successful natural resource conservation. However, managers and scientists do not know to what extent human versus environmental factors influence current species composition and population sizes. Though many agencies are charged with management and conservation of species that transcend virtually all levels of the food chain, there are still significant concerns about the ecosystem, about how individual species interact, and about the extent to which management can play a decisive role in protecting the health of the ecosystem. No one can forecast the future fluctuations in the many diverse components of the ecosystem. Therefore, managers and scientists cannot know with confidence the level of human activities that will inadvertently transpose the ecosystem from a healthy to an unhealthy condition.

The overarching problem is that there has been no commitment by nations and their resource management agencies to long-term ecosystem research and monitoring. More often than not, funding comes in response to emergencies rather than as proactive management strategies. This approach often can lead to crisis management, when it would have been much better to be able to anticipate the problems and take proactive mitigating actions. For many U.S. institutions, the Bering Sea is considered part of the Arctic yet it commands a very small percentage of the funds appropriated for Arctic research. Consequently, insufficient attention has been focused on the Bering Sea, particularly from an ecosystem perspective or for those resources without a clear economic value. Because of the minor role given to Bering Sea issues, there is no cohesive research legacy in place to provide the badly needed information to manage this significant, complex ecosystem. Delays in closing the research gap could leave natural resource managers in a bind and crisis management could become the norm as in other marine ecosystems throughout the world.

Though there is a strong collective will, and indeed a tremendous need, for managers and scientists to move toward ecosystem-based management of the diverse resources of the Bering Sea, they simply do not have sufficient information to do so. This lack of information likely will be a main finding of the National Academy of Sciences, which recently was tasked by Congress to report on the extent to which ecosystem principles are being used in fisheries management decisions throughout the nation. The urgency of the congressional request came on the heels of general pervasive public opinion that natural resource management, whether for fish, marine mammals, or seabirds, needs to be more holistic in its approach. Identification of information gaps, formulation of research strategies, and funding for scientific research on the Bering Sea ecosystem are badly needed if the system is to be maintained in a healthy condition.

The Proposed Solution: What is Needed and Why

Organize the North Pacific Research Board to Prioritize and Fund Bering Sea Ecosystem Research

As stated above, management agencies have limited resources for the Bering Sea and, necessarily, must confine their efforts to research and management of a few, but important, species. There are existing coordinating entities with responsibilities for various aspects of the Bering Sea ecosystem. Though none is responsible for the entire ecosystem, there are many federal and state agencies that have substantial and interdependent responsibilities to manage certain aspects of the Bering Sea.

The NPFMC, for example, in its advisory role to NMFS, comes closest to having the most overarching responsibility for ecosystem-based management, but it is mainly for the commercial fisheries in the region. The NPFMC has established an Ecosystem Committee to advise it on how to move toward ecosystem-based management of the fisheries. By law the Council must also consider the impacts of its management activities on marine mammals and any threatened or endangered species.

NMFS, as noted above, manages and conducts research on Bering Sea fisheries and marine mammals. NMFS manages the valuable groundfish fisheries and cooperates with the State of Alaska in management of the offshore crab and scallop fisheries in the Bering Sea and the Aleutians. Other components of NMFS' parent organization, NOAA, have responsibilities for meteorology and physical oceanography.

ADFG manages nearshore fisheries for herring, salmon, and groundfish, and offshore fisheries for crabs, scallops, and other shellfish. ADFG also conducts substantial fishery research and collaborates with federal agencies on marine mammal research.

The Marine Mammal Commission (MMC) advises the Secretary of Commerce on that agency's scientific research program to monitor the health and stability of Bering Sea marine mammals and to resolve causes for their declines. The Department of State (DOS) supports the MMC in cooperative, international marine mammal efforts. DOS also chairs an Arctic Policy Group that advises the federal government on international policies that include certain aspects of the Bering Sea Region.

FWS has responsibility for the well being of migratory birds, sea otters, polar bears *(Ursus maritimus)*, Pacific walrus *(Odobenus rosmarus)* and the trust resources of the national wildlife refuges located around the Bering Sea. The FWS has direct input to the NPFMC process and may offer advice and comments as necessary to ensure the protection of seabirds, these marine mammals and listed species.

Finally, the Arctic Research and Policy Act of 1984 established an Arctic Research Commission (ARC) and an Interagency Research Policy Committee each with coordinating roles concerning research in the entire Arctic Region.

An opportunity for integrated, collaborative research exists through the Title IV Environmental Improvement and Restoration Fund of the FY 1998 Interior Appropriations Bill that provides for a North Pacific Research Board (NPRB) as currently structured. The NPRB will provide the Secretary of Commerce with recommendations to determine research priorities for funding grants to federal, state, private or foreign organizations or individuals to conduct marine research in the north Pacific, Bering Sea and Arctic Ocean. The NPRB will consist of two co-chairs, the Secretary of Commerce and the Alaska Commissioner of Fish and Game. Other members will be: the Secretary of State, the Secretary of the Interior, the Commandant of the Coast Guard, the Director of the Office of Naval Research, the Chairman of the NPFMC, the Chairman of the ARC, the Director of the Oil Spill Recovery Institute, the Director of the Alaska SeaLife Center, or their designees. In addition, there will be five representatives nominated by the Governor of Alaska, three members nominated by the Governor of Washington and one member nominated by the Governor of Oregon. The Board will require a small administrative staff.

The NPRB will provide a forum for identifying and addressing information gaps, ensuring that adequate funding is available for long term ecosystems research, and coordinating integrative research efforts. Some of these gaps do not fall within any one agency's legislated mandate and funding often is not available to address research needs to fill in those gaps that do fall within an agency's jurisdiction. Many information gaps require more quantified data on life histories, species inventory, technology development and ecosystem dynamics. The NPRB would advise on all aspects of these research issues. Following its organizational meeting, a comprehensive science plan can assist the NPRB to help set the agenda, scope of work, and to prioritize short- and long-term research needs for the Bering Sea ecosystem.

Address Critical Research Gaps Through a Comprehensive Science Plan

A significant amount of research has been done on the Bering Sea ecosystem. In general the research has underpinned sustainable management of fish and wildlife, but it has raised many questions that need to be answered concerning the general health of the Bering Sea and the ecological changes that are taking place. A Bering Sea Ecosystem Workshop was organized by the Department of Commerce, the Department of the Interior and the ADFG in Anchorage, Alaska December 4-5, 1997. Bering Sea researchers from government agencies and academia, and other interested researchers, were invited to present research findings in their fields of expertise and to submit data gap charts to reveal research needs.

There are many excellent science plans from groups conducting biological and environmental studies. For example, the University of Alaska Fairbanks is investigating the effects of global climate change on the Bering Sea ecosystem and on the resource-based communities along its rim. NOAA's Southeast Bering Sea Carrying Capacity program addresses information gaps in climate variability, limits to population growth, the effects of oceanographic conditions and basic productivity. NMFS conducts extensive surveys of the fisheries resources throughout the Bering Sea. FWS has an ongoing monitoring program focusing on predators in upper trophic levels as indicators of the health of the food web that supports them.

A collaborative, interagency steering committee has consolidated science plans from these research groups and others into an integrated draft science plan for the Bering Sea. Agency staff met with university scientists, fishing industry members, environmentalists, coastal residents, and other members of the public on June 2-3, 1998, to discuss the draft plan and provide comments and suggestions to make it comprehensive. The science plan will address ecosystem science and management issues, present research needs, implementation strategies.

Traditional knowledge (TK) is important for creating biological models and will be included in the science plan. The indigenous peoples around the Bering Sea have lived and hunted in this and other Arctic ecosystems over thousands of years and their knowledge of the ecosystems has become a part of their oral tradition. For example, TK data on seal abundance have already been gathered. At the December 4-5, 1997, workshop, a representative of coastal indigenous people of Alaska presented proposals for combining the vested interests and abilities of Bering Sea coastal communities with managers, environmentalists, and other users of the area resources. Another representative suggested protocols and approaches to accessing traditional knowledge. One of the more challenging aspects of finding ways to better understand the Bering Sea ecosystem is to reconcile the apparent dichotomy of western scientific disciplines with traditional knowledge of indigenous people. Another complication of integrating TK into ecosystem research and management is the notion that such information is regarded as sacred and proprietary to indigenous people.

In addition to TK, efforts are under way by the Minerals Management Service and the NPFMC to compile local knowledge of marine resource events that reflect unusual environmental conditions. Such information would contribute additional observations for ecosystem studies.

Collaboration between western science and traditional knowledge must continue. Various agencies have already invested significant resources into the integration of TK into its process through collaboration between indigenous people and agency researchers/managers. Consideration should also be given to empowering various coastal communities to carry out environmental monitoring projects as a component and/or adjunct of a Bering Sea Ecosystem Science Plan.

Coordinate Research and Communication

Some research data collected a decade ago have not been fully analyzed or published in scientific journals. Likewise, some researchers remain isolated from others performing similar research. To facilitate and speed communication between scientists and other interested parties, a metadatabase has been established. The most efficient means of sharing data is to develop a metadatabase on the World Wide Web. This project has already been initiated by NOAA (by the Alaska Fisheries Science Center and the Pacific Marine Environmental Laboratory [PMEL]) and has been in operation for several months. There is no need to initiate another database project since the database already resides in a web site http://www.pmel.noaa.gov/bering/mdb/ maintained by NOAA's PMEL.

The interagency steering committee recommended that work continue on development of NOAA's Bering Sea Ecosystem Biophysical Metadatabase project and that all research projects link their existing databases into it. NOAA should remain the lead agency to facilitate the coordination, development, and maintenance of the metadatabase.

The web site will also provide an opportunity for all Bering Sea researchers to coordinate their field sampling plans. A 3-step process can facilitate the coordination of agency research projects and cross-placement of scientific personnel on field sampling programs: (i) conduct a research coordination meeting to discuss field plans early each year, (ii) develop an interactive web site to share research plans, and (iii) annually compile and publish agency research plans in a compendium. The intent of this process is to integrate the current work of multiple agencies, reduce redundancies, and maximize the research benefit of existing agency resources.

The interagency steering committee recommended that a web site be used initially to facilitate coordination. Such a web site can be integrated with the metadatabase project to build a "living document" on the World

Wide Web for sharing information on research planning on a continuing basis. NOAA should be designated the leader to undertake the project.

Other near-term possibilities might include a series of meetings focusing on the Bering Sea ecosystem and its significance to the United States and the international community. Congress would be encouraged to send representatives to these meetings. It is expected that additional funding will be needed to fill Bering Sea information gaps and begin coordinated, adequate long-term ecosystem research. Possibly, some duplication of research efforts may be uncovered, and then reduced to produce a leaner, more cost-effective Bering Sea research program.

Engage in Additional International Partnering

Because of its location, long-term protection of the Bering Sea's extraordinary ecosystem can only be accomplished through an international effort. Multinational research cooperation and agreements must be forged with all active participants that have an interest in and an impact on the Bering Sea ecosystem. Much of the Bering Sea is within Russian waters and Russia should be a major player in this endeavor. Other interested parties for partnering include Japan, China, Taiwan, Canada, and possibly Korea and Poland. There are a number of existing international organizations with marine scientific research issues in the Arctic, the North Pacific, and the Bering Sea. The NPRB would be in a position to fund research and enhance coordination of international efforts in the Bering Sea, with the advice of other organizations and institutions involved.

Conclusions

The Bering Sea ecosystem demands a long-term commitment to assure its protection. It is an extremely valuable and ecologically complex ecosystem which the United States must act assertively and wisely to safeguard. Its long-term protection will require comprehensive and coordinated research as well as secure funding for research. The proposed NPRB would help to provide the funding and focus needed to sustain this highly productive, showcase ecosystem of North America.

Participants

Ackley, David
Alaska Dept. of Fish and Game
P.O. Box 25526
Juneau AK 99802-5526
Ph: 907/465-6108; Fax: 907/465-2604
e-mail: davida@fishgame.state.ak.us

Adkison, Milo
University of Alaska Fairbanks
Fisheries Division
11120 Glacier Highway
Juneau AK 99801
Ph: 907/465-6251; Fax: 907/465-6447
e-mail: ffmda@aurora.alaska.edu

Ahlgren, Molly
Sheldon Jackson College
801 Lincoln Street
Sitka AK 99835

Aydin, Kerim
University of Washington
Fisheries Research Institute
Box 357980
Seattle WA 98195-7980
Ph: 206/685-2724; Fax: 206/685-7471
e-mail: kerim@fish.washington.edu

Bax, Nicholas
CSIRO Division of Marine Research
Castray Esplanade
GPO Box 1538
Hobart TAS 7001
AUSTRALIA
Ph: 61 3 62325341;
Fax: 61 3 62325485
e-mail: bax@marine.csiro.au

Beamish, Richard
Dept. of Fisheries and Oceans
Pacific Biological Station
Nanaimo BC V9R 5K6
CANADA

Blackburn, Jim
Alaska Dept. of Fish and Game
211 Mission Road
Kodiak AK 99615
e-mail: JimB1@fishgame.state.ak.us

Brannian, Linda
Alaska Dept. of Fish and Game
333 Raspberry Road
Anchorage AK 99518-1599
Ph: 907/267-2118
e-mail: lindab@fishgame.state.ak.us

Brown, Evelyn
University of Alaska Fairbanks
Institute of Marine Science
P.O. Box 757220
Fairbanks AK 99775-7220
Ph: 907/474-5801; Fax: 907/474-1943
e-mail: ebrown@ims.uaf.edu

Bue, Brian
Alaska Dept. of Fish and Game
333 Raspberry Road
Anchorage AK 99518
Ph: 907/267-2123
e-mail: BrianB@fishgame.state.ak.us

Bulgakova, Tatiana
VNIRO
V. Krasnoselskaya 17
Moscow 107140
RUSSIA
Ph: 095 264 1633; Fax: 095 264 9187
e-mail: frol@vniro.msk.su

Bundy, Alida
Department of Fisheries and Oceans
P.O. Box 1006
Dartmouth NS B2Y 1A2
CANADA
e-mail: bundy@mar.dfo-mpo.zc.ca

Burkey Jr., Charles
Alaska Dept. of Fish and Game
P.O. Box 1467
Bethel AK 99559-1467
e-mail: lsceb@uaf.edu

Byerly, Mike
University of Alaska Fairbanks
P.O. Box 210751
Juneau AK 99821
e-mail: ftmmb@uaf.edu

Byrne, Larry
Alaska Dept. of Fish and Game
211 Mission Road
Kodiak AK 99615
Ph: 907/486-1875; Fax: 907/486-1824
e-mail: lbyrne@fishgame.state.ak.us

Cannon, Dave
U.S. Fish and Wildlife Service
Bethel AK 99559

Cannon, Richard
Alaska Dept. of Fish and Game
333 Raspberry Road
Anchorage AK 99518

Cartwright, Meg
Alaska Dept. of Fish and Game
304 Lake Street, Room 103
Sitka AK 99835
e-mail: megac@fishgame.state.ak.us

Cheslak, Ed
Uribe and Associates
2930 Lakeshore Avenue
Oakland CA 94610-3614
e-mail: ed@uribe.com

Collie, Jeremy
University of Rhode Island
Graduate School of Oceanography
Narragansett Bay Campus
Narragansett RI 02882
Ph: 401/874-6859; Fax: 401/874-6240
e-mail: jcollie@limanda.gso.uri.edu

Cundy, Terry
Potlatch Corporation
805 Mill Road
Lewiston ID 83501
Ph: 208/799-4135; Fax: 208/799-1707
e-mail: twcundy@potlatchcorp.com

Cyr, Ned
NOAA/NMFS/STZ
1315 East-West Highway
Silver Spring MD 20910
Ph: 301/713-2319; Fax: 301/713-0376
e-mail: Ned_Cyr@noaa.gov

Dambacher, Jeffrey
Oregon Dept. of Fish and Wildlife
28655 Highway 34
Corvallis OR 97333
Ph: 541/757-4263
e-mail: dambachj@ucs.orst.edu

Danchy, Bob
Boise-Cascade
P.O. Box 50
Boise ID 83729
e-mail: bdanchy@bc.com

Dressel, Sherri
University of Alaska Fairbanks
Institute of Marine Science
P.O. Box 757220
Fairbanks AK 99775-7220
Ph: 907/474-7918
e-mail: dressel@ims.alaska.edu

Faris, Tamra
NMFS Alaska Region
P.O. Box 21668
Juneau AK 99802

Fetterolf, Carlos
8200 Pine Cross Lane
Ann Arbor MI 48103-9701

Fluharty, David
University of Washington
School of Marine Affairs HF-05
Seattle WA 98195
Ph: 206/685-2518

Focht, Rick
Douglas Island Pink and Chum
2697 Channel Drive
Juneau AK 99803
e-mail: dipacops@pobox.alaska.net

Ford, Tim
Turlock Irrigation District
333 E. Canal Drive
Turlock CA 95380
e-mail: tford@ainet.com

Fowler, Charles
National Marine Fisheries Service
National Marine Mammal Laboratory
7600 Sand Point Way, NE Bldg 4
Seattle WA 98115-0070
Ph: 206/526-4031
e-mail: Charles.Fowler@noaa.gov

Foy, Bob
University of Alaska Fairbanks
Institute of Marine Science
P.O. Box 7220
Fairbanks AK 99775-7220
Ph: 907/474-5801
e-mail: foy@ims.uaf.edu

Fried, Steven
Alaska Dept. of Fish and Game
333 Raspberry Road
Anchorage AK 99518
e-mail: stephenf@fishgame.state.ak.us

Fritz, Lowell
National Marine Fisheries Service
Alaska Fisheries Science Center
7600 Sand Point Way NE Bldg 4
Seattle WA 98115
Ph: 206/526-4246; Fax: 206/526-4004
e-mail: lfritz@noaa.afsc.gov

Fu, Caihong
University of Alaska Fairbanks
Fisheries Division
11120 Glacier Highway
Juneau AK 99801
Ph: 907/465-6372; Fax: 907/465-6320
e-mail: ftcf@uaf.edu

Fuerstenberg, Robert
King County Water and Land Resources
700 5th Avenue, Suite 2200
Seattle WA 98104

Furness, Robert
University of Glasgow
Graham Kerr Building
Glasgow G12 8QQ
UNITED KINGDOM
Ph: 44 141 330 3560;
Fax: 44 141 330 5971
e-mail: r.furness@bio.gla.ac.uk

Gallaway, Benny
LGL Alaska Research Associates
1410 Cavitt
Bryan TX 77801
Ph: 409/775 2000; Fax: 409/775-2002
e-mail: 73447.3635@compuserve.com

Ganey, Steve
Alaska Marine Conservation Council
P.O. Box 101145
Anchorage AK 99510
Ph: 907/277-5357; Fax: 907/277-5975

Gardner, Lee Ann
RWJ Consulting
P.O. Box 672302
Chugiak AK 99567-2302
Ph: 907/688-1400; Fax: 907/688-1499
e-mail: rwjconsult@corecom.net

Geiger, Harold
Alaska Dept. of Fish and Game
P.O. Box 25526
Juneau AK 99802
e-mail: halg@adfg.state.ak.us

Gislason, Henrik
University of Copenhagen
Danish Inst. for Fisheries Research
Charlottenlund Castle
Copenhagen DK 2920
DENMARK
Ph: 33963361
e-mail: hg@dfu.min.dk

Grabacki, Stephen
GRAYSTAR Pacific Seafoods Ltd.
P.O. Box 100506
Anchorage AK 99510
Ph: 907/272-5600; Fax: 907/272-5603

Grossman, Andrew
NMFS Alaska Region
P.O. Box 20431
Juneau AK 99802
Ph: 907/586-7358
e-mail: Andrew_Grossman@noaa.gov

Gunther, Andrew
Applied Marine Sciences, Inc.
4740 Bennett Drive #L
Livermore CA 94550-0100
Ph: 510/451-7936; Fax: 510/451-3631
e-mail: gunther@amarine.com

Hagen, Peter
Alaska Dept. of Fish and Game
P.O. Box 25526
Juneau AK 99802-5526
Ph: 907/465-3054; Fax: 907/465-2765
e-mail: peterh@fishgame.state.ak.us

Haggan, Nigel
University of British Columbia
Fisheries Centre, 2204 Main Mall
Vancouver BC V6T 1Z4
CANADA
Ph: 604/822-6939; Fax: 604/822-8934
e-mail: nhaggan@fisheries.com

Hansson, Sture
Stockholm University
Dept. of Systems Ecology
Stockholm S-106 91
SWEDEN
Ph: +46 816 42 48;
Fax: +46 (0) 8 158417
e-mail: sture_h@system.ecology.su.se

Harrington, Gretchen
NMFS Alaska Region
P.O. Box 21668
Juneau AK 99802
Ph: 907/780-6741; Fax: 907/780-6223
e-mail: gretchen.harrington@noaa.gov

Hartmann, Cindy
NMFS Alaska Region
P.O. Box 21668
Juneau AK 99802-1668
Ph: 907/586-7585
e-mail: Cindy.Hartmann@noaa.gov

Hauser, Bill
Alaska Dept. of Fish and Game
Habitat and Restoration
333 Raspberry Road
Anchorage AK 99518

Heizer, Steve
Dept. of Fisheries and Oceans
3225 Stephenson Pt. Road
Nanaimo BC V9T 1K3
CANADA
Ph: 604/756-7270; Fax: 604/756-7162

Hjerne, Olle
Stockholm University
Dept. of Systems Ecology
Stockholm S-106 91
SWEDEN
Ph: +46 867 475 51
e-mail:
Olle.Hjerne@system.ecology.su.se

Hodges, James
USDA Forest Service
Cordova Ranger District
P.O. Box 280
Cordova AK 99574

Hollowed, Anne
NMFS Alaska Fisheries Science Ctr.
7600 Sand Point Way NE
Building 4
Seattle WA 98115
Ph: 206/526-4223; Fax: 206/526-6723
e-mail: Anne.Hollowed@noaa.gov

Honnold, Steven
Alaska Dept. of Fish and Game
211 Mission Road
Kodiak AK 99615
e-mail: stevehon@fishgame.state.ak.us

Ianelli, Jim
NMFS Alaska Fisheries Science Ctr.
7600 Sand Point Way NE
Building 4
Seattle WA 98115
Ph: 206/526-6510; Fax: 206/526-6723
e-mail: jim.ianelli@noaa.gov

Jasper, Jim
National Park Service
P.O. Box 237
Gustavus AK 99826
e-mail: Jim_Jasper@nps.gov

Johnston, Samuel
Hydoacoustic Technology, Inc.
715 EN Northlake Way
Seattle WA 98105
e-mail: support@htisonar.com

Kitchell, Jim
University of Wisconsin
Center for Limnology
680 North Park Street
Madison WI 53706
Ph: 608/262-9512; Fax: 608/262-2340
e-mail: kitchell@macc.wisc.edu

Kline, Tom
Prince William Sound Science Center
P.O. Box 705
Cordova AK 99574
Ph: 907/424-5800; Fax: 907/424-5820
e-mail: tkline@grizzly.pwssc.gen.ak.us

Knudsen, Eric
USGS - BRD, Alaska Science Center
1011 East Tudor Road
Anchorage AK 99503
Ph: 907/786-3842; Fax: 907/786-3636
e-mail: Eric_Knuksen@usgs.gov

Kruse, Gordon
Alaska Dept. of Fish and Game
P.O. Box 25526
Juneau AK 99802
Ph: 907/465-6106; Fax: 907/465-2604
e-mail: GordonK@fishgame.state.ak.us

Krygier, Earl
Alaska Dept. of Fish and Game
P.O. Box 25526
Juneau AK 99802
Ph: 907/465-6112; Fax: 907/465-2604
e-mail: earlk@fishgame.state.ak.us

Link, Jason
NMFS Northeast Fisheries Science Ctr.
166 Water Street
Woods Hole MA 02543
Ph: 508/495-2340; Fax: 508/495-2258
e-mail: jlink@whsun1.wh.whoi.edu

Link, Michael
LGL Alaska Research Assoc., Inc.
4175 Tudor Centre Drive #202
Anchorage AK 99508
e-mail: mlink@lgl.com

Livingston, Patricia
NMFS Alaska Fisheries Science Ctr.
7600 Sand Point Way NE
Building 4
Seattle WA 98115
Ph: 206/526-4242; Fax: 206/526-6723
e-mail: Pat.Livingston@noaa.gov

Lorenz, Mitch
NMFS Auke Bay Lab
11305 Glacier Highway
Juneau AK 99801
e-mail: mitch.lorenz@noaa.gov

Lucey, Bill
USDA Forest Service
P.O. Box 327
Yakutat AK 99689

Lynch, Kristine
Michigan State University
Dept. of Fisheries and Wildlife
13 Natural Resources Bldg.
East Lansing MI 48824-1222
Ph: 517/353-6697; Fax: 517/432-1699
e-mail: dawsonkr@pilot.msu.edu

Marmulla, Gerd
FAO
Via Delle Terme Drive
Caracalla
Rome I-00100
ITALY
e-mail: gerd.marmulla@fao.org

Marotz, Brian
Montana Fish, Wildlife and Parks
490 North Meridian
Kalispell MT 59901
Ph: 406/751-4546; Fax: 406/357-0349
e-mail: marotz@digisys.net

Martin, Doug
Martin Environmental
2103 N 62nd Street
Seattle WA 98103
e-mail: martin1696@aol.com

Mathisen, Ole
University of Alaska Fairbanks
Fisheries Division
11120 Glacier Highway
Juneau AK 99801
Ph: 907/465-6442
e-mail: ffoam@uaf.edu

Mazumber, Asit
University of Montreal
CP 6128
POB Centre Ville
Montreal Quebec H3C 3J7
CANADA
e-mail: mazumdea@ere.umontreal.ca

McCallum, Charles
Chignik Regional Aquaculture Assoc.
1155 N State Street, #426
Bellingham WA 98225
e-mail: craa@premier1.net

McCullough, Jim
Alaska Dept. of Fish and Game
211 Mission Road
Kodiak AK 99615

McIntosh, Bruce
University of Alaska Fairbanks
P.O. Box 80071
Fairbanks AK 99708
e-mail: fsbcm1@uaf.edu

McKelvey, Robert
University of Montana
Depts. of Economics and Math Sci.
1740 Madeline Avenue
Missoula MT 59801
Ph: 406/543-5996
e-mail: ma_rwm@selway.umt.edu

Meadows, Steve
Quileute Natural Resources
P.O. Box 187
LaPush WA 98350

Megrey, Bern
NMFS Alaska Fisheries Science Ctr.
7600 Sand Point Way NE
Building 4
Seattle WA 98115
Ph: 206/526-4147; Fax: 206/526-6190
e-mail: bmegrey@afsc.noaa.gov

Menard, Jim
Alaska Dept. of Fish and Game
P.O. Box 1467
Bethel AK 99559-1467
e-mail: bethel@fishgame.state.ak.us

Merritt, Peggy
Alaska Dept. of Fish and Game
1300 College Road
Fairbanks AK 99701
Ph: 907/459-7296
e-mail: pmerritt@fishgame.state.ak.us

Moffitt, Steven
Alaska Dept. of Fish and Game
Box 669
Cordova AK 99574
e-mail: stevemo@fishgame.state.ak.us

Moloney, Coleen
Sea Fisheries Research Institute
Private Bag X2
Rogge Bay 8012
SOUTH AFRICA
Ph: 27-21-402-3171; Fax: 27-21-21-7406
e-mail: cmoloney@sfri.wcape.gov.za

Molyneaux, Doug
Alaska Dept. of Fish and Game
333 Raspberry Road
Anchorage AK 99518
e-mail: dougm@fishgame.state.ak.us

Mortensen, Don
NMFS Auke Bay Lab
11305 Glacier Highway
Juneau AK 99801
e-mail: don.mortensen.@noaa.gov

Mueter, Franz
University of Alaska Fairbanks
Institute of Marine Science
P.O. Box 757220
Fairbanks AK 99775
Ph: 907/474-7839
e-mail: mueter@ims.alaska.edu

Mundy, Phillip
Fisheries and Aquatic Sciences
1015 Sher Lane
Lake Oswego OR 97034-1744
e-mail: mundy@teleport.com

Murphy, James
NMFS Auke Bay Lab
11305 Glacier Highway
Juneau AK 99802
e-mail: Jim.Murphy@noaa.gov

Murphy, Peggy
Alaska Dept. of Fish and Game
P.O. Box 25526
Juneau AK 99802
Ph: 907/465-6107; Fax: 907/465-2604
e-mail: PeggyM@fishgame.state.ak.us

Newman, Kurt
Michigan State University
Dept. of Fisheries and Wildlife
13 Natural Resources Building
East Lansing MI 48824-1222
Ph: 517/353-6697; Fax: 517/432-1699
e-mail: newmanku@pilot.msu.edu

Norris, Jim
Marine Resources Consultants
P.O. Box 816
Port Townsend WA 98368
Ph: 206/385-4486

O'Brien, Chris
Ministry of Fisheries
P.O. Box 1020
Wellington
NEW ZEALAND
e-mail: obrienc@fish.govt.nz

Okey, Tom
University of British Columbia
Fisheries Centre
2204 Main Mall
Vancouver BC V6T 1Z4
CANADA
Ph: 604/822-7231; Fax: 604/822-8934
e-mail: tokey@fisheries.com

Olito, Carmen
Alaska Dept. of Fish and Game
333 Raspberry Road
Anchorage AK 99518
e-mail: carmeno@fishgame.state.ak.us

Olson, Robert
Inter-American Tropical Tuna Comm.
Scripps Institution of Oceanography
8604 La Jolla Shores Drive
La Jolla CA 92037-1508
Ph: 619/546-7160; Fax: 619/546-7133
e-mail: rolson@iattc.ucsd.edu

Orlov, Alexei
Russian Federal Research Inst. of
Fisheries and Oceangraphy (VNIRO)
17a, V. Krasnoselskaya
Moscow 107140
RUSSIA
Ph: (095)264-94-54; Fax: (095)264-91-78
e-mail: akrovnin@mx.iki.rssi.ru

Overholtz, William
NMFS Northeast Fisheries Center
166 Water Street
Woods Hole MA 02543
Ph: 508/495-2000
e-mail: william.overholtz@noaa.gov

Parker, James
Alaska Dept. of Fish and Game
P.O. Box 605
Delta Jct. AK 99737
e-mail: fparker@fishgame.state.ak.us

Patino, Enrique
Quileute Natural Resources
Box 189
LaPush WA 98350
e-mail: epatino@olypen.com

Patrick-Riley, Kent
Alaska Dept. of Environ. Conservation
555 Cordova Street
Anchorage AK 99501
e-mail: kriley@environcom.state.ak.us

Paul, A.J.
University of Alaska Fairbanks
Seward Marine Station
P.O. Box 730
Seward AK 99664
Ph: 907/224-5261; Fax: 907/224-3392
e-mail: ffajp@aurora.alaska.edu

Pauly, Daniel
University of British Columbia
Fisheries Centre, 2204 Main Mall
Vancouver BC V6T 1Z4
CANADA
Ph: 604/822-1201; Fax: 604/822-8934
e-mail: pauly@fisheries.com

Pavlychev, Mikhail
Pacific Research Fisheries Centre
TINRO-centre
4 Shevchenko Alley
Vladivostok 690600
RUSSIA
e-mail: MPAVL@nt.pin.dvgu.ru

Pavlychev, Vadim
Pacific Fisheries Research Centre
TINRO-centre
4 Shevchenko Alley
Vladivostok 690600
RUSSIA
e-mail: MPAVL@pub.marine.su

Perez-Fuentetaja, Alicia
Alaska Dept. of Fish and Game
211 Mission Road
Kodiak AK 99615
e-mail: aperez@fishgame.state.ak.us

Perry, Ian
Dept. of Fisheries and Oceans
Pacific Biological Station
Hammond Bay Road
Nanaimo BC V9R 5K6
CANADA
Ph: 250/756-7137; Fax: 250/756-7053
e-mail: perryi@pac.dfo-mpo.gc.ca

Peterman, Randall
Simon Fraser University
Natural Resource Mgmt. Program
Burnaby BC V5A 1S6
CANADA
Ph: 604/291-4659; Fax: 604/291-4968
e-mail: peterman@sfu.ca

Pinnix, William
University of Washington
School of Fisheries
Box 357980
Seattle WA 98195-7980
Ph: 206/616-9418; Fax: 206/685-7471
e-mail: pinnix@fish.washington.edu

Pitcher, Tony
University of British Columbia
Fisheries Centre
2204 Main Mall
Vancouver BC V6T 1Z4
CANADA
Ph: 604/822-2731; Fax: 604/822-8934
e-mail: TPitcher@fisheries.com

Poe, Patrick
Bonneville Power Administration
P.O. Box 3621
Portland OR 97208
e-mail: patrickhpoe@bpa.gov

Polovina, Jeffrey
NMFS Southwest Fisheries Center
2570 Dole Street
Honolulu HI 96822-2396
Ph: 808/943-1218; Fax: 808/943-1290
e-mail:
jpolovin@honlab.nmfs.hawaii.edu

Power, Melanie
University of British Columbia
Fisheries Centre
2204 Main Mall
Vancouver BC V6T 1Z4
CANADA
Ph: 604/822-1864; Fax: 604/822-8934
e-mail: melanie@fisheries.com

Preikshot, Dave
University of British Columbia
Fisheries Centre, 2204 Main Mall
Vancouver BC V6T 1Z4
CANADA
Ph: 604/822-0618; Fax: 604/822-8934
e-mail: preikshot@fisheries.com

Quinn, Terry
University of Alaska Fairbanks
Fisheries Division
11120 Glacier Highway
Juneau AK 99801
Ph: 907/465-5389; Fax: 907/465-6320
e-mail: fftjq@aurora.alaska.edu

Radchenko, Vladimir
Pacific Fisheries Research Centre
TINRO-centre
4 Shevchenko Alley
Vladivostock 690600
RUSSIA
Ph: 7 4232 25 95 04
e-mail: root@tinro.marine.su

Rice, Stanley
National Marine Fisheries Service
Auke Bay Lab
11305 Glacier Highway
Juneau AK 99801-8626
e-mail: jeep.rice@noaa.gov

Roach, Don
Alaska Dept. of Fish and Game
1300 College Road
Fairbanks AK 99701
Ph: 907/459-7229
e-mail: droach@fishgame.state.ak.us

Roley, Dennis
Bio-Oregon
P.O. Box 429
Warrenton OR 97146
e-mail: roley@seafurf.com

Roni, Phil
National Marine Fisheries Service
2725 Montlake Boulevard East
Seattle WA 98112
e-mail: phil.roni@noaa.gov

Roseman, Edward
Michigan State University
Dept. of Fisheries and Wildlife
East Lansing MI 48824-1222
Ph: 517/4353-6697; Fax: 517/432-1699
e-mail: rosemane@pilot.msu.edu

Sandone, Gene
Alaska Dept. of Fish and Game
1800 Glenn Highway
Palmer AK 99654
e-mail: gsandone@fishgame.state.ak.us

Sands, Norma Jean
Alaska Dept. of Fish and Game
P.O. Box 25526
Juneau AK 99802-5526
Ph: 907/465-4256; Fax: 907/465-2332
e-mail: normas@fishgame.state.ak.us

Schiewe, Michael
National Marine Fisheries Service
2725 Montlake Boulevard East
Seattle WA 98112
e-mail: michael.schiewe@noaa.gov

Schmidt, Dana
RL&L Environmental Services, Ltd.
235 Fairview Drive
Castlegar BC V1N 3S8
CANADA
e-mail: dschmidt@insidenet.com

Seeb, Lisa
Alaska Dept of Fish and Game
Division of Commercial Fisheries
333 Raspberry Road
Anchorage AK 99518

Seitz, William
U.S. Geological Survey
1011 E Tudor Road
Anchorage AK 99503
Ph: 907/786-3385; Fax: 907/786-3636
e-mail: william_seitz@usgs.gov

Setter, Ann
Oregon Dept. of Fish and Wildlife
65495 Alder Slope Road
Enterprise OR 97828
e-mail: asetter@orednet.org

Shannon, L.J.
Sea Fisheries Research Institute
Private Bag X2
Rogge Bay 8012
SOUTH AFRICA

Shaw, Linda
National Marine Fisheries Service
P.O. Box 21668
Juneau AK 99802
e-mail: linda.shaw@noaa.gov

Shima, Michiyo
University of Washington
School of Fisheries
P.O. Box 357980
Seattle WA 98195
Ph: 206/526-4330; Fax: 206/526-6723
e-mail: mshima@u.washington.edu

Shin, Yunne-Jai
IRD, Laboratory LIA
32 avenue Henri-Varagnat
93143 Bondy cedex
FRANCE
e-mail: shin@bondy.ird.fr

Shirakihara, Kunio
Mei University
1515 Kamihama
Tsu 514-8507
JAPAN
e-mail: shirak@bio.mei-u.ac.jp

Sisson, Joyce
National Marine Fisheries Service
8604 La Jolla Shore Drive
La Jolla CA 92037
e-mail: joyce@CA.liban.ucsd

Smoker, Bill
University of Alaska Fairbanks
Fisheries Division
11120 Glacier Highway
Juneau AK 99801

Soiseth, Chad
National Park Service
P.O. Box 140
Gustavus AK 99826
Ph: 907/697-2230
e-mail: chad_soiseth@nps.gov

Spies, Bob
Applied Marine Sciences
4749 Bennet Drive #L
Livermore CA 94550-0100
Ph: 510/373-7142
e-mail: spies@amarinc.com

Steer, Greg
Fisheries and Oceans Canada
360-555 W Hastings Street
Vancouver BC V6B 5G3
CANADA
e-mail: steerg@dfo-MPO.gc.ca

Stuby, Lisa
Alaska Dept. of Fish and Game
1300 College Road
Fairbanks AK 99701
e-mail: lstuby@fishgame.state.ak.us

Sullivan, Joe
12300 Audubon Drive
Anchorage AK 99516

Tarbox, Kenneth
Alaska Dept. of Fish and Game
34828 K-Beach Road, #B
Soldotna AK 99669
e-mail: kent@fishgame.state.ak.us

Tol, Dennis
BLM
222 W 7th #13
Anchorage AK 99513
e-mail: dtol@ak.blm.gov

Tracy, Donn
Alaska Dept. of Fish and Game
211 Mission Road
Kodiak AK 99615
Ph: 907/486-1822
e-mail: donnt@fishgame.state.ak.us

Trites, Andrew
University of British Columbia
NPUMMRC Room 18, Hut B-3
6248 Biological Sciences Road
Vancouver BC V6T 1Z4
CANADA
Ph: 604/822-8182; Fax: 604/822-8180
e-mail: trites@bcu.ubc.ca

Tweten, Randy
National Marine Fisheries Service
525 NE Roegon St., Suite 500
Portland OR 97232
e-mail: randy.tweten@noaa.gov

Tyler, Al
University of Alaska Fairbanks
SFOS
P.O. Box 757220
Fairbanks AK 99775-7220

Urban, Dan
Alaska Dept. of Fish and Game
211 Mission Road
Kodiak AK 99615
e-mail: danu@fishgame.state.ak.us

Vining, Ivan
Alaska Dept. of Fish and Game
211 Mission Road
Kodiak AK 99615
e-mail:
IVAN_Vining@fishgame.state.ak.us

Wedemeyer, Kate
2CES/CEVP 22040 Maple Street
Elmendorf AK 99506-3240

Wiedmer, Michael
Alaska Dept. of Fish and Game
333 Raspberry Road
Anchorage AK 99518
e-mail: mikewi@fishgame.state.ak.us

Wietkamp, Laurie
National Marine Fisheries Service
2725 Montlake Boulevard East
Seattle WA 98112
e-mail: laurie-wietkamp@noaa.gov

Wilcock, John
Alaska Dept. of Fish and Game
P.O. Box 669
Cordova AK 99574
Ph: 907/424-3212; Fax: 907/424-3235
e-mail: johnwi@state.ak.us

Willette, Mark
Alaska Dept. of Fish and Game
P.O. Box 669
Cordova AK 99574
Ph: 907/424-3214; Fax: 907/424-3235
e-mail: markw@fishgame.state.ak.us

Witherell, Dave
N. Pac. Fishery Management Council
605 W 4th Avenue
Anchorage AK 99501
Ph: 907/271-2809; Fax: 907/271-2817

Witteveen, Mark
Alaska Dept. of Fish and Game
211 Mission Road
Kodiak AK 99615
e-mail: mwitteve@fishgame.state.ak.us

Wood, Mark
Native Village of Quinhagak
P.O. Box 149
Quinhagak AK 99655
e-mail: asmcw8@uaa.alaska.edu

Wright, Bruce
EVOS Trustee Council
National Marine Fisheries Service
11305 Glacier Highway
Juneau AK 99801
Ph: 907/789-6601

Wunderlich, Bob
U.S. Fish and Wildlife Service
510 Desmond Drive, Suite 102
Lacey WA 98503
e-mail: Bob_Wunderlich@fws.gov

Yang, Yufeng
Shantou University
Science Center
Shantou 515063
P.R. CHINA
e-mail: YFYang@mailserv.stu.edu.cn

Zheng, Jie
Alaska Dept. of Fish and Game
P.O. Box 25526
Juneau AK 99802-5526
Ph: 907/465-6102; Fax: 907/465-2604
e-mail: jiez@fishgame.state.ak.us

Index

A

B

D

E

F

G

H

I

J

K

L

M

N

O

Q

R

S